Basic Mathematics for Economics, Business, and Finance

T0221412

This book can help overcome the widely observed math-phobia and math-aversion among undergraduate students in economics, business, and finance. The book can also help them understand why they have to learn different mathematical techniques, how these techniques can be applied, and how the techniques will equip the students in their further studies.

The book provides a thorough but lucid exposition of most of the mathematical techniques applied in the fields of economics, business, and finance. The book deals with topics from high school mathematics to relatively advanced areas of integral calculus, also covering subjects such as linear algebra, differential calculus, classical optimization, linear and nonlinear programming, and game theory.

Though the book directly caters to the needs of undergraduate students in economics, business, and finance, graduate students in these subjects will also find the book an invaluable text for supplementary reading. The website of the book – www.emeacollege.ac.in/bmebf – provides supplementary materials and further chapters on difference equations, differential equations, elements of Mathematica®, and graphics in Mathematica®. The website also provides materials on the applications of Mathematica® throughout the book, as well as teacher and student manuals.

E. K. Ummer graduated with a Ph.D. degree from the University of London. He is currently an Assistant Professor at the Department of Economics of E.M.E.A. College of Arts and Science. Prior to joining E.M.E.A. College, he was with the faculty of Economics and Management Studies at International Islamic University, Malaysia. His current research interests are quantitative methods, economic theory, and economics of development and growth.

Basic Mathematics for Economics, Business, and Finance

E. K. Ummer

Routledge
Taylor & Francis Group

LONDON AND NEW YORK

First published 2012
by Routledge
4 Park Square, Milton Park, Abingdon, Oxon OX14 4RN

Simultaneously published in the USA and Canada
by Routledge
605 Third Avenue, New York, NY 10017

Routledge is an imprint of the Taylor & Francis Group, an informa business

British Library Cataloguing in Publication Data
A catalogue record for this book is available from the British Library

Library of Congress Cataloging-in-Publication Data
Ummer, E. K.
Basic mathematics for economics, business, and finance / E. K. Ummer.
   p. cm.
Includes bibliographical references and index.
1. Economics–Mathematical models. 2. Programming (Mathematics) I. Title.
HB135.U44 2011
330.01'51–dc23   2011019188

ISBN: 978-0-415-66419-6 (hbk)
ISBN: 978-0-415-66420-2 (pbk)
ISBN: 978-0-203-81732-2 (ebk)

Typeset in Times New Roman by Cenveo Publisher Services

Dedicated to my parents, my wife, my sons, and all others in my family

Contents

Preface

This book grew out of the lecture notes I gave to students in undergraduate programs in economics, business, and finance for more than a decade at different institutions in different countries. The book can be adopted either wholly or partially for undergraduate or beginning graduate programs in these subjects. The only prerequisite, I assume, to follow the topics covered in this book is a *bit of patience*.

It will not be an exaggeration if one states that mathematics has become the language of economics. The states of affairs in related subjects such as business and finance are not much different. Most of the beginning undergraduate programs in these subjects mainly apply geometric tools for the exposition of relationships and theories. But, as the courses progress, the inherent limitations of the geometric tools necessitate a shift from them to more general algebraic forms. This shift calls for a training in some of the techniques and tools of mathematics.

I have found from experience that an alarmingly large proportion of the students who enroll in undergraduate programs in economics, business, and finance in particular (and social science in general) possess some degree of "math-phobia" and "math-aversion." These feelings, I believe, have their root in the unpropitious presentation of the subject to them. Although the books that have been written on mathematics for these subjects are all excellent in their own respects, many of them still continue the unpropitious form of presentation. Many of these books follow either a notoriously technical or oversimplified approach making the subject esoteric or humdrum. My aim, through an intermediate approach, is an auspicious presentation of the subject so that the feelings of phobia and aversion can be replaced by passion and appreciation. Therefore, I attempt to present to undergraduates in these subjects through this book why they need to learn all the mathematical techniques expected of them; the importance of these techniques and their interrelationships; and how these techniques are applied in their subjects. I believe that this approach will make them appreciate mathematics and, thereby, help them understand their subjects properly.

Most of the graduate programs in economics, business, and finance apply mathematical techniques and tools that are far beyond the levels of those covered in this book. Similarly, most journals (even those considered to be *applied* in nature) appear with articles that contain high-level mathematics. I do not claim either that this book will be sufficient for graduate mathematical requirements or that it will prepare students to read and understand the said articles. But, I do claim that this book is sufficient for undergraduate mathematical requirements and it can build a strong foundation for graduate studies in these subjects, which will eventually help in the reading and understanding of the journal articles mentioned above.

One important feature of the book is that the complete presentation of different topics is based on intuition. Since I believe that visual aids such as graphs help students learn faster,

I have included them throughout the book. Though proofs of theorems and propositions are important and necessary for a proper understanding of mathematics, I believe that it will be inauspicious and counterproductive to impose these proofs on already math-phobic and math-averse students. They can learn these proofs once they understood the basics and if they are interested in them. Therefore, I have deliberately omitted the proofs of most of the theorems and propositions. Another feature is that most subsections in every chapter of the book contain a number of numerical examples. Moreover, most major sections of the book contain application examples and exercises from different branches of economics, business, and finance. Although the examples in the book are drawn primarily from these subjects, the main body of the book can be successfully adopted (through suitable selection of examples) in similar programs in subjects such as political science, psychology, life sciences, etc.

The book is organized into eight chapters. I have attempted to include in the book most of the mathematical techniques and tools that are normally taught in undergraduate programs in economics, business, and finance throughout the world. I believe that a review of some of the necessary mathematics learned in school will help students much and, therefore, the *first* chapter of the book is devoted to this purpose. It covers most of the important topics that students learned in school including basics of sets; number system; exponents; logarithms; equations; inequalities; intervals; absolute values; functions; limits; continuity; sequences; series; and sum and product symbols.

The *second* chapter covers linear algebra. This chapter explores most of the topics in vectors and matrices that are required by undergraduate programs in the subjects mentioned above. Specifically, it discusses the basics of vectors and matrices; vector spaces; vector and matrix operations; determinants; inverse; rank; solutions to systems of linear equations; and some special matrices and determinants. Differential calculus is discussed in the *third* chapter. It explores differentiation and derivatives; differentiability of functions; rules of differentiation; higher-order differentiation; curvature of curves; convex sets; transformation of functions and Maclaurin and Taylor series; partial derivatives; differentials; total derivatives; implicit differentiation; etc.

Since static optimization is crucial in the subjects of our interest, chapters four through seven are set apart for this topic. The *fourth* chapter is on classical optimization, which is primarily concerned with the application of linear algebra and differential calculus covered in the second and third chapters, respectively. It begins with a discussion of optima and extrema of univariate functions and progresses through their differential versions; optima of multivariate functions and their extensions; and optimization with equality constraints and its extensions. The *fifth* chapter is devoted to linear programming. The topics covered in this chapter include graphical approach; the tabular and matrix approaches of the simplex method; the revised simplex method; duality and sensitivity analyses; the dual simplex method; transportation and assignment problems; etc. The nonlinear programming approach to optimization is covered in the *sixth* chapter. Topics such as geometric forms of nonlinear objective functions and constraints; geometric and algebraic solutions to nonlinear programming problems; and concave, quasiconcave, and quadratic programming are dealt with in this chapter. Another important topic of static optimization, namely game theory, is dealt with in the *seventh* chapter. It presents topics including static games of complete and perfect information; dominant and dominated strategies; Nash equilibrium; mixed and maximin strategies; dynamic games of complete and perfect information; extensive form representations; subgame perfect Nash equilibrium; repeated games; etc.

The last, and the *eighth*, chapter of the book is devoted to the presentation of one of the tools of dynamic analysis, namely integral calculus. This chapter introduces the meaning of integration; the relationship between integration and differentiation; indefinite integrals; rules of integration; initial value problems; partial and multiple integrals; definite integrals and the fundamental theorem of integral calculus; areas under and between curves; definite partial and multiple integrals; and improper integrals.

I had planned to include in the book, along with a few supplementary topics in the existing chapters, exclusive chapters on difference equations and differential equations. Two issues compelled me to exclude them from the book. One was, of course, the space constraint. The other was the fact that these excluded topics are not widely covered in most undergraduate programs in the subjects of our interest. However, I have prepared these supplementary topics and the two chapters on difference equations and differential equations, which can be found at the book's website: www.emeacollege.ac.in/bmebf. Interested readers can access and use them for learning purposes.

I had also planned to integrate Mathematica®, one of the world's most versatile software packages, into the book. Mathematica is a highly advanced computational software package. It is beyond doubt that it can facilitate students' learning of mathematics. All the figures in this book are generated in Mathematica. But, again, space constraints forced me to exclude it from the book. These materials include exclusive chapters on elements of Mathematica and graphics in Mathematica. They also include the applications of Mathematica in most of the topics covered in the book and in difference and differential equations. These materials can also be found at the website mentioned above and interested readers can use them for learning purposes.

The website also contains teaching aids such as PowerPoint and overhead projector slides; an instructor's manual; and a student solution manual.

E. K. Ummer
ummerek@emeacollege.ac.in

Acknowledgments

To a certain extent, this book is an outcome of cooperation. Several people have cooperated in many ways during the past few years by providing support, encouragement, guidance, advice, and love. It is certainly a pleasure to acknowledge my debt to them.

This book grew out of the lecture notes I gave to undergraduate students in mathematics for economics, business, and finance at different institutions in different countries. Knowingly or unknowingly, many of these students have contributed greatly to the contents of the book. I am thankful to them all.

My institution, EMEA College of Arts and Science, and my colleagues have been a constant source of support and inspiration. Special thanks are due to Professor K. Abdul Hameed (College Principal); Professor K. Hamza and my other colleagues at the Department of Economics; T.V. Zacharia, Assistant Professor and Head, Department of Political Science; and V. Abdul Muneer, Assistant Professor and Head, Department of Journalism. I am also thankful to A.M. Riyad, Assistant Professor and Head, Department of Computer Science; and to K. Muhammedali, College Superintendent.

Professor P. Mohammed, Head, Department of Mathematics, has patiently reviewed the entire manuscript of the book. His suggestions and comments have certainly improved its contents. I am indebted to him.

It was a great pleasure to deal with the concerned officials at Routledge, Taylor & Francis Group, all the way from the proposal of the book to its present form. In particular, I mention Ms. Lam Yongling, Associate Editor, Routledge. She has been highly cooperative and considerate, and I am grateful to her and all others concerned at Routledge.

My wife, Ani, and my sons, Kittu and Kuttu, have extended me unstinted encouragement and support right from the inception of the idea of writing the book until now. They missed me a lot while I was working away on the book. Although I cannot repay them what is lost, I love them and dedicate the book to them and to all others in my family.

1 Review of basics

1.1 Introduction

Economic activities have played an important role in the lives of humans for centuries past. We now know that they have an even greater influence on our modern lives. The economic agents in the old civilizations too possessed some perception, though not as sophisticated as we do today, of some of the economic phenomena that affected their lives. But the difference is that they needed only the rudiments of mathematics to analyze and comprehend these phenomena. It was under these circumstances that some of the earliest writers on economics communicated their misty visions.

However, events such as the Renaissance and the Industrial Revolution resulted in radical transformations in production, consumption, trade, and economic management. These transformations are now bolstered by the advent of information technology. These events and the accompanying transformations have made modern economic life highly complex. This suggests that we can no longer be complacent about the rudimentary mathematics that was sufficient until about the beginning of the twentieth century.

One simple example can illuminate the argument we made above. Assume that a consumer wishes to purchase a good offered for sale. But, we are aware of the fact that the consumer's demand for the good depends, *ceteris paribus*, on the price of the good. We know that this is a highly simplified version of reality. In fact, the consumer's demand for the good is also influenced by factors such as the price of related goods (determined in the markets for the related goods); the consumer's income (determined in the factor market); events taking place in the government sector; and so on. Although we started with the simple proposition that a consumer's demand for a good depends on the price of the good, we ended up with a complex situation involving many markets or sectors of the economy.

It would be difficult to analyze such a complex structure as the one presented above without mathematics. The reason is that mathematics can reduce the complexity to manageable limits. Mathematics can help define the elements of a theory precisely; can help generate new insights; and can help in the applicability of the theory. The following view of Fisher (1925: 119), a celebrated American economist, is a testimony to our above statements (italics added):

The economic world is a misty region. The first explorers used unaided vision. Mathematics is the lantern by which what before was dimly visible now looms up in firm, bold outlines. The old phantasmagoria[1] disappears. We see better. We also see further.

The above presented necessity generated by the complexity of the economic world paved the way for the advent of mathematics in economic sciences. Mathematics has, in fact, become the language of modern economics, business, and finance. Students of these subjects require a wide variety of mathematical tools of varying degrees of complexity. Since several of the mathematical tools used in these subjects are far beyond the scope of a basic book such as this, we include here only those necessary tools that are required by students for the successful completion of undergraduate programs, and to prepare them for graduate programs, in these subjects.

In this chapter we review some of the essential topics that we will use later. This review will include the basics of topics such as set theory; the number system; exponents; logarithms; equations; inequalities, intervals, and absolute values; relations and functions; limits and continuity; sequences and series; and summation and product notations.

Section 1.2 discusses the fundamental concepts in set theory. This is followed by the number system and the associated properties in Section 1.3. Exponents and their laws are covered in Section 1.4. Section 1.5 reviews logarithms and their properties. A review of the basics of equations is provided in Section 1.6. Section 1.7 presents inequalities, intervals, and absolute values. A review of the fundamental ideas of relations and functions is given in Section 1.8. Limits and continuity are dealt with in Section 1.9. Sequences and series are covered in Section 1.10. We introduce some of the sum and product notations in Section 1.11.

1.2 Set Theory

1.2.1 Meaning of sets

Sets play a crucial role in almost all branches of mathematics and are being increasingly used in economics, business, and finance. It is sometimes convenient to consider many items together. Such a collective entity is called a *set*. A set is defined as any well-defined list, collection, or class of objects. The objects in a set can be anything: students, numbers, vehicles, countries, trees, or anything else. Examples of sets include:

> The people living in the city of New York.
> The even numbers between 0 and 10.
> The odd numbers between 0 and 10.
> The numbers 1, 2, 3, 4, and 5.

1.2.2 Set notations

Sets are usually denoted by uppercase letters such as A, B, C, X, Y, Z, etc. The objects in a set are called the *elements* or *members* of the set. These objects are usually denoted by lowercase letters such as a, b, c, x, y, z, etc. If x is an object in the set A, then x is called an element of the set and is denoted as

> $x \in A$, and is read "x belongs to A" or "x is a member of A"

If x is not an object in A, then we may write it as

> $x \notin A$, and is read "x does not belong to A" or "x is not a member of A"

We can represent a set by listing its elements and using { } notation. Assume that the set A consists of numbers 2, 4, 6, 8, and 10. Then we may write the set A as

$$A = \{2, 4, 6, 8, 10\}$$

Notice that in the set A above we separated the elements by commas and enclosed them in curly brackets. We call this form of representation of a set the *tabular form*. Sets can also be represented by stating properties that its elements must satisfy. Assume that we want a set B of even numbers. Then we may write it as

$$B = \{x \mid x \text{ is even}\}$$

which we read as "B is the set of numbers x such that x is even." This form of representation of a set is called the *set-builder form*.

1.2.3 Equality of sets and subsets

Two sets A and B are said to be equal if they have the same elements; that is, if every element in A also belongs to B and if every element in B also belongs to A. Let $A = \{9, 8, 7, 6\}$ and $B = \{8, 7, 9, 6\}$. Then $A = B$. Notice that a set does not change if its elements are rearranged. Notice also that the set $\{1, 2, 3, 3, 4\} = \{1, 2, 3, 4\}$.

Let there be two sets A and B. If every element in A is also an element of B, then A is called a *subset* of B. In other words, A is a subset of B if $a \in A$ and $a \in B$, and is denoted as $A \subseteq B$. For example, let $A = \{1, 2, 3\}$ and $B = \{1, 2, 3, 4, 5\}$. Since the elements 1, 2, and 3 appear in both sets and since B contains more elements than A does, then $A \subseteq B$. Notice that if $A = B$, $A \subseteq B$ and $B \subseteq A$. Assume that $A \subseteq B$. Then, we may also write $B \supseteq A$, which we read "B is a *superset* of A."

Another term widely used is the *proper subset*. Let there be two sets A and B. Then A is called a proper subset of B if $A \subseteq B$ and $A \neq B$, and is denoted as $A \subset B$. As an example, if $A = \{1, 2, 3\}$ and $B = \{1, 2, 3, 4, 5\}$, then $A \subset B$.

1.2.4 Types of sets

There are a number of different types of sets. One of the basic types of sets is the *null set* or *empty set*, which is denoted by the Greek letter Φ (phi).[2] As an example, let A be a set of people who are neither dead nor alive. We can write this set using the set-builder for as $A = \{x \mid x \text{ is a person who is neither dead nor alive}\}$. We know that this set is a null or empty set. Notice that Φ is considered to be a subset of all other sets.

Sets can be finite or infinite. A set is said to be a *finite set* if it contains a finite number of different elements. Otherwise the set is called an *infinite set*. The set of months in a year, the set of hours in a day, etc., are examples of finite sets. The set of stars in the sky, the set of real numbers, etc., are the examples of infinite sets.

Two other important sets widely used are *universal set* and *complementary set*. The universal set consists of all the objects that are being considered in a particular situation. It is generally denoted by U. The complementary set is the set of all elements that are not the elements of a particular set (say A) but are of U. The complementary set of, say, B is denoted by B'.

Sometimes two or more sets may not have common elements. Such sets are called *disjoint sets*. For example, if $A = \{1, 2, 3, 4\}$ and $B = \{5, 6, 7, 8\}$, then A and B are called disjoint sets. Another important type of set is the *power set*. The power set is defined as the set of all the subsets that can be generated from a given set A. It can be shown that if A has n elements, then the power set will contain 2^n elements and is usually denoted as $2^{n(A)}$. For example, let $A = \{1, 2\}$. Then $2^{n(A)} = \{\{1, 2\}, \{1\}, \{2\}, \phi\}$.

1.2.5 Set operations

There are three basic set operations: *union*, *intersection*, and *difference*. We shall review each of them below. The union of two sets A and B is defined as the set of all elements which belong to A, or to B, or to both A and B. We denote the union of sets A and B by $A \cup B$, which is read "A union B." Let $A = \{1, 2, 3, 4\}$ and $B = \{4, 3, 5, 6\}$. Then $A \cup B = \{1, 2, 3, 4, 5, 6\}$.

The intersection of two sets A and B is defined as the set of elements that are common to A and B, and is denoted by $A \cap B$, which is read "A intersection B." In our last example, $A \cap B = \{3, 4\}$.

The difference of two sets A and B is defined as the set of elements which belong to A but not to B and is noted by $A - B$, which is read "A difference B" or "A minus B." In our last example, $A - B = \{1, 2\}$. Notice that $B - A = \{5, 6\}$.

A useful way of representing sets and their operations is the *Venn diagram*, named after the English logician and mathematician John Venn. In a Venn diagram, the universal set U is represented by a square or a rectangle within which individual sets are shown as circles. The Venn diagram representations of union, intersection, difference, and complement are illustrated by the shaded areas in Figures 1.2.1(A)–(D), respectively.

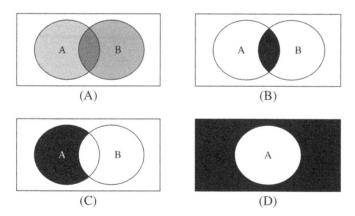

Figure 1.2.1

1.2.6 Laws of set operations

The basic laws of set operations are

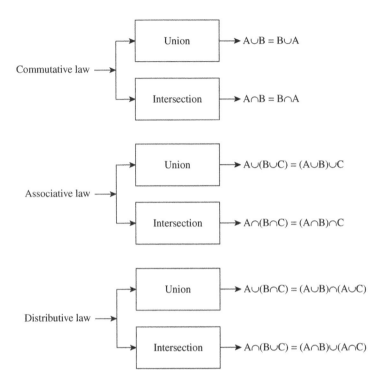

1.2.7 Application examples

Example 1.   Assume that a company wanted to frame a marketing strategy. The company randomly chose 100 students from the hostels of a university and asked them three questions: (1) Do you have a computer in your room? (2) Do you have a TV in your room? (3) Do you have a computer and a TV in your room? Assume also that 60 of them answered yes to (1), 40 answered yes to (2), and 25 answered yes to (3). (i) How many students have either a computer or a TV in their rooms? (ii) How many students do not have a either a computer or a TV in their rooms? (iii) How many students do have a computer but not a TV in their rooms? (iv) How many students do not have both a computer and a TV in their rooms?

Solution.   If we use the Venn diagram, it is easy to solve this problem. But, for this, we need to use specifications such as $U =$ the set of students in the sample (100), $C =$ the set of students who have computers in their rooms (60), $T =$ the set of students who have TV in their rooms (40), and $T \cap C =$ the set of students who have computers and TV in their rooms (25). Now we can use the Venn diagram illustrated in Figure 1.2.2. (i) The number of students who have either a computer or a TV in their rooms is the number of students in the set $T \cup C$. As can be seen from Figure 1.2.2, this number is $35 + 25 + 15 = 75$. (ii) This is equal to the number of students in the set $(T \cup C)'$; that is, $100 - 75 = 25$. (iii) The number of

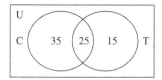

Figure 1.2.2

students who have a computer in their rooms but not a TV is $C - T = 35$. (iv) This is equal to $(T \cap C)' = 75$.

Example 2. Assume that four managers of a company, denoted by the set $\{M_1, M_2, M_3, M_4\}$, wish to select a committee of two people from among themselves. In how many ways can this committee be formed? Or, in other words, how many two-person subsets can be formed from a set of four people?

Solution. Since the elements of the set are M_1, M_2, M_3, and M_4, the subsets with exactly two elements are $\{M_1, M_2\}$, $\{M_1, M_3\}$, $\{M_1, M_4\}$, $\{M_2, M_3\}$, $\{M_2, M_4\}$, and $\{M_3, M_4\}$. This shows that there are six different ways of forming a committee of two managers from among four managers or there are six different subsets of two elements each in a set of four elements.

1.2.8 Exercises

1. Write the following using the tabular form of sets:
    (a) The days in a week. (b) The numbers 1, 2, 3, 4, and 5. (c) The vowels of English alphabet. (d) The South Asian countries India, Pakistan, Sri Lanka, Bangladesh, and Nepal.
2. Continue with exercise 1 above. Write the following using the set-builder form of sets:
    (i) (a); (ii) (b); (iii) (c); (iv) (d).
3. Continue with exercise 1 above. Write the following statements using set notations:
    (i) Sunday is an element of (a); (ii) 6 does not belong to (b); (iii) "b" is not a subset of (c); (iv) India is a subset of (d); (v) Nepal is a proper subset of (d).
4. Let $A = \{a, b, c\}$. Decide whether the following statements are true or false:
    (i) $a \notin A$; (ii) $\{c\} \subseteq A$; (iii) $\{b\} \in A$; (iv) $\{a\} \subset A$; (v) $2^A = 7$.
5. Given the sets $A_1 = \{1, 2, 3\}$, $A_2 = \{5, 1, 3\}$, $A_3 = \{2, 1, 3\}$, and $A_4 = \{3, 1\}$, find:
    (i) $A_1 \cup A_2$; (ii) $A_1 \cap A_2$; (iii) $A_2 \cup A_3$; (iv) $A_2 \cap A_3$; (v) $A_1 \cup A_3$; (vi) $A_1 \cap A_3$;
    (vii) $A_1 \cup A_2 \cup A_3$; (viii) $A_1 \cap A_2 \cap A_3$.
6. Given $A = \{1, 2, \{3, 4\}, 5\}$, which of the following statements are true and why?
    (i) $\{3, 4\} \subseteq A$; (ii) $\{3, 4\} \in A$; (iii) $\{\{3, 4\}\} \subset A$; (iv) $\{\{3, 4\}, 5\} \not\subset A$.
7. Which of the following statements are valid?
    (i) $A \cup A = A$; (ii) $A \cap A = A$; (iii) $A \cup \phi = A$; (iv) $A \cup U = U$; (v) $A \cap \phi = \phi$; (vi) $A \cap U = U$;
    (vi) $(A')' = A$.
8. *Application exercise.* A marketing survey of 100 people found that 70 people watch TV news, 40 people listen to radio news, and 30 people both watch TV news and listen to radio news. Find:
    (i) The set of people who watch either TV news or listen to radio news. (ii) The set of people who both watch TV news and listen to radio news. (iii) The set of people who do

not watch either TV news or listen to radio news. (iv) The set of people who do not both watch TV news and listen to radio news. (v) The set of people who watch TV news but do not listen to radio news. (vi) The set of people who listen to radio news but do not watch TV news.

 Web supplement: S1.2.9 Mathematica applications

1.3 Number system

Many of the models in the subjects of our interest often use numbers. Moreover, most commercial and financial transactions involve the use of numbers. Therefore, students of economics, business, and finance require knowledge of the fundamental operations involving numbers. Besides, a reasonable understanding of the classification of numbers is also required by these students for further study of mathematics.

1.3.1 Classification of numbers

Numbers are classified into different sets according to certain characteristics. We shall discuss here these sets and their characteristics. Let us begin the classification with *natural numbers*. The natural numbers are also called the *counting numbers*, and we denote them by N. Natural numbers constitute the set of *positive whole numbers*. Therefore, the set of natural numbers is $N = \{1, 2, 3, 4, 5, \ldots\}$. Notice that the natural numbers are closed only under the operations of addition and multiplication. What this means is that when we add or multiply two natural numbers we obtain another natural number. This also means that the difference or quotient of two natural numbers need *not* be a natural number.

Another set of numbers, which is close to the set of natural numbers, is the set of *prime numbers*. The prime numbers are those natural numbers that are only divisible by 1 and by the number itself. We represent the set of prime numbers by P. The set of prime numbers is, therefore, $P = \{2, 3, 5, 7, 11, 13, 17, 19, 23, 29, \ldots\}$.

When we add *negative whole numbers* and zero to the set of natural numbers, we obtain what is called the set of *integers*, denoted by Z. Therefore, the set of integers is written as $Z = \{\ldots, -3, -2, -1, 0, 1, 2, 3, \ldots\}$. The integers are also referred to as *whole numbers*. Notice that the integers are closed under the operations of addition, subtraction, and multiplication. This means that the sum, difference, or product of two integers is an integer.

Assume that we divide one integer by another integer (except zero). Then the quotient may or may not be an integer. Such a number is called a *rational number* and is denoted by Q. Therefore, we write the set of rational numbers as $Q = \{x | x = z_1/z_2\}$, where $z_1 \in Z$ and $z_2 \in Z$. It should be noticed that each integer is also a rational number; since, for example, $2/1 = 2$ and so $Z \subset Q$ (i.e. Z is a proper subset of Q). It should also be noticed that rational numbers are closed under all *arithmetic operations*; that is, under addition, subtraction, multiplication, and division. This means that the sum, difference, product, or quotient (except under division by 0) of two rational numbers is a rational number.

Figure 1.3.1

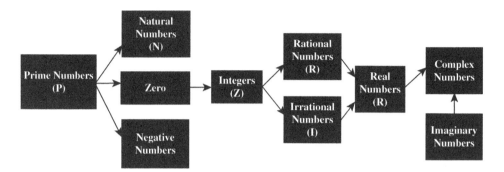

Figure 1.3.2

Can we write every number as the quotient of two numbers? In other words, is every number a rational number? The answer is no. The reason is that some numbers like $\sqrt{2}$, $\sqrt{3}$, e $(= 2.71828\ldots)$, and the value denoted by the Greek letter $\pi (= 3.1415\ldots)$ cannot be written as ratios of integers. The numbers that cannot be written as ratios of integers or the numbers that are not rational numbers are called *irrational numbers*, and we denoted them by I.

One of the most important sets of numbers is the set of *real numbers* denoted by R. The set of all rational and irrational numbers is the set of real numbers. They contain all possible decimal representations. One of the important properties of the real numbers is that they can be represented by points on a straight line. As can be seen from Figure 1.3.1, we choose a point called the *origin* to represent 0 and another point, to the right of 0, to represent 1. Similarly, we choose a point to the left of 0 to represent -1. Then each point will represent a unique real number, and vice versa. We call this line the *real line*. Those numbers to the right of 0 are called the *positive numbers* and those numbers to the left of 0 are called the *negative numbers*. The number 0 is neither positive nor negative.

There is still another set of numbers called *imaginary numbers*. These are the numbers whose squares are negative numbers. $i = \sqrt{-1}$, which implies $i^2 = -1$, is an imaginary number. The last category of numbers is the set of *complex numbers*. Complex numbers have both real and imaginary components and are written in the form $a + bi$, where a and b are real numbers: a is the *real part* and bi is the *imaginary part*. Examples of complex numbers are $2 + 3i$, $10 - 3i$, etc. The above classification of numbers can be represented by a *tree diagram* as illustrated in Figure 1.3.2.

1.3.2 Properties of real numbers

Given any real numbers a, b, c, and 0, the following properties are valid:

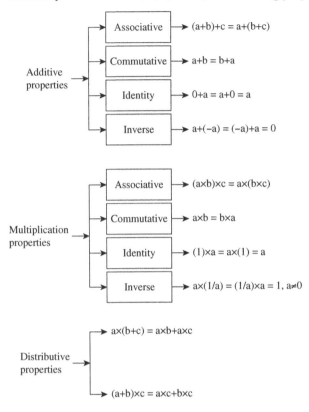

1.3.3 Exercises

1. Given the sets of numbers N (natural numbers), Z (integers), Q (rational numbers), and R (real numbers), indicate to which set(s) each of following numbers belongs:
   (i) 5; (ii) $\sqrt{3}$; (iii) -3.59; (iv) $-5/3$; (v) 0; (vi) -1.
2. State whether each of the following is true or false:
   (i) $\sqrt{-1} \in R$; (ii) $5 \in N$; (iii) $-1 \in N$; (iv) $4 \in P$; (v) $(1/2) \in Z$; (vi) $2 \in Q$; (vii) $1 \in R$; (viii) $2.5 \in I$; (ix) $0 \notin N$; (x) $5 \in N$; (xi) $\pi \in R$; (xii) $e \notin R$.
3. Give an example of each of the following:
   (i) A prime number; (ii) a natural number; (iii) an integer; (iv) a rational number; (v) an irrational number; (vi) a real number; (vii) a complex number; (viii) an imaginary number.

 Web supplement: S1.3.4 Mathematica applications

1.4 Exponents

In mathematical calculations we often make use of expressions such as x^2, x^3, x^n, etc. We shall now undertake a closer consideration of such expressions. We have learned in school mathematics that the product $x \times x \times x$ is x^3. In general, for a positive integer n, x^n is the short form for the product of n expressions of x's. The letter n in x^n is called the *exponent* and x is called the *base*. x^n is called the nth power of x. Notice that $x^0 = 1$ for $x \neq 0$, and 0^0 is undefined.

1.4.1 Rules of exponents

There exist a number of rules related to the use of exponents. The first rule of exponents is the *product rule*. Let there be two exponents x^n and x^m. Then their product $x^n \times x^m$ is defined as

$$x^n \times x^m = x^{n+m} \qquad\qquad (1.4.1)$$

Equation (1.4.1) can be confirmed by using specific values for n and m. Let $n = 4$ and $m = 3$. Therefore, using equation (1.4.1), we may write $x^4 \times x^3 = (x \times x \times x \times x) \times (x \times x \times x) = (x \times x \times x \times x \times x \times x \times x) = x^7 = x^{4+3}$.

The second rule is called the *power rule*. Assume that x is raised to the power n as x^n. Assume again that the last expression is again raised to the power m. Then we may write these statements as

$$(x^n)^m = x^{n \times m} \qquad\qquad (1.4.2)$$

As in the case of equation (1.4.1), equation (1.4.2) can also be verified by using specific values for n and m. Let $n = 4$ and $m = 3$. Then we may write $(x \times x \times x \times x) \times (x \times x \times x \times x) \times (x \times x \times x \times x) = x^{12} = (x^4)^3 = x^{4 \times 3}$.

Sometimes algebraic expressions may contain more complicated forms than those in equations (1.4.1) or (1.4.2). One such expression is $\{[(x^n)]^m\}^p$. We may use the power rule in equation (1.4.2) to simplify this expression. The result will be $x^{n \times m \times p}$. Another complicated expression is $(x.y)^n$, which involves two bases, x and y. This expression can be simplified, as before, by applying the power rule, and the simplified form is $x^n \times y^n$. A combination of the last two expressions is $(x^n \times y^m)^p$. The simplified form of this is $x^{n \times p} \times y^{m \times p}$.

The third rule is called the *quotient rule*. Assume that there exists an expression such as x^n/x^m. Then the quotient rule says that this expression is equivalent to

$$x^n/x^m = x^n \times x^{-m} = x^{n-m} \qquad\qquad (1.4.3)$$

As an example, suppose that the expression is $x^4/x^2 = (x.x.x.x)/(x.x)$. This expression can be written as $x^4/x^2 = x^{4-2} = x^2$. In the case of $n = m$, we have $x^0 = 1$.

1.4.2 Rational and irrational exponents

In this section we shall discuss the concepts, in the context of exponents, of rational and irrational numbers that we reviewed in Section 1.3.1. We shall first consider *rational exponents*. One of the simplest examples of a rational exponent is $x^{1/2}$. Notice that in this expression the numerator of the exponent is 1 and the denominator (2) is an integer. This is a simple way to represent the *square root* of x and it includes both the positive and the negative square roots.[3]

Let there be an integer n. If we raise x to the power $1/n$, then the result is called the nth root of x and is represented as $x^{1/n}$ or $\sqrt[n]{x}$. Now suppose, instead of 1, we use a number (m)

for the numerator of the exponent. Then the result is called the *m*th root of the exponent $x^{1/n}$. This is represented as $x^{m/n}$ or $\sqrt[n]{x^m}$. It should be noticed that all the three rules of exponents we stated above are applicable to rational exponents.

One can also encounter *irrational exponents*. In these cases, the exponents will be irrational numbers. Let the exponent be π. Then the product of the same two bases with powers π is $x^\pi \times x^\pi = x^{\pi+\pi} = x^{2.\pi}$. As an example, assume that the exponent is $\sqrt{2}$. Then we have $x^{\sqrt{2}} \times x^{\sqrt{2}} = x^{\sqrt{2}+\sqrt{2}} = x^{2\times\sqrt{2}} = x^{2\times(1.4142)} = x^{2.8284}$.

1.4.3 Exercises

1. Simplify the following expressions using the rules of exponents:
   (i) $x^n.x^{3n}.x^{2n}$; (ii) (x/x^2); (iii) $x^5.y^3.z^{-2}.x^{-4}.y^0.z^{-1}$; (iv) $(x^{-n}.y^m.z^2)/(x^n.y^{-m}.z^{-1})$.
2. Use the rules of exponents to simplify the following exponents:
   (i) $(((x)^2)^3)^4$; (ii) $((((x^2))^2)^2)^2$; (iii) $(x^{1/2})^{1/2}$; (iv) $(((x^{1/2}/x^{1/2})^{1/2})^{1/2})^{1/2}$.

 Web supplement: S1.4.4 Mathematica applications

1.5 Logarithms

Logarithms are widely used for both computational and analytical or theoretical purposes in economics, business, and finance. A reasonable understanding of logarithms is necessary for students aspiring for graduation, and for further studies, in these fields. Therefore, we shall review in this section the fundamentals of logarithmic expressions.

1.5.1 Meaning of logarithm

We discussed some of the fundamental ideas of exponents in Section 1.4. Logarithms are closely related to exponents. Assume that we have two numbers, 5 and 25. We know that these two are related by the equation $5^2 = 25$. Now we may define the exponent 2 as the *logarithm* of 25 to the base of 5. We write this as $\log_5 25 = 2$ and is read "logarithm to the base 5 of 25 is 2." Similarly, we may write $\log_5 125 = 3$, $\log_2 4 = 2$, $\log_3 27 = 3$, and so on. In general, if $b^x = y$ then one may state that the exponent x is the logarithm of y to the base b, and is written as

$$\log_b y = x \tag{1.5.1}$$

It should be clear by now that logarithm is the power (x) to which a base (b) must be raised to obtain y. This is precisely what we did in the case of the equation $5^2 = 25$ above. Hence one may write the expression that shows the relationship between exponents and logarithms as

$$y = b^x \text{ is equivalent to } \log_b y = x \tag{1.5.2}$$

1.5.2 Common and natural logarithms

One may use any positive number for the base, b. But, in practice, the widely used bases are 10 and 'e'. Logarithm to the base 10 is called *common logarithm* and is denoted by "log" or "log$_{10}$." Logarithm to the base e is called *natural logarithm* and is denoted by "ln" or "log$_e$." Finding logarithms is not a time-consuming problem in our digital age as most scientific calculators have specific buttons for both common and natural logarithms.

Students of economics, business, and finance might have noticed that common logarithms are mainly used for computational purposes whereas natural logarithms are frequently applied in analytical or theoretical areas. Common logarithms usually used for computational purposes include

$$\log_{10} 0 = \text{undefined}, \quad \log_{10} 0.001 = -3, \quad \log_{10} 0.01 = -2, \quad \log_{10} 0.1 = -1,$$

$$\log_{10} 1 = 0, \quad \log_{10} 10 = 1, \quad \log_{10} 100 = 2, \quad \log_{10} 1000 = 3, \text{ and so on} \qquad (1.5.3)$$

Notice that common logarithm of any number between 0 and 1 is negative and that of any number above 1 is positive. Notice also that logarithm of a number below 0 is not defined.

Similarly, natural logarithms usually used for analytical purposes include

$$\log_e e^{-1} = \ln \frac{1}{e} = -1, \quad \log_e 0 = \text{undefined}, \quad \log_e e^0 = \ln 1 = 0, \quad \log_e e^1 = \ln e^1 = 1,$$

$$\log_e e^2 = \ln e^2 = 2, \quad \text{and so on} \qquad (1.5.4)$$

There are a few important points to be noted here. First, like common logarithm, natural logarithm is not defined for 0 and the numbers below 0. Second, natural logarithm of an expression e^n, where n is any real number, is negative if n is negative, is 0 if n is 0, and is positive if n is positive. Third, natural logarithm of an expression e^n, where n (as before) is any real number, is n.

1.5.3 Properties of logarithms

We list below some of the important properties of logarithms.[4]

Property I.   Logarithm of a product is equal to the sum of the logarithm of the factors:

$$\ln(x \times y) = \ln x + \ln y \quad \text{where } x, y > 0 \qquad (1.5.5)$$

Property II.   Logarithm of a ratio is the difference between the logarithms of its numerator and denominator:

$$\ln(x/y) = \ln x - \ln y \quad \text{where } x, y > 0 \qquad (1.5.6)$$

Property III.   Logarithm of a power or exponent is the power or exponent times the logarithm of the base:

$$\ln x^n = n \times \ln x \quad \text{where } x > 0 \qquad (1.5.7)$$

Property IV.   Logarithm of the reciprocal of a number is the negative of the logarithm of the number:

$$\ln(1/n) = -\ln n \quad \text{where } n > 0 \qquad (1.5.8)$$

Property V.   Logarithm of 1 is always 0:

$$\ln 1 = 0 \tag{1.5.9}$$

Property VI.   Logarithm of a number to the same base is equal to 1:

$$\ln e = 1 \tag{1.5.10}$$

Property VII.   Logarithm of any number to which e is raised is equal to that number:

$$e^{\ln x} = x \quad \text{where } x > 0 \tag{1.5.11}$$

Property VIII.   Let there be two bases a and b such that $a, b > 0$. Then

$$\log_b n = (\log_a n)/(\log_a b) \text{ or } \log_a b \times \log_b n = \log_a n \tag{1.5.12}$$

Property IX.   Let $a > 0$. Then

$$\log_a e = 1/\log_e a \tag{1.5.13}$$

Equations (1.5.12) and (1.5.13) are, respectively, called the *change of base* and the *inversion of base* formulas of logarithms.

1.5.4 Exercises

1. Convert the following logarithmic (natural logarithmic) forms into their equivalent exponential (natural exponential) forms:
    (i) $\log_3 9 = 2$; (ii) $\log_{125} 5 = 1/3$; (iii) $\log_{81} 9 = 1/2$; (iv) $\log_b x = 2$; (v) $\ln 15 = 2.71$;
    (vi) $\ln x = 2.71$; (vii) $\ln y = (1/5)$; (viii) $\ln x = y - w$.
2. Convert the following exponential (natural exponential) forms into their equivalent logarithmic (natural logarithmic) forms:
    (i) $25 = 5^2$; (ii) $625 = 5^4$; (iii) $3 = 27^{1/3}$; (iv) $2 = 64^{1/6}$; (v) $2.66 = e^{0.98}$; (vi) $x = e^{0.98}$;
    (vii) $1.131 = e^{0.123}$; (viii) $y = e^{x+2}$.
3. Find the values of the following logarithms:
    (i) $\log_{10} 1000$; (ii) $\log_{10} 0.001$; (iii) $\log_2 32$; (iv) $\log_3 81$; (v) $e^{\ln 5}$; (vi) $\ln e^{\ln 1}$;
    (vii) $e^{\ln 5} - e^{\ln 5}$; (viii) $\ln(e^{\ln e})$.
4. Evaluate (or simplify) the following expressions using the properties of logarithms:
    (i) $\log(5x^2)$; (ii) $\log(5x/y^2)$; (iii) $\log \sqrt[3]{125}$; (iv) $\log(10 \times 1/10)$; (v) $\log 100^{1/2}$;
    (vi) $\ln e^{\ln e}$; (vii) $\ln e^{\ln e^0}$; (viii) $\ln(x.y^{1/2})$; (ix) $\ln(e^2 - e)$; (x) $\ln(1/e)$; (xi) $(\log_5 e)(\log_e 25)$;
    (xii) $(\ln 5)(\log_{25} e)$; (xiii) $\log 10^x = \log 1$; (xiv) $3\ln(1/2)$.

  Web supplement: S1.5.5 Mathematica applications

1.6 Equations

Many of the relationships in economics, business, and finance are quantitative in nature. This nature helps these subjects apply various mathematical tools in their analysis of relationships.

One such tool is the concept of equations. Economists and the students of economics deal with a variety of equations such as the demand equation, supply equation, price equation, national income equation, etc. Business managers often engage with the profit equation, revenue equation, cost equation, etc. Students of finance also make use of equations such as the discount and compound formulas, break-even equations, etc. Therefore, students of these subjects require a good understanding of the meaning of equations, their types, their geometric forms, their determinations and solutions, etc. This section helps build this understanding.

1.6.1 Meaning of equations

Let there be two expressions: $y - 5$ and $2x$. An equation is defined as a statement that two expressions are equal, and the two statements are separated by an equals the sign '$=$'. Therefore, if we write $y - 5 = 2x$ it is an equation. The left-hand side of this equation $(y - 5)$ is called the LHS and the right-hand side $(2x)$ is called the RHS. A few examples of equations are $2x - 5 = 9$, $y_i = 5 + 2x_i$, $z_i = 2 + 3x_i + 4y_i$, $y_i = a + bx_{i1}$, and $z_i = a + bx_i + cy_i$. Notice the subscript "i" (or, subsequently, letters or numbers) in these equations. These subscript notations are temporarily used to distinguish equations from functions, which will be clear from the discussion at the end of Section 1.8.1.

We shall first review some terms related to equations. For example, consider the equation $2x - 5 = 9$. In this equation, there is only one *unknown* quantity x. Therefore, this is an equation with one unknown. Now consider the equation $y_i = 5 + 2x_i$. There are two unknown quantities in this equation: x and y. Notice that in this equation when x changes, y also changes. Therefore, x and y are called *variables*. Since there is only one variable on the RHS (x), this is called a *one-variable equation*, or a *single-variable equation*, or a *univariate equation*. The equation $z_i = 2 + 3x_i + 4y_i$ is a *multivariable equation* because there are two variables on the RHS. Notice that in the last two equations there is only one variable on the LHS. This variable on the LHS is called the *dependent variable*. The reason for this name is that it depends on the value(s) of the variable(s) on the RHS. The variable (s) on the RHS is (are) called the *independent variable(s)* since it (they) does (do) not depend on any other variable in the equation.

When we work with equations, our aim is, among others, to solve the equations or to find the solution(s) to the equations. When we say "to solve" an equation or "to find the solution" to an equation what we mean is to determine a number(s) or value(s) for the unknown quantity (quantities) that satisfies (satisfy) the equation. The solution(s) to an equation is (are) called the root(s) of that equation. Consider, for example, $2x - 5 = 9$. This equation can be written as $2x = 9 + 5 = 14$. Dividing both the LHS and RHS of this equation by 2, we obtain $x = 14/2 = 7$. This means that there is only one number or value that satisfies this equation and it is 7. Therefore, 7 is the solution to this equation.

Sometimes, even in the case of an equation with one unknown, there may be more than one number that satisfies or solves the equation. For example, consider the equation $x^2 = 4$. This means that $x = \pm\sqrt{4} = \pm 2$. That is, both $+2$ and -2 satisfy the equation, and there are two solutions. The set of all solutions that satisfy an equation is called the *solution set*.

There are a few other terms related to equations that need to be touched upon. One is the *constant* of an equation. A constant term is the fixed term in an equation. Consider the equation $y_1 = 5 + 2x_1$. In this equation 5 and 2 are constants. In the equation $y_i = 2x_i$, there is only one constant and it is 2. But, in $z_i = 2 + 3x_i + 4y_i$, there are three constants and they are 2, 3, and 4. These examples suggest that an equation may contain one or more than one constant.

Another important term is the *intercept* of an equation. There are two intercept terms in a univariate equation: x-intercept and y-intercept. The former is the point where the graph of the equation crosses the x-axis (that is, when $y = 0$) and the latter is the point where the graph crosses the y-axis (that is, when $x = 0$). We often work with the y-intercept. In the equation $y_i = 5 + 2x_i$, the y-intercept is 5 (when $x = 0$); and in the equation $z_i = 2 + 3x_i + 4y_i$, the z-intercept is 2 (when $x_i = y_i = 0$). Notice that some equations such as $y = 2x$ will not have a positive y-intercept and some equations such as $y_i = -5 + 3x_i$ will have a negative y-intercept. Notice also that the graph of the former will pass through the origin (that is, the value zero) and the graph of the latter will pass through the negative value -5 on the y-axis.

The last term in connection with equations often found in the literature is the *parameter(s)*. This can be explained with an example. Consider two equations: $y_i = 5 + 2x_i$ and $y_i = a + bx_i$. If we substitute 5 for a and 2 for b into the second equation, we obtain the first equation. This means that the latter equation is more general than the former equation. That is, the latter general equation can represent a number of specific equations like the former one. Notice that in this general equation the values of the constants a and b are not specified. Such constants are called parameters. One of the important objectives in a branch of applied mathematics called *statistics* is to determine the values of these parameters.

1.6.2 Types of single-variable or univariate equations

Equations can be classified as shown in Figure 1.6.1. We shall discuss each of these equations below. Let us begin with polynomial equations. A *polynomial equation* is defined as an equation in which there is more than one term, with different powers, of an independent variable. An example of a general polynomial equation is

$$y_n = b_0 x^0 + b_1 x^1 + b_2 x^2 + b_3 x^3 + \cdots + b_n x^n \tag{1.6.1}$$

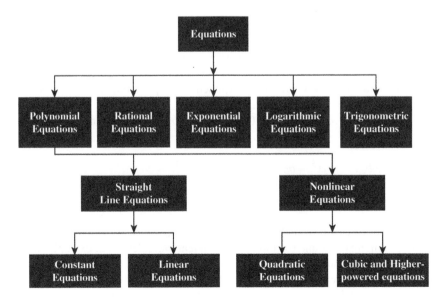

Figure 1.6.1

Notice that in equation (1.6.1) we have only one independent variable x. But, x appears with different powers: from 0, through 1, to n. Depending on the value of n (the highest power in equation (1.6.1)), we obtain two subsets of the polynomial equation. The first subset comprises *constant equations* and *linear equations*, and the second subset comprises *nonlinear equations* of *quadratic equations* and *cubic equations* (and *higher-powered equations*). We will later discuss the distinguishing features of linear and nonlinear equations. Notice that the polynomial equation (1.6.1) consists of a number of terms on the RHS and each term is a *monomial*. The degree of the polynomial is the highest value of the exponent (n) in it.

A constant equation can be obtained if we set the powers to zero and $b_0 \neq 0$ in equation (1.6.1), and is given as

$$y_0 = b_0 x^0 = b_0 \tag{1.6.2}$$

Notice that equation (1.6.2) contains no independent variable on the RHS, but only the constant b_0. What this means is that the dependent variable will be equal to this value of the constant irrespective of the value taken by the independent variable. Notice also that this equation is a *zero-degree polynomial* because $n = 0$.

A linear equation can be derived from equation (1.6.1) if we set $n = 1$, $b_0 \neq 0$, and $b_1 \neq 0$, and can be written as

$$y_1 = b_0 x^0 + b_1 x^1 = b_0 + b_1 x \tag{1.6.3}$$

Notice that in equation (1.6.3) there is one independent variable (x) with power 1 and two constants b_0 and b_1. Since there are two terms on the RHS of this equation, it is a *binomial*. And since the highest exponent in the equation is 1, it is called a *first-degree polynomial*.

We can derive a *quadratic equation* if we set $n = 2$, $b_0 \neq 0$, $b_1 \neq 0$, and $b_2 \neq 0$ in equation (1.6.1) as

$$y_2 = b_0 x^0 + b_1 x^1 + b_2 x^2 = b_0 + b_1 x + b_2 x^2 \tag{1.6.4}$$

Notice that, since there are three terms on the RHS of equation (1.6.4), this equation is a *trinomial*. Besides, since the highest value of the exponents in the equation is 2, a quadratic equation is also called a *second-degree polynomial*. A second-degree polynomial is also a nonlinear equation. If the highest value of the exponents in a polynomial is n, then that polynomial is called an *nth-degree polynomial*.

Similarly, if we specify $n = 3$, $b_0 \neq 0$, $b_1 \neq 0$, $b_2 \neq 0$, and $b_3 \neq 0$ in equation (1.6.1), we obtain the *cubic equation* as

$$y_3 = b_0 x^0 + b_1 x^1 + b_2 x^2 + b_3 x^3 = b_0 + b_1 x + b_2 x^2 + b_3 x^3 \tag{1.6.5}$$

The reader must have noticed that a cubic equation is a *third-degree polynomial* and that it is a nonlinear equation.

Another important set of equations is the set of *rational equations*. A rational equation is the ratio of two polynomials in one (or more) independent variable(s). An example of a rational equation in one independent variable is $y_i = (3 + 2x_i)/(2 - 5x_i)$. One of the simplest cases of rational equations is the equation $y_i = b/x_i$.

Exponential equation and *logarithmic equation* are two of the other important types of equations widely used in economics, business, and finance. We have already discussed the exponential and logarithmic expressions in Sections 1.4 and 1.5, respectively. We provide below only the general forms of exponential and logarithmic equations, respectively, in equations (1.6.6) and (1.6.7). Notice that these equations are nonlinear equations:

$$y_i = a^{x_i} \tag{1.6.6}$$

$$y_i = \log_a x_i \tag{1.6.7}$$

1.6.3 Solution of equations

In most applications, we need to solve equations. The reader will have noticed that there is no need to solve a constant equation as the RHS is a constant term and its value is given. But, in the case of all other equations, we may need to solve for the unknowns. We begin this discussion with a linear equation. For example, assume that our linear equation is of the form $2x + 7 = 17 - 8x$. In order to solve this equation, collect the like terms on one side applying the required changes in the signs. After doing this we obtain $2x + 8x = 17 - 7$. Now add the like terms, and the result is $10x = 10$. Then divide both sides of this equation by 10 and the result will be $x = 1$. Notice that the last equation is now in the form of a constant equation.

We can now consider the solution of a quadratic equation. The quadratic equation (1.6.4), by treating $y_2 = 0$, $b_0 = c$, $b_1 = b$, and $b_2 = a$, is generally written as

$$ax^2 + bx + c = 0 \tag{1.6.8}$$

where $a(= b_2)$, $b(= b_1)$, and $c(= b_0)$ are constants and $a \neq 0$. Notice that a quadratic equation may possess no *real solution*, one real solution, or two different real solutions while a linear equation possesses only one solution. A quadratic equation can be solved either by *factoring* or by applying the *quadratic formula*.

For example, assume that we have a quadratic equation of the form $x^2 - 3x - 10 = 0$. Let us first solve this equation by factoring. The LHS of this can be written as the product of two factors: $(x + 2)$ and $(x - 5)$ or $x^2 - 3x - 10 = (x + 2) \times (x - 5) = 0$. The RHS is still the same, 0. For the LHS to equal the RHS, at least one term or both terms on the LHS must be zero. This is possible only when $x = -2$ or $x = +5$, or $x = -2, +5$. Therefore, the roots of (or the solutions to) the above quadratic equation are -2 and 5, and the solution set is $x = \{-2, +5\}$.

We can apply the quadratic formula

$$x = \frac{-b \pm \sqrt{b^2 - 4ac}}{2a} \tag{1.6.9}$$

to find the roots of a quadratic equation. One advantage of the quadratic formula is that it is useful when factoring is difficult. One can derive equation (1.6.9) as follows. Notice that equation (1.6.8) may be written as $a[x^2 + (b/a)x + (c/a)] = 0$. Since $a \neq 0$, the last equation has the same solution as that of $x^2 + (b/a)x + (c/a) = 0$, which may also be written as $x^2 + (b/a)x = -(c/a)$. Adding $(b/2a)^2$ to both sides of the last equation yields $x^2 + (b/a)x + (b/2a)^2 = -(c/a) + (b/2a)^2$, or $[x + (b/2a)]^2 = [b^2 - 4ac]/4a^2$, or $x = [-b \pm \sqrt{b^2 - 4ac}]/2a$, which is the same as equation (1.6.9).

We shall now solve the quadratic equation discussed above, $x^2 - 3x - 10 = 0$, using the quadratic formula. Notice that, in our example, $a = 1$, $b = -3$, and $c = -10$. Substituting these values into equation (1.6.9) we obtain, after simplification, the set of solutions $x = [-(-3) \pm \sqrt{3^2 - 4 \times 1 \times (-10)}]/2 \times 1 = [3 \pm \sqrt{49}]/2 = [3 \pm 7]/2 = \{10/2 = 5, -4/2 = -2\}$, which are precisely the solutions we obtained through factoring.

We now consider the solution of a cubic equation. Since a cubic equation is a third-degree polynomial as in equation (1.6.5), it is quite natural that it is relatively more difficult to solve such equations. As an example, we consider the simple cubic equation with $y_3 = 0$, $b_0 = 0$, $b_1 = 1$, and $b_2 = -2$, and $b_3 = 1$: $x^3 - 2x^2 + x = 0$.

As the first step in solving this equation, we may factor one x out of it. Then we obtain $x(x^2 - 2x + 1) = 0$. For the LHS to be zero, either $x = 0$ or $(x^2 - 2x + 1) = 0$, or both must be equal to zero. Therefore, one root of the above cubic equation is 0. We can now use either factoring or the quadratic equation to find the roots of $(x^2 - 2x + 1) = 0$. Factoring the LHS of the equation yields $(x - 1)(x - 1)$. This suggests that the remaining two roots are $+1$ and $+1$. Therefore, the solution set of the above cubic equation is $x = \{0, 1, 1\}$.

It is quite simple to solve rational equations such as $y_i = b/x_i$. Assume that $b = 10$ and $y_i = 5$, then the rational equation is of the form $5 = 10/x$. This rational equation can be solved by cross multiplying and dividing both sides of the equation by x and 5, respectively, and then simplifying the resulting expression as $x = 10/5 = 2$. It is needless to mention that more complicated forms of rational equations will be harder to solve. Since we have already considered the solutions of exponential and logarithmic equations in detail in Sections 1.4 and 1.5, respectively, we do not repeat them here.

1.6.4 Determination of the equation of a straight line

Students of economics, business, and finance often work with straight lines whose general, algebraic form is given in equation (1.6.3). But, how do we determine the equation of a straight line in the first place? Generally, there are four ways one can determine the equation of a straight line.

The first and the simplest way of determining the equation of a straight line is the use of *literal description*. To explain this let us consider the description of a firm's supply of a commodity. Assume that the firm requires a minimum amount of $1 to begin the supply of the commodity. In addition to this, the firm decides to charge a constant amount of $2 on each unit of the commodity sold. How can we write the equation of the price set by the firm (or, in other words, the *supply equation*)?

In order to sell the first unit of the commodity the firm has to receive $1 + $2 \times 1 = $3. The firm has to receive $1 + $2 \times 2 = $5 to sell the second unit. It has to receive $1 + $2 \times 3 = $7 to sell the third unit. This means that the price (y) increases by $2 when the firm sells each additional unit of the commodity (x). This implies that we can write the supply equation as

$$y = 1 + 2x \tag{1.6.10}$$

Notice that we have dropped the subscript notations for convenience. A comparison of the general linear equation (1.6.3) with equation (1.6.10) shows that $b_0 = 1$ and $b_1 = 2$. Notice also that equation (1.6.10) is a binomial and a first-degree polynomial equation.

The other methods of the determination of the equation of a straight line require that we understand the meaning of the term *slope*. One of the important features of a straight line (or any other curve) is its steepness. The concept of slope represents the measure of this steepness. We discuss here the concept of slope with the help of the graph of equation (1.6.10) as shown in Figure 1.6.2. As can be seen from the figure, the line representing the linear equation $y = 1 + 2x$ goes upward from left to right. Two points on this line are

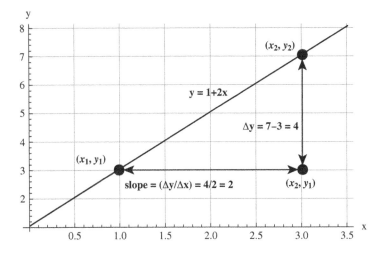

Figure 1.6.2

represented by the points (1, 3) and (3, 7), which for convenience we denote by (x_1, y_1) and (x_2, y_2), respectively. Notice that (x_1, y_1) represents the initial (or the original) point and (x_2, y_2) represents the new (or the changed) point. The difference between the new $y(= y_2)$ and the initial $y(= y_1)$ points is called the *vertical change* denoted by $\Delta y = y_2 - y_1$. Similarly, the difference between the new $x(= x_2)$ and the initial $x(= x_1)$ points is called the *horizontal change* denoted by $\Delta x = x_2 - x_1$. We now define slope as the *ratio of vertical change to horizontal change*. In mathematical notation this definition takes the form

$$\text{Slope} = b_1 = \frac{\text{Vertical change}}{\text{Horizontal change}} = \frac{\Delta y}{\Delta x} = \frac{y_2 - y_1}{x_2 - x_1} \tag{1.6.11}$$

Substituting $y_2 = 7, y_1 = 3, x_2 = 3$, and $x_1 = 1$ into equation (1.6.11) yields

$$\text{Slope} = b_1 = \frac{\text{Vertical change}}{\text{Horizontal change}} = \frac{\Delta y}{\Delta x} = \frac{y_2 - y_1}{x_2 - x_1} = \frac{7 - 3}{3 - 1} = \frac{4}{2} = 2$$

This means that the slope of the straight line in Figure 1.6.2 is 2. Notice that the coefficient of x in equation (1.6.10) is also 2. Therefore, we may generalize that the slope of a straight line is the same as the coefficient of the independent variable in the equation that represents the straight line.

The second way to determine the equation of a straight line is to use the notion of *point–slope form*. Assume that there exists a straight line L. Assume also that the slope of this line is b_1 and that two points on this line are (x_1, y_1) and (x, y). We can now find the equation for L if we know b_1 and (x_1, y_1). This is the idea behind the notion of the point–slope form. The procedure can be explained as follows. We know from equation (1.6.11) that $b_1 = (y - y_1)/(x - x_1)$ or $y - y_1 = (x - x_1)b_1$. This equation is the point–slope form of L. We can use this idea to find the equation of a line that passes through $(x_1, y_1) = (1, 3)$ and $b_1 = 2$. Substituting these values into the above point–slope form, we obtain the required equation as $y - 3 = (x - 1) \times 2 = 2x - 2$ or $y = 2x - 2 + 3 = 1 + 2x$.

The third way is useful when we have only two points instead of a point and slope as above. How can we then find the equation of the line? Let the two points on the

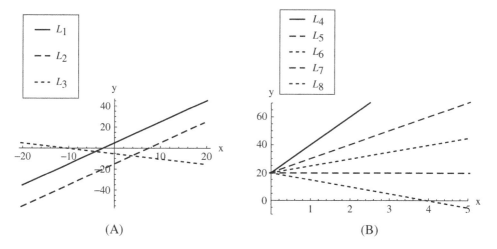

Figure 1.6.3

straight line be $(x_1, y_1) = (1, 3)$ and $(x_2, y_2) = (3, 7)$. Then using equation (1.6.11) we obtain $b_1 = (y_2 - y_1)/(x_2 - x_1) = (7 - 3)/(3 - 1) = 4/2 = 2$. Having obtained the slope $b_1 = 2$, we may apply the point–slope form in equation (1.6.11) with the points $(x_1, y_1) = (1, 3)$ and $(x_2, y_2) = (x, y)$. Thus, we obtain $y - 3 = (x - 1) \times 2$. This simplifies to $y = 2x - 2 + 3 = 1 + 2x$, which is the same as equation (1.6.10).

The last method of determining the equation of a straight line is to use the *slope–intercept form*. We know that b_0 represents the y-intercept of the linear equation (1.6.3). If we are given b_0 and the slope b_1, how can we find the equation of the line? The answer is straightforward: apply equation (1.6.3). Let $b_0 = 1$ and $b_2 = 2$. Then application of equation (1.6.3) yields $y = b_0 + b_1 x = 1 + 2x$, which the required equation of the line.

Notice that there are two types of straight lines in the literature: *parallel lines* and *perpendicular lines*. Let there be two straight lines L_1 and L_2, and let their respective slopes be $b_{1,1}$ and $b_{1,2}$. These two lines are parallel to each other if and only if $b_{1,1} = b_{1,2}$. And they are perpendicular to each other if and only if $b_{1,1} = -1/b_{1,2}$ or $b_{1,2} = -1/b_{1,1}$. As an example of parallel and perpendicular lines, suppose that $L_1 = 5 + 2x$, $L_2 = -15 + 2x$, and $L_3 = -5 - 0.5x$. As the graphs of these equations in Figure 1.6.3(A) show, L_2 is parallel to L_1 and L_3 is perpendicular to L_3 (or, to L_2).

By this time the reader might have guessed that the steepness of a straight line, for a given intercept, depends upon the slope of the line. The higher the slope, the steeper the line will be, and vice versa. This can be verified by the graphs of equations $L_4 = 20 + 20x$, $L_5 = 20 + 10x$, $L_6 = 20 + 5x$, $L_7 = 20$, and $L_8 = 20 - 5x$ illustrated in Figure 1.6.3(B). Notice that these lines are drawn with the same intercept (5).

1.6.5 Systems of linear equations

The study of most branches of economics, business, and finance often starts with models. Many of these models contain more than one linear equation in two or more unknowns. These equations constitute, when taken together, a *system of simultaneous linear equations* (SSLEs). A system of equations containing n variables and m equations is called an $n \times m$ (read "n by m") system or $n \times m$ SSLEs.

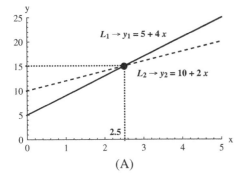
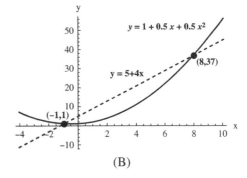

Figure 1.6.4

We have already discussed the meaning of the solution of an equation in Section 1.6.3. An $n = m$ system may have a *unique solution* (that is, only one solution for each of the variables), an *infinite solution*, or no solution. An $m < n$ system will give *multiple solutions* (unlimited) or no solution (and never a unique solution). An $n < m$ system may give solutions as those of an $n = m$ system.

One can present many examples of an $n = m = 2$ system. But, consider the general 2×2 system of two equations in two unknowns

$$y = b_0 + b_1 x \quad \text{and} \quad y = b_2 + b_3 x \tag{1.6.12}$$

It can be shown that if $b_0 \neq b_2$ and $b_1 \neq b_3$ in the system (1.6.12), there will be a unique solution to the system; if $b_0 = b_2$ and $b_1 \neq b_3$, there is no unique solution; and if $b_0 = b_2$ and $b_1 = b_3$, the two equations coincide (that is, they are equivalent) and there is an infinite number of solutions.

Let us now attempt to solve a 2×2 SSLEs. As an example, we consider the simple system $y = 10 + 2x$ and $y = 5 + 4x$. Generally there are four methods to solve this system.[5] The first one we discuss is the *graphical method*. The first step in this method is to draw the graphs representing the equations. These are given as L_1 and L_2 in Figure 1.6.4(A). The next step is to draw a straight line from the point of intersection of L_1 and L_2 to the x-axis to obtain the *x-coordinate*. The third, and the last, step is to draw a straight line from the same intersection point to the y-axis to obtain the *y-coordinate*. These coordinates will be the solution of the system. In our example, as shown in Figure 1.6.4(A), the solutions are $x = 2.5$ and $y = 15$.

The second method of solving a system of linear equations is the *substitution method*. The first step in this method of solving a 2×2 linear system, with x and y as the unknowns, is to solve for one of the unknowns (e.g. x). Then this solution of x is substituted back to solve for y. Notice that this method will become more and more cumbersome as the size of the system increases. In order to carry out the procedure just outlined, we again consider the example of our previous 2×2 system: $y = 10 + 2x$ and $y = 5 + 4x$.

As stated above, solving for x from the first equation gives $2x = y - 10$ or $x = (y/2) - 5$. Substituting this solution of x into the second equation yields $y = 5 + 4[(y/2) - 5] = 5 + (4y/2) - 20 = (4y/2) - 15$, or $y - (4y/2) = -15$. Taking the *least common denominator* (LCD) of the last equation we get $[(2y - 4y)/2] = -15$, and by cross multiplying it by 2 we obtain $2y - 4y = -30$. Simplifying this yields $-2y = -30$ or $y = 15$. The solution $y = 15$ can be substituted back to the solution for $x = (y/2) - 5$ to obtain $x = (5/2) - 5 = 7.5 - 5 = 2.5$, which are precisely the solutions we obtained from the graphical method.

The third method of solving a 2×2 SSLEs we discuss here is the *elimination method*. This method, as the name suggests, eliminates one of the unknowns in the system. Once again, for exposition, we use the above system and eliminate x. The first step in this method is to make equal the coefficients of x in the two equations by multiplying every term in one equation by the coefficient of x in the other equation and, then, from the resulting equations, subtract one equation from the other. The procedure is as follows:

$$\begin{array}{llll} y = 10 + 2x & 4y = 40 + 8x & 4y = 40 + 8x & 4y - 2y = 30 \\ \quad\quad\quad \text{or} & \quad\quad\quad \text{or} & \quad\quad\quad \text{or} & \\ y = 5 + 4x & 2y = 10 + 8x & 2y = 10 + 8x & 2y = 30 \end{array}$$

Notice that we obtain $y = 15$ from the second equation in the last column above. The next and the last step in this method is to substitute the solution $y = 15$ back into one of the original equations, say $y = 5 + 4x$. This gives us $15 = 5 + 4x$, or $15 - 5 = 4x = 10$, which simplifies to $x = (10/4) = 2.5$. Notice, again, that these are the same solutions as those we obtained from both graphical and substitution methods.

So far in this section we have dealt with the 2×2 SSLEs. There are plenty of instances in the subjects of our interest that a student encounters models involving *higher-order systems*. Now imagine a system with $n = 3$ and $m = 3$ (that is, three variables and three equations). The general form of a 3×3 SSLEs is

$$z = b_0 + b_1 x + b_2 y, \quad z = b_3 + b_4 x + b_5 y, \quad \text{and} \quad z = b_6 + b_7 x + b_8 y \tag{1.6.13}$$

As a concrete example of a 3×3 SSLEs, consider the system

$$z = 6 - 2x - y, \quad z = 2 - x - 2y, \quad \text{and} \quad z = 2 - 0.5x - 0.5y \tag{1.6.14}$$

which can be solved to obtain $x = 3$, $y = -1$, and $z = 1$.

1.6.6 Nonlinear equations and their systems

Many of the models in economics, business, and finance involve *nonlinear equations* in two or more variables or unknowns. Therefore, students of these subjects must possess at least a basic knowledge of these equations, their systems, and their solutions. An equation is a nonlinear equation if it is a logarithmic or exponential equation, or if the power(s) of the independent variable(s) is other than 0 or 1. We have already introduced some of these equations earlier and we will deal with them in greater detail in Chapter 3.

A *nonlinear system of equations* is defined as a system of equations in which either all or at least one equation is nonlinear. The general form of a nonlinear system of two equations in two variables is given as

$$y = b_0 + b_1 x + b_2 x^2 \quad \text{and} \quad y = b_3 + b_4 x \tag{1.6.15}$$

Notice that the first equation in this system is a quadratic equation and the second one is a linear equation. Therefore, the system is a nonlinear system of equations. As a specific example, consider the simple system $x^2 + x + 2 = 2y$ and $4x + 5 = y$. The first step in solving this system is to substitute the value of y from the second equation into the first equation. The resulting expression will be $x^2 + x + 2 = 2(4x + 5)$ or $x^2 + x + 2 = 8x + 10$ or $x^2 - 7x - 8 = 0$. The last expression is a quadratic equation. As we discussed earlier in Section 1.6.3, we can resort either to factorization or to the quadratic formula to find the solution of this equation. Without repeating the algebraic manipulations, we simply write that the roots of this equation are 8 and -1. Therefore, the solution set is $x = \{8, -1\}$. The next, and the last step is to substitute these values in any one of the original equations. We choose the simplest equation for this: $4x + 5 = y$. The result, after simplification, will be the solution set $y = \{37, 1\}$. Therefore, the solutions to this nonlinear system are $x = (8, -1)$ and $y = (37, 1)$. The graphs of this nonlinear system are illustrated in Figure 1.6.4(B) which confirms our result.

The system in equation (1.6.15) involves only one independent variable. How about a single equation that involves more than one unknown? Assume that an equation involves unknowns x, y, and z, and is given by $z = x^2 - y^2$. This equation has two independent variables, x and y, and is nonlinear. Now assume that we have a nonlinear system of three equations in three unknowns, x, y, and z, and is given by

$$z = x^2 - y^2, \quad z = x^2 - y, \quad \text{and} \quad z = x - y^2. \tag{1.6.16}$$

The above system of nonlinear equations can be solved to obtain, among others, $(x = 1, y = 1,$ and $z = 0)$ and $(x = 0, y = 0,$ and $z = 0)$.

1.6.7 Application examples

Example 1. Assume that in a market a firm can sell any quantity of a commodity produced by it at the existing price of \$5 per unit. This price is equal to the *average revenue* (AR) and the *marginal revenue* (MR) to the firm from the sale of the commodity. Show these two statements in a graph.

Solution. Since the price (\$5) remains the same irrespective of the quantity of the commodity sold, the graph of the price (= AR = MR) must be a straight line parallel to the x-axis and with slope equal to zero. This graph is illustrated in Figure 1.6.5(A).

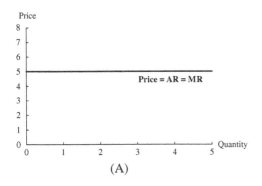

(A)

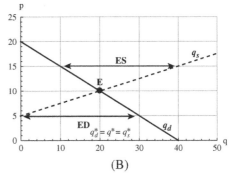

(B)

Figure 1.6.5

Example 2.   Suppose that a consumer purchases 20 units of apples when price (p) is zero and reduces the quantity of apples purchased (q) by 2 units for every unit increase in price. Derive the consumer's *demand equation* for apples.

Solution.   We are given that when $p = 0$, $q = 20$ and q diminishes by 2 units as p increases by 1 unit. This implies that when $p = 1$, $q = 18$; when $p = 2$, $q = 16$; when $p = 3$, $q = 14$; and so on. Therefore, the required demand equation can be written as $q = 20 - 2p$, or the *inverse demand equation* as $p = 10 - 0.5q$.

Example 3.   Assume that a straight line supply equation for a good has a slope of 5 and passes through point (5, 60). Find the supply equation.

Solution.   Let y denote the price of the good and x the quantity of the good supplied. We may use the point–slope form $[(y - y_1) = b_1(x - x_1)]$ to solve this problem. Assuming that $y_1 = 60$, $x_1 = 5$, and substituting 5 for b_1, we can write the point–slope form as $y - 60 = 5(x - 5)$, which simplifies to $y = 35 + 5x$. This is the equation of the supply curve, or the required supply equation.

Example 4.   Suppose that the straight line of a demand equation passes through points (10, 20) and (5, 30). Find the demand equation.

Solution.   Let p denote the price of the good and q the quantity demanded of the good. Once again we can use the point–slope form to solve this problem. Since the points are given, first we have to find out the slope to use the point–slope form. This can be found from the equation $b_1 = (p - p_1)/(q - q_1)$. Substituting $p = 20$, $p_1 = 30$, $q = 10$, and $q_1 = 5$ into this equation, we obtain slope $b_1 = -2$. Now using b_1, p_1, and q_1, we obtain the demand equation as $q = 20 - 0.5p$ or the inverse demand equation as $p = 40 - 2q$.

Example 5.   Assume that the demand and supply equations of a good are given, respectively, by $q_d = 40 - 2p$ and $q_s = -20 + 4p$, where p represents the price per unit of the good in dollars, q_d stands for the quantity of the good demanded, and q_s denotes the quantity sold. Find the *equilibrium price* and *equilibrium quantity*, and the *excess supply* (ES) or *surplus* and *excess demand* (ED) or *shortage*. Show these equilibrium and *disequilibrium* conditions with the help of a figure.

Solution.   *Equilibrium* is a state in which the variables (p, q_d, and q_s in the present example) do not have a tendency to change. In other words, this is the point in a graph where the lines of demand and supply equations intersect. The price, quantity demanded, and quantity supplied at this point of intersection are the equilibrium price, equilibrium quantity demanded, and equilibrium quantity supplied and are denoted by p^*, q_d^*, and q_s^*, respectively. Notice that at this point $q_d^* = q_s^*$. This quantity is called the equilibrium quantity denoted by q^* and, therefore, we have $q^* = q_d^* = q_s^*$ at the equilibrium. Therefore, we can obtain p^* and q^* by equating $q_d = 40 - 2p$ and $q_s = -20 + 4p$, and solving for p and $q^* = q_d^* = q_s^*$. The results will be $p^* = \$10$ and $q^* = q_d^* = q_s^* = 20$. This equilibrium is shown in Figure 1.6.5(B).

In our present example, disequilibrium occurs when $q_s > q_d$ or $q_d > q_s$ which happens when $p > p^*$ or $p < p^*$, respectively. That is, any price above the equilibrium price $(p > p^*)$ causes excess supply or surplus $(q_s > q_d)$, and any price below the equilibrium price $(p < p^*)$

causes excess demand or shortage ($q_d > q_s$). For example, when $p = \$15 > p^* = \$10$, $q_s = 40 > q_d = 10$. Therefore, the ES is equal to $q_s - q_d = 40 - 10 = 30$ units. And when $p = \$5 < p^* = \$10$, the ED is equal to $q_d - q_s = 30 - 0 = 30$ units. These disequilibrium states are also shown in Figure 1.6.5(B).

Example 6. Assume that the demand and supply equations of a product of a seller are the same as those in example 5: $q_d = 40 - 2p$ and $q_s = -20 + 4p$, respectively. Also assume that the government imposes an *excise tax* of \$3 on every unit of the product sold. What will be the new equilibrium price and quantity? How much of the per unit tax is borne by the buyer and how much of the per unit tax is borne by the seller? Show the initial and the new equilibrium positions with the help of a figure.

Solution. In the last example we saw that the original equilibrium price and quantity were $p^* = \$10$ and $q^* = 20$ units, respectively. Two points must be noted here. First, an excise tax affects the supply equation only. Second, the price the seller receives after a tax of \$3 per unit of the product will be $p - 3$. Therefore, the new supply equation will be $q_s^t = -20 + 4(p - 3) = -32 + 4p$. Equating this with the original demand equation and simplifying gives $q^{t*} = 16$ and $p^{t*} = \$12$, where q^{t*} and p^{t*} denote the equilibrium quantity demanded and supplied and the equilibrium price after the tax, respectively. Notice that the equilibrium quantity demanded and supplied decreased from 20 units to 16 units and the equilibrium price increased from \$10 to \$12 per unit after the tax. Notice also that though the tax on the seller was \$3, she could pass a burden of only \$2 to the buyer and the rest (\$1) is borne by her. The two equilibrium positions are shown in Figure 1.6.6(A). In this figure, E and E^t represent the pre-tax and post-tax equilibrium positions, respectively.

If the government imposed a sales tax, instead of an excise tax, the impact would again be on the supply equation. In this event, the supply equation would change to $q_s = -20 + 4(1 - t)p$, where t represents the rate of tax. One can equate this new supply equation with the original demand equation and solve for q^{t*} and p^{t*}. It is evident that the impacts of an excise tax and a sales tax are on the intercept and the slope of the supply equation, respectively.

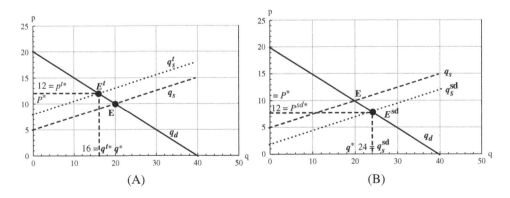

(A)　　　　　　　　　　　　　　　(B)

Figure 1.6.6

Example 7.   Once again assume that the demand and supply equations of a product of a seller are the same as those in example 5: $q_d = 40 - 2p$ and $q_s = -20 + 4p$, respectively. Also assume that the government gives a *subsidy* to the seller of $3. What will be the new equilibrium price and quantity? How much of the subsidy is passed to the buyer and how much of the subsidy is retained by the seller? Show the initial and new equilibrium positions with the help of a figure.

Solution.   In the last two examples we saw that the original equilibrium price and quantity were $p^* = \$10$ and $q^* = 20$ units, respectively. As in the case of tax, two points must be noticed here too. First, a subsidy affects the supply equation only. Second, the price the seller receives after a subsidy of $3 per unit of the product will be $p + 3$. Therefore, the new supply equation will be $q_s^{sd} = -20 + 4(p + 3) = -8 + 4p$. Equating this with the original demand equation and simplifying gives $q^{sd*} = 24$ and $p^{sd*} = \$8$, where q^{sd*} and p^{sd*} represent the equilibrium quantity demanded and sold and the equilibrium price after the subsidy, respectively. Notice that the equilibrium quantity demanded and supplied increased from 20 units to 24 units and the equilibrium price decreased from $10 to $8 per unit after the subsidy. Notice also that though the subsidy to the seller was $3 per unit, she passed only $2 to the buyer and the rest ($1) is retained by her. The two equilibrium positions are shown in Figure 1.6.6(B). In this figure, E and E^{sd} represent the pre-subsidy and post-subsidy equilibrium positions, respectively.

Example 8.   Assume that the *inverse demand curve* for a good is nonlinear and is given by $p = 4/q_d$, and the *inverse supply curve* is linear and is given by $p = 2 + 2q_s$. Find the equilibrium quantity (q^*) and price (p^*), and illustrate this equilibrium in a figure.

Solution.   If we equate the demand and supply equations and simplify (assuming that $q^* = q_d = q_s = q$), we get $2q^2 + 2q - 4 = 0$. Now using the quadratic formula from equation (1.6.9) yields the roots $(-2, 1)$. If we substitute first root (-2) into any of the original equations, we obtain the corresponding solution for price, -2. Since the quantity and price cannot be negative, we discard the negative root, and choose the positive root, 1 (i.e. $q^* = 1$). Substitution of $q^* = 1$ into any of the two original equations yields $p^* = 4$. This equilibrium state is illustrated in Figure 1.6.7(A).

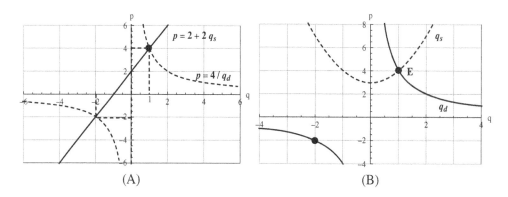

(A)   (B)

Figure 1.6.7

Example 9. Assume, as in example 8, that a good's inverse demand curve is nonlinear and is given by $p = 4/q_d$, and (unlike in example 8) that its inverse supply curve also is nonlinear and is given by $p = 3 + q_s^2$. Find the equilibrium quantity (q^*) and price (p^*), and illustrate this equilibrium in a figure.

Solution. If we equate the demand and supply equations and simplify (assuming that $q^* = q_d = q_s = q$), we obtain $q^3 + 3q - 4 = 0$, which is a cubic equation. Now factoring and using the quadratic formula (equation (1.6.9)) yields the roots $(-0.5 + 1.94i, -0.5 - 1.94i, 1)$. As can be seen from this, except the solution 1, the other solutions are complex numbers. Therefore, we choose the real, positive root, 1 ($q^* = 1$). Substitution of $q^* = 1$ into any of the two original equations yields $p^* = 4$. This equilibrium state is illustrated in Figure 1.6.7(B). Notice that we obtained the same solution in example 8 even when we used a linear supply equation.

1.6.8 Exercises

1. Solve the following equations:
   (i) $8x - 5 = -7 + 10x$; (ii) $x = 1/x$; (iii) $2x^2 + x - 2 = 0$; (iv) $x^3 + x^2 = 0$; (v) $(x/1) + (x^2/2) = 5 - 5$; (vi) $(1/x) - (2/x^2) = 10 - 10$; (vii) $x(x - 5) = 0$.

2. Find the equations of the straight lines that pass through the following points:
   (i) $(4, 3), (5, 2)$; (ii) $(5, 2), (8, 4)$; (iii) $(-3, 4), (8, 6)$; (iv) $(4, -4), (4, 6)$.

3. Find the equations of the straight lines that pass through the following points and have the following slopes:
   (i) $(10, 5)$, and slope $= 2$; (ii) $(-10, 5)$, and slope $= -2$; (iii) $(-3, -5)$, and slope $= 4$; (iii) $(3, 5)$, and slope $= -4$.

4. Find the equations of the straight lines that have the following features:
   (i) slope $= 3$, and x-intercept $= 5$; (ii) slope $= -3$, and y-intercept $= 5$; (iii) slope $= 0$, and y-intercept $= 4$; (iv) slope $= \infty$, and x-intercept $= 4$.

5. Solve the following SSLEs by graphical, substitution, and elimination methods:
   (i) $\begin{aligned} 2x + y &= 5 \\ x + 2y &= 2 \end{aligned}$; (ii) $\begin{aligned} 2x - y &= -5 \\ x - 2y &= -2 \end{aligned}$; (iii) $\begin{aligned} x + 0.5y &= 0 \\ 3x + y &= 4 \end{aligned}$; (iv) $\begin{aligned} 3x + y - 2 &= 0 \\ x - 2y &= 1 \end{aligned}$.

6. Solve the following systems of nonlinear equations:
   (i) $\begin{aligned} x^2 &= y \\ 2x + y &= 0 \end{aligned}$; (ii) $\begin{aligned} x^2 + y &= 0 \\ x + y &= 0 \end{aligned}$; (iii) $\begin{aligned} y &= 2x - x^2 \\ y &= x^2 - 2x \end{aligned}$; (iv) $\begin{aligned} x^2 + y &= 2 \\ x + y^2 &= 2 \end{aligned}$.

7. *Application exercise.* When the per unit price of a good was $6 a consumer purchased 2 units of the good, and when the per unit price was $2 the consumer purchased 6 units. Determine the demand equation.

8. *Application exercise.* Assume that a supply equation passes through points (2, 6) and has a slope of 2. Determine the equation.

9. *Application exercise.* Assume that the market price of a good supplied by a company increased $3 every month. Its price after 10 months was $220. Find the supply equation of this company for this good.

10. *Application exercise.* Suppose that the per unit profit from the sale of a good (x) by a company is $15, its *total fixed cost* is $50, and its *total variable cost* is $5x$. Find the *breakeven* (that is, no profit and no loss) quantity of x sold.

11. *Application exercise.* Assume that the inverse demand and supply equations of a good sold by a firm are $P = 20 - 2Q$ and $P = 2 + Q$, respectively. Also assume that the

government imposes a tax of \$2 on every unit of the good sold. Determine the pre- and post-tax equilibrium prices and quantities, and the portions of the tax borne by the seller and by the buyer after the tax.

12. *Application exercise.* Assume that the inverse demand and supply equations of a good sold by a firm are as those in application exercise 11. Also assume that, instead of imposing a tax, the government now gives a subsidy of \$1 for every unit sold to the seller of the good. Determine the pre- and post-subsidy equilibrium prices and quantities, and the portions of the subsidy received by the seller and the buyer.

13. *Application exercise.* Assume that the total cost and total revenue equation of a firm are $C = 6 - x^2$ and $R = x^2$, respectively, where C is the *total cost*, R is the *total revenue*, and x is the amount of the good produced. Determine the breakeven quantity. Now assume that the government imposes an excise tax of \$2 on every unit of the good produced. Determine how this affects the above results. How will a subsidy (instead of an excise tax) of \$1 on every unit of the good produced change the original results?

 Web supplement: S1.6.9 Mathematica applications

1.7 Inequalities, intervals, and absolute values

It is always convenient to express relationships in terms of equations. But this is often an extreme case. Students of the subjects of our interest are frequently required to manipulate relationships that are not equations, rather *inequalities*. This is particularly so in the study of topics like *optimization* involving *inequality constraints*, which we will discuss later in the book. In addition to inequalities, these students are often required to use ideas of *intervals* and *absolute values* in algebraic manipulations. Therefore, it is important that students of these subjects possess a reasonable understanding of inequalities, intervals, and absolute values. Although there are both *linear inequalities* and *nonlinear inequalities*, our discussion here is confined only to linear inequalities only.

1.7.1 Linear inequalities

An inequality is a statement that one expression is not equal to (i.e. less than or greater than) another expression. We begin the discussion of linear inequalities with ideas of the number system we briefly discussed in Section 1.3.1. We understand from the discussion of real numbers that -2 is less than 0 and $+2$ is greater than 0. How do we write this? One way to write these is in words: -2 is less than 0, and $+2$ is greater than 0. This can be seen from Figure 1.7.1.

Imagine that, instead of numbers, we use letters to represent points (or, in other words, numbers) on a real line as shown in Figure 1.7.2. A visual inspection of Figure 1.7.2 indicates that a is less than b and c is greater than b; and these may be written with inequality signs as $a < b$ and $c > b$, respectively. These are sometimes read as "a is strictly less than b" and

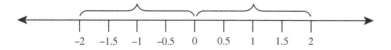

Figure 1.7.1

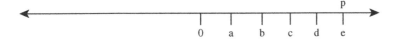

Figure 1.7.2

Table 1.7.1

Notation	=	≠	>	<	≤ or <=	≥ or >=
Meaning	Equal to	Not equal to	Greater than or strictly greater than	Less than or strictly less than	Less than or equal to	Greater than or equal to

"c is strictly greater than b," respectively. The figure also indicates that since p is equal to e, we write it as $p = e$; and since a is not equal to b, we write it as $a \neq b$. If there is a point or number which is *very* close to another point or number, then we state that the former is approximately equal to the latter, and denote by the sign $\approx$; for example, $2.99987 \approx 3$. Sometimes we may state that "b is greater than or equal to a" and "c is less than or equal to d." These are written as $b \geq a$ and $c \leq d$, respectively. These notations and their meanings are summarized in Table 1.7.1.

Students of economics, business, and finance may at times also encounter *double inequalities*. Consider Figure 1.7.2. We know that $a < b$ and $b < c$. Therefore, we may write $a < b < c$ or $c > b > a$, which are examples of double inequalities.

1.7.2 Properties of linear inequalities[6]

Some of the important properties of linear inequalities are the following.

Property I. If $a < b$, then $a + c < b + c$, and $a - c < b - c$.
Property II. If $a < b$ and $c > 0$, then $a \times c < b \times c$, and $(a/c) < (b/c)$.
Property III. If $a < b$ and $c > 0$, then $a(-c) > b(-c)$, and $[a/(-c)] > [b/(-c)]$.
Property IV. If $a < b$, and $a \neq 0$ and $b \neq 0$, then $(1/a) > (1/b)$.

Notice that the third and the fourth properties above are tricky, and the reader has to be clear about their meanings. The third property implies that if both sides of an inequality are multiplied by a negative number, then the inequality sign will be reversed. Similarly, the fourth property implies that if we take the reciprocals of both sides of an inequality, then the inequality sign will be reversed.

1.7.3 Solution of inequalities

As an example, consider the inequality $3(2 - a) < 5$. Then, by applying the second property of linear inequalities, we can divide both sides of the inequality by $1/3$ and get $(3/3)(2 - a) < 5/3$, which simplifies to $2 - a < 5/3$. Now, by applying the first property, we can write $2 - a - 2 < (5/3) - 2 = -a < -1/3$. And, by using the third property, we obtain $(-1)(-a) > (-1/3)(-1)$. This simplifies to the solution $a > 1/3$.

As another example, consider the double inequality $0 < 2/a < 5$. Using the second property, we get $(0/2) < (2/2a) < (5/2) = 0 < (1/a) < 5/2$. Again using the second property, we obtain $0 \times a < (1/a) \times a < (5/2) \times a = 0 < 1 < (5/2)a$. Applying the second property for the third time yields $0 \times (2/5) < 1 \times (2/5) < (5/2) \times (2/5) \times a = 0 < 2/5 < a$. Therefore, we have the solution $a > 2/5$.

1.7.4 Linear inequalities in two variables

Some of our earlier examples of inequalities used numbers or symbols that were constants. All the properties of inequalities discussed in Section 1.7.2 are equally applicable if we use a variable (say, x) instead of a constant. As an example, an inequality with the variable x can be written as $x + 1 < 2$. We know that the solution of this inequality is $x < 1$. The graph of this will represent the set of all values on the real line to the left of the thick vertical line at $x = 1$ (the complete area except that on the thick vertical line) in Figure 1.7.3(A). If our inequality were $x + 1 \leq 2$, then the solution would be $x \leq 1$. The graph of this would represent the set of all values on the real line to the left and on the thick vertical line at $x = 1$ (the complete area) in Figure 1.7.3(A). If the inequalities were $x + 1 > 2$ (with solution $x > 1$) and $x + 1 \geq 2$ (with solution $x \geq 1$), their graphs would represent the set of all values on the real line to the right of the thick vertical line at $x = 1$ (the complete area except that on the thick vertical line) and to the right and on the thick vertical line at $x = 1$ (the complete area) in Figure 1.7.3(A), respectively.

Sometimes we are required to use two variables in a single inequality. For example, consider the inequality $x + y < 1$. We know that when $y = 0$, $x < 1$. This gives the values that x can take: all values on the horizontal axis below 1. Similarly, if $x = 0$, $y < 1$, which gives the values that y can take: all values on the vertical line below 1. If we draw a thick straight line, as the white line in Figure 1.7.3(B), connecting points (0, 1) and (1, 0), then all combinations of x and y below this line ($x + y = 1$) will be the set of solutions to this inequality, as shown by the lower area in Figure 1.7.3(B). If the inequality were $x + y \leq 1$, then all combinations of x and y on and below the white line would be the set of solutions to this inequality. If the inequality were $x + y > 1$ (or $x + y \geq 1$), the set of solutions would

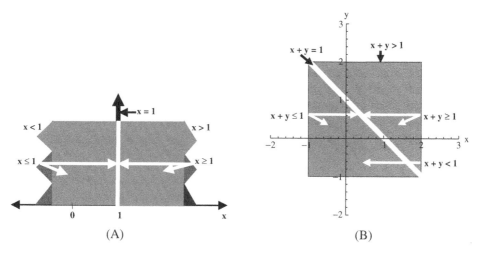

Figure 1.7.3

consist of all combinations of x and y above the white line (or on and above the white line) in Figure 1.7.3(B).

Some of the optimization methods in the subjects of our interest make use of *systems of linear inequalities* in more than one variable. Therefore, students of these subjects are expected to possess at least a working knowledge of these systems. For the time being, we shall consider the systems of linear inequalities in two variables.

As an example, consider the system of two linear inequalities in two variables: $2x+y < 10$ and $x + 2y < 10$. Let us first attempt to solve this system and, then, graph the solution set. Consider $2x + y < 10$ first. If we treat $x = 0$, we get $y < 10$, and if $y = 0$, we get $x < 5$. This gives us the y-intercept $(0, <10)$ and the x-intercept $(< 5, 0)$. Therefore, the solution of this inequality is the set of combinations of all points below the line $2x + y = 10$ in Figure 1.7.4(A). Now consider the second inequality: $x+2y < 10$. Following the same logic, we get the y-intercept $(0, < 5)$ and the x-intercept $(<10, 0)$. The solution of this inequality is the set of combinations of all points below the line $x + 2y = 10$ in the Figure 1.7.4(A). It can be seen from Figure 1.7.4(A) that the set of combinations of x and y that satisfy both inequalities simultaneously is given by the set of points below the thick, solid line. But, if our inequalities were $2x+y \le 10$ and $x+2y \le 10$, then the solution set would contain the set of all combinations of x and y that lie on and below the thick, solid line in Figure 1.7.4(A).

Now consider the same inequalities $2x+y < 10$ and $x+2y < 10$, but with the constraint that both x and y must be nonnegative (that is, $x, y \ge 0$). This constraint is called the *nonnegativity constraint*. This means that the solution set must include only nonnegative values of x and y. Given the inequalities and the nonnegative constraints $x, y \ge 0$, the solution set would be the set of all combinations of x and y within the solid quadrilateral as in Figure 1.7.4(B). If, instead, the inequalities were such as $2x+y \le 10$ and $x+2y \le 10$, and with the constraints, the solution set would be the set of all combinations of x and y within and on the thick quadrilateral as in Figure 1.7.4(B).

As another example, consider the system of linear inequalities $2x + y > 10$ (with y-intercept $(0, >10)$ and x-intercept $(>5, 0)$) and $x + 2y > 10$ (with y-intercept $(0, >5)$ and x-intercept $(>10, 0)$). In the case of $2x +y > 10$ and $x + 2y > 10$, the solution set will be the set of points above the thick line in Figure 1.7.5(A). If the system of inequalities

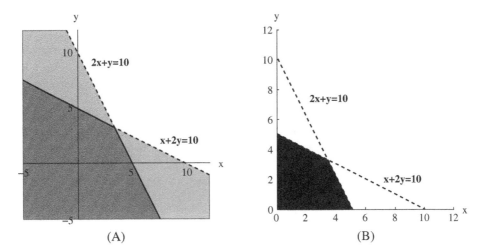

(A) (B)

Figure 1.7.4

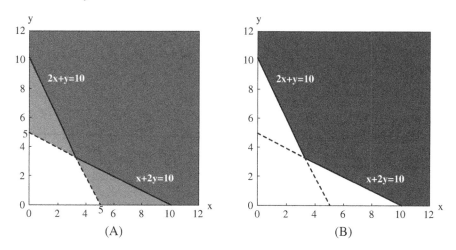

Figure 1.7.5

were $2x + y \geq 10$ and $x + 2y \geq 10$, the solutions set would be the set of all points on and above the thick line in Figure 1.7.5(A). The solutions set for inequalities $2x + y > 10$ and $x + 2y > 10$ with nonnegative constraint $x, y \geq 0$ will be the set of points above the thick line in Figure 1.7.5(B), and the solutions set for inequalities $2x + y \geq 10$ and $x + 2y \geq 10$ with nonnegative constraint $x, y \geq 0$ will be the set of points on and above the thick line.

1.7.5 Intervals

A concept closely related to inequalities is that of *intervals*. An interval represents the set of all numbers that lie between two numbers or points (called *end points*) on a real line, either including or excluding the points. If the set contains the end points, it is called a *closed interval*, and if the set does not contain the end points, it is called an *open interval*. We can use Figure 1.7.6 to explain the ideas of intervals. If an interval is written with square brackets, [...], we call it a closed interval, and if it is written with parentheses, (...), it is called an open interval. If an interval is written with one parenthesis and one square bracket, (...] or [...), it is called a *half-open interval* or a *half-closed interval*. Therefore, with reference to Figure 1.7.6, $[e, f]$ represents a closed interval; (j, k) represents an open interval; and $(c, d]$ and $[l, m)$ represent half-open (or half-closed) intervals.

Notice that all the intervals discussed above possess upper and lower end points. Such intervals are called *bounded intervals* (with both *upper bound* and *lower bound*). How about the intervals representing the points b and $a = -\infty$, and n and $o = +\infty$? Since we cannot specify a value for ∞ (or infinity), we may write the first interval as $(-\infty, b]$ and the second interval as $[n, +\infty)$. Therefore, we say that $(-\infty, b]$ has no lower bound and $[n, +\infty)$ has

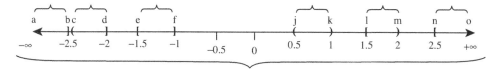

Figure 1.7.6

Table 1.7.2

Notation	$[e, f]$	$(c, d]$ and $[l, m)$	(j, k)	$(-\infty, +\infty)$
Name	Closed	Half-open	Open	Open
Example	$e \leq x \leq f$	$c < x \leq d, l \leq x < m$	$j < x < k$	$-\infty < x < +\infty$

no upper bound. Notice that all the real numbers that lie on the interval $(a = -\infty, o = +\infty)$, or $(-\infty, +\infty)$, are represented by the downward-pointing brace.

Let us now pick a number, say x, which lies in the interval $[e, f]$. The inequality representing this can be written as $e \leq x \leq f$. If x lies between (j, k), $(c, d]$, and $[l, m)$, we represent the position of x in the form of inequalities as $j < x < k$; $c < x \leq d$; and $l \leq x < m$, respectively. If x lies in the interval $(-\infty, +\infty)$, we represent it by the inequality $-\infty < x < +\infty$. The above presentation of interval notations, their names, and related examples (of a number, x, lying within the interval) are summarized in Table 1.7.2. It must be clear by now the connection between intervals and inequalities.

1.7.6 Absolute values

Sometimes, the sign of a number is important. Consider the annual profit or loss of a company. The number that represents the annual loss of the company in its accounts is always preceded by the negative sign; for example, $-\$10$ million, $-\$15$ million, etc. Similarly, the annual profit of the company in its accounts is always preceded by the positive sign (although it is ignored in actual practice); for example, $+\$5$ (or $\$5$) million, $+\$20$ (or $\$20$) million, etc.

At some other times, the sign of a number is immaterial. Consider the example of the distance run by an athlete within a specified time. Here the sign is immaterial. The magnitude or size of a number irrespective or regardless of its sign is called the *modulus* or *absolute value* of that number. Consider the real line in Figure 1.7.7. On the real line, the distance from the point of origin (0) to a point or value, say x_0, is called the modulus or absolute value of x_0 and is denoted by $|x_0|$. Similarly, the distance from 0 to x_1 is called the modulus or absolute value of x_1 and is denoted by $|x_1|$. For example, the distance from 0 to 2 is the modulus of 2 and is written as $|2| = 2$; and the distance from 0 to -2 is the modulus of -2 and is written as $|-2| = 2$. This implies that the distances from 0 to 2 and from 0 to -2 are equal; that is, $|2| = |-2| = 2$. Notice that irrespective of the fact that the point or the number lies to the left or to the right of the origin, the modulus or absolute value of that number considers only the magnitude of the number.

We shall now define the absolute value. The absolute value of a real number, x, written as $|x|$, is defined as

$$|x| = \begin{cases} x, & \text{if } x \geq 0 \\ -x, & \text{if } x < 0 \end{cases} \tag{1.7.1}$$

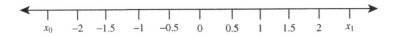

Figure 1.7.7

Sometimes one may find in the literature expressions called *absolute value equations* such as $|x - 2| = 3$. How do we solve this equation? This equation states that the distance from 0 to the number $|x - 2|$ is equal to 3, in either direction (that is, $x \geq 0$ or $x < 0$). Then, applying the definition in equation (1.7.1) yields $x - 2 = 3$ or $x - 2 = -3$. These two equations simplify to $x = 5$ and $x = -1$, respectively. Therefore, the solution set is $x = \{5, -1\}$.

One may also find expressions called *absolute value inequalities* in the literature. Examples of such inequalities are $|x| < 2$; $|x + 2| < 2$; $|x - 2| < 2$; and so on. How do we solve such inequalities? Before we do this, we need to be a bit clearer about such inequalities.

Consider, for example, the absolute value inequality $|x| < 2$. This inequality says that $|x|$ is less than 2 units away from 0 or the distance from 0 to $|x|$ is less than 2. Applying the definition of absolute value shows that x lies between -2 and 2; that is, on the interval $-2 < x < 2$. If $|x| > 2$, then $x < -2$ or $x > 2$. Notice that these results are valid even if we use $\leq$ and $\geq$ signs instead of $<$ and $>$ signs. Let us now generalize the above results. Assume that we now use a general value, b, instead of specific values such as 2 or 3. Then, we may write the solutions of the following absolute value inequalities:

$$|x| > b \quad \text{as} \quad x < -b \text{ or } x > b, \quad |x| \geq b \quad \text{as} \quad x \leq -b \text{ or } x \geq b, \quad |x| < b \quad \text{as}$$

$$-b < x < b, \quad \text{and} \quad |x| \leq b \quad \text{as} \quad -b \leq x \leq b \tag{1.7.2}$$

Let us now consider specific examples of absolute value inequalities. Consider, for example, $|x - 2| < 6$. We need to solve this inequality for x. Following the general solutions given in inequalities (1.7.2), we can write the last inequality as $-6 < x - 2 < 6$, or $-6 + 2 < x < 6 + 2$, or $-4 < x < 8$. This shows that x lies between -4 and 8. In terms of intervals discussed in the last section, this can be written as $(-4, 8)$.

As another example, consider the absolute value inequality $|x - 2| \geq 6$ for solving for x. Following again the general solutions in inequalities (1.7.2), we may write the last inequality as $x - 2 \leq -6$ or $x - 2 \geq 6$. This implies that the solution is $x \leq -4$ or $x \geq 8$. In terms of the intervals discussed in the last section, these can be written as $(-\infty, -4]$ and $[8, \infty)$.

We now state the properties of absolute values. Let there be two points, a and b, on the real line. Then, the following properties are valid.

Property I.   The absolute value of the product of a and b is equal to the product of their individual absolute values: $|a.b| = |a| . |b|$.

Property II.   The absolute value of the ratio of a to b is equal to the ratio of their individual absolute values: $|a/b| = |a| / |b|$.

Property III. a lies between the negative absolute value of a and the positive absolute value of a: $-|a| \leq a \leq |a|$.

Property IV.   The absolute value of the difference of a and b is equal to the absolute value of the difference of b and a: $|a - b| = |b - a|$.

1.7.7 Application examples

Example 1.   Let that a consumer has \$270 to spend on two goods (x and y), and that the per unit price of x is \$5 and the per unit price of y is \$6. If the consumer buys 16 units of x, what is the maximum units of y that can be bought?

Solution.   Given the information, we may write the inequality as $5x + 6y \le 270$, or $516 + 6y$ ≤ 270, or $80 + 6y \le 270$, or $6y \le 190$. This implies that the consumer can buy a maximum of 31.67 ($y \le 190/6 = 31.67$) units of y.

Example 2.   Let that a company expects to sell a minimum of 8000 units and a maximum of 9000 units of its product every month for the next 9 months. Also let that the actual sale (S) lies between the minimum and maximum expected sales. Express the actual sales of the company with the help of a double inequality.

Solution.   Since the company expects to sell a minimum of 8000 units every month, the minimum sales for 9 months will be $9 \times 8000 = 72000$ units. And since the company expects to sell a maximum of 9000 units every month, the maximum sales for 9 months will be $9 \times 9000 = 81000$. Therefore, the actual sales S will be between 72 000 units and 81 000 units: $72\,000 < S < 81\,000$.

Example 3.   Suppose that a consumer has $1000 to spend on two goods x and y. Suppose also that the per unit price of x is $4 and the per unit price of y is $8. Construct an inequality that describes the above statements if the consumer spends the complete money or less than that. Show the solution sets, using a graph, when the consumer spends $1000 completely and when the consumer spends less than that.[7]

Solution.   Since the unit prices of x and y are $4 and $8, respectively, the expenditures on the two goods are $4 $\times x$ and $8 $\times y$. Therefore, the required inequalities are $4x + 8y < 1000$ and $4x + 8y \le 1000$ if the consumer spends less than and equal to the available money, respectively. In order to find the solution sets we shall proceed as follows. Notice that the x-intercept and y-intercept for the inequality $4x + 8y < 1000$ are (≤ 250, 0) and (0, ≤ 125), respectively. Connecting these two intercepts we obtain the thick straight line Figure 1.7.8(A). The set of all the combinations of x and y below this line (shaded area) in Figure 1.7.8(A) will be the set of solutions of the inequality $4x + 8y < 1000$. The set of all the combinations of x

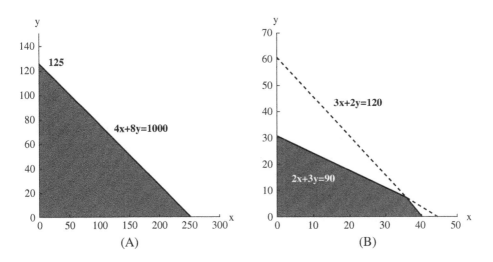

Figure 1.7.8

Table 1.7.3

	Output x requirement	Output y requirement	Input availability
Input A	2	3	90
Input B	3	2	120

and y below and on the thick line will be the set of solutions of the equality $4x + 8y \leq 1000$. The latter set is called the *budget set*, and the equation $4x + 8y = 1000$ is called the *budget line* or *budget equation*.

Example 4.   Assume that a firm uses two inputs x and y in order to produce two goods A and B. Also assume that to produce one unit of x the firm has to use 2 units of A and 3 units of B, and to produce one unit of y it has to use 3 units of A and 2 units of B. Assume again that the maximum amount of A available is 90 units and the maximum amount of B available is 120 units. Show this information with the help of a system of inequalities, and determine the solution set (with $x, y \geq 0$).

Solution.   Let us first construct Table 1.7.3 using the above information. The next step is to convert the information in Table 1.7.3 into inequalities. Then we obtain

$$2x + 3y \leq 90 \quad \text{and} \quad 3x + 2y \leq 120 \tag{1.7.3}$$

From the first inequality we obtain $(0 \leq 30)$ and $(\leq 45, 0)$ as the x-intercept and y-intercept, respectively. Similarly, from the second inequality, we obtain $(0, \leq 60)$ and $(\leq 40, 0)$ as the x-intercept and y-intercept, respectively. Connecting these respective intercepts will give us the straight lines in Figure 1.7.8(B). The set of solutions satisfying both inequalities simultaneously is given by the set of all combinations lying on and inside the quadrilateral in the figure (the shaded area).

1.7.8 Exercises

1.   Solve the following inequalities:
     (i) $2 + 2x \leq 10$; (ii) $-x \geq -2$; (iii) $[(2x + 1)/(x - 3)] < 1$; (iv) $4x - 2 > 0$;
     (v) $[(x+1)/(2y+1)] > 1$; (vi) $3x + 2y + 3 \leq x + y + 5$; (vii) $(5/2)x > (2/5)$.
2.   Find the solution sets of the following inequalities graphically:
     (i) $x < y$; (ii) $x > y$; (iii) $x < y < 2 + x$; (iv) $x < 5$; (v) $x > 5$; (vi) $4 < x < 6$;
     (vii) $4 < y < 6, 4 < x < 6$; (viii) $0 < x < 2, 0 < y < 2$; (ix) $2x + 2y \leq 10$; (x) $2x + 2y \geq 10$;
     (xi) $2x + 2y \leq 10, (x, y) \geq 0$; (xii) $5x + 2y \geq 10, 40x + 10y \geq 70$; (xiii) $4x + 2y \leq 20$,
     $x + y \leq 4, (x, y) \geq 0$.
3.   Solve the following equations:
     (i) $|10/x| = 5$; (ii) $|x/10| = 5$; (iii) $|x| = 5$; (iv) $|-x| = 5$; (v) $|2 + 4x| = 8$.
4.   Solve the following inequalities:
     (i) $|x| < 10$; (ii) $|x| > 10$; (iii) $|x| \leq 10$; (iv) $|x| \geq 10$; (v) $|4x - 2| < 8$; (vi) $|4x - 2| \leq 8$;
     (vii) $|2 + 4x| > 8$; (viii) $|2 + 4x| \geq 8$.
5.   *Application exercise.* Assume that a marketing executive expected a bonus (B) between $4000 and $6000 every month over the next 5 months. But, the actual monthly B turned out to be between these two values. Express these statements in terms of an inequality.

6. *Application exercise.* Assume that a person has to eat two types of food, say x and y, as a part of a diet prescribed by her doctor. The doctor advised that the food must contain two types of nutrients, say A and B. Each unit of x needs at least 10 units of A and 20 units of B, and each unit of y needs at least 4 units of A and 2 units of B. The food must contain a total minimum of 80 units of A and 20 units of B. Find the solution set graphically such that $x, y \geq 0$.

7. *Application exercise.* Assume that a firm produces two goods, say x and y, using three inputs, say A, B, and C. Also assume that the production of one unit of x requires one unit of A, the production of one unit of y requires one unit of B, and one unit of x and y each requires one and two units, respectively, of C. Again assume that the maximum quantities of A, B, and C available are 10, 15, and 24, respectively. Find the solution set graphically such that $x, y \geq 0$.

8. *Application exercise.* Suppose that a producer has \$10 000 to spend on two factors l and k. Suppose also that the unit price of l is \$2 and the unit price of k is \$4. Construct an inequality that describes the above statements if the producer spends the complete money or less than that. Show the solution sets, using a graph, when the producer spends \$10 000 completely and when less is spent.

 Web supplement: S1.7.9 Mathematica applications

1.8 Functions

Functions lie at the heart of the science of economics. A student of economics has to study a multitude of functions: *demand function, supply function, cost function, revenue function, production function, profit function, consumption function,* and so on. Business and finance students also deal with some of these or related functions. Therefore, it is extremely important that students in these branches possess a reasonably good understanding of the meaning, generation, nature, and types of functions.

1.8.1 Univariate functions

Before we get into functions, we need to explore a few concepts such as the *Cartesian coordinate system* or *rectangular coordinate system*; *ordered pair*; *Cartesian product*; *relation*; and *mapping* for a proper understanding of functions. Therefore, we present below a brief exposition of each of these concepts.

Assume that we pick a real line and place it horizontally. Assume also that we pick another real line and place it such that it is perpendicular to the real line already placed horizontally and that each line passes through the middle of the other. Then the former will be called the horizontal axis or the x-axis and the latter will be called the vertical axis or the y-axis. Such a system of lines is called a Cartesian or a rectangular coordinate system, as shown in Figure 1.8.1. As can be seen, the two axes divide the *plane* into four areas called *quadrants*. It is important to note that the signs of the values taken by the variables (x and y) are different in different quadrants. Notice that the axes cross each other at point 0, and this point is called the *origin*. Both variables have positive signs in the first quadrant and have negative signs in the third quadrant. But, x has negative and positive signs in the second and fourth quadrants, respectively; and y has positive and negative signs in the second and fourth quadrants, respectively.

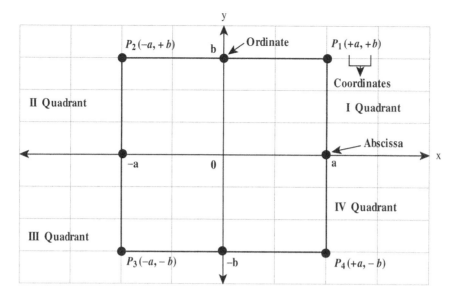

Figure 1.8.1

Assume now that we drop a straight line from a specific point to the *x*-axis and another one from the same point to the *y*-axis (such as point P_1, P_2, P_3, or P_4). Then the value at which this line touches the *x*-axis is called *abscissa* and the value at which the second line touches the *y*-axis is called *ordinate*. These two values together are called *coordinates*. Notice that there is an innumerable number of coordinates in the *x*, *y* plane. Hence the name *coordinate system*.

We mentioned that, to draw Figure 1.8.1, we first picked a real line that represented the *x*-axis. This means that we have one real line and that all the numbers on this line constitute a set of real numbers. Therefore, the points or numbers on this real line form a *one-dimensional space* denoted by $\Re^1$, or simply by $\Re$. If we have two axes (for two variables), as in Figure 1.8.1, then we have two sets of complete real numbers. Then the plane (e.g. quadrant I, II, III, or IV in the figure) defined by these two sets constitutes a *two-dimensional space* denoted by $\Re^2$. Assume that, instead of two, we have three sets of complete real numbers represented by three real lines (for three variables). Then the plane defined by these three sets jointly constitutes a *three-dimensional space* denoted by $\Re^3$. In short, if we have *n* real lines to represent *n* variables, then we will have *n* sets of real numbers and the plane defined by these sets jointly is called an *n-dimensional space* represented by $\Re^n$. Notice that if $n > 3$, a geometric interpretation of space is not possible.

Let us now consider the concept of an ordered pair. An ordered pair is a grouping of elements in a specific order. Let there be two elements *a* and *b*. Also let that we group them in a specific order, for example *a* first and then *b*, and denote it by (a, b). This is called an ordered pair of *a* and *b*. Another ordered pair of the same elements is (b, a). One may wonder here whether (a, b) is equal to (b, a). The answer is no; because they are based on different orders. This can be confirmed from Figure 1.8.3 because the ordered pairs represented by points *E* and *F* are different.

An astute reader may now wonder how an ordered pair differs from a set. To explain the difference between them, let us take two sets $\{a, b\}$ and $\{b, a\}$. We know from our discussion in Section 1.2 that $\{a, b\} = \{b, a\}$. But, as discussed above, the ordered pair (a, b) is not equal to the ordered pair (b, a). The reason for this is that orders are immaterial among sets, while they are important among ordered pairs.

Assume now that we have a set of four points on the x-axis, namely a, b, c, and d: $x = \{a, b, c, d\}$. Assume also that we have another set of four similar points on the y-axis: $y = \{a, b, c, d\}$. The total number of ordered pairs that can be generated from two sets of four elements is 16 ($4^2 = 16$). The set of all these ordered pairs is called the *Cartesian product* or *direct product* of the sets x and y, and is denoted by $x \times y$ (read "x cross y"). Notice that even if x and y are sets of numbers, $x \times y$ will represent a set of ordered pairs of numbers. However, how do we get 16 ordered pairs from two sets of 4 letters or elements? The answer can be shown through Figures 1.8.2(A)–(D). As shown, we associate each element of x to every element of y. Thus, we can construct the Cartesian product generated from the two sets x and y as

$$x \times y = \{(a, a), (a, b), (a, c), (a, d), (b, a), (b, b), (b, c), (b, d), (c, a), (c, b), (c, c), (c, d),$$
$$(d, a), (d, b), (d, c), (d, d)\}.$$

Graphing these cross products in a diagram gives us the Cartesian system or rectangular system as illustrated in Figure 1.8.3(A).

As an example, consider two small sets: $x = \{1, 2, 3\}$ and $y = \{1, 2, 3\}$. We know that we can generate $3^2 = 9$ ordered pairs from these sets. They are: (1, 1), (1, 2), (1, 3), (2, 1), (2, 2), (2, 3), (3, 1), (3, 2), and (3, 3). Therefore, the Cartesian or cross product in the present case is $x \times y = C = \{(1, 1), (1, 2), (1, 3), (2, 1), (2, 2), (2, 3), (3, 1), (3, 2), (3, 3)\}$. The graph of this rectangular system is illustrated in Figure 1.8.3(B).

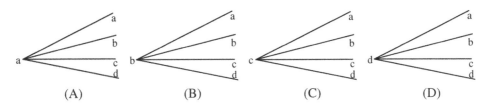

| (A) | (B) | (C) | (D) |

Figure 1.8.2

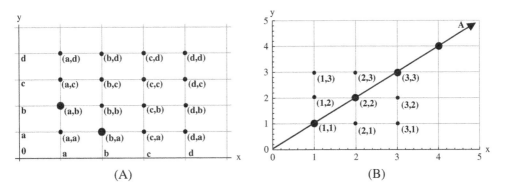

| (A) | (B) |

Figure 1.8.3

The next concept we need to explain is relation. To do this, consider the statement that y is greater than x. This statement can be expressed using the inequality $y > x$. What will be the solution set of this inequality? This problem can be easily solved with the help of the graph in Figure 1.8.3(B). Notice that we have drawn a straight line from the origin to the point A $(0A)$ that passes through points representing values of x and y such that $x = y$. Therefore, all points above this line must satisfy the inequality $y > x$. These points are, among others, $(1, 2)$, $(1, 3)$, and $(2, 3)$. We denote the set of these points by D such that $D = \{(1, 2), (1, 3), (2, 3)\}$. Notice that D is a subset of the cross product C; that is, $D \subseteq C$. A relation is defined as a subset of the Cartesian or cross product. Therefore, D constitutes a relation.

The graph in Figure 1.8.3(B) can be used to represent other examples of relations. Consider the two statements: y is equal to x, and y is less than x. These can be written as $y = x$ and $y < x$, respectively. All the points lying on the line $(0A)$ satisfy the first statement and all the points below that line satisfy the second statement. Assume that the set of all the points on the line $0A$ is denoted by the set E; that is, $E = \{(1, 1), (2, 2), (3, 3)\}$. This means that $E \subseteq C$. Therefore, E constitutes a relation. Similarly, we denote the set of all points below the line $0A$ by the set F. We know that $F \subseteq C$. Therefore, F defines another relation.

We shall now explain the meaning of the term mapping with the help of two sets (x and y), which we used to derive the Cartesian product illustrated in Figure 1.8.3(B). The term mapping refers to the process of associating the elements of one set (x) to those of another set (y). Notice that we picked up one element from x to associate it with an element from y such that the statement $y > x$, or $y < x$, or $y = x$ is satisfied. This process of associating the elements of two or more sets is called mapping.

We are now ready to define a function.[8] Let there be two sets, x and y. A function is a special type of relation where for each (or more) element(s) in set x, there is *only one* corresponding element in set y. Formally, a function is a relation or rule denoted by f that produces a *correspondence* between the elements of two sets such that to each (or more) element(s) in the first set (x) there corresponds *one and only one* element in the second set (y) denoted by $f(x)$, and read "f of x" (or $y = f(x)$, read "y is a function of x"). The first set is called the *domain* of the function and the second set is called the *range* of the function.

We may again use the graph in Figure 1.8.3(B) to explain the above definition of a function. We discussed above the statement that $y = x$ and found some points in the x, y plane that satisfied this statement. These points were represented by the straight line $0A$ in the figure. We stated that the set of points on this line described a relation. And, therefore, this relation constituted a function. The reason is that for one (or more) value(s) of the variable x, we have one and only one value of the variable y.

As another example, consider a statement in $\Re^2$: y is half of x. This statement can be written as $y = (1/2)x$. Assume that $x \in \Re$ and $y \in \Re$. When we map the different values of x into y we obtain the relation that constitutes a function $y = 0.5 \times x$ as illustrated in Figure 1.8.4(A). In fact what we have done in Figure 1.8.4(A) is simply a mapping of different values of x onto a unique value of y, such that $y = 0.5 \times x$ is satisfied, as shown in Figure 1.8.4(B).

It is important to understand whether a graph represents a function or not. One can judge whether a graph represents a function or not using a test called the *vertical line test*. Imagine that we have a function with dependent variable y and independent variable x. We know from the definition of a function given above that for a graph to represent a function there should be only one point on the graph that intersects with a line drawn vertically to the graph, or there should be only one value of y for one (or more) value(s) of x. This can be checked by drawing a vertical line through the graph to the x-axis at, say, $x = x_1$. If the vertical line

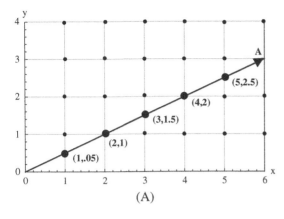

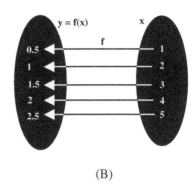

(A)   (B)

Figure 1.8.4

intersects the graph at exactly one point, then the graph represents a function, otherwise it does not represent a function.

Let us now use the vertical line test to check whether the graphs illustrated in Figures 1.8.5(A)–(G) satisfy the definition of a function. In the graph in Figure 1.8.5(A) there is only one y value for each value of x and, therefore, it represents a function. In Figure 1.8.5(B), there is no unique y value for each value of x and, therefore, it does not represent a function. In Figures 1.8.5(C), (F), and (G), there is only one y value for each value of x and, therefore, these graphs represent functions. And there is no unique y value for each value of x in Figures 1.8.5(D) and (E) and, therefore, they do not represent functions.

So far we have defined functions and graphically illustrated them using concepts such as cross products, relations, and mappings. However, the reader must have noticed that the domains of all these functions were defined over the elements of a single set. In other words, all the functions had only one independent variable (in our examples, x). Such functions are called *univariate functions*. Notice that one can find plenty of functions whose domains are defined over the elements (values) of more than one set (variable), or multivariate functions. We will deal with such functions in Section 1.8.5.

We discussed equations in detail in Section 1.6. Let us now take a closer look at equations and, thereby, their dissimilarity with functions. The reader must have noticed that when we mapped the set x onto the set y, as illustrated in Figure 1.8.4(B), using the statement "y is half of x" we obtained some specific values in the set y. Therefore, for every value $x_i \in x$ we have a unique value $y_i \in y$. If we pick $y_i = 2$ and reframe the statement as "half of x_i is equal to y_i" we may write the statement algebraically as $0.5 \times x_i = y_i$ or $0.5 \times x_i = 2$. The last expression is an equation whose solution is $x_i = 4$ as confirmed by the mapping in Figure 1.8.4(B). In short, a function purports a *sequence* of associations between two or more sets but an equation purports only one (few) *specific* association(s). This was the reason why we used subscripts in Sections 1.6.1 and 1.6.2.

However, in most practical cases we discard the subscripts for convenience and write the expressions as $0.5 \times x = y = 2$. Our above exposition suggests that functions are defined over the entire domain set(s). Therefore, if we replace the specific elements of the domain (say x_i) or of the range (say y_i) by the domain (x) or by the range (y) itself, an equation becomes a function. This implies that we if replace x_i and y_i by x and y, respectively, and use the notation $f(x)$ then an equation becomes a function.

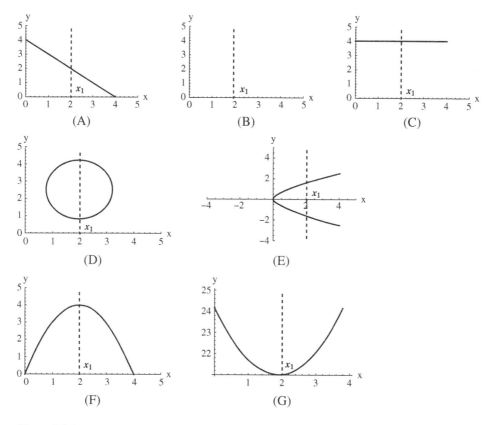

Figure 1.8.5

Two important points are worth mentioning here. First, a general function such as $y = f(x)$ may take many specific forms depending upon the constants and coefficients involved in it and upon the powers of the independent variable, x. In other words, a general function $y = f(x)$ may take many specific forms depending upon the rule f applied to the function. For example, $y = f(x) = x$, $y = f(x) = 1 + x$; $y = f(x) = 2 + 2x^2$; etc. Second, one has to be clear about the equality of functions. Let there be two functions f and g. Assume that both are the functions of x. Then these two functions are equal, i.e. $f = g$, only if they have the same domain (the set of values that x can take).

1.8.2 Combination of functions; composite and compound functions

Functions obey the four fundamental arithmetic rules of addition, subtraction, multiplication, and division. One can generate a new function by combining two or more functions through the four arithmetic rules if the functions' domains are the same. This process of combining functions, or the function so generated, is called a *combination of functions*. Assume that we have two functions, f and g, of x. Then we can do the above four operations that generate new functions as

$$F(x) = (f + g)(x) = f(x) + g(x) \tag{1.8.1}$$

$$G(x) = (f - g)(x) = f(x) - g(x) \tag{1.8.2}$$

$$H(x) = (f \times g)(x) = f(x) \times g(x) \tag{1.8.3}$$

$$J(x) = (f/g)(x) = [f(x)/g(x)], \quad g(x) \neq 0 \tag{1.8.4}$$

These rules suggest that we may express a function as a combination of two or more functions, and vice versa. As an example, consider the two functions $f(x) = 3 + 2x$ and $g(x) = 1 + 4x$. Then the combinations of functions corresponding to equations (1.8.1)–(1.8.4) will respectively be

$$F(x) = (f + g)x = f(x) + g(x) = (3 + 2x) + (1 + 4x) = 4 + 6x$$

$$G(x) = (f - g)(x) = f(x) - g(x) = (3 + 2x) - (1 + 4x) = 2 - 2x$$

$$H(x) = (f \times g)(x) = f(x) \times g(x) = (3 + 2x) \times (1 + 4x) = 3 + 14x + 8x^2$$

$$J(x) = (f/g)(x) = [f(x)/g(x)] = [(3 + 2x)/(1 + 4x)]$$

The next concept that we need to understand is *composite function* and the process of forming it, called *composition of functions*. We shall discuss composition of functions with the help of an example. We know that the quantity of output (z) produced of paddy, other things remaining the same, depends on the quantity of water (y) in the paddy field, that is $z = g(y)$; and the quantity of water, again other things remaining the same, depends on the amount of rainfall (x), that is $y = f(x)$. Therefore, in this example, we have z depending on y, which in turn depends on x; that is, we have a *function-of-function*. This means that z is a direct function of y and an indirect function of x. Such a function is called a composite function, and we denote it by $(g \circ f)(x) = g(f(x))$. This is illustrated in Figure 1.8.6.

In composition of two functions, $z = g(y)$ and $y = f(x)$, we first apply one function (in our example $y = f(x)$) to a number and then apply the other function $z = g(y)$ to the result. We use a specific example to show how this works. Assume that $z = g(y) = y - 2y^2$ and $y = f(x) = 6x^2$. Then the composite function $(g \circ f)(x) = g[f(x)]$ can be obtained through the substitution of $f(x) = y$ for every occurrence of y in $g(y)$. Therefore, we obtain the result $g[f(x)] = 6x^2 - 2(6x^2)^2 = 6x^2 - 2 \times 6^2 x^4 = 6x^2 - 2 \times 36 \times x^4 = 6x^2 - 72x^4$.

The last concept we consider in this section is *compound functions*. They are also called *piecewise functions*. This may be better illustrated with an example. Suppose that a saleswoman of a firm receives her remuneration (y) in two forms. The first is that she receives \$10 000 a month irrespective of sales. The second is that, in addition, she receives a commission of \$1 for every unit sold beyond the target level of 10 000 units. How do we represent her remuneration in the form of a function? One minute's thought would suggest that one simple function would not suffice to represent her remuneration. This can be seen from Figure 1.8.7. The question now is how we can write the information given in the

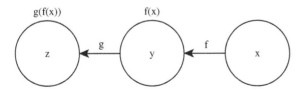

Figure 1.8.6

y = Remuneration

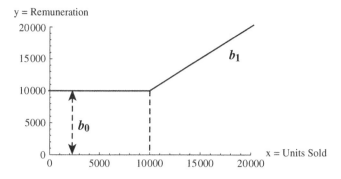

Figure 1.8.7

remuneration example and its graphic illustration in Figure 1.8.7 in the form of a function. Following the figure, we can write it as

$$y = f(x) = \begin{cases} 10000 & 0 \le x < 10000 \\ x & 10000 \le x \le 20000 \end{cases}$$

1.8.3 Types of functions and their graphs

Having considered the meaning and generation of functions, we now turn to the types of functions. We presented a detailed discussion of equations in Section 1.6.1. And we found the link between equations and functions at the end of Section 1.8.1. Following these, we may write equations (1.6.1) through (1.6.7) as

$$y = f(x) = b_0 x^0 + b_1 x^1 + b_2 x^2 + b_3 x^3 + \cdots + b_n x^n \tag{1.8.5}$$

$$y = f(x) = b_0 x^0 = b_0 \tag{1.8.6}$$

$$y = f(x) = b_0 x^0 + b_1 x^1 = b_0 + b_1 x \tag{1.8.7}$$

$$y = f(x) = b_0 x^0 + b_1 x^1 + b_2 x^2 = b_0 + b_1 x + b_2 x^2 \tag{1.8.8}$$

$$y = f(x) = b_0 x^0 + b_1 x^1 + b_2 x^2 + b_3 x^3 = b_0 + b_1 x + b_2 x^2 + b_3 x^3 \tag{1.8.9}$$

$$y = f(x) = a^x \tag{1.8.10}$$

$$y = f(x) = \log_a x \tag{1.8.11}$$

which are called general *polynomial function, constant function, linear function, quadratic function, cubic function, exponential function,* and *logarithmic function,* respectively. The graphs of the second to the seventh of these functions, with the given values for constants, are illustrated in Figures 1.8.8(A)–(F).

1.8.4 Other types of functions and their graphs

One type of function that we have not discussed so far is a *continuous function.* All of the functions we have considered so far were continuous functions. A function is said to be

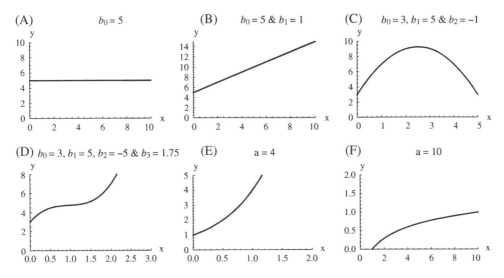

Figure 1.8.8

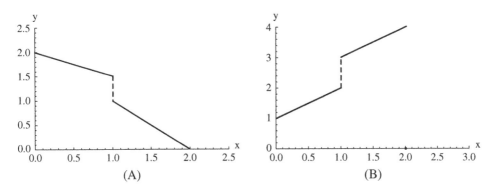

Figure 1.8.9

continuous if its graph does not contain gaps (either vertical or horizontal). In other words, a continuous function is one whose graph can be drawn without lifting the pencil from the paper. However, it may contain corners. The difference between a continuous function and a *smooth function* is that the former may contain corners while the latter does not. The opposite of a continuous function is a *discontinuous function*, which contains gaps as shown in Figures 1.8.9(A) and (B).

Another type of function that we have yet to consider is a *step function*. In this function, for each interval in the domain, there will be a specific constant value in the range. But the value in the range jumps up (or down) as we move forward (or backward) from one interval to another interval in the domain. Graphs of such functions are shown in Figures 1.8.10(A) and (B).

Another important function frequently used in the subjects of our interest is an *inverse function*. Let $y = f(x)$. If f is a *one-to-one function*,[9] then the inverse of f, denoted by $x = f^{-1}(y)$, read as "x is an inverse function of y," is the function formed by interchanging

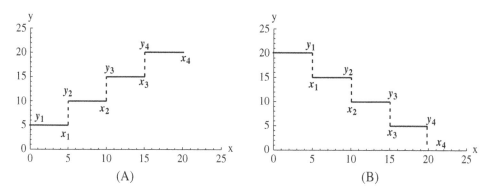

Figure 1.8.10

the independent and dependent variables of f. If f is not a one-to-one function (that is, an *onto function*), then f does not have an inverse. Inverse functions are mirror images of each other. Since the exponential and logarithmic functions are mirror images of each other, they are inverse functions as their graphs in Figures 1.8.8(E) and (F) suggest.

As an example, let our original function be $y = f(x) = 3 + 2x$. Then interchanging the independent and dependent variables, we get $x = f^{-1}(y) = -1.5 - 0.5y$, which is the inverse of the original function. The graphs of these two functions are illustrated in Figure 1.8.11(A).

An *implicit function* is another important type of function. Let our function be $y^{1/2} = f(x) = n/(x^{1/2})$, where n is a positive real number. Functions such as these are called *explicit functions*. In fact all the functions we have dealt with so far in the present chapter are explicit functions. Notice that from the last function we can judge that y is the dependent variable and x is the independent variable. Suppose that we write this function as $f(x, y) = y^{1/2}x^{1/2} = n$. But now we cannot distinguish between the dependent and independent variables. Functions such as these are called implicit functions. The graph of this function is illustrated in Figure 1.8.11(B).

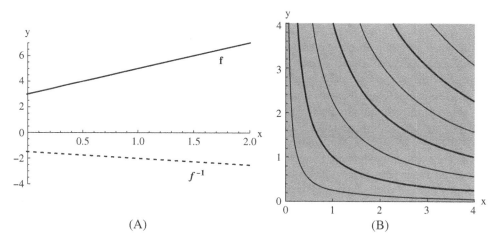

Figure 1.8.11

There is another type of function that is often used in the literature. This is called a *monotonic function*. Sometimes a function will always be increasing or decreasing. Functions like these are called monotonic functions. If the function is always increasing, it is called an *isotonic function*; if the function is always decreasing, it is called an *antitonic function*. Examples of isotonic and antitonic functions are the supply and the demand functions of a good (assuming other things remain the same), respectively.

1.8.5 Multivariate functions

Our discussion so far in the present section has been limited to univariate functions. This is, in fact, a simplified form of reality. Students of economic sciences normally deal with a large number of functions that have more than one independent variable. A function with more than one independent variable is called a *multivariable function* or *multivariate function*. Suppose that we have a function with the dependent variable z and the independent variables x and y. Then this multivariate function is written as $z = f(x, y)$.

Normally the decision to cultivate most agricultural commodities is determined not just by their respective current prices, but by their past prices as well. A company's revenue from the sale of a good is determined by many factors including the price of the good and the amount of advertising expenditure. The price of the share of a company is influenced by a host of forces including the performance of the company and government policies. A person's current year's consumption expenditure depends not only on their current year's income but also on their past year's income and consumption. A company's decision to invest in a project depends, among others, on the *cost of capital* and the *expected return* from the project. Examples of multivariate functions like these are plenty in our fields of interest. Therefore, it is important that students of these subjects posses a reasonable understanding of such functions.

As an example of a multivariate function, consider the *Cobb–Douglas production function*. This *production function* is written as $Q = A.K^{\alpha}.L^{1-\alpha}$, where Q represents the quantity of output produced, A denotes the efficiency parameter reflecting technology, K represents the amount of capital employed, L denotes the quantity of labor employed, and the Greek letter α represents the *elasticity of output* with respect to K. The graph of this function, for $A = 1$ and $\alpha = 0.5$, is illustrated in Figure 1.8.12(A).

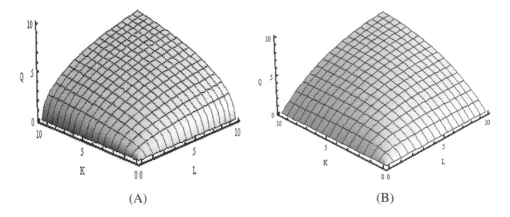

(A)   (B)

Figure 1.8.12

Another example of a multivariable function in economics is the *constant elasticity of substitution production function* (also called the *CES production function*). This function is written as $Q = A[\beta.K^{-\lambda} + (1-\beta).L^{-\lambda}]^{-1/\lambda}$, where Q, A, K, and L are the same as in the Cobb–Douglas production function. The Greek letters β and λ represent the *distribution parameter* and the *substitution parameter*, respectively. The graph of this function, for $A = 1$, $\beta = 0.5$, and $\lambda = 0.9$, is illustrated in Figure 1.8.12(B).

1.8.6 Application examples

Example 1.   Assume that a company's total revenue (R) and total cost (C) are functions of output produced (q). If $R = f(q) = 0.3q$ and $C = f(q) = 1 - 0.5q$, find the function for the company's total profit (Π).

Solution.   We know that total profit is the difference between total revenue and cost: $\Pi = (q) = R(q) - C(q)$. Therefore, by applying equation (1.8.2), we obtain $\Pi(q) = R(q) - C(q) = (0.3q) - (1 - 0.5q) = 0.3q - 1 + 0.5q = -1 + 0.8q$.

Example 2.   Suppose that a firm's total variable cost function is given by TVC $= (q)0.5q$ and its total fixed cost is given by TFC $= 20$, where q is the quantity of the output produced. Find the total cost (C) function.

Solution.   We know that the total cost is the sum of TVC and TFC: $C = (q) = \text{TVC}(q) + \text{TFC}(q)$. Therefore, application of equation (1.8.1) yields $C = (q) = \text{TFC}(q) + \text{TVC}(q) = 20 + 0.5q$.

Example 3.   Assume that a firm's total cost is given by $C = 20 + 0.5q$, where q is the quantity of the output produced. Find the firm's *average cost* (AC) function.

Solution.   We know that AC is the ratio of total cost to the quantity of the good produced: $\text{AC} = f(q) = C(q)/q$. Therefore, application of equation (1.8.4) gives $\text{AC} = f(q) = C(q)/q = (20 + 0.5q)/q = 0.5 + 20/q$.

Example 4.   Suppose that a firm's total cost C per day is a function of the amount of output (q) that it produces on that day, and it is given by $C = f(q) = 10 + 2q$. Also assume that output per day is given by $q = g(l) = 3 + 100l$, where l is the number of workers employed. Find the firm's total cost as a function of the number of workers employed.

Solution.   To solve this problem, we can use the equation for the generation of a composite function discussed in Section 1.8.2; that is, $C = g(f(x))$. Therefore, we can write the total cost as $C = g(f(l)) = 10 + 2(3 + 100l) = 10 + 6 + 200l = 16 + 200l$.

Example 5.   Assume that Ms. Stella was employed by a company. The company offered her a remuneration (y) package that involved three slabs. The first slab was that every month she would be paid a monthly consolidated salary of \$15 000 if the profit (x) is above 0 but below \$100 000. The second slab is that she would be paid a commission of 15 percent of the company's profit every month if the profit is between \$100 000 and \$150 000. The last slab is that she would be given a monthly consolidated salary of \$22 500 if the profit is above \$150 000. Write the function for her remuneration (assuming $x > 0$).

Solution. In this example, $y = f(x)$. Given the information we have, we can write $y = 15\,000$ if $0 \le x < 100\,000$, $y = 0.15x$ if $100\,000 < x < 150\,000$, and $y = 22\,500$ if $x > 150\,000$. This can be written in functional form as

$$y = f(x) = \begin{cases} 15\,000 & 0 \le x < 100\,000 \\ 0.15x & 100\,000 \le x < 150\,000 \\ 22\,500 & x \ge 150\,000 \end{cases}$$

Example 6. Suppose that the demand function for a good is given by $Q_d = f(P) = 10 - 2P$, where Q_d denotes the quantity demanded of the good and P denotes it's unit price. Find the *inverse demand function*.

Solution. The inverse demand function can be found by solving the demand equation for P. Therefore, the required inverse demand function is $P = f^{-1}(Q_d) = 5 - 0.5Q_d$.

1.8.7 Exercises

1. Find the domain of the following functions:
   (i) $y = 1/(3 - x)$; (ii) $y = 1/(\sqrt{5 - x})$; (iii) $y = 2x + 6$; (iv) $y = \sqrt{5 - x}$; (v) $y = 1/x$.
2. Find the range of the following functions:
   (i) $y = [1/(1 + x^2)]$; (ii) $y = x^2 - 5$; (iii) $y = x^2$; (iv) $y = x^3$; (v) $y = 1/x$.
3. If $y = f(x) = 10 + 5x$ and $y = g(x) = 2 + x$, find
   (i) $(f + g)(x)$; (ii) $(f - g)(x)$; (iii) $(f.g)(x)$; (iv) $(f/g)(x)$; (v) $(f + g)(5)$; (vi) $(f - g)(6)$;
   (vii) $(f \circ g)(2)$; (viii) $(f/g)(0)$.
4. Find $(g \circ f)(x)$ and $(f \circ g)(x)$ if
   (i) $g(y) = 2 + 3y$ and $f(x) = 1 + x$; (ii) $g(y) = 2x^2$ and $f(x) = 2 + x$; (iii) $g(y) = 2y^2 + y + 2$
   and $f(x) = x$; (iv) $g(y) = y$ and $f(x) = 2x^2 - x - 2$.
5. Find the inverse functions of the following functions:
   (i) $y = 50 - 5x$; (ii) $y = 1/x$; (iii) $y = x^2$; (iv) $y = \sqrt{x}$; (v) $y = x$.
6. *Application exercise.* Assume that a company sells a good in two markets. The revenue from market one is given by the function $R_1 = f(x) = 0.5x$ and that from market two is given by $R_2 = g(x) = 0.6x$, where x denotes the quantity of the good sold. Find the function for the total revenue.
7. *Application exercise.* Assume that a company's average revenue is given by $AR = f(q) = 5 + q^2 - q$, where q represents the amount output sold. Find the firm's total revenue $R(q)$ function.
8. *Application exercise.* Suppose that a firm's total profit is given by $\Pi = f(x) = 3x^3 - (1/x^2) - (1/x) + 10$, where x denotes the quantity of output produced. Find the firm's profit per unit of output.
9. *Application exercise.* If the price of a good produced by a firm is given by $P = 1/Q$, find the total revenue function of the firm.
10. *Application exercise.* Suppose that a person's income (Y) is a function of the amount of human capital (h) that the person possesses and is given by $Y = f(h) = 5000 + 500h$. Also suppose that the amount of human capital depends on the years of schooling, s, and is given by the function $h = g(s) = 2 + s$. Find the function for the person's income in terms of years of schooling.
11. *Application exercise.* Suppose that an insurance agent receives no commission if she does not sell any policies. The commission will be 5 percent of the value of the policy

if she sells policies below 10 and the commission will be 7 percent if she sells policies equal to and above 10. Express her commission in the form of a suitable function.

12. *Application exercise.* Suppose that a company obtains a profit of $3 per unit from the sale of up to 1000 units of a good, $2 per unit from the sale of the next 1000 units, and $1 per unit from the sale of units above 2000. Illustrate this information in a suitable graph.

13. *Application exercise.* Suppose that the inverse demand function for a good is given by $P = f^{-1}(Q_d) = 90 - 3Q_d$, where P is price and Q_d is quantity demanded. Find the demand function.

1.9 Limits and continuity

The concept of *limit* is fundamental to the branch of *calculus* in mathematics. Therefore, it is important that one understands the meaning and rules of limits for a proper understanding of calculus, a topic that we will deal with later. We shall discuss the concept of *continuity* after the discussion of limits.

1.9.1 Meaning of limits

Limits are closely related to functions. In fact, the limit of a function is the value in the range of the function when the independent variable assumes a particular value in the domain. Consider, for example, the function $y = f(x) = (x^2 - 1)/(x - 1)$. Suppose that x takes values (in the domain) from 0 to 1 such that x will never be equal to 1 but closer and closer to 1. Then the values taken by the function (in the range) will come closer and closer to 2. These movements of x and $f(x)$ are presented in the first two rows, respectively, of Table 1.9.1. Similarly, the values taken by $f(x)$ as x moves from 2 to 1 (again, never equal to 1) are presented in the last two rows of the table.

Table 1.9.1 suggests that $f(x)$ converges to 2 irrespective of the movement of x: either when x comes down from 2 to 1 (from the right) or when x goes up from 0 to 1 (from the left). Therefore, $f(x)$ settles down at 2. This is the limit of the function $f(x)$, and is written $\lim_{x \to 1} f(x) = 2$ and read "the limit of $f(x)$ as x approaches, or tends, to 1 is 2." This can be illustrated as in Figure 1.9.1(A).

Notice that when x approaches 1 from the right, the limit of the function $f(x)$ is 2. This is called *right-side limit* (or *one-sided limit*). When x approaches 1 from the left also the limit of the function $f(x)$ is 2, and it is called *left-side limit*. For the limit to exist, the left- and right-side limits must be the same. We illustrate in Figures 1.9.2(A)–(D) some of the graphical examples of limits (of functions $f(x) = 0.25 + 0.5x - 0.25x^2$; $g(x) = 0.5x$ for $0 \leq x < 1$ and $g(x) = 1 - 0.5x$ for $1 \leq x \leq 2$; $h(x) = 1 - x + 0.5x^2$; and $j(x) = 1 - 0.5x$ for $0 \leq x < 1$ and $j(x) = 0.5x$ for $1 \leq x \leq 2$) as x approaches $a = 1$.

Table 1.9.1

$x < 1$	0	0.1	0.2	0.3	0.4	0.5	0.6	0.7	0.8	0.9
$f(x)$	1	1.1	1.2	1.3	1.4	1.5	1.6	1.7	1.8	1.9
$x > 1$	2	1.9	1.8	1.7	1.6	1.5	1.4	1.3	1.2	1.1
$f(x)$	3	2.9	2.8	2.7	2.6	2.5	2.4	2.3	2.2	2.1

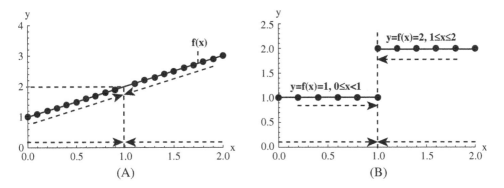

Figure 1.9.1

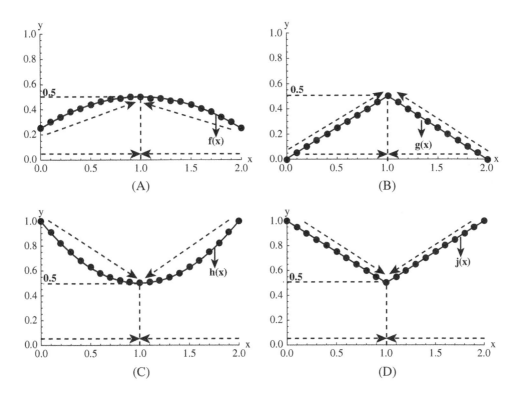

Figure 1.9.2

We now formally define limit. Suppose that we have a function $y = f(x)$. Then the limit of $f(x)$ as x approaches a is the number b and is written as $\lim_{x \to a} f(x) = b$. Notice that the limit is independent of the way x approaches a. Therefore, the limit will exist if and only if both right-side and left-side limits exist and are equal. Notice also that $b = 0.5$ in the examples of the graphs illustrated in Figure 1.9.2.

Now consider, as an example, the piecewise function $y = f(x) = 1$ for $0 \leq x < 1$ and $y = f(x) = 2$ for $1 \leq x \leq 2$, the graph of which is illustrated in Figure 1.9.1(B). It is clear

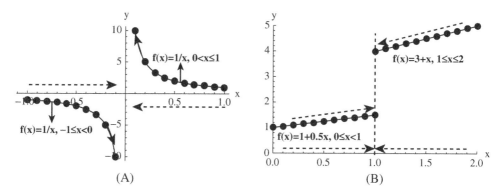

Figure 1.9.3

that $f(x)$ approaches two values when $x \to 1$ from either directions. When $x \to 1$ from the left of $x = 1$, the limit is 1 and when $x \to 1$ from the right of $x = 1$, the limit is 2. Since the left-side and right-side limits are different, as per the definition of limit above, the limit of the piecewise function $y = f(x) = 1$ for $0 \le x < 1$ and $y = f(x) = 2$ for $1 \le x \le 2$ does not exist at $x = 1$.

We illustrate in Figures 1.9.3(A) and (B) two graphical examples of limits that do not exist at $x = a$. In Figure 1.9.3(A), the left-side limit of the function $y = f(x) = 1/x$ is $-\infty$ and the right-side limit of the same function is $+\infty$ as $x \to a = 0$. Similarly, in Figure 1.9.3(B) the left-side limit of the piecewise function ($y = f(x) = 1 + 0.5x$ for $0 \le x < 1$ and $y = f(x) = 3 + x$ for $1 \le x \le 2$) is 1.5 and the right-side limit of the same function is 4 as $x \to a = 1$. Since the left-side and right-side limits of these functions are different, the functions that define the graphs in Figures 1.9.3(A) and (B) do not have limits at $x \to a = 0$ and $x \to a = 1$, respectively.

1.9.2 Properties of limits

Limits exhibit some important properties. Suppose that we have two functions $f(x)$ and $g(x)$. Also let that both $\lim_{x \to a} f(x)$ and $\lim_{x \to a} g(x)$ exist. Then we have the properties given below.

Property I. Let $f(x) = k$, then $\lim_{x \to a} f(x) = \lim_{x \to a} k = k$, where k is a constant.

Property II. Let $f(x) = x^n$, then $\lim_{x \to a} f(x) = \lim_{x \to a} x^n = a^n$, where n is any positive integer.

Property III. $\lim_{x \to x} k \times f(x) = k \times \lim_{x \to a} f(x)$, where k is a constant.

Property IV. $\lim_{x \to a} [f(x) \pm g(x)] = \lim_{x \to a} f(x) \pm \lim_{x \to a} g(x)$.

Property V. $\lim_{x \to a} [f(x) \times g(x)] = \lim_{x \to a} f(x) \times \lim_{x \to a} g(x)$.

Property VI. $\lim_{x \to a} [f(x)/g(x)] = \lim_{x \to a} f(x) / \lim_{x \to a} g(x)$.

Property VII. $\lim_{x \to a} [f(x)]^n = [\lim_{x \to a} f(x)]^n$.

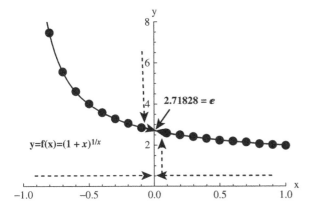

Figure 1.9.4

1.9.3 Evaluation of limits

Students of economics, businesses, and finance are frequently required to evaluate the limits of a variety of functions. We shall explain in the present section how one can evaluate the limits of some of the important functions using the properties listed in Section 1.9.2.

Assume that we have the function $f(x) = 2 + 2x$. What is the limit of $f(x)$ when $x \to 2$? Applying Properties III and IV listed in Section 1.9.2, we can show that the limit of this function is 6. How did we obtain this value? Although x only approaches 2 and is never equal to 2, we obtained the limit 6 through the substitution of 2 for x into the function $f(x)$.

There may be many instances where the method of substitution does not work. For example, consider the function $f(x) = (x^2 - 1)/(x - 1)$. What is the limit of this function as $x \to 1$? Suppose that we apply the substitution method to find the limit of this function. If we substitute $x = 1$ into the function we get a meaningless quantity: zero divided by zero. This shows that the substitution method is not always advisable. Then, how do we find the limit of the function? One way is to convert the original function into another form and then do the substitution. We may write the numerator of the given function as $(x^2 - 1) = (x^2 - x + x - 1) = (x + 1)(x - 1)$. Thus, we may convert the original function $f(x) = (x^2 - 1)/(x - 1)$ into the form $f(x) = (x^2 - 1)/(x - 1) = [(x - 1)(x + 1)]/(x - 1) = (x + 1)$. Now, substitution of $x = 1$ yields $\lim_{x \to 1} f(x) = \lim_{x \to 1} (x + 1) = 1 + 1 = 2$. Therefore the limit of the function as $x \to 1$ is 2. Notice that this was the result we obtained in Section 1.9.1.

Now consider an important function that often appears in the subjects of our interest: $f(x) = (1 + x)^{1/x}$. What is the limit of this function as $x \to 0$? It is easy to see the limit of this function as $x \to 0$ with help of the graph of this function as illustrated in Figure 1.9.4. We see from this graph that the limit of the function as $x \to 0$ from either side is 2.71828. This is the value of the term e, the base of natural logarithm, we mentioned in Section 1.5. Therefore, we can write $\lim_{x \to 0} [(1 + x)^{1/x}] = 2.71828 = e$.

1.9.4 Continuity

Informally, a continuous function is a function whose graph can be drawn without lifting the pencil from the paper (i.e. the graph does not contain gaps). When looking from this angle, many of the graphs that we have constructed so far are the graphs of continuous functions. Similarly, a function whose graph is broken or contains gap, say at $x = a$, is a discontinuous function. Graphs in Figures 1.8.9, 1.8.10, 1.9.1(B), and 1.9.3(B) are examples of discontinuous functions.

However, we need a formal definition of continuity. We state it here. Let there be a function $f(x)$ and a point $x = a$ in the domain of $f(x)$. Then $f(x)$ is said to be continuous at $x = a$ if

1. $\lim_{x \to a} f(x)$ exists,   2. $f(a)$ exists,   and 3. $\lim_{x \to a} f(x) = f(a)$

If one or more of these three requirements are not met by the function $f(x)$ at $x = a$, then the function is said to be discontinuous at $x = a$. We also state that a function is continuous on the open interval (a, b) if it is continuous at each point on the interval, and that a polynomial function is continuous at every point in its domain.

As an example, we consider the function $f(x) = 3 + x$ to check whether it is continuous at $x \to a = 1$. Since $\lim_{x \to a = 1} f(x) = \lim_{x \to a = 1} (3 + x) = 3 + 1 = 4$, the first condition above is satisfied. Since $f(x = a = 1) = f(1) = 4$, the second condition is satisfied. And since $\lim_{x \to a = 1} f(x) = f(x = a = 1) = 4$, the third condition also is satisfied. Therefore, the function $f(x) = 3 + x$ is continuous at $x \to a = 1$.

As another example, we check whether the function $f(x) = x^2 - 1$ is continuous at $x = 3$. First, we know that $\lim_{x \to 3} f(x) = \lim_{x \to 3} (x^2 - 1) = 8$. Second, we obtain $f(3) = (3^2 - 1) = 8$ at $x = 8$. Third, we have $\lim_{x \to 3} f(x) = f(3) = 8$. Since all the requirements of a continuous function are satisfied, then $f(x) = x^2 - 1$ is continuous at $x = 3$.

Now consider the function $f(x) = 1/(1 - x)$. Is this function continuous at $x = a = 1$? First, we check its limit at $x = a = 1$. At $x = 1$, the function approaches infinity $(f(1) = 1/(1 - 1) = \infty)$. Therefore, the limit of the function does not exist at $x = a = 1$. This means that the first condition of continuity is violated. Since $f(1) = 1/(1 - 1) = \infty$, the second condition is also violated. And since $f(a = 1) = \lim_{x \to a = 1} f(x) = \infty$, the third condition is satisfied. Since the first two conditions are violated, the function $f(x) = 1/(1 - x)$ is not continuous at $x = a = 1$.

1.9.5 Properties of continuity

The important properties of continuity are the following.

Property I.   Let there be two continuous functions, $f(x)$ and $g(x)$, on an interval. Then $f(x) + g(x), f(x) - g(x), f(x).g(x)$, and $f(x)/g(x)$ are also continuous on the same interval, except for those values of x that make a denominator zero.

Property II.   A constant function $f(x) = a$, where a is a constant, is continuous for all x. For example, $f(x) = 1$ is continuous for all x.

Property III.   A power function $f(x) = x^n$, where n is a positive integer, is continuous for all x. For example, $f(x) = x^2$ is continuous for all x.

Property IV.   A polynomial function is continuous for all x. For example, the function $f(x) = x^2 + x - 1$ is continuous for all x.

Property V.   A rational function is continuous for all x except for those values of x that make the denominator zero. For example, $f(x) = [(x^2 - 1)/(x - 1)]$ is continuous for all $x \neq 1$.

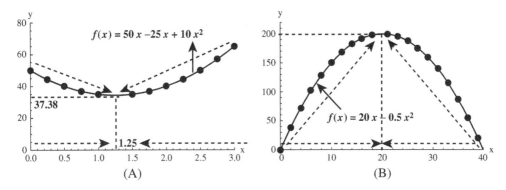

Figure 1.9.5

Property VI. Let n be a positive integer greater than 1. Then, the function $\sqrt[n]{f(x)}$ is continuous whenever $f(x)$ is continuous. For example, $\sqrt[5]{x^3}$ is continuous for all x.

1.9.6 Application examples

Example 1. Suppose that a company's total average cost is given by the function $f(x) = 10x^2 - 25x + 50$, where x represents the quantity of the good produced. Draw a graph of this function and evaluate its limit as $x \to 1.25$.

Solution. The function approaches its minimum value of 37.38 as $x \to 1.25$. The graph of the function is illustrated in Figure 1.9.5(A).

Example 2. Assume that the total profit of a company is given by $f(x) = 20x - 0.5x^2$, where x denotes the quantity of the good produced. Show the limit of the function graphically as $x \to 20$.

Solution. The graph of the function is illustrated in Figure 1.9.5(B). It can be seen from the graph that the function approaches its maximum value 200 as $x \to 20$.

1.9.7 Exercises

1. Find the limits, if they exist, of the following functions:
   (i) $\lim_{x \to 3} 5$; (ii) $\lim_{x \to 3}(14x)$; (iii) $\lim_{x \to 3}(14x - 20)$; (iv) $\lim_{x \to 3}(14x + 20)$; (v) $\lim_{x \to 3}[(1 + 3x)/(2 + 2x)]$.
2. Find the limits, if they exist, of the following functions:
   (i) $\lim_{x \to 0}(1/x)$; (ii) $\lim_{x \to 0}[(x^2 - 1)/(x + 1)]$; (iii) $\lim_{x \to 0}[(x^2 + 2x + 1)x]$; (iv) $\lim_{x \to 0}[(2x + 1)^2/x]$;
   (v) $\lim_{x \to \infty}[x^2/(1 + x^2)]$.
3. Determine whether the following functions are continuous or discontinuous at the specified points:
   (i) $f(x) = 3$ at $x = 1$; (ii) $f(x) = 3x + 3$ at $x = 1$; (iii) $f(x) = (1 - x)/2$ at $x = 1$;
   (iv) $f(x) = (x - 3)/(x^2 - 9)$ at $x = 3$; (v) $f(x) = (x^2 - 9)/(x - 3)$ at $x = 1$.

⚠️ **Web supplement: S1.9.8 Mathematica applications**

1.10 Sequences and series

Many of the relationships in the subjects of our interest follow some specific patterns. Sequences and series are among the features of these patterns. Therefore, students of these subjects require a reasonable understanding of sequences and series. We present a discussion of these in this section.

1.10.1 Meaning of a sequence

Suppose that someone asked us to list all odd integer numbers greater than 0. Then our list would be like this

$$1, 3, 5, 7, 9, 11, 13, 15, 17, 19, \ldots$$

Informally, a list of numbers such as the list above is called a *sequence*. We now attempt to formalize this informal definition of a sequence. The first number (1) in the above list can be written as $2 \times 0 + 1 = 1$. The second number (3) can be written as $2 \times 1 + 1 = 3$. The third number (5) can be written as $2 \times 2 + 1 = 5$. This process can be continued indefinitely. In general, the nth odd number in the above sequence will be equal to

$$2 \times n + 1 = f(n) = s_n \quad \text{or} \quad s_n = f(n) = 1 + 2 \times n \tag{1.10.1}$$

Notice that we have a set of positive integers represented by the set $n = \{1, 2, 3, \ldots\}$. What we did was that we mapped the set n into a set of odd integers given in the above list by the function in equation (1.10.1). We may now formally define a sequence as a function whose domain is a set of positive integers. An entry, for example s_n, is called a *term* of the sequence. Notice also that the sequence represented by equation (1.10.1) has, or any other sequence would have, four distinct essentials: an equation or function (as equation (1.10.1)), a domain (as $n = \{1, 2, 3, \ldots\}$), a range (as the above list), and a sequence of entries (as in the above list). If n is finite or if the domain of a sequence is a finite set of numbers, we have a *finite sequence*. If, instead, n is infinite, we have an *infinite sequence*.

As an example, consider the function $s_n = f(n) = 1 + n$. Assume that the domain is the set $n = \{1, 2, 3, 4, \ldots\}$, the set of all positive integers. When we map the values in the domain using the function $s_n = 1 + n$, we obtain the range. This set includes all positive integers greater than one. This list of numbers, (2, 3, 4, 5, \ldots), constitutes a sequence.

1.10.2 Converging and diverging sequences

Sequences may generally be divided into two groups: *convergent sequence* (or finite sequence) and *divergent sequence* (or infinite sequence). We explain each with examples. Consider the function $f(n) = s_n = 4 - n$. Assume that $f(n)$ represents the price of a good and n represents the quantity of the good that is demanded at price $f(n)$. Then this is an example of a linear inverse demand function introduced earlier. As is evident from the function, s_n decreases as n increases. And in the limit, s_n will converge to 0 (ruling out the case of negative price) as $n \to 4$. Such a sequence is called a convergent sequence. The graph of this convergent sequence and the graph of the function are illustrated in Figure 1.10.1(A).

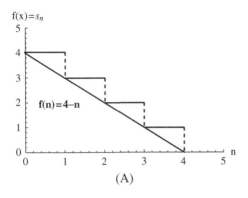
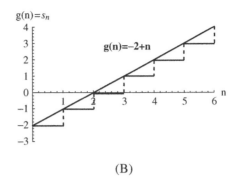

Figure 1.10.1

As another example, consider the function $g(n) = -2 + n$. Assume that $g(n)$ represents the price of a good and n represents the quantity of the good supplied at price $g(n)$. This is an example of a linear inverse supply function introduced earlier. It is evident from this function that $g(n)$ increases without bounds, as n increases. Such a sequence is called a divergent sequence. The graph of this divergent sequence and the graph of the function are illustrated in Figure 1.10.1(B).

1.10.3 Series

The sum of the terms of a sequence is called a *series*. For example, assume that we have a sequence $s_1, s_2, s_3, s_4, \ldots, s_n, \ldots$ Then the sum of this sequence $s_1 + s_2 + s_3 + s_4 + \cdots + s_n + \cdots$ is called a series. If the sequence is finite, the associated series is called a *finite series* or a *convergent series*; if the sequence is infinite, the associated series is called an *infinite series* or a *divergent series*. An example of a finite series is $2 + 4 + 6 + 8$ if the associated sequence is 2, 4, 6, and 8; an example of an infinite series is $1 + 3 + 5 + 7 + \ldots$ if the associated sequence is $1, 3, 5, 7, \ldots$.

1.10.4 Arithmetic progression

Before we begin the discussion of *arithmetic progression*, we need to understand the meaning of the term *progression*. The term progression in mathematics is closely related to a sequence. Progression is an arrangement of a number of terms in a definite order for which there is a pattern or rule that allows the identification of successive terms. The fundamental difference between a sequence and a progression is that the former may or may not exhibit a definite order between terms whereas in the latter there will always be an order between terms. Therefore, a sequence with definite order between terms is called a progression. We will be concerned mainly with progressions.

For example, consider the following sequences of numbers: 1, 11, 21, 31, 41, ...; 1, 2, 3, 4, 5, ...; and 10, 9, 8, 7, ... Consider the first set of numbers to begin with. In this set, the first number is 1 and each subsequent number is obtained by adding 10 to the previous number. In the second set, the first number is again 1 but each subsequent number is obtained by adding 1 to the previous number. In the third set, the first number is 10 and each subsequent number is obtained by subtracting 1 from (or adding -1 to) the previous number. These are examples of *arithmetic sequence* or arithmetic progression; sequence or progression obtained through addition. Arithmetic progression is one in which there is a constant difference between any two consecutive terms.

We now define arithmetic progression formally. Let there be a sequence $s_1, s_2, s_3, s_4, \ldots,$ $s_n, \ldots$. This sequence is called an arithmetic progression if there is a constant term a such that $s_n - s_{n-1} = a$ or $s_n = a + s_{n-1}$. The term a is called the *common difference*. Notice that $a = 10$, $a = 1$, and $a = -1$, respectively, in the last three sets of numbers.

Notice that if we are given a and the first term, we can find the nth term of the progression. However, we can state the formal method of finding the nth term of an arithmetic progression. Suppose that we have an arithmetic progression $s_1, s_2, s_3, s_4, \ldots, s_n, \ldots$. We know that $s_2 = s_1 + a$; $s_3 = s_2 + a = s_1 + a + a = s_1 + 2a$; $s_4 = s_3 + a = s_2 + a + a = s_1 + a + a + a = s_1 + 3a$; $\ldots$. This implies that the nth term of an arithmetic progress is given by

$$s_n = s_1 + (n-1) \times a, \quad \text{for all } n > 1 \tag{1.10.2}$$

Sometimes we may need to find the sum of an arithmetic progression. Let $s_1, s_2, s_3, s_4, \ldots, s_n$ be a *finite arithmetic sequence*. Then the associated series is given by $s_1 + s_2 + s_3 + s_4 + \cdots + s_n$ and is called a *finite arithmetic series*. How can we find this sum? We may find it with the help of a formula, which can be derived through the following procedure.

We know from above that $s_2 = s_1 + a$; $s_3 = s_2 + a = s_1 + a + a$; $s_4 = s_3 + a = s_2 + a + a = s_1 + a + a + a = s_1 + 3a$; and so on. It follows that $s_n = s_1 + (n-1)a$. Therefore, the sum of this finite arithmetic series is given by S_n as: $S_n = s_1 + (s_1 + a) + (s_1 + 2a) + (s_1 + 3a) + \cdots + [s_1 + (n-1)a]$, or $S_n = ns_1 + a + 2a + 3a + \cdots + (n-1)a = ns_1 + (n-1)a + (n-2)a + \cdots + 3a + 2a + a$. Adding the last two equations gives us $2S_n = 2ns_1 + na + na + na + \cdots + na = 2ns_1 + (n-1)na$. Therefore, we have

$$S_n = (n/2) \times [2s_1 + (n-1)a] \tag{1.10.3}$$

Using equation (1.10.2), equation (1.10.3) can be written as

$$S_n = (n/2) \times [s_1 + s_n] \tag{1.10.4}$$

1.10.5 Geometric progression

Consider the sequence of numbers 1, 2, 4, 8, 16, 32, 64, $\ldots$. The important feature of this sequence is that each successive term is 2 times the previous term. That is, $2 = 1 \times 2$; $4 = 1 \times 2 \times 2 = 1 \times 2^2$; $8 = 1 \times 2 \times 2 \times 2 = 1 \times 2^3$; and so on. This is an example of a *geometric sequence* or *geometric progression*. A geometric progression is one in which there is a constant ratio between any two consecutive terms. The general form of any geometric progression is $s_1, s_1 k, s_1 k^2, s_1 k^3, s_1 k^4, \ldots, s_1 k^n, \ldots$, where s_1 is the first term and k is the constant *common ratio*.

We now define geometric progression formally. Let there be a sequence $s_1, s_2, s_3, s_4, \ldots,$ $s_n, \ldots$. This sequence is called a geometric progression if there is a constant common ratio k such that $k = s_n/s_{n-1}$ or $s_n = k \times s_{n-1}$. Notice that $s_1 = 1$ and $k = 2$ in the above geometric sequence.

How can we find the nth term of the geometric progression? We know that $s_2 = s_1.k$; $s_3 = s_2.k = s_1.k.k = s_1k^2$; $s_4 = s_3.k = s_1k^2.k = s_1k^3$; and so on. This implies that the nth term of a geometric progression is given by

$$s_n = s_1.k^{n-1} \quad \text{for all } n > 1 \tag{1.10.5}$$

Equation (1.10.5) implies that once we are given s_1 and k, we can easily find the nth term. However, as in the case of arithmetic progression, we may need to find out the sum of a *finite geometric sequence* or a *finite geometric progression*. We can find it with the help of a formula. But, we shall first derive the formula through the following procedure. We know from above that our finite geometric progression is $s_1, s_1k, s_1k^2, s_1k^3, s_1k^4, \dots, s_1k^{n-1}$. Our aim is to find the sum $s_1 + s_1k + s_1k^2 + s_1k^3 + \cdots + s_1k^{n-1}$, which is called a *geometric series*. In other words, we need to find $S_n = s_1 + s_1k + s_1k^2 + s_1k^3 + \cdots + s_1k^{n-1}$. Multiplying both sides of this equation by k yields $kS_n = s_1k + s_1k + s_1k^3 + s_1k^4 + \cdots + s_1k^n$. Now subtracting kS_n from S_n yields $S_n(1-k) = s_1 - s_1k^n = s_1(1-k^n)$, or

$$S_n = [s_1(1-k^n)/(1-k)] = s_1(k^n-1)/(k-1), \quad \text{where } k \neq 1 \tag{1.10.6}$$

which can be written in an alternative form. We have from equation (1.10.5) that $s_n = s_1k^{n-1}$, which may be written as $ks_n = s_1k^n$. Therefore, equation (1.10.6) may be transformed into

$$S_n = s_1(1-k^n)/(1-k) = (s_1 - s_1k^n)/(1-k)$$
$$= (s_1 - ks_n)/(1-k) = (ks_n - s_1)/(k-1), \quad \text{where } k \neq 1 \tag{1.10.7}$$

One pertinent question now is how one can find the sum of an *infinite geometric progression* such as $s_1, s_2, s_3, s_4, \dots, s_n, \dots$ called an *infinite geometric series*. This question may be answered by rewriting equation (1.10.6) as

$$S_n = s_1(k^n-1)/(k-1) = [s_1k^n/(k-1)] - [s_1/(k-1)] \tag{1.10.8}$$

Notice a special case of $-1 < k < 1$. In this case, to find the series we may rewrite equation (1.10.6) once again as

$$S_n = s_1(1-k^n)/(1-k) = (s_1 - s_1k^n)/(1-k) = [s_1/(1-k)] - [s_1k^n/(1-k)] \tag{1.10.9}$$

It can be shown that, given $-1 < k < 1$, k^n will approach 0 as n becomes infinitely large (or, $k^n \to 0$ as $n \to \infty$). Therefore, the geometric series with $-1 < k < 1$ will converge to $S_n = s_1/(1-k)$. Hence we have the result

$$S_n = s_1/(1-k) \tag{1.10.10}$$

1.10.6 The binomial theorem

Before we begin a discussion of the *binomial theorem*, we need to explain three concepts used in the development of this theorem. The first is the concept of *factorials*. Let there be a natural number n. If we write the product of all the integers between (and including) n and 1, it is called *n factorial*, and is denoted by $n!$. That is, $n! = n \times (n-1) \times (n-2) \times (n-3) \times 2 \times 1$.

For example, $4! = 4 \times 3 \times 2 \times 1 = 24$. It should be noted that both one factorial and zero factorial are defined to be 1 (that is, $1! = 0! = 1$).

The second concept is that of *permutation*. If r items are selected (without replacement) from a set of n items, any particular sequence of these r items is called a permutation. As an example, we know that ab is a permutation of two of the first three letters of the English alphabet. The other permutations are ba, ac, ca, bc, and cb. In general, the number of permutations of r items that can be selected from n items is $P_{n,r} = n \times (n-1) \times (n-2) \times \cdots \times (n-r+1)$. This equation can be expressed in another way. Let us multiply and divide the RHS of the equation by $(n-r) \times (n-r-1) \times \cdots \times (2) \times (1)$ to yield $P_{n,r} = [n \times (n-1) \times (n-2) \times \cdots \times (n-r+1) \times (n-r) \times (n-r-1) \times \cdots \times (2) \times (1)]/[(n-r) \times (n-r-1) \times \cdots \times (2) \times (1)]$, where the numerator is $n!$ and the denominator is $(n-r)!$. Therefore, the last equation can be written as

$$P_{n,r} = n!/(n-r)! \tag{1.10.11}$$

In our example of letters, the number of permutations, using equation (1.10.11), of $r = 2$ items one can select from a set of $n = 3$ items is $P_{3,2} = 3!/(3-2)! = 6/1 = 6$.

The last concept is *combination*. It is clear that the order of the items matters in permutations. A combination is a selection of items where the order does not matter. The number of combinations of r items that can be selected from n items is

$$C_{n,r} = [n \times (n-1) \times (n-2) \times \ldots \ldots \times (n-r+1)]r!$$

$$= [n \times (n-1) \times (n-2) \times \ldots \times (n-r+1) \times (n-r) \times (n-r-1) \times \cdots$$

$$\times (2) \times (1)]/[(n-r) \times (n-r-1) \times \cdots \times (2) \times (1) \times r!]$$

$$= n!/(n-r)!r! \tag{1.10.12}$$

where the second step has been multiplied and divided by $(n-r) \times (n-r-1) \times \cdots \times (2) \times (1)$. The reader will have noticed the connection between permutation and combination: $P_{n,r} = r! \times C_{n,r}$. As an example, suppose that we want to select three balls from five balls. How many ways can we do this? In this example, $n = 5$ and $r = 3$. Therefore, applying equation (1.10.12), the answer is $C_{5,3} = [5!/(5-3)!3!] = [\{5 \times 4 \times 3 \times 2 \times 1\}/\{(2 \times 1) \times (3 \times 2 \times 1)\}] = 20/2 = 10$. Notice that equation (1.10.12) is the basis of the solution we obtained for example 2 in Section 1.2.7.

We shall now begin the discussion of the binomial theorem. One might wonder why we need a theorem such as this. The rationale is as follows. Suppose that we want to expand the expression $(a+b)^2$. We can do this easily by factoring $(a+b)^2$ and expanding the factors through multiplication. The result is $(a+b)^2 = (a+b) \times (a+b) = a^2 + 2ab + b^2$. Similarly, we can obtain the expansion of $(a+b)^3$ as $(a+b)^3 = (a+b)(a+b)^2 = (a+b) \times (a^2 + 2ab + b^2) = (a^3 + 3a^2b + 3ab^2 + b^3)$. However, the job involved becomes tedious as the power of the expression increases. The question, therefore, is how one can find the expansion of an expression such as $(a+b)^n$. The binomial theorem comes in handy here. It states that

$$(a+b)^n = C_{n,0}a^n + C_{n,1}a^{n-1}b^1 + C_{n,2}a^{n-2}b^2 + C_{n,3}a^{n-3}b^3 + \cdots + C_{n,n}a^{n-n}b^n$$

$$= C_{n,0}a^n + C_{n,1}a^{n-1}b + C_{n,2}a^{n-2}b^2 + C_{n,3}a^{n-3}b^3 + \cdots + C_{n,n}b^n \tag{1.10.13}$$

which may be derived as follows. We know from above that $(a+b)^1 = a+b$; $(a+b)^2 = a^2 + 2ab + b^2$; $(a+b)^3 = a^3 + 3a^2b + 3ab^2 + b^3$; $(a+b)^4 = a^4 + 4a^3b + 6a^2b^2 + 4ab^3 + b^4$;

$(a+b)^5 = a^5 + 5a^4b + 10a^3b^2 + 10a^2b^3 + 5ab^4 + b^5$. Notice that every expansion here is a particular series. Also notice some of the features of the above expansions. Firstly, there are $(n+1)$ terms in the expansion of $(a+b)^n$. Secondly, the power of a (b) decreases (increases) as we move rightward in each expansion. Thirdly, the sum of powers of a and b in each term is equal to n. Fourthly, and lastly if we are given a term, we can get the coefficient of the next term by multiplying the coefficient of the given term by the exponent of a and dividing by the number that represents the position of the term in the series of terms. Using these features, we can write the expansion of the general case as

$$(a+b)^n = a^n + \frac{n}{1}a^{n-1}b + \frac{n(n-1)}{1 \times 2}a^{n-2}b^2 + \frac{n(n-1)(n-2)}{1 \times 2 \times 3}a^{n-3}b^3 + \cdots + b^n$$

$$= \frac{n!}{0!(n-0)!}a^n + \frac{n!}{1!(n-1)!}a^{n-1}b + \frac{n!}{2!(n-2)!}a^{n-2}b^2$$

$$+ \frac{n!}{3!(n-3)!}a^{n-3}b^3 + \cdots + \frac{n!}{n!(n-n)!}b^n$$

$$= C_{n,0}a^n + C_{n,1}a^{n-1}b + C_{n,2}a^{n-2}b^2 + C_{n,3}a^{n-3}b^3 + \cdots + C_{n,n}b^n$$

which is equation (1.10.13).

One application of the binomial theorem is in the determination of the vale of e, the base of natural logarithm. Remember that we stated in Section 1.3.1 that value $e = 2.71828$ and that we obtained this value in Section 1.9.3 when we geometrically evaluated $\lim_{x \to 0}[(1+x)^{1/x}]$. One can also apply the binomial theorem to obtain the value $e = 2.71828$. The procedure is as follows. Let us first convert $\lim_{x \to 0}[(1+x)^{1/x}]$ into the equivalent form $\lim_{n \to \infty}[(1+1/n)^n]$, where $n = 1/x$ or $x = 1/n$. Our problem now is to evaluate $\lim_{n \to \infty}[(1+1/n)^n]$. But, before this we need to expand the expression $(1+1/n)^n$. Using the binomial theorem in equation (1.10.13) with $a = 1$ and $x = b = 1/n$, the expression $(1+1/n)^n$ can be expanded as

$$\left(1+\frac{1}{n}\right)^n = C_{n,0}a^{n-0}b^0 + C_{n,1}a^{n-1}b^1 + C_{n,2}a^{n-2}b^2 + C_{n,3}a^{n-3}b^3 + \cdots + C_{n,n}a^{n-n}b^n$$

$$= 1 \times 1 \times 1 + n \times 1 \times \frac{1}{n} + \frac{n(n-1)}{2!} \times 1 \times \frac{1}{n^2} + \frac{n(n-1)(n-2)}{3!}$$

$$\times 1 \times \frac{1}{n^3} + \cdots + 1 \times 1 \times \frac{1}{n^n}$$

$$= 1 + 1 + \left(\frac{n}{n}\right)\left(\frac{n-1}{n}\right)\frac{1}{2!} + \left(\frac{n}{n}\right)\left(\frac{n-1}{n}\right)\left(\frac{n-2}{n}\right)\frac{1}{3!} + \cdots + \frac{1}{n^n}$$

Taking limits on both sides of the last equation as $n \to \infty$ we obtain

$$\lim_{n \to \infty}\left(1+\frac{1}{n}\right)^n = \lim_{n \to \infty}\left[1 + 1 + \left(\frac{n}{n}\right)\left(\frac{n-1}{n}\right)\frac{1}{2!} + \left(\frac{n}{n}\right)\left(\frac{n-1}{n}\right)\left(\frac{n-2}{n}\right)\frac{1}{3!} + \cdots + \frac{1}{n^n}\right]$$

$$e = \lim_{x \to 0}(1+x)^{1/x} = \lim_{n \to \infty}\left(1+\frac{1}{n}\right)^n = 1 + 1 + \frac{1}{2!} + \frac{1}{3!} + \frac{1}{4!} + \frac{1}{5!} + \frac{1}{6!} + \frac{1}{7!} + \frac{1}{8!} + \cdots$$

$$(1.10.14)$$

This expansion of e is called the *series expansion* of e, which can be used to find the value of e to any decimal. The value of e when we sum the first 9 terms of the above expansion is 2.71828, which was the value we mentioned in Section 1.3.1, and geometrically obtained in Section 1.9.3.

Another closely related application of the binomial theorem lies in the evaluation of $\lim_{n \to \infty} [(1 + x/n)^n]$. How do we find this limit? The procedure is similar to the one above. Let us first find $(1 + x/n)^n$. Letting $a = 1$, $b = x/n$ and using the binomial theorem, we have

$$\left(1 + \frac{x}{n}\right)^n = C_{n,0} a^{n-0} b^0 + C_{n,1} a^{n-1} b^1 + C_{n,2} a^{n-2} b^2 + C_{n,3} a^{n-3} b^3 + \cdots + C_{n,n} a^{n-n} b^n$$

$$= 1 \times 1 \times 1 + \frac{n}{1!} \times 1 \times \frac{x}{n} + \frac{n(n-1)}{2!} \times 1 \times \frac{x^2}{n^2} + \frac{n(n-1)(n-2)}{3!}$$

$$\times 1 \times \frac{x^3}{n^3} + \cdots + 1 \times 1 \times \frac{x^n}{n^n}$$

$$= 1 + \frac{x}{1!} + \left(\frac{n-1}{n}\right) \frac{x^2}{2!} + \left(\frac{n-1}{n}\right) \left(\frac{n-2}{n}\right) \frac{x^3}{3!} + \cdots + \frac{1}{n^n}$$

Taking limits on both sides of the last equation as $n \to \infty$ we obtain

$$\lim_{n \to \infty} \left(1 + \frac{x}{n}\right)^n = \lim_{n \to \infty} \left[1 + \frac{x}{1!} + \left(\frac{n-1}{n}\right) \frac{x^2}{2!} + \left(\frac{n-1}{n}\right) \left(\frac{n-2}{n}\right) \frac{x^3}{3!} + \cdots + \frac{1}{n^n}\right]$$

$$\lim_{n \to \infty} \left(1 + \frac{x}{n}\right)^n = 1 + \frac{x}{1!} + \frac{x^2}{2!} + \frac{x^3}{3!} + \cdots \tag{1.10.15}$$

Notice that $\lim_{n \to \infty} (1 + x/n)^n = \lim_{m \to \infty} (1 + 1/m)^{m \times x}$, where $m = n/x$. But, as per equation (1.10.14), $\lim_{m \to \infty} (1 + 1/m)^m = e$. This implies that $\lim_{n \to \infty} (1 + x/n)^n = \lim_{m \to \infty} (1 + 1/m)^{m \times x} = e^x$. Therefore, equation (1.10.15) can be written as

$$\lim_{n \to \infty} \left(1 + \frac{x}{n}\right)^n = e^x = 1 + \frac{x}{1!} + \frac{x^2}{2!} + \frac{x^3}{3!} + \cdots \tag{1.10.16}$$

which can be used to evaluate e^x for any value of x. The expansion of e^x in equation (1.10.16) is called the *exponential series*.

1.10.7 Application examples

Example 1. Suppose that the inverse demand function for a good is $f(x) = 1/x$, where $f(x)$ represents the price of the good and x represents the quantity demanded of the good. Determine whether the sequence represented by this demand function is converging or diverging.

Solution. If we plug positive integers for x in $f(x) = 1/x$, we get $f(1) = 1/1 = 1$, $f(2) = 1/2 = 0.5, f(3) = 1/3 = 0.333, f(4) = 1/4 = 0.25, f(5) = 1/5 = 0.2$, and so on. This suggests that $f(x)$ decreases as x increases and in the limit $f(x) \to 0$ as $x \to \infty$. Therefore, the sequence represented by the above function is a converging sequence. The graph of this function is illustrated in Figure 1.10.2(A).

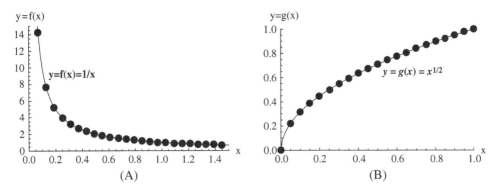

Figure 1.10.2

Example 2. Assume that the total profit of a company is given by the function $g(x) = x^{1/2}$, where $g(x)$ represents the total profit of the company and x represents the quantity of the good produced. Determine whether the sequence represented by this profit function is converging or diverging.

Solution. If we plug positive integers for x in $g(x) = x^{1/2}$, we obtain $g(1) = 1^{1/2} = 1$, $g(2) = 2^{1/2} = 1.414$, $g(3) = 3^{1/2} = 1.732$, $g(4) = 4^{1/2} = 2$, and so on. This suggests that $g(x)$ increases as x increases and in the limit $g(x) \to \infty$ as $x \to \infty$. Therefore, the sequence represented by this profit function is a diverging sequence. The graph of this function is illustrated in Figure 1.10.2(B).

Example 3. Suppose that a person receives a monthly salary of \$15 000, which increases by \$1000 every year. What will be the person's salary in the tenth year and what will be the total salary the person received after ten years?

Solution. This is an example of an arithmetic progression with the initial value (s_1) of \$15 000 and the common difference (a) of \$1000. In order to find the person's salary in the tenth year we can apply equation (1.10.2). Therefore, the person's salary in the tenth year will be $s_{10} = s_1 + (n-1)a = \$15000 + (10-1)\$1000 = \$15000 + \$9000 = \$24000$. To find the total salary the person received after 10 years we can use equation (1.10.3). Therefore, the total salary will be $S_n = (n/2).[2s_1 + (n-1)a] = (10/2).[2 \times \$15000 + (10-1)\$1000] = \$195\,000$.

Example 4. Suppose that a small company expects to achieve a profit of \$10 000 in the first year, which is expected to double every year after that. What will be the company's fifth year's profit and what will be the total profit after the first five years?

Solution. This is an example of a geometric progression with the initial value (s_1) of \$10 000 and the common ratio (k) of 2. To find out the company's profit in the fifth year we can apply equation (1.10.5). Therefore, the fifth year's profit will be $s_5 = s_1 k^{n-1} = \$10000 \times 2^{5-1} = \$10000 \times 16 = \$160000$. To find out the company's total profit after the first five years we may apply equation (1.10.6). Therefore, the total profit of the company after the fifth year will be $S_5 = s_1(k^n - 1)/(k - 1) = \$10000(2^5 - 1)/(2 - 1) = \$310000$.

Example 5.   Suppose that the initial change in the investment in an economy ($\Delta I = s_1$) is \$2 billion. If the *marginal propensity to consume* (MPC, the fraction of a dollar that consumers wish to spend) in the economy is 0.75, what will be the total amount of income generated ($\Delta Y = S_n$) in the economy as the economic transactions become infinitely large (that is, $n \to \infty$)?

Solution.   This is an example of an infinite geometric series. In order to find the changed total income ($\Delta Y = S_n$) we have to apply equation (1.10.10) because the common ratio in this example lies between 0 and 1 ($0 < k < 1$). Therefore, the economy's changed total income will be $\Delta Y = S_n = s_1/(1-k) = \$2 \text{ billion}/(1-0.75) = \$8$ billion. Since $\Delta I = s_1$, we can write $\Delta Y/\Delta I = 1/(1-k)$. This ratio is called *investment multiplier*, which measures the change in national income per unit change in investment in the economy. In the present example, $\Delta Y/\Delta I = 1/(1-k) = 1/(1-0.75) = 4$. Notice that the total change in national income ($\Delta Y = S_n = \$8$ billion) is equal to the initial change in investment ($\Delta I = s_1 = \$2$ billion) times the investment multiplier (4).

1.10.8 Exercises

1.   Determine which sequences represented by the following functions converge (or diverge):
     (i) $f(x) = [(x)^4]^{1/2}$; (ii) $f(x) = 1/x^2$; (iii) $f(x) = 10 + x^2$; (iv) $f(x) = 10 - 1/x^2$.
2.   Give the fifteenth terms (and determine the sum of the first fifteen terms) of the following progressions:
     (i) 1, 2, 3, 4, 5,...; (ii) 50, 100, 150, 200, 250,...; (iii) 10, 9, 8, 7, 6, 5,....
3.   Determine the tenth term (and the sum of the first ten terms) of the following progressions:
     (i) 5, 25 125, 625,...; (ii) 2, 4, 8, 16,...; (iii) 6, 36, 216, 1296,....
4.   *Application exercise.* Suppose that a company received a profit of \$500 000 in the first year of its operation. The company expects that this profit will increase by \$50 000 every year after the first year. What will the company's profit at the end of the seventh year and what will be its total profit after the first seven years?
5.   *Application exercise.* Assume that a company expects to achieve a profit of \$50 000 in the first year of its operation, which is expected to halve every year after the first year. What will be the company's profit at the end of the fifth year and what will be the total profit of the company after the first five years?

 Web supplement: S1.10.9 Mathematica applications

1.11 Sums and products

Sum and product symbols, and their rules, are widely used in mathematics and economic sciences. A working knowledge of these symbols and the associated rules is necessary for students in these branches. Therefore, we provide here a fairly comprehensive treatment of these symbols and their rules.

1.11.1 Sums

Assume that we wish to find out the total population of five cities in a country. Also assume that the population of a city is denoted by the letter P_i, where the subscript i denotes the *i*th city. Then the sum of populations of the five cities can be shown as $P_1+P_2+P_3+P_4+P_5$. Every number in the last expression is called an entry.

It was easy to write the above expression as there were only five entries. The above expression of sum can be written in an alternative, abbreviated form with the Greek letter Σ (read "sigma") as $\sum_{i=1}^{i=5} P_i$, where i represents the ith city and is called the *index of summation*. Notice that one can use any index. This expression is read "sigma i equals 1 to i equals 5 P_i," and it gives the sum of P (population) for city $i=1$ to city $i=5$.

But, if we have a list of populations of all the cities and towns in a vast country like the USA, India, China, or Russia (or any other similar country), it will be a laborious task to write the list. In order to write such a long expression that involves n entries, one can use the "sigma" notation introduced above. Suppose that we have an expression that involves n entries such as $P_1+P_2+P_3+P_4+P_5+\cdots+P_n$. Then we can write this long expression in the abbreviated form as $\sum_{i=1}^{i=n} P_i$.

It should be noticed that the abbreviation from $i=1$ to $i=n$ is over positive integer values. It must also be noticed that one can begin the abbreviation of the sum from any positive integer (not just 1) to any other positive integer (not just n). However, one condition must be satisfied: the upper limit or upper positive integer (that is, $i=n$) must be greater than or equal to the lower limit or lower positive integer ($i=1$). For example, suppose that we wish to write the abbreviated form of the sum $n_{10}+n_{11}+n_{12}+\cdots+n_{25}$. Then it can be written as $\sum_{i=10}^{i=25} n_i$ or $\sum_{i=10}^{25} n_i$. If we sum from $i=1$ to $i=1$ then we get the first entry itself: $\sum_{i=10}^{i=10} n_i = \sum_{i=10}^{10} n_i = n_{10}$. Some of the examples of abbreviation of sums are: $\sum_{i=5}^{10} x^i = x^5+x^6+x^7+x^8+x^9+x^{10}$; $\sum_{i=5}^{10}(x+1)^i = (x+1)^5+(x+1)^6+(x+1)^7+$

$(x+1)^8+(x+1)^9+(x+1)^{10}$; $\sum_{i=5}^{10}[x/y]^i = [x/y]^5+[x/y]^6+[x/y]^7+[x/y]^8+[x/y]^9+$

$[x/y]^{10}$; $\sum_{i=1}^{4} 1/i = (1/1)+(1/2)+(1/3)+)(1/4)=1+0.5+0.33+0.25=2.08$; $\sum_{i=5}^{10} x^{10-i}y^{i-5}=$

$x^{10-5}y^{5-5}+x^{10-6}y^{6-5}+x^{10-7}y^{7-5}+x^{10-8}y^{8-5}+x^{10-9}y^{9-5}+x^{10-10}y^{10-5}=x^5+x^4y+$

$x^3y^2+x^2y^3+xy^4+y^5$; and $\sum_{i=5}^{10} 1^i = 1^5+1^6+1^7+1^8+1^9+1^{10}=1+1+1+1+1+1=6$.

Notice that abbreviated sums obey two rules. They are $\sum_{i=1}^{n}(x_i+y_i)=\sum_{i=1}^{n}x_i+\sum_{i=1}^{n}y_i$ and $\sum_{i=1}^{n} kx_i = k.\sum_{i=1}^{n}x_i$.

There are occasions for one to use *double sums* or *multiple sums*. We consider a simple example to demonstrate the use of double sums. Assume that a company sells its product in different regions ($i=1$ to $i=m$) of a country during different weeks ($j=1$ to $j=n$). Then the quantities of the product sold in different regions during different weeks can be written in the form of a rectangular array called a *matrix* (discussed in detail in Chapter 2) as

$$
\begin{bmatrix}
a_{11} & a_{12} & a_{13} & \cdots & a_{1j} & \cdots & a_{1n} \\
a_{21} & a_{22} & a_{23} & \cdots & a_{2j} & \cdots & a_{21} \\
a_{31} & a_{32} & a_{33} & \cdots & a_{3j} & \cdots & a_{3n} \\
\cdots & \cdots & \cdots & \cdots & \cdots & \cdots & \cdots \\
a_{i1} & a_{i2} & a_{i3} & \cdots & a_{ij} & \cdots & a_{in} \\
\cdots & \cdots & \cdots & \cdots & \cdots & \cdots & \cdots \\
a_{m1} & a_{m2} & a_{m3} & \cdots & a_{mj} & \cdots & a_{mn}
\end{bmatrix}
$$

Notice that this matrix has m rows and n columns. The notation a_{ij} in the matrix is called an *element* of the matrix. The first subscript i of each element denotes the row and the second subscript j denotes the column in which the element appears in the matrix. It is much easier to represent the rows and columns with the help of letters. Therefore, we write a_{ij} to represent the element a in the ith row and the jth column. This implies that we can write the sums of each row as $\sum_{j=1}^{n} a_{1j}, \sum_{j=1}^{n} a_{2j}, \sum_{j=1}^{n} a_{3j}, \ldots \sum_{j=1}^{n} a_{mj}$. The sum of these row sums can then be written as $\sum_{j=1}^{n} a_{1j} + \sum_{j=1}^{n} a_{2j} + \sum_{j=1}^{n} a_{3j} + \cdots + \sum_{j=1}^{n} a_{mj} = \sum_{i=1}^{m} (\sum_{j=1}^{n} a_{ij})$. Similarly, the sums of each column can be written as $\sum_{i=1}^{m} a_{i1}, \sum_{i=1}^{m} a_{i2}, \sum_{i=1}^{m} a_{i3}, \ldots \sum_{i=1}^{m} a_{in}$; and the sum of these column sums can then be written as $\sum_{i=1}^{m} a_{i1} + \sum_{i=1}^{m} a_{i2} + \sum_{i=1}^{m} a_{i3} + \cdots + \sum_{i=1}^{m} a_{in} = \sum_{j=1}^{n} (\sum_{i=1}^{m} a_{ij})$. Notice that $\sum_{i=1}^{m} (\sum_{j=1}^{n} a_{ij}) = \sum_{j=1}^{n} (\sum_{i=1}^{m} a_{ij})$.

1.11.2 Products

We now consider the product notation, which is similar to the sum notation we discussed above. Suppose that we have a list of numbers $x_1, x_2, x_3, x_4, \ldots x_n$. How do we write the product of these numbers? If n is small, we can easily write the product. But if n is large, one has to go through time-consuming work. One can save a lot of time if one adopts the generally followed method of using a special notation for the product. This notation is the uppercase Greek letter $\prod$ (read "pi"), and can be used in our example as $\prod_{i=1}^{n} x_i = x_1.x_2.x_3.x_4.\cdots .x_n$. Notice that, as in the case of the sum notation, multiplication is over the positive integer values, and the upper limit ($i = n$) must be greater than or equal to the lower limit ($i = 1$). Examples of abbreviated product notation are

$$\prod_{i=1}^{5} x^i = x^1.x^2.x^3.x^4.x^5 = x^{15}; \quad \prod_{i=1}^{5} x^i y^i = x^1 y^1 .x^2 y^2 .x^3 y^3 .x^4 y^4 .x^5 y^5 = x^{15} y^{15}; \quad \prod_{i=1}^{5} x^{i-1} y^{1-i} =$$

$$x^0 y^0 .x^1 y^{-1} .x^2 y^{-2} .x^3 y^{-3} .x^4 y^{-4} = x^{10} y^{-10} = x^{10}/y^{10}; \text{ and } \prod_{1=1}^{2} (x+y)^i = (x+y)^1 .(x+y)^2 =$$

$$(x+y)(x^2 + 2xy + y^2) = x^3 + 3x^2 y + 3xy^2 + y^3.$$

1.11.3 Application examples

Example 1.   Assume that a company employs different numbers of workers to work for four days at different wages. Write the total wages paid by the company to the four workers using sum notations.

Solution.   For convenience, we denote the number of workers employed on the ith day by n_i and the wages paid on the ith day by w_i, where $1 \le i \le 4$. The total wages (W) paid can be written in the expanded form as $W = w_1.n_1 + w_2.n_2 + w_3.n_3 + w_4 n_4$. In abbreviated form using the sum notation, we may write it as $W = \sum_{i=1}^{4} w_i n_i$.

Example 2.   Suppose that a consumer buys different units of five commodities at their respective prices. Write the consumer's total expenditure using sum notations.

Solution.   For convenience, we denote the units of commodities by q_i and prices by p_i, where $1 \le i \le 5$. The total expenditure (E) can be written in the expanded form as

$E = q_1.p_1 + q_2.p_2 + q_3.p_3 + q_4p_4 + q_5p_5$. In abbreviated form using the sum notation, we may write it as $E = \sum_{i=1}^{5} q_i p_i$.

Example 3. Suppose that the consumers in an economy buy different goods (q_i) at their respective prices (p_i) during a particular year (say 0, called the *base year*), where $1 \leq i \leq n$. Also assume that they buy the same goods at their corresponding prices during year t (where $t > 0$). We know that the sum of $q_i \times p_i$ gives the *cost of living* in an economy during a particular year. Show, with the help of sum notations, the percentage change in the cost of living in the economy between the two time periods.

Solution. The sum $\sum_{i=1}^{n} q_i^0 p_i^t$ gives the cost of living in the economy during year t. Similarly, the cost of living for the base year is given by $\sum_{i=1}^{n} q_i^0 p_i^0$. If we divide the former by the latter and multiply the quotient by 100 we obtain the percentage change in the cost of living in the economy between the two time periods. That is, the percentage change in the cost of living in the economy between the two time periods is given by $(\sum_{i=1}^{n} q_i^0 p_i^t) / \sum_{i=1}^{n} q_i^0 p_i^0) \times 100$. This is called a *price index* for the year t. Specifically, it is called the *Laspeyres price index*. Instead, if we write the index as $(\sum_{i=1}^{n} q_i^t p_i^t) / \sum_{i=1}^{n} q_i^t p_i^0) \times 100$, it is called the *Paasche price index*. The other two important indices are $\sqrt{[(\sum q_i^0 p_i^t)/(\sum q_i^0 p_i^0)] \times [\sum q_i^t p_i^t)/(\sum q_i^t p_i^0)]} \times 100$ and $[(\sum q_i^0 p_i^t + \sum q_i^t p_i^t)/(\sum q_i^0 p_i^0 + \sum q_i^t p_i^0)] \times 100$. The former is called the *Fisher index number* and the latter is called the *Marshall–Edgeworth index number*.

Example 4. Suppose that we have n observations whose values are denoted by x_i, where $1 \leq i \leq n$. Find the *arithmetic mean* (or simply the *mean*) of these values.

Solution. The arithmetic mean (of *ungrouped data*), denoted by $\bar{x}$, is defined as the sum of the values of the observations divided by the number of observations n. Therefore, the mean is $\bar{x} = (1/n) \times \sum_{i=1}^{n} x_i$.

Example 5. Suppose that we have n observations whose values are denoted by x_i, where $1 \leq i \leq n$. Find the *variance* of these values.

Solution. The variance (of ungrouped data), denoted by σ^2, is defined as the sum of squares of the *mean deviations* (*deviation* of the *observed value* from the mean). Therefore, the variance is $\sigma^2 = \sum_{i=1}^{n} (x_i - \bar{x})^2$. Notice that the *standard deviation* of the values is obtained by the square root of the variance. Therefore, the standard deviation of the values in our present example is given by $\sigma = \sqrt{\sum_{i=1}^{n} (x_i - \bar{x})^2}$.

1.11.4 Exercises

1. Write the following expressions in abbreviated form:
   (i) $1 + 1 + 1 + 1 + 1$; (ii) $1 + 1/2 + 1/3 + 1/4 + 1/5$; (iii) $1 + 2 + 4 + 8 + 16$; (iv) $a1 + a1 + a1 + a1 + a1$; (v) $x_{1j}.y_{1j} + x_{2j}.y_{2j} + x_{3j}.y_{3j} + x_{4j}.y_{4j}$; (vi) $x_{i1}.y_{i1} + x_{i2}.y_{i2} + x_{i3}.y_{i3} + x_{i4}.y_{i4}$.

2. Write the following expressions in abbreviated form:
    (i) $2 \times 4 \times 8 \times 16$; (ii) $(1/2) \times (1/4) \times (1/8) \times (1/16)$; (iii) $(x+y)(x+y)^2(x+y)^3$;
    (iv) $k^0 \times k^1 \times k^2 \times k^3 \times k^4 \times k^5$; (v) $[(a+b)^1/(x+y)^0] \times [(a+b)^2/(x+y)^1]$
    $\times [(a+b)^3/(x+y)^2]$.
3. Expand the following expressions:
    (i) $\sum_{i=0}^{5} 2^i$; (ii) $\sum_{i=0}^{5} (5i-1)$; (iii) $\sum_{i=0}^{5} (1/i^2)$; (iv) $\sum_{i=2}^{5} [(1-i)/(i-1)]$; (v) $\sum_{i=1}^{5} 6$.
4. Expand the following expressions:
    (i) $\prod_{k=1}^{5} x^{k-1}/y^{1-k}$; (ii) $\prod_{k=1}^{5} x^{1/k}$; (iii) $\prod_{k=1}^{5} 5^{k-1}$; (iv) $\prod_{k=1}^{5} 5^{1-k}$; (v) $\prod_{k=1}^{5} x^{1-k} \cdot y^{k-1}$.
5. *Application exercise.* Suppose that a company sold $q_1 = 50$, $q_2 = 40$, $q_3 = 30$, and $q_4 = 20$ units of goods numbered 1, 2, 3, and 4, respectively. The unit prices of these goods were $10, $8, $6, and $4, respectively. Find the company's total and average revenues using abbreviation notations.
6. *Application exercise.* Suppose that a firm purchased $r_1 = 500$, $r_2 = 400$, $r_3 = 300$, and $r_4 = 200$ units of raw materials numbered 1, 2, 3, and 4, respectively. The unit prices of these raw materials were $100, $80, $60, and $40, respectively. Find the firm's total and average costs using abbreviation notations.
7. *Application exercise.* Assume that we have the base year prices (p) and commodities (q) in an imaginary economy as $p_1^0 = 10$, $q_1^0 = 50$, $p_2^0 = 15$, $q_2^0 = 60$, $p_3^0 = 12$, and $q_3^0 = 55$; and the tth year prices and quantities as $p_1^t = 12$, $q_1^t = 35$, $p_2^t = 16$, $q_2^t = 50$, $p_3^t = 11$, and $q_3^t = 60$. Find the Laspeyres price index and the Paasche price index.
8. Suppose that the total monthly sales of a company for one year are $m_1 = 10$, $m_2 = 15$, $m_3 = 25$, $m_4 = 30$, $m_5 = 30$, $m_6 = 35$, $m_7 = 45$, $m_8 = 50$, $m_9 = 55$, $m_{10} = 60$, $m_{11} = 70$, and $m_{12} = 80$, where m_i denotes the ith month and the amounts are in thousands of dollars. Find the arithmetic mean, variance, and standard deviation of sales using abbreviation notations.

 Web supplement: S1.12 Trigonometry

 Web supplement: S1.12.7 Mathematica applications

2   Linear algebra

Vectors and matrices

2.1 Introduction

Matrices are widely applied in various branches of economics. They are also applied in business and finance. Important applications of matrices can be found extensively in areas of *econometrics*, statistics, *input–output analysis*, solutions of *system(s) of simultaneous linear equations* (SSLEs), *optimization* topics, and so on. Therefore, it is important that students of economics, business, and finance possess a good understanding of matrices and related topics.

Our main objective in this chapter is to learn some of the basic ideas of matrices and related topics. However, a meaningful study of matrices must be preceded by a discussion of *vectors*. Therefore, we begin this chapter with an introduction to the basics of vectors.

2.2 Vectors

2.2.1 Meaning, geometric representation, and types of vectors

A vector is defined as an ordered set of numbers, parameters, or variables usually enclosed in square brackets. We can use a simple example to elucidate this definition of a vector. Assume that a consumer buys x_1 and y_1 units of two goods x and y, respectively. We can write these units either in a column as $\begin{bmatrix} x_1 \\ y_1 \end{bmatrix}$ or in a row as $\begin{bmatrix} x_1 & y_1 \end{bmatrix}$. This column or row of numbers is called a vector. If the vector appears in the form of a column, it is called a *column vector*; if it appears in the form of a row, it is called a *row vector*. Each entry inside the brackets is called a *component* of the vector. Column vectors are generally denoted by bold, lowercase letters such as **u**, **v**, etc. And row vectors are denoted by notations **u'**, **v'**, etc. Since both of the above vectors have two components each, we say that each is a *2-vector* or each has *dimension* 2. If a vector has n components, then we can say that it is an *n-vector* or it has dimension n.

Let us now consider the geometric representation of vectors. Suppose that $x_1 = 2$ and $y_1 = 2$ in our example above. Then we have a row vector **u'** of dimension 2 as $\mathbf{u'} = \begin{bmatrix} 2 & 2 \end{bmatrix}$. This vector can be represented by an arrow from the origin to the coordinate point $P = (2,2)$ in a two-dimensional space as illustrated in Figure 2.2.1(A).

Assume now that we have two row 2-vectors: $\mathbf{u'} = \begin{bmatrix} x_1 & y_1 \end{bmatrix}$ and $\mathbf{v'} = \begin{bmatrix} x_2 & y_2 \end{bmatrix}$. Also assume that $x_2 = 2$, $y_1 = 0$, $x_2 = 0$, and $y_2 = 2$. Then these two row 2-vectors can be written as $\mathbf{u'} = \begin{bmatrix} 2 & 0 \end{bmatrix}$ and $\mathbf{v'} = \begin{bmatrix} 0 & 2 \end{bmatrix}$, respectively. These two vectors can be illustrated geometrically as Figure 2.2.1(B).

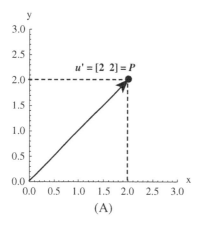
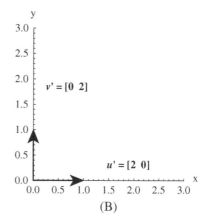

(A)   (B)

Figure 2.2.1

There exist various types of vectors. One type is called *unit vector*. A unit 2-vector is of the form $\mathbf{u'} = \begin{bmatrix} 1 & 0 \end{bmatrix}$ or $\mathbf{v'} = \begin{bmatrix} 0 & 1 \end{bmatrix}$. Notice that the geometric forms of these unit vectors coincide with the horizontal and the vertical axes, respectively, as Figure 2.2.1(B). Another type is called *zero vector* or *null vector*. A null vector will contain only zeros as its components such as $\mathbf{v'} = \begin{bmatrix} 0 & 0 \end{bmatrix} = \mathbf{0}$. This null vector represents the point of origin. The third type is called *equal vectors*. Two vectors are said to be equal vectors only if they have the same dimension and if their corresponding components are equal. For example, the two row 2-vectors $\mathbf{u'} = \begin{bmatrix} u_1 & u_2 \end{bmatrix}$ and $\mathbf{v'} = \begin{bmatrix} v_1 & v_2 \end{bmatrix}$ are said to be equal vectors only if $u_1 = v_1$ and $u_2 = v_2$. The fourth type comprises of vectors called *like vectors* and *unlike vectors*. Vectors having the same direction are called like vectors (such as $0P$ and $0P'$ in Figure 2.2.2(A)), and vectors with opposite directions are called unlike vectors (such as $0P$ and $0P''$ in Figure 2.2.2(A)). A fifth type of vector is *collinear vectors*. If two vectors lie on the same line (again, as in Figure 2.2.2(A)) or on parallel lines, they are called collinear vectors. The last type we mention here is *coplanar vectors*, which are vectors lying on the same or parallel planes.

2.2.2 Vector operations

Operations on vectors include addition, subtraction, and scalar multiplication. We shall discuss each of these here. Suppose that we have two row 2-vectors: $\mathbf{u'} = \begin{bmatrix} 1 & 2 \end{bmatrix}$ and $\mathbf{v'} = \begin{bmatrix} 2 & 1 \end{bmatrix}$. Then the sum $\mathbf{u'} + \mathbf{v'}$, called the *addition of vectors*, is obtained by adding each component of $\mathbf{u'}$ to the corresponding component of $\mathbf{v'}$. Therefore, we have $\mathbf{u'} + \mathbf{v'} = \begin{bmatrix} 1+2 & 2+1 \end{bmatrix} = \begin{bmatrix} 3 & 3 \end{bmatrix}$. This new vector is illustrated in Figure 2.2.2(B). The reader must have noticed that for addition of two vectors, the number of components in the given vectors (or the dimension of the vectors) must be the same.

We now consider the *subtraction of vectors*. Assume that we have two vectors $\mathbf{u'} = \begin{bmatrix} 2 & 1 \end{bmatrix}$ and $\mathbf{v'} = \begin{bmatrix} 1 & 2 \end{bmatrix}$. The difference of these two vectors, called the subtraction of vectors, is obtained by subtracting each component of $\mathbf{v'}$ from the corresponding component of $\mathbf{u'}$. Therefore, we have $\mathbf{u'} - \mathbf{v'} = \begin{bmatrix} 2 & 1 \end{bmatrix} - \begin{bmatrix} 1 & 2 \end{bmatrix} = \begin{bmatrix} 2 & 1 \end{bmatrix} + \begin{bmatrix} -1 & -2 \end{bmatrix} = \begin{bmatrix} 2-1 & 1-2 \end{bmatrix} = \begin{bmatrix} 1 & -1 \end{bmatrix}$. This is illustrated in Figure 2.2.3(A). Notice that $\mathbf{u'} + (-\mathbf{u'})$ or $\mathbf{v'} + (-\mathbf{v'})$ will

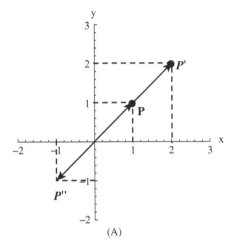
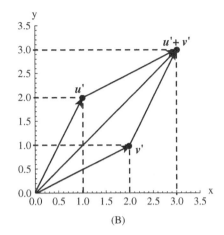

(A)   (B)

Figure 2.2.2

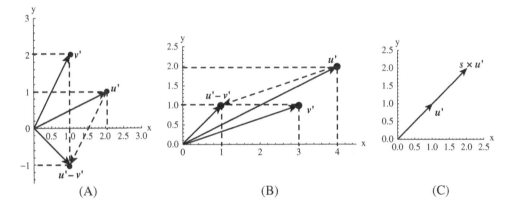

(A)   (B)   (C)

Figure 2.2.3

yield a zero vector (the point of origin). If our vectors were $\mathbf{u'} = \begin{bmatrix} 4 & 2 \end{bmatrix}$ and $\mathbf{v'} = \begin{bmatrix} 3 & 1 \end{bmatrix}$, then the difference would be $\mathbf{u'} - \mathbf{v'} = \begin{bmatrix} 1 & 1 \end{bmatrix}$, which is illustrated in Figure 2.2.3(B).

The last operation we consider here is *scalar multiplication* of vectors. A *scalar* is any real number. Suppose that s is a scalar and that $s = 2$. Also let $\mathbf{u'} = \begin{bmatrix} 1 & 1 \end{bmatrix}$. Then, in scalar multiplication, we multiply every component of $\mathbf{u'}$ by s, denoted by $s\,\mathbf{u'}$, giving us a new vector which is s times the vector $\mathbf{u'}$. Therefore, given our 2-vector $\mathbf{u'} = \begin{bmatrix} 1 & 1 \end{bmatrix}$, the scalar product is given by $s \times \mathbf{u'} = 2 \times \begin{bmatrix} 1 & 1 \end{bmatrix} = \begin{bmatrix} 2 & 2 \end{bmatrix}$. This scalar multiplication is illustrated in Figure 2.2.3(C). Notice that the direction of the newly generated vector will be reversed if s is negative. Also notice that one can do scalar multiplication in combination with addition and subtraction of vectors. This implies that given two vectors of the same dimension $\mathbf{u'}$ and $\mathbf{v'}$ and the scalar s, $s(\mathbf{u'} \pm \mathbf{v'}) = s\,\mathbf{u'} \pm s\mathbf{v'}$.

The properties of vector addition and scalar multiplication are presented below. The first four properties are the *properties of vector addition* and the last three are the *properties of scalar multiplication*.[1]

Property I.   Vector addition is commutative. For any two vectors $\mathbf{u}$ and $\mathbf{v}$, $\mathbf{u} + \mathbf{v} = \mathbf{v} + \mathbf{u}$.

Property II.   Vector addition is associative. For any three vectors $\mathbf{u}$, $\mathbf{v}$, and $\mathbf{z}$, $(\mathbf{u} + \mathbf{v}) + \mathbf{z} = \mathbf{u} + (\mathbf{v} + \mathbf{z})$.

Property III.   Existence of *additive identity*. For any vector $\mathbf{u}$, there exists a zero vector $\mathbf{0}$, called the additive identity, such that $\mathbf{u} + \mathbf{0} = \mathbf{u}$.

Property IV.   Existence of *additive inverse*. For any vector $\mathbf{u}$, there exists a vector called $-\mathbf{u}$, such that $\mathbf{u} + (-\mathbf{u}) = \mathbf{0}$.

Property V.   Scalar multiplication is associative. If there are scalars s_1 and s_2 and if $\mathbf{u}$ is a vector, then $(s_1 \times s_2)\,\mathbf{u} = s_1 \times (s_2 \times \mathbf{u})$.

Property VI.   Scalar multiplication is distributive. If s_1 and s_2 are two scalars and $\mathbf{u}$ and $\mathbf{v}$ are two vectors of the same direction, then $(s_1 + s_2)\,\mathbf{u} = s_1 \times \mathbf{u} + s_2 \times \mathbf{u}$, and $s_1 \times (\mathbf{u} + \mathbf{v}) = s_1 \times \mathbf{u} + s_1 \times \mathbf{v}$.

Property VII.   Existence of *multiplicative identity*. For any vector $\mathbf{u}$, $1 \times \mathbf{u} = \mathbf{u}$.

2.2.3 Linear combination and linear dependence of vectors

The first concept we discuss in this section is the *linear combination of vectors*. Suppose that we have n n-vectors $\mathbf{v}_1$, $\mathbf{v}_2$, $\mathbf{v}_3$, ..., $\mathbf{v}_n$; and n scalars $s_1, s_2, s_3, \ldots, s_n$. Then we can generate a new n-vector by multiplying the components of each original vector by the corresponding scalars. The new vector generated in this way is called a *linear combination* of the original vectors and is given by

$$s_1\mathbf{v_1} + s_2\mathbf{v_2} + s_3\mathbf{v_3} + \cdots + s_n\mathbf{v_n} \tag{2.2.1}$$

As an example of linear combination of vectors, consider the two 2-vectors $\mathbf{v}_1 = \begin{bmatrix} 1 \\ 2 \end{bmatrix}$ and $\mathbf{v}_2 = \begin{bmatrix} 2 \\ 1 \end{bmatrix}$, and scalars s_1 and s_2. Then the linear combination of the given vectors with the given scalars can be written as

$$\mathbf{v} = s_1 \begin{bmatrix} 1 \\ 2 \end{bmatrix} + s_2 \begin{bmatrix} 2 \\ 1 \end{bmatrix} = \begin{bmatrix} s_1.1 + s_2.2 \\ s_1.2 + s_2.1 \end{bmatrix}$$

The next concept we are concerned with here is the *linear dependence of vectors* or the *linear independence of vectors*. Suppose that we have n n-vectors $\mathbf{v}_1$, $\mathbf{v}_2$, $\mathbf{v}_3$, ..., $\mathbf{v}_n$; and n scalars $s_1, s_2, s_3, \ldots, s_n$, *not all zeros*. Given these vectors and scalars, if we can write $s_1\mathbf{v_1} + s_2\mathbf{v_2} + s_3\mathbf{v_3} + \cdots + s_n\mathbf{v_n} = \mathbf{0}$, then the vectors are said to be linearly dependent. On the other hand, if there do not exist scalars $s_1, s_2, s_3, \ldots, s_n$, again not all zeros, such that $s_1\mathbf{v_1} + s_2\mathbf{v_2} + s_3\mathbf{v_3} + \cdots + s_n\mathbf{v_n} = \mathbf{0}$, then the vectors are said to be linearly independent.

As an example, consider the two vectors $\mathbf{v_1} = \begin{bmatrix} 2 \\ 2 \end{bmatrix}$, and $\mathbf{v}_2 = \begin{bmatrix} 1 \\ 1 \end{bmatrix}$. Now, using scalars $s_1 = 1$ and $s_2 = -2$, we can write $1\mathbf{v_1} - 2\mathbf{v_2} = 1\begin{bmatrix} 2 \\ 2 \end{bmatrix} - 2\begin{bmatrix} 1 \\ 1 \end{bmatrix} = \begin{bmatrix} 2 \\ 2 \end{bmatrix} - \begin{bmatrix} 2 \\ 2 \end{bmatrix} = \begin{bmatrix} 0 \\ 0 \end{bmatrix} = \mathbf{0}$. This equation of vectors can also be written as $1\mathbf{v_1} + (-2)\mathbf{v_2} = \mathbf{0}$ so that it corresponds with the above definition of linear dependence. Therefore, we say that the vectors $\mathbf{v_1}$ and $\mathbf{v_2}$ are linearly dependent. Notice that the geometric forms of these vectors are similar to those of the vectors illustrated in Figure 2.2.3(C).

As another example, consider the vectors $\mathbf{v}_1 = \begin{bmatrix} 1 \\ 3 \end{bmatrix}$ and $\mathbf{v}_2 = \begin{bmatrix} 1 \\ 1 \end{bmatrix}$. Now, using scalars s_1 and s_2, we can write $s_1\mathbf{v}_1 - s_2\,\mathbf{v}_2 = s_1 \begin{bmatrix} 1 \\ 3 \end{bmatrix} + s_2 \begin{bmatrix} 1 \\ 1 \end{bmatrix} = \begin{bmatrix} s_1 \\ 3s_2 \end{bmatrix} - \begin{bmatrix} s_1 \\ s_2 \end{bmatrix} = \begin{bmatrix} 0 \\ 0 \end{bmatrix} = \mathbf{0}$. It can be shown that this SSLEs will be satisfied only if $s_1 = s_2 = 0$. Therefore, the two vectors $\mathbf{v}_1$ and $\mathbf{v}_2$ are linearly independent. One can see that these two vectors have different directions if one represents them geometrically.

2.2.4 Product of vectors

We now consider the *multiplication of vectors* or the *product of vectors*. The product of two or more vectors is also called the *inner product*, or the *scalar product*, or the *dot product* of vectors. Suppose that we have the two 3-vectors $\mathbf{q}' = [q_1 \quad q_2 \quad q_3]$ and $\mathbf{p}' = [p_1 \quad p_2 \quad p_3]$, where q_i denotes the quantity of the ith good purchased by a consumer and p_i represents the price of the ith good, and $i = 1, 2, 3$. When we multiply the quantity of the ith good in $\mathbf{q}'$ by the price of the ith good in $\mathbf{p}$, we obtain the expenditure on the ith good. Then, when we sum the amounts spent on the three goods, we obtain the total expenditure of the consumer on these three goods. That is, the total expenditure will be equal to $\sum_{i=1}^{3} q_i p_i$. This is exactly what we get when we multiply the vector $\mathbf{q}$ by the vector $\mathbf{p}$. This product is called the inner product, or scalar product, or dot product, or simply the product of vectors $\mathbf{q}'$ and $\mathbf{p}$. Let us now generalize the last result. Suppose that we have two n-vectors $\mathbf{u}' = [u_1 \quad u_2 \quad \ldots \quad u_n]$ and $\mathbf{v}' = [u_1 \quad u_2 \quad \ldots \quad u_n]$. Then the scalar product of $\mathbf{u}'$ and $\mathbf{v}$, denoted by $\mathbf{u}'.\mathbf{v}$, is defined as

$$\mathbf{u}'.\mathbf{v} = [u_1.v_1 + u_2.v_2 + u_3.v_3 + \cdots + u_n.v_n] = \sum_{i=1}^{n} u_i.v_i \qquad (2.2.2)$$

Notice that the product in equation (2.2.2) is not a vector but a real number. This is the reason why it is called a scalar product. Also notice that in order for the scalar product to exist, $\mathbf{u}'$ and $\mathbf{v}'$ must be of the same dimension. We list below the properties of the scalar product of any three n-vectors $\mathbf{u}$, $\mathbf{v}$, and $\mathbf{z}$, and the scalar s_1.

Property I.   Scalar product is commutative: $\mathbf{u}'.\mathbf{v} = \mathbf{v}'.\mathbf{u}$.
Property II.   Scalar product is distributive: $\mathbf{u}'.(\mathbf{v} + \mathbf{z}) = \mathbf{u}'.\mathbf{v} + \mathbf{u}'.\mathbf{z}$.
Property III.   Scalar product is associative: $(s_1.\mathbf{u}').\mathbf{v} = \mathbf{u}'.(s_1.\mathbf{v}) = s_1.(\mathbf{u}'.\mathbf{v})$.

2.2.5 Vector spaces

In Section 1.8.1 we introduced some of the rudiments of spaces and dimensions. We shall extend them here a bit further. Let us begin with the geometric forms of vectors in Figure 2.2.4(A). We mentioned in our discussion in Section 1.8.1 that each ordered pair that lies on a real line, say a *2-tuple* such as (1, 0) on the x-axis or (0, 1) on the y-axis, is an ordered pair generated from two sets $x \in \Re$ and $y \in \Re$. We also mentioned that each real line represents a one-dimensional space (or a *one-space*) denoted by $\Re$ or $\Re^1$. This pointed to the fact that $\Re$ constitutes the set of all 2-tuple ordered pairs that lie on a real line such as those on the x-axis or y-axis in Figure 2.2.4(A).

Similarly, we know from our discussion so far in this chapter that every point on a real line can be represented by a vector such as $\mathbf{C}$ on the x-axis or $\mathbf{F}$ on the y-axis of Figure 2.2.4(A). Notice that these vectors can be generated by scalar multiplication of the unit vectors such as $\mathbf{B}$ or $\mathbf{E}$ in the same figure. Therefore, $\Re$ also represents the set of all

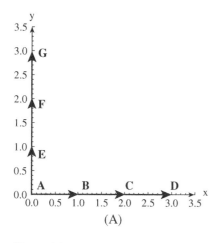 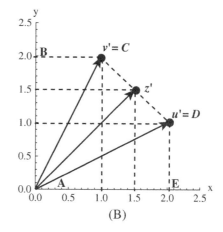

(A)   (B)

Figure 2.2.4

the 2-vectors generated by scalar multiplications of a single unit 2-vector. This implies that these newly generated 2-vectors (like **C**, **D**, **F**, **G**, etc.) *span* a *one-space* or $\Re$. Therefore, a *one-dimensional vector space* is defined as the set of all the unit 2-vectors generated by scalar multiplications of a single unit vector. Notice that all the vectors that span a one-space are linearly dependent vectors.

Now consider the geometric representations of vectors in Figure 2.2.4(B). These 2-vectors are $\mathbf{u'} = \begin{bmatrix} 2 & 1 \end{bmatrix}$; $\mathbf{v'} = \begin{bmatrix} 1 & 2 \end{bmatrix}$; and $\mathbf{z'} = \begin{bmatrix} 1.5 & 1.5 \end{bmatrix}$. Notice that we generated the vector $\mathbf{z'}$ through linear combination of $\mathbf{u'}$ and $\mathbf{v'}$ with scalars $s_1 = 0.5$ and $s_2 = 0.5$ (that is, $\mathbf{z'} = s_1 \times \mathbf{u'} + s_2 \times \mathbf{v'}$). We can generate such infinite number of linear combinations of $\mathbf{u}$ and $\mathbf{v}$ with different values for s_1 and s_2. Such linear combinations of independent vectors span the entire two-dimensional space ($\Re^2$ or *two-space*) such as the area ABCDE in Figure 2.2.4(B). Therefore, we define the *two-dimensional vector space* as the set of all the 2-vectors generated by linear combinations of two independent 2-vectors.

The next question is how we can define a *three-dimensional vector space* ($\Re^3$ or *3-space*). For this consider the geometric representation of three linearly independent row 3-vectors $\mathbf{u'} = \begin{bmatrix} 0.5 & 0 & 0 \end{bmatrix}$, $\mathbf{v'} = \begin{bmatrix} 0 & 0.5 & 0 \end{bmatrix}$, and $\mathbf{w'} = \begin{bmatrix} 0 & 0 & 0.5 \end{bmatrix}$ illustrated in Figure 2.2.5(A). The elements of each of these vectors represent the values taken by three variables z, x, and y, respectively. This implies that we will have three axes as can be seen in Figure 2.2.5(A). Notice that we generated the vector $2\mathbf{u'} + 2\mathbf{v'} + 2\mathbf{w'}$ through linear combination of $\mathbf{u}$, $\mathbf{v}$, and $\mathbf{w}$ with scalars $s_1 = 2$, $s_2 = 2$, and $s_3 = 2$ (that is, $2\mathbf{u} + 2\mathbf{v} + 2\mathbf{w}$). We can generate such infinite number of linear combinations of $\mathbf{u}$, $\mathbf{v}$, and $\mathbf{w}$ using different values for s_1, s_2 and s_3. Such linear combinations of independent vectors span the entire three-dimensional vector space illustrated in Figure 2.2.5(A). Therefore, we define the three-dimensional vector space as the set of all the 3-vectors generated through linear combinations of three independent 3-vectors. The condition here is that, as in the definition of the two-space, the three vectors must be independent and have the same dimension.

It is needless to mention that, although it is not possible to show geometrically, we can define an *n-dimensional vector space* (*n-space* or $\Re^n$) as the set of all the *n*-vectors generated through linear combinations of *n* independent *n*-vectors. Notice that, since each point in an *n*-space is an ordered *n-tuple*, each *n*-vector represents a point in the *n*-space. Since the *n*-space or $\Re^n$ contains all the real numbers, it is also known as *Euclidian n-space*.

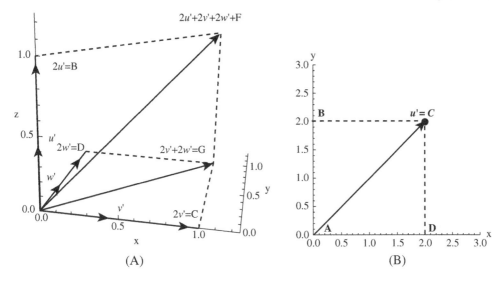

Figure 2.2.5

2.2.6 Lengths and orthogonality of vectors

We may sometimes need to find the *lengths of vectors* in different dimensions or spaces. Consider the simplest problem first: the problem of finding the length of a vector (also called the *norm of a vector*) in a one-space. Two unit vectors are illustrated in Figure 2.2.4(A): AB and AE. Consider the vector AB in this figure. For convenience we denote it by $\mathbf{u'} = \begin{bmatrix} 1 & 0 \end{bmatrix}$. What is the length of this vector? We know that this vector originates at point A = 0 and terminates at point B = 1. Therefore, we obtain the length of the vector $\mathbf{u'} = \begin{bmatrix} 1 & 0 \end{bmatrix}$ by subtracting its terminal value from its initial value; that is, the length of the vector $\mathbf{u'} = \begin{bmatrix} 1 & 0 \end{bmatrix}$ is equal to $1 - 0 = 1$. In this way we can find the *length* of any vector in one-space.

Let us now consider the length of a vector in a 2-space or the length of a row 2-vector. For this we can use the geometric form of the row 2-vector $\mathbf{u'} = \begin{bmatrix} 2 & 2 \end{bmatrix}$ illustrated in Figure 2.2.5(B). The graph of the vector shows that its initial point is A = 0 and its terminal point is C. Notice that the vector makes a right-angled *triangle* CAD. Our problem is to find the length of the vector $\mathbf{u'} = \begin{bmatrix} 2 & 2 \end{bmatrix}$; that is, the length of the line segment AC or the *distance* from A to C. We can apply *Pythagoras' theorem* to find this distance. Applying this theorem, we obtain the distance as $\|AC\| = \sqrt{(DA)^2 + (CD)^2}$. Since $DA = 2 - 0 = 2$ and $CD = 2 - 0 = 2$, we have $\|AC\| = \sqrt{2^2 + 2^2} = \sqrt{4+4} = \sqrt{8} = 2.8284$.

Assume that the elements of the vector $\mathbf{u'}$ are now general: $\mathbf{u'} = [x_1 \ y_1]$. Then the length of the vector $\mathbf{u'}$, following our specific case above, can be written as

$$\|\mathbf{u'}\| = \sqrt{x_1^2 + y_1^2} \tag{2.2.3}$$

Following a line of thought similar to the above, we can find the lengths of vectors in 3-space (or row 3-vectors) and in n-space (or row n-vectors). Suppose that the row 3-vector and the row n-vectors are $\mathbf{u'} = [x_1 \ y_1 \ z_1]$ and $\mathbf{v'} = [x_1 \ x_2 \ \ldots \ x_n]$, respectively. Then the lengths

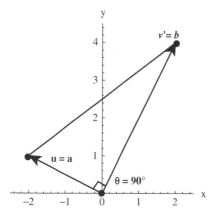

Figure 2.2.6

of these vectors can be given as

$$||\mathbf{u}'|| = \sqrt{x_1^2 + y_1^2 + z_1^2} \tag{2.2.4}$$

and

$$||\mathbf{v}'|| = \sqrt{x_1^2 + x_2^2 + \cdots + x_n^2} \tag{2.2.5}$$

respectively.

Let us now turn our attention to the concept of *orthogonality of vectors*. For this we may use the geometric representation of two vectors $\mathbf{u}' = \begin{bmatrix} -2 & 1 \end{bmatrix}$ and $\mathbf{v}' = \begin{bmatrix} 2 & 4 \end{bmatrix}$ as illustrated in Figure 2.2.6. We know from Pythagoras' theorem that the angle θ between the vectors $\mathbf{u}'$ and $\mathbf{v}'$ is equal to 90° (that is, $\theta = 90°$) only if it is a right angle. In other words, $\theta = 90°$ only if $||\mathbf{b} - \mathbf{a}||^2 = ||\mathbf{a}||^2 + ||\mathbf{b}||^2$ or $||\mathbf{b}||^2 - ||2.\mathbf{a}.\mathbf{b}|| + ||\mathbf{a}||^2 = ||\mathbf{a}||^2 + ||\mathbf{b}||^2$. The last equation implies that $||2\mathbf{a}.\mathbf{b}|| = 0$ or $\mathbf{a}.\mathbf{b} = 0$, which suggests that two vectors (such as $\mathbf{u}'$ and $\mathbf{v}'$ in Figure 2.2.6) are *orthogonal vectors* (or they are perpendicular or the angle between them is equal to 90°) only if their inner product is zero. Although we defined orthogonality of vectors with respect to 2-space, the definition can be generalized to *n*-space.

2.2.7 Lines, planes, and hyperplanes

We begin the discussion of a line with the geometric forms of two linearly independent row 2-vectors $\mathbf{u}' = \begin{bmatrix} x_3 & y_1 \end{bmatrix} = \begin{bmatrix} 3 & 1 \end{bmatrix}$ and $\mathbf{v}' = \begin{bmatrix} x_1 & y_3 \end{bmatrix} = \begin{bmatrix} 1 & 3 \end{bmatrix}$, as illustrated in Figure 2.2.7(A). Notice that the vector $\mathbf{z}' = \begin{bmatrix} x_2 & y_2 \end{bmatrix} = \begin{bmatrix} 2 & 2 \end{bmatrix}$ in this figure is generated through the linear combination of vectors $\mathbf{u}'$ and $\mathbf{v}'$ with the scalar $s = 0.5$. That is, we can write $\mathbf{z}' = \mathbf{v}' + s(\mathbf{u}' - \mathbf{v}') = s\,\mathbf{u}' + (1 - s)\mathbf{v}'$. Notice also that the scalar s is such that $0 \leq s \leq 1$. Now using different values for s yields the set $\mathbf{z}'_i = \begin{bmatrix} x_i & y_i \end{bmatrix}$ (where $i = 1, 2, 3, \ldots, n$), such as $\mathbf{z}'_1$, $\mathbf{z}'_2$, etc. It can be seen from Figure 2.4.7(A) that this set is the same as the set of points lying on the line L. Therefore, $\mathbf{z}'_i$ describes the line L. We shall now formally define a line L through vectors $\mathbf{u}' = \begin{bmatrix} x_1 & y_1 \end{bmatrix}$ and $\mathbf{v}' = \begin{bmatrix} x_2 & y_2 \end{bmatrix}$ in $\mathfrak{R}^2$ as the set of all $\mathbf{z}'_i$ satisfying the equation $\mathbf{z}' = s\,\mathbf{u}' + (1 - s)\mathbf{v}'$.

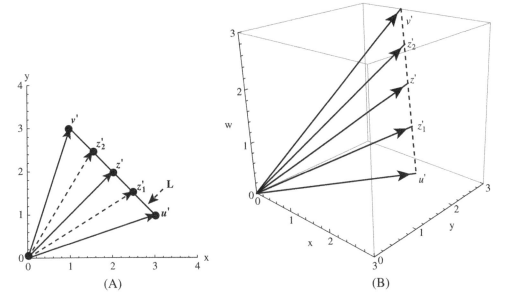

Figure 2.2.7

Similarly, we can define a line in $\mathfrak{R}^3$ using geometric forms of two linearly independent row 3-vectors $\mathbf{u'} = \begin{bmatrix} x_1 & y_1 & w_1 \end{bmatrix} = \begin{bmatrix} 3 & 1 & 1 \end{bmatrix}$ and $\mathbf{v'} = \begin{bmatrix} x_2 & y_2 & w_2 \end{bmatrix} = \begin{bmatrix} 1 & 3 & 3 \end{bmatrix}$. Notice that, as in Figure 2.2.7(A), the vector $\mathbf{z'}$ in Figure 2.2.7(B) is generated through linear combination of vectors $\mathbf{u'}$ and $\mathbf{v'}$: $\mathbf{z'} = s\,\mathbf{u'} + (1-s)\mathbf{v'}$, where s is a scalar such that $0 \le s \le 1$. Now using different values for s, as before, yields the set $\mathbf{z'}_i = \begin{bmatrix} x_i & y_i & w_i \end{bmatrix}$ (where $i = 1, 2, 3, \ldots, n$), such as $\mathbf{z'}_1$, $\mathbf{z'}_2$, etc. It can be seen, from Figure 2.2.7(B), that this set is the same as the set of points lying on the line L. Therefore, $\mathbf{z'}$ describes the line L. We can now formally define, again as before, a line L through $\mathbf{u'} = \begin{bmatrix} x_1 & y_1 & w_1 \end{bmatrix}$ and $\mathbf{v'} = \begin{bmatrix} x_2 & y_2 & w_2 \end{bmatrix}$ in $\mathfrak{R}^3$ as the set of all $\mathbf{z'}_i$ satisfying the equation $\mathbf{z'} = s\,\mathbf{u'} + (1-s)\mathbf{v'}$.

Continuing in an analogous fashion, we can define a line in $\mathfrak{R}^n$. Suppose that we have two linearly independent row n-vectors $\mathbf{u'} = \begin{bmatrix} u_1 & u_2 & u_3 & \ldots & u_n \end{bmatrix}$ and $\mathbf{v'} = \begin{bmatrix} v_1 & v_2 & v_3 & \ldots & v_n \end{bmatrix}$ and a scalar s such that $0 \le s \le 1$. Now using different values for s, as above, yields the set $\mathbf{z'}_i = \begin{bmatrix} z_1 & z_2 & z_3 & \ldots & z_n \end{bmatrix}$. It can be shown that this set is the same as the set of points lying on a specific line in $\mathfrak{R}^n$. Thus $\mathbf{z'}$ describes this specific line. Therefore, we shall formally define a line through $\mathbf{u'} = \begin{bmatrix} u_1 & u_2 & u_3 & \ldots & u_n \end{bmatrix}$ and $\mathbf{v'} = \begin{bmatrix} v_1 & v_2 & v_3 & \ldots & v_n \end{bmatrix}$ in $\mathfrak{R}^n$ as the set of all $\mathbf{z'}_i$ satisfying the equation $\mathbf{z'} = s\,\mathbf{u'} + (1-s)\mathbf{v'}$.

The above description of a line using a *parameter* (s) is called the *parametric representation of a line*. Notice that if $s = 0$, we have the point represented by the vector $\mathbf{z'} = \mathbf{v'}$; if $s = 1$, we have the point represented by the vector $\mathbf{z'} = \mathbf{u'}$. When s lies between 0 and 1 (that is, when $0 \le s \le 1$) we are at vector points, represented by $\mathbf{z'}_i$, between $\mathbf{u'}$ and $\mathbf{v'}$. Therefore, any point in this intermediate range can be shown through an equation called the *parametric equation of a line*:

$$L\,(\mathbf{u'}, \mathbf{v'}) = \{s\,\mathbf{u'} + (1-s)\mathbf{v'} \mid \quad 0 \le s \le 1\} \tag{2.2.6}$$

So far we have been concerned with lines in $\mathfrak{R}^1$, $\mathfrak{R}^2$, $\mathfrak{R}^3$, ..., and $\mathfrak{R}^n$. We now consider *planes* in these respective dimensions. But, before this, notice that while lines are one

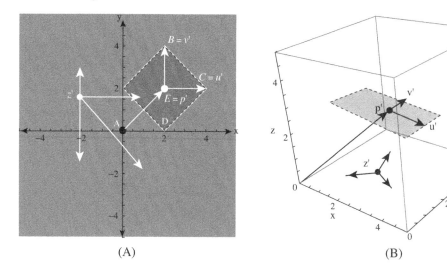

(A)   (B)

Figure 2.2.8

dimensional (irrespective of whether they lie in $\Re^1$, $\Re^2$, $\Re^3$, or $\Re^n$) and can be described by one parameter (s in our case), planes are two dimensional. This is illustrated in Figure 2.2.8(A).

The two-dimensional character of planes suggests that we require two parameters to describe them. Therefore, we now use two scalars (parameters) to define a plane. Assume that we have three linearly independent row 2-vectors $\mathbf{p'} = \begin{bmatrix} p_1 & p_2 \end{bmatrix}$, $\mathbf{u'} = \begin{bmatrix} x_1 & y_1 \end{bmatrix}$, and $\mathbf{v'} = \begin{bmatrix} x_2 & y_2 \end{bmatrix}$ and two scalars s and r, where the vector $\mathbf{p'}$ is a *normal vector* (or a vector perpendicular) to the plane $\mathbf{z'}$ in Figure 2.2.8(A). Then we can use the above method of *parameterization* to define a plane in $\Re^2$, such as the plane $\mathbf{z'}$, which can be given by the equation $\mathbf{z'} = \mathbf{p'} + s\,\mathbf{u'} + r\,\mathbf{v'}$. One can show that this set $\mathbf{z'}_j$ is the same as the set of points lying in the plane $\Re^2$. Notice that the vectors $\mathbf{u'} - \mathbf{p'}$ and $\mathbf{v'} - \mathbf{p'}$ are *displacement vectors* from the normal vector $\mathbf{p'}$. Therefore, the parametric equation given above for a two-dimensional plane can be expressed as $\mathbf{z'} = \mathbf{p'} + s(\mathbf{u'} - \mathbf{p'}) + r(\mathbf{v'} - \mathbf{p'})$, which simplifies to $\mathbf{z'} = (1 - s - r)\mathbf{p'} + s\,\mathbf{u'} + r\,\mathbf{v'}$. Notice also that if $\mathbf{p'} = \begin{bmatrix} 0 & 0 \end{bmatrix}$, then the plane will pass through the origin. In such cases, the parametric equation becomes $\mathbf{z'} = s\,\mathbf{u'} + r\,\mathbf{v'}$.

The last equation is called the *parametric equation of a plane*. It implies that the set of vectors $\mathbf{z}_j$ generated through linear combinations of $\mathbf{u'}$ and $\mathbf{v'}$ with scalars s and r, and with the normal vector $\mathbf{p'}$, will span the entire two-dimensional plane (including the darker one) in Figure 2.2.8(A). This suggests that, just as two points (represented by two linearly independent 2-vectors) determine a line, three points (represented by three linearly independent 3-vectors) determine a plane.

The above parametric equation can be used to describe a plane in $\Re^3$. This is illustrated in Figure 2.2.8(B). Let us now attempt to define a plane in $\Re^3$. Suppose that we have three linearly independent 3-vectors $\mathbf{p'} = \begin{bmatrix} p_1 & p_2 & p_3 \end{bmatrix}$, $\mathbf{u'} = \begin{bmatrix} x_1 & y_1 & z_1 \end{bmatrix}$, and $\mathbf{v'} = \begin{bmatrix} x_2 & y_2 & z_2 \end{bmatrix}$; and two scalars s and r. Notice that $\mathbf{p'}$ is still a normal vector. Following the above procedure, we define a plane in $\Re^3$ as a three-dimensional area that satisfies the same equation as the one that defines a plane in $\Re^2$: $\mathbf{z'} = \mathbf{p'} + s(\mathbf{u'} - \mathbf{p'}) + r(\mathbf{v'} - \mathbf{p'})$. The only difference here is that in the former the vectors $\mathbf{p'}$, $\mathbf{u'}$, and $\mathbf{v'}$ are 3-vectors, while in the latter they are 2-vectors.

Continuing in an analogous fashion, we can define a plane in $\Re^n$. Suppose that we have three linearly independent row n-vectors $\mathbf{u'} = \begin{bmatrix} u_1 & u_2 & u_3 & \ldots & u_n \end{bmatrix}$, $\mathbf{v'} = \begin{bmatrix} v_1 & v_2 & v_3 & \ldots & v_n \end{bmatrix}$, and $\mathbf{p'} = \begin{bmatrix} p_1 & p_2 & p_3 & \ldots & p_n \end{bmatrix}$; and two scalars s and r. Also suppose that we have another vector $\mathbf{z'} = \mathbf{p'} + s\,\mathbf{u'} + r\,\mathbf{v'} = \begin{bmatrix} z_1 & z_2 & z_3 & \ldots & z_n \end{bmatrix}$, which is generated through a linear combination of the given vectors and using the given scalars. It can be shown that this set $\mathbf{z'}_j$ is the same as the set of points lying in a plane in $\Re^1$. Therefore, a plane in $\Re^n$, like a plane in $\Re^2$ or $\Re^3$, is again the set of all $\mathbf{z'}$ satisfying the equation

$$\mathbf{z'} = (1 - s - r)\mathbf{p'} + s\mathbf{u'} + r\mathbf{v'} \tag{2.2.7}$$

Equation (2.2.7) represents the parametric equation of the plane that contains points $\mathbf{p'}$, $\mathbf{u'}$, and $\mathbf{v'}$. Notice that if $\mathbf{p'} = 0$ and $r = 1 - s$, then equation (2.2.7) reduces to equation (2.2.6). This suggests that the equation that describes a plane (equation (2.2.7)) is more general than the equation that describes a line (equation (2.2.6)). That is, one can derive the latter from the former. What this suggests is that equation (2.2.7) can describe both lines and planes. Therefore, a line in $\Re^1$ such as AB in Figure 2.2.4(A) (or in $\Re^2$ such as L or in $\Re^3$ such as L in Figures 2.2.7(A) and (B), respectively, or in $\Re^n$) or a plane in $\Re^2$ such as the shaded area in Figure 2.2.8(A) (or in $\Re^3$ such as the shaded area in Figure 2.2.8(B), or in $\Re^n$) are called *hyperplanes*.

2.2.8 Application examples

Example 1. Assume that two combinations of two goods (x and y) purchased by a consumer are given by the two row 2-vectors $\mathbf{u'} = \begin{bmatrix} x_1 & y_1 \end{bmatrix}$ and $\mathbf{v'} = \begin{bmatrix} x_2 & y_2 \end{bmatrix}$, and the prices of the two goods are given by the row 2-vector $\mathbf{p'} = [p_x \quad p_y]$. The consumer's income is given by I. Show the consumer's *budget line* and *commodity space* or *commodity plane*. Also show that the price vector is orthogonal to the budget line (or to the commodity plane).

Solution. Suppose that we use the combination given by the vector $\mathbf{u'} = \begin{bmatrix} x_1 & y_1 \end{bmatrix}$. Then the total expenditure of the consumer is given by the dot product of two vectors: $\mathbf{p'}.\mathbf{u} = p_x.x_1 + p_y y_1$. Since the consumer's income must be less than or equal to the expenses, we have the inequality $p_x.x_1 + p_y.y_1 \leq I$. If we write $p_x.x_1 + p_y.y_1 = I$, then it gives us the consumer's budget line as shown by the line BC (a line in a 2-space) in Figure 2.2.9(A).

Notice that the line BC in Figure 2.2.9(A) conforms to the definition of a line in a two-dimensional space given in equation (2.2.6) if we use the parameter s such that $0 \leq s \leq 1$: $L(\mathbf{u'}, \mathbf{v'}) = s\,\mathbf{u'} + (1-s)\mathbf{v'}$. However, using parameters s and r such that $0 \leq s \leq 1$ and $0 \leq r \leq 1$, we may express $\mathbf{z'} = (1 - s - r)\mathbf{p'} + s\,\mathbf{u'} + r\,\mathbf{v'}$ (as in equation (2.2.7)). Then the set of points represented by the plane $\mathbf{z'}$ is the same as the set of points lying within (or on) the shaded triangle in Figure 2.2.9(A). This space is called the commodity space or the commodity plane.

Let us now show that the price vector $\mathbf{p'} = [p_x \quad p_y]$ is orthogonal to the budget line or the commodity space. For this, assume that $\mathbf{u'} = [0 \quad I/p_y]$ and $\mathbf{v'} = \begin{bmatrix} I/p_x & 0 \end{bmatrix}$. Therefore, we have $\mathbf{z'} = \mathbf{u'} - \mathbf{v'} = [0 \quad I/p_y] - \begin{bmatrix} I/p_x & 0 \end{bmatrix} = [I/p_y \quad -I/p_x]$. We can now multiply the vector $\mathbf{z'}$ by the vector $\mathbf{p'}$ to obtain $\mathbf{p'}.\mathbf{z'} = [p_x \quad p_y][I/p_y \quad -I/p_x] = 0$. This shows that the vector that represents the budget line or the commodity plane and the price vector in Figure 2.4.9(A) are orthogonal.

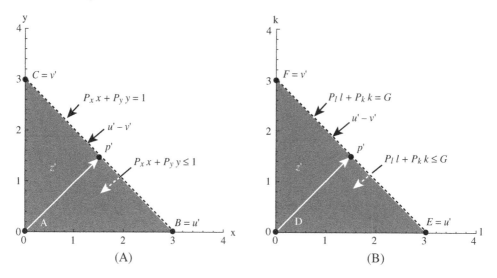

Figure 2.2.9

Example 2. Assume that the two combinations of the two inputs (l and k) that a firm employs to produce a good are given by the two row 2-vectors $\mathbf{u'} = \begin{bmatrix} l_1 & k_1 \end{bmatrix}$ and $\mathbf{v'} = \begin{bmatrix} l_2 & k_2 \end{bmatrix}$, and the prices of the two inputs are given by the row 2-vector $\mathbf{p'} = \begin{bmatrix} p_l & p_k \end{bmatrix}$. The firm's budget is given by G. Show the producer's *iso-cost line* and *input space* or *input plane*. Also show that the price vector is orthogonal to the iso-cost line (or to the input plane).

Solution. Following the same procedure as in the solution of example 1 above, we can draw the firm's iso-cost line (EF) and input plane (the shaded area) as illustrated in Figure 2.2.9(B). Let us now show that the price vector $\mathbf{p'} = \begin{bmatrix} p_l & p_k \end{bmatrix}$ is orthogonal to the iso-cost line or the input space. For this, assume that $\mathbf{v'} = \begin{bmatrix} 0 & G/p_k \end{bmatrix}$ and $\mathbf{u'} = \begin{bmatrix} G/p_l & 0 \end{bmatrix}$. Therefore, we have $\mathbf{z'} = \mathbf{u'} - \mathbf{v'} = \begin{bmatrix} G/p_l & 0 \end{bmatrix} - \begin{bmatrix} 0 & G/p_k \end{bmatrix} = \begin{bmatrix} G/p_l & -G/p_k \end{bmatrix}$. We can now multiply the vector $\mathbf{p'}$ by the vector $\mathbf{z}$ to obtain $\mathbf{p'}.\mathbf{z} = \mathbf{0}$. This shows that the vector that represents the iso-cost line or the input plane and the price vector in Figure 2.2.9(B) are orthogonal.

2.2.9 Exercises

1. Let $\mathbf{u'} = [1 \ 3 \ 5]$ and $\mathbf{v'} = [2 \ 4 \ 6]$. Find the following:
    (i) $\mathbf{u'} + \mathbf{v'}$; (ii) $\mathbf{u'} - \mathbf{v'}$.
2. Let $\mathbf{u'} = [1 \ 3]$ and $\mathbf{v'} = [2 \ 4]$. Find, and illustrate, the following:
    (i) $\mathbf{u'} + \mathbf{v'}$; (ii) $\mathbf{u'} - \mathbf{v'}$.
3. Let $\mathbf{u'} = [1 \ 3]$ and $\mathbf{v'} = [2 \ 4]$. Find, and illustrate, the following:
    (i) $2(\mathbf{u'} + \mathbf{v'})$; (ii) $2(\mathbf{u'} - \mathbf{v'})$.
4. Which of the following vectors are linearly independent (or dependent), and why?
    (i) $\mathbf{u'} = [1 \ 3]$ and $\mathbf{v'} = [2 \ 6]$; (ii) $\mathbf{u'} = [1 \ 3]$ and $\mathbf{v'} = [2 \ 4]$; (iii) $\mathbf{u'} = [1 \ 3 \ 5]$ and $\mathbf{v'} = [2 \ 4 \ 6]$; (iv) $\mathbf{u'} = [5 \ 15 \ 25]$ and $\mathbf{v'} = [1 \ 3 \ 5]$.
5. Find the inner products of the following vectors:
    (i) $\mathbf{u'} = [1 \ 3]$ and $\mathbf{v'} = [2 \ 6]$; (ii) $\mathbf{u'} = [2 \ 6]$ and $\mathbf{v'} = [1 \ 3]$; (iii) $\mathbf{u'} = [1 \ 3 \ 5]$ and $\mathbf{v'} = [2 \ 4 \ 6]$; (iv) $\mathbf{u'} = [5 \ 15 \ 25]$ and $\mathbf{v'} = [1 \ 3 \ 5]$.

6. Draw the graph of a line in $\mathfrak{R}^2$ using the vectors $\mathbf{u'} = [1\ \ 3]$ and $\mathbf{v'} = [2\ \ 4]$ and the scalar $s = 0.5$.

7. Draw the graph of a plane in $\mathfrak{R}^2$ using the vectors $\mathbf{u'} = [1\ \ 3]$ and $\mathbf{v'} = [2\ \ 4]$ and the scalars $s = 0.4$ and $r = 0.4$.

8. Draw the graph of a line in $\mathfrak{R}^3$ using the vectors $\mathbf{u'} = [1\ \ 3\ \ 5]$ and $\mathbf{v'} = [2\ \ 4\ \ 8]$ and the scalar $s = 0.5$.

9. Draw the graph of a plane in $\mathfrak{R}^3$ using the vectors $\mathbf{u'} = [2\ \ 0\ \ 0]$ and $\mathbf{v'} = [0\ \ 2\ \ 0]$ and $\mathbf{v'} = [0\ \ 0\ \ 2]$, and the scalars $s_1 = 1.5$, $s_2 = 1.5$, and $s_3 = 1.5$.

 Web supplement: S2.2.10 Mathematica applications

2.3 Matrices

We are now ready to begin the discussion of matrices. In Section 1.11.1 we constructed a matrix based on the sales data of a company. The indices i and j in that matrix denoted the ith month and jth market, respectively. Therefore, the elements in the ith row represented the sales of the company in all the markets during the ith month and the elements in the jth column represented the sales of the company in the jth market during all the months. And, the element a_{ij} denoted the sales of the company during the ith month in the jth market.

Suppose now that the company sells its product in only three markets ($j = 1, 2, 3$) during only three months ($i = 1, 2, 3$). Suppose also that we denote these sales by three row 3-vectors $\mathbf{u'} = [a_{11}\ \ a_{12}\ \ a_{13}]$, $\mathbf{v'} = [a_{21}\ \ a_{22}\ \ a_{23}]$, and $\mathbf{z'} = [a_{31}\ \ a_{32}\ \ a_{33}]$. Now imagine that we write these three vectors together by stacking one on the top of the other:

$$\mathbf{A} = \begin{bmatrix} u' \\ v' \\ z' \end{bmatrix} = \begin{bmatrix} a_{11} & a_{12} & a_{13} \\ a_{21} & a_{22} & a_{23} \\ a_{31} & a_{32} & a_{33} \end{bmatrix} \qquad (2.3.1)$$

We call this row-wise stack of vectors in equation (2.3.1) a matrix. If our vectors were $\mathbf{u'} = [a_{11}\ \ a_{12}\ \ a_{13}\ \ \dots\ \ a_{1j}\ \ \dots\ \ a_{1n}]$, $\mathbf{v'} = [a_{21}\ \ a_{22}\ \ a_{23}\ \ \dots\ \ a_{2j}\ \ \dots\ \ a_{2n}]$, $\mathbf{w'} = [a_{31}\ \ a_{32}\ \ a_{33}\ \ \dots\ \ a_{3j}\ \ \dots\ \ a_{3n}]$,..., $\mathbf{x'} = [a_{i1}\ \ a_{i2}\ \ a_{i3}\ \ \dots\ \ a_{ij}\ \ \dots\ \ a_{in}]$,..., and $\mathbf{z'} = [a_{m1}\ \ a_{m2}\ \ a_{m3}\ \ \dots\ \ a_{mj}\ \ \dots\ \ a_{mn}]$, then the matrix would become

$$\mathbf{B} = \begin{bmatrix} u' \\ v' \\ w' \\ \dots \\ x' \\ \dots \\ z' \end{bmatrix} = \begin{bmatrix} a_{11} & a_{12} & a_{13} & \dots & a_{1j} & \dots & a_{1n} \\ a_{21} & a_{22} & a_{23} & \dots & a_{2j} & \dots & a_{2n} \\ a_{31} & a_{32} & a_{33} & \dots & a_{3j} & \dots & a_{3n} \\ \dots & \dots & \dots & \dots & \dots & \dots & \dots \\ a_{i1} & a_{i2} & a_{i3} & \dots & a_{ij} & \dots & a_{in} \\ \dots & \dots & \dots & \dots & \dots & \dots & \dots \\ a_{m1} & a_{m2} & a_{m3} & \dots & a_{mj} & \dots & a_{mn} \end{bmatrix} \qquad (2.3.2)$$

Let us now formally define a matrix, such as $\mathbf{A}$ or $\mathbf{B}$ above, as a rectangular array of numbers (or parameters or variables). Every number (or parameter or variable) appearing inside the matrix (such as a_{12}, or a_{21}, or a_{ij}) is called an element of the matrix. Equation (2.3.2) gives the general form of a matrix, and its one specific form is given in equation (2.3.1).

Matrices are usually represented by bold, uppercase letters, such as **A** in equation (2.3.1) or **B** in equation (2.3.2). The *size of a matrix* (or the *dimension of a matrix* or the *order of a matrix*) is determined by the number of rows and columns it contains. Since matrix **A** has 3 rows and 3 columns, it is of order 3×3 (read "three by three") and is written as $\mathbf{A}_{3 \times 3}$. Similarly, matrix **B** is of order $m \times n$ and is written as $\mathbf{B}_{m \times n}$.

2.3.1 Types of matrices

One of the most frequently used types of matrices is a *square matrix*. A square matrix is a matrix in which the number of rows is equal to the number of columns; that is, $m = n$. The matrices **A** and **B** we discussed earlier are examples of square matrices. Another example of a square matrix is $\mathbf{C}_{2 \times 2} = \begin{bmatrix} 1 & 2 \\ 3 & 4 \end{bmatrix}$. If in a matrix $m \neq n$, then that matrix is called a *rectangular matrix*. An example of a rectangular matrix is $\mathbf{D}_{2 \times 3} = \begin{bmatrix} 1 & 2 & 3 \\ 4 & 5 & 6 \end{bmatrix}$.

Another important type of matrix is a *null matrix*. A null matrix is a matrix in which all the elements are zeros. A null matrix is normally denoted by **0**. Examples of a square null matrix and a rectangular null matrix are $\mathbf{E}_{2 \times 2} = \begin{bmatrix} 0 & 0 \\ 0 & 0 \end{bmatrix} = \mathbf{0}_{2 \times 2}$ and $\mathbf{F}_{2 \times 3} = \begin{bmatrix} 0 & 0 & 0 \\ 0 & 0 & 0 \end{bmatrix}$, respectively.

Another type of matrix is a *triangular matrix*. Before we explain the meaning of a triangular matrix, we need to know the meaning of the *diagonal elements* of a matrix. Consider, for example, matrix $\mathbf{C}_{2 \times 2}$ mentioned above. If we draw a line from the upper left corner of this matrix to its lower right corner, the line we obtain is called the *diagonal* of the matrix or the *main diagonal* of the matrix. The diagonal of this matrix will pass through elements 1 and 4. These elements are called the diagonal elements of matrix **C**.

We now define a triangular matrix as a matrix whose elements off the main diagonal are all zeros. This implies that a diagonal matrix is a matrix in which $a_{ij} \neq 0$ for all $i = j$, and $a_{ij} = 0$ for all $i \neq j$. If $a_{ij} = 0$ for all $i < j$ in a matrix (that is, if the elements in a matrix above the main diagonal are all zeros), then that matrix is called a *lower triangular matrix*. If $a_{ij} = 0$ for all $i > j$ in a matrix (that is, if the elements in a matrix below the main diagonal are all zeros), then that matrix is called an *upper triangular matrix*. Notice that for a matrix to be either triangular, or lower or upper triangular, the matrix must be a square matrix. The following matrices, **A**, **B**, and **C**, are examples of triangular, lower triangular, and upper triangular matrices, respectively:

$$\mathbf{A}_{3 \times 3} = \begin{bmatrix} 1 & 0 & 0 \\ 0 & 2 & 0 \\ 0 & 0 & 3 \end{bmatrix}, \quad \mathbf{B}_{3 \times 3} = \begin{bmatrix} 1 & 0 & 0 \\ 2 & 3 & 0 \\ 4 & 5 & 6 \end{bmatrix}, \quad \text{and } \mathbf{C}_{3 \times 3} = \begin{bmatrix} 1 & 2 & 3 \\ 0 & 4 & 5 \\ 0 & 0 & 6 \end{bmatrix}$$

Another important type of matrix is called an *identity matrix* or a *unit matrix*. Imagine what happens if, in a triangular matrix, the elements on the main diagonal are all 1's. Such a matrix is called an identity or a unit matrix and is normally denoted by **I**. Examples of identity matrices are

$$\mathbf{I}_{2 \times 2} = \begin{bmatrix} 1 & 0 \\ 0 & 1 \end{bmatrix}, \quad \mathbf{I}_{3 \times 3} = \begin{bmatrix} 1 & 0 & 0 \\ 0 & 1 & 0 \\ 0 & 0 & 1 \end{bmatrix}, \quad \text{etc.}$$

The last three types of matrices we discuss here are *row matrix, column matrix,* and *equal matrices*. A row matrix is a matrix in which there is only one row. A column matrix is a matrix in which there is only one column. Examples of row matrix and column matrix are $\mathbf{A}_{1\times 2} = \begin{bmatrix} 1 & 2 \end{bmatrix}$ and $\mathbf{B}_{2\times 1} = \begin{bmatrix} 3 \\ 4 \end{bmatrix}$, respectively. Notice that these row and column matrices are nothing other than the row and column vectors, respectively, we discussed and used in Section 2.2. Two matrices are said to be equal only if their corresponding elements are equal and they are of the same order. For example, suppose we have four matrices: $\mathbf{A} = \begin{bmatrix} 1 & 2 \\ 3 & 4 \end{bmatrix}$, $\mathbf{B} = \begin{bmatrix} 1 & 2 \\ 3 & 4 \end{bmatrix}$, $\mathbf{C} = \begin{bmatrix} 5 & 6 \\ 7 & 8 \end{bmatrix}$, and $\mathbf{D} = \begin{bmatrix} 1 & 2 \end{bmatrix}$. Since $\mathbf{A}$ and $\mathbf{B}$ have the same order (2×2) and since their corresponding elements are equal, $\mathbf{A} = \mathbf{B}$. Although $\mathbf{A}$, $\mathbf{B}$, and $\mathbf{C}$ are of the same order, their corresponding elements are not equal. Therefore, $\mathbf{A} = \mathbf{B} \neq \mathbf{C}$. Since the orders of $\mathbf{A}$ and $\mathbf{B}$ are different from that of $\mathbf{D}$, $\mathbf{A} = \mathbf{B} \neq \mathbf{D}$.

2.3.2 Matrix operation: scalar multiplication

We often need to use *scalar multiplication* and its properties in other matrix operations. Therefore, we first discuss the multiplication of one or more matrices by one or more scalars. Suppose that we have a matrix $\mathbf{A} = \begin{bmatrix} a_{11} & a_{12} \\ a_{21} & a_{22} \end{bmatrix}$ and a scalar s. What we mean by scalar multiplication is that we multiply every element of the matrix $\mathbf{A}$ by s, and we write it as $s\mathbf{A} = s \begin{bmatrix} a_{11} & a_{12} \\ a_{21} & a_{22} \end{bmatrix} = \begin{bmatrix} s.a_{11} & s.a_{12} \\ s.a_{21} & s.a_{22} \end{bmatrix}$. Notice that, for scalar multiplication, the matrix need not be a square matrix. Notice also that the reverse is also possible. Two of the important *properties of scalar multiplication* of matrices are the following.

Property I.   Suppose that we have two scalars s and t, and a matrix $\mathbf{A}$. Then $(s \pm t) \times \mathbf{A} = s \times \mathbf{A} \pm t \times \mathbf{A}$.

Property II.   Suppose that we have a scalar s and two matrices $\mathbf{A}$ and $\mathbf{B}$ of the same order. Then $s \times (\mathbf{A} + \mathbf{B}) = s \times \mathbf{A} + s \times \mathbf{B}$.

2.3.3 Matrix operations: addition and subtraction

Assume that we have two matrices of the same order: $\mathbf{A}_{ij} = [a_{ij}]$ and $\mathbf{B}_{ij} = [b_{ij}]$. Then a new matrix $\mathbf{C}_{ij} = [c_{ij}]$, where $[c_{ij}] = [a_{ij} + b_{ij}]$ for all i and j, and $i = 1, 2, 3, \ldots, m$ and $j = 1, 2, 3, \ldots, n$, can be generated. What this means is that we add the (i, j)th element of $\mathbf{A}$ to the corresponding (i, j)th element of $\mathbf{B}$ to obtain the (i, j)th of $\mathbf{C}$. As an example consider two matrices of order 2×2: $\mathbf{A}_{2\times 2} = \begin{bmatrix} 1 & 2 \\ 3 & 4 \end{bmatrix}$ and $\mathbf{B}_{2\times 2} = \begin{bmatrix} 5 & 6 \\ 7 & 8 \end{bmatrix}$. Then the sum, or addition, of the two matrices $\mathbf{A}_{ij}$ and $\mathbf{B}_{ij}$ is given as

$$\mathbf{C}_{2\times 2} = \mathbf{A}_{2\times 2} + \mathbf{B}_{2\times 2} = \begin{bmatrix} 1 & 2 \\ 3 & 4 \end{bmatrix} + \begin{bmatrix} 5 & 6 \\ 7 & 8 \end{bmatrix} = \begin{bmatrix} 1+5 & 2+6 \\ 3+7 & 4+8 \end{bmatrix} = \begin{bmatrix} 6 & 8 \\ 10 & 12 \end{bmatrix}$$

Subtraction of two or more matrices is similar to the addition of two or more matrices. Suppose that, as in the case of addition, we have two matrices of the same order: $\mathbf{A}_{ij} = [a_{ij}]$ and $\mathbf{B}_{ij} = [b_{ij}]$. Then a new matrix $\mathbf{C}_{ij} = [c_{ij}]$, where $[c_{ij}] = [a_{ij} - b_{ij}]$ for all i and j, and

$i = 1, 2, 3, \ldots, m$ and $j = 1, 2, 3, \ldots, n$, can be generated. What this means is that we subtract the (i, j)th element of **B** from the corresponding (i, j)th element of **A** to obtain the (i, j)th of **C**. Notice that, in subtraction, we are in fact adding the negative of **B** (which is nothing but a scalar multiple of **B** with the scalar being -1) to the corresponding element of **A**. As an example, consider our last two matrices $\mathbf{A}_{2 \times 2}$ and $\mathbf{B}_{2 \times 2}$. Then the difference between **A** and **B** is given as

$$\mathbf{C}_{2 \times 2} = \mathbf{A}_{2 \times 2} - 1 \times \mathbf{B}_{2 \times 2} = \mathbf{A}_{2 \times 2} - \mathbf{B}_{2 \times 2} = \begin{bmatrix} 1 & 2 \\ 3 & 4 \end{bmatrix} - \begin{bmatrix} 5 & 6 \\ 7 & 8 \end{bmatrix} = \begin{bmatrix} 1-5 & 2-6 \\ 3-7 & 4-8 \end{bmatrix}$$

$$= \begin{bmatrix} -4 & -4 \\ -4 & -4 \end{bmatrix} = (-4) \begin{bmatrix} 1 & 1 \\ 1 & 1 \end{bmatrix}$$

Notice that for addition (or subtraction) of two matrices, the matrices must be of the same order. This is called the *conformability condition* for addition (or subtraction) of matrices. Notice also that the resulting matrix **C** will be of the same order as those of the original matrices. The important *properties of matrix addition* (or of *matrix subtraction*) are presented below, supposing that we have four matrices of the same order: **A**, **B**, **C**, and **0** (null matrix).

Property I.   Matrix addition is commutative: $\mathbf{A} + \mathbf{B} = \mathbf{B} + \mathbf{A}$.
Property II.   Matrix addition is associative: $(\mathbf{A} + \mathbf{B}) + \mathbf{C} = \mathbf{A} + (\mathbf{B} + \mathbf{C})$.
Property III.   Existence of additive inverse: $\mathbf{A} + (-\mathbf{A}) = \mathbf{0}$.
Property IV.   Existence of additive identity: $\mathbf{A} + \mathbf{0} = \mathbf{A}$.

2.3.4 Matrix operation: multiplication

Let there be two matrices: $\mathbf{A}_{m \times p} = [a_{ij}]_{m \times p}$ and $\mathbf{B}_{p \times n} = [b_{ij}]_{p \times n}$. Then the product of **A** and **B**, denoted by **C**, is given by $\mathbf{AB} = \mathbf{C} = [c_{ij}]_{m \times n}$, where $c_{ij} = a_{i1}.b_{1j} + a_{i2}.b_{2j} + a_{i3}.b_{3j} + \ldots\ldots + a_{ip}.b_{pj}$, $i = 1, 2, 3, \ldots, m$ and $j = 1, 2, 3, \ldots, n$. This implies that the (i, j)th element of **C** is obtained by multiplying the ith row of **A** and the jth column of **B** and summing the result. Suppose that the two matrices are: $\mathbf{A} = \begin{bmatrix} a_{11} & a_{12} & a_{13} \\ a_{21} & a_{22} & a_{23} \end{bmatrix}_{2 \times 3}$ and $\mathbf{B} = \begin{bmatrix} b_{11} & b_{12} \\ b_{21} & b_{22} \\ b_{31} & b_{32} \end{bmatrix}_{3 \times 2}$.

Assume that we multiply every element in the first row of **A** by the corresponding elements in the first column of **B**, multiply every element in the first row of **A** by the corresponding elements in the second column of **B**, multiply every element in the second row of **A** by the corresponding elements in the first column of **B**, and multiply every element in the second row of **A** by the corresponding elements in the second column of **B**. When we take the sums of these four products and write it in the form of a matrix, we obtain matrix **C** (which is the product of **A** and **B**):

$$\mathbf{C} = \mathbf{A}.\mathbf{B} = \begin{bmatrix} a_{11} \times b_{11} + a_{12} \times b_{21} + a_{13} \times b_{31} & a_{11} \times b_{12} + a_{12} \times b_{22} + a_{13} \times b_{32} \\ a_{21} \times b_{11} + a_{22} \times b_{21} + a_{23} \times b_{31} & a_{21} \times b_{12} + a_{22} \times b_{22} + a_{23} \times b_{32} \end{bmatrix}_{2 \times 2}$$

Notice that these are nothing but the scalar products of vectors contained in the matrices **A** and **B**.

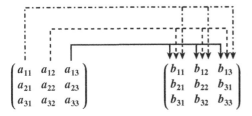

Figure 2.3.1

As an example of finding the products of two matrices, we shall use the matrices **A** and

B: $\mathbf{A} = \begin{bmatrix} 1 & 2 & 2 \\ 1 & 2 & 2 \end{bmatrix}_{2\times 3}$ and $\mathbf{B} = \begin{bmatrix} 2 & 1 \\ 1 & 2 \\ 2 & 1 \end{bmatrix}_{3\times 2}$. Therefore, the product of **A** and **B** is

$$\mathbf{A.B} = \begin{bmatrix} 1\times 2 + 2\times 1 + 2\times 2 & 1\times 1 + 2\times 2 + 2\times 1 \\ 1\times 2 + 2\times 1 + 2\times 2 & 1\times 1 + 2\times 2 + 2\times 1 \end{bmatrix}_{2\times 2} = \begin{bmatrix} 8 & 7 \\ 8 & 7 \end{bmatrix}_{2\times 2}$$

Suppose that our matrices are $\mathbf{A} = \begin{bmatrix} a_{11} & a_{12} & a_{13} \\ a_{21} & a_{22} & a_{23} \\ a_{31} & a_{32} & a_{33} \end{bmatrix}_{3\times 3}$ and $\mathbf{B} = \begin{bmatrix} b_{11} & b_{12} & b_{13} \\ b_{21} & b_{22} & b_{23} \\ b_{31} & b_{32} & b_{33} \end{bmatrix}_{3\times 3}$. Then,

following a procedure similar to the one above or as illustrated in Figure 2.3.1, we obtain the product of **A** and **B** as

$$\mathbf{AB} = \begin{bmatrix} a_{11}\times b_{11} + a_{12}\times b_{21} + a_{13}\times b_{31} & a_{11}\times b_{12} + a_{12}\times b_{22} + a_{13}\times b_{32} & a_{11}\times b_{13} + a_{12}\times b_{23} + a_{13}\times b_{33} \\ a_{21}\times b_{11} + a_{22}\times b_{21} + a_{23}\times b_{31} & a_{21}\times b_{12} + a_{22}\times b_{22} + a_{23}\times b_{32} & a_{31}\times b_{13} + a_{32}\times b_{23} + a_{33}\times b_{33} \\ a_{31}\times b_{11} + a_{32}\times b_{21} + a_{33}\times b_{31} & a_{31}\times b_{12} + a_{32}\times b_{22} + a_{33}\times b_{32} & a_{31}\times b_{13} + a_{32}\times b_{23} + a_{33}\times b_{33} \end{bmatrix}_{3\times 3}$$

As an example of finding the product of two 3×3 matrices, suppose that the matrices are

$\mathbf{A} = \begin{bmatrix} 2 & 2 & 1 \\ 2 & 1 & 1 \\ 2 & 2 & 1 \end{bmatrix}_{3\times 3}$ and $\mathbf{B} = \begin{bmatrix} 1 & 1 & 2 \\ 1 & 2 & 2 \\ 1 & 1 & 2 \end{bmatrix}_{3\times 3}$. Then their product will be

$$\mathbf{A.B} = \begin{bmatrix} 2\times 1 + 2\times 1 + 1\times 1 & 2\times 1 + 2\times 2 + 1\times 1 & 2\times 2 + 2\times 2 + 1\times 2 \\ 2\times 1 + 1\times 1 + 1\times 1 & 2\times 1 + 1\times 2 + 1\times 1 & 2\times 2 + 1\times 2 + 1\times 2 \\ 2\times 1 + 2\times 1 + 1\times 1 & 2\times 1 + 2\times 2 + 1\times 1 & 2\times 2 + 2\times 2 + 1\times 2 \end{bmatrix}_{3\times 3} = \begin{bmatrix} 5 & 7 & 10 \\ 4 & 5 & 8 \\ 5 & 7 & 10 \end{bmatrix}_{3\times 3}$$

The reader would have noticed that in the multiplication of **A** by **B**, the number of columns of **A** was equal to the number of rows of **B**. This is called the *conformability condition for the multiplication of matrices*. This implies that two matrices (**A** and **B**) are said to be conformable for multiplication (if we multiply **A** by **B**) only if the number of columns in **A** is equal to the number of rows in **B**. Otherwise the product will not exist. This points to the fact that the order of **A.B** will be equal to the number of rows of **A** and the number of columns of **B**. In our first example above, the order of **A** was 2×3 and the order of **B** was 3×2 giving us the product **A.B** whose order was 2×2. In the second example, **A** and **B** were of the same order (3×3) yielding us **A.B** whose order was 3×3.

Similarly, if we multiply **B** by **A**, then the number of rows in **B** must be equal to the number of columns in **A**, and the order of the resulting matrix will be equal to the number of rows of **B** and the number of columns of **A**. Therefore, one needs to check whether the matrices are conformable for multiplication before one carries out multiplication. Notice that those two square matrices are always (in both ways) conformable for multiplication. We state below the *properties of matrix multiplication*, assuming that we have five matrices, **A**, **B**, **C**, an identity matrix (**I**), and a null matrix (**0**), such that they are conformable for the operations indicated below.

Property I.   Matrix multiplication is associative: $(\mathbf{AB})\mathbf{C} = \mathbf{A}(\mathbf{BC})$.

Property II.   Matrix multiplication is distributive over addition and subtraction: $\mathbf{A}(\mathbf{B} \pm \mathbf{C}) = \mathbf{AB} \pm \mathbf{AC}$ and $(\mathbf{B} \pm \mathbf{C})\mathbf{A} = \mathbf{BA} \pm \mathbf{CA}$.

Property III.   Matrix multiplication is *not always* commutative: $\mathbf{AB} = \mathbf{BA}$ or $\mathbf{AB} \neq \mathbf{BA}$ (that is, **AB** *may or may not be equal to* **BA**, even if these two operations are conformable or even if the products exist).

Property IV. $\mathbf{AI} = \mathbf{A} = \mathbf{IA}$.

Property V. $\mathbf{A0} = \mathbf{0} = \mathbf{0A}$.

2.3.5 *Transpose of a matrix and powers of square matrices*

Suppose that we have the matrix $\mathbf{A} = [a_{ij}]_{m \times n}$. Then the transpose of this matrix, denoted by $\mathbf{A}^\mathbf{T}$, is defined as $\mathbf{A}^\mathbf{T} = [b_{ji}]_{n \times m}$, where $b_{ji} = a_{ij}$ for all $i = 1, 2, 3, \ldots, m$ and $j = 1, 2, 3, \ldots, n$. What this means is that we obtain $\mathbf{A}^\mathbf{T}$ by interchanging the rows and columns of **A**; that is, the rows of $\mathbf{A}^\mathbf{T}$ will be columns of **A**. This implies that the transposition of a matrix reverses the order of the matrix; that is, if **A** is of order $m \times n$, then $\mathbf{A}^\mathbf{T}$ will be of order $n \times m$. Notice that the transposes of matrices **I** and **0** will be the same; i.e., $\mathbf{I} = \mathbf{I}^\mathbf{T}$ and $\mathbf{0} = \mathbf{0}^\mathbf{T}$.

As an example, consider the matrices $\mathbf{A} = \begin{bmatrix} 1 & 2 & 3 \\ 4 & 5 & 6 \end{bmatrix}_{2 \times 3}$ and $\mathbf{B} = \begin{bmatrix} 1 & 2 & 3 \\ 4 & 5 & 6 \\ 7 & 8 & 9 \end{bmatrix}_{3 \times 3}$. Then,

$\mathbf{A}^\mathbf{T} = \begin{bmatrix} 1 & 4 \\ 2 & 5 \\ 3 & 6 \end{bmatrix}_{3 \times 2}$ and $\mathbf{B}^\mathbf{T} = \begin{bmatrix} 1 & 4 & 7 \\ 2 & 5 & 8 \\ 3 & 6 & 9 \end{bmatrix}_{3 \times 3}$. The reader will have noticed that if a square

matrix (such as **B** above) is transposed, the diagonal elements remain unchanged.

Two matrices related to the transpose of a matrix are *symmetric matrix* and *skew symmetric matrix*. A square matrix, say **C**, is said to be a symmetric matrix if $\mathbf{C}^\mathbf{T} = \mathbf{C}$. Notice that **I** and **0** are two examples of a symmetric matrices. Other examples include $\mathbf{C}^\mathbf{T}_1 =$

$\mathbf{C}_1 = \begin{bmatrix} a & b \\ b & c \end{bmatrix}$, $\mathbf{C}^\mathbf{T}_2 = \mathbf{C}_2 = \begin{bmatrix} 1 & 2 \\ 2 & 4 \end{bmatrix}$, $\mathbf{C}^\mathbf{T}_3 = \mathbf{C}_3 = \begin{bmatrix} a & b & c \\ b & d & e \\ c & e & f \end{bmatrix}$, and $\mathbf{C}^\mathbf{T}_4 = \mathbf{C}_4 = \begin{bmatrix} 1 & 2 & 3 \\ 2 & 4 & 5 \\ 3 & 5 & 6 \end{bmatrix}$.

A square matrix, say **C**, is said to be a skew symmetric matrix if $\mathbf{C}^\mathbf{T} = -\mathbf{C}$. Examples of a skew symmetric matrix include $\mathbf{C}^\mathbf{T}_1 = \begin{bmatrix} 0 & -b \\ b & 0 \end{bmatrix}$ given $\mathbf{C}_1 = \begin{bmatrix} 0 & b \\ -b & 0 \end{bmatrix}$, and $\mathbf{C}^\mathbf{T}_2 =$

$\begin{bmatrix} 0 & b & c \\ -b & 0 & e \\ -c & -e & 0 \end{bmatrix}$ given $\mathbf{C}_2 = \begin{bmatrix} 0 & -b & -c \\ b & 0 & -e \\ c & e & 0 \end{bmatrix}$.

The important properties of matrix transposition are the following.

Property I. Let there be a matrix $\mathbf{A}$ and scalar s, then $(s.\,\mathbf{A})^{\mathrm{T}} = s.\,(\mathbf{A})^{\mathrm{T}}$.

Property II. For any matrix $\mathbf{A}$, then $(\mathbf{A}^{\mathrm{T}})^{\mathrm{T}} = \mathbf{A}$.

Property III. Let there be two matrices, $\mathbf{A}$ and $\mathbf{B}$, of the same order, then $(\mathbf{A} \pm \mathbf{B})^{\mathrm{T}} = \mathbf{A}^{\mathrm{T}} \pm \mathbf{B}^{\mathrm{T}}$.

Property IV. Let there be two matrices, $\mathbf{A}$ and $\mathbf{B}$, and their product $\mathbf{A}.\mathbf{B}$ exists, then $(\mathbf{A}.\mathbf{B})^{\mathrm{T}} = \mathbf{B}^{\mathrm{T}}.\mathbf{A}^{\mathrm{T}}$.

We now consider the *powers of a square matrix*. Let $\mathbf{A}$ be a square matrix. Then, we define $\mathbf{A}.\mathbf{A} = \mathbf{A}^2$; $\mathbf{A}.\mathbf{A}^2 = \mathbf{A}^3$; $\mathbf{A}.\mathbf{A}^3 = \mathbf{A}^4$;...; and $\mathbf{A}.\mathbf{A}^{n-1} = \mathbf{A}^n$, where n is any positive integer. Notice that $\mathbf{A}^m.\mathbf{A}^n = \mathbf{A}^{m+n}$, and $(\mathbf{A}^m)^n = \mathbf{A}^{m.n}$.

2.3.6 Matrices and SSLEs

There are plenty of models in the subjects of our interest that use matrices to abbreviate SSLEs. Therefore, we consider here the topic of abbreviation of SSLEs using matrices. We discussed SSLEs in detail in Section 1.6.5. Consider, for example, the 2×2 system we dealt with in equation (1.6.12). Let us reproduce that system: $-b_1 x + y = b_0$ and $-b_3 x + y = b_2$. Using, for convenience, $a_{11} = -b_1$, $a_{12} = 1$, $a_{21} = -b_3$, $a_{22} = 1$, $x_1 = x$, $x_2 = y$, $d_1 = b_0$, and $d_2 = b_2$ we may rewrite the last system as

$$a_{11} x_1 + a_{12} x_2 = d_1$$
$$a_{21} x_1 + a_{22} x_2 = d_2 \tag{2.3.3}$$

The question now is how one can abbreviate the system in equation (2.3.3) using matrices. Notice that a_{11}, a_{12}, a_{21}, and a_{22} are called coefficients, x_1 and x_2 are called variables, and d_1 and d_2 are called constants. One can write the coefficients, the variables, and the constants in the above system using matrices as $\mathbf{A} = \begin{bmatrix} a_{11} & a_{12} \\ a_{21} & a_{22} \end{bmatrix}$, $\mathbf{x} = \begin{bmatrix} x_1 \\ x_2 \end{bmatrix}$, and $\mathbf{d} = \begin{bmatrix} d_1 \\ d_2 \end{bmatrix}$. Therefore, the system in equation (2.3.3) can be written in matrix form as $\mathbf{Ax} = \mathbf{d}$. Notice that $\mathbf{x}$ and $\mathbf{d}$ in this equation are nothing but two column 2-vectors.

Consider now, for example, the 3×3 system we dealt with in equation (1.6.13). Using, for convenience, $a_{11} = -b_1$, $a_{12} = -b_2$, $a_{13} = 1$, $d_1 = b_0$, $a_{21} = -b_4$, $a_{22} = -b_5$, $a_{23} = 1$, $d_2 = b_3$, $a_{31} = -b_7$, $a_{32} = -b_8$, $a_{33} = 1$, $d_3 = b_6$, $x_1 = x_2 = y$, and $x_3 = z$, we may write the system (1.6.13) as

$$a_{11} x_1 + a_{12} x_2 + a_{13} x_3 = d_1$$
$$a_{21} x_1 + a_{22} x_2 + a_{23} x_3 = d_2 \tag{2.3.4}$$
$$a_{31} x_1 + a_{32} x_2 + a_{33} x_3 = d_3$$

One can now write the coefficients, the variables, and the constants in the above system using matrices as $\mathbf{A} = \begin{bmatrix} a_{11} & a_{12} & a_{13} \\ a_{21} & a_{22} & a_{23} \\ a_{31} & a_{32} & a_{33} \end{bmatrix}$, $\mathbf{x} = \begin{bmatrix} x_1 \\ x_2 \\ x_3 \end{bmatrix}$, and $\mathbf{d} = \begin{bmatrix} d_1 \\ d_2 \\ d_3 \end{bmatrix}$. Therefore, the system in equation (2.3.4) can be written in matrix form as $\mathbf{Ax} = \mathbf{d}$. Suppose now that we have an $m \times n$ system of the form

$$a_{11}x_1 + a_{12}x_2 + a_{13}x_3 + \cdots + a_{1n}x_n = d_1$$

$$a_{21}x_1 + a_{22}x_2 + a_{23}x_3 + \cdots + a_{2n}x_n = d_2$$

$$a_{31}x_1 + a_{32}x_2 + a_{33}x_3 + \cdots + a_{3n}x_n = d_3 \qquad (2.3.5)$$

$$\cdots\cdots\cdots\cdots\cdots\cdots\cdots\cdots\cdots\cdots\cdots$$

$$a_{m1}x_1 + a_{m2}x_2 + a_{m3}x_3 + \cdots + a_{mn}x_n = d_m$$

The above $m \times n$ system can be written, using matrices, as $\mathbf{Ax} = \mathbf{d}$, where

$$\mathbf{A} = \begin{bmatrix} a_{11} & a_{12} & a_{13} & \cdots & a_{1n} \\ a_{21} & a_{22} & a_{21} & \cdots & a_{2n} \\ a_{31} & a_{32} & a_{33} & \cdots & a_{3n} \\ \cdots & \cdots & \cdots & \cdots & \cdots \\ a_{m1} & a_{m2} & a_{m3} & \cdots & a_{mn} \end{bmatrix}, \mathbf{x} = \begin{bmatrix} x_1 \\ x_2 \\ x_3 \\ .. \\ x_n \end{bmatrix}, \text{ and } \mathbf{d} = \begin{bmatrix} d_1 \\ d_2 \\ d_3 \\ .. \\ d_m \end{bmatrix}$$

Consider now an $m \times n$ system whose constants are all zeros (that is, $d_i = 0$, where $i = 1, 2, 3, \ldots, m$). Such a system is of the form

$$a_{11}x_1 + a_{12}x_2 + a_{13}x_3 + \cdots + a_{1n}x_n = 0$$

$$a_{21}x_1 + a_{22}x_2 + a_{23}x_3 + \cdots + a_{2n}x_n = 0$$

$$a_{31}x_1 + a_{32}x_2 + a_{33}x_3 + \cdots + a_{3n}x_n = 0 \qquad (2.3.6)$$

$$\cdots\cdots\cdots\cdots\cdots\cdots\cdots\cdots\cdots\cdots\cdots$$

$$a_{m1}x_1 + a_{m2}x_2 + a_{m3}x_3 + \cdots + a_{mn}x_n = 0$$

We can write the system in equation (2.3.6) in matrix form as $\mathbf{Ax} = \mathbf{d} = \mathbf{0}$, where $\mathbf{A}$ denotes the $m \times n$ matrix of coefficients, $\mathbf{x}$ denotes the $m \times 1$ column matrix (or vector) of variables, and $\mathbf{d} = \mathbf{0}$ denotes the $m \times 1$ column matrix (or vector) of constants. A SSLEs whose constants are all zeros (as in equation (2.3.6)) is called a *homogenous SSLEs*.

2.3.7 *Augmented matrices, elementary row operations, echelon forms, and the solution of SSLEs*

Consider the 3×3 SSLEs we used in the system in equation (1.6.14). Letting $x_1 = x$, $x_2 = y$, and $x_3 = z$, we may rewrite that system as

$$2x_1 + x_2 + x_3 = 6$$

$$x_1 + 2x_2 + x_3 = 2 \qquad (2.3.7)$$

$$0.5x_1 + 0.5x_2 + x_3 = 2$$

We know that the coefficient matrix, the column vector of the variable, and the column vector of the constant of the system, respectively, are $\mathbf{A} = \begin{bmatrix} 2 & 1 & 1 \\ 1 & 2 & 1 \\ 0.5 & 0.5 & 1 \end{bmatrix}$, $\mathbf{x} = \begin{bmatrix} x_1 \\ x_2 \\ x_3 \end{bmatrix}$, and

$\mathbf{d} = \begin{bmatrix} 6 \\ 2 \\ 2 \end{bmatrix}$. We also know that the matrix form of the system is $\mathbf{Ax} = \mathbf{d}$. Notice that if we can transform the system $\mathbf{Ax} = \mathbf{d}$ into $\mathbf{Ix} = \mathbf{e}$ (where we transformed the coefficient matrix ($\mathbf{A}$) into identity matrix ($\mathbf{I}$) and the constant matrix $\mathbf{d}$ into another column matrix of constants ($\mathbf{e}$)), then we can be sure that $\mathbf{Ix} = \mathbf{x} = \mathbf{e}$ would be the solution to the above system.

In order to obtain the matrix equation $\mathbf{Ix} = \mathbf{x} = \mathbf{e}$, we proceed as follows. Suppose now that we augment $\mathbf{A}$ by adding the column matrix $\mathbf{d}$ inside $\mathbf{A}$; that is, as $[\mathbf{A}|\mathbf{d}]$ and denote it by $\mathbf{A_{aug}d}$. Then the result will be $\mathbf{A_{aug}d} = \begin{bmatrix} 2 & 1 & 1 & | & 6 \\ 1 & 2 & 1 & | & 2 \\ 0.5 & 0.5 & 1 & | & 2 \end{bmatrix}$, which is called the *augmented matrix* based on equation (2.3.7).

Suppose again that we multiply or divide the rows of the augmented matrix by a nonzero constant; or multiply one of the rows of the augmented matrix by a nonzero constant and add the result to another row of the same augmented matrix; or interchange any of the two rows of the augmented matrix. These three operations, when applied to an augmented matrix $\mathbf{A_{aug}d}$ of an original system of equations, are called the *elementary row operations*. Therefore, the elementary row operations include any one, or all, of the following:

I. Multiply or divide one row of the augmented matrix by a nonzero constant. When we multiply the first row by 3 (or by 1/3), we denote it as $R_1 \rightarrow 3R_1$ (or $R_1 \rightarrow (1/3)R_1$). Notice that when we do this operation on R_1 of the augmented matrix, only R_1 will change and all other rows will remain unchanged.

II. Add a multiple of one row of the augmented matrix to another row of the same matrix. When we multiply the second row by 3 and add the result to (or subtract the result from) the third row, we denote it as $R_3 \rightarrow R_3 \pm 3R_2$. As above, when we do this operation, only R_3 will change and all other rows remain unchanged.

III. Interchange two rows of the augmented matrix. When we interchange the first and the second rows, we denote it as $R_1 \leftrightarrow R_2$. When we do this operation, the positions of R_1 and R_2 will be interchanged while the positions of all other rows remain unchanged.

We shall now carry out these elementary row operations on our augmented matrix $\mathbf{A_{aug}d}$ above. Notice that one can carry out individual elementary row operations, or one or more of the elementary row operations at the same time.

$$\begin{bmatrix} 2 & 1 & 1 & | & 6 \\ 1 & 2 & 1 & | & 2 \\ 0.5 & 0.5 & 1 & | & 2 \end{bmatrix} \xrightarrow{R_3 \rightarrow 2R_3} \begin{bmatrix} 2 & 1 & 1 & | & 6 \\ 1 & 2 & 1 & | & 2 \\ 1 & 1 & 2 & | & 4 \end{bmatrix} \xrightarrow{R_3 \rightarrow 2R_3 - R_1} \begin{bmatrix} 1 & 1 & 1 & | & 6 \\ 1 & 2 & 1 & | & 2 \\ 0 & 1 & 3 & | & 2 \end{bmatrix}$$

$$\xrightarrow{R_1 \rightarrow R_1 - R_2} \begin{bmatrix} 1 & -1 & 0 & | & 4 \\ 1 & 2 & 1 & | & 2 \\ 0 & 1 & 3 & | & 2 \end{bmatrix} \xrightarrow{R_2 \rightarrow R_2 - R_3} \begin{bmatrix} 1 & -1 & 0 & | & 4 \\ 1 & 1 & -2 & | & 0 \\ 0 & 1 & 3 & | & 2 \end{bmatrix} \xrightarrow[R_2 - R_1]{R_2 \rightarrow} \begin{bmatrix} 1 & -1 & 0 & | & 4 \\ 0 & 2 & -2 & | & -4 \\ 0 & 1 & 3 & | & 2 \end{bmatrix}$$

$$\xrightarrow[R_2 \rightarrow R_2 - 2R_3]{R_1 \rightarrow R_1 + R_3} \begin{bmatrix} 1 & 0 & 3 & | & 6 \\ 0 & 0 & -8 & | & -8 \\ 0 & 1 & 3 & | & 2 \end{bmatrix} \xrightarrow[R_2 + R_3]{R_2 \rightarrow} \begin{bmatrix} 1 & 0 & 3 & | & 6 \\ 0 & 1 & -5 & | & -6 \\ 0 & 1 & 3 & | & 2 \end{bmatrix} \xrightarrow[R_3 - R_2]{R_3 \rightarrow} \begin{bmatrix} 1 & 0 & 3 & | & 6 \\ 0 & 1 & -5 & | & -6 \\ 0 & 0 & 8 & | & 8 \end{bmatrix}$$

$$\xrightarrow{R_3 \rightarrow \frac{1}{8}R_3} \begin{bmatrix} 1 & 0 & 3 & | & 6 \\ 0 & 1 & -5 & | & -6 \\ 0 & 0 & 1 & | & 1 \end{bmatrix} \xrightarrow[R_2 \rightarrow R_2 + 5R_3]{R_1 \rightarrow R_1 - 3R_3} \begin{bmatrix} 1 & 0 & 0 & | & 3 \\ 0 & 1 & 0 & | & -1 \\ 0 & 0 & 1 & | & 1 \end{bmatrix}$$

Notice that the last matrix contains two matrices: the identity matrix $\mathbf{I} = \begin{bmatrix} 1 & 0 & 0 \\ 0 & 1 & 0 \\ 0 & 0 & 1 \end{bmatrix}$ and

the column matrix $\mathbf{e} = \begin{bmatrix} 3 \\ -1 \\ 1 \end{bmatrix}$. We can now write these two matrices backward, with the

matrix $\mathbf{x}$, as we did before (that is, $\mathbf{Ix} = \mathbf{e}$). Now, what this new system $\mathbf{Ix} = \mathbf{e}$ means is the solution of our 3×3 SSLEs in equation (2.3.7). In other words we have the solutions $x_1 = 3$, $x_2 = -1$, and $x_3 = 1$. Therefore, the system possesses a unique solution. Notice that these were exactly the solutions we mentioned at the end of Section 1.6.5.

As another example, consider the 2×2 system in Section 1.6.5. This system can be written as

$$-2x_1 + x_2 = 10$$
$$-4x + x_2 = 5$$
(2.3.8)

where $x_1 = x$ and $x_2 = y$. We can write this system in matrix form as $\mathbf{Ax} = \mathbf{d}$, where $\mathbf{A} = \begin{bmatrix} -2 & 1 \\ -4 & 1 \end{bmatrix}$, $\mathbf{x} = \begin{bmatrix} x_1 \\ x_2 \end{bmatrix}$, and $\mathbf{d} = \begin{bmatrix} 10 \\ 5 \end{bmatrix}$. Then, as we did above, the augmented matrix of this

2×2 system is $\mathbf{A_{aug}d} = \begin{bmatrix} -2 & 1 & | & 10 \\ -4 & 1 & | & 5 \end{bmatrix}$. We shall now carry out the elementary row operations on $\mathbf{A_{aug}d}$ as

$$\begin{bmatrix} -2 & 1 & | & 10 \\ -4 & 1 & | & 5 \end{bmatrix} \xrightarrow{R_1 \to -\frac{1}{2}R_1} \begin{bmatrix} 1 & -1/2 & | & -5 \\ -4 & 1 & | & 5 \end{bmatrix} \xrightarrow{R_2 \to R_2 + 4R_1} \begin{bmatrix} 1 & -1/2 & | & -5 \\ 0 & -1 & | & -15 \end{bmatrix}$$

$$\xrightarrow{R_2 \to (-1)R_2} \begin{bmatrix} 1 & -1/2 & | & -5 \\ 0 & 1 & | & 15 \end{bmatrix} \xrightarrow{R_1 \to R_1 + \frac{1}{2}R_2} \begin{bmatrix} 1 & 0 & | & 5/2 \\ 0 & 1 & | & 15 \end{bmatrix}$$

Notice that, as before, the last matrix contains two matrices: the identity matrix $\mathbf{I} = \begin{bmatrix} 1 & 0 \\ 0 & 1 \end{bmatrix}$

and the column matrix $\mathbf{e} = \begin{bmatrix} 5/2 \\ 15 \end{bmatrix}$. We can write these two matrices backward, with the

matrix $\mathbf{x}$, as we did before (that is, as $\mathbf{Ix} = \mathbf{e}$). Now, what this new system $\mathbf{Ix} = \mathbf{e}$ means, as before again, is the solution of the 2×2 SSLEs in equation (2.3.8). In other words we have the solutions $x_1 = 5/2 = 2.5$ and $x_2 = 15$. Therefore, the system possesses a unique solution. Notice that these were exactly the solutions we obtained in Section 1.6.5.

There may be occasions where one might end up with, instead of an identity matrix, a

matrix such as $\begin{bmatrix} 1 & 0 & 0 \\ 0 & 1 & 0 \\ 0 & 0 & 0 \end{bmatrix}$ or $\begin{bmatrix} 1 & 1 \\ 0 & 0 \end{bmatrix}$ for a 3×3 or for a 2×2 system, respectively. In these

cases, there will be an infinite number of solutions.

All that we did in the elementary row operations was that we converted the coefficient matrix $\mathbf{A}$, contained in the augmented matrix, into an identity matrix. But, as a byproduct of this, we obtained a new column vector that replaced the original constant vector ($\mathbf{d}$) and we called this new column vector $\mathbf{e}$, which is nothing but the vector of the solution of the system. Therefore, to solve a SSLEs all that one has to do is to convert the coefficient part of the augmented matrix into an identity matrix using the elementary row operations. This process of converting the coefficient part of the augmented matrix into an identity matrix is called *row reduction*.

Suppose that, as above, we transform or reduce a coefficient matrix (or any square matrix) into any other matrix of similar order containing only 1's and 0's. Then this latter matrix is called the *reduced row echelon form* of the original coefficient matrix. Let us now define the reduced row echelon form of a matrix. A matrix is in reduced row echelon form if the following four conditions are satisfied:

I.   All rows that contain only zeros, if any, must appear at the bottom of the matrix.

II.   Any row that contains nonzero elements must begin with 1 as its first nonzero element.

III.   If two consecutive rows contain nonzero elements, then the first nonzero element (that is, 1) in the lower row must be to the right of the first nonzero element (again, 1) in the upper row.

IV.   Any column that contains the first 1 (as the first nonzero element of the concerned row) must contain zeros as other elements.

The last four matrices are among the examples of row reduced echelon forms. However, the fourth condition may not be satisfied in certain cases. If the first three conditions given above are satisfied and if the fourth condition is not satisfied in a specific row reduction, then we call such matrices *row echelon forms*. Examples of such matrices include $\begin{bmatrix} 1 & 2 \\ 0 & 1 \end{bmatrix}$,

$\begin{bmatrix} 1 & 0 & 3 \\ 0 & 0 & 0 \end{bmatrix}$, and $\begin{bmatrix} 1 & 2 & 3 \\ 0 & 1 & 4 \\ 0 & 0 & 1 \end{bmatrix}$.

2.3.8 Solution of SSLEs: the Gauss–Jordan and the Gauss methods

We continue our discussion of the topic of solution of SSLEs in this section. We discussed reduced row echelon forms and row echelon forms in the previous section. However, one can find two methods of solution of SSLEs, using row reduction, in the literature.

One of these two methods is called the *Gauss–Jordan method* or the *Gauss–Jordan elimination method*. This is the same method as the one we used in the solution of the systems in equations (2.3.7) and (2.3.8). Therefore, we will not repeat this method here. However, we state that the crux of the Gauss–Jordan elimination method lies in finding the reduced row echelon form from the augmented form of the coefficient matrix of the given system. This reduced row echelon form will give us the solutions to the system. The Gauss–Jordan method is used extensively in finding the *inverse* of a square matrix, a topic we will consider later.

The *Gauss method* is closely related to the Gauss–Jordan method. However, in the Gauss method we first find the row echelon form, which will give us the solution for the last unknown. This solution is used to substitute backward until we obtain the solutions for all other unknowns. As an example of the use of the Gauss method, consider the following simple system:

$$2x_1 + x_2 = 4$$
$$x_1 + 2x_2 = 2$$

(2.3.9)

The system in equation (2.3.9) implies that the coefficient matrix, the variable matrix, and the constant matrix respectively are $\mathbf{A} = \begin{bmatrix} 2 & 1 \\ 1 & 2 \end{bmatrix}$, $\mathbf{x} = \begin{bmatrix} x_1 \\ x_2 \end{bmatrix}$, and $\mathbf{d} = \begin{bmatrix} 4 \\ 2 \end{bmatrix}$. Therefore, we write the augmented matrix as $\mathbf{A_{aug}d} = \begin{bmatrix} 2 & 1 & | & 4 \\ 1 & 2 & | & 2 \end{bmatrix}$. One can carry out the elementary row operations given below to obtain the row echelon form:

$$\begin{bmatrix} 2 & 1 & | & 4 \\ 1 & 2 & | & 2 \end{bmatrix} \xrightarrow{R_1 \rightarrow R_1 - R_2} \begin{bmatrix} 1 & -1 & | & 2 \\ 1 & 2 & | & 2 \end{bmatrix} \xrightarrow{R_2 \rightarrow R_2 - R_1} \begin{bmatrix} 1 & -1 & | & 2 \\ 0 & 3 & | & 2 \end{bmatrix}$$

Notice that the last matrix gives the required row echelon form. From the last row of this matrix it is clear that $x_2 = 0$. From its first row we get $x_1 - x_2 = 2$. Substitution of $x_2 = 0$ into $x_1 - x_2 = 2$ yields $x_1 - 0 = x_1 = 2$. Therefore, the solutions are $x_1 = 2$ and $x_2 = 0$.

As another example of the use of the Gauss method, consider the following 3×3 system:

$$2x_1 + x_2 + x_3 = 10$$
$$x_1 + 2x_2 + x_3 = 5 \qquad\qquad (2.3.10)$$
$$x_1 + x_2 + 2x_3 = 5$$

The system in equation (2.3.10) implies that the coefficient matrix, the variable matrix, and the constant matrix, respectively, are $\mathbf{A} = \begin{bmatrix} 2 & 1 & 1 \\ 1 & 2 & 1 \\ 1 & 1 & 2 \end{bmatrix}$, $\mathbf{x} = \begin{bmatrix} x_1 \\ x_2 \\ x_3 \end{bmatrix}$, and $\mathbf{d} = \begin{bmatrix} 10 \\ 5 \\ 5 \end{bmatrix}$. Therefore, we write the augmented matrix as $\mathbf{A_{aug}d} = \begin{bmatrix} 2 & 1 & 1 & | & 10 \\ 1 & 2 & 1 & | & 5 \\ 1 & 1 & 2 & | & 5 \end{bmatrix}$. One can carry out the elementary row operation given below, as before, to obtain the row echelon form:

$$\begin{bmatrix} 2 & 1 & 1 & | & 10 \\ 1 & 2 & 1 & | & 5 \\ 1 & 1 & 2 & | & 5 \end{bmatrix} \begin{array}{c} R_3 \rightarrow R_3 - R_2 \\ \xrightarrow{\hspace{1cm}} \\ R_1 \rightarrow R_1 - R_2 \end{array} \begin{bmatrix} 1 & -1 & 0 & | & 5 \\ 1 & 2 & 1 & | & 5 \\ 0 & -1 & 1 & | & 0 \end{bmatrix} \begin{array}{c} R_2 \rightarrow R_2 - R_1 \\ \xrightarrow{\hspace{1cm}} \\ R_3 \rightarrow (-1)R_3 \end{array} \begin{bmatrix} 1 & -1 & 0 & | & 5 \\ 0 & 3 & 1 & | & 0 \\ 0 & 1 & -1 & | & 0 \end{bmatrix}$$

$$\xrightarrow{R_1 \rightarrow R_1 + R_3} \begin{bmatrix} 1 & 0 & -1 & | & 5 \\ 1 & 2 & 1 & | & 5 \\ 0 & 1 & -1 & | & 0 \end{bmatrix} \xrightarrow{R_2 \rightarrow R_2 - R_1} \begin{bmatrix} 1 & 0 & -1 & | & 5 \\ 0 & 2 & 2 & | & 0 \\ 0 & 1 & -1 & | & 0 \end{bmatrix} \xrightarrow{R_2 \rightarrow R_2 - R_3} \begin{bmatrix} 1 & 0 & -1 & | & 5 \\ 0 & 1 & 3 & | & 0 \\ 0 & 1 & -1 & | & 0 \end{bmatrix}$$

$$\xrightarrow{R_3 \rightarrow R_3 - R_2} \begin{bmatrix} 1 & 0 & -1 & | & 5 \\ 0 & 1 & 3 & | & 0 \\ 0 & 1 & -4 & | & 0 \end{bmatrix} \xrightarrow{R_3 \rightarrow R_3 - R_2} \begin{bmatrix} 1 & 0 & -1 & | & 5 \\ 0 & 1 & 3 & | & 0 \\ 0 & 0 & -7 & | & 0 \end{bmatrix}$$

$$\xrightarrow{R_3 \rightarrow \left(\frac{-1}{7}\right)R_3} \begin{bmatrix} 1 & 0 & -1 & | & 5 \\ 0 & 1 & 3 & | & 0 \\ 0 & 0 & 1 & | & 0 \end{bmatrix} \xrightarrow{R_1 \rightarrow R_1 + R_3} \begin{bmatrix} 1 & 0 & 0 & | & 5 \\ 0 & 1 & 3 & | & 0 \\ 0 & 0 & 1 & | & 0 \end{bmatrix}$$

Notice that the last augmented matrix gives the required row echelon form. From the first and the last rows of this matrix it is clear that $x_1 = 5$ and $x_3 = 0$, respectively. From the second row we get $x_2 + 3x_3 = 0$. Substitution of $x_3 = 0$ into $x_2 + 3x_3 = 0$ yields $x_2 = 0$. Therefore, the solutions are $x_1 = 5$, $x_2 = 0$, and $x_3 = 0$.

We now consider homogeneous SSLEs, as that in equation (2.3.6). As we will describe in Section 2.7, a general SSLEs may possess a unique solution, an infinite number of solutions, or no solution at all. However, in the case of a homogeneous SSLEs there are only two possibilities: either one *trivial solution* (also called *zero solution*) or infinite *nontrivial solutions*. The former happens when all the x_s are zeros (that is, when $x_1 = x_2 = x_3 = \cdots\cdots = x_n = 0$).

2.3.9 Application examples

Example 1. Suppose that the demand and the supply functions of a good are given respectively by $x_1 = 4 - 4x_2$ and $x_1 = 2 + 2x_2$, where x_1 denotes the quantity demanded and supplied of the good, and x_2 denotes the price of the good in dollars. Find the quantity demanded (and supplied) and the price that solve the system.

Solution. To find out the quantity and price, we need to solve the given system. For convenience, we write the system as

$$x_1 + 4x_2 = 4$$

$$x_1 - 2x_2 = 2$$

The coefficient, variable, constant, and augmented matrices of this system, respectively, are $\mathbf{A}$ $= \begin{bmatrix} 1 & 4 \\ 1 & -2 \end{bmatrix}$, $\mathbf{x} = \begin{bmatrix} x_1 \\ x_2 \end{bmatrix}$, $\mathbf{d} = \begin{bmatrix} 4 \\ 2 \end{bmatrix}$, and $\mathbf{A_{aug}d} = \begin{bmatrix} 1 & 4 & | & 4 \\ 1 & -2 & | & 2 \end{bmatrix}$. We can now use the Gauss–Jordan method. Application of this method yields the reduced row echelon form $\begin{bmatrix} 1 & 0 & | & 2.67 \\ 0 & 1 & | & 0.33 \end{bmatrix}$. This suggests that the quantity demanded and supplied is 2.67 units (that is, $x_1^* = 2.67$ units) and the price is \$0.33 (that is, $x_2^* = \$0.33$). These are the quantity and price that solve the given system.

Example 2. Assume that income and interest rate in an *IS-LM model* of an economy are related by the following SSLEs:

$$0.2x_1 + 200x_2 = 40$$

$$0.2x_1 - 800x_2 = -60$$

where x_1 denotes income in billions of dollars and x_2 denotes interest rate. Find the level of income and interest rate that solve the system.

Solution. We represent this system in matrix form as $\mathbf{Ax} = \mathbf{d}$ and $\mathbf{A_{aug}d}$, where $\mathbf{A} = \begin{bmatrix} 0.2 & 200 \\ 0.2 & -800 \end{bmatrix}$, $\mathbf{x} = \begin{bmatrix} x_1 \\ x_2 \end{bmatrix}$, $\mathbf{d} = \begin{bmatrix} 40 \\ -60 \end{bmatrix}$, and $\mathbf{A_{aug}d} = \begin{bmatrix} 0.2 & 200 & | & 40 \\ 0.2 & -800 & | & -60 \end{bmatrix}$. We can now apply the Gauss–Jordan method to yield the reduced row echelon form $\begin{bmatrix} 1 & 0 & | & 100 \\ 0 & 1 & | & 0.1 \end{bmatrix}$, which suggests that the level of income is \$100 billion (or $x_1^* = \$100$ billion) and the equilibrium interest rate is 0.1 or 10 percent (or $x_2^* = 0.1$ or 10%). These are the values that solve the given system.

2.3.10 Exercises

1. Let $\mathbf{A} = [a_{ij}]$, where $i, j = 1, 2, 3, 4$. Construct a matrix using $\mathbf{A}$ such that $a_{ij} = 1$ for all $i = j$ and $a_{ij} = 0$ for all $i \neq j$. What do you call this matrix? Construct a second matrix using $\mathbf{A}$ such that $a_{ij} = 0$ for all $i = j$ and $a_{ij} = 0$ for all $i \neq j$. What do you call this second matrix? Construct a third matrix using $\mathbf{A}$ such that $a_{ij} = 0$ for all $i < j$ and $a_{ij} = 1$ for all $i > j$. What do you call this third matrix? Construct a fourth matrix using $\mathbf{A}$ such that $a_{ij} = 0$ for all $i > j$ and $a_{ij} = 1$ for all $i < j$. What do you call this forth matrix?

2. Let $\mathbf{A} = \begin{bmatrix} 1 & 2 & 3 \\ 4 & 5 & 6 \\ 7 & 8 & 9 \end{bmatrix}$, $\mathbf{B} = \begin{bmatrix} 9 & 8 \\ 7 & 6 \\ 5 & 4 \end{bmatrix}$, $\mathbf{C} = \begin{bmatrix} 9 & 8 & 7 \\ 6 & 5 & 4 \\ 3 & 2 & 1 \end{bmatrix}$, and $\mathbf{D} = \begin{bmatrix} 1 & 2 \\ 3 & 4 \\ 5 & 6 \end{bmatrix}$. Which of these matrices are conformable for addition and subtraction? Find the sums and differences of those matrices that are conformable for addition and subtraction.

3. Suppose that we have an equation of matrices given by $4\mathbf{A} - 2\mathbf{B} = \mathbf{C}$. Let $\mathbf{A} = \begin{bmatrix} 1 & 2 & 3 \\ 4 & 5 & 6 \\ 7 & 8 & 9 \end{bmatrix}$ and $\mathbf{B} = \begin{bmatrix} 9 & 8 & 7 \\ 6 & 5 & 4 \\ 3 & 2 & 1 \end{bmatrix}$. Find $\mathbf{C}$.

4. Let $\mathbf{A} = [a_{ij}]$ and $\mathbf{B} = [b_{ij}]$, where $i, j = 1, 2, 3$ and $a_{ij} = 1$ for all $i = j$ and $a_{ij} = 0$ for all $i \neq j$ in $\mathbf{A}$; and where $i, j = 1, 2, 3$ and $b_{ij} = 0$ for all $i = j$ and $b_{ij} = 1$ for all $i \neq j$ in $\mathbf{B}$. Find their product $\mathbf{AB}$.

5. Let $\mathbf{A} = \begin{bmatrix} 1 & 2 & 3 \\ 4 & 5 & 6 \\ 7 & 8 & 9 \end{bmatrix}$, $\mathbf{B} = \begin{bmatrix} 9 & 8 \\ 7 & 6 \\ 5 & 4 \end{bmatrix}$, $\mathbf{C} = \begin{bmatrix} 5 & 6 & 7 \\ 8 & 9 & 10 \end{bmatrix}$, and $\mathbf{D} = \begin{bmatrix} 1 & 2 \\ 3 & 4 \end{bmatrix}$. Which of these matrices are conformable for multiplication? Find the products of those matrices that are conformable for multiplication.

6. Let $\mathbf{a'} = [5]$, $\mathbf{b'} = [2]$, $\mathbf{c'} = \begin{bmatrix} 2 & 3 \end{bmatrix}$, $\mathbf{d} = \begin{bmatrix} 4 \\ 5 \end{bmatrix}$, $\mathbf{e} = \begin{bmatrix} 6 \\ 7 \\ 8 \end{bmatrix}$, and $\mathbf{f} = \begin{bmatrix} 9 \\ 10 \\ 11 \end{bmatrix}$. Which of these matrices are conformable for multiplication? Find the products of those matrices that are conformable for multiplication.

7. Find the transpose of the following matrices, and state which of them are symmetric or skew symmetric matrices:

    $\mathbf{A} = \begin{bmatrix} a & b \\ c & d \end{bmatrix}$, $\mathbf{B} = \begin{bmatrix} 1 & 0 & 0 \\ 0 & 1 & 0 \\ 0 & 0 & 1 \end{bmatrix}$, $\mathbf{C} = \begin{bmatrix} 0 & -1 & -2 \\ 1 & 0 & -3 \\ 2 & 3 & 0 \end{bmatrix}$, $\mathbf{D} = \begin{bmatrix} 0 & 5 \\ -5 & 0 \end{bmatrix}$, $\mathbf{e} = \begin{bmatrix} 1 \\ 2 \\ 3 \end{bmatrix}$,

    $\mathbf{f'} = \begin{bmatrix} 1 & 2 & 3 \end{bmatrix}$, and $\mathbf{g'} = [1]$.

8. Suppose that we have three matrices $\mathbf{A} = [a_{ij}]$, $\mathbf{x} = [x_i]$, and $\mathbf{d} = [d_i]$ where $i, j = 1, 2, 3$. Construct a matrix equation of the form $\mathbf{Ax} = \mathbf{d}$ using the above three matrices.

9. Write the following SSLEs as matrix equation(s):

    (i) $a_{11}x_1 + a_{12}x_2 = d$; (ii) $\begin{aligned} 4x_1 &= d_1 \\ 2x_2 &= d_2 \end{aligned}$; (iii) $\begin{aligned} 4x_1 + 2x_2 &= d_1 \\ 2x_2 &= d_2 \end{aligned}$; (iv) $\begin{aligned} 4x_1 &= d_1 \\ x_1 + 2x_2 &= d_2 \end{aligned}$;

    (v) $\begin{aligned} 2x_1 - 3x_2 &= d_1 \\ 4x_2 + x_3 &= d_2 \\ x_1 + x_3 &= d_3 \end{aligned}$; (vi) $\begin{aligned} 2x_1 - 3x_2 - d_1 &= 0 \\ 4x_2 + x_3 &= d_2 \\ x_1 + x_3 &= d_3 \end{aligned}$; (vii) $\begin{aligned} 2x_1 - 3x_2 + x_3 &= d_1 \\ x_1 + 4x_2 + x_3 &= d_2 \\ x_1 + x_2 - x_3 &= d_3 \end{aligned}$.

10. Let $\mathbf{A} = \begin{bmatrix} 1 & 2 \\ 3 & 4 \end{bmatrix}$. Find $\mathbf{A}^n$ if $n = 1, 2, 3$. Also find $(\mathbf{A}^2)^2$ and $(\mathbf{A}^2).(\mathbf{A}^3)$.

11. Let $\mathbf{A} = \begin{bmatrix} 1 & 2 & 3 \\ 4 & 5 & 6 \\ 7 & 8 & 9 \end{bmatrix}$. Find $\mathbf{A}^n$ if $n = 1, 2, 3$. Also find $(\mathbf{A}^2)^2$ and $(\mathbf{A}^2).(\mathbf{A}^3)$.

12. Solve the following SSLEs using the Gauss method and the Gauss–Jordan method:

   (i) $\begin{aligned} x_1 + x_2 &= 10 \\ 2x_1 + x_2 &= 4 \end{aligned}$; (ii) $\begin{aligned} x_1 + 2x_2 &= 8 \\ 2x_1 + x_2 &= 4 \end{aligned}$; (iii) $\begin{aligned} x_1 + x_2 + x_3 &= 1 \\ 2x_1 + 2x_2 &= 2 \\ x_2 + x_3 &= 1 \end{aligned}$; (iv) $\begin{aligned} 2x_1 + 2x_2 + 2x_3 &= 1 \\ 2x_1 + 2x_2 &= 2 \\ 2x_2 + 2x_3 &= 1 \end{aligned}$.

13. *Application exercise.* Suppose that the demand and the supply functions of a good are given respectively by $x_1 = 10 - 0.5x_2$ and $x_1 = 1 + x_2$, where x_1 denotes the quantity demanded and supplied of the good, and x_2 denotes the price of the good in dollars. Find the quantity demanded and supplied and the price that solve the system using the Gauss–Jordan method.

14. *Application exercise.* Assume that income and interest rate in an IS-LM model of an economy are related by the SSLEs $0.1x_1 - 400x_2 = -30$ and $0.1x_1 + 100x_2 = 10$, where x_1 denotes income in billions of dollars and x_2 denotes the interest rate. Find the level of income and interest rate that solve the system using the Gauss–Jordan method.

15. *Application exercise.* Suppose that the prices of three goods sold by a firm are interrelated by the SSLEs $5x_1 - 2x_2 - 2x_3 = 5$, $-2x_1 + 5x_2 - 2x_3 = 5$, and $-2x_1 + -2x_2 + 5x_3 = 5$, where x_1, x_2, and x_3 denote prices in dollars of goods 1, 2, and 3, respectively. Find the prices that solve the system using the Gauss–Jordan elimination method.

 Web supplement: S2.3.11 Mathematica applications

2.4 Determinants

Determinants are used extensively in economics, business, and finance. One of the most important uses of determinants lies in the solution of SSLEs. Therefore, it is important that students of economics, business, and finance possess a reasonably good knowledge of determinants, their evaluations, and their properties. This is attempted in this section.

2.4.1 Meaning and evaluation of determinants

So far we were dealing with vectors and matrices. Some students may wonder if we can associate a number with every square matrix. Yes, we can, and it is called its *determinant*. But, how can one find such a number? We discuss below the methods of finding the values that we associate with square matrices of different orders (that is, *evaluating determinants*).

Suppose that we have a 2×2 matrix given by $\mathbf{A} = \begin{bmatrix} 1 & 2 \\ 3 & 4 \end{bmatrix}$. Then the determinant of $\mathbf{A}$, denoted by $| \mathbf{A} |$, is written as $|\mathbf{A}| = \begin{vmatrix} 1 & 2 \\ 3 & 4 \end{vmatrix}$. We can find the value we associate with $\mathbf{A}$ (that is, $|\mathbf{A}|$) by finding the products of the elements on the *main diagonal* and subtracting from this the product of the *off-diagonal* elements. The product of the elements on the main diagonal in the present example is $1 \times 4 = 4$ and the product of the off-diagonal elements is $2 \times 3 = 6$. Therefore, the determinant is $|\mathbf{A}| = \begin{vmatrix} 1 & 2 \\ 3 & 4 \end{vmatrix} = 4 - 6 = -2$.

In general, if we have a 2×2 matrix $\mathbf{B} = \begin{bmatrix} a_{11} & a_{12} \\ a_{21} & a_{22} \end{bmatrix}$, then the determinant of $\mathbf{B}$ is given by

$$|\mathbf{B}| = \begin{vmatrix} a_{11} & a_{12} \\ a_{21} & a_{22} \end{vmatrix} = (a_{11} \times a_{22}) - (a_{21} \times a_{12}) \tag{2.4.1}$$

Now suppose we have a matrix of order 3×3 given by

$$\mathbf{A} = \begin{bmatrix} a_{11} & a_{12} & a_{13} \\ a_{21} & a_{22} & a_{23} \\ a_{31} & a_{32} & a_{33} \end{bmatrix} \longrightarrow \begin{matrix} a_{11} & a_{12} & a_{13} & a_{11} & a_{12} \\ a_{21} & a_{22} & a_{23} & a_{21} & a_{22} \\ a_{31} & a_{32} & a_{33} & a_{31} & a_{32} \end{matrix}$$

One easy way of evaluating the determinant of this third-order matrix is to write it as given alongside the matrix $\mathbf{A}$. Notice that the last two columns above are the first two columns of $\mathbf{A}$. The next step is to find the product of the elements on the downward sloping arrows, and sum these products, which is given by $a_{11} \times a_{22} \times a_{33} + a_{12} \times a_{23} \times a_{31} + a_{13} \times a_{21} \times a_{32}$. The third step is to find the product of the elements on the upward sloping arrows, and sum them too, which is given by $a_{31} \times a_{22} \times a_{13} + a_{32} \times a_{23} \times a_{11} + a_{33} \times a_{21} \times a_{12}$. The final step is to subtract the latter from the former. Therefore, the determinant of $\mathbf{A}$ is given by

$$|\mathbf{A}| = (a_{11} \times a_{22} \times a_{33} + a_{12} \times a_{23} \times a_{31} + a_{13} \times a_{21} \times a_{32})$$
$$- (a_{31} \times a_{22} \times a_{13} + a_{32} \times a_{23} \times a_{11} + a_{33} \times a_{21} \times a_{12}) \tag{2.4.2}$$

As an example, consider the following 3×3 matrix:

$$\mathbf{A} = \begin{bmatrix} 1 & 2 & 3 \\ 4 & 5 & 6 \\ 7 & 8 & 9 \end{bmatrix} \longrightarrow \begin{matrix} 1 & 2 & 3 & 1 & 2 \\ 4 & 5 & 6 & 4 & 5 \\ 7 & 8 & 9 & 7 & 8 \end{matrix}$$

As before, we write the matrix as given above alongside the matrix $\mathbf{A}$. First, we find the product of elements on the downward sloping arrows, and sum these products. This sum is equal to $1 \times 5 \times 9 + 2 \times 6 \times 7 + 3 \times 4 \times 8 = 225$. Second, we obtain the product of elements on the upward sloping arrows, which is equal to $7 \times 5 \times 3 + 8 \times 6 \times 1 + 9 \times 4 \times 2 = 225$. The difference between these two sums is $225 - 225 = 0$. Therefore, the determinant is zero; that is, $|\mathbf{A}| = 0$. If the determinant of a matrix is zero, then the determinant is said to *vanish*. It is important to notice that the above method using arrows, popularly called *Sarrus' rule*, works *only in the case of* 3×3 *matrices*. Notice that the determinant of the matrix $\mathbf{A} = [3]$ is given by $|\mathbf{A}| = 3$. Also notice that determinants are defined only for square matrices. We will consider later the evaluation of determinants of square matrices of order greater than 3.

2.4.2 Properties of determinants

Determinants obey some important properties. These properties are the following.

Property I.   The determinant of a square matrix, $\mathbf{A}$, and the determinant of its transpose are equal, that is, $|\mathbf{A}| = |\mathbf{A}^{\mathrm{T}}|$. This implies that the determinant of a matrix remains unchanged if the rows and columns of that matrix are interchanged.

Property II.   If two rows (or columns) of a determinant, $|\mathbf{A}|$, are interchanged, then the sign of the determinant changes but the determinant remains the same.

Property III.   If two rows (or columns) of a determinant, $|\mathbf{A}|$, are identical, then the determinant is zero.

Property IV.   If any one row (or one column) of a determinant is multiplied by a scalar s, then the determinant will be multiplied by the scalar s.

Property V.   If a multiple of one row (or column) is added to any row (or column), then the determinant remains unchanged.

Property VI.   The determinant of an identity matrix, $\mathbf{I}$, of any order is equal to 1.

Property VII.   Suppose we have two matrices, $\mathbf{A}$ and $\mathbf{B}$. Also suppose that the product $\mathbf{AB}$ exists. Then $|\mathbf{AB}| = |\mathbf{A}|.|\mathbf{B}|$.

2.4.3 Sub-matrices, minors, co-factors, Laplace expansion, and determinants of matrices of order greater than 3

So far we were evaluating the determinants of square matrices of order less than or equal to 3 and we applied Sarrus' rule for this. Since Sarrus' rule is not applicable to evaluate determinants of order greater than 3, we need to find out a method that can deal with such determinants. There exists another method, called the method of *expansion by co-factors*, to evaluate determinants of any order. However, to discuss this method, we need to explain few concepts such as *sub-matrices*, *minors*, and *co-factors*. We shall do this first.

For simplicity, we begin our exposition of the concepts with a 3×3 matrix $\mathbf{A}$ we used in the previous section. Suppose, now, that we discard the first row and the first column of $\mathbf{A}$ (one can discard any row and any column); that is, we discard $i = 1$ and $j = 1$. Then the matrix we obtain is called a sub-matrix (of order 2×2) and we denote it by $\mathbf{S}_{11} = \begin{bmatrix} 5 & 6 \\ 8 & 9 \end{bmatrix}$, where $\mathbf{S}$ represents the *sub-matrix*, and the two subscripts (1's) represent the numbers of discarded rows and columns. One can generate a number of sub-matrices like this depending upon the rows and columns discarded. For example, two other sub-matrices of $\mathbf{A}$ are $\mathbf{S}_{12} = \begin{bmatrix} 4 & 6 \\ 7 & 9 \end{bmatrix}$ and $\mathbf{S}_{13} = \begin{bmatrix} 4 & 5 \\ 7 & 8 \end{bmatrix}$.

We are now ready to define a sub-matrix formally. Assume that we have a square matrix $\mathbf{A} = (a_{ij})_{n \times n}$. The square matrix that can be generated if we discard the ith row and jth column of $\mathbf{A}$, denoted by $\mathbf{S}_{ij}$, is called the ij^{th} sub-matrix of $\mathbf{A}$. We can generate, in total, nine sub-matrices from a 3×3 matrix such as $\mathbf{A}$ above. Therefore, all the sub-matrices of order two that can be generated from $\mathbf{A}$ above are $\mathbf{S}_{11}$, $\mathbf{S}_{12}$, $\mathbf{S}_{13}$, $\mathbf{S}_{21}$, $\mathbf{S}_{22}$, $\mathbf{S}_{23}$, $\mathbf{S}_{31}$, $\mathbf{S}_{32}$, and $\mathbf{S}_{33}$.

Now suppose that we take the determinant of a sub-matrix, say $\mathbf{S}_{ij}$, and denote it by $|\mathbf{S}_{ij}|$. This determinant is called the ij^{th} *minor* of $\mathbf{A}$, and we denote it by $\mathbf{M}_{ij} = |\mathbf{S}_{ij}|$. For example, $\mathbf{M}_{11} = |\mathbf{S}_{11}| = (5 \times 9 - 8 \times 6) = 45 - 48 = -3$. Similarly, $\mathbf{M}_{12} = |\mathbf{S}_{12}| = (4 \times 9 - 7 \times 6) = 36 - 42 = -6$, and $\mathbf{M}_{13} = |\mathbf{S}_{13}| = (4 \times 8 - 7 \times 5) = 32 - 35 = -3$. Therefore, given $\mathbf{A}$, we will have nine minors in total: $\mathbf{M}_{11} = -3$, $\mathbf{M}_{12} = -6$, $\mathbf{M}_{13} = -3$, $\mathbf{M}_{21} = -6$, $\mathbf{M}_{22} = -12$, $\mathbf{M}_{23} = -6$, $\mathbf{M}_{31} = -3$, $\mathbf{M}_{32} = -6$, and $\mathbf{M}_{33} = -3$.

Assume now that we multiply $\mathbf{M}_{11}$ by $(+1)$, $\mathbf{M}_{12}$ by (-1), $\mathbf{M}_{13}$ by 1, and so on. Notice that we are multiplying each minor by the scalar $(+1)$ or the scalar (-1) depending upon the sum of the subscripted numbers of the corresponding minor. If the sum is even, the scalar is $+1$; if the sum is odd, the scalar is (-1). In general, we can write the scalar as $(-1)^{(i+j)}$. If we multiply $\mathbf{M}_{ij}$ by $(-1)^{(i+j)}$, then it is called the *signed minor* of $\mathbf{M}_{ij}$. The signed minor of $\mathbf{M}_{ij}$ (that is, $(-1)^{(i+j)} \mathbf{M}_{ij}$) is also called the *co-factor* of $\mathbf{M}_{ij}$, and is denoted by C_{ij}. Therefore, the co-factor of M_{11} is $C_{11} = (-1)^{(i+j)} M_{11} = (-1)^{(1+1)} M_{11} = (-1)^2$ $M_{11} = (1) M_{11} = M_{11} = -3$; of M_{12} is $C_{12} = (-1) M_{12} = 6$; and of M_{13} is $C_{13} = (1)$ $M_{13} = -3$. Similarly, $C_{21} = (-1) M_{21} = 6$; $C_{22} = (1) M_{22} = -12$; $C_{23} = (-1) M_{23} = 6$; $C_{31} = (1) M_{31} = -3$; $C_{32} = (-1) M_{32} = 6$; and $C_{33} = (1) M_{33} = -3$.

We are now ready to use the concepts of co-factors to evaluate determinants of order 3×3. Consider, for example, our last 3×3 matrix $\mathbf{A}$ to find $|\mathbf{A}|$ using co-factors. Then $|\mathbf{A}|$ is defined as $|\mathbf{A}| = 1 \times C_{11} + 2 \times C_{12} + 3 \times C_{13}$. Since $C_{11} = (-1)^{(i+j)} M_{11} = (-1)^{(1+1)}$ $M_{11} = (-1)^2 M_{11} = (1) M_{11} = (1) M_{11}$; $C_{12} = (-1) M_{12}$; and $C_{13} = (1) M_{13}$, we can write $|\mathbf{A}| = 1 \times (1) M_{11} + 2 \times (-1) M_{12} + 3 \times (1) M_{13}$. This implies that $|\mathbf{A}| = 1 \times (1) \times$

$$\begin{vmatrix} 5 & 6 \\ 8 & 9 \end{vmatrix} + 2 \times (-1) \times \begin{vmatrix} 4 & 6 \\ 7 & 9 \end{vmatrix} + 3 \times (1) \times \begin{vmatrix} 4 & 5 \\ 7 & 8 \end{vmatrix} = 1 \times \begin{vmatrix} 5 & 6 \\ 8 & 9 \end{vmatrix} - 2 \times \begin{vmatrix} 4 & 6 \\ 7 & 9 \end{vmatrix} + 3 \begin{vmatrix} 4 & 5 \\ 7 & 8 \end{vmatrix}.$$

Since $\begin{vmatrix} 5 & 6 \\ 8 & 9 \end{vmatrix} = -3$, $\begin{vmatrix} 4 & 6 \\ 7 & 9 \end{vmatrix} = -6$, and $\begin{vmatrix} 4 & 5 \\ 7 & 8 \end{vmatrix} = -3$, the last result simplifies to $|\mathbf{A}| = 1 \times -3 + -2 \times -6 + 3 \times -3 = -3 + 12 - 9 = 12 - 12 = 0$. This was precisely the result we obtained when we used Sarrus' rule in the previous section.

Let us now generalize the idea with the help of a general matrix, $\mathbf{A} = \begin{bmatrix} a_{11} & a_{12} & a_{13} \\ a_{21} & a_{22} & a_{23} \\ a_{31} & a_{32} & a_{33} \end{bmatrix}$.

Then we define $|\mathbf{A}|$ as

$$|\mathbf{A}| = a_{11} \times (1) \times \begin{vmatrix} a_{22} & a_{23} \\ a_{32} & a_{33} \end{vmatrix} + a_{12} \times (-1) \times \begin{vmatrix} a_{21} & a_{23} \\ a_{31} & a_{33} \end{vmatrix}$$

$$+ a_{13} \times (1) \times \begin{vmatrix} a_{21} & a_{22} \\ a_{31} & a_{32} \end{vmatrix} = a_{11}C_{11} + a_{12}C_{12} + a_{13}C_{13} \tag{2.4.3}$$

Equation (2.4.3) implies that $|\mathbf{A}| = \sum_j^3 a_{ij}C_{ij}$. If we evaluate (that is, expand) the last matrix using the jth column, then we can write $|\mathbf{A}| = \sum_i^3 a_{ij}C_{ij}$. Notice that if the matrix is an $n \times n$ matrix, then we have $|\mathbf{A}| = \sum_j^n a_{ij}C_{ij}$ if we use the jth column and $|\mathbf{A}| = \sum_i^n a_{ij}C_{ij}$ if we use the ith row. Also notice that $\sum_j^n a_{ij}C_{ij} = \sum_i^n a_{ij}C_{ij}$; that is, the determinant remains the same irrespective of the row or column of the matrix that is used to evaluate it. This method of evaluating (or expanding) a determinant, using co-factors, is called the *Laplace expansion* of a determinant.

We now consider evaluating the determinant of a 4×4 matrix, $\mathbf{A} = \begin{bmatrix} 1 & 2 & 3 & 4 \\ 5 & 6 & 7 & 8 \\ 9 & 10 & 11 & 12 \\ 13 & 14 & 15 & 16 \end{bmatrix}$.

Following exactly the process of Laplace expansion as detailed above, we obtain

$$|\mathbf{A}| = 1 \times (1) \times \begin{vmatrix} 6 & 7 & 8 \\ 10 & 11 & 12 \\ 14 & 15 & 16 \end{vmatrix} + 2 \times (-1) \times \begin{vmatrix} 5 & 7 & 8 \\ 9 & 11 & 12 \\ 13 & 15 & 16 \end{vmatrix}$$

$$+ 3 \times (1) \times \begin{vmatrix} 5 & 6 & 8 \\ 9 & 10 & 12 \\ 13 & 14 & 16 \end{vmatrix} + 4 \times (-1) \times \begin{vmatrix} 5 & 6 & 7 \\ 9 & 10 & 11 \\ 13 & 14 & 15 \end{vmatrix} = 0.$$

2.4.4 Co-factors and adjoint matrices

We described in the last section concepts like sub-matrices, minors, and co-factors. Let us now construct a matrix using the co-factors called the *co-factor matrix*. Given the 3×3 matrix $\mathbf{A} = \begin{bmatrix} 1 & 2 & 3 \\ 4 & 5 & 6 \\ 7 & 8 & 9 \end{bmatrix}$, the co-factors we obtained in the previous section were $C_{11} = -3$, $C_{12} = 6$, $C_{13} = -3$, $C_{21} = 6$, $C_{22} = -12$, $C_{23} = 6$, $C_{31} = -3$, $C_{32} = 6$, and $C_{33} = -3$. Therefore, the co-factor matrix denoted by $\mathbf{C_{FA}}$, where $\mathbf{A}$ represents the original matrix in our present example, will be

$$\mathbf{C_{FA}} = \begin{bmatrix} C_{11} & C_{12} & C_{13} \\ C_{21} & C_{22} & C_{23} \\ C_{31} & C_{32} & C_{33} \end{bmatrix} = \begin{bmatrix} -3 & 6 & -3 \\ 6 & -12 & 6 \\ -3 & 6 & -3 \end{bmatrix} \tag{2.4.4}$$

Notice that if our original square matrix were of order 2×2, its co-factor matrix would be of order 2×2; if it were of order 4×4, its co-factor matrix would be of order 4×4; and so on. If we transpose the co-factor matrix, we obtain what is called the *adjoint matrix*, denoted by $\mathbf{A_{adj}}$. Therefore, the adjoint matrix in the case of our present example will be

$$\mathbf{A_{adj}} = (\mathbf{C_{FA}})^{\mathrm{T}} = \begin{bmatrix} -3 & 6 & -3 \\ 6 & -12 & 6 \\ -3 & 6 & -3 \end{bmatrix} \tag{2.4.5}$$

Notice that in this case $\mathbf{C_{FA}} = (\mathbf{C_{FA}})^{\mathrm{T}}$ because they are symmetric matrices.

2.4.5 Solution of SSLEs: Cramer's rule

In Section 1.6.5 we used graphical, substitution, and elimination methods to solve SSLEs. In Sections 2.3.7 and 2.3.8 we used elementary row operations or the Gauss–Jordan elimination method and the Gauss elimination method, respectively, to solve SSLEs. Notice that each method has its own merits and demerits.

However, in this section, we discuss another popular method called *Cramer's rule*. For this, we use the $m \times n$ (where $m = n$) SSLEs given in equation (2.3.5). We expressed there the system as the matrix equation $\mathbf{Ax} = \mathbf{d}$. Now suppose that we replace the first column of $\mathbf{A}$ by $\mathbf{d}$ and denote the resulting matrix by $\mathbf{A_1}$. When we continue the replacement of the other columns of $\mathbf{A}$ by $\mathbf{d}$, we obtain n such new matrices (including $\mathbf{A_1}$) each denoted by $\mathbf{A_1}, \mathbf{A_2}, \mathbf{A_3}, \ldots$, and $\mathbf{A_n}$. We know that the determinants of these matrices will be $|\mathbf{A_1}|, |\mathbf{A_2}|, |\mathbf{A_3}|, \ldots$, and $|\mathbf{A_n}|$, respectively. Notice that the determinant of the original coefficient matrix is $|\mathbf{A}|$. It can now be shown that

$$x_1 = |\mathbf{A_1}| \div |\mathbf{A}|, x_2 = |\mathbf{A_2}| \div |\mathbf{A}|, x_3 = |\mathbf{A_3}| \div |\mathbf{A}|, \ldots, x_j = |\mathbf{A_j}| \div |\mathbf{A}|, \ldots, x_n = |\mathbf{A_n}| \div |\mathbf{A}| \tag{2.4.6}$$

Let us now state Cramer's rule formally. Let $\mathbf{A}$ be an $n \times n$ matrix and $|\mathbf{A}| \neq 0$. Then the unique solution to the system $\mathbf{Ax} = \mathbf{d}$ is given by $x_1 = |\mathbf{A_1}| \div |\mathbf{A}|$, $x_2 = |\mathbf{A_2}| \div |\mathbf{A}|$, $x_3 = |\mathbf{A_3}| \div |\mathbf{A}|, \ldots, x_j = |\mathbf{A_j}| \div |\mathbf{A}|, \ldots, x_n = |\mathbf{A_n}| \div |\mathbf{A}|$, where $|\mathbf{A_j}|$ denotes the determinant of the adjoint matrix obtained by replacing the jth column of the coefficient matrix $\mathbf{A}$ by the vector of the constants (**d**). Notice that we can use either Sarrus' rule if **A** (or **A**$_j$)is of order 3×3 or Laplace expansion if **A** is of order larger than 3×3 to find the determinants of **A** (or of **A**$_j$).

As an example, consider the SSLEs in equation (2.3.7). We expressed this system in matrix form as $\mathbf{Ax} = \mathbf{d}$, where $\mathbf{A} = \begin{bmatrix} 2 & 1 & 1 \\ 1 & 2 & 1 \\ 1/2 & 1/2 & 1 \end{bmatrix}$, $\mathbf{x} = \begin{bmatrix} x_1 \\ x_2 \\ x_3 \end{bmatrix}$, and $\mathbf{d} = \begin{bmatrix} 6 \\ 2 \\ 2 \end{bmatrix}$. We now generate

$\mathbf{A_1} = \begin{bmatrix} 6 & 1 & 1 \\ 2 & 2 & 1 \\ 2 & 1/2 & 1 \end{bmatrix}$, $\mathbf{A_2} = \begin{bmatrix} 2 & 6 & 1 \\ 1 & 2 & 1 \\ 1/2 & 2 & 1 \end{bmatrix}$, and $\mathbf{A_3} = \begin{bmatrix} 2 & 1 & 6 \\ 1 & 2 & 2 \\ 1/2 & 1/2 & 2 \end{bmatrix}$ by replacing the first,

the second, and the third columns of **A**, respectively, by **d**. We can compute that $|\mathbf{A_1}| = 6$, $|\mathbf{A_2}| = -2$, $|\mathbf{A_3}| = 2$, and $|\mathbf{A}| = 2$. Therefore, we obtain the solutions $x_1 = |\mathbf{A_1}| \div |\mathbf{A}| = 6/2 = 3$, $x_2 = |\mathbf{A_2}| \div |\mathbf{A}| = -2/2 = -1$, and $x_3 = |\mathbf{A_3}| \div |\mathbf{A}| = 2/2 = 1$. Notice that these are exactly the same solutions as those we obtained for the system in Section 2.3.7.

As another example, consider the SSLEs given in equation (2.3.8). We expressed this system in matrix form as $\mathbf{Ax} = \mathbf{d}$, where $\mathbf{A} = \begin{bmatrix} -2 & 1 \\ -4 & 1 \end{bmatrix}$, $\mathbf{x} = \begin{bmatrix} x_1 \\ x_2 \end{bmatrix}$, and $\mathbf{d} = \begin{bmatrix} 10 \\ 5 \end{bmatrix}$. As before, we generate $\mathbf{A_1} = \begin{bmatrix} 10 & 1 \\ 5 & 1 \end{bmatrix}$ and $\mathbf{A_2} = \begin{bmatrix} -2 & 10 \\ -4 & 5 \end{bmatrix}$ by replacing the first and the second columns of **A**, respectively, by **d**. We can obtain the determinants $|\mathbf{A_1}| = 5$ and $|\mathbf{A_2}| = 30$. The determinant of the original coefficient matrix is $|\mathbf{A}| = 2$. Therefore, we obtain the solutions $x_1 = |\mathbf{A_1}| \div |\mathbf{A}| = 5/2 = 2.5$, and $x_2 = |\mathbf{A_2}| \div |\mathbf{A}| = 30/2 = 15$. Notice, again, that these solutions are identical to the solutions we obtained for the system in Section 2.3.7.

2.4.6 Application examples

Example 1. Assume that the demand for and supply of a good are given by the system $x_1 + 5x_2 = 10$ and $x_1 - 5x_2 = 5$, where x_1 denotes the quantity demanded and supplied of the good and x_2 denotes the price of the good in dollars. Solve the system for the quantity and price using Cramer's rule.

Solution. We can write the above system in matrix form as $\mathbf{Ax} = \mathbf{d}$, where $\mathbf{A} = \begin{bmatrix} 1 & 5 \\ 1 & -5 \end{bmatrix}$, $\mathbf{x} = \begin{bmatrix} x_1 \\ x_2 \end{bmatrix}$, and $\mathbf{d} = \begin{bmatrix} 10 \\ 5 \end{bmatrix}$. According to Cramer's rule, $x_j = |\mathbf{A}_j| \div |\mathbf{A}|$ where $|\mathbf{A}_j|$ denotes the determinant of the coefficient matrix whose jth column is replaced by the constant vector **d**. This means that $|\mathbf{A_1}| = -75$ and $|\mathbf{A_2}| = -5$. The determinant of the coefficient matrix is $|\mathbf{A}| = -10$. Therefore, $x_1 = |\mathbf{A_1}| \div |\mathbf{A}| = -75/-10 = 7.5$ and $x_2 = |\mathbf{A_2}| \div |\mathbf{A}| = -5/-10 = 0.5$. That is, the quantity and price that solve the given SSLEs are 7.5 units and \$0.5, respectively.

Example 2. Suppose that the IS-LM model of a two-sector economy is given by the SSLEs $0.15x_1 + 500x_2 = 150$ and $0.25x_1 - 500x_2 = 150$, where x_1 represents income in billions of dollars and x_2 represents interest rate. Find the levels of income and interest rate that solve the system using Cramer's rule.

Solution. We can write the above system in matrix form as $\mathbf{Ax} = \mathbf{d}$, where $\mathbf{A} = \begin{bmatrix} 0.15 & 500 \\ 0.25 & -500 \end{bmatrix}$, $\mathbf{x} = \begin{bmatrix} x_1 \\ x_2 \end{bmatrix}$, and $\mathbf{d} = \begin{bmatrix} 150 \\ 150 \end{bmatrix}$. Following Cramer's rule, we have $x_j = |\mathbf{A}_j| \div |\mathbf{A}|$, where $|\mathbf{A}_j|$ denotes the determinant of the coefficient matrix whose jth column is replaced by the constant vector **d**. We can get that $|A_1| = -150\,1000$ and $|A_2| = -15$. The determinant of the coefficient matrix is $|A| = -200$. Therefore, $x_1 = |A_1| \div |A| = -150\,000/-200 = 750$ and $x_2 = |A_2| \div |A| = -15/-200 = 0.075 = 7.5$ percent. That is, the levels of income and interest rate that solve the given SSLEs are \$750 billion and 7.5 percent, respectively.

Example 3. Suppose that prices of three goods sold by a company are interrelated by the system $-4x_1 + 3x_2 + 3x_3 = 10$, $3x_1 - 4x_2 + 3x_3 = 10$, and $3x_1 + 3x_2 - 4x_3 = 10$, where x_1, x_2, and x_3 denote price in dollars of goods 1, 2, and 3, respectively. Find the prices that solve the system using Cramer's rule.

Solution. We first write the above system in matrix form as $Ax = d$, where

$$A = \begin{bmatrix} -4 & 3 & 3 \\ 3 & -4 & 3 \\ 3 & 3 & -4 \end{bmatrix}, \; x = \begin{bmatrix} x_1 \\ x_2 \\ x_3 \end{bmatrix}, \text{ and } d = \begin{bmatrix} 10 \\ 10 \\ 10 \end{bmatrix}.$$ Following Cramer's rule, we have x_j

$= |A_j| \div |A|$, where $|A_j|$ denotes the determinant of the coefficient matrix whose jth column is replaced by the constant vector **d**. As above, we can obtain that $|A_1| = 490$, $|A_2| = 490$, and $|A_3| = 490$. The determinant of the coefficient matrix is $|A| = 98$. Therefore, $x_1 = |A_1| \div |A| = 490/98 = 5$, $x_2 = |A_2| \div |A| = 490/98 = 5$, and $x_3 = |A_3| \div |A| = 490/98 = 5$. Therefore, the prices that solve the system are $x_1 = \$5$, $x_2 = \$5$, and $x_3 = \$5$.

2.4.7 Exercises

1. Find the determinants of the following matrices:

   (i) $\begin{bmatrix} 1 & 3 \\ 4 & 2 \end{bmatrix}$; (ii) $\begin{bmatrix} 1 & 3 \\ 2 & 6 \end{bmatrix}$; (iii) $\begin{bmatrix} a & b \\ a & b \end{bmatrix}$; (iv) $\begin{bmatrix} 1 & 3 \\ 0 & 0 \end{bmatrix}$; (v) $\begin{bmatrix} 0 & 3 \\ 0 & 6 \end{bmatrix}$.

2. Find the determinant of the following matrices using Sarrus' rule:

   (i) $\begin{bmatrix} 1 & 4 & 3 \\ 2 & 5 & 4 \\ 3 & 5 & 1 \end{bmatrix}$; (ii) $\begin{bmatrix} 1 & 2 & 3 \\ 2 & 5 & 4 \\ 3 & 5 & 1 \end{bmatrix}$; (iii) $\begin{bmatrix} 2 & 8 & 9 \\ 2 & 5 & 4 \\ 3 & 5 & 1 \end{bmatrix}$; (iv) $\begin{bmatrix} a & b & c \\ d & e & f \\ g & h & i \end{bmatrix}$.

3. Find the determinants of the following matrices using co-factors:

   (i) $\begin{bmatrix} 5 & 4 & 3 \\ 6 & 5 & 4 \\ 3 & 5 & 5 \end{bmatrix}$; (ii) $\begin{bmatrix} 1 & 2 & 3 \\ 2 & 1 & 4 \\ 3 & 5 & 1 \end{bmatrix}$; (iii) $\begin{bmatrix} 2 & 8 & 9 \\ 6 & 5 & 4 \\ 3 & 4 & 1 \end{bmatrix}$; (iv) $\begin{bmatrix} a & d & g \\ b & e & h \\ c & f & i \end{bmatrix}$.

4. Find the determinants of the following matrices:

   (i) $\begin{bmatrix} 16 & 15 & 14 & 13 \\ 12 & 11 & 10 & 9 \\ 8 & 7 & 6 & 5 \\ 4 & 3 & 2 & 1 \end{bmatrix}$; (ii) $\begin{bmatrix} 1 & 5 & 4 & 1 \\ 2 & 1 & 0 & 3 \\ 2 & 3 & 6 & 5 \\ 4 & 1 & 2 & 0 \end{bmatrix}$; (iii) $\begin{bmatrix} a & b & c & d \\ e & f & g & h \\ a & b & c & d \\ i & j & k & l \end{bmatrix}$; (iv) $\begin{bmatrix} a & 1 & b & 1 \\ d & 9 & e & 9 \\ f & 5 & g & 5 \\ h & 1 & i & 1 \end{bmatrix}$;

   (v) $\begin{bmatrix} 2a & 2b & 2c & 2d \\ e & f & g & h \\ i & j & k & l \\ m & n & o & p \end{bmatrix}$; (vi) $\begin{bmatrix} a+2e & b+2f & c+2g & d+2h \\ e & f & g & h \\ a & b & c & d \\ i & j & k & l \end{bmatrix}$; (vii) $\begin{bmatrix} e & f & g & h \\ a & b & c & d \\ a & b & c & d \\ i & j & k & l \end{bmatrix}$;

$$\text{(viii)} \begin{bmatrix} a & b & c & d \\ 0 & f & g & h \\ 0 & 0 & c & d \\ 0 & 0 & 0 & l \end{bmatrix}; \text{(ix)} \begin{bmatrix} 1 & 0 & 0 & 0 \\ 2 & 3 & 0 & 0 \\ 4 & 5 & 6 & 0 \\ 7 & 8 & 9 & 10 \end{bmatrix}; \text{(x)} \begin{bmatrix} 1 & 0 & 0 & 0 \\ 0 & 1 & 0 & 0 \\ 0 & 0 & 1 & 0 \\ 0 & 0 & 0 & 1 \end{bmatrix}; \text{(xi)} \begin{bmatrix} 1 & 2 & 3 & 4 \\ 0 & 5 & 6 & 7 \\ 0 & 0 & 8 & 9 \\ 0 & 0 & 0 & 10 \end{bmatrix}.$$

5. Solve the following SSLEs using Cramer's rule:

(i) $\begin{aligned} x_1 + 4x_2 &= 2 \\ 2x_1 + 3x_2 &= 4 \end{aligned}$; (ii) $\begin{aligned} 2x_1 + 4x_2 &= 1 \\ 2x_1 + 3x_2 &= 4 \end{aligned}$; (iii) $\begin{aligned} -x_1 + 4x_2 &= 25 \\ 2x_1 - 3x_2 &= 40 \end{aligned}$; (iv) $\begin{aligned} -2x_1 - 3x_2 &= 20 \\ 2x_1 - x_2 &= 30 \end{aligned}$;

(v) $\begin{aligned} 3x_1 + x_2 - x_3 &= 2 \\ x_1 - 3x_2 - x_3 &= 2 \\ -x_1 - x_2 - 3x_3 &= 2 \end{aligned}$; (vi) $\begin{aligned} 3x_1 - x_2 - x_3 &= 1 \\ -x_1 + 3x_2 - x_3 &= 1 \\ -x_1 - x_2 + 3x_3 &= 1 \end{aligned}$; (vii) $\begin{aligned} x_1 + x_3 &= 2 \\ x_2 + x_3 &= 4 \\ x_1 + x_3 &= 1 \end{aligned}$.

6. *Application exercise.* Suppose that the demand for and the supply of a good offered for sale in a market are given by functions $x_1 = 10 - 4x_2$ and $x_1 = 4 + 2x_2$, respectively, where x_1 denotes the quantity of the good demanded and supplied and x_2 denotes the price of the good in dollars. Find the quantity and price that solve the system using Cramer's rule.

7. *Application exercise.* Assume that the IS-LM model of a two-sector economy is given by the SSLEs $0.3x_1 + 10x_2 = 25$ and $0.15x_1 - 700x_2 = -50$, where x_1 denotes income in billions of dollars and x_2 denotes interest rate. Find the levels of income and interest rate in the economy that solve the system using Cramer's rule.

8. *Application exercise.* Suppose that the prices in three markets of a good are interrelated as shown by the SSLEs $5x_1 - 0.5x_2 - 0.5x_3 = 2$, $0.5x_1 + 5x_2 - 0.5x_3 = 2$, and $0.5x_1 - 0.5x_2 + 5x_3 = 2$, where x_1, x_2, and x_3 denote prices in dollars in markets 1, 2, and 3, respectively. Find the prices that solve the system using Cramer's rule.

 Web supplement: S2.4.8 Mathematica applications

2.5 Inverse of a matrix

2.5.1 Meaning of inverse

We know from elementary algebra that $a \times (1/a) = a \times a^{-1} = a^{1-1} = a^0 = 1$. Here we call a^{-1} the *multiplicative inverse* of a. We shall now attempt to extend this result in algebra to matrices. Suppose that we have two square matrices of the same order: matrix **A** and identity matrix **I**. Can we now find a matrix **B** such that $\mathbf{A}.\mathbf{B} = \mathbf{I}$? If we can, then **B** is called the *inverse* of **A**, and is written as $\mathbf{A}^{-1}$ (that is, $\mathbf{B} = \mathbf{A}^{-1}$). Then **A** is said to be *invertible*.

Since $\mathbf{A}.\mathbf{B} = \mathbf{I}$, **A** is the inverse of **B** and **B** is the inverse of **A**; that is, **A** and **B** are the inverses of each other. Notice that inverses are defined only for square matrices. However, how does one find the inverse in the first place? Two methods are generally used to find the inverse of a square matrix. One of these methods uses the determinant and the adjoint of the given matrix. The other method uses the Gauss–Jordan method discussed in Section 2.3.8. Let us discuss each of these below.

2.5.2 Finding inverse using determinant and adjoint matrix

Before finding the inverse, we need to check whether the inverse of a matrix exists or not. Above, we used the matrix equation $\mathbf{AB} = \mathbf{I}$. We need to know whether **A** is invertible or not. We know, from the properties of determinants, that $|\mathbf{AB}| = |\mathbf{A}|.|\mathbf{B}| = |\mathbf{I}|$. It is clear from this equation that if $|\mathbf{A}| = 0$, then we will obtain a contradictory result that $|\mathbf{A}|.|\mathbf{B}| = |\mathbf{I}| = 0.|\mathbf{B}| = 0 = |\mathbf{I}|$. The only way to avoid this contradiction is that $|\mathbf{A}| \neq 0$. This suggests that the condition for $\mathbf{A}$ to have an inverse is that its determinant, $|\mathbf{A}|$, must be nonzero or nonvanishing. Therefore, one can check whether a matrix is invertible or not by checking its determinant. Notice that a square matrix will have only one inverse.

We are now ready to apply the determinant of a matrix and its adjoint to find the inverse of that matrix. We defined the adjoint of matrix $\mathbf{A}$ (denoted by $\mathbf{A_{adj}}$) as the transpose of the co-factor matrix, $\mathbf{C_{FA}}$. That is, $\mathbf{A_{adj}} = (\mathbf{C_{FA}})^{\mathrm{T}}$. One can show that the product of $\mathbf{A}$ and the adjoint of $\mathbf{A}$ is equal to the product of the determinant of $\mathbf{A}$ and the identity matrix $\mathbf{I}$ of the same order; that is, $\mathbf{A}.\mathbf{A_{adj}} = |\mathbf{A}|.\mathbf{I}$. Now dividing both sides of this equation by $|\mathbf{A}|$ yields $(\mathbf{A}.\mathbf{A_{adj}})/|\mathbf{A}| = |\mathbf{A}|.\mathbf{I}/|\mathbf{A}| = \mathbf{I}$. And pre-multiplying both sides of the last equation by $\mathbf{A}^{-1}$ gives $\mathbf{A}^{-1}(\mathbf{A}.\mathbf{A_{adj}})/|\mathbf{A}| = \mathbf{A}^{-1}\mathbf{I}$. Since $\mathbf{A}^{-1}\mathbf{A} = \mathbf{I}$, $(\mathbf{I}.\mathbf{A_{adj}})/|\mathbf{A}| = \mathbf{A_{adj}}/|\mathbf{A}|$ and $\mathbf{A}^{-1}\mathbf{I} = \mathbf{A}^{-1}$, the last equation can be written as $\mathbf{A_{adj}}/|\mathbf{A}| = \mathbf{A}^{-1}$. Therefore, the inverse of $\mathbf{A}$ is given by

$$\mathbf{A}^{-1} = \mathbf{A_{adj}} \div |\mathbf{A}| \tag{2.5.1}$$

As an example, consider the 3×3 matrix $\mathbf{A} = \begin{bmatrix} 1 & 2 & 1 \\ 1 & 1 & 2 \\ 2 & 2 & 1 \end{bmatrix}$. The determinant of this matrix is $|\mathbf{A}| = 3$. The co-factor matrix of $\mathbf{A}$ is $\mathbf{C_{FA}} = \begin{bmatrix} C_{11} & C_{12} & C_{13} \\ C_{21} & C_{22} & C_{23} \\ C_{31} & C_{32} & C_{33} \end{bmatrix} = \begin{bmatrix} -3 & 3 & 0 \\ 0 & -1 & 2 \\ 3 & -1 & -1 \end{bmatrix}$, which implies that $\mathbf{A_{adj}} = (\mathbf{C_{FA}})^{\mathrm{T}} = \begin{bmatrix} -3 & 0 & 3 \\ 3 & -1 & -1 \\ 0 & 2 & -1 \end{bmatrix}$. Therefore, $\mathbf{A}^{-1} = \mathbf{A_{adj}} \div |\mathbf{A}| = \frac{1}{3}\begin{bmatrix} -3 & 0 & 3 \\ 3 & -1 & -1 \\ 0 & 2 & -1 \end{bmatrix} = \begin{bmatrix} -3/3 & 0/3 & 3/3 \\ 3/3 & -1/3 & -1/3 \\ 0/3 & 2/3 & -1/3 \end{bmatrix} = \begin{bmatrix} -1 & 0 & 1 \\ 1 & -1/3 & -1/3 \\ 0 & 2/3 & -1/3 \end{bmatrix}$.

2.5.3 *Finding inverse using Gauss–Jordan elimination method*

We shall now use the Gauss–Jordan elimination method to find the inverse of a square matrix. But, before this, we shall recap this method. Suppose that the SSLEs we deal with now is the general SSLEs in equation (2.3.5). We expressed this SSLEs, in Section 2.3.6, as the matrix equation $\mathbf{Ax} = \mathbf{d}$, where $\mathbf{A}$ represented the coefficient matrix, $\mathbf{x}$ represented the column vector of variables, and $\mathbf{d}$ represented the column vector of constants. If the number of equations in the system is equal to the number of the variables, then we have $m = n$. We can generate, following the procedure outlined in Section 2.3.7, the augmented matrix $\mathbf{A_{aug}d}$. If we carry out the elementary row operations on this augmented matrix and convert the coefficient part of the augmented matrix into an identity matrix, we obtain the reduced row echelon form, and that process of conversion was called row reduction. We know that, in the row reduction process, the original column vector $\mathbf{d}$ would be replaced by a new column vector $(\mathbf{e})$ which would represent the solution to the system. We called this method of finding the solution to the system the Gauss–Jordan elimination method.

We can now attempt to use the same method, with a slight difference, to find the inverse of a matrix. The difference here is that, instead of using the column vector **d** to augment the coefficient matrix **A**, we augment **A** by an identity matrix of the same order as

$$\mathbf{A_{aug}I} = \begin{bmatrix} a_{11} & a_{12} & ... & a_{1n} & 1 & 0 & ... & 0 \\ a_{21} & a_{22} & ... & a_{2n} & 0 & 1 & ... & 0 \\ ... & ... & ... & ... & 0 & 0 & ... & 0 \\ a_{m1} & a_{m2} & ... & a_{mn} & 0 & 0 & ... & 1 \end{bmatrix}$$

That is, we form $\mathbf{A_{aug}I} = [\mathbf{A}|\mathbf{I}]$. We now carry out the same elementary row operations to convert the coefficient part of $\mathbf{A_{aug}I}$ into an identity matrix (the reduced echelon form of **A**). This conversion process will replace, as before, the identity part of $\mathbf{A_{aug}I}$ into another matrix, which will be the inverse of the coefficient matrix. This suggests that a square matrix **A** is invertible only if its reduced echelon form is an identity matrix of the same order as **A**. This method is called the *Gauss–Jordan elimination method of finding inverse* of a square matrix. Notice that in this reduction process, we may end up with, instead of the reduced echelon form of **A**, a row that contains only zeros (on the left of the vertical line). If this happens, **A** is not invertible. This is similar to the condition, in finding inverse with determinant and adjoint discussed in the previous section, that |**A**| should not be zero for **A** to have an inverse.

Let us now consider examples of finding inverse using the Gauss–Jordan elimination method. Consider first the 2×2 matrix $\mathbf{A} = \begin{bmatrix} 3 & 4 \\ 3 & 2 \end{bmatrix}$. We can now augment this matrix by a 2×2 identity matrix to obtain $\mathbf{A_{aug}I} = \begin{bmatrix} 3 & 4 & 1 & 0 \\ 3 & 2 & 0 & 1 \end{bmatrix}$. We may then carry out the elementary row operations as follows:

$$\begin{bmatrix} 3 & 4 & 1 & 0 \\ 3 & 2 & 0 & 1 \end{bmatrix} \begin{matrix} R_1 \to (1/3)R_1 \\ R_2 \to (1/3)R_2 \end{matrix} \begin{bmatrix} 1 & 4/3 & 1/3 & 0 \\ 1 & 2/3 & 0 & 1/3 \end{bmatrix} R_2 \to R_2 - R_1 \begin{bmatrix} 1 & 4/3 & 1/3 & 0 \\ 0 & -2/3 & -1/3 & 1/3 \end{bmatrix}$$

$$R_1 \to R_1 + 2R_2 \begin{bmatrix} 1 & 0 & -1/3 & 2/3 \\ 0 & -2/3 & -1/3 & 1/3 \end{bmatrix} R_2 \to (-3/2)R_2 \begin{bmatrix} 1 & 0 & -1/3 & 2/3 \\ 0 & 1 & 1/2 & -1/2 \end{bmatrix}$$

Notice that the last augmented matrix is in reduced row echelon form. Therefore, as we stated above, the inverse of **A** is $\mathbf{A}^{-1} = \begin{bmatrix} -1/3 & 2/3 \\ 1/2 & -1/2 \end{bmatrix}$.

Consider now the 3×3 matrix we inverted in the previous section using determinant and adjoint, $\mathbf{A} = \begin{bmatrix} 1 & 2 & 1 \\ 1 & 1 & 2 \\ 2 & 2 & 1 \end{bmatrix}$. Then, the corresponding augmented matrix of this matrix is $\mathbf{A_{aug}I} = \begin{bmatrix} 1 & 2 & 1 & 1 & 0 & 0 \\ 1 & 1 & 2 & 0 & 1 & 0 \\ 2 & 2 & 1 & 0 & 0 & 1 \end{bmatrix}$. We can carry out the elementary row operations on

$A_{aug}I$ as follows:

$$R_3 \rightarrow R_3 - R_2 \begin{bmatrix} 1 & 2 & 1 & | & 1 & 0 & 0 \\ 1 & 1 & 2 & | & 0 & 1 & 0 \\ 1 & 1 & -1 & | & 0 & -1 & 1 \end{bmatrix} R_3 \rightarrow (-1)R_3 \begin{bmatrix} 1 & 2 & 1 & | & 1 & 0 & 0 \\ 1 & 1 & 2 & | & 0 & 1 & 0 \\ -1 & -1 & 1 & | & 0 & 1 & -1 \end{bmatrix}$$

$$R_3 \rightarrow R_3 + R_2 \begin{bmatrix} 1 & 2 & 1 & | & 1 & 0 & 0 \\ 1 & 1 & 2 & | & 0 & 1 & 0 \\ 0 & 0 & 3 & | & 0 & 2 & -1 \end{bmatrix}$$

$$R_3 \rightarrow (1/3)R_3 \begin{bmatrix} 1 & 2 & 1 & | & 1 & 0 & 0 \\ 1 & 1 & 2 & | & 0 & 1 & 0 \\ 0 & 0 & 1 & | & 0 & 2/3 & -1/3 \end{bmatrix} R_2 \rightarrow R_2 - R_1 \begin{bmatrix} 1 & 2 & 1 & | & 1 & 0 & 0 \\ 0 & -1 & 1 & | & -1 & 1 & 0 \\ 0 & 0 & 1 & | & 0 & 2/3 & -1/3 \end{bmatrix}$$

$$R_1 \rightarrow R_1 - R_3 \begin{bmatrix} 1 & 2 & 0 & | & 1 & -2/3 & 1/3 \\ 0 & -1 & 1 & | & -1 & 1 & 0 \\ 0 & 0 & 1 & | & 0 & 2/3 & -1/3 \end{bmatrix} R_2 \rightarrow (-1)R_2 \begin{bmatrix} 1 & 2 & 0 & | & 1 & -2/3 & 1/3 \\ 0 & 1 & -1 & | & 1 & -1 & 0 \\ 0 & 0 & 1 & | & 0 & 2/3 & -1/3 \end{bmatrix}$$

$$R_2 \rightarrow R_2 + R_3 \begin{bmatrix} 1 & 2 & 0 & | & 1 & -2/3 & 1/3 \\ 0 & 1 & 0 & | & 1 & -1/3 & -1/3 \\ 0 & 0 & 1 & | & 0 & 2/3 & -1/3 \end{bmatrix} R_1 \rightarrow R_1 - 2R_2 \begin{bmatrix} 1 & 0 & 0 & | & -1 & 0 & 1 \\ 0 & 1 & 0 & | & 1 & -1/3 & -1/3 \\ 0 & 0 & 1 & | & 0 & 2/3 & -1/3 \end{bmatrix}.$$

Notice that the last augmented matrix is in reduced row echelon form. Therefore, as we stated above, the inverse of A is $A^{-1} = \begin{bmatrix} -1 & 0 & 1 \\ 1 & -1/3 & -1/3 \\ 0 & 2/3 & -1/3 \end{bmatrix}$. Notice that this is exactly the same as the result we obtained when we used determinant and adjoint to find the inverse of A in the previous section.

2.5.4 Properties of inverse

The inverse of a square matrix obeys some important properties. These properties are the following.

Property I.   If A is an invertible square matrix, then $(A^{-1})^{-1} = A$.

Property II.   If s is a scalar not equal to zero and A is an invertible square matrix, then $(sA)^{-1} = s^{-1}A^{-1}$.

Property III.   If A is a square matrix, and if A^T is invertible, then $(A^T)^{-1} = (A^{-1})^T$.

Property IV.   If A and B are two square matrices of the same order, and if $(AB)^{-1}$ exists, then $(AB)^{-1} = B^{-1}A^{-1}$.

Property V.   If A is a square matrix and I is an identity matrix of the same order as that of A, and if A is invertible, then $AA^{-1} = I$.

2.5.5 Solution of SSLEs using inverse

In Section 2.4.5 we applied Cramer's rule to solve two SSLEs. In this section we attempt to solve the same systems with the help of inverse – the last method of solution of SSLEs we consider in this book.

Suppose, as in Section 2.4.5, we have an $m \times n$ (with $m = n$) SSLEs represented by the matrix equation $\mathbf{Ax} = \mathbf{d}$, where $\mathbf{A}$ represents the $n \times n$ coefficient matrix, $\mathbf{x}$ represents the $n \times 1$ variable vector, and $\mathbf{d}$ represents the $n \times 1$ constant vector. Let us now *post-multiply* $\mathbf{A}$ and *pre-multiply* $\mathbf{d}$ of the equation $\mathbf{Ax} = \mathbf{d}$ by the inverse of $\mathbf{A}$, $\mathbf{A}^{-1}$. These yield $\mathbf{AA}^{-1}\mathbf{x} = \mathbf{A}^{-1}\mathbf{d}$. Since $\mathbf{A}\,\mathbf{A}^{-1} = \mathbf{I}$, the last equation simplifies to

$$\mathbf{Ix} = \mathbf{x} = \mathbf{A}^{-1}\mathbf{d} \tag{2.5.2}$$

Equation (2.5.2) is a very important result in linear algebra. It states that the solution of a SSLEs is given by the product of the inverse of the matrix of coefficients and the vector of the constants of the system. This is called the *method of solving SSLEs using inverse*.

As an example of the application of this method, consider again the 3×3 SSLEs we solved in Section 2.4.5. In this SSLEs we had $\mathbf{A} = \begin{bmatrix} 2 & 1 & 1 \\ 1 & 2 & 1 \\ 1/2 & 1/2 & 1 \end{bmatrix}$, $\mathbf{x} = \begin{bmatrix} x_1 \\ x_2 \\ x_3 \end{bmatrix}$, and $\mathbf{d} = \begin{bmatrix} 6 \\ 2 \\ 2 \end{bmatrix}$.

Therefore, following the method of inverse discussed above, we can solve for $\mathbf{x}$ (that is, find $\mathbf{x} = \mathbf{A}^{-1}\mathbf{d}$). But, for this, we need to find the inverse of $\mathbf{A}$, $\mathbf{A}^{-1}$. We know how to find $\mathbf{A}^{-1}$ using determinant and adjoint of $\mathbf{A}$ (using equation (2.5.1)) or using the Gauss–Jordan elimination method (as we did in Section 2.5.3). Following any one of these two methods, we can obtain $\mathbf{A}^{-1} = \begin{bmatrix} 3/4 & -1/4 & -1/2 \\ -1/4 & 3/4 & -1/2 \\ -1/4 & -1/4 & 3/2 \end{bmatrix}$. Therefore, applying equation (2.5.2) yields

$$\mathbf{x} = \begin{bmatrix} x_1 \\ x_2 \\ x_3 \end{bmatrix} = \mathbf{A}^{-1}\mathbf{d} = \begin{bmatrix} 3/4 & -1/4 & -1/2 \\ -1/4 & 3/4 & -1/2 \\ -1/4 & -1/4 & 3/2 \end{bmatrix} \begin{bmatrix} 6 \\ 2 \\ 2 \end{bmatrix} = \begin{bmatrix} 3 \\ -1 \\ 1 \end{bmatrix}.$$ This implies that $x_1 = 3$,

$x_2 = -1$, and $x_3 = 1$. Notice that these solutions are identical with the solutions we obtained when we used Cramer's rule in Section 2.4.5.

As another example, consider the second SSLEs we solved in Section 2.4.5. In this SSLEs we had $\mathbf{A} = \begin{bmatrix} -2 & 1 \\ -4 & 1 \end{bmatrix}$, $\mathbf{x} = \begin{bmatrix} x_1 \\ x_2 \end{bmatrix}$, and $\mathbf{d} = \begin{bmatrix} 10 \\ 5 \end{bmatrix}$. In this case, we can find that

$\mathbf{A}^{-1} = \begin{bmatrix} 1/2 & -1/2 \\ 2 & -1 \end{bmatrix}$. Therefore, application of equation (2.5.2) gives $\mathbf{x} = \begin{bmatrix} x_1 \\ x_2 \end{bmatrix} = \mathbf{A}^{-1}\mathbf{d}$

$= \begin{bmatrix} 1/2 & -1/2 \\ 2 & -1 \end{bmatrix} \begin{bmatrix} 10 \\ 5 \end{bmatrix} = \begin{bmatrix} 5/2 \\ 15 \end{bmatrix}.$ This implies that $x_1 = 5/2 = 2.5$ and $x_2 = 15$. Notice again that these solutions and the solutions we obtained when we used Cramer's rule in Section 2.4.5 are identical.

2.5.6 Application examples

Example 1. Suppose that the equilibrium in the goods market of an economy is given by $0.30x_1 + 100x_2 = 180$ and that the equilibrium in the money market of the economy is given by $0.30x_1 - 200x_2 = 150$, where x_1 represents income in billions of dollars and x_2 represents interest rate. Find the levels of income and interest rate that solve the system using inverse.

Solution. We first write the system in matrix form as $\mathbf{Ax} = \mathbf{d}$, where $\mathbf{A} = \begin{bmatrix} 0.3 & 100 \\ 0.3 & -200 \end{bmatrix}$,

$\mathbf{x} = \begin{bmatrix} x_1 \\ x_2 \end{bmatrix}$, and $\mathbf{d} = \begin{bmatrix} 180 \\ 150 \end{bmatrix}$. The solution to this system can be obtained using equation (2.5.2): $\mathbf{x} = \mathbf{A}^{-1}\mathbf{d}$. But, to apply this equation, we need to find $\mathbf{A}^{-1}$. We can find that $\mathbf{A}^{-1} = \begin{bmatrix} 2.22 & 1.11 \\ 0.0033 & -0.0033 \end{bmatrix}$. Now application of equation (2.5.2) gives $\mathbf{x} = \mathbf{A}^{-1}\mathbf{d}$

$= \begin{bmatrix} 2.22 & 1.11 \\ 0.0033 & -0.0033 \end{bmatrix}\begin{bmatrix} 180 \\ 150 \end{bmatrix} = \begin{bmatrix} 566 \\ 0.1 \end{bmatrix}$. Therefore, the levels of income and the interest rate that solve the system are $x_1 = \$566$ billion and $x_2 = 0.1$ or 10 percent, respectively.

Example 2. Suppose that the three markets in an economy are related by the prices x_1, x_2, and x_3. Also suppose that the relationships among these prices are given by $-2x_1 + 2x_2 + 2x_3 - 25 = 0$, $2x_1 - 2x_2 + 2x_3 - 25 = 0$, and $2x_1 + 2x_2 - 2x_3 - 25 = 0$. Find the prices in dollars that solve the system using inverse.

Solution. As above, we first write the system in matrix form as $\mathbf{Ax} = \mathbf{d}$, where $\mathbf{A} = \begin{bmatrix} -2 & 2 & 2 \\ 2 & -2 & 2 \\ 2 & 2 & -2 \end{bmatrix}$, $\mathbf{x} = \begin{bmatrix} x_1 \\ x_2 \\ x_3 \end{bmatrix}$, and $\mathbf{d} = \begin{bmatrix} 25 \\ 25 \\ 25 \end{bmatrix}$. The solution to this system can be obtained by applying equation (2.5.2): $\mathbf{x} = \mathbf{A}^{-1}\mathbf{d}$. But, for this we need to find $\mathbf{A}^{-1}$. We can find that $\mathbf{A}^{-1} = \begin{bmatrix} 0 & 1/4 & 1/4 \\ 1/4 & 0 & 1/4 \\ 1/4 & 1/4 & 0 \end{bmatrix}$. Then, application of equation (2.5.2) yields $\mathbf{x} = \mathbf{A}^{-1}\mathbf{d} = \begin{bmatrix} 0 & 1/4 & 1/4 \\ 1/4 & 0 & 1/4 \\ 1/4 & 1/4 & 0 \end{bmatrix}\begin{bmatrix} 25 \\ 25 \\ 25 \end{bmatrix} = \begin{bmatrix} 12.5 \\ 12.5 \\ 12.5 \end{bmatrix}$. Therefore, the prices that solve the system are $x_1 = x_2 = x_3 = \$12.5$.

2.5.7 Exercises

1. Find the inverses, if they exist, of the following matrices using their determinants and adjoint matrices:

   (i) $\begin{bmatrix} 2 & 5 \\ 1 & 3 \end{bmatrix}$; (ii) $\begin{bmatrix} 2 & 1 \\ 1 & 2 \end{bmatrix}$; (iii) $\begin{bmatrix} 1 & 0 \\ 0 & 1 \end{bmatrix}$; (iv) $\begin{bmatrix} 1 & 2 \\ 2 & 1 \end{bmatrix}$; (v) $\begin{bmatrix} 3 & 1 \\ 1 & 3 \end{bmatrix}$; (vi) $\begin{bmatrix} 1 & 2 & 2 \\ 2 & 1 & 2 \\ 2 & 2 & 1 \end{bmatrix}$;

   (vii) $\begin{bmatrix} 2 & 1 & 1 \\ 1 & 2 & 1 \\ 1 & 1 & 2 \end{bmatrix}$; (viii) $\begin{bmatrix} 1 & 0 & 0 \\ 0 & 1 & 0 \\ 0 & 0 & 1 \end{bmatrix}$; (ix) $\begin{bmatrix} 1 & 2 & 2 \\ 2 & 1 & 2 \\ 2 & 2 & 1 \end{bmatrix}$; (x) $\begin{bmatrix} 1 & 2 & 2 \\ 2 & 1 & 2 \\ 2 & 2 & 1 \end{bmatrix}$; (xi) $\begin{bmatrix} 1 & 2 & 3 \\ 0 & 4 & 5 \\ 0 & 0 & 6 \end{bmatrix}$;

   (xii) $\begin{bmatrix} 1 & 0 & 0 \\ 2 & 3 & 0 \\ 4 & 5 & 6 \end{bmatrix}$.

2. Find the inverses, if they exist, of the matrices in exercise 1 above using the Gauss–Jordan elimination method.

3. Solve the following SSLEs using inverse:

   (i) $\begin{matrix} 5x_1 + 2x_2 = 4 \\ 3x_1 + x_2 = 6 \end{matrix}$; (ii) $\begin{matrix} 3x_1 + x_2 = 4 \\ x_1 + 3x_2 = 6 \end{matrix}$; (iii) $\begin{matrix} 2x_1 + x_2 = 4 \\ x_1 + 2x_2 = 6 \end{matrix}$; (iv) $\begin{matrix} x_1 + 2x_2 = 4 \\ 2x_1 + x_2 = 6 \end{matrix}$;

   (v) $\begin{matrix} 2x_1 + x_2 + x_3 = 2 \\ x_1 + 2x_2 + x_3 = 4 \\ x_1 + x_2 + 2x_3 = 6 \end{matrix}$; (vi) $\begin{matrix} x_1 = 2 \\ 2x_1 + 3x_2 = 4 \\ 4x_1 + 5x_2 + 6x_3 = 6 \end{matrix}$; (vii) $\begin{matrix} x_1 + 2x_2 + 3x_3 = 2 \\ 4x_2 + 5x_3 = 4 \\ 6x_3 = 6 \end{matrix}$.

4. *Application exercise.* Assume that the demand and supply functions of a good are given respectively by $x_1 = 4 - 2x_2$ and $x_1 = 8 + 0.5x_2$, where x_1 represents the quantity of the good demanded and supplied and x_2 represents the price in dollars of the good. Find the quantity and price that solve the system using inverse.

5. *Application exercise.* Suppose that the equilibrium in the goods market of an economy is given by $0.4x_1 + 200x_2 = 360$ and that the equilibrium in the money market of the economy is given by $0.15x_1 - 100x_2 = 100$, where x_1 represents income in billions of dollars and x_2 represents interest rate. Find the levels of income and interest rate that solve the system.

 Web supplement: S2.5.8 Mathematica applications

2.6 Rank of a matrix

In Section 2.4 we associated a numerical value with a square matrix, and we called it the determinant of that matrix. In this section we associate another numerical value with a matrix (square or rectangular). We call this value the *rank of the matrix*. Now the question is: how does one associate such a value to a square or a rectangular matrix? We take up this problem in the present section.

2.6.1 Meaning of the rank of a matrix

We have already discussed the meaning of linear independence among vectors in Section 2.2.3. We defined linear independence as follows. Suppose that we have n n-vectors $\mathbf{v_1}$, $\mathbf{v_2}$, $\mathbf{v_3}$,..., $\mathbf{v_n}$; and n scalars $s_1, s_2, s_3, \ldots, s_n$, not all zeros. Given these vectors and scalars, if we can write $s_1\mathbf{v_1} + s_2\mathbf{v_2} + s_3\mathbf{v_3} + \cdots + s_n\mathbf{v_n} = \mathbf{0}$, then the vectors are said to be linearly dependent. On the other hand, if there do not exist scalars $s_1, s_2, s_3, \ldots, s_n$, again not all zeros, such that $s_1\mathbf{v_1} + s_2\mathbf{v_2} + s_3\mathbf{v_3} + \cdots + s_n\mathbf{v_n} = \mathbf{0}$, then the vectors are said to be linearly independent.

Notice that matrices are nothing but the stacking of vectors (see equation (2.3.2)). Suppose we have three row 3-vectors $\mathbf{v_1}' = [1\ 2\ 3]$, $\mathbf{v_2}' = [3\ 2\ 1]$, and $\mathbf{v_3}' = [2\ 3\ 1]$. We now generate the matrix, $\mathbf{A}$, by stacking these row vectors horizontally $\mathbf{A} = \begin{bmatrix} v_1' \\ v_2' \\ v_3' \end{bmatrix} = \begin{bmatrix} 1 & 2 & 3 \\ 3 & 2 & 1 \\ 2 & 3 & 1 \end{bmatrix}$.

One may now ask if the rows (or columns) of $\mathbf{A}$ are linearly dependent or independent. It can be shown that given $\mathbf{v_1}'$, $\mathbf{v_2}'$, and $\mathbf{v_3}'$, and scalars s_1, s_2, and s_3, not all zeros, no linear combination such as $s_1\mathbf{v_1}' + s_2\mathbf{v_2}' + s_3\mathbf{v_3}' = \mathbf{0}$ can be generated. Such a linear combination, in the case of $\mathbf{A}$, is possible only when the scalars are all zeros. This means that the rows or columns in $\mathbf{A}$ are linearly independent.

Consider other three row 3-vectors: $\mathbf{u_1}' = [1\ 2\ 3]$, $\mathbf{u_2}' = [2\ 4\ 6]$, and $\mathbf{u_3}' = [2\ 3\ 1]$. Using these three vectors we can generate the matrix $\mathbf{B} = \begin{bmatrix} u_1' \\ u_2' \\ u_3' \end{bmatrix} = \begin{bmatrix} 1 & 2 & 3 \\ 2 & 4 & 6 \\ 2 & 3 & 1 \end{bmatrix}$. We know that the second row in $\mathbf{B}$ is, in fact, two times the first row. Notice that a linear combination such as $s_1\mathbf{u_1}' + s_2\mathbf{u_2}' + s_3\mathbf{u_3}' = \mathbf{0}$ with $s_1 = -2$, $s_2 = 1$, and $s_3 = 0$ can be generated. This means that the rows (specifically the first and the second) in $\mathbf{B}$ are linearly dependent.

We are now ready to define the rank of a matrix. The rank of a matrix $\mathbf{A}$, usually denoted by $\mathbf{r}(\mathbf{A})$ (read "$\mathbf{r}$ of $\mathbf{A}$"), is defined as the maximum number of *linearly independent* rows or columns in $\mathbf{A}$. If $\mathbf{A}$ has 3 linearly independent rows or columns, then its rank is $\mathbf{r}(\mathbf{A}) = 3$. If $\mathbf{A}$ is of order 3×3, then $\mathbf{r}(\mathbf{A}) \leq 3$. Notice that if $\mathbf{A}$ is of order $m \times n$, then $\mathbf{r}(\mathbf{A}) \leq \min\{m, n\}$, that is, the rank will be equal to or less than the minimum of the set of m and n. Notice also that $\mathbf{r}(\mathbf{0}) = 0$, where $\mathbf{0}$ is any null matrix. However, how can one find the rank in the first place? This is dealt with in the next section.

2.6.2 Finding the rank of a matrix

One can find the rank of a matrix using the determinant(s) of the given matrix (or its square sub-matrices). The procedure is as follows. Suppose that we have a 3×3 matrix $\mathbf{A}$ as in the previous section. We first find $|\mathbf{A}|$. If we find that $|\mathbf{A}| \neq 0$, then the rank of $\mathbf{A}$ is said to be 3; that is, $\mathbf{r}(\mathbf{A}) = 3$. But, if it happens that $|\mathbf{A}| = 0$, then $\mathbf{r}(\mathbf{A}) < 3$. In this event, in order to find the exact rank we need to evaluate the determinants of the largest square sub-matrices ($\mathbf{S}_{ij}$) of the given matrix or find minors $\mathbf{M}_{ij} = |\mathbf{S}_{ij}|$. If the minors $\mathbf{M}_{ij}$ are different from zero, then $\mathbf{r}(\mathbf{A}) = i, j$. If it happens that $\mathbf{M}_{ij} = 0$ again, then we need to find $\mathbf{M}_{i-1, j-1}$. If we obtain that $\mathbf{M}_{i-1, j-1} \neq 0$, then $\mathbf{r}(\mathbf{A}) = i - 1, j - 1$. But, if we obtain that $\mathbf{M}_{i-1, j-1} = 0$, then we need to evaluate $\mathbf{M}_{i-2, j-2}$, and so on.

We stated in the previous section that the rank of a matrix is equal to the number of linearly independent rows or columns of that matrix. We found there that the matrix $\mathbf{B}$ has only two linearly independent rows. Therefore, the rank of $\mathbf{B}$ is equal to 2: $\mathbf{r}(\mathbf{B}) = 2$. This confirms our statement in the previous section that $\mathbf{r}(\mathbf{B}) \leq 3$.

The same conclusion can be arrived at by checking the determinant of $\mathbf{B}$ or of the determinants of its square sub-matrices. We can find that the determinant of the matrix in the previous section equals zero: $|\mathbf{B}| = 0$. Now, as outlined above, we have to find the minors of $\mathbf{B}$: $\mathbf{M}_{ij}$ with $i, j = 2$. One can verify that $\mathbf{M}_{11} = -14 \neq 0$. Notice that we do not need to check other minors. This confirms that $\mathbf{r}(\mathbf{B}) = 2$.

Our discussion above shows that it is much easier to use the minors of a matrix (or the determinants of its square sub-matrices) to find the rank of that matrix. Therefore, we redefine the rank of a matrix as the largest-order nonzero minor of that matrix, or the order of the nonzero determinant of the largest square sub-matrix that can be generated from that matrix.

2.6.3 Exercises

1. Find the rank of each of the following matrices:

(i) $\begin{bmatrix} 1 & 5 \\ 3 & 4 \end{bmatrix}$; (ii) $\begin{bmatrix} 1 & 1 \\ 1 & 1 \end{bmatrix}$; (iii) $\begin{bmatrix} 1 & 0 \\ 0 & 1 \end{bmatrix}$; (iv) $\begin{bmatrix} 0 & 1 \\ 1 & 0 \end{bmatrix}$; (v) $\begin{bmatrix} 0 & 0 \\ 0 & 0 \end{bmatrix}$; (vi) $\begin{bmatrix} 1 & 0 \\ 1 & 0 \end{bmatrix}$; (vii) $\begin{bmatrix} 0 & 1 \\ 0 & 1 \end{bmatrix}$;

(viii) $\begin{bmatrix} 1 & 2 \\ 3 & 6 \end{bmatrix}$; (ix) $\begin{bmatrix} 3 & 5 \\ 9 & 15 \end{bmatrix}$; (x) $\begin{bmatrix} 1 & 5 & 7 \\ 3 & 2 & 1 \end{bmatrix}$; (xi) $\begin{bmatrix} 1 & 3 \\ 5 & 2 \\ 7 & 1 \end{bmatrix}$; (xii) $\begin{bmatrix} 1 & 5 & 7 \\ 4 & 20 & 28 \end{bmatrix}$;

(xiii) $\begin{bmatrix} 1 & 3 \\ 5 & 15 \\ 7 & 21 \end{bmatrix}$.

2. Find the rank of each of the following matrices:

(i) $\begin{bmatrix} 1 & 2 & 3 \\ 4 & 5 & 6 \\ 7 & 8 & 9 \end{bmatrix}$; (ii) $\begin{bmatrix} 9 & 8 & 7 \\ 6 & 5 & 4 \\ 3 & 2 & 1 \end{bmatrix}$; (iii) $\begin{bmatrix} 2 & 3 & 1 \\ 5 & 1 & 1 \\ 3 & 5 & 2 \end{bmatrix}$; (iv) $\begin{bmatrix} 2 & 3 & 1 \\ 4 & 6 & 2 \\ 3 & 5 & 2 \end{bmatrix}$; (v) $\begin{bmatrix} 2 & 3 & 1 \\ 8 & 5 & 7 \\ 3 & 5 & 2 \end{bmatrix}$.

⚠ **Web supplement: S2.6.4 Mathematica applications**

2.7 Solution of SSLEs: consistency, existence, and uniqueness

We have used different methods to solve SSLEs. In Section 1.6.5 we used graphical, substitution, and elimination methods. The Gauss–Jordan elimination method was used in Section 2.3.7. In Section 2.3.8 we used the Gauss elimination method. The methods using determinants (Cramer's rule) and inverse were employed in Sections 2.4.5 and 2.5.5, respectively.

However, before one can employ any of these methods one needs to know whether a SSLEs is consistent or not. One might also need to know whether a solution exists or not for a SSLEs. If a solution exists, one might wonder whether it is unique or not. Therefore, the following three sections are devoted to the consistency of SSLEs; and the existence, and uniqueness of their solutions.

2.7.1 SSLEs: consistency

Even before checking for the existence of solutions, we need to know whether a SSLEs is a *consistent system of equations* or an *inconsistent system of equations*. The reason is that it will be futile to spend time on checking the existence of solutions for a system if it is an inconsistent system. We use the system $5x_1 + 3x_2 = d_1$ and $15x_1 + 9x_2 = d_2$ to illustrate the meaning of consistency of a SSLEs. Notice that we write the matrix form of this system as $\mathbf{Ax} = \mathbf{d}$, where $\mathbf{A} = \begin{bmatrix} 5 & 3 \\ 15 & 9 \end{bmatrix}$, $\mathbf{x} = \begin{bmatrix} x_1 \\ x_2 \end{bmatrix}$, and $\mathbf{d} = \begin{bmatrix} d_1 \\ d_2 \end{bmatrix}$.

We have two situations here depending upon the values that d_1 and d_2 may take. Notice that the second equation of the above system is perfectly three times the first equation if $d_2 = 3d_1$. Since the second row of $\mathbf{A}$ is three times the first row, the two rows are linearly dependent. This implies that one equation of the system, say the second, is redundant leaving us with only one equation with two unknowns: $5x_1 + 3x_2 = d_1$. Solving this equation for x_1 gives us $x_1 = (d_1/50) - (3/5)x_2$, which implies that there are an infinite number of solutions (each value of x_1 depends on the value that x_2 takes, and vice versa). Therefore, we state that a SSLEs is a consistent system if there exists at least one solution for the system. Hence, given $d_2 = 3d_1$, the last system is a consistent SSLEs.

Suppose now, instead of $d_2 = 3d_1$, we treat $d_1 = 6$. Then the first equation will be $5x_1 + 3x_2 = 6$. Suppose also that $d_2 = 10$. Then the second equation will become $15x_1 + 9x_2 = 10$. Given these two values, the system will be $5x_1 + 3x_2 = 6$ and $15x_1 + 9x_2 = 10$. Notice carefully that the two equations in this system represent parallel lines. We know that parallel lines do not intersect and that a solution to a SSLEs exists only when the graphs of the equations in the system intersect at some point. This means that the last system does not possess a solution. Therefore, we state that a SSLEs is an inconsistent system if it has no solution.

2.7.2 SSLEs: existence of solutions

We found in the previous section that one could judge whether a SSLEs is consistent or not by checking whether the system has a solution or not. Whether a system has a solution or not can be checked through visual inspection (as in the previous section) or through geometric representation of the system (as in Section 1.6.5). But, these procedures are inefficient, or even impossible particularly when the system has a number of equations or has more than three unknowns. This calls for an efficient procedure that can deal with an $m \times n$ system. This is explained below.

Instead of using specific systems, we shall use a general system such as the one in equation (2.3.5). We have expressed this system in Section 2.3.6 as $\mathbf{Ax} = \mathbf{d}$, where $\mathbf{A}$ denoted the $m \times n$ coefficient matrix, $\mathbf{x}$ denoted the $m \times 1$ column vector of variables or unknowns, and $\mathbf{d}$ denoted the $m \times 1$ column vector of constants. We shall now generate an augmented matrix by appending the column vector $\mathbf{d}$ to the coefficient matrix $\mathbf{A}$ and denote it, as before, by $\mathbf{A_{aug}d}$. We know that $\mathbf{A}$ is of order $m \times n$. This implies that $\mathbf{A_{aug}d}$ must be of order $m \times [n + 1]$. Assume now that $\mathbf{r(A)} = r$. Since $\mathbf{d}$ is a linear combination of the elements of $\mathbf{A}$, the rank of $\mathbf{A_{aug}d}$ must be equal to r: $\mathbf{r(A_{aug}d)} = r$. Recall that the rank of a matrix is the number of linearly independent rows or columns of that matrix. Recall also that our statement in the previous section that the graphs of linearly dependent equations will be parallel (when the constants are different) and will coincide (when the constants are equal), while the graphs of linearly independent equations will intersect leading to solution(s).

The above discussion suggests that the rank of the coefficient matrix of a system and the rank of its augmented form together determine whether the system has a solution or not. Therefore, we have the following theorem: a system of m linear equations in n unknowns such as $\mathbf{Ax} = \mathbf{d}$ has a solution if and only if $\mathbf{r(A)} = \mathbf{r(A_{aug}d)}$. Notice that $\mathbf{r(A)} = 1$ and $\mathbf{r(A_{aug}d)} = 2$ in the case of the example with $\mathbf{d'} = [\,6\ 10\,]$ in the previous section. Therefore, the system with $\mathbf{d'} = [\,6\ 10\,]$ in that example does not have a solution.

We stated in the previous section that a SSLEs such as $\mathbf{Ax} = \mathbf{d}$ is consistent only if it has at least one solution. But, we found above that the existence of solution to a the system $\mathbf{Ax} = \mathbf{d}$ depends on the equality of $\mathbf{r(A)}$ and $\mathbf{r(A_{aug}d)}$. Therefore, the consistency of the system and the existence of its solution boil down to the same condition: $\mathbf{r(A)} = \mathbf{r(A_{aug}d)}$. The system will be inconsistent if $\mathbf{r(A)} < \mathbf{r(A_{aug}d)}$. Notice that the system in the example with $\mathbf{d'} = [\,6\ 10\,]$ in the previous section is inconsistent because $\mathbf{r(A)} = 1 < \mathbf{r(A_{aug}d)} = 2$, and is consistent with $\mathbf{d'} = [\,2\ 6\,]$ because $\mathbf{r(A)} = \mathbf{r(A_{aug}d)} = 1$.

2.7.3 SSLEs: uniqueness of solutions

We discussed the inverse of a matrix and the methods to find the inverse in Section 2.5. We found from our discussion that for a matrix to have an inverse, it must be a square matrix. But, the squareness of a matrix alone does not guarantee its inverse. An example of such a matrix is the coefficient matrix $\mathbf{A}$ in Section 2.7.1. If a square matrix has no inverse, it is called a *singular matrix*; if it has an inverse, it is called a *nonsingular matrix*.

Consider two matrices $\mathbf{A}$ and $\mathbf{B}$ (those dealt with in Sections 2.7.1 and 2.5.3, respectively): $\mathbf{A} = \begin{bmatrix} 5 & 3 \\ 15 & 9 \end{bmatrix}$ and $\mathbf{B} = \begin{bmatrix} 3 & 4 \\ 3 & 2 \end{bmatrix}$. One can show that $\mathbf{A}$ does not have an inverse and we have seen in Section 2.5.3 that $\mathbf{B}$ has an inverse. An important question now is: why does the matrix $\mathbf{A}$ not have inverse, and why does the matrix $\mathbf{B}$ have inverse? The answer is simple: the rows in the former and in the latter are linearly dependent and independent, respectively. Therefore, for a matrix to have inverse, its rows or columns must be linearly independent. But, we know from the previous section that when the rows or columns of the coefficient matrix of a system are linearly dependent as in $\mathbf{A}$, the system may have either multiple solutions or no solution. Therefore, for a unique solution, the rows or columns of the coefficient matrix must be linearly independent; or, in other words, the coefficient matrix must have an inverse.

However, as we saw in Section 2.5.2 that, for a square matrix (say $\mathbf{A}$) to have inverse, its determinant must be nonvanishing or a nonzero number: that is, $|\mathbf{A}| \neq 0$. This is evident from equation (2.5.1): $\mathbf{A}^{-1} = \mathbf{A_{adj}} \div |\mathbf{A}|$. Moreover, as can be seen from Cramer's rule in equation (2.4.6), the solution to the jth variable of a linear system of equations depends on the determinant of the system's coefficient matrix, $\mathbf{A}$: $x_j = |\mathbf{A}_j| \div |\mathbf{A}|$. These imply that our whole enquiry boils down to a single value: the determinant of the coefficient matrix of a SSLEs. In other words, if $|\mathbf{A}| \neq 0$, $\mathbf{A}$ is nonsingular, $\mathbf{A}^{-1}$ exists, and a unique solution exists for the system.

We now sum up our findings so far in the present section. A SSLEs such as $\mathbf{Ax} = \mathbf{d}$ is consistent if $\mathbf{r(A)} = \mathbf{r(A_{aug}d)}$ and inconsistent if $\mathbf{r(A)} < \mathbf{r(A_{aug}d)}$. The system will have a solution if $\mathbf{r(A)} = \mathbf{r(A_{aug}d)}$. The solution to the system will be unique if $\mathbf{A}^{-1}$ exists or if $|\mathbf{A}| \neq 0$. If $\mathbf{r(A)} = \mathbf{r(A_{aug}d)}$ and if $|\mathbf{A}| = 0$, the system will have multiple or infinite number of solutions.

2.7.4 SSLEs: homogeneous case

Assume that we have a general SSLEs as in equation (2.3.5). If in this system $\mathbf{d} \neq \mathbf{0}$, that is, if the constant vector is not a vector of zeros, then the system is called a *nonhomogeneous SSLEs*. So far our analyses of the SSLEs have been concerned with these nonhomogeneous cases. However, there are situations where one needs to work with homogeneous SSLEs. We defined a homogeneous SSLEs, in equation (2.3.6), as a system in which $\mathbf{d} = \mathbf{0}$ and represented it in matrix form as $\mathbf{Ax} = \mathbf{d} = \mathbf{0}$. We mentioned at the end of Section 2.3.8 that a homogeneous system has either a trivial solution (or a zero solution) or an infinite number of nontrivial solutions.

Suppose that we have a *homogeneous SSLEs* such as equation (2.3.6) whose matrix representation is given by $\mathbf{Ax} = \mathbf{d} = \mathbf{0}$; where, as before, $\mathbf{A}$ denotes the $m \times n$ coefficient matrix, $\mathbf{x}$ denotes the $m \times 1$ column vector of unknowns or variables, $\mathbf{d}$ denotes the $m \times 1$ column vector of constants, and $\mathbf{0}$ denotes the $m \times 1$ column, null vector. Notice that when $\mathbf{d} = \mathbf{0}$, we know from equation (2.5.2) that the solution of the system, if $\mathbf{A}^{-1}$ exists or if $|\mathbf{A}| \neq 0$, is $\mathbf{x} = \mathbf{A}^{-1}\mathbf{d} = \mathbf{A}^{-1}\mathbf{0} = \mathbf{0}$. This means, when $\mathbf{d} = \mathbf{0}$, we have the solution $x_1 = x_2 = x_3 = \cdots = x_n = 0$. This is the trivial or zero solution that we mentioned in Section 2.3.8. The existence of a trivial solution is a feature of all homogeneous SSLEs.

If the system is $\mathbf{Ax} = \mathbf{d} = \mathbf{0}$ or $\mathbf{x} = \mathbf{A}^{-1}\mathbf{0}$, then the system may yield a set of infinite nontrivial solutions of which the trivial solution will be an element. This can be verified as follows. If $\mathbf{d} = \mathbf{0}$, $\mathbf{x} = \mathbf{A}^{-1}\mathbf{0} = [|(\mathbf{A_{aug}d})| \div |\mathbf{A}|] \times \mathbf{0} = \mathbf{0}$. Moreover, we have seen in the previous section that the system $\mathbf{Ax} = \mathbf{d} = \mathbf{0}$ may have infinite solutions if $|\mathbf{A}| = 0$. Since $|\mathbf{A_{adj}}| = 0$ when $\mathbf{d} = \mathbf{0}$, we may obtain $\mathbf{x} = \mathbf{0}/\mathbf{0}$. This suggests that, instead of a unique solution, we obtain an infinite number of solutions. One can check this using the example of the simple linearly dependent system we used in Section 2.7.1 by treating $d_1 = d_2 = 0$ (that is, $\mathbf{d} = \mathbf{0}$).

2.7.5 Solution of SSLEs: summary of results

As can be seen widely in the literature, we present a summary of all the ideas related to the solution of SSLEs we have discussed so far in the form of a table. The advantage of presenting them in a compact, table form is that the reader can quickly understand which SSLEs have unique solutions (trivial or nontrivial, or multiple solutions) and why they have

Table 2.7.1

	$\mathbf{d} \neq 0$	$\mathbf{d} = 0$
$|\mathbf{A}| \neq 0$ (nonsingular)	Unique, nontrivial solution, i.e. $\mathbf{x} \neq \mathbf{0}$	Unique, trivial solution, i.e. $\mathbf{x} = \mathbf{0}$
$|\mathbf{A}| = 0$ (singular; linear dependence, but consistent)	Infinite solutions, not including $\mathbf{x} = \mathbf{0}$ (trivial)	Infinite solutions, including $\mathbf{x} = \mathbf{0}$ (trivial)
$|\mathbf{A}| = 0$ (singular; inconsistent)	No solution	Not applicable

those solutions. Suppose that we have a SSLEs represented in matrix form $\mathbf{Ax} = \mathbf{d}$, then the entries in Table 2.7.1 are valid.

2.7.6 Exercises

1. Determine whether each of the following SSLEs has a solution, a unique solution, or multiple solutions:

    (i) $\begin{aligned} 2x_1 + x_2 &= 10 \\ 3x_1 + x_2 &= 10 \end{aligned}$; (ii) $\begin{aligned} 2x_1 + x_2 &= 0 \\ 3x_1 + x_2 &= 0 \end{aligned}$; (iii) $\begin{aligned} 2x_1 + x_2 &= 0 \\ 3x_1 + x_2 &= 10 \end{aligned}$; (iv) $\begin{aligned} 2x_1 + x_2 &= 10 \\ 3x_1 + x_2 &= 0 \end{aligned}$;

    (v) $\begin{aligned} 2x_1 + x_2 &= 10 \\ 4x_1 + 2x_2 &= 10 \end{aligned}$; (vi) $\begin{aligned} 2x_1 + 6x_2 &= 10 \\ 2x_1 + 6x_2 &= 10 \end{aligned}$.

2. Determine whether each of the following SSLEs has a solution, a unique solution, or multiple solutions:

    (i) $\begin{aligned} x_1 + 2x_2 + 3x_3 &= 2 \\ x_1 + 2x_2 + 3x_3 &= 2; \\ x_1 + 2x_2 + 3x_3 &= 2 \end{aligned}$ (ii) $\begin{aligned} x_1 + 2x_2 + 3x_3 &= 2 \\ x_1 + 2x_2 + 3x_3 &= 0; \\ x_1 + 2x_2 + 3x_3 &= 2 \end{aligned}$ (iii) $\begin{aligned} x_1 + 2x_2 + 3x_3 &= 4 \\ x_1 + x_2 + 1.5x_3 &= 2; \\ x_1 + 2x_2 + 3x_3 &= 2 \end{aligned}$

    (iv) $\begin{aligned} 3x_1 + 2x_2 + 3x_3 &= 2 \\ x_1 + 2x_2 + x_3 &= 2 \\ x_1 + 2x_2 + 3x_3 &= 2 \end{aligned}$.

2.8 Linear algebra: extensions

In this section we discuss some of the matrices and determinants that we will make use of in the applications of differential calculus (Chapter 3) and in optimization using differential calculus (Chapter 4). Since most of the topics in this section are dependent on the topics in Chapters 4 and 5, readers are encouraged to do Chapters 4 and 5 before they do the following sections. The topics we cover in this section are *Jacobian matrices* and *Jacobian determinants*; *quadratic forms* and *discriminants*; *Hessian matrices, Hessian determinants, bordered Hessian matrices*, and *bordered Hessian determinants*; and *characteristic equations* and *characteristic roots*. We begin with Jacobian matrices and determinants.

2.8.1 Jacobian matrices and determinants

So far we were using the determinant of the coefficient matrix, of a SSLEs, to test whether there existed linear dependence among the rows and columns of that matrix. This test of linear dependence among equations helped us see whether a particular SSLEs had a unique solution or multiple solutions.

Sometimes one may come across functions, linear or nonlinear, that are dependent or independent. Such functions may not have a solution, have a unique solution, or have multiple solutions. How do we know whether a *system of simultaneous functions* (SSF) possesses a solution(s) or not? Therefore, we need a technique to test this; or in other words, to test whether such a SSFs is dependent or not. For this we use the Jacobian matrices and their determinants as discussed below.

Suppose that we have a SSFs of x_1 and x_2 as

$$y_1 = f^1(x_1, x_2)$$
$$y_2 = f^2(x_1, x_2)$$

(2.8.1)

where f^i represents the ith function and y_i represents the ith dependent variable. The *partial derivative* of each of these functions with respect to x_1 and x_2 are $\partial y_1/\partial x_1$ and $\partial y_1/\partial x_2$; and $\partial y_2/\partial x_1$ and $\partial y_2/\partial x_2$, respectively. Assume now that we arrange these partial derivatives in the form of a matrix $\mathbf{J}$:

$$\mathbf{J} = \begin{bmatrix} \dfrac{\partial y_1}{\partial x_1} & \dfrac{\partial y_1}{\partial x_2} \\ \dfrac{\partial y_2}{\partial x_1} & \dfrac{\partial y_2}{\partial x_2} \end{bmatrix}$$

(2.8.2)

The matrix $\mathbf{J}$ in equation (2.8.2) is called the Jacobian matrix of the partial derivatives of the system in equation (2.8.1). Notice that the Jacobian matrix in equation (2.8.2) may also be written as

$$\mathbf{J} = [\partial y_1, \partial y_2] / [\partial x_1, \partial x_2].$$

(2.8.3)

Let us now take the determinant of the Jacobian matrix $\mathbf{J}$ in equation (2.8.2) or (2.8.3) to obtain

$$|\mathbf{J}| = \begin{vmatrix} \partial y_1/\partial x_1 & \partial y_1/\partial x_2 \\ \partial y_2/\partial x_1 & \partial y_2/\partial x_2 \end{vmatrix}$$

(2.8.4)

which can also be written as $|\mathbf{J}| = |\partial y_1, \partial y_2/\partial x_1, \partial x_2|$. The determinant in equation (2.8.4) is called the Jacobian determinant or, in short, the *Jacobian*. If $|\mathbf{J}| = 0$, then, the equations are said to be functionally dependent; and if $|\mathbf{J}| \neq 0$, the equations are said to be functionally independent.

We can generalize the above results into an $m \times n$ system of functions. Let y_i be a function of x_j, where $i = 1, 2, 3, \ldots, m$ and $j = 1, 2, 3, \ldots, n$, as

$$y_1 = f^1(x_1, x_2, x_3, \ldots, x_n)$$
$$y_2 = f^2(x_1, x_2, x_3, \ldots, x_n)$$
$$y_3 = f^3(x_1, x_2, x_3, \ldots, x_n)$$
$$\ldots\ldots\ldots\ldots\ldots\ldots\ldots\ldots\ldots$$
$$y_m = f^m(x_1, x_2, x_3, \ldots, x_n)$$

(2.8.5)

Then the associated Jacobian matrix and Jacobian determinants, respectively, are

$$\mathbf{J} = [\partial y_1, \partial y_2, \partial y_3, \dots, \partial y_n / \partial x_1, \partial x_2, \partial x_3, \dots, \partial x_n]$$

$$= \begin{bmatrix} \partial y_1/\partial x_1 & \partial y_1/\partial x_2 & \partial y_1/\partial x_3 & \dots & \partial y_n/\partial x_1 \\ \partial y_2/\partial x_1 & \partial y_2/\partial x_2 & \partial y_2/\partial x_3 & \dots & \partial y_2/\partial x_n \\ \partial y_3/\partial x_1 & \partial y_3/\partial x_2 & \partial y_3/\partial x_3 & \dots & \partial y_3/\partial x_n \\ \dots & \dots & \dots & \dots & \dots \\ \partial y_m/\partial x_1 & \partial y_m/\partial x_2 & \partial y_m/\partial x_3 & \dots & \partial y_m/\partial x_n \end{bmatrix} \tag{2.8.6}$$

and

$$|\mathbf{J}| = |\partial y_1, \partial y_2, \partial y_3, \dots, \partial y_n / \partial x_1, \partial x_2, \partial x_3, \dots, \partial x_n|$$

$$= \begin{vmatrix} \partial y_1/\partial x_1 & \partial y_1/\partial x_2 & \partial y_1/\partial x_3 & \dots & \partial y_n/\partial x_1 \\ \partial y_2/\partial x_1 & \partial y_2/\partial x_2 & \partial y_2/\partial x_3 & \dots & \partial y_2/\partial x_n \\ \partial y_3/\partial x_1 & \partial y_3/\partial x_2 & \partial y_3/\partial x_3 & \dots & \partial y_3/\partial x_n \\ \dots & \dots & \dots & \dots & \dots \\ \partial y_m/\partial x_1 & \partial y_m/\partial x_2 & \partial y_m/\partial x_3 & \dots & \partial y_m/\partial x_n \end{vmatrix} \tag{2.8.7}$$

As an example, consider the SSFs: $y_1 = f^1 = 2x_1 + x_2$ and $y_2 = f^2 = 4x_1^2 + 4x_1x_2 + x_2^2$. The associated partial derivatives are $\partial y_1/\partial x_1 = 2$, $\partial y_1/\partial x_2 = 1$, $\partial y_2/\partial x_1 = 8x_1 + 4x_2$, and $\partial y_2/\partial x_2 = 4x_1 + 2x_2$. Then, we obtain the Jacobian matrix and Jacobian, respectively, as $\mathbf{J} = \begin{bmatrix} 2 & 1 \\ 8x_1 + 4x_2 & 4x_1 + 2x_2 \end{bmatrix}$ and $|\mathbf{J}| = \begin{vmatrix} 2 & 1 \\ 8x_1 + 4x_2 & 4x_1 + 2x_2 \end{vmatrix}$. Therefore, $|\mathbf{J}| = [2(4x_1 + 2x_2) - 1(8x_1 + 4x_2)] = (8x_1 + 4x_2) - (8x_1 + 4x_2) = 0$. This implies that the last two functions are dependent. Notice that the second function is simply the square of the first function: $y_2 = (y_1)^2$. Notice also that if the above system were a SSLEs, then we would have used the determinant of its coefficient matrix to test for *singularity* (that is, to test whether the system has a solution). In which case, we would have obtained the same result. This implies that the determinantal test of the existence of solution in the case of a SSLEs (or SSFs) is just an application of the *Jacobian test* (which can handle both linear and nonlinear functions) we discussed above.

As another example, consider the system of functions $y_1 = f^1 = 2x_1 + x_2$ and $y_2 = f^2 = 3x_1 + 4x_2$. The associated partial derivatives are $\partial y_1/\partial x_1 = 2$, $\partial y_1/\partial x_2 = 1$, $\partial y_2/\partial x_1 = 3$, and $\partial y_2/\partial x_2 = 4$. Then, we have $\mathbf{J} = \begin{bmatrix} 2 & 1 \\ 3 & 4 \end{bmatrix}$ and $|\mathbf{J}| = \begin{vmatrix} 2 & 1 \\ 3 & 4 \end{vmatrix}$. Therefore, $|\mathbf{J}| = 2 \times 4 - 3 \times 1 = 8 - 3 = 5$. This implies that the two functions are independent. Notice that we can write the above system in matrix notation as $\mathbf{Ax} = \mathbf{y}$, where $\mathbf{A} = \mathbf{J}$ denotes the coefficient matrix, $\mathbf{x}$ denotes the vector of x's, and $\mathbf{y}$ denotes the vector of y's. Then we have the determinant of the coefficient matrix $|\mathbf{A}| = |\mathbf{J}| = 2 \times 4 - 3 \times 1 = 8 - 3 = 5$, which is the same as the Jacobian of the system. This confirms our statement above that the Jacobian will be equal to the determinant of the coefficient matrix in the case of linear SSFs (or SSLEs).

As the last example, consider the functions $y_1 = f^1 = 2x_1 + x_1x_2$ and $y_2 = f^2 = 3x_1x_2$. The partial derivatives of this system are $\partial y_1/\partial x_1 = 2$, $\partial y_1/\partial x_2 = x$, $\partial y_2/\partial x_1 = 3x_2$, and $\partial y_2/\partial x_2 = 3x_1$. Then we obtain $\mathbf{J} = \begin{bmatrix} 2 & x_1 \\ 3x_2 & 3x_1 \end{bmatrix}$ and $|\mathbf{J}| = \begin{vmatrix} 2 & x_1 \\ 3x_2 & 3x_1 \end{vmatrix}$. Therefore, $|\mathbf{J}| = 2 \times 3x_1 - 3x_2 \times x_1 = 6x_1 - 3x_1x_2$. This suggests that the two functions are independent.

2.8.2 Quadratic forms, discriminants, and sign definiteness

Students of economics, business, and finance usually deal with optimization problems involving two or more independent variables. In solving these problems, both the *first-order condition* (FOC) and the *second-order condition* (SOC) are to be satisfied. The SOCs involve the determination of the sign of the *second partial derivatives*. This task can be simplified considerably if one uses the concerned *quadratic forms* and the associated *discriminants*. However, we first explain a few related concepts before we go on to the determination of the sign of the quadratic forms using discriminants.

We presented a general polynomial in equation (1.6.1): $y = b_0 x^0 + b_1 x^1 + b_2 x^2 + b_3 x^3 + \cdots + b_n x^n = \sum_{i=0}^{n} b_i x^i$. Notice that in this equation there is only one independent variable, x. If we limit $i = 2$, we obtain the quadratic equation $y = b_0 x^0 + b_1 x^1 + b_2 x^2$, as in equation (1.6.4).

Assume now that we have a polynomial equation with more than one independent variable. Then each term of the new polynomial equation may contain one, or more than one, variable with different powers. Notice that $z = b_0 + b_1 x_1 + b_2 x_2$ is such a polynomial equation. This equation is called a *linear form* (of two independent variables, x_1 and x_2) because the sum of powers of the variables in each term is equal to 1. If the polynomial is $z = b_0 + b_1 x_1^2 + b_2 x_1 x_2 + b_3 x_2^2$, then this is called a *quadratic form* (as before, of two independent variables) because the sum of powers of the variables in each term is equal to 2. If we have a polynomial such as $z = b_0 + b_1 x_1^3 + b_2 x_1^2 x_2 + b_3 x_1 x_2^2 + b_4 x_2^3$, then it is called a *cubic form* since the sum of powers of the variables in each term is equal to 3. We are mainly concerned here with quadratic forms.

Suppose that we have a function of two independent variables, such as $U = f(x_1, x_2)$, to be optimized. Its first and second partial derivatives are given by $f_{x_1} = f_1, f_{x_2} = f_2, f_{x_1 x_1} = f_{11}$, and $f_{x_2 x_2} = f_{22}, f_{x_1 x_2} = f_{12}, f_{x_2 x_1} = f_{21}$, respectively. Moreover, its first and second *total differentials* are given, respectively, by

$$dU = f_1 dx_1 + f_2 dx_2 \tag{2.8.8}$$

and

$$d^2 U = f_{11} dx_1^2 + 2 f_{12} dx_1 dx_2 + f_{22} dx_2^2, \text{ since } f_{12} = f_{21} \tag{2.8.9}$$

Notice that equation (2.8.8) is a special case of equation (3.8.4) with $n = 2$ and equation (2.8.9) is identical with equation (3.8.7). We can now use either partial derivatives or total differential to find the *optimum* (either *maximum* or *minimum*) of U. We will show in Chapter 4 that the FOCs for an optimum of U are $f_1 = f_2 = 0$ in terms of partial derivatives and $dU = 0$ in terms of total differential. The SOCs for a minimum of U are $f_{11}, f_{22} > 0$ and $f_{11}.f_{22} > (f_{12})^2$ in terms of partial derivatives and $d^2 U > 0$ in terms of total differential. Similarly, the SOCs for a maximum of U are $f_{11}, f_{22} < 0$ and $f_{11}.f_{22} > (f_{12})^2$ in terms of partial derivatives and $d^2 U < 0$ in terms of total differential.

Notice that in the case of optimization involving two independent variables, as in our example so far, we can use the partial derivatives (f_{11}, f_{22}, and f_{12}) directly to determine the sign as shown by the above conditions. But, when the number of independent variables increases beyond two, the use of partial derivatives becomes tedious. In such cases, we need to use an alternative method that may simplify the computation. Such a method can be found by putting the total differential ($d^2 U$) into further analysis. We present this below.

The above exposition shows that we need to determine the sign of d^2U as the SOC when we use the total differential to decide whether U has an optimum. Notice that the equation $d^2U = f_{11}dx_1^2 + 2f_{12}dx_1dx_2 + f_{22}dx_2^2$ is a quadratic form because dx_1 and dx_2 are the variables in the present case and the sum of their powers in each term is equal to 2. If this quadratic form is negative, that is, if $d^2U < 0$, then d^2U is said to be *negative definite*. If $d^2U > 0$, then d^2U is said to be *positive definite*. If $d^2U \leq 0$ or $d^2U \geq 0$, d^2U is said to be *negative semidefinite* or *positive semidefinite*, respectively. What all these mean is that for U to have a minimum, d^2U must be positive definite ($d^2U > 0$); for U to have a maximum, d^2U must be negative definite ($d^2U < 0$). In the case where d^2U is indeterminate, then we have what is called a *saddle point*. However, how does one test the sign of d^2U? The procedure is outlined below.

Notice that the quadratic form in equation (2.8.9) can be expressed as a product of matrices as

$$d^2U = \begin{bmatrix} dx_1 & dx_2 \end{bmatrix} \begin{bmatrix} f_{11} & f_{12} \\ f_{21} & f_{22} \end{bmatrix} \begin{bmatrix} dx_1 \\ dx_2 \end{bmatrix} \tag{2.8.10}$$

In equation (2.8.10), as mentioned above, the *differentials* (dx_1 and dx_2) are considered to be variables while the second partial derivatives are considered to be constants. One restriction we impose here is that both differentials cannot be zero at the same time. The determinant of the matrix of the second partial derivatives is called the *discriminant* of the quadratic form d^2U and we denote it by $\mathbf{D_R}$. Therefore,

$$\mathbf{D_R} = \begin{vmatrix} f_{11} & f_{12} \\ f_{21} & f_{22} \end{vmatrix} \tag{2.8.11}$$

Notice that, in equation (2.8.11), f_{11} is the first element on the principal diagonal of $\mathbf{D_R}$. The determinant of this first element (that is, a *sub-determinant* of $\mathbf{D_R}$), which we denote by $\mathbf{D_{R1}}$, is given by $\mathbf{D_R} = |f_{11}| = f_{11}$ and is called the *first principal minor* of $\mathbf{D_R}$. Also notice that $\mathbf{D_{R2}}$, which is equal to $\mathbf{D_R}$, is a sub-determinant of $\mathbf{D_R}$. $\mathbf{D_{R2}}$ is called the *second principal minor* of $\mathbf{D_R}$. Since there are only two sub-determinants in the present example ($\mathbf{D_R}$ and $\mathbf{D_{R2}} = \mathbf{D_R}$), we need to consider them only. If the quadratic form is a function of three independent variables, then the associated discriminant will have three sub-determinants (such as $\mathbf{D_{R1}}$, $\mathbf{D_{R2}}$, and $\mathbf{D_{R3}} = \mathbf{D_R}$); if the quadratic form is a function of four independent variables, then the associated discriminant will have four sub-determinants (such as $\mathbf{D_{R1}}$, $\mathbf{D_{R2}}$, $\mathbf{D_{R3}}$, and $\mathbf{D_{R4}} = \mathbf{D_R}$); and so on.

It is easy to see that d^2U is positive definite ($d^2U > 0$) if, and only if, $\mathbf{D_{R1}} = f_{11} > 0$ and $\mathbf{D_{R2}} = \mathbf{D_R} > 0$; d^2U is negative definite ($d^2U < 0$) if, and only if, $\mathbf{D_{R1}} = f_{11} < 0$ and $\mathbf{D_{R2}} = \mathbf{D_R} > 0$. To validate this statement we use the quadratic form $d^2U = f_{11}dx_1^2 + 2f_{12}dx_1dx_2 + f_{22}dx_2^2$ by adding to it, and subtracting from it, the expression $f_{12}^2dx_2^2/f_{11}$. After doing this and after factoring out, we obtain $d^2U = f_{11}[dx_1^2 + 2f_{12}dx_1dx_2/f_{11} + f_{12}^2dx_2^2/f_{11}^2] + [f_{22} - f_{12}^2/f_{11}]dx_2^2$. The expression in the first set of brackets of this equation is the square of $[dx_1 + f_{12}dx_2/f_{11}]$. Therefore, we can write the equation as $d^2U = f_{11}[dx_1 + f_{12}dx_2/f_{11}]^2 + [(f_{11}f_{22} - f_{12}^2)/f_{11}]dx_2^2$. Since the squared terms in the equation cannot be negative, the sign of d^2U depends on f_{11}, f_{22}, and $f_{12} = f_{21}$. If $f_{11} > 0$ and $f_{11}f_{22} - f_{12}^2 > 0$, then $d^2U > 0$ (or positive definite); if $f_{11} < 0$ and $f_{11}f_{22} - f_{12}^2 > 0$, then $d^2U < 0$ (or negative definite). But, these are the same statements we made at the beginning of this paragraph. Notice that the condition for the inequality $\mathbf{D_{R2}} = f_{11}.f_{22} - (f_{12})^2 > 0$ is that both f_{11} and f_{22} must be of the same sign. Notice also that these are the same conditions as those stated earlier.

As an example, consider the quadratic form $d^2U = 2dx_1^2 + 0dx_1dx_2 + 2dx_2^2$. Our aim here is to check whether d^2U is positive or negative definite. We know, from this quadratic form, that $f_{11} = 2$, $f_{22} = 2$, and $f_{12} = f_{21} = 0$. Therefore, we can construct the discriminant $\mathbf{D_R} = \mathbf{D_{R2}} = 4 - 0 = 4$. The first principal minor, $\mathbf{D_{R1}}$, of the discriminant is $\mathbf{D_{R1}} = |2| = 2$. Since the first and the second principal minors are positive, we conclude that d^2U is positive definite.

As another example, consider the quadratic form $d^2U = -2dx_1^2 + 0dx_1dx_2 - 2dx_2^2$. Then we can obtain $f_{11} = -2$, $f_{22} = -2$, and $f_{12} = f_{21} = 0$. Therefore, the discriminant is $\mathbf{D_R} = \mathbf{D_{R2}} = 4$. Notice that $\mathbf{D_{R1}} = |-2| = -2$. Since the second principal minor is positive and the first principal minor is negative, we conclude that d^2U is negative definite.

Consider now the quadratic form of three independent variables

$$d^2U = f_{11}dx_1^2 + f_{22}dx_2^2 + f_{33}dx_3^2 + f_{12}dx_1dx_2 + f_{13}dx_1dx_3 + f_{23}dx_2dx_3 \tag{2.8.12}$$

When the quadratic form is a function of three independent variables (as in equation (2.8.12)) one can show, following a similar line of arguments as earlier, that the conditions for d^2U to be positive definite are $\mathbf{D_{R1}} > 0$, $\mathbf{D_{R2}} > 0$, and $\mathbf{D_R} = \mathbf{D_{R3}} > 0$, and that the conditions for d^2U to be negative definite are $\mathbf{D_{R1}} < 0$, $\mathbf{D_{R2}} > 0$, and $\mathbf{D_R} = \mathbf{D_{R3}} < 0$. The discriminant of the last quadratic form can be written as

$$\mathbf{D_R} = \begin{vmatrix} f_{11} & f_{12} & f_{13} \\ f_{21} & f_{22} & f_{23} \\ f_{31} & f_{32} & f_{33} \end{vmatrix} \tag{2.8.13}$$

Assume now that $f_{11} = 2, f_{22} = 2, f_{33} = 2, f_{12} = f_{21} = 0, f_{13} = f_{31} = 0$, and $f_{23} = f_{32} = 0$. Then the discriminant of the last quadratic form becomes $\mathbf{D_R} = \begin{vmatrix} f_{11} & f_{12} & f_{13} \\ f_{21} & f_{22} & f_{23} \\ f_{31} & f_{32} & f_{33} \end{vmatrix} = \begin{vmatrix} 2 & 0 & 0 \\ 0 & 2 & 0 \\ 0 & 0 & 2 \end{vmatrix}$, from which we obtain that the third, the second, and the first principal minors, respectively, are $\mathbf{D_{R3}} = \mathbf{D_R} = \begin{vmatrix} f_{11} & f_{12} & f_{13} \\ f_{21} & f_{22} & f_{23} \\ f_{31} & f_{32} & f_{33} \end{vmatrix} = \begin{vmatrix} 2 & 0 & 0 \\ 0 & 2 & 0 \\ 0 & 0 & 2 \end{vmatrix} = 8$; $\mathbf{D_{R2}} = \begin{vmatrix} f_{11} & f_{12} \\ f_{21} & f_{22} \end{vmatrix} = \begin{vmatrix} 2 & 0 \\ 0 & 2 \end{vmatrix} = 4$; and $\mathbf{D_R} = |f_{11}| = |2| = 2$. Since the first and second principal minors and the discriminant are all positive, we conclude that d^2U is positive definite.

As the last example, consider the quadratic form of three independent variables as in equation (2.8.12). Assuming that $f_{11} = -2, f_{22} = -2, f_{33} = -2, f_{12} = f_{21} = 0, f_{13} = f_{31} = 0$, and $f_{23} = f_{32} = 0$, we get the discriminant as $\mathbf{D_R} = \begin{vmatrix} f_{11} & f_{12} & f_{13} \\ f_{21} & f_{22} & f_{23} \\ f_{31} & f_{32} & f_{33} \end{vmatrix} = \begin{vmatrix} -2 & 0 & 0 \\ 0 & -2 & 0 \\ 0 & 0 & -2 \end{vmatrix}$, from which we obtain that the third, the second, and the first principal minors, respectively, are $\mathbf{D_{R3}} = \mathbf{D_R} = \begin{vmatrix} f_{11} & f_{12} & f_{13} \\ f_{21} & f_{22} & f_{23} \\ f_{31} & f_{32} & f_{33} \end{vmatrix} = \begin{vmatrix} -2 & 0 & 0 \\ 0 & -2 & 0 \\ 0 & 0 & -2 \end{vmatrix} = -8$; $\mathbf{D_{R2}} = \begin{vmatrix} f_{11} & f_{12} \\ f_{21} & f_{22} \end{vmatrix} = \begin{vmatrix} -2 & 0 \\ 0 & -2 \end{vmatrix} = 4$;

and $\mathbf{D_R} = |f_{11}| = |-2| = -2$. Since the first and the third principal minors are negative and since the second principal minor is positive, we conclude that $d^2 U$ is negative definite.

We now extend the application of discriminantal test of sign definiteness to quadratic forms that involve n independent variables. Before we do this extension, we can convert the earlier examples of the quadratic forms with two (and three) independent variables into more convenient expressions. This is presented first.

Our quadratic form with two independent variables in equation (2.8.9) was $d^2 U = f_{11}dx_1^2 + 2f_{12}dx_1dx_2 + f_{22}dx_2^2$. This quadratic form can be written in matrix form as

$$d^2 U = \mathbf{x^T D x} \tag{2.8.14}$$

where $\mathbf{x} = \begin{bmatrix} dx_1 \\ dx_2 \end{bmatrix}$ and $\mathbf{D} = \begin{bmatrix} f_{11} & f_{12} \\ f_{21} & f_{22} \end{bmatrix}$. The associated discriminant is $\mathbf{D_R} = \begin{vmatrix} f_{11} & f_{12} \\ f_{21} & f_{22} \end{vmatrix}$.

Notice that the first principal minor is $\mathbf{D_{R1}} = |f_{11}| = f_{11}$ and the second principal minor is $\mathbf{D_R} = \mathbf{D_{R2}}$.

Similarly, our quadratic form with three independent variables in equation (2.8.12) was $d^2 U = f_{11}dx_1^2 + f_{22}dx_2^2 + f_{33}dx_3^2 + 2f_{12}dx_1dx_2 + 2f_{13}dx_1dx_3 + 2f_{23}dx_2dx_3$. As above, this quadratic form can be written in matrix form, following equation (2.8.14), as $d^2 U = \mathbf{x^T D x}$,

where $\mathbf{x} = \begin{bmatrix} dx_1 \\ dx_2 \\ dx_3 \end{bmatrix}$ and $\mathbf{D} = \begin{bmatrix} f_{11} & f_{12} & f_{13} \\ f_{21} & f_{22} & f_{23} \\ f_{31} & f_{32} & f_{33} \end{bmatrix}$ with $\mathbf{D_R} = \begin{vmatrix} f_{11} & f_{12} & f_{13} \\ f_{21} & f_{22} & f_{23} \\ f_{31} & f_{32} & f_{33} \end{vmatrix}$. Notice that the

first, the second, and the third principal minors of the discriminant here, respectively, are:

$$\mathbf{D_{R1}} = |f_{11}| = f_{11}; \ \mathbf{D_{R2}} = \begin{vmatrix} f_{11} & f_{12} \\ f_{21} & f_{22} \end{vmatrix}; \text{ and } \mathbf{D_R} = \mathbf{D_{R3}} = \begin{vmatrix} f_{11} & f_{12} & f_{13} \\ f_{21} & f_{22} & f_{23} \\ f_{31} & f_{32} & f_{33} \end{vmatrix}.$$ Instead of quadratic

forms with two or three independent variables, suppose that we have a quadratic form with n independent variables as

$$d^2 U = f_{11}dx_1^2 + f_{22}dx_2^2 + f_{33}dx_3^2 + \cdots + f_{nn}dx_n^2$$
$$+ 2f_{12}dx_1dx_2 + 2f_{13}dx_1dx_3 + \cdots + 2f_{1n}dx_1dx_n + 2f_{23}dx_2dx_3 + 2f_{24}dx_2dx_4$$
$$+ \cdots + 2f_{2n}dx_2dx_n + \cdots + 2f_{n(n-1)}dx_ndx_{n-1} \tag{2.8.15}$$

which is identical to equation (3.8.9) and which in matrix notation can be written as $d^2 U = \mathbf{x^T D x}$, where $\mathbf{x} = \begin{bmatrix} dx_1 \\ dx_2 \\ .. \\ dx_n \end{bmatrix}$ and $\mathbf{D} = \begin{bmatrix} f_{11} & f_{12} & .. & f_{1n} \\ f_{21} & f_{22} & .. & f_{2n} \\ .. & .. & .. & .. \\ f_{n1} & f_{n2} & .. & f_{nn} \end{bmatrix}$ with $\mathbf{D_R} = \begin{vmatrix} f_{11} & f_{12} & .. & f_{1n} \\ f_{21} & f_{22} & .. & f_{2n} \\ .. & .. & .. & .. \\ f_{n1} & f_{n2} & .. & f_{nn} \end{vmatrix}$. Notice

that the first, the second, the third, ..., and the nth principal minors of the discriminant,

respectively, are: $\mathbf{D_{R1}} = |f_{11}| = f_{11}; \ \mathbf{D_{R2}} = \begin{vmatrix} f_{11} & f_{12} \\ f_{21} & f_{22} \end{vmatrix}; \ \mathbf{D_{R3}} = \begin{vmatrix} f_{11} & f_{12} & f_{13} \\ f_{21} & f_{22} & f_{23} \\ f_{31} & f_{32} & f_{33} \end{vmatrix};...$ and $\mathbf{D_R}$

$= \mathbf{D_{Rn}} = \begin{vmatrix} f_{11} & f_{12} & .. & f_{1n} \\ f_{21} & f_{22} & .. & f_{2n} \\ .. & .. & .. & .. \\ f_{n1} & f_{n2} & .. & f_{nn} \end{vmatrix}$. Therefore, one can show (as in the case of the quadratic form

with two or three variables), that $d^2 U$ will be positive definite if $\mathbf{D_{R1}}, \mathbf{D_{R2}}, \mathbf{D_{R3}}, ..., \mathbf{D_R}$ $= \mathbf{D_{Rn}}$ are all positive. Similarly, $d^2 U$ will be negative definite if $\mathbf{D_{R1}} < 0$, $\mathbf{D_{R2}} > 0$,

$\mathbf{D_{R3}} < 0, \ldots, (-1)^n \, \mathbf{D_{Rn}} > 0$; that is, when the principal minors of the discriminant alternate in sign starting with $\mathbf{D_{R1}} < 0$. These are exactly the same results as those we employed when we considered specific examples earlier.

2.8.3 Hessian and bordered Hessian matrices and determinants

Students of economics, business, and finance usually come across optimization problems that involve *equality constraints*. Examples of these include *utility maximization* by a consumer who faces a budget constraint, *cost minimization* by a firm that faces a constraint in the form of an isocost line, etc. However, in order to find the optimum in all these problems we need to use *bordered Hessian matrices* and *bordered Hessian determinants*. Therefore, it is necessary that students of these subjects possess a good understanding of these matrices and their determinants, and of using them in solving the said optimization problems. But, before this, we need to be familiar with the concepts of Hessian matrices and Hessian determinants, which will be dealt with first.

We carried out the sign definiteness test of the second total differentials in the previous section. The n independent variable, general second total differential, or quadratic form we used was given in equation (2.8.15). We then presented this second total differential in matrix form as $d^2U = \mathbf{x^T D x}$. The square matrix $\mathbf{D}$ that we obtained, which is in fact the matrix of the partial derivatives in the quadratic form d^2U, is also called the Hessian matrix, and we denote it by $\mathbf{H}$. And the determinant of $\mathbf{D}$ (or the discriminant $\mathbf{D_R}$) is called the Hessian determinant (or, simply, the Hessian), and we denote it by $|\mathbf{H}|$. This shows that $\mathbf{D}$ and $\mathbf{H}$, and $|\mathbf{D}|$ (or $\mathbf{D_R}$) and $|\mathbf{H}|$ are identical. Moreover, the conditions of sign definiteness using $|\mathbf{D}|$ (or $\mathbf{D_R}$) are equally applicable when we use $|\mathbf{H}|$. Specifically, d^2U is positive definite (or U has a minimum) when $|\mathbf{H_1}| > 0$, $|\mathbf{H_2}| > 0$, $|\mathbf{H_3}| > 0, \ldots, |\mathbf{H_n}| > 0$, and d^2U is negative definite (or U has a maximum) when $|\mathbf{H_1}| < 0$, $|\mathbf{H_2}| > 0$, $|\mathbf{H_3}| < 0, \ldots, (-1)^n \, |\mathbf{H_n}| > 0$; that is, when the principal minors of $|\mathbf{H}|$ alternate in sign. This implies that we may use either discriminant or Hessian to test the sign definiteness of the quadratic form.

However, Hessian matrices and their determinants become particularly useful when we do optimization problems involving equality constraints. We shall consider here the general case. Suppose that we have an objective function with two independent variables given by $U = f(x_1, x_2)$ and a constraint with the same independent variables given by $g(x_1, x_2) = c$, where c is a constant. We now construct a new function

$$L = f(x_1, x_2) + \lambda[c - g(x_1, x_2)] \tag{2.8.16}$$

where L is called the *Lagrangian function*, and λ is called the *Lagrangian multiplier*. Then the SOC for an optimum, as shown in Chapter 4 and as before, amounts again to determining the sign of d^2U. Notice that d^2U here is a *constrained quadratic form*, with the constraint $g(x_1, x_2) = c$. Therefore, the constraint quadratic form can be written as

$$d^2U = L_{11}dx_1^2 + 2L_{12}dx_1dx_2 + L_{22}dx_2^2 \tag{2.8.17}$$

where $L_{11} = L_{x_1x_1} = f_{11} - \lambda g_{11}$, $L_{12} = L_{x_1x_2} = f_{12} - \lambda g_{12}$, and $L_{22} = L_{x_2x_2} = f_{22} - \lambda g_{22}$. As discussed in Section 4.4.5, the quadratic form in equation (2.8.17) can be written as equation (4.4.15): $d^2U = [L_{11}g_2^2 - 2L_{12}g_1g_2 + L_{22}g_1^2][dx_1^2/g_2^2]$. It is easy to see from this equation that d^2U is positive (negative) definite if and only if the sum of the terms in the brackets is positive (negative). But, the reader must have noticed that the determinant

$$|\mathbf{H}| = \begin{vmatrix} 0 & g_1 & g_2 \\ g_1 & L_{11} & L_{12} \\ g_2 & L_{21} & L_{22} \end{vmatrix} = -\left[L_{11}g_2^2 - 2L_{12}g_1g_2 + L_{22}g_1^2\right] \text{ is the negative of the last bracketed}$$

sum. Therefore, we can state that $d^2U > 0$ ($d^2U < 0$), or positive (negative) definite, if and only if $|\mathbf{H}| < 0$ ($|\mathbf{H}| > 0$).

The matrix corresponding to the determinant $|\mathbf{H}|$ is written as $\mathbf{H} = \begin{bmatrix} 0 & g_1 & g_2 \\ g_1 & L_{11} & L_{12} \\ g_2 & L_{21} & L_{22} \end{bmatrix}$ and

is called a bordered Hessian matrix, because its borders (the top and the left) are the partial derivatives of the constraint with respect to each independent variable and its constant. The determinant of $\mathbf{H}$ above, $|\mathbf{H}|$, is called a bordered Hessian determinant (or, simply, a bordered Hessian).

Let us now generalize the above results to the case of an objective function with n independent variables, $U = f(x_1, x_2, \ldots, x_n)$, subject to one constraint of the same variables, $g(x_1, x_2, \ldots, x_n) = c$. The associated Lagrangian function can be written as $L = f(x_1, x_2, \ldots, x_n) + \lambda[c - g(x_1, x_2, \ldots, x_n)]$. If one follows a similar line of argument as the one we used above, one can also derive the quadratic form in the present case of the n-variable function. Once again the sign of the present quadratic form is dependent upon the associated bordered Hessian. The associated bordered Hessian matrix and the bordered Hessian, respectively, are

$$\mathbf{H} = \begin{bmatrix} 0 & g_1 & g_2 & \cdots & g_n \\ g_1 & L_{11} & L_{12} & \cdots & L_{1n} \\ g_2 & L_{21} & L_{22} & \cdots & L_{2n} \\ \cdots & \cdots & \cdots & \cdots & \cdots \\ g_n & L_{n1} & L_{n2} & \cdots & L_{nn} \end{bmatrix} \text{ and } |\mathbf{H}| = \begin{vmatrix} 0 & g_1 & g_2 & \cdots & g_n \\ g_1 & L_{11} & L_{12} & \cdots & L_{1n} \\ g_2 & L_{21} & L_{22} & \cdots & L_{2n} \\ \cdots & \cdots & \cdots & \cdots & \cdots \\ g_n & L_{n1} & L_{n2} & \cdots & L_{nn} \end{vmatrix} \qquad (2.8.18)$$

Notice that, as in the case of discriminants and Hessians, we also have *bordered principal minors* for the bordered Hessians. In our general case the *second bordered principal minor*, the *third bordered principal minor*, . . ., and the *nth bordered principal minor*, respectively, are

$$|\mathbf{H}_2| = \begin{vmatrix} 0 & g_1 & g_2 \\ g_1 & L_{11} & L_{12} \\ g_2 & L_{21} & L_{22} \end{vmatrix}, |\mathbf{H}_3| = \begin{vmatrix} 0 & g_1 & g_2 & g_3 \\ g_1 & L_{11} & L_{12} & L_{13} \\ g_2 & L_{21} & L_{22} & L_{23} \\ g_3 & L_{31} & L_{32} & L_{33} \end{vmatrix}, \ldots,$$

$$\text{and } |\mathbf{H}| = |\mathbf{H}_n| = \begin{vmatrix} 0 & g_1 & g_2 & \cdots & g_n \\ g_1 & L_{11} & L_{12} & \cdots & L_{1n} \\ g_2 & L_{21} & L_{22} & \cdots & L_{2n} \\ \cdots & \cdots & \cdots & \cdots & \cdots \\ g_n & L_{n1} & L_{n2} & \cdots & L_{nn} \end{vmatrix} \qquad (2.8.19)$$

Having explained the meaning and derivation of $|\mathbf{H}|$ and having discussed the bordered principal minors of $|\mathbf{H}|$, we are now ready to state the SOCs for the sign definiteness of d^2U. One can show that d^2U is positive definite (then the objective function, U, has a minimum), given $dg = 0$, if and only if $|\mathbf{H}_2|, |\mathbf{H}_3|, |\mathbf{H}_4|, \ldots, |\mathbf{H}|$ $(= |\mathbf{H}_n|) < 0$; that is, if all the bordered principal minors are less than zero. Similarly, d^2U is negative definite (then the objective function, U, has a maximum), given $dg = 0$, if and only if $|\mathbf{H}_2| > 0$, $|\mathbf{H}_3| < 0$, $|\mathbf{H}_4| > 0, \ldots, (-1)^n |\mathbf{H}| (= (-1)^n |\mathbf{H}_n|) > 0$; that is, if the bordered principal minors alternate in sign.

As an example, let us consider an objective function with two independent variables and a constraint. Suppose the objective function is given by $U = f(x_1, x_2) = x_1^{1/2} + x_2^{1/2}$ and the constraint is given by $2x_1 + 4x_2 = 40$. Then we construct the Lagrangian function as

$$L = f(x_1, x_2) + \lambda[c - g(x_1, x_2)] = x_1^{1/2} + x_2^{1/2} + \lambda[40 - 2x_1 - 4x_2] \tag{2.8.20}$$

We know that, given the Lagrangian function, the associated quadratic form is given by

$$d^2U = L_{11}dx_1^2 + L_{22}dx_2^2 + L_{12}dx_1dx_2 + L_{21}dx_2dx_1 \tag{2.8.21}$$

The Hessian matrix and the bordered Hessian associated with the quadratic form in equation (2.8.21) can be written, respectively, as $\mathbf{H} = \begin{bmatrix} 0 & g_1 & g_2 \\ g_1 & L_{11} & L_{12} \\ g_2 & L_{21} & L_{22} \end{bmatrix}$ and $|\mathbf{H}| = |\mathbf{H}_2| = \begin{vmatrix} 0 & g_1 & g_2 \\ g_1 & L_{11} & L_{12} \\ g_2 & L_{21} & L_{22} \end{vmatrix}$. We now need to check the sign of $|\mathbf{H}_2|$. To do this, we substitute the values (we will show in Section 4.4.5 how we obtain them) $L_{11} = [-(13.33)^{-3/2}]/4$, $L_{22} = [-(3.33)^{-3/2}]/4$, $L_{12} = 0$, $L_{21} = 0$, $g_1 = 2$, and $g_2 = 4$ into $|\mathbf{H}_2|$. Then, upon evaluating this determinant we obtain $|\mathbf{H}_2| = 0.25 > 0$. Since the bordered principal minor $|\mathbf{H}| = |\mathbf{H}_2|$ is positive, we conclude that the quadratic form is negative definite. This suggests that the objective function U has a maximum.

As another example, consider the function $C = h(x_1, x_2) = q_1^2 + q_2^2 - q_1q_2$ with the constraint $q_1 + q_2 = 10$. Then we construct the Lagrangian function as

$$L = h(q_1, q_2) + \lambda[c - g(q_1, q_2)] = q_1^2 + q_2^2 + \lambda[10 - q_1 - q_2] \tag{2.8.22}$$

We know that, given the Lagrangian function, the associated quadratic form is given by

$$d^2C = L_{11}dq_1^2 + L_{22}dq_2^2 + L_{12}dq_1dq_2 + L_{21}dq_2dq_1 \tag{2.8.23}$$

We can find the Hessian matrix and the bordered Hessian associated with the quadratic form in equation (2.8.23). We then need to check, as above, the sign of the bordered Hessian $|\mathbf{H}_2|$. To do this, we substitute the values (we will show in Section 4.4.5 how we obtain them) $L_{11} = 2$, $L_{22} = 2$, $L_{12} = 0$, $L_{21} = 0$, $g_1 = 1$, and $g_2 = 1$ into $|\mathbf{H}_2|$. Then, upon evaluating this determinant we obtain $|\mathbf{H}_2| = -4 < 0$. Since the bordered principal minor $|\mathbf{H}| = |\mathbf{H}_2|$ is negative, we conclude that the quadratic form is positive definite. This suggests that the objective function C has a minimum.

Some of the constrained optimization problems may contain more than one constraint. How do we know that the SOC for an optimum of the objective functions in such problems is satisfied or not? We take up this problem here. Suppose that we have an objective function with n independent variables as $U = f(x_1, x_2, x_3, \ldots, x_n)$. Also suppose that there are

m constraints (such that $m < n$) with the same n variables given by $g^i(x_1, x_2, x_3, \ldots, x_n) = c_i$, where $i = 1, 2, 3, \ldots, m$. Then the Lagrangian function can be written as

$$L = f(x_1, x_2, x_3, \ldots, x_n) + \sum_{i=1}^{m} \lambda_i [c_i - g^i(x_1, x_2, x_3, \ldots, x_n)] \tag{2.8.24}$$

From equation (2.8.24), we can obtain the second partial derivatives as L_{ij}, where $i = 1, 2, 3, \ldots, m$ and $j = 1, 2, 3, \ldots, n$. From the constraints we can get the partial derivatives of each independent variable (that is, g_j^i, where $i = 1, 2, 3, \ldots, m$ and $j = 1, 2, 3, \ldots, n$). These partial derivatives can be used to write the quadratic form of the second total differential, $d^2 U$. This quadratic form can be expressed as an equation, as we did before, involving matrices. The matrix in the middle of these matrices will be the bordered Hessian matrix. When we take the determinant of the bordered Hessian matrix, exactly as we did before, we obtain the bordered Hessian. As can be seen widely in the literature, this bordered Hessian of the n-variable, m-constraint optimization problem can be given as in the following equation:

$$|\mathbf{H}| = |\mathbf{H}_n| = \begin{vmatrix} 0 & 0 & 0 & \cdots & 0 & g_1^1 & g_2^1 & g_3^1 & \cdots & g_n^1 \\ 0 & 0 & 0 & \cdots & 0 & g_1^2 & g_2^2 & g_3^2 & \cdots & g_n^2 \\ 0 & 0 & 0 & \cdots & 0 & g_1^3 & g_2^3 & g_3^3 & \cdots & g_n^3 \\ \cdots & \cdots & \cdots & \cdots & \cdots & \cdots & \cdots & \cdots & & \cdots \\ 0 & 0 & 0 & \cdots & 0 & g_1^m & g_2^m & g_3^m & \cdots & g_n^m \\ g_1^1 & g_1^2 & g_1^3 & \cdots & g_1^m & L_{11} & L_{12} & L_{13} & \cdots & L_{1n} \\ g_2^1 & g_2^2 & g_2^3 & \cdots & g_2^m & L_{21} & L_{22} & L_{23} & \cdots & L_{2n} \\ g_3^1 & g_3^2 & g_3^3 & \cdots & g_3^m & L_{31} & L_{32} & L_{33} & \cdots & L_{3n} \\ \cdots & \cdots & \cdots & \cdots & \cdots & \cdots & \cdots & \cdots & & \cdots \\ g_n^1 & g_n^2 & g_n^3 & \cdots & g_n^m & L_{n1} & L_{n2} & L_{n3} & \cdots & L_{nn} \end{vmatrix} \tag{2.8.25}$$

Notice that we can construct $(n - m)$ bordered principal minors, from the above $|\mathbf{H}| = |\mathbf{H}_n|$, depending upon n and m. If $n = 2$ and $m = 1$, $n = 3$ and $m = 2$, and so on, we have

$$|\mathbf{H}_{m+1}| = |\mathbf{H}_2| = \begin{vmatrix} 0 & g_1^1 & g_2^1 \\ g_1^1 & L_{11} & L_{12} \\ g_2^1 & L_{21} & L_{22} \end{vmatrix}, |\mathbf{H}_{m+1}| = |\mathbf{H}_3| = \begin{vmatrix} 0 & 0 & g_1^1 & g_2^1 & g_3^1 \\ 0 & 0 & g_1^2 & g_2^2 & g_3^2 \\ g_1^1 & g_1^2 & L_{11} & L_{12} & L_{13} \\ g_2^1 & g_2^2 & L_{21} & L_{22} & L_{23} \\ g_3^1 & g_3^2 & L_{31} & L_{32} & L_{33} \end{vmatrix} \tag{2.8.26}$$

and so on, respectively. Notice that the number of rows (which is equal to the number of columns) of the null square matrix inside $|\mathbf{H}|$ is equal to the number of constraints (m) in the problem. Therefore, if $m = 2$ we have $\mathbf{0}_{2 \times 2}$ inside $|\mathbf{H}|$; if $m = 3$ we have $\mathbf{0}_{3 \times 3}$ inside $|\mathbf{H}|$; and so on.

Our next task in this n-variable, m-constraint optimization problem is to check the signs of the $(n - m)$, from the largest to the smallest, bordered principal minors as the SOC for U to have an optimum. We state that $d^2 U$ is positive definite (so that U will have a minimum) if all the bordered principal minors take the same sign; that is, if they take the sign as the sign of $(-1)^m$. If these bordered principal minors alternate in sign, starting with the sign of $|\mathbf{\bar{H}}_{m+1}|$ equals the sign of $(-1)^{m+1}$, then d^2U is said to be negative definite (so that U will have a maximum).

2.8.4 Characteristic equations, characteristic roots, and characteristic vectors

In Section 2.8.2 we used the discriminant $\mathbf{D_R}$ and its principal minors to test the sign definiteness of a quadratic form such as $d^2U = \mathbf{x^T D x}$. In Section 2.8.3 we applied the bordered Hessian $|\mathbf{\bar{H}}|$ and its bordered principal minors to test the sign definiteness of a quadratic form such as $d^2U = \mathbf{x^T \bar{H} x}$.

Suppose now that we still have a quadratic form as before: $d^2U = \mathbf{x^T D x}$. We can use another method to test its sign definiteness. This method uses the concept of the *characteristic roots* of the matrix $\mathbf{D}$, where $\mathbf{D}$ is the matrix of second partial derivatives. We know that $\mathbf{D}$ is a square matrix. The question now is: can we find, given $\mathbf{D}$, a vector $\mathbf{v}$ ($\mathbf{v} \neq 0$) and a scalar v such that $\mathbf{D}_{n \times n}\mathbf{v}_{n \times 1} = v\mathbf{v}_{n \times 1}$? If yes, then the vector $\mathbf{v}$ is called the *characteristic vector* (or *latent vector* or *eigenvector*) of the matrix $\mathbf{D}$; and the constant v is called the *characteristic root* (or *latent root* or *eigenvalue*) of the matrix $\mathbf{D}$.

Notice that the matrix equation $\mathbf{D}_{n \times n}\mathbf{v}_{n \times 1} = v\mathbf{v}_{n \times 1}$ can be written as $\mathbf{D}_{n \times n}\mathbf{v}_{n \times 1} - v\mathbf{I}_{n \times n}\mathbf{v}_{n \times 1} = \mathbf{0}_{n \times 1}$ or as $[\mathbf{D} \quad -v\mathbf{I}]\,\mathbf{v} = 0$, where $[\mathbf{D} \quad -v\mathbf{I}]$ is called the *characteristic matrix* of $\mathbf{D}$. Also notice that the determinant of this characteristic matrix must be a singular matrix. Since the previous matrix equation represents a homogeneous SSLEs, the condition for a nontrivial solution to the system is that the characteristic matrix must be singular (that is, $[\mathbf{D} \quad -v\mathbf{I}]$ must be singular or its inverse should not exist). Again, the condition for singularity of the matrix $[\mathbf{D} \quad -v\mathbf{I}]$ is that its determinant must be zero (in other words, its determinant must vanish). Therefore, we have the equation

$$|\mathbf{D} \quad -v\mathbf{I}| = \begin{vmatrix} a_{11} - v & a_{12} & a_{13} & \ldots & a_{1n} \\ a_{21} & a_{22} - v & a_{23} & \ldots & a_{2n} \\ a_{31} & a_{32} & a_{33} - v & \ldots & a_{3n} \\ \ldots & \ldots & \ldots & \ldots & \ldots \\ a_{n1} & a_{n2} & a_{n3} & \ldots & a_{nn} - v \end{vmatrix} = 0 \qquad (2.8.27)$$

This determinantal equation is called the *characteristic equation* (or the *characteristic polynomial*) of the matrix $\mathbf{D}$. Notice that equation (2.8.27) represents an nth-degree polynomial equation because we can obtain from it an nth-degree polynomial equation using Laplace expansion. This implies that there will be n roots for (or solutions to) this nth-degree polynomial equation. We denote these roots by $v_1, v_2, v_3, \ldots, v_n$.

As a specific case, suppose that $\mathbf{D}_{2 \times 2} = \begin{bmatrix} a_{11} & a_{12} \\ a_{21} & a_{22} \end{bmatrix}$. Then the characteristic equation associated with this matrix, following equation (2.8.27), can be written as $|\mathbf{D} \quad -v\mathbf{I}| = \begin{vmatrix} a_{11} - v & a_{12} \\ a_{21} & a_{22} - v \end{vmatrix} = v^2 - (a_{11} + a_{22})v + (a_{11}a_{22} - a_{12}a_{21}) = 0$, which is a second-degree polynomial and can be solved to obtain the characteristic roots or eigenvalues using the quadratic formula in equation (1.6.9) as $v_1, v_2 = [(a_{11} + a_{22}) \pm \sqrt{(a_{11} + a_{22})^2 - 4(a_{11}a_{22} - a_{12}a_{21})}]/2$. Notice that the eigenvalues v_1 and v_2 are real if $(a_{11} + a_{22})^2 \geq 4(a_{11}a_{22} - a_{12}a_{21})$ or if $(a_{11} - a_{22})^2 + 4a_{12}a_{21} \geq 0$. In other words, the eigenvalues

v_1 and v_2 are real if $\mathbf{D}_{2\times2}$ is a symmetric matrix in which case $a_{12} = a_{21}$. Suppose that the eigenvalues are v_1 and v_2. Then the expression $v^2 - (a_{11} + a_{22})v + (a_{11}a_{22} - a_{12}a_{21})$ can be written as $v^2 - (a_{11} + a_{22})v + (a_{11}a_{22} - a_{12}a_{21}) = (v - v_1)(v - v_2) = v^2 - (v_1 + v_2)v + v_1 v_2$. This implies that $v_1 + v_2 = a_{11} + a_{22}$, or that the sum of the eigenvalues is equal to the sum of the diagonal elements of the matrix $\mathbf{D}$ (also referred to as the *trace* of the matrix $\mathbf{D}$, denoted by tr($\mathbf{D}$)). It also implies that $v_1 \times v_2 = a_{11} \times a_{22} - a_{12} \times a_{21} = |\mathbf{D}|$, or the product of the eigenvalues is equal to the determinant of the matrix. Then we can infer that both eigenvalues, v_1 and v_2, are positive if and only if tr($\mathbf{D}$) > 0 and $|\mathbf{D}| > 0$; both eigenvalues are negative if and only if tr($\mathbf{D}$) < 0 and $|\mathbf{D}| > 0$; the two eigenvalues possess different signs if and only if $|\mathbf{D}| < 0$; and one of the two eigenvalues will be zero if $|\mathbf{D}| = 0$ and the other will be equal to $a_{11} + a_{22}$. One can extend this procedure to determine the signs of the eigenvalues of the matrix $\mathbf{D}$ with any order.

We are now ready to state the conditions for the sign definiteness of the quadratic form $d^2U = \mathbf{x}^T\mathbf{D}\mathbf{x}$. Our above discussion shows that the sign of this quadratic form is determined by the signs of the eigenvalues of the matrix $\mathbf{D}$ if it is a symmetric matrix (as in all our examples). Therefore, d^2U is negative definite (so that U will have a maximum) if all the eigenvalues (v_i, where $i = 1, 2, \ldots, n$) are negative. Similarly, d^2U is positive definite (so that U will have a minimum) if all the eigenvalues are positive. d^2U is negative semidefinite or positive semidefinite if all the eigenvalues are nonpositive (and at least one is equal to zero) and if all the eigenvalues are nonnegative (and at least one is equal to zero), respectively. If some eigenvalues are positive and others are negative, then d^2U will be indefinite.

As an example, assume that $\mathbf{D}$ in the quadratic form $d^2U = \mathbf{x}^T\mathbf{D}\mathbf{x}$ and its associated characteristic matrix are given, respectively, by $\mathbf{D}_{2\times2} = \begin{bmatrix} 2 & 1 \\ 1 & 2 \end{bmatrix}$ and $[\mathbf{D} - v\mathbf{I}] = \begin{bmatrix} 2-v & 1 \\ 1 & 2-v \end{bmatrix}$. Then the characteristic equation is $|\mathbf{D} - v\mathbf{I}| = (2-v)(2-v) - 1 = 4 - 4v + v^2 - 1 = v^2 - 4v + 3 = 0$. This equation can be written as $(v - 3)(v - 1) = 0$. Therefore, the two roots of this quadratic equation are $v_1 = 3$ and $v_2 = 1$. This means that the eigenvalues of the characteristic equation $|\mathbf{D} - v\mathbf{I}|$, or of $\mathbf{D}$, are $v_1 = 3$ and $v_2 = 1$. This suggests that the quadratic form $d^2U = \mathbf{x}^T\mathbf{D}\mathbf{x}$ with the given $\mathbf{D}$, following the above conditions, is positive definite.

As another example, assume now that $\mathbf{D}$ in the quadratic form $d^2U = \mathbf{x}^T\mathbf{D}\mathbf{x}$ and its associated characteristic matrix are given, respectively, by $\mathbf{D}_{2\times2} = \begin{bmatrix} -2 & 1 \\ 1 & -2 \end{bmatrix}$ and $[\mathbf{D} - v\mathbf{I}] = \begin{bmatrix} -2-v & 1 \\ 1 & -2-v \end{bmatrix}$. Then the characteristic equation is given by $|\mathbf{D} - v\mathbf{I}| = (-2-v)(-2-v) - 1 = 4 + 4v + v^2 - 1 = v^2 + 4v + 3 = 0$. This equation can be written as $(v + 3)(v + 1) = 0$. Therefore, the two roots of this quadratic equation are $v_1 = -3$ and $v_2 = -1$. This means that the eigenvalues of the characteristic equation $|\mathbf{D} - v\mathbf{I}|$, or of $\mathbf{D}$, are $v_1 = -3$ and $v_2 = -1$. This suggests that, the quadratic form $d^2U = \mathbf{x}^T\mathbf{D}\mathbf{x}$ with the given $\mathbf{D}$, is negative definite.

Let us now consider the derivation of eigenvectors. Notice that when we substitute one of the above roots (say, v_i) into the equation $[\mathbf{D} - v\mathbf{I}]\mathbf{v} = \mathbf{0}$, we will obtain a particular vector (say, $\mathbf{v}_i$). Since the system $[\mathbf{D} - v\mathbf{I}]\mathbf{v} = \mathbf{0}$ is homogeneous, we will, in fact, obtain a set with as many vectors as the degree of the polynomial (each vector obtained will correspond to a particular root). This implies that there will be n eigenvectors ($\mathbf{v_1}, \mathbf{v_2}, \ldots, \mathbf{v_n}$) in the set if there are (or corresponding to each of the) n roots or eigenvalues ($v_1, v_2, v_3, \ldots, v_n$).

We stated that when we use a particular eigenvalue (v_i) we obtained a corresponding eigenvector, and we denoted it by $\mathbf{v}_i$. But, to do this we follow a procedure called *normalization*. Normalization here is nothing but making the sum of squares of all the eigenvalues in a particular eigenvector $\mathbf{v}_i$ equal to unity; that is, $\sum_{i=1}^{n} v_i^2 = 1$.

Consider, for example, the eigenvalues in our first example above: $v_1 = 3$ and $v_2 = 1$. When we use the first eigenvalue, the characteristic equation $[\mathbf{D} - v\mathbf{I}]\ \mathbf{v} = \mathbf{0}$ becomes

$$\begin{bmatrix} 2-3 & 1 \\ 1 & 2-3 \end{bmatrix} \begin{bmatrix} v_1 \\ v_2 \end{bmatrix} = \begin{bmatrix} -1 & 1 \\ 1 & -1 \end{bmatrix} \begin{bmatrix} v_1 \\ v_2 \end{bmatrix} = \mathbf{0} = \begin{bmatrix} 0 \\ 0 \end{bmatrix}.$$ Notice that the rows and columns

of the matrix $[\mathbf{D} - v\mathbf{I}]$ are linearly dependent yielding an infinite number of solutions. Therefore, we choose one of these equations, $-1v_1 + v_2 = 0$, which gives $v_1 = v_2$. Then, normalizing $v_1 = v_2$ yields $v_1^2 + v_2^2 = v_1^2 + v_1^2 = 1$. This means that $2v_1^2 = 1$ or $v_1^2 = 1/2$, or $v_1 = \sqrt{0.5}$ if we take the positive square root only. Since $v_1 = v_2$, $v_2 = \sqrt{0.5}$. Therefore,

the eigenvector using the eigenvalue $v_1 = 3$ is $\mathbf{v_1} = \begin{bmatrix} \sqrt{0.5} \\ \sqrt{0.5} \end{bmatrix}$. When we use the second

eigenvalue, the characteristic equation $[\mathbf{D} - v\mathbf{I}]\mathbf{v} = \mathbf{0}$ becomes $\begin{bmatrix} 2-1 & 1 \\ 1 & 2-1 \end{bmatrix} \begin{bmatrix} v_1 \\ v_2 \end{bmatrix} =$

$\begin{bmatrix} 1 & 1 \\ 1 & 1 \end{bmatrix} \begin{bmatrix} v_1 \\ v_2 \end{bmatrix} = \mathbf{0} = \begin{bmatrix} 0 \\ 0 \end{bmatrix}$. Notice that, as above, the rows and columns of the matrix

$[\mathbf{D} - v\mathbf{I}]$ are linearly dependent. Therefore, we choose one of these equations: $v_1 + v_2 = 0$. This gives us $v_1 = -v_2$. Then normalizing $v_1 = -v_2$ yields $v_1^2 + (v_2)^2 = v_1^2 + (-v_1)^2 = 1$ or $v_1^2 + v_1^2 = 2v_1^2 = 1$, or $v_1^2 = 1/2 = 0.5$, which yields $v_1 = \sqrt{0.5}$ (as before, we take the positive square root only). Since $v_2 = -v_1$, $v_2 = -v_1 = -\sqrt{0.5}$. Therefore, the eigenvector

using the eigenvalue $v_2 = 1$ is $\mathbf{v_2} = \begin{bmatrix} \sqrt{0.5} \\ -\sqrt{0.5} \end{bmatrix}$.

2.8.5 Exercises

1. Find the Jacobian matrices and the Jacobians associated with the following SSFs. Use these Jacobians to test for functional dependence in each of these SSFs.

(i) $\begin{aligned} y_1 &= f^1 = 2x_1 + x_2 \\ y_2 &= f^2 = 4x_1^2 + 4x_1x_2 + x_2^2 \end{aligned}$; (ii) $\begin{aligned} y_1 &= f^1 = x_1 + x_2 \\ y_2 &= f^2 = 4x_1 + 5x_2 \end{aligned}$;

(iii) $\begin{aligned} y_1 &= f^1 = x_1 + x_2 \\ y_2 &= f^2 = x_1^4 + 4x_1^3x_2 + 6x_1^2x_2^2 + 4x_1x_2^3 + x_2^4 \end{aligned}$; (iv) $\begin{aligned} y_1 &= f^1 = x_1^3 - 3x_1^2x_2 - x_2^3 \\ y_2 &= f^2 = x_1 - x_2 \end{aligned}$;

(v) $\begin{aligned} y_1 &= f^1 = x_1 + x_2 \\ y_2 &= f^2 = x_1^3 + 3x_1^2x_2 + 3x_1x_2^2 + x_2^3 \\ y_3 &= f^3 = x_1^2 + 2x_1x_2 + x_2^2 \end{aligned}$; (vi) $\begin{aligned} y_1 &= f^1 = x_1 + x_2 \\ y_2 &= f^2 = 3x_1 + 2x_2 \\ y_3 &= f^3 = 5x_1 + 3x_2 \end{aligned}$.

2. Find the discriminants of the following quadratic forms. Use these discriminants to test the sign definiteness of the quadratic forms.
(i) $d^2U = -2dx_1^2 + 3dx_1dx_2 - 6dx_2^2$; (ii) $d^2U = 2dx_1^2 - 3dx_1dx_2 + 6dx_2^2$; (iii) $d^2U = 2dx_1^2 + 3dx_2^2 + 6dx_3^2 + dx_1dx_2 + 2dx_1dx_3 + 6dx_2dx_3$; (iv) $d^2U = -2dx_1^2 - 3dx_2^2 - 6dx_3^2 + dx_1dx_2 + 2dx_1dx_3 + 6dx_2dx_3$.

3. Given the following quadratic forms, construct the bordered Hessian matrices and bordered Hessians. Use these bordered Hessians to test the sign definiteness of the quadratic forms.

(i) $d^2U = 2dx_1^2 + 2dx_2^2 + dx_1dx_2 + dx_2dx_1, g_1 = g_2 = 2$; (ii) $d^2U = 0.5dx_1^2 + 0.5dx_2^2 + dx_1dx_2 + dx_2dx_1, g_1 = g_2 = 2$; (iii) $d^2U = 5dx_1^2 + 5dx_2^2 + 5dx_3^2 + dx_1dx_2 + dx_2dx_1 + dx_1dx_3 + dx_2dx_3, g_1 = g_2 = g_3 = 2$; (iv) $d^2U = -5dx_1^2 - 5dx_2^2 - 5dx_3^2 + dx_1dx_2 + dx_2dx_1 + dx_1dx_3 + dx_2dx_3, g_1 = g_2 = g_3 = 2$.

4. Given the following quadratic forms, find the characteristic matrices, characteristic equations, characteristic roots, and characteristic vectors. Test the sign definiteness of the quadratic forms using characteristic roots.
   (i) $d^2U = -2dx_1^2 + 2dx_1dx_2 - 2x_2^2$; (ii) $d^2U = 2dx^2 - 2dx_1dx_2 + 2dx_2^2$; (iii) $d^2U = 2dx_1^2 + 2dx_2^2$; (iv) $d^2U = 2dx_1dx_2$; (v) $d^2U = 2dx_1^2 + 3dx_2^2 + 6dx_3^2 + dx_1dx_2 + 2dx_1dx_3 + 6dx_2dx_3$; (vi) $d^2U = -2dx_1^2 - 3dx_2^2 - 6dx_3^2 + dx_1dx_2 + 2dx_1dx_3 + 6dx_2dx_3$.

⚠ **Web supplement: S2.8.6 Mathematica applications**

⚠ **Web supplement: S2.8.7 *Diagonalization of matrices***

⚠ **Web supplement: S2.9 *Application of Linear Algebra*: *Regression Analysis***

⚠ **Web supplement: S2.9.7 Mathematica applications**

⚠ **Web supplement: S2.10 *Application of Linear Algebra*: *Input–Output Analysis***

⚠ **Web supplement: S2.10.7 Mathematica applications**

3 Differential calculus

3.1 Introduction

Students of economics come across a large number of functions such as the functions of *marginal cost, marginal product, marginal revenue, marginal utility, marginal social benefit, marginal social cost, marginal return,* etc. Students of business and finance also encounter some of these or similar functions. One common feature of these functions is that they all involve the word "marginal." Some students wonder what this word marginal means. Although we will deal with *marginal analysis* in greater detail later, we present below a simple example to illustrate the meaning of this word and its importance in the subjects of our interest.

Before we begin the example, we need to be clear about the meaning of a change in a variable. Suppose that y is the dependent variable and x is the independent variable in a function. Also suppose that y changes from y_0 to y_1 in response to a change in x from x_0 to x_1. In the former, the initial value is y_0 and the new value is y_1; in the latter, they are x_0 and x_1, respectively. We now define the change in a variable as the difference between the new value and the initial value of the variable, and it is usually denoted by the Greek letter Δ. This means that the change in y is given by $\Delta y = y_1 - y_0$ and that in x is given by $\Delta x = x_1 - x_0$.

Assume now that a small firm produces a good called x. At present the firm produces 4 units of x. The total profit (y) from producing these 4 units of x is assumed to be $16. Also assume that, expecting an increase in demand for x, the firm decides to increase the production of x by two more units. Then the firm expects the total profit to increase to $36. The question now is: what is the impact on total profit of producing the additional two units of good x? We will answer this question in the next section.

Notice that, in our example above, we assumed that x is indivisible; that is, we treated the quantity produced of the good as a *discrete variable*, a variable that takes only whole numbers over a specific interval on the real line. Suppose now that we treat it as a *continuous variable*, a variable that can take any value over a specific interval on the real line. Then the second question that arises is: what will be the impact on the total profit if the firm increases production by an infinitesimally small amount (rather than one or two discrete units). We will answer this question also in the next section as we need to discuss some additional concepts.

We will see in the next section that the answers to the above two questions are straightforwardly related to the topic called *differential calculus*, one of the most fundamental branches of mathematics. As most of the models and theories in the subjects of our interest make heavy use of the topic of differential calculus, a detailed study of that topic is indispensable for a meaningful understanding of the models and theories that students of economics, business, and finance need to learn. We undertake this task in this chapter beginning with the next section. We first consider differential calculus involving univariate functions, before moving to differential calculus involving multivariate functions.

3.2 Difference quotient, derivatives of univariate functions, and notations

3.2.1 Difference quotient, and secant and tangent lines

Consider our previous example of the profit of a firm that produces the good x. Assume that the firm's profit is given by the univariate function $y = f(x)$. Also assume, for convenience, that this general function takes the particular form $y = f(x) = x^2$ as illustrated by its graph in Figure 3.2.1(A).

Applying the equations for the changes in y and x, we obtain $\Delta y = y_1 - y_0 = 36 - 16 = 20$ and $\Delta x = x_1 - x_0 = 6 - 4 = 2$. We now divide Δy by Δx to obtain $\Delta y/\Delta x = [\{f(x_0 + \Delta x) - f(x_0)\}/\{(x_0 + \Delta x) - x_0\}] = [\$36 - \$16]/[4 + 2 - 4] = \$20/2 = \$10$. This ratio (that is, the ratio of the total change in the dependent variable to the total change in the independent variable) is called the *difference quotient*, or the *average rate of change*, or the *rate of change* in y per unit change in x. Therefore, the average rate of change in the total profit when the firm produces two more units of the good is $10. This is the answer to our first question in the last section.

Notice that, from the last equation, we can find $\Delta y/\Delta x$ if we have $f(x_0), f(x_0 + \Delta x)$, and Δx. Assume now, for example, that $y_0 = f(x_0) = x_0^2$ and $y_1 = f(x_1) = f(x_0 + \Delta x) = (x_0 + \Delta x)^2$. Then we obtain $\Delta y/\Delta x = [\{(x_0 + \Delta x)^2 - (x_0^2)\}/\Delta x] = [x_0^2 + 2x_0\Delta x + (\Delta x)^2 - x_0^2]/\Delta x$. This equation can be simplified to yield

$$\Delta y/\Delta x = [2x_0\Delta x + (\Delta x)^2]/\Delta x = 2x_0 + \Delta x \tag{3.2.1}$$

We can now obtain, by plugging $x_0 = 4$ and $\Delta x = 2$ in equation (3.2.1), $\Delta y/\Delta x = 2 \times 4 + 2 = \$10$. This means that the rate of change of y per unit change in x is 10. Notice that this is exactly the same as the value we obtained above.

We now consider *secant lines* and *tangent lines*. These lines can be illustrated through Figure 3.2.1(B). Assume that we have a point, such as P, on the graph of a function $f(x)$.

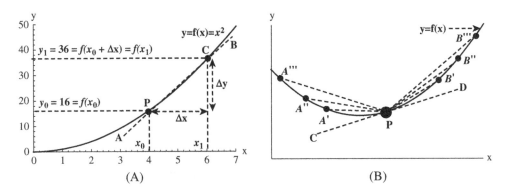

(A)   (B)

Figure 3.2.1

Also assume that we draw few straight lines from the point P rightwards and leftwards. The rightward lines are PB′, PB″, and PB‴, and the leftward lines are PA′, PA″, and PA‴. These lines are called *secant lines*. Secant lines are those lines that *intersect* a curve at one or more points. Notice that, in Figure 3.2.1(B), we have drawn another line CD that touches, but does not intersect, the graph of the function $f(x)$ at point P. This line is called the *tangent line* to the point P. One can draw as many tangent lines as the number of points on the graph of a function.

3.2.2 Derivative, slope, differentiation, and notations

We mentioned in the above example that x was indivisible; that is, we treated the quantity produced of the good as a discrete variable. Suppose now that we treat it as a continuous variable, taking every possible value over an interval on the horizontal real line. Then the question that arises is: what would be the impact on the total profit if the firm continuously reduced (to an *infinitesimally small* amount) the additional quantity produced of the good? In other words, what will be the change in profit if the change in x approaches zero (or, $\Delta x \to 0$), but never equal to zero (or, $\Delta x \neq 0$)? This question can be answered by taking the limit, if it exists, of equation (3.2.1) when $\Delta x \to 0$. Following the discussion of limits presented in Section 1.9, we can write

$$\lim_{\Delta x \to 0}\left[\frac{\Delta y}{\Delta x}\right] = \lim_{\Delta x \to 0}\left[\frac{2x_0\Delta x + (\Delta x)^2}{\Delta x}\right] = \lim_{\Delta x \to 0}[2x_0 + \Delta x] = 2x_0 \qquad (3.2.2)$$

Equation (3.2.2) is an important result in differential calculus. It states that the limit of the average rate of change, or of the difference quotient, when $\Delta x \to 0$ is the *derivative* of the function $y = f(x)$ with respect to x at $x = x_0$. Notice that the derivative is a function of x_0 only and not of Δx, whereas the difference quotient in equation (3.2.1) is a function of both x_0 and Δx. Notice also that equation (3.2.2) is a *derived function*; derived from the original or *primitive function* ($y = f(x)$). This is the reason why equation (3.2.2) is called the derivative. Equation (3.2.2) is also called the *instantaneous rate of change* because of the fact that we treated Δx as infinitesimally small.

Notice that the line AB in Figure 3.2.1(A) is similar to the line PB‴ (or any other secant line) in Figure 3.2.1(B). Therefore, we can use Figure 3.2.1(B) to answer the above question. If we reduce Δx continuously, the rightward sloping secant lines (PB‴, PB″, PB′, etc.) become closer and closer to the tangent CD to the point P on the graph (so do the leftward sloping secant lines). And, in the limit (or, when Δx becomes infinitesimally small) the secant lines coincide with the tangent line. This is illustrated in Figure 3.2.2 for the case of rightward sloping secant lines.

We now explain the meaning of the *slope of a curve*. The slope of a curve at a point (such as P in Figure 3.2.2) is defined as the slope, if it exists, of the tangent line at that point. The slope of a straight line such as CD in Figure 3.2.2 is equal to the ratio of the vertical distance (DE or Δy) to the horizontal distance (EC or Δx); that is, it is equal to DE/EC or $\Delta y/\Delta x$. Since the point P is common to both the graph of the function and the straight line CD, the slope of the graph at P is equal to $\Delta y/\Delta x$ as $\Delta x \to 0$. This implies that the slope of the graph at point P corresponds to the value of the derivative at point P.

We are now ready to answer the question that we posed at the beginning of this section: what is the change in total profit if the change in x approaches zero (or, $\Delta x \to 0$), but never equal to zero (or, $\Delta x \neq 0$)? Using equation (3.2.2) the answer to this question is $2x_0$.

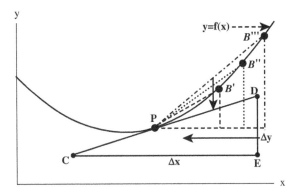

Figure 3.2.2

Moreover, we can see from Figure 3.2.1(A), corresponding to point P, $x_0 = 4$. Substitution of this value in equation (3.2.2) yields $\lim_{\Delta x \to 0} [\Delta y/\Delta x] = 2 \times 4 = \$8$. Therefore, the derivative of the function $y = f(x) = x^2$ when $x = x_0 = 4$ is 8. Similarly, when $x = x_0 = 5$, the derivative of the function is 10; when $x = x_0 = 8$, it is 16; and so on. Notice that the derivative of a function refers to the change in its dependent variable when the independent variable changes *at the margin* (or by an infinitesimally small amount). This is the reason why we used the word "marginal" with the names of some of the functions at the beginning of this chapter. Therefore, the marginal functions we referred to are nothing but the derivatives of (or functions derived from) some functions when their independent variables change at the *margin*.

One can find in the literature different notations that represent the derivative of a function ($f(x)$) with respect to one independent variable (x in our example). We present few of them in the following equation:

$$\frac{dy}{dx} = y' = y'(x) = f' = f'(x) = \lim_{\Delta x \to 0} \left[\frac{\Delta y}{\Delta x} \right] \tag{3.2.3}$$

Using equation (3.2.3) we can write the derivative of the function $y = f(x)$ in our example above when $x = x_0 = 4$ as $\left. \frac{dy}{dx} \right|_{x_0=4} = y'|_{x_0=4} = y'(x)|_{x_0=4} = f'|_{x_0=4} = f'(x)|_{x_0=4} = \lim_{\Delta x \to 0} \left[\frac{\Delta y}{\Delta x} \right]\Big|_{x_0=4} = 2 \times 4 = 8$, where the superscript is read "f prime (or y prime)" or "f dash (or y dash)." Although we can represent the derivative of $y = f(x)$ by the notations in equation (3.2.3), we will mainly use either dy/dx or $f'(x)$ in this book.

Let us now formally define the concepts of the derivative of a function and *differentiation* of a function. The derivative of a function $y = f(x)$ is the function denoted by dy/dx or $f'(x)$ and is given by

$$\frac{dy}{dx} = f'(x) = \lim_{\Delta x \to 0} \left[\frac{f(x + \Delta x) - f(x)}{\Delta x} \right] \tag{3.2.4}$$

given that the limit exists. If we can find dy/dx at, say, $x = x_0$, then the function $y = f(x)$ is said to be *differentiable* at $x = x_0$; and dy/dx is called the derivative of $y = f(x)$ at $x = x_0$ or the derivative of $y = f(x)$ with respect to x at $x = x_0$. The process of finding the derivative is called *differentiation*. Differential calculus deals with differentiation of various types of functions and their applications.

3.2.3 Differentiability of a univariate function

Having demonstrated the meaning of the derivative of a function and differentiation, we now turn our attention to the concept called the *differentiability* of a function. We pose few questions to begin with: are all functions differentiable, or is a function differentiable at every point on its graph? In other words, how does one know if a function is differentiable or not? We will answer these questions below.

Since the differentiability of a function is closely related to the continuity of that function, we recap here our previous discussion of the continuity of a function. In Section 1.8.4 we stated that a function is continuous if its graph does not contain gaps. In Section 1.9.4 we presented the formal definition of continuity of a function. As per this definition, a function, say $y = f(x)$, is said to be continuous at a point $x = x_0$ if: (1) $\lim_{x \to x_0} f(x)$ exists; (2) $f(x_0)$ exists; and (3) $\lim_{x \to x_0} f(x) = f(x_0)$. If these three conditions are satisfied, then the function $y = f(x)$ is said to be continuous at $x = x_0$.

Now consider equation (3.2.4). Notice that the derivative $f'(x)$ in this equation will exist only if the limit exists. But, we know from our discussion of limits in Section 1.9 that for the function $y = f(x)$ to have a limit at $x = x_0$, it must be continuous at $x = x_0$. If $y = f(x)$ is discontinuous at $x = x_0$, then the limit of $y = f(x)$ will not exist. This suggests that for the derivative of $y = f(x)$ to exist at $x = x_0$, $y = f(x)$ must be continuous at $x = x_0$.

However, the *continuity condition* for differentiability of a function is only a *necessary condition* for differentiability. The *sufficient condition* is that the graph of the function must be "smooth" (or the function must be a *smooth function*) at the chosen point. For a function to be a smooth function, its graph must not only be continuous but must also be free of sharp points. Recall that the derivative of a function at a particular point on the graph of the function is equal to the slope of the tangent line drawn to that point. If the graph of the function contains a sharp point, then we cannot define a tangent line to that point implying that the derivative of the function at that point is not defined. Therefore, the condition for the function $y = f(x)$ to be differentiable and to have a derivative at a point ($x = x_0$), it must be a smooth function or its graph must be a smooth graph at that point.

As examples of differentiable and nondifferentiable functions, consider the graphs of various functions in Figures 3.2.3(A)–(F). Figures 3.2.3(A)–(D) illustrate the graphs of two functions each, and Figures 3.2.3(E) and (F) show the graphs of four functions each. The graphs of functions in Figures 3.2.3(A) and (B) are continuous and discontinuous, respectively. But, they have sharp points and gaps, respectively, at $x = x_0$. Therefore, following the condition of differentiability stated above, the functions of these graphs are not differentiable at $x = x_0$. The graphs of the two functions in Figure 3.2.3(C) are smooth and, therefore, are differentiable at $x = x_0$. The graphs of the two functions in Figure 3.2.3(D) and those of the four functions in Figure 3.2.3(E) are continuous but do not possess limits at $x = x_0$. Therefore, these functions are not differentiable at $x = x_0$. Lastly, the graphs of the four functions in Figure 3.2.3(F) are smooth and, therefore, are differentiable at $x = x_0$.

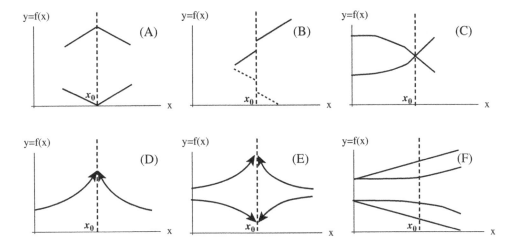

Figure 3.2.3

3.2.4 Application examples

Example 1. Assume that the utility (y) that a consumer obtains from the consumption of different units of a good (x) is given by the function $y = f(x) = 1/x$. Find the marginal utility function using the definition of the derivative. Also find the marginal utility when $x = x_0 = 5$.

Solution. Substituting $y = f(x) = 1/x$ into equation (3.2.4), we obtain

$$
\begin{aligned}
dy/dx &= \lim_{\Delta x \to 0} [\{f(x+\Delta x) - f(x)/\Delta x\}] \\
&= \lim_{\Delta x \to 0} \left[\left\{ \frac{1}{(x+\Delta x)} - \frac{1}{x} \right\} / \Delta x \right] \\
&= \lim_{\Delta x \to 0} [\{-\Delta x / x(x+\Delta x)\} / \Delta x] \\
&= \lim_{\Delta x \to 0} [-\Delta x/(x^2 + \Delta x)\Delta x] = [-\Delta x/(x^2 + \Delta x)\Delta x] \\
&= \lim_{\Delta x \to 0} [-1/x^2 + \Delta x] = -1/x^2
\end{aligned}
$$

Therefore, the marginal utility function is $dy/dx = -1/x^2$. Marginal utility when $x = x_0 = 5$ can be found by substituting $x = x_0 = 5$ into dy/dx. This yields $dy/dx|_{x=x_0=5} = -1/x^2{}_{x=x_0=5} = -1/5^2 = -1/25$. Therefore, marginal utility when $x = x_0 = 5$ is $-1/25$.

Example 2. Assume that the total cost (y) in dollars of producing x units of a good to a firm is given by the function $y = f(x) = 1 + 0.5x$. Find the marginal cost function using the definition of the derivative. Also find the marginal cost when $x = x_0 = 10$.

Solution. Substituting $y = f(x) = 1 + 0.5x$ into equation (3.2.4), we obtain $dy/dx = \lim_{\Delta x \to 0} [\{f(x+\Delta x) - f(x)\}/\Delta x] = \lim_{\Delta x \to 0} [\{(1+0.5(x+\Delta x)) - (1 - 0.5x)\}/\Delta x] = \lim_{\Delta x \to 0} [\{1 + 0.5x$

$+0.5\Delta x - 1 - 0.5x\}/\Delta x] = \lim_{\Delta x \to 0} [0.5\Delta x/\Delta x] = \lim_{\Delta x \to 0} 0.5 = 0.5$. Therefore, the marginal cost function is $dy/dx = 0.5$, a constant function. Notice that marginal cost is the same, \$0.5, irrespective of the value that x takes.

Example 3. Suppose that the total output produced of good y by a firm using l units of labor is given by the function $y = f(l) = 2 + l^2$. Find the marginal product function using the definition of the derivative. Also find the marginal product when $l = l_0 = 20$.

Solution. Substituting $y = f(l) = l + l^2$ into equation (3.2.4), we obtain $dy/dl = \lim_{\Delta l \to 0} [\{f(l + \Delta l) - f(l)\}/\Delta l] = \lim_{\Delta l \to 0} [\{(2 + (l + \Delta l)^2) - (2 + l^2)\}/\Delta l] = \lim_{\Delta l \to 0} [\{(2 + l^2 + 2l\Delta l + (\Delta l)^2) - (2 + l^2)\}/\Delta l] = \lim_{\Delta l \to 0} [\{2l\Delta l + (\Delta l)^2\}/\Delta l] = \lim_{\Delta l \to 0} (2l + \Delta l) = 2l$. Therefore, the marginal product function is $dy/dl = 2l$. The marginal product when $l = l_0 = 20$ is $dy/dl|_{l=l_0=20} = 2l|_{l=l_0=20} = 2 \times 20 = 40$. Therefore, the marginal product when $x = x_0 = 20$ is 40 units.

3.2.5 Exercises

1. Find the difference quotient of each of the following functions:
    (i) $y = f(x) = 5 - x$; (ii) $y = f(x) = x^3$; (iii) $y = f(x) = 2 + x^2$; (iv) $y = f(x) = 1/(1 - x^2)$.
2. Find the derivatives of the following functions, using equation (3.2.4), at $x = x_0 = 2$:
    (i) $y = f(x) = 5 - x$; (ii) $y = f(x) = x^3$; (iii) $y = f(x) = 2 + x^2$; (iv) $y = f(x) = 1/(1 - x^2)$.
3. *Application exercise.* Suppose that the monthly consumption expenditure of a household is given by the function $y = f(x) = 5 + 0.7x$, where y denotes the monthly consumption expenditure and x denotes the monthly disposable income (both in dollars). Find the function for *marginal propensity to consume* (MPC) using equation (3.2.4). Also find the MPC when $x = x_0 = \$3500$.
4. *Application exercise.* Suppose that the monthly savings of a household are given by the function $y = f(x) = 5 + 0.3x$, where y denotes the amount of money saved a month and x denotes the monthly disposable income (both in dollars). Find the function for *marginal propensity to save* (MPS) using equation (3.2.4). Also find the MPS when $x = x_0 = \$3500$.
5. *Application exercise.* Suppose that the total social benefit of a public project is given by the function $y = f(x) = x^3$, where y denotes the monetary value of the social benefit and x denotes the amount of money invested in the project (both in dollars). Find the function for *marginal social benefit* of the project using equation (3.2.4). Also find the marginal social benefit when $x = x_0 = \$5\,000\,000\,000$.

  Web supplement: S3.2.6 Mathematica applications

3.3 Rules of differentiation of univariate functions

In the last section we differentiated few simple functions using the definition of a derivative given in equation (3.2.4). We found that the application of this equation was a tedious job even when the function was a simple one. It is needless to say that finding the derivative of a complicated function will be a difficult task if we use equation (3.2.4). Another problem with this equation is that we need to check the limit every time we differentiate a function.

Therefore, we need a set of routines that can be used straightforwardly to find the derivative of different types of functions. These routines are given by the *rules of differentiation*, which we discuss in the following sections. We use these rules, either individually or in combination, throughout the book. Moreover, these rules lie at the heart of differential calculus. A good grasp of these rules is necessary for a proper understanding of differential calculus.

3.3.1 Differentiation of constant functions or constants

The first rule we discuss is the derivative of a constant function. Suppose that our differentiable function is $y = f(x) = c$, where c is a constant. Then

$$\frac{dy}{dx} = \frac{d[c]}{dx} = f'[c] = 0 \tag{3.3.1}$$

which is popularly called the *constant function rule* of differentiation. The constant function rule of differentiation states that the derivative of a constant or of a constant function is equal to zero.

The rule in equation (3.3.1) can be validated using equation (3.2.4). Substitution of $y = f(x) = c$ into equation (3.2.4) yields $dy/dx = f'(x) = \lim_{\Delta x \to 0} [\{f(x + \Delta x) - f(x)\}\Delta x] = \lim_{\Delta x \to 0} [\{c - c\}/\Delta x] = \lim_{\Delta x \to 0} [0/\Delta x] = \lim_{\Delta x \to 0} 0 = 0$, which checks with equation (3.3.1).

As an example, consider the function $y = f(x) = 10$. Since the power of x on the RHS is zero, $y = f(x) = 10$ is a constant function. Therefore, $dy/dx = f'(10) = 0$.

3.3.2 Differentiation of power functions

Let our differentiable function be $y = f(x) = x^n$. Then

$$\frac{dy}{dx} = \frac{d[y]}{dx} = f'(x) = n.x^{n-1} \tag{3.3.2}$$

which is popularly called the *power function rule* of differentiation. The power function rule states that the derivative of a function raised to a power is equal to the power times the independent variable raised to the power minus one.

The power function rule can be substantiated using equation (3.2.4). Substitution of the function into equation (3.2.4) yields $dy/dx = f'(x) = \lim_{\Delta x \to 0} [\{f(x + \Delta x) - f(x)\}/\Delta x] = \lim_{\Delta x \to 0} [\{(x + \Delta x)^n - x^n\}/\Delta x]$. The first term in the numerator, $(x + \Delta x)^n$, can be expanded using the binomial theorem given in equation (1.10.13) as $(x + \Delta x)^n = C_{n,0} x^n (\Delta x)^0 + C_{n,1} x^{n-1}(\Delta x)^1 + C_{n,2} x^{n-2}(\Delta x)^2 + C_{n,3} x^{n-3}(\Delta x)^3 + \cdots + C_{n,n-n=0} x^{n-n}(\Delta x)^n = x^n + n.x^{n-1} \Delta x + C_{n,2} x^{n-2}(\Delta x)^2 + C_{n,3} x^{n-3}(\Delta x)^3 + \cdots + (\Delta x)^n$. Therefore, $dy/dx = \lim_{\Delta x \to 0} [\{f(x + \Delta x) - f(x)\}/\Delta x] = \lim_{\Delta x \to 0} [\{(x + \Delta x)^n - x^n\}/\Delta x] = \lim_{\Delta x \to 0} [\{x^n + n.x^{n-1} \Delta x + C_{n,2} x^{n-2}(\Delta x)^2 + C_{n,3} x^{n-3}(\Delta x)^3 + \ldots + (\Delta x)^n - x^n\}/\Delta x] = \lim_{\Delta x \to 0} [n.x^{n-1} + C_{n,2} x^{n-2}(\Delta x) + C_{n,3} x^{n-3}(\Delta x)^2 + \cdots + (\Delta x)^{n-1}] = n.x^{n-1}$, which checks with equation (3.3.2).

As an example, consider the function $y = f(x) = x^3$. In this example, $n = 3$. Therefore, applying the rule in equation (3.3.2), we obtain $dy/dx = f'(x) = 3.x^{3-1} = 3x^2$.

3.3.3 Differentiation of functions with a constant

Suppose that our differentiable function is $y = f(x) = c.x^n$, where c is a constant. Then

$$\frac{dy}{dx} = \frac{d[y]}{dx} = f'[x] = c(n.x^{n-1})$$ (3.3.3)

which is popularly called the *function-with-constant rule* of differentiation. This rule states that the derivative of a function with a constant is equal to the constant times the derivative of the function.

As earlier, the rule in equation (3.3.3) can be verified using equation (3.2.4). Substitution of the function into equation (3.2.4) yields $dy/dx = f'(x) = \lim_{\Delta x \to 0} [c.\{f(x + \Delta x) - f(x)\}/\Delta x] = c.[\lim_{\Delta x \to 0} [\{(x + \Delta x)^n - x^n\}/\Delta x]$. Then following exactly the same arguments as in the verification of the derivative of the power function, we obtain $dy/dx = f'[x] = c.n.x^{n-1}$, which checks with equation (3.3.3). As an example, consider the function $y = f(x) = [2/3].x^3$. Then, applying the rule in equation (3.3.3), we obtain $dy/dx = f'(x) = f[(2/3).x^3] = (2/3)[3.x^{3-1}] = 2x^2$. Notice that $c = 2/3$ and $n = 3$ in the present example.

3.3.4 Differentiation of sums or differences (or combinations) of functions

Assume that we have two differentiable functions: $f(x)$ and $g(x)$. Since $f(x)$ and $g(x)$ are assumed to be differentiable, their sum $F(x) = f(x) + g(x)$ and their difference $G(x) = f(x) - g(x)$ are also differentiable. Consider first the sum: $F(x) = f(x) + g(x)$. Then

$$\frac{d[F(x)]}{dx} = F'(x) = \frac{d[f(x) + g(x)]}{dx} = f'(x) + g'(x)$$ (3.3.4)

which is called the *sum rule* of differentiation. The sum rule of differentiation states that the derivative of the sum of two or more functions is the sum of the derivatives of individual functions.

We can validate the rule in equation (3.3.4) using equation (3.2.4). Application of equation (3.2.4) yields $d[F(x)]/dx = F'(x) = \lim_{\Delta x \to 0} [\{F(x + \Delta x) - F(x)\}/\Delta x] = \lim_{\Delta x \to 0} [\{f(x + \Delta x) + g(x + \Delta x) - [f(x) + g(x)]\}/\Delta x] = \lim_{\Delta x \to 0} [\{[f(x + \Delta x) - f(x)] + [g(x + \Delta x) - g(x)]\}/\Delta x] = \lim_{\Delta x \to 0} [\{f(x + \Delta x) - f(x)\}/\Delta x] + \lim_{\Delta x \to 0} [\{g(x + \Delta x) - g(x)\}/\Delta x]$. Since the first term and the second term in the last step are $f'(x)$ and $g'(x)$, respectively, we can write $d[F(x)]/dx = F'(x) = f'(x) + g'(x)$, which checks with equation (3.3.4).

Following exactly the same reasoning as above, one can show that the derivative of the difference of two or more functions is the difference of the derivatives of individual functions; that is

$$\frac{d[G(x)]}{dx} = G'(x) = \frac{d[f(x) - g(x)]}{dx} = f'(x) - g'(x)$$ (3.3.5)

where $G(x) = f(x) - g(x)$. Equation (3.3.5) gives the *difference rule* of differentiation. Notice that the rules in equations (3.3.4) and (3.3.5) can be generalized to the sum and difference, respectively, of any number of functions.

As an example, consider $F(x) = f(x) + g(x)$, where $f(x) = 2 + x^2$ and $g(x) = 1 + x^3$. This means that $F(x) = f(x) + g(x) = 2 + x^2 + 1 + x^3 = 3 + x^2 + x^3$. Therefore, using the rules we have discussed so far, $F'(x) = 2.x^{2-1} + 3.x^{3-1} = F'(x) = 2.x + 3.x^2 = x(2+3x)$. We can check this result by finding the derivatives of $f(x)$ and $g(x)$ individually and summing them up. That is, $f'(x) = 2x$ and $g'(x) = 3x^2$. Therefore, $f'(x) + g'(x) = 2x + 3x^2 = x(2+3x) = F'(x)$. This confirms the rule $d[F(x)]/dx = F'(x) = f'(x) + g'(x)$.

As another example, consider the function $G(x) = f(x) - g(x)$, where $f(x) = 5 + x$ and $g(x) = 1 + x^4$. This implies that $G(x) = f(x) - g(x) = 5 + x - 1 - x^4 = 4 + x - x^4$. Therefore, again using the rules we have discussed so far, $G'(x) = 1.x^{1-1} - 4.x^{4-1} = 1 - 4.x^3 = 1 - 4x^3$. As above, we can check this result by finding the derivatives of $f(x)$ and $g(x)$ individually and subtracting the latter from the former. That is, $f'(x) = 1$ and $g'(x) = 4x^3$. Therefore, $f'(x) + g'(x) = 1 - 4x^3 = F'(x)$. This confirms the rule $d[G(x)]/dx = G'(x) = f'(x) - g'(x)$.

3.3.5 Differentiation of the product of two functions (product rule)

Suppose that we have two functions: $f(x)$ and $g(x)$. We assume that $f(x)$ and $g(x)$ are differentiable. Therefore, their product $H(x) = f(x).g(x)$ is also differentiable. Then

$$\frac{dy}{dx} = H'(x) = \frac{d[f(x) \times g(x)]}{dx} = f'(x).g(x) + f(x).g'(x) \tag{3.3.6}$$

which is called the *product rule* of differentiation. The product rule of differentiation states that the derivative of the product of two functions is equal to the derivative of the first function times the second function plus the derivative of the second function times the first function.

Let us substantiate the product rule using equation (3.2.4). Substitution of $H(x) = f(x).g(x)$ into equation (3.2.4) yields $H'(x) = \lim_{\Delta x \to 0} [\{H(x+\Delta x) - H(x)\}/\Delta x] = \lim_{\Delta x \to 0} [\{f(x+\Delta x).g(x+\Delta x) - f(x).g(x)\}/\Delta x]$. Now adding and subtracting $[f(x).g(x+\Delta x)]/\Delta x$ from the RHS of the last equation yields $H'(x) = \lim_{\Delta x \to 0} [\{f(x+\Delta x).g(x+\Delta x) - f(x).g(x)\}/\Delta x + \{f(x).g(x+\Delta x) - f(x).g(x+\Delta x)\}/\Delta x] = \lim_{\Delta x \to 0} [\{f(x+\Delta x).g(x+\Delta x) - f(x).g(x+\Delta x)\}/\Delta x + \{f(x).g(x+\Delta x) - f(x).g(x)\}/\Delta x] = \lim_{\Delta x \to 0} [\{f(x+\Delta x) - f(x)\}.g(x+\Delta x)\}/\Delta x + \{f(x)[g(x+\Delta x) - g(x)]\}/\Delta x] = \lim_{\Delta x \to 0} [\{[f(x+\Delta x) - f(x)].g(x+\Delta x)\}\Delta x] + \lim_{\Delta x \to 0} [\{f(x)[g(x+\Delta x) - g(x)]\}/\Delta x] = \lim_{\Delta x \to 0} [\{f(x+\Delta x) - f(x)\}/\Delta x] \times \lim_{\Delta x \to 0} g(x+\Delta x) + \lim_{\Delta x \to 0} f(x) \times \lim_{\Delta x \to 0} [\{g(x+\Delta x) - g(x)\}/\Delta x] = f'(x).g(x) + f(x).g'(x)$, which checks with equation (3.3.6).

As an example, consider the functions $H(x) = f(x).g(x)$, where $f(x) = 2 + x^2$ and $g(x) = 1 + x^3$. This means that $H(x) = (2+x^2).(1+x^3)$. Therefore, applying the product rule yields $dy/dx = H'(x) = d[f(x) + g(x)]/dx = f'(x).g(x) + f(x).g'(x) = (2x) \times (1+x^3) + (2+x^2) \times (3x^2)$. After simplification we obtain $dy/dx = H'(x) = 2x + 6x^2 + 5x^4$. This result can be verified by multiplying $f(x) = 2 + x^2$ and $g(x) = 1 + x^3$ and finding the derivative of the resulting expression using the rules we have discussed so far. Multiplication of $(2+x^2)$ by $(1+x^3)$ gives us $2 + x^2 + 2x^3 + x^5$. The derivative of this expression with respect to x is $d(2+x^2+2x^3+x^5)/dx = 2x + 6x^2 + 5x^4$, which checks with the result we obtained through the product rule.

The product rule we outlined above can be extended to three functions of the same independent variable. Let $f(x)$, $g(x)$, and $h(x)$ be three differentiable functions. Then, it can be shown that $d[f(x).g(x).h(x)]/dx = f'(x).g(x).h(x) + f(x).g'(x).h(x) + f(x).g(x).h'(x)$.

3.3.6 *Differentiation of rational functions (quotient rule)*

Assume that we have two differentiable functions: $f(x)$ and $g(x)$, where $g(x) \neq 0$. Since these two functions are assumed to be differentiable, their quotient $K(x) = f(x)/g(x)$ is also differentiable. Then

$$\frac{d[K(x)]}{dx} = K'(x) = \frac{g(x).f'(x) - f(x).g'(x)}{[g(x)]^2} \qquad (3.3.7)$$

which is called the *quotient rule* of differentiation. The quotient rule of differentiation states that the derivative of the quotient of two functions is the denominator times the derivative of the numerator minus the numerator times the derivative of the denominator, all divided by the square of the denominator.

The quotient rule can be verified as follows. Given $K(x) = f(x)/g(x)$, we have $f(x) = K(x).g(x)$. By applying the product rule in equation (3.3.6), we obtain $f'(x) = K'(x).g(x) + K(x).g'(x)$. This equation can be solved for $K'(x)$ to obtain $K'(x) = [f'(x) - K(x).g'(x)]/g(x)$. Since $K(x) = f(x)/g(x)$, we can write $K'(x)$ as $K'(x) = [\{f'(x) - f(x)/g(x)\}.g'(x)]/g(x)$. This equation can be simplified to obtain $K'(x) = [g(x).f'(x) - f(x).g'(x)]/[g(x)]^2$, which checks with equation (3.3.7). As an example, consider the functions $f(x) = 2 + x^2$ and $g(x) = 1 + x^3$. Therefore, we define $K(x) = f(x)/g(x) = (2 + x^2)/(1 + x^3)$. Then, applying equation (5.3.7), we obtain $K'(x) = [g(x).f'(x) - f(x).g'(x)]/[g(x)]^2 = [(1 + x^3).(2x) - (2 + x^2).(3x^2)]/(1 + x^3)^2 = [2x - 6x^2 - x^4]/(1 + x^3)^2$.

3.3.7 *Differentiation of composite functions (chain rule)*

Assume that we have two differentiable functions: $y = f(u)$ and $u = g(x)$. Notice that in the former function, y depends on u; and in the latter function, u depends on x. This means that we have two independent variables. These functions together imply that changes in x cause changes in u which, in turn, will change y.

Now assume that y and u are differentiable functions of u and x, respectively. Then

$$\frac{dy}{dx} = \frac{dy}{du} \times \frac{du}{dx} \qquad (3.3.8)$$

which is called the *chain rule* or the *composite function rule* of differentiation. The chain rule states that the derivative of the first function ($y = f(u)$) with respect to the independent variable in the second function (x in $u = g(x)$) is equal to the derivative of the first function with respect to its independent variable (u) times the derivative of the second function with respect to its independent variable (x).

The chain rule can be validated as follows. Given $y = f(u)$ and $u = g(x)$, we know that $\Delta y/\Delta u = [f(u + \Delta u) - f(u)]/\Delta u$ and $\Delta u/\Delta x = [f(x + \Delta x) - f(x)]/\Delta x$. Therefore, $\Delta y/\Delta x = [\Delta y/\Delta u].[\Delta u/\Delta x] = [f(u + \Delta u) - f(u)/\Delta u].[f(x + \Delta x) - f(x)]/\Delta x$. Now taking limits on both sides of this equation when $\Delta x \to 0$, we obtain $dy/dx = \lim_{\Delta x \to 0}[\Delta y/\Delta x] = (\lim_{\Delta u \to 0}\Delta y/\Delta u).(\lim_{\Delta x \to 0}\Delta u/\Delta x) = (\lim_{\Delta u \to 0}\{f(u + \Delta u) - f(u)\}/\Delta u) \times \lim_{\Delta x \to 0}\{f(x + \Delta x) - f(x)\}/\Delta x = [dy/du].[du/dx]$, which checks with equation (3.3.8).

As an example, consider the functions $y = f(u) = 2 + u^2$ and $u = g(x) = 2x + 3x^2$. Then, by applying the chain rule, we obtain $dy/dx = [dy/du].[du/dx] = (2u).(2 + 6x)$. Since $u = g(x) = 2x + 3x^2$, $dy/dx = [dy/du].[du/dx] = (2u).(2 + 6x) = 2(2x + 3x^2).(2 + 6x) = 2(4x + 12x^2 + 6x^2 + 18x^3) = 8x + 36x^2 + 36x^3$.

Notice that the chain rule can be extended to include any number of functions. Suppose, for example, we have three functions $y = f(u)$, $u = g(v)$, and $v = h(x)$. Then, following exactly the same arguments as those we used in the validation of equation (3.3.8), we can obtain

$$\frac{dy}{dx} = \frac{dy}{du} \times \frac{du}{dv} \times \frac{dv}{dx} \qquad (3.3.9)$$

3.3.8 Differentiation of composite function with power (power rule)

Suppose that we have a differentiable composite function $y = f(u) = u^n$, where $u = g(x)$. Then the derivative of y with respect to x, if y and u are differentiable, is defined as

$$\frac{dy}{dx} = n.u^{n-1} \times \frac{du}{dx} \qquad (3.3.10)$$

which is called the *power rule* of differentiation. Notice that this rule is, in fact, the chain rule applied to the power of a function.

As an example, let $y = f(u) = (2x + 5)^3$. We now define $u = g(x) = 2x + 5$. This implies that we can write $y = f(u) = (2x + 5)^3 = u^3$. Then, by applying equation (3.3.10), we obtain $dy/dx = n.u^{n-1}.du/dx = 3.(2x + 5)^2.(2) = 6(2x + 5)^2$.

3.3.9 Differentiation of logarithmic functions

Assume that our function is a differentiable natural logarithmic function given by $y = f(x) = \ln x$. Then the derivative of $y = (x)$ with respect to x is given by

$$\frac{dy}{dx} = f'(x) = \frac{d[\ln x]}{dx} = \frac{1}{x} \qquad (3.3.11)$$

which is called the *logarithmic rule* of differentiation. This rule implies that the derivative of a logarithmic function of x is the inverse of x.

As before, the logarithmic function rule can be substantiated by applying equation (3.2.4). Substituting the function $y = f(x) = \ln x$ into equation (3.2.4) yields $dy/dx = f'(x) = d(\ln x)/dx = \lim_{\Delta x \to 0} [\{f(x + \Delta x) - f(x)\}/\Delta x] = \lim_{\Delta x \to 0} [\{\ln(x + \Delta x) - \ln x\}/\Delta x]$. Using equation (3.5.6), we can write the last equation as $dy/dx = d[\ln x]/dx = \lim_{\Delta x \to 0} [\ln\{(x + \Delta x)/x\}/\Delta x] = \lim_{\Delta x \to 0} (1/\Delta x).\ln[(x + \Delta x)/x] = \lim_{\Delta x \to 0} (1/\Delta x).\ln[1 + \Delta x/x] = \lim_{\Delta x \to 0} (1/x).(x/\Delta x).\ln[1 + \Delta x/x]$. Using equation (1.5.7), we can write the last equation as $dy/dx = d[\ln x]/dx = \lim_{\Delta x \to 0} [(1/x).\ln[1 + \Delta x/x]^{x/\Delta x} = (1/x). \lim_{\Delta x \to 0} \ln[1 + \Delta x/x]^{x/\Delta x} = (1/x).$ $\ln[\lim_{\Delta x \to 0} [1 + \Delta x/x]^{x/\Delta x}] = (1/x).\ln[\lim_{\Delta x \to 0} (1 + 1/n)^n]$, where $n = x/\Delta x$, which implies that as $\Delta x \to 0$, $n \to \infty$. Therefore, we can write $dy/dx = d[\ln x]/dx = (1/x).\ln[\lim_{n \to \infty} (1 + 1/n)^n]$. Notice that the term $\lim_{n \to \infty} (1 + 1/n)^n$ in the last equation is identical with the term that we found in the derivation of equation (1.10.14). We found in Section 1.10.6 that this term was equal to the value $e = 2.71828$. Therefore, we have $dy/dx = d[\ln x]/dx = (1/x).\ln e$. We know from equation (1.5.4) that $\ln e = 1$, which shows that $dy/dx = d[\ln x]/dx = (1/x)\ln e = (1/x).1 = 1/x$, which checks with equation (3.3.11).

As an example, assume that our logarithmic function is given by $y = \ln x^3$. Notice that we can write it as $y = 3\ln x$. We shall now apply equations (3.3.11) and (3.3.3) to obtain the derivative as $dy/dx = d[3.\ln x]/dx = 3.(1/x) = 3/x$.

Assume now that our function is a *logarithmic composite function* such as $y = f(u) = \ln u$, where $u = g(x)$. Also assume that y and u are differentiable functions. How do we differentiate this function? The answer is now simple: apply equations (3.3.11) and (3.3.8) together; that is, use the logarithmic rule and the chain rule together. The result will be

$$\frac{dy}{dx} = \frac{dy}{du} \times \frac{du}{dx} = \frac{d[\ln u]}{du} \times \frac{du}{dx} = \frac{1}{x} \times \frac{du}{dx} \tag{3.3.12}$$

As an example, assume that $y = f(u) = \ln u$, where $u = g(x) = 2 + x^2$. Then, applying equation (3.3.12) gives us $dy/dx = d[\ln u]/dx = (dy/du).(du/dx) = \{d[\ln u]/du\}.(du/dx) = (1/u).(du/dx) = (2x)/(2 + x^2)$.

Suppose that our function is a logarithmic composite function with base other than e such as $y = f(u) = \log_b u$, where $u = g(x)$. How do we differentiate this function if y and u are differentiable functions? Before we differentiate $y = f(u) = \log_b u$, we convert it into $y = \ln u/\ln b$ using equation (1.5.12). Notice that, in the last equation, $\ln b$ is a constant. This implies that we can write it as $y = (1/\ln b).\ln u$. We can now apply equation (3.3.12) to obtain

$$\frac{dy}{dx} = \frac{d[\log_b u]}{dx} = \frac{d(\ln u/\ln b)}{dx} = \frac{1}{\ln b} \times \frac{d(\ln u)}{dx} . = \frac{1}{\ln b} \times \frac{d(\ln u)}{du} \times \frac{du}{dx} = \frac{1}{\ln b} \times \frac{1}{u} \times \frac{du}{dx} \tag{3.3.13}$$

As an example, consider the function $y = \log_4(2 + x^2)$. Assume that $u = g(x) = 2 + x^2$ and $b = 4$. Notice that we can write $y = \log_4(2 + x^2)$ as $y = \log_4 u$, and $y = \log_4 u$ as $y = \ln u/\ln 4 = (1/\ln 4).\ln u$, where $u = 2 + x^2$. We can now apply $y = (1/\ln 4).\ln u$ to equation (3.3.13) to yield $dy/dx = (1/\ln b).\{d[\ln u]/dx\}.(du/dx) = (1/\ln b).(1/u).(du/dx) = (1/\ln 4).[1/((2 + x^2)].(2x)$.

3.3.10 Differentiation of exponential functions

Suppose that our differentiable exponential function is given by $y = f(x) = e^x$. The derivative of this function with respect to x is given by

$$\frac{dy}{dx} = f'(x) = e^x \tag{3.3.14}$$

which is called the *exponential rule* of differentiation. Notice that derivative of the exponential function $y = f(x) = e^x$ is the function itself.

The exponential rule can be verified as follows. Let us first convert the function $y = f(x) = e^x$ into the form $\ln y = x.\ln e = x$. Differentiating both sides of $\ln y = x$ with respect to x, we obtain $(1/y).(dy/dx) = d[x]/dx = dx/dx = 1$. Rearranging this equation, we get $dy/dx = y$. Since $y = f(x) = e^x$, we obtain $dy/dx = e^x$, which checks with equation (3.3.14).

As an example consider the function $y = f(x) = 5e^x$. Then the derivative of this function with respect to x is given by $dy/dx = f'(x) = 5e^x$.

Suppose that the differentiable exponential function is a composite function such as $y = f(u) = e^u$, where $u = g(x)$. We can find the derivative of $y = f(u)$ with respect to x through the combined application of equations (3.3.14) and (3.3.8). Then, following the above procedure by using u (instead of x), we obtain

$$\frac{dy}{dx} = f'(x) = \frac{d[e^u]}{dx} = e^u \times \frac{du}{dx} \tag{3.3.15}$$

As an example, consider the function $y = f(u) = e^u$, where $u = g(x) = 2 + x^2$. Then, using equation (3.3.15), we obtain the derivative of $y = f(u)$ with respect to x as $dy/dx = f'(x) = d[e^u]/dx = (e^u).(du/dx) = (e^u)(2x) = 2xe^u = 2xe^{2+x^2}$.

So far we have used e as the base of the exponential functions. What will be the derivative of an exponential function if its base is a value other than e; say, for example, a? Then the function will be of the form $y = f(x) = a^x$. Notice that we can write, using equation (1.5.11), $y = f(x) = a^x$ as $y = f(x) = (e^{\ln a})^x = e^{\ln a^x} = e^{x \ln a}$. Then the derivative of the last function with respect to x is given by

$$\frac{dy}{dx} = f'(x) = \frac{d[a^x]}{dx} = \frac{d[e^{x \ln a}]}{dx} = (e^{x \ln a}) \times (\ln a) \times \frac{dx}{dx} = (a^x) \times (\ln a) \tag{3.3.16}$$

Consider, as an example, the function $y = f(x) = 5^x$. We shall now treat $a = 5$. Then the derivative of this function with respect to x, following equation (3.3.16), is $dy/dx = d[5^x]/dx = d[e^{x \ln 5}]/dx = (e^{x \ln 5}).(\ln 5).(dx/dx) = (5^x).(\ln 5)$.

As a last case, what will be the derivative of the differentiable function $y = f(u) = a^u$ with respect to x, where $u = g(x)$? This can be found through the combined application of equations (3.3.16) and equation (3.3.8). That is

$$\frac{dy}{dx} = \frac{d[a^u]}{dx} = \frac{d[e^{u \ln a}]}{dx} = (e^{u \ln a}) \times (\ln a) \times \frac{du}{dx} = (a^u) \times (\ln a) \times \frac{du}{dx} \tag{3.3.17}$$

As an example, consider the function $y = f(x) = 5^u$, where $u = g(x) = x^2$. This implies that $y = f(x) = 5^{x^2}$. We now treat $a = 5$. Therefore, applying equation (3.3.17) yields $dy/dx = d[5^u]/dx = d[e^{u \ln 5}]/dx = (e^{u \ln 5}).(\ln 5).(du/dx) = (5^u).(\ln 5).(2x) = 5^{x^2}.(\ln 5).(2x)$.

3.3.11 Differentiation of inverse functions (the inverse function rule)

Suppose that we have a one-to-one function, $y = f(x)$. We know from Section 1.8.4 that the inverse function of this original function is defined as $x = f^{-1}(y)$. Then the derivative of this inverse function is

$$\frac{dx}{dy} = \frac{1}{dy/dx} = \frac{1}{f'(x)}, \text{ where } dy/dx = f'(x) \neq 0 \tag{3.3.18}$$

which is called the *inverse function rule* of differentiation. Notice that this rule states that the derivative of the inverse function is the inverse of the derivative of the original function.

The inverse function rule can be validated as follows. Our aim is to find dy/dx from $x = f^{-1}(y)$. Since it is difficult to differentiate $x = f^{-1}(y)$ explicitly, we take a shortcut here. The shortcut is to write $x = f^{-1}(y)$ as $x = f(y)$. Then we can differentiate both sides of this function with respect to x. This gives us $d[x]/dx = \{d[f(y)]/dy\}.(dy/dx) = dx/dx = 1$, or $\{d[f(y)]/dy\}.(dy/dx) = 1$, which implies that $d[f(y)]/dy = 1/(dy/dx)$. Since $x = f(y)$, the last result can be written as $dx/dy = 1/(dy/dx) = 1/f'(x)$, which checks with equation (3.3.18). Notice the condition: $dy/dx = f'(x) \neq 0$.

As an example of the application of the inverse function rule, assume that $y = f(x) = 2 + x^2$. From this we obtain $dy/dx = f'(x) = 2x$. Therefore, using equation (3.3.18), we obtain $dx/dy = 1/(dy/dx) = 1/f'(x) = 1/2x$. Notice that $dy/dx = f'(x) \neq 0$ for all values of $x \neq 0$.

3.3.12 Differentiation of implicit functions (the implicit function rule)

Suppose that our implicit function is $y^3 + x^2 - 4 = 0$. Notice that in an implicit function such as $y^3 + x^2 - 4 = 0$, we are unable to say whether y or x is the dependent variable or not. How do we differentiate such a function and find dy/dx? The method of differentiating such functions involves four steps as outlined below.

Firstly, assume that y is a differentiable function of x. Secondly, differentiate both sides of the equation with respect to x. Thirdly, collect all the terms that contain dy/dx, and factor it. Fourthly, solve the result for dy/dx. Let us carry out these operations for the above implicit function to yield $d[y^3]/dx + d[x^2]/dx - d[4]/dx = d[0]/dx$; $3y^2(dy/dx) + 2x(dx/dx) - 0 = 0$; and $3y^2(dy/dx) = -2x$. This equation can be simplified to obtain $dy/dx = -2x/3y^2$.

3.3.13 Logarithmic differentiation

Sometimes one may come across functions that are difficult to be differentiated using the rules stated so far. In such a situation, one can use the method called *logarithmic differentiation*. Logarithmic differentiation is not a rule of differentiation such as the product or the quotient rule, but a technique or method of differentiating complicated functions. However, this technique makes use of the rule of differentiation of logarithmic functions and other rules we discussed earlier.

Suppose that we have a differentiable function $y = f(x)$. Notice that $f(x)$ may be a complicated expression involving products, sums, differences, quotients, exponents, and so on. How do we differentiate such a function and find dy/dx? The procedure is as follows. Firstly, take the natural logarithm of both sides of the function, that is, $\ln y = \ln[f(x)]$, and simplify the resulting expression using the properties of logarithms given in Section 1.5.3. Secondly, differentiate both sides of the result in the last step with respect to x, and then solve for dy/dx.

As an example, suppose that our differentiable function is $y = f(x) = x^{\ln x}$. We know that it is difficult to differentiate this function with respect to x through the direct application of the rules of differentiation we discussed so far. Therefore, we follow the procedure outlined above. Let us first convert the function, using the rules of logarithms, into the form $\ln y = \ln x^{\ln x} = \ln x . \ln x = \ln(x+x) = \ln(2x)$. Notice that we can write this equation as $\ln y = \ln(2x) = \ln u$, where $u = g(x) = 2x$ is assumed to be a differentiable function of x. Differentiation of both sides of this equation with respect to x yields $(1/y).(dy/dx) = (d[\ln u]/du).(du/dx) = (1/u).(du/dx) = (1/2x).2 = (2/2x) = 1/x$. Since $y = x^{\ln x}$, the last result can be restated as $dy/dx = (1/x).y = x^{\ln x}/x = x^{\ln x}.x^{-1} = x^{\ln x - 1}$.

3.3.14 Relative and percentage rates of change

Before we present the application examples of the rules of differentiation, we need to explain two important concepts. We know that the derivative of the function $y = f(x)$, denoted by $dy/dx = f'(x)$, is, in fact, the *instantaneous rate of change* or the *rate of change* of the dependent variable with respect to an infinitesimally small change in the independent variable. Suppose that we divide the derivative (or the derived function) by the original function to obtain the ratio $f'(x)/f(x)$. This ratio is called the *relative rate of change* of $f(x)$. Therefore, we have the equation

$$\text{Relative rate of change of } f(x) = f'(x)/f(x) \tag{3.3.19}$$

If we multiply the relative rate of change by 100, we obtain the *percentage rate of change* of $f(x)$. Therefore, we have the equation

$$\text{Percentage rate of change of } f(x) = [f'(x)/f(x)] \times 100 \tag{3.3.20}$$

3.3.15 Application examples

Example 1. Assume that the inverse demand function for a firm's product is given by $p = f(q) = 25/q$, where p denotes the per unit price in dollars and q denotes the quantity of the good demanded. Find the rate of change of price with respect to quantity. How fast will the price change when $q = 10$?

Solution. We can find the rate of change of price with respect to quantity by differentiating the function with respect to q. Therefore, applying the quotient rule yields $dp/dq = \{[q.d(25)/dq] - 25.d(q)/dq\}/(q)^2 = (0 - 25)/(q)^2 = -25/(q)^2$. This means that the rate of change of price with respect to quantity demanded is $dp/dq = -25/(q)^2$. This implies that when $q = 10$, $dp/dq|_{q=10} = -25/(q)^2|_{q=10} = -125/10^2 = -25/100 = -0.25$. This suggests that an increase of one unit of the good demanded when $q = 10$ reduces the price by 25 cents. Notice that, to have any economic sense out of this result, we have to interpret it oppositely.

Example 2. Suppose that a company's total revenue is given by the function $R = f(q) = 50q - 0.2q^2$, where R denotes the total revenue in dollars and q denotes the total quantity of the good sold by the company. Find the marginal revenue when $q = 20$. Also find the percentage rate of change of revenue with respect to the quantity of the good sold when $q = 20$ and interpret the result.

Solution. Marginal revenue is obtained by differentiating the total revenue function with respect to the quantity of the good sold. Therefore, using the sum–difference rule and the power rule of differentiation jointly we obtain $dR/dq = 50 - 2(0.2q) = 50 - 0.4q$. Therefore, marginal revenue when $q = 20$ is $dR/dq|_{q=20} = 50 - 0.4q|_{q=20} = 50 - (0.4)20 = 50 - 8 = \$42$. The percentage rate of change of revenue with respect to the quantity of the good sold when $q = 20$ can be found by applying equation (3.3.20). This gives us $[f'(q)/f(q)] \times 100 = 42/f(q)|_{q=20} \times 100 = \{42/[50(20) - 0.2(20^2)]\} \times 100 \cong 0.05 \times 100 = 5\%$. This means that the total revenue of the company increases by 5 percent when an additional unit of the good beyond 20 units is sold.

Example 3. Suppose that a firm's total cost is given by the function $C = f(q) = 8 + 2q + 0.25q^2$, where C denotes the total cost in dollars and q denotes the total quantity of the good produced. Find the marginal cost of the firm when $q = 10$, and the percentage rate of change of total cost with respect to q when $q = 10$ and interpret the result.

Solution. Marginal cost is obtained by differentiating the total cost function with respect to the quantity of the good produced. Therefore, using the sum–difference rule we obtain $dC/dq = f'(q) = q - 2(0.25q) = q - 0.5q = 0.5q$. Therefore, marginal cost when $q = 10$ is $dC/dq|_{q=10} = 0.5q|_{q=10} = (0.5)10 = \$5$. The percentage rate of change of cost with respect to the quantity of the good produced when $q = 10$ can be found by applying equation (3.3.20). This gives us $f'(q)/f(q) \times 100 = [5/f(q)|_{q=10}] \times 100 = \{5/[8+2(10)+0.25(10^2)]\} \times 100 \cong 0.09 \times 100 = 9\%$. This implies that the total cost of the firm increases by 9 percent when an additional unit of the good beyond 10 units is produced.

Example 4. Assume that the total utility that an individual obtains from the consumption of different units of a good is given by the function $U = f(x) = \ln(2+x^2)$, where U denotes the total utility obtained and x denotes the units of the good consumed. Find the individual's marginal utility when $x = 10$; and find the percentage rate of change of the individual's total utility when $x = 10$ and interpret the result.

Solution. Marginal utility is obtained by differentiating the total utility function with respect to the quantity of the good consumed. Notice that we can write the utility function as $U = \ln u$, where $u = 1 + x^2$. Therefore, using the logarithmic rule of differentiation and the chain rule jointly, we obtain $dU/dx = \{d[\ln u]/du\}.(du/dx) = [1/(2+x^2)].2x = 2x/(2+x^2)$. Thus, marginal utility when $x = 10$ is $dU/dx|_{x=10} = 2x/(2+x^2)|_{x=10} = 20/102 = 20/102 \cong 0.198$. The percentage rate of change of utility with respect to the quantity of the good consumed when $x = 10$ can be found by applying equation (3.3.20). This gives us $[f'(x)/f(x)] \times 100 = [0.19/\ln(2 + 100)] \times 100 = [0.19/4.6] \times 100 \cong 0.04 \times 100 = 4.3\%$. This implies that the total utility that the individual obtains increases by 4.3 percent when an additional unit of the good beyond 10 units is consumed.

Example 5. Suppose that the inverse demand for a good is given by the function $p = f(q)$, where p denotes price and q denotes the quantity demanded. Find the equation for the *point elasticity of demand* (Φ) using the definition of derivative in equation (3.2.4).

Solution. Point elasticity of demand is defined as the ratio of the percentage change in quantity demand to the percentage change in price. The former can be written as $[\{(q + \Delta q) - q\}/q] \times 100 = [\Delta q/q] \times 100$ and the latter can be written as $[\{f(q + \Delta q) - f(q)\}/f(q)] \times 100$. Assume that $f(q)$ is differentiable. Then we have $\Phi = \lim_{\Delta q \to 0} [\{(\Delta q/q) \times 100\}/\{[f(q+\Delta q)-f(q)]/f(q)\} \times 100 = \lim_{\Delta q \to 0} [\{f(q).100/q.100\}.\{\Delta q/[f(q+\Delta q)-f(q)]\}] = \lim_{\Delta q \to 0} [\{f(q)/q\}/\{[f(q+\Delta q)-f(q)]/\Delta q\}]$. Using Properties I and VI of limits in Section 1.9.2, we can write the last equation as $\Phi = [\lim_{\Delta q \to 0} f(q)/q]/\lim_{\Delta q \to 0} [\{f(q + \Delta q) - f(q)\}/\Delta q] = [f(q)/q]/\lim_{\Delta q \to 0} [\{f(q + \Delta q) - f(q)\}/\Delta q]$. From equation (3.2.4) we know that the limit of $[f(q + \Delta q) - f(q)]/\Delta q$, as $\Delta q \to 0$, is the derivative of $f(q)$ with respect to q; that is, it is $f'(q) = dp/dq$. Therefore, we obtain $\Phi = [f(q)/q]/[dp/dq]$. Since $f(q) = p$, we have

$$\Phi = \frac{f(q)}{q} \div \frac{dp}{dq} = \frac{p}{q} \div \frac{dp}{dq} \tag{3.3.21}$$

Since dp and dq move in opposite directions (which we call the *law of demand*) and since p and q are assumed to be nonnegative, $\Phi < 0$; that is, point elasticity of demand is always negative. But, for convenience, we consider only the absolute value of elasticity, discarding the negative sign. The absolute value of elasticity of demand can be less than one, equal to one, or greater than one. If it is less than one, equal to one, and greater than one, we say that demand is *inelastic, unitary elastic*, and *elastic*, respectively.

Example 6. Find the elasticity of demand if the inverse demand function is $p = f(x) = 10 - 0.5q$ and when $q = 10$.

Solution. Given $p = f(x) = 10 - 0.5q$ and $q = 10$, we have $p = f(x) = 10 - 0.5(10) = 5$, and $dp/dq = -0.5$. Therefore, using equation (3.3.21), we obtain $\Phi = (p/q) \div (dp/dq) = (5/10) \div -0.5 = -1$. This implies that the demand is unitary elastic when $p = 5$.

Example 7. Suppose that the aggregate consumption (C), in billions of dollars, in an economy is given implicitly by the equation $C^2 + 0.2Y^2 = CY + Y$, where Y denotes national income of the economy in billions of dollars. Find the marginal propensities to consume and to save when $C = \$70$ billion and $Y = \$100$ billion and interpret the results.

Solution. Marginal propensity to consume (MPC) is defined as the rate of change of the *aggregate consumption* with respect to *national income*, and is obtained by differentiating aggregate consumption, C, with respect to national income, Y. But, notice that the given equation represents an implicit function. Assuming C as a function of Y, and differentiating C in $C^2 + 0.2Y^2 = CY + Y$ with respect to Y using the implicit function rule, we obtain $2.C.(dC/dY) + 0.2.(2).Y(dY/dY) = C.(dY/dY) + Y.(dC/dY) + (dY/dY)$. This expression can be simplified to obtain $2.C.(dC/dY) + 0.4Y = C + Y.(dC/dY) + 1$, or $(2.C - Y)(dC/dY) = C + 1 - 0.4Y$, or $dC/dy = [C + 1 - 0.4Y]/[2.C - Y]$. Therefore, the MPC, when $C = \$70$ billion and $Y = \$100$ billion, can be found by substituting the given values in $dC/dy = [C + 1 - 0.4Y]/[2.C - Y]$. This gives us $dC/dY|_{C=70, Y=100} = [C + 1 - 0.4Y]/[2.C - Y]|_{C=70, Y=100} = 31/40 = \$0.775$ billion. This implies that when the national income increases by \$1 billion, the aggregate consumption will increase by \$0.775 billion.

Marginal propensity to save (MPS) is defined as the rate of change of *aggregate saving* with respect to national income, and is obtained by differentiating aggregate saving, S, with respect to national income, Y. MPS is also defined as one minus MPC; that is, MPS $= dS/dy = 1 - \text{MPC} = 1 - dC/dY$. Therefore, MPS $= \$1\text{billion} - \$0.775 \text{ billion} = \$0.225$ billion. This value of MPS implies that when national income increases by \$1 billion, the aggregate saving increases by \$0.225 billion. Notice that MPC + MPS $= 1$.

Example 8. Suppose that a company's total revenue from the sale of a good is given by the function $y = f(x) = x^{1/x}$, where y denotes the total revenue in dollars and x denotes the quantity of the good sold. Find the marginal revenue of the company when $x = 10$.

Solution. Marginal revenue can be found by differentiating the function with respect to x. Since it is difficult to apply the rules of differentiation to this function, we use the method of logarithmic differentiation. For this, we first take natural logarithm of both sides of the function to obtain $\ln y = f(x) = \ln x^{1/x} = (1/x)\ln x$. We can now differentiate $\ln y$ with respect to x. This process yields $d[\ln y]/dx = (1/x).\{d[\ln x]/dx\} + (\ln x).\{d[1/x]/dx\}$, which can be simplified to yield $(1/y)(dy/dx) = (1/x)(1/x) + \ln x.(-1/x^2) = (1/x^2) + \ln x.(-1/x^2) = (1/x^2). - (\ln x/x^2) = (1 - \ln x)/x^2$, or $dy/dx = [(1 - \ln x)/x^2] \times y$. Since $y = x^{1/x}$, we obtain $dy/dx = [(1 - \ln x)/x^2].x^{1/x}$. Substituting $x = 10$ into this equation gives us $dy/dx|_{x=10} = [(1 - \ln x)/x^2].x^{1/x}|_{x=10} = -0.0164$. Therefore, the company's marginal revenue is $-\$0.0164$.

Example 9. Suppose that the quantity demanded of a good is given by the function $q = f(p) = 5 - 0.5p$, where p denotes the price in dollars and q denotes the quantity demanded of the good at price p. Find the derivative of the inverse demand function: dp/dq.

Solution. From $q = f(p) = 5 - 0.5p$, we obtain $dq/dp = -0.5$. Applying equation (3.3.18), we obtain the derivative of the inverse demand function as $dp/dq = 1/-0.5 = -2$. Notice that this result can be verified by writing the inverse demand function explicitly and then differentiating it with respect to q. That is, the inverse demand function can be written as $p = 10 - 2q$. Now differentiating this function with respect to q yields $dp/dq = -2$, which confirms the result we obtained.

Example 10. Equation (3.3.21) implies that the point elasticity of demand varies when p or q varies (or when both vary). Show that this is true in the case of a linear inverse demand function.

Solution. Assume that the linear inverse demand function is $p = f(q) = a + bq$, where we assume that (to make economic sense) $a > 0$ and $b < 0$. Applying equation (3.3.21), we obtain the point elasticity of the linear inverse demand function: $\Phi = (p/q) \div (dp/dq) = (p/q)/b = p/bq = p/(p - a)$. Notice now, using the quotient rule of differentiation, that $d\Phi/dp = [\{(p - a).(-1.p)\}/(p - a)^2] = [-a/(p - a)^2] = -[a/(p - a)^2]$. This implies that the point elasticity of demand is a decreasing function of price; that is, $[d\Phi/dp] < 0$. In other words, as price increases, point elasticity of demand decreases. But, as can be seen from Figure 3.3.1, p lies between 0 and b. At the middle of this interval, we have $p = b/2$. Substituting this in the last equation, we get $\Phi = p/(p - a) = (b/2)/[(b/2) - b] = -1$. Therefore, at the middle of the linear demand curve, the point elasticity of demand is equal to -1 (or $\Phi = -1$). But, notice that $\Phi > -1$ when $p < b/2$ and $\Phi < -1$ when $p > b/2$. Since Φ is always negative, we take its absolute value. Therefore, we have three cases: when $p = b/2$, $\Phi = 1$; when $p < b/2$, $\Phi < 1$; and when $p > b/2$, $\Phi > 1$. These results can be seen in Figure 3.3.1.

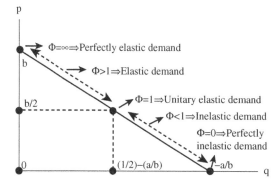

Figure 3.3.1

Example 11. Assume that the inverse demand function of a firm's output is given by $p = f(q)$ and its total revenue function is given by $R = p.q$, where p denotes the price of the good and q denotes the quantity of the good. Show that marginal revenue is a function of both price and the point elasticity of demand, and interpret the result.

Solution. We know that marginal revenue is dR/dq. Therefore, differentiating R with respect to q using the product rule of differentiation yields $dR/dq = p + q.(dp/dq) = p[1 + (q/p).(dp/dq)]$. Notice that the second term inside the brackets is the inverse of the point elasticity of demand (Φ). Therefore, marginal revenue can be written as

$$\frac{dR}{dq} = p\left[1 + \frac{1}{\Phi}\right] \tag{3.3.22}$$

We know that $p > 0$. Therefore, for given $p > 0$, equation (3.3.22) implies that demand is elastic if $[1 + (1/\Phi)] > 0$ (or when $\Phi < -1$) and demand is inelastic if $[1 + (1/\Phi)] < 0$ (or when $\Phi > -1$). This suggests that if the firm reduces (increases) the price of the good when demand is elastic (inelastic), its total revenue will increase (decrease).

3.3.16 Exercises

1. Use the constant function and the function-with-constant rules to find $dy/dx = f'(x)$ given
   (i) $y = f(x) = 1/2$; (ii) $y = f(x) = 2^{1/2}$; (iii) $y = f(x) = 0$; (iv) $y = f(x) = -0.5x$;
   (v) $y = f(x) = 3x/4$; (vi) $y = f(x) = (1/2) - (2x/3)$.

2. Use the sum-or-difference and power rules to find $dy/dx = f'(x)$ given
   (i) $y = f(x) = 0.5 + 3x + 2x^2$; (ii) $y = f(x) = 0.5 - 2x^2$; (iii) $y = f(x) = 0.5 - 3x - 2x^2$;
   (iv) $y = f(x) = 0.5 + 3x - 2x^2$; (v) $y = f(x) = 0.5 + 3x + 2x^2 - 3x^3$; (vi) $y = f(x) = \sqrt{x}$;
   (vii) $y = f(x) = (1 + x^2)^5$; (viii) $y = f(x) = (1 + x^2)^{1/2}$; (ix) $y = f(x) = (\sqrt{x})^{1/2}$.

3. Use the product and quotient rules to find $dy/dx = f'(x)$ given
   (i) $y = f(x) = (1 + x)(2 + x^2)$; (ii) $y = f(x) = \sqrt{x}(\sqrt[3]{1 + x^2})$; (iii) $y = f(x) = (x + 2x^2 + 3x^3)(-x - 2x^2 - 3x^3)$; (iv) $y = f(x) = (2x + x^3)/(x^2 - x^3)$; (v) $y = f(x) = (10 - x^2)/(x^2 - 10)$; (vi) $y = f(x) = 1/x^{1/n}$; (vii) $y = f(x) = (x - 2)/(x + 2).(x - 2)$.

4. Use the chain rule to find $dy/dx = f'(x)$ given
   (i) $y = 2u + 1/u$, where $u = 1 + x$; (ii) $y = 2u^2$, where $u = 2x$; (iii) $y = u^3$, where $u = 1 + x$;
   (iv) $y = 1/u^3$, where $u = 1 + x$; (v) $y = \sqrt[3]{u - 1}$, where $u = 1 + x$; (vi) $y = u^3/u^2$, where $u = 1 + x$.

5. Use the logarithmic and exponential function rules to find $dy/dx = f'(x)$ given
   (i) $y = \ln[(1 - x)/(1 + x)]$; (ii) $y = \ln(2x + 3x^2 + 4x^3)$; (iii) $y = \log_5 x$; (iv) $y = e^{10 - (1/x)}$;
   (v) $y = (e^x + 1)/(e^x - 1)$; (vi) $e^{x.\ln(1/x)}$.

6. Find $dy/dx = f'(x)$, using implicit differentiation, given
   (i) $x^{1/3} + y^{1/3} = 3$; (ii) $x^3 + y^2 + xy = 4$; (iii) $x^2 + y^2 = 9$.

7. Use the technique of logarithmic differentiation to find $dy/dx = f'(x)$ given
   (i) $y = \sqrt{(2 + x^2).(x^2 + 2)}$; (ii) $y = (2 + x^2)(3 + x^3)$; (iii) $y = x^{\sqrt{x}}$.

8. *Application exercise.* Assume that the inverse demand function for a good produced by a firm is given by $p = e^{\ln q^2}$, where p denotes the price per unit of the good in dollars and q denotes the quantity demanded of the good. Find the firm's marginal revenue when $q = 20$.

9. *Application exercise.* We know that P amount of money invested for t years when interest rate r is compounded continuously would have grown to $F = f(t) = P.e^{r \times t}$. What is the relative rate of change of F with respect to t?

10. *Application exercise.* Suppose that the population of a city from now until time t is given by the function $P = 500\,000e^{0.08.t}$, where P denotes the population and t denotes the number of years. Find the rate of change of population when $t = 10$.

11. *Application exercise.* Assume that the quantity of output (q) produced by a firm using l units of labor is given by the function $q = (2l^2)/(2 + l^{1/2})$. Also assume that the demand for the firm's product is given by the function $p = 20 - 0.5q$, where p denotes the price per unit of the good in dollars. Find the *marginal revenue product* when $l = 10$.

12. *Application exercise.* Suppose that the inverse demand for a good is given by $p = 10q^{(-4/\sqrt{q})}$, where p denotes the price per unit of the good in dollars and q denotes the quantity demanded of the good. Find the point elasticity of demand when $q = 4$, and interpret the result. Should the producer of the good increase or decrease the price, or keep the price constant, to increase the revenue?

13. *Application exercise.* Assume that the inverse demand function for a good is given by $p = f(q) = 25/(2 + q^2)$, where p denotes the price per unit of the good in dollars and q denotes the quantity demanded of the good. Find the rate of change of quantity with respect to price.

 Web supplement: S3.3.17 *Differentiation of trigonometric functions*

3.4 Higher-order differentiation of univariate functions

3.4.1 Meaning and notations of higher-order differentiation

In Section 3.2.2 we found that the derivative of a primitive, univariate function was in fact another function. We called this function the derived function. A question that arises here is: can we differentiate again the derived functions with respect to the independent variable x? Yes, we can differentiate them again. If we differentiate for a second time, then the result is called the *second-order derivative* or the *second derivative* of $y = f(x)$ with respect to x. If we differentiate the second derivative for another time, it is called the *third-order derivative* or the *third derivative* of $y = f(x)$ with respect to x. And, if we differentiate the nth derivative, then the result is called the nth-*order derivative* or the nth *derivative* of $y = f(x)$ with respect to x. Notice that the higher derivatives are also derived functions.

We know that the first derivative, or simply the derivative, of $y = f(x)$ is denoted by $dy/dx = f'(x) = y'$. Similarly, the second derivative of this function is denoted by $d^2y/dx^2 = f''(x) = y''$; the third derivative is denoted by $d^3y/dx^3 = f'''(x) = y'''$; and the nth derivative is denoted by $d^ny/dx^n = f^n(x) = y^n$.

3.4.2 Finding the higher derivatives of univariate functions

Suppose that we have a univariate function given by $y = f(x) = 2x - 3x^2$. We know that the first derivative of this function with respect to x is $dy/dx = f'(x) = 2 - 6x$. Notice that dy/dx is a function of x. We can now differentiate dy/dx with respect to x to obtain the second derivative of $y = f(x)$. This second derivative is $d^2y/dx^2 = f''(x) = -6$.

Notice that all the higher derivatives of the given function beyond the second derivative are equal to zero.

We now consider the second derivative of an exponential function. Suppose that the exponential function is $y = f(x) = e^x$. Then we know from Section 3.3.10 that $dy/dx = e^x$. Since the RHS of the derived function (dy/dx) is the same as that of the primitive function, all the higher derivatives of an exponential function with base e will be equal to the original, primitive function. Therefore, $d^n y/dx^n = f^n(x) = e^x$. However, the results will be different if the exponential function has a base different from e. For example, consider the function $y = f(x) = a^x$. Then, following equation (3.3.16), we obtain $dy/dx = a^x.(\ln a)$. The second derivative is $d^2 y/dx^2 = (\ln a).a^x.(\ln a) = (\ln a)^2 a^x$; and continuing like this, we can obtain the nth derivative of $y = f(x) = a^x$ as $d^n y/dx^n = f^n(x) = (\ln a)^n a^x$. Similarly, one can show that the higher derivatives will be different from the first derivative if the exponential function is $y = f(u) = e^u$ or if $y = f(u) = a^u$, where $u = g(x)$.

The last important function that we consider here is the implicit function. Consider the implicit function $3x^2 + 5y^2 = 2$. Following our discussion in Section 3.3.12, we obtain the first derivative $3 \times 2 \times x.(dx/dx) + 5 \times 2 \times y.(dy/dx) = 0$, which can be written as $6x + 10y.(dy/dx) = 0$, which can be simplified to obtain $10y.(dy/dx) = -6x$, or $dy/dx = (-6x/10y) = (-3x/5y)$. We can now use the quotient rule to obtain the second derivative: $d^2 y/dx^2 = \{(5y).[d(-3x)/dx] - [(-3x).d(5y)/dx]\}/(5y)^2\} = [(5y)(-3) - (-3x)4.(dy/dx)]/(5y)^2] = [\{(-15y) + 12x(dy/dx)\}/(5y)^2]$. Substituting $dy/dx = (-3x)/5y$ into the last equation we obtain $d^2 y/dx^2 = [\{(-15y) + 12x(dy/dx)\}/(5y)^2] = [\{(-15y) + 12x(-3x/5y)\}/(5y)^2] = [(-15y) + (-26x^2/5y)]/(5y)^2] = -(75y + 26x^2)(5y)$. Notice that one can continue like this to find the other higher derivatives.

3.4.3 Derivatives and limits: l'Hôpital's rule

We discussed limits of functions and their properties in Section 1.9. In our discussion we found that limits might not exist for some functions. Assume that our function is $y = f(x) = g(x)/h(x)$ and we want to find the limit of this function as $x \rightarrow x_0$. Notice that when $x \rightarrow x_0$, the limit of the function does not exist if one of these cases happen: (1) $g(x) \rightarrow 0$ and $h(x) \rightarrow 0$ (giving us the meaningless expression $0/0$); (2) $g(x) \rightarrow +\infty$ and $h(x) \rightarrow -\infty$ (giving us the expression $+\infty/-\infty$); (3) $g(x) \rightarrow -\infty$ and $h(x) \rightarrow -\infty$ (giving us the expression $-\infty/-\infty$); or (4) $g(x) \rightarrow +\infty$ and $h(x) \rightarrow +\infty$ (giving us the expression $+\infty/+\infty$).

As an example, consider the function we used in Section 1.9.3. The function we used there, to find the limit when $x \rightarrow x_0 = 1$, was $y = f(x) = g(x)/h(x)$, where $g(x) = x^2 - 1$ and $h(x) = x - 1$. When we substituted the value $x = 1$ in this function, we ended up with case (1) noted above. However, we factored the function $g(x) = x^2 - 1$ into $g(x) = (x + 1)(x - 1)$ and rewrote the original function $y = f(x) = (x^2 - 1)/(x - 1)$ as $y = f(x) = [(x + 1)(x - 1)]/(x - 1) = x + 1$. Then we evaluated the limit of the function when $x \rightarrow x_0 = 1$ and obtained the limit 2: $\lim_{x \to 1} f(x) = \lim_{x \to 1} (x + 1) = (1 + 1) = 2$.

Notice that, in the above operation, we factored the numerator $g(x)$ (one can also factor the denominator if needed and possible) and then carried out the operation. Factoring may not always be advisable in the case of some functions. Suppose that $g(x) = 10x + 1$ and $h(x) = x - 1$ so that $y = f(x) = g(x)/h(x) = (10x + 1)/(x - 1)$. Also suppose that we want to find the limit of this function when $x \rightarrow x_0 = \infty$. If we substitute $x \rightarrow x_0 = \infty$ in this function, we end up with case (4) noted above. Notice that factoring does not work in this example. How, then, do we find the limit of this function when $x \rightarrow \infty$? In situations like these *l'Hôpital's rule* can be of help. This rule states that

$$\lim_{x \to x_0} f(x) = \lim_{x \to x_0} \left[\frac{g(x)}{h(x)} \right] = \lim_{x \to x_0} \left[\frac{g'(x)}{h'(x)} \right] \tag{3.4.1}$$

We now apply l'Hôpital's rule to the above example. We know that $g'(x) = 10$ and $f'(x) = 1$. Therefore, the limit of $y = f(x) = g(x)/h(x) = (10x + 1)/(x - 1)$ when $x \to x_0 = \infty$ is $\lim_{x \to x_0 = \infty} y = \lim_{x \to x_0 = \infty} f(x) = \lim_{x \to x_0 = \infty} [g(x)/h(x)] = \lim_{x \to x_0 = \infty} [(10x + 1)/(x - 1)] = \lim_{x \to x_0 = \infty} [g'(x)/h'(x)] = \lim_{x \to x_0 = \infty} (10/1) = 10$.

As another example, consider our previous function $y = f(x) = (x^2 - 1)/(x - 1)$, where we assumed that $g(x) = x^2 - 1$ and $h(x) = x - 1$, whose limits we wanted to find when $x \to x_0 = 1$. Then we obtain $g'(x) = 2x$ and $h'(x) = 1$. Substituting these values in equation (3.4.1) yields $\lim_{x \to x_0 = 1} y = \lim_{x \to x_0 = 1} f(x) = \lim_{x \to x_0 = 1} [(x^2 - 1)/(x - 1)] = \lim_{x \to x_0 = 1} [g(x)/h(x)] = \lim_{x \to x_0} [g'(x)/h'(x)] = \lim_{x \to x_0 = 1} [2x/1] = 2 \times 1 = 2$. Notice that this is identical with the result we obtained when we used the method of factoring above and in Section 1.9.3.

However, in some cases one single application of l'Hôpital's rule may still produce indeterminate limits. In such cases one can apply l'Hôpital's rule more than once. This implies that $\lim_{x \to x_0} f(x) = \lim_{x \to x_0} [g(x)/h(x)] = \lim_{x \to x_0} [g'(x)/h'(x)] = \lim_{x \to x_0} [g''(x)/h''(x)] = \dots$. As an example of this, consider the function $y = f(x) = g(x)/h(x)$, where $g(x) = 4x^2 - 5x$ and $h(x) = 3x^2 - 16$. Then, we obtain $g'(x) = 8x^2 - 5$ and $h'(x) = 6x$. When we substitute this in equation (3.4.1), we get $\lim_{x \to x_0 = \infty} f(x) = \lim_{x \to x_0 = \infty} [g(x)/h(x)] = \lim_{x \to x_0 = \infty} [g'(x)/h'(x)] = \lim_{x \to x_0 = \infty} [(8x - 5)/6x] = \infty/\infty$, which is the same as case (4) noted earlier. Therefore, we apply l'Hôpital's rule again to yield $\lim_{x \to x_0 = \infty} f(x) = \lim_{x \to x_0 = \infty} [g(x)/h(x)] = \lim_{x \to x_0 = \infty} [g'(x)/h'(x)] = \lim_{x \to x_0 = \infty} [g''(x)/h''(x)] = \lim_{x \to x_0 = \infty} [16/6] = 16/6 = 8/3$.

3.4.4 Application examples

Example 1.   Assume that the total utility that an individual obtains from the consumption of different units of a good is given by the function $U = f(x) = \ln x$, where U denotes the total utility obtained and x denotes the units of the good consumed. Find the individual's marginal utility, and the rate of change of marginal utility or determine how marginal utility behaves.

Solution.   Marginal utility is obtained by differentiating the total utility function with respect to x. Using equation (3.3.11) we obtain marginal utility as $dU/dx = d[\ln x]/dx = 1/x$. This shows that marginal utility of the individual diminishes as x increases. This phenomenon is called the *law of diminishing marginal utility* in the *theory of demand* in economics. The rate of change of marginal utility is given by the second derivative of the total utility function. Differentiating dU/dx again with respect to x yields $d^2 U/dx^2 = (x.0 - 1 \times 1)/x^2 = -1/x^2$. This implies that the total utility that the individual obtains increases at a diminishing rate as x consumed increases (because $dU/dx = (1/x) > 0$ and $d^2 U/dx^2 = (-1/x^2) < 0$ for $x > 0$).

Example 2. Suppose that the total cost (assuming that there is no fixed cost) of producing x units of output of a firm is given by $C = f(x) = 40x - 9x^2 + x^3$. Determine the rate of change of marginal cost of the firm or determine how marginal cost behaves as x increases.

Solution. Marginal cost is obtained by differentiating the total cost function with respect to x. We can obtain marginal cost as $dC/dx = 40 - 18x + 3x^2$. A plot of this will show that marginal cost of the firm diminishes in the beginning, and then increases as the quantity of the good produced increases. The rate of change of marginal cost is given by the second derivative of the total cost function. Differentiating dC/dx again with respect to x yields $d^2C/dx^2 = -18 + 6x$. This implies that the total cost of the firm increases at a diminishing rate until $x < 3$ (because $(dC/dx) > 0$ and $(d^2C/dx^2) < 0$ at $x < 3$) and increases at an increasing rate when $x > 3$ (because $(dC/dx) > 0$ and $(d^2C/dx^2) > 0$ for $x > 3$).

Example 3. Assume that a firm's total output of producing q units of output using l units of labor is given by the function $q = f(l) = 50l^2 - 5l^3$. Determine the rate of change of *marginal product of labor* (or determine how marginal product of labor behaves) as l increases.

Solution. Marginal product of labor is obtained by differentiating the total output function with respect to l. We can obtain it as $dq/dl = 100l - 15l^2$. A plot of this will show that the marginal product of labor increases in the beginning (until $l = 3.3$), and then diminishes as l increases beyond 3.3 units. This phenomenon is called the *law of diminishing returns* in the *theory of production* in economics. The rate of change of marginal product of labor is given by the second derivative of the total output function. Differentiating dq/dl again with respect to l yields $d^2q/dl^2 = 100 - 30l$. This implies that the total product of the firm increases at an increasing rate until $l < 3.3$ (because $(dq/dl) > 0$ and $(d^2q/dl^2) > 0$ for $l < 3.3$) and increases at a diminishing rate when $l > 3.3$ (because $(dq/dl) > 0$ and $(d^2q/dl^2) < 0$ for $l > 3.3$).

Example 4. Suppose that the total revenue (R) of a firm from the production of q units of output is given by $R = f(q) = 50q - 5q^2 - 0.1q^3$. Determine the rate of change of marginal revenue (or determine how marginal revenue behaves) as q increases.

Solution. Marginal revenue of the firm is obtained by differentiating the total revenue function with respect to q. We can obtain it as $dR/dq = f'(q) = 50 - 10q - 0.3q^2$. A plot of this will show that marginal revenue decreases as q increases. The rate of change of marginal revenue is given by the second derivative of the total revenue function. Differentiating dR/dq again with respect to q yields $d^2R/dq^2 = -10 - 0.6q$. This implies that the total revenue increases at a diminishing rate until $q = 4.42$ (because $(dR/dq) > 0$ and $(d^2R/dl^2) < 0$ for $q < 4.42$).

3.4.5 *Exercises*

1. Find the third derivatives of the following functions:
   (i) $y = f(x) = 2x - 3x^2$; (ii) $y = f(x) = (2x + 1)/x^2$; (iii) $y = f(u) = \ln u$, $u = g(x) = (1 + x^2)$; (iv) $y = f(u) = e^u$, $u = g(x) = (1 + x^2)$.
2. Find the third derivatives of the following functions:
   (i) $y = f(x) = 10^{2x}$; (ii) $y = f(x) = 10^{2x/(1+x)}$; (iii) $y = f(x) = e^{\ln x^2}$.
3. Find the second derivatives given the following implicit functions:
   (i) $y^2 = 2x^2$; (ii) $y^2 - 2x^2 = 4$; (iii) $y - xy + 4 = 0$; (iv) $y^2 + xy + x^2 = 8$.

4. *Application exercise.* Suppose that the total cost (C) of producing q units of output is given by $C = f(q) = 50q - 10q^2 + q^3$. Determine how marginal cost behaves as q increases. Interpret the results.

5. *Application exercise.* Let the total output (q) that a firm produces using k units of capital be given by $q = f(k) = 10 + k + 5k^2 - k^3$. Determine how *marginal product of capital* $(= dq/dk)$ behaves as k increases. Interpret the results.

6. *Application exercise.* Suppose that the total utility (U) that a consumer obtains from the consumption of x units of a good is given by $U = f(x) = x^2$. Determine how marginal utility behaves as x increases. Interpret the results.

7. *Application exercise.* Assume that the total profit (Π) that a company obtains from the production of q units of its good is given by $\Pi = f(q) = q^{1/2}$. Determine how marginal profit behaves as q increases. Interpret the results.

 Web supplement: S3.4.6 *Higher derivatives of trigonometric functions*

 Web supplement: S3.4.7 Mathematica applications

3.5 Derivatives and curvature of curves; concave and convex functions; convex sets

Undergraduate courses in the fields of our interest often make use of curves as geometric representations of relationships among variables. One frequently needs to determine whether a particular curve always increases, always decreases, alternates in nature, or remains as a straight line. Therefore, a reasonable understanding of the *curvature of curves*, or the nature of curves, is a must for these students.

Similarly, students in these fields frequently deal with curves such as *indifference curves*, *isoquants*, *production possibility curves*, etc. These curves are the geometric representations of either *concave functions* or *convex functions*. This suggests that a study of these functions will certainly enrich students' understanding of the geometric representations mentioned above. It will also equip them with a powerful tool which can be used later, particularly in solving optimization problems.

Another important topic that we consider in this section is the concept of *convex sets*. A good understanding of convex sets is essential for a meaningful understanding of the topic of optimization, particularly *linear programming*. We first consider the topic of the curvature of curves followed by concave and convex functions, and then convex sets.

3.5.1 Derivatives and curvature of curves

One important question is: can we say, without plotting, whether the graph, or the curve, of a function is always increasing, is always decreasing, is alternating in nature, or is remaining constant? Is there any way to judge the nature, or the curvature, of a curve from its function? The answers to these questions are the same: yes, one can use the derivatives of a function to judge the nature or curvature of a curve defined by its function. We discuss the use of derivatives in determining the curvature of curves below.

Assume that we have two functions: $y = f(x)$ and $y = g(x)$. We know, from the definition of the derivative of $f(x)$ at $x = x_0$ (or of the slope of the tangent to $f(x)$ corresponding to $x = x_0$), that the derivative or slope $(f'(x))$ measures the rate of change of $f(x)$ when x changes by an infinitesimally small amount. If a positive (negative) infinitesimal change in x causes a

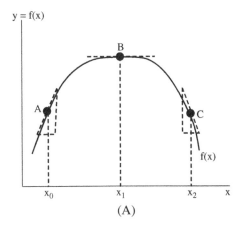
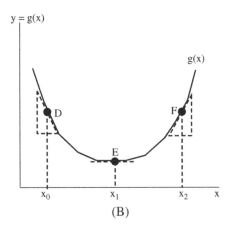

Figure 3.5.1

positive (negative) change in y, then these two changes move in the same direction. Therefore, the ratio of the latter to the former must be positive (that is, $f'(x) > 0$). This implies that when the derivative or the slope of the tangent is positive at $x = x_0$ (if $f'(x)\big|_{x=x_0} > 0$), then the function or the curve representing the function must be increasing at $x = x_0$. This conclusion is applicable if $g'(x)\big|_{x=x_0} > 0$.

Similarly, if a positive (negative) infinitesimal change in x causes a negative (positive) change in y, then these two changes move in the opposite direction. Therefore, the ratio of the latter to the former must be negative (that is, $f'(x) < 0$). This implies that when the derivative or the slope of the tangent is negative at $x = x_2$ (if $f'(x)\big|_{x=x_2} < 0$), then the function or the curve representing the function must be decreasing at $x = x_2$. This conclusion is applicable if $g'(x)\big|_{x=x_2} < 0$.

Another possibility is that a positive (negative) infinitesimal change in x may cause no change in y. Then the ratio of the latter to the former must be zero (that is, $f'(x) = 0$). This implies that when the derivative or the slope of the tangent is zero at $x = x_1$ (if $f'(x)\big|_{x=x_1} = 0$), then the function or the curve representing the function must be constant at $x = x_1$. Again, the conclusion is applicable if $g'(x)\big|_{x=x_1} = 0$. The above three arguments can be best understood with the help of Figures 3.5.1(A) and (B). The first argument corresponds to points A and F, the second argument corresponds points C and D, and the third argument corresponds points B and E in the figure.

We found above that one could use the sign of the derivative of (the slope of the tangent to) $y = f(x)$ at $x = x_0$ to judge whether $y = f(x)$ or its graph is increasing, decreasing, or remains constant at $x = x_0$. But, a question that arises now is: if $y = f(x)$ is increasing (decreasing) at $x = x_0$, does the function increase (decrease) at an increasing (decreasing) rate at $x = x_0$? The answer to this question depends on the sign of the second derivative of $y = f(x)$ with respect to x or of the slope of the slope of the tangent at $x = x_0$.

We know that the second derivative of $y = f(x)$, or $d^2y/dx^2 = f''(x)$, at a point on its graph (say, at $x = x_0$) measures the rate change of the rate of change of $y = f(x)$ at that point. Therefore, the sign of $f''(x)\big|_{x=x_0}$ shows whether the curve of $y = f(x)$ increases (or decreases) at an increasing (or decreasing) rate at $x = x_0$. If $f'(x)\big|_{x=x_0} > 0$ and if $f''(x)\big|_{x=x_0} < 0$, then the function will be increasing at a decreasing rate at $x = x_0$.

Table 3.5.1

$y = f(x)$	$dy/dx = f'(x)$	$d^2y/dx^2 = f''(x)$	Point on the graph in Figure 3.5.1(A)	Curvature of $y = f(x)$
$x = x_0$	> 0	< 0	A	Increasing at decreasing rate
$x = x_1$	$= 0$	< 0	B	Constant
$x = x_2$	< 0	< 0	C	Decreasing at decreasing rate
$y = g(x)$	$dy/dx = g'(x)$	$d^2y/dx^2 = g''(x)$	Point on the graph in Figure 3.5.1(B)	Curvature of $y = g(x)$
$x = x_0$	< 0	> 0	D	Decreasing at increasing rate
$x = x_1$	$= 0$	> 0	E	Constant
$x = x_2$	> 0	> 0	F	Increasing at increasing rate

This corresponds to point A in Figure 3.5.1(A). If $f'(x)\big|_{x=x_2} < 0$ and $f''(x)\big|_{x=x_2} < 0$, then the function or its curve will be decreasing at a decreasing rate at $x = x_2$. This corresponds to point C in Figure 3.5.1(A). If $g'(x)\big|_{x=x_0} < 0$ and $g''(x)\big|_{x=x_0} > 0$, then the function will be decreasing at an increasing rate at $x = x_0$. This corresponds to point D in Figure 3.5.1(B). If $g'(x)\big|_{x=x_2} > 0$ and $g''(x)\big|_{x=x_2} > 0$, then the function will be increasing at an increasing rate. This corresponds to point F in Figure 3.5.1(B). If $f'(x)\big|_{x=x_1} = f''(x)|_{x=x_1} = 0$ and $g'(x)\big|_{x=x_1} = g''(x)|_{x=x_1} = 0$, then the function will be constant or its graph will be a straight line parallel to the horizontal axis as represented by points B and E in Figures 3.5.1(A) and (B), respectively.

As an example, consider the function $y = f(x) = 5 - x + x^2$. The first derivative of this function at $x = 5$ is $f'(x)\big|_{x=5} = -1 + 2x|_{x=5} = -1 + 2 \times 5 = 9 > 0$. The second derivative of this function at $x = 5$ is $f''(x)\big|_{x=5} = 2 > 0$. Since the first derivative is positive at $x = 5$, the function is increasing at $x = 5$. And since the second derivative is positive, the function is increasing at an increasing rate at $x = 5$.

As another example, consider the function $y = g(x) = 5 + x - x^2$. The first derivative of this function at $x = 5$ is $g'(x)\big|_{x=5} = 1 - 2x|_{x=5} = 1 - 2 \times 5 = -9 < 0$. The second derivative of this function at $x = 5$ is $g''(x)\big|_{x=5} = -2 < 0$. Since the first derivative is negative, the function is decreasing at $x = 5$. And since the second derivative is negative at $x = 5$, the function is decreasing at a decreasing rate at $x = 5$. Let us now condense all the results that we derived in this section for easy reference. This is presented in Table 3.5.1.

3.5.2 Univariate concave and convex functions

There are two types of concave functions: *strictly concave functions* and *weakly concave functions* (or, simply, concave functions). From now on we call a weakly concave function a concave function. Let us first explain the meaning of a strictly concave function geometrically. We will then present the algebraic representation of such a function.

Suppose that we a have a function $y = f(x)$ whose graph resembles that in Figure 3.5.1(A). Also suppose that we choose two points (G and H) on the graph of this function and connect them by a straight line GH as shown in Figure 3.5.2(A). Notice that the line GH lies entirely

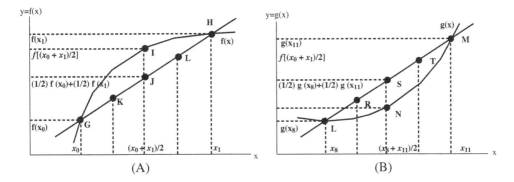

Figure 3.5.2

below, except at points G and H, the graph of the function. Such a curve is called a *strictly concave curve* or the function of that curve is called a strictly concave function. This suggests that that a strictly concave curve is like an *inverted U-shaped curve*.

We now provide a formal, algebraic description of a strictly concave function. Notice that point J on the line GH lies exactly halfway between G and H. Therefore, the vertical coordinate of J is $(1/2)f(x_0) + (1/2)f(x_1)$ and its horizontal coordinate is $(x_0 + x_1)/2$. The horizontal coordinate of point I on the graph of the function is the same as that of point J. But the vertical coordinate of I is $f[(x_0 + x_1)/2]$, which is different from that of J. This means that the vertical coordinate of I is higher than that of J. This argument gives us the inequality $f[(x_0 + x_1)/2] > (1/2)f(x_0) + (1/2)f(x_1)$. One can repeat the above argument for any other point on the line GH yielding inequalities similar to the last one. This is exactly the description of a strictly concave function we geometrically stated above. Therefore, the last inequality gives the definition of a strictly concave univariate function.

We used the value 1/2 in the above inequality because we chose point J that lay exactly in the middle of the line GH. But when we choose point K, then we will be using 1/4 in the inequality, or 3/4 if we use point L. Instead of specific values such as 1/2, 1/4, or 3/4, we now use a general constant α such that $0 < \alpha < 1$. Using this constant, and using i and j as indices of x, we can rewrite the above definition of a strictly concave function as

$$f\left(\alpha.x_i + (1-\alpha).x_j\right) > \alpha.f(x_i) + (1-\alpha).f(x_j), \text{ where } i \neq j \tag{3.5.1}$$

Let us now consider a concave function. As we defined a strictly concave function geometrically, we can define a concave function geometrically. Suppose that our function is $y = f(x)$ whose graph resembles that in Figure 3.5.3(A). As we drew a straight line connecting two points of a strictly concave function, we now draw a straight line connecting points B and C on the graph. But this line coincides with the graph of the function between points B and C. However, if we draw a straight line connecting points A and B (or points C and D), we obtain a straight line that lies completely below the graph of the function (except at the end points). Therefore, the straight lines we drew lie either below the graph or coincide with it. This implies that, following the definition given in inequality (3.5.1), the function is strictly concave between points A and B (and points C and D), and concave between points B and C. Therefore, the function whose graph is illustrated in Figure 3.5.3(A) is both strictly concave and concave.

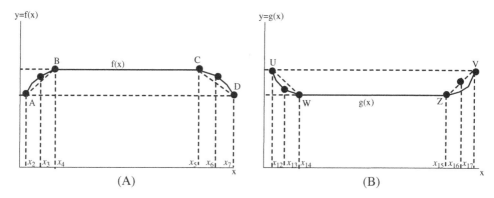

Figure 3.5.3

The above arguments imply that we can define a concave function by a slight modification of the definition of a strictly concave function given in inequality (3.5.1). The required modification relates to replacing the ">" sign by the "≥" sign. Therefore, a function is a concave function if the following inequality is satisfied:

$$f\left(\alpha.x_i + (1-\alpha).x_j\right) \geq \alpha.f(x_i) + (1-\alpha).f(x_j), \text{ where } i \neq j \tag{3.5.2}$$

We shall now consider a *strictly convex function*. Suppose that we have the function $y = g(x)$ whose graph resembles that in Figure 3.5.2(B). Also suppose that we choose two points (L and M) on the graph of this function and connect them by the straight line LM. Notice that the line LM lies entirely above (except at points L and M) the graph of the function. Such a curve is called a *strictly convex curve* or the function of that curve is called a strictly convex function. This suggests that that a strictly convex curve is like a *U-shaped curve*.

As in the case of a strictly concave function, we now provide a formal, algebraic description of a strictly convex function. Notice that point S on the line LM lies exactly halfway between L and M. Therefore, the vertical coordinate of S is $(1/2)g(x_8) + (1/2)g(x_{11})$ and its horizontal coordinate is $(x_8 + x_{11})/2$. The horizontal coordinate of point N on the graph of the function is the same as that of point S. But the vertical coordinate of N is $g[(x_8 + x_{11})/2]$, which is different from that of point S. This means that the vertical coordinate of S is higher than that of N. This argument gives us the inequality $g[(x_8 + x_{11})/2] < (1/2)g(x_8) + (1/2)g(x_{11})$. One can repeat the above argument for any other point on the line LM yielding inequalities similar to the one above. This is exactly the description of a strictly convex function we geometrically stated above. Therefore, the last inequality gives the definition of a strictly convex univariate function. Again as in the case of a strictly concave function, we use a general constant β (such that $0 < \beta < 1$) instead of specific values. Using this constant, and using i and j as indices of x, we can rewrite the definition of a strictly convex function as

$$g\left(\beta.x_i + (1-\beta).x_j\right) < \beta.g(x_i) + (1-\beta).g(x_j), \text{ where } i \neq j \tag{3.5.3}$$

We shall now consider a *weakly convex function* or *convex function*. From now on we call a weakly convex function a convex function. As for a strictly convex function, we can define a convex function geometrically. Suppose that our function is $g = f(x)$ whose graph resembles that in Figure 3.5.3(B). We now draw a straight line connecting points W and Z. But this line coincides with the graph of the function between points W and Z. However, if we draw a straight line connecting points U and W (and points Z and V), we obtain a straight line that lies completely above the graph of the function (except at the end points). Therefore, the straight lines we drew lie either above the graph or coincide with it. This implies that, following the definition given in equation (3.5.3), the function is strictly convex between points U and W (and Z and V), and convex between points W and Z. Therefore, the function whose graph is illustrated in Figure 3.5.3(B) is both strictly convex and convex.

The above arguments imply that, as before, we can define a convex function by a slight modification of the definition of a strictly convex function given in inequality (3.5.3). The required modification relates to replacing the "<" by the "≤" sign. Therefore, a function is a convex function if the following inequality is satisfied:

$$g\left(\beta.x_i + (1 - \beta).x_j\right) \leq \beta.g(x_i) + (1 - \beta).g(x_j), \text{ where } i \neq j \tag{3.5.4}$$

Notice that the graph of the function $y = f(x)$ between points B and C in Figure 3.5.3(A) is a straight line. As per the definition in inequality (3.5.2), this straight line is a concave function. Similarly, the graph of the function $g = f(x)$ between points W and Z in Figure 3.5.3(B) is a straight line. And, as per the definition in inequality (3.5.4), this straight line is a convex function. What these imply is that a straight line represents both a concave as well as a convex function. At the same time, a straight line represents neither a strictly concave nor a strictly convex function. Inequalities (3.5.1)–(3.5.4) are called *Jensen's inequalities* for univariate functions.

Also notice an important feature of strictly concave and strictly convex functions. We found that function $y = f(x)$ in Figure 3.5.2(A) was a strictly concave function, which resembles the function $y = f(x)$ in Figure 3.5.1(A). An important feature of this function is that its second derivative is negative for all x in its domain. Therefore, we can infer that, if the second derivative is negative, then the function must be a strictly concave function. Similarly, the strictly convex function $y = g(x)$ in Figure 3.5.2(B) is similar to the function $y = g(x)$ illustrated in Figure 3.5.1(B), whose feature is that its second derivative is positive for all x in its domain. Therefore, we can infer again that, if the second derivative is positive, then the function must be a strictly convex function.

We shall now sum up the above results. If the inequalities $f\left(\alpha.x_i + (1 - \alpha).x_j\right) > \alpha.f(x_i) + (1 - \alpha).f(x_j)$ and $f\left(\alpha.x_i + (1 - \alpha).x_j\right) \geq \alpha.f(x_i) + (1 - \alpha).f(x_j)$ with i and j being the indices of x and $i \neq j$ are satisfied, then the function $y = f(x)$ is said to be a strictly concave function and a concave function, respectively. If the inequalities $g\left(\beta.x_i + (1 - \beta).x_j\right) < \beta.g(x_i) + (1 - \beta).g(x_j)$ and $g\left(\beta.x_i + (1 - \beta).x_j\right) \leq \beta.g(x_i) + (1 - \beta).g(x_j)$ with i and j being the indices of x and $i \neq j$ are satisfied, then the function $y = g(x)$ is said to be a strictly convex function and a convex function, respectively. Moreover, $y = f(x)$ is also said to be a strictly concave function if $f''(x) < 0$, and $y = g(x)$ is also said to be a strictly convex function if $g''(x) > 0$.

Concave and convex functions obey few important properties. These properties are the following.

Property I.   Assume that $y = f(x)$ is a concave (strictly concave) function. Then, $-f(x) [= g(x)]$ is a convex (strictly convex) function, and vice versa. This property implies that concave (strictly concave) and convex (strictly convex) functions are the mirror images of each other (see the graphs in Figures 3.5.1(A) and (B)).

Property II.   Assume that $y = f(x)$ (or $g(x)$) are both concave (convex) functions. Then, $f(x) + g(x)$ is a concave (convex) function. Assume again that $f(x)$ and

$g(x)$ are both concave (convex) functions and either one or both of them is a strictly concave (strictly convex) function, then $f(x) + g(x)$ is a strictly concave (strictly convex) function.

Property III.   Assume that $f(x)$ (or $g(x)$) is a linear function. Then $f(x)$ (or $g(x)$) is a concave function and a convex function at the same time (but is neither a strictly concave nor a strictly convex function). Notice that we have already mentioned this property.

3.5.3 Differentiable univariate functions, and concave and convex functions

Notice that in the definitions of $f(x)$ and $g(x)$ as strictly concave, concave, strictly convex, and convex functions given in inequalities (3.5.1)–(3.5.4) we did not use derivatives of the functions $f(x)$ and $g(x)$. In other words, we did not impose the conditions in those definitions that the functions $f(x)$ and $g(x)$ are differentiable functions. Suppose now that these functions are differentiable. Then, one can give alternative definitions to concave and convex (including strictly concave and strictly convex) functions using the derivatives $f'(x)$ and $g'(x)$. We present these below.

Consider the graph of the function $f(x)$ illustrated in Figure 3.5.4(A). In this graph, we have drawn a tangent to the point A. The vertical coordinate of point A is $f(x_0)$ and its horizontal coordinate is x_0. We have drawn another straight line from point A to point B on the same graph. The vertical coordinate of point B is $f(x_1)$ and its horizontal coordinate is x_1. Notice that the change in x is $x_1 - x_0$. Also notice that the slopes of the tangent line AC and the secant line AB are CD/AD and BD/AD, respectively. A visual inspection of the graph shows that the slope of the secant line is smaller than that of the tangent line; that is, (BD/AD) < (CD/AD). But the slope of the secant line is given by $[\{f(x_1) - f(x_0)\}/(x_1 - x_0)]$ and the slope of the tangent is given by $f'(x_0)$. Therefore, we can write the inequality as $[\{f(x_1) - f(x_0)\}/(x_1 - x_0)] < f'(x_0)$. Multiplying both sides of this inequality by $(x_1 - x_0)$ yields $f(x_1) - f(x_0) < f'(x_0)(x_1 - x_0)$, and rearranging gives $f(x_1) < f(x_0) + f'(x_0)(x_1 - x_0)$. This gives us the definition of a differentiable, univariate, strictly

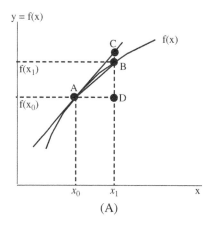

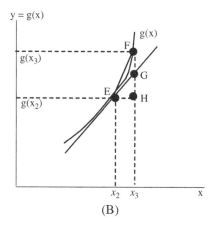

Figure 3.5.4

concave function. Therefore, a differentiable, univariate function is said to be a differentiable, univariate, strictly concave function if the following inequality is satisfied:

$$f(x_1) < f(x_0) + f'(x_0)(x_1 - x_0) \tag{3.5.5}$$

We know from the previous section that if we replace the strong inequality sign by the weak inequality sign, we obtain the definition of a concave function. Therefore, a differentiable, univariate function is said to be a differentiable, univariate, concave function if the following inequality is satisfied:

$$f(x_1) \leq f(x_0) + f'(x_0)(x_1 - x_0) \tag{3.5.6}$$

We now attempt to define a differentiable, univariate, strictly convex function. For this consider the graph of the function $g(x)$ illustrated in Figure 3.5.4(B). In this figure, we have drawn a tangent to the point E. The vertical coordinate of point E is $g(x_2)$ and its horizontal coordinate is x_2. We have drawn a straight line from point E to point F on the same graph. The vertical coordinate of point F is $g(x_3)$ and its horizontal coordinate is x_3. Notice that the change in x is $x_3 - x_2$. Also notice that the slope of the tangent line EG is GH/EH, and the slope of the secant line EF is FH/EH. A visual inspection of the graph shows that the slope of the secant line is larger than that of the tangent line; that is, (FH/EH) > (GH/EH). But the slope of the secant line is given by $[\{g(x_3) - g(x_2)\}/(x_3 - x_2)]$, and the slope of the tangent is given by $g'(x_2)$. Therefore, we can write the above inequality as $[\{g(x_3) - g(x_2)\}/(x_3 - x_2)] > g'(x_2)$. Multiplying both sides of the last inequality by $(x_3 - x_2)$ yields $g(x_3) - g(x_2) > g'(x_2)(x_3 - x_2)$, and rearranging gives $g(x_3) > g(x_2) + g'(x_2)(x_3 - x_2)$. This gives us the definition of a differentiable, univariate, strictly convex function. Therefore, a differentiable, univariate function is said to be a differentiable, univariate, strictly convex function if the following inequality is satisfied:

$$g(x_3) > g(x_2) + g'(x_2)(x_3 - x_2) \tag{3.5.7}$$

As in the case of inequality (5.5.6), if we replace ">" by "≥" in inequality (3.5.7), we obtain the definition of a convex function. Therefore, a differentiable, univariate function is said to be a differentiable, univariate, convex function if the following inequality is satisfied:

$$g(x_3) \geq g(x_2) + g'(x_2)(x_3 - x_2) \tag{3.5.8}$$

3.5.4 Inflection points and the curvature of curves

We found in Section 3.5.1 that a function $y = f(x)$ would increase at $x = x_0$ if $f'(x)|_{x=x_0} > 0$ and would decrease at $x = x_0$ if $f'(x)|_{x=x_0} < 0$. We also saw that the function would increase at an increasing rate if $f''(x)|_{x=x_0} > 0$ and would decrease at a decreasing rate if $f''(x)|_{x=x_0} < 0$.

Suppose that we have a continuous function $y = f(x)$. Then, one might ask whether this function is always strictly concave or strictly convex. The answer is that $y = f(x)$ may or may not always be a strictly concave or a strictly convex function. In fact, the students of the subjects of our interest often deal with functions that are strictly concave (for some values in the domain of the functions) as well as strictly convex (for some other values in the domain of the functions).

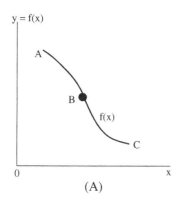
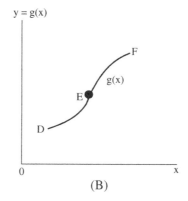

(A)   (B)

Figure 3.5.5

As an example, assume that we have two functions $y = f(x)$ and $y = g(x)$ whose graphs are illustrated, respectively, in Figures 3.5.5(A) and (B). We know, following our previous discussions, that the portion of the graph between points A and B and the portion of the graph between points B and C in Figures 3.5.5(A) represent strictly concave and strictly convex portions, respectively, of the function $y = f(x)$. Similarly, the portion of the graph between points D and E and the portion of the graph between points E and F in Figures 3.5.5(B) represent strictly convex and strictly concave portions, respectively, of the function $y = g(x)$.

How does one know whether a function is always strictly concave, strictly convex, or both strictly concave and convex for different values in the domain of the function? One simple condition, as we saw at the end of Section 3.5.2, for $y = f(x)$ to be a strictly concave function is $f''(x) < 0$, and to be a strictly convex function is $f''(x) > 0$, both evaluated at $x = x_0$. Therefore, one can use the sign of the second derivative of a function to determine whether that function is always strictly concave or always strictly convex.

Before we apply the second derivative test, we need to explain a concept called *inflection point*. An inflection point on the graph of a function, if it exists, is a point after which the second derivative of the function changes its sign from either negative to positive or from positive to negative. In other words, the inflection point on the graph of a function, if it exists, is a point before which the graph of the function will be strictly concave (or strictly convex) and after which the graph of the function will be strictly convex (or strictly concave). The graph of $y = f(x)$ and the graph of $y = g(x)$ in Figure 3.5.5 are the geometric illustrations of strictly concave (and strictly convex) and strictly convex (and strictly concave) functions, respectively. The reason is that $f''(x) < 0$ between points A and B, and $f''(x) > 0$ between points B and C; and $g''(x) > 0$ between points D and E, and $g''(x) < 0$ between points E and F.

The above discussion leads us to a question: what will be the value of the second derivative of the function at the inflection point? Since $f''(x)$ changes sign at the inflection point, its value must be zero (that is, $f''(x) = 0$) or not defined at the inflection point. Therefore, if $f''(x) = 0$ is solved for $x = x^*$, then the function will have an inflection point corresponding to $x = x^*$.

As an example, consider the function $y = f(x) = 18x - 5x^2 + 0.5x^3$. Therefore, $f'(x) = 18 - 10x + 1.5x^2$ and $f''(x) = -10 + 3x$. Setting $f''(x) = -10 + 3x = 0$, and solving for x yields $x = x^* = 3.33$. For values $x = x^* < 3.33$, $f''(x) < 0$ and, therefore, the function is

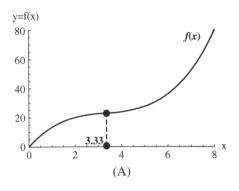 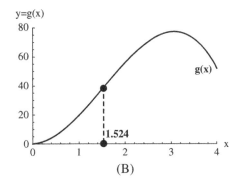

Figure 3.5.6

strictly concave. For values $x = x^* > 3.33$, $f''(x) > 0$ and, therefore, the function is strictly convex. The graph of this function is illustrated in Figure 3.5.6(A).

As another example, consider the function $y = g(x) = x + 24x^2 - 5.25x^3$. Therefore, $g'(x) = 1 + 48x - 15.75x^2$ and $g''(x) = 48 - 31.5x$. Setting $g''(x) = 48 - 31.5x = 0$, and solving for x yields $x = x^* = 1.524$. For values $0 < x = x^* < 1.524$, $g''(x) > 0$ and, therefore, the function is strictly convex. For values $x = x^* > 1.524$, $g''(x) < 0$ and, therefore, the function is strictly concave. The graph of this function is illustrated in Figure 3.5.6(B).

3.5.5 Convex sets

Let us now consider the concept of *convex set*. Convex sets are used extensively in optimization problems, including linear programming problems, which we will consider in the next few chapters. Therefore, we present here an introduction to the concept of convex sets. We first present a geometric definition of convex sets followed by an algebraic definition.

Assume that we have a set of points in a space in $\Re^2$. Suppose that we pick two points in the set and connect them by a straight line. If the resulting line lies completely within the set, then the set is said to be a convex set. All the sets of points in the two-dimensional spaces of Figures 3.5.7(A)–(H) are examples of convex sets. Other examples of convex sets include a straight line, a set with a single point, and a null set with no points.

But, if the resulting line does *not* lie completely within the set, then the set is said to be a *nonconvex set*. All the sets of points in the two-dimensional spaces of Figure 3.5.8(A)–(D) are examples of nonconvex sets.

We can now define a convex set algebraically. Assume that we have two linearly independent row 2-vectors: $\mathbf{u}' = [\, x_1 \; y_1 \,]$ and $\mathbf{v}' = [\, x_2 \; y_2 \,]$. Also assume that $x_1 = 1$, $y_1 = 2$, $x_2 = 2$, and $y_2 = 1$. Then we have $\mathbf{u}' = [\, 1 \; 2 \,]$ and $\mathbf{v}' = [\, 2 \; 1 \,]$. We now choose a scalar s such that $0 \leq s \leq 1$, and generate the linear combination of $\mathbf{u}$ and $\mathbf{v}$ to obtain $s. \begin{bmatrix} x_1 \\ y_1 \end{bmatrix} + (1-s). \begin{bmatrix} x_2 \\ y_2 \end{bmatrix}$.

This is a special type of linear combination, the specialty being $0 \leq s \leq 1$. This special linear combination is called a *convex combination* of vectors $\mathbf{u}$ and $\mathbf{v}$. In terms of our numerical values and using $s = 0.5$, the convex combination of vectors $\mathbf{u}$ and $\mathbf{v}$ can be written as

$$\mathbf{u} + \mathbf{v} = 0.5. \begin{bmatrix} 1 \\ 2 \end{bmatrix} + (1-0.5). \begin{bmatrix} 2 \\ 1 \end{bmatrix} = \begin{bmatrix} 0.5 \\ 1 \end{bmatrix} + \begin{bmatrix} 1 \\ 0.5 \end{bmatrix} = \begin{bmatrix} 1.5 \\ 1.5 \end{bmatrix} = \mathbf{w}.$$ Notice that this convex combination of vectors $\mathbf{u}$ and $\mathbf{v}$ gives us another vector, which lies between the original

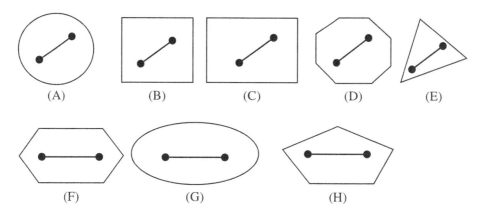

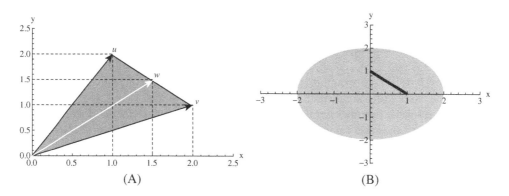

Figure 3.5.8

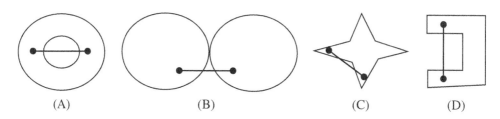

Figure 3.5.9

vectors **u** and **v**. We denote this new vector, generated through the convex combination, by **w**. The geometric illustration of this convex combination is shown in Figure 3.5.9(A).

Similarly, we can generate an infinite number of new vectors like **w** simply by changing the value of s such that $0 \leq s \leq 1$. This way we obtain the entire plane represented by the shaded triangle. Notice an important feature of the convex combination that the new vector(s) that we generate lies (lie) completely with in the triangle. This was the geometric definition of a convex set that we provided earlier. Notice also that this shaded triangle is similar to the triangle in Figure 3.5.7(E).

We can now state the algebraic definition of a convex set. Let there be a set A of points in a plane. Then the set A is said to be a convex set if, for any two vector points **u** and **v** in A and for the scalar s such that $0 \leq s \leq 1$, the following equation holds:

$$\mathbf{w} = s\mathbf{u} + (1 - s)\mathbf{v} \tag{3.5.9}$$

Notice that the vectors we used above (**u** and **v**) can contain values, as components, representing any number of variables, not just two. Therefore, equation (3.5.9) can be used to define a convex set in any dimension. Notice also the similarity between equation (3.5.9) and the parametric equation (2.2.6), which defines a line in $\Re^n$.

As an example, consider the implicit function $2x^2 + 4y^2 = 8$. How can we say whether this function represents a convex set or not? One easy way to answer this question is to plot the graph of the function. This is illustrated in Figure 3.5.9(B). This graph shows that the function $2x^2 + 4y^2 = 8$ gives us an *oval*, which is similar to the graph in Figure 3.5.7(G), which (as per our definition) represents a convex set. Therefore, the set of points represented by the implicit function $2x^2 + 4y^2 = 8$ defines a convex set. Notice that if we pick two points inside the oval, say $x = 1$ and $y = 1$, then the line joining these two points lies completely within the oval.

Let us now consider some related concepts. Suppose that a point (x_0, y_0) exists in a set of points, denoted by A, represented by a plane and that there exists a circle whose centre is the point (x_0, y_0), such as the circle in Figure 3.5.10(A). Then this point is called the *interior point* of the set A if all the points inside the circle lie within the set. If the circle's centre (x_0, y_0) lies on the boundary of the set such that the circle contains points that are included both in the set A and in its complement, as can be seen in Figure 3.5.10(A), then the point (x_0, y_0) is called the *boundary point* of the set A. A set is called an *open set*, such as the set of points represented by Figure 3.5.10(B), if it consists of only the interior points (and does not include the boundary points). If the set contains all its boundary points, such as the one in Figure 3.5.10(C), then the set is called a *closed set*. Notice that the set represented by the plane in Figure 3.5.10(D) is neither an open set nor a closed set.

Notice that the set of points in the planes defined by inequalities $p_x x + p_y y \leq I$ and $p_l l + p_k k \leq G$ and illustrated, respectively, in Figures 2.2.9(A) and (B) are examples of closed sets. The sets defined by these inequalities include the boundary points. However, if we replace the *weak inequalities* (such as $\leq$ or $\geq$) by *strong inequalities* (such as $<$ or $>$) the boundary points are excluded, and, therefore, we obtain open sets.

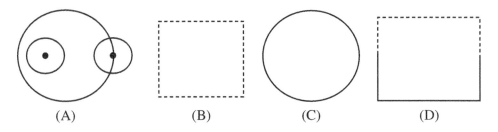

(A)   (B)   (C)   (D)

Figure 3.5.10

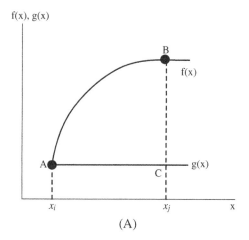

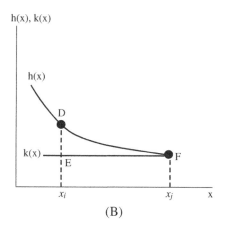

Figure 3.5.11

Three other related terms are *bounded set*, *unbounded set*, and *compact set*. Imagine that we place a set in a large circle. If the set is contained fully within the circle, then that set is called a bounded set. All the sets represented by planes in Figures 3.5.7–3.5.9 (except the one in Figure 3.5.8(A)), and those represented by the triangles in Figures 2.2.9(A) and (B) are examples of bounded sets. If a set is both closed and bounded, then the set is called a compact set. But, the set of points in the plane represented by the inequalities $x \geq 1$ or $x \leq 1$ (the graph of which is illustrated in Figure 1.7.3) are examples of unbounded (closed) sets. Notice that the set of points represented by the plane in Figure 3.5.8(A) is neither closed nor open, but is unbounded.

3.5.6 Univariate quasiconcave and quasiconvex functions

In Section 3.5.2 we presented the meaning, nature, and properties of univariate concave and convex functions. We extended this presentation to include univariate differentiable functions in Section 3.5.3. And in Section 3.5.5 we discussed convex sets and related topics. We are now ready to present functions called *quasiconcave functions* and *quasiconvex functions*.

We begin with a geometric illustration of a quasiconcave function followed by its algebraic representation. Consider the graphs of two univariate functions, $f(x)$ and $g(x)$, illustrated in Figure 3.5.11(A).

As *concavity* is subdivided into *strict concavity* and *weak concavity* (or, simply, concavity), *quasiconcavity* is also subdivided into *strict quasiconcavity* and *weak quasiconcavity* (or, simply, quasiconcavity). Notice that both functions illustrated in Figure 3.5.11(A) have the same domain, which is a convex set. Suppose that we pick two different points x_i and x_j such that $x_i < x_j$ in this convex domain. Also suppose that the function $f(x)$ forms an arc between x_i and x_j such that the point on the arc corresponding to x_i is A and the point on the arc corresponding to x_j is B. The feature of the arc AB is that point B is higher than point A. Then the function $f(x)$ is said to be a *strictly quasiconcave function*. Now consider the graph of the function $g(x)$, which forms a straight line between points A and C corresponding, respectively, to points x_i and x_j, respectively. The feature of this line is that point C is higher than or equal to point A (in fact, they are equal). Then the function $g(x)$ is said to be a *weak quasiconcave function* or, simply, a quasiconcave function. This suggests that a linear function is a quasiconcave function.

Let us now provide an algebraic definition of quasiconcave functions. Although the definition we present here pertains to univariate functions, it can be easily generalized to multivariate functions. Assume that f is a function of x. Then, for any two points x_i and x_j in the convex domain of the function such that $x_i < x_j$ and for $0 < \alpha < 1$, the function is said to be a strictly quasiconcave function if the following inequality is satisfied:

$$f(x_j) > f(x_i) \Rightarrow f\left(\alpha x_i + (1 - \alpha)x_j\right) > f(x_i) \tag{3.5.10}$$

If we replace the strict inequality (3.5.10) with weak inequality in the case of a function ($g(x)$ in our example), we have the definition for a weak quasiconcave function as

$$g(x_j) \geq g(x_i) \Rightarrow g\left(\alpha x_i + (1 - \alpha)x_j\right) \geq g(x_i) \tag{3.5.11}$$

Let us now present the geometric illustration of a quasiconvex function. Consider the graphs of the two univariate functions, $h(x)$ and $k(x)$, illustrated in Figure 3.5.11(B). As in the case of *convexity*, *quasiconvexity* is also subdivided into *strict quasiconvexity* and *weak quasiconvexity* (or, simply, quasiconvexity). Notice that the domain of both functions illustrated in Figure 3.5.11(B) is a convex set. Suppose that we pick two different points, x_i and x_j, in this convex domain such that $x_i < x_j$. Also suppose that the function $h(x)$ forms an arc between x_i and x_j such that the point on the arc corresponding to x_i is D and the point on the arc corresponding to x_j is F. The feature of the arc DF is that point D is higher than point F. Then the function $h(x)$ is said to be a *strictly quasiconvex function*. Now consider the graph of the function $k(x)$, which forms a straight line between points E and F corresponding to points x_i and x_j, respectively. The feature of this graph is that point F is higher than or equal to point E (in fact, they are equal). Then the function $k(x)$ is said to be a *weak quasiconvex function* or, simply, a quasiconvex function. This suggests that a linear function is quasiconcave as well as quasiconvex.

We can now present an algebraic definition of a quasiconvex function. Although the definition we present here pertains to univariate functions, it can be generalized to multivariate functions. Assume that h is a function of x. Then, for any two points x_i and x_j in the convex domain of the function such that $x_i < x_j$ and for $0 < \alpha < 1$, the function is said to be a strictly quasiconvex function if the following inequality is satisfied:

$$h(x_i) > h(x_j) \Rightarrow h(x_i) > h\left(\alpha x_i + (1 - \alpha)x_j\right) \tag{3.5.12}$$

If we replace the strict inequality (3.5.12) with weak inequality in the case of a function ($k(x)$ in our example), we have the definition for a weak quasiconcave function as

$$k(x_i) \geq k(x_j) \Rightarrow k(x_i) \geq k\left(\alpha x_i + (1 - \alpha)x_j\right) \tag{3.5.13}$$

So far we have not imposed the condition of differentiability on all the functions ($f(x)$, $g(x)$, $h(x)$, and $k(x)$) we have considered in the present section. Now assume that these functions are differentiable. Then these functions are strictly quasiconcave, quasiconcave, strictly quasiconvex, and quasiconvex if, for two points x_i and x_j on their convex domain such that $x_i < x_j$, the following inequalities, respectively, are satisfied:

$$f(x_j) > f(x_i) \Rightarrow f'(x_i)[x_j - x_i] > 0 \tag{3.5.14}$$

$$g(x_j) \geq g(x_i) \Rightarrow g'(x_i)[x_j - x_i] \geq 0 \tag{3.5.15}$$

$$h(x_i) > h(x_j) \Rightarrow h'(x_i)[x_j - x_i] < 0 \tag{3.5.16}$$

and

$$k(x_i) \geq k(x_j) \Rightarrow k'(x_i)[x_j - x_i] \leq 0 \tag{3.5.17}$$

Quasiconcave (or quasiconvex) functions obey three important properties. These properties are:

Property I.   As we stated before, a linear function is both quasiconcave and quasiconvex.

Property II.   All concave (convex) functions, strict or not strict, are quasiconcave (quasiconvex), strict or not strict, but the opposite is not valid.

Property III.   If a function is quasiconcave, strict or not strict, then the negative of that function is quasiconvex, strict or not strict.

3.5.7 Determinantal tests for quasiconcavity and quasiconvexity

We discussed in the previous section the nature of quasiconcave and quasiconvex functions. But, how does one know whether a function is quasiconcave or quasiconvex in the first place? One can rely on a test called a *determinantal test* to determine whether the function is quasiconcave or quasiconvex. This test uses the determinant of the partial derivatives (see Section 3.7) of the function.

Suppose that we have a multivariate function, say, $f(x_1, x_2, x_3, \ldots, x_n)$. Also suppose that this function is twice differentiable. Then, one can show that the function $f(x_1, x_2, x_3, \ldots, x_n)$ is quasiconcave if the principal minors of the matrix

$$\mathbf{Q} = \begin{bmatrix} 0 & f_1 & f_2 & f_3 & \cdots & f_n \\ f_1 & f_{11} & f_{12} & f_{13} & \cdots & f_{1n} \\ f_2 & f_{21} & f_{22} & f_{23} & \cdots & f_{2n} \\ f_3 & f_{31} & f_{32} & f_{33} & \cdots & f_{3n} \\ \cdots & \cdots & \cdots & \cdots & \cdots & \cdots \\ f_n & f_{n1} & f_{n2} & f_{n3} & \cdots & f_{nn} \end{bmatrix}$$

alternate in sign, beginning with nonpositive (that is, for even n, $|\mathbf{Q}_n| \geq 0$; and for odd n, $|\mathbf{Q}_n| \leq 0$). In other words, the function $f(x_1, x_2, \ldots, x_n)$ is quasiconcave if

$$|\mathbf{Q}_1| = \begin{vmatrix} 0 & f_1 \\ f_1 & f_{11} \end{vmatrix} \leq 0, |\mathbf{Q}_2| = \begin{vmatrix} 0 & f_1 & f_2 \\ f_1 & f_{11} & f_{12} \\ f_2 & f_{21} & f_{22} \end{vmatrix} \geq 0, |\mathbf{Q}_3| = \begin{vmatrix} 0 & f_1 & f_2 & f_3 \\ f_1 & f_{11} & f_{12} & f_{13} \\ f_2 & f_{21} & f_{22} & f_{23} \\ f_3 & f_{31} & f_{32} & f_{33} \end{vmatrix} \leq 0, \ldots$$

Table 3.5.2

Function	Quasiconcave	Quasiconvex
$f(x_1, x_2, x_3, \ldots, x_n)$	$\|Q_1\| \leq 0, \|Q_2\| \geq 0, \|Q_3\| \leq 0, \ldots$	$\|Q_1\| \leq 0, \|Q_2\| \leq 0, \|Q_3\| \leq 0, \ldots,$ $\|Q_n\| = \|Q\| \leq 0$
	That is, $\|Q_n\| \leq 0$ if n is odd and $\|Q_n\| \geq 0$ if n is even	That is, $\|Q_n\| \leq 0$ for all n

One can also show that the function $f(x_1, x_2, x_3, \ldots, x_n)$ is quasiconvex if the principal minors of the matrix Q are all nonpositive (that is, $\|Q_n\| \leq 0$, for all n). In other words, the function $f(x_1, x_2, \ldots, x_n)$ is quasiconvex if

$$|Q_1| = \begin{vmatrix} 0 & f_1 \\ f_1 & f_{11} \end{vmatrix} \leq 0, |Q_2| = \begin{vmatrix} 0 & f_1 & f_2 \\ f_1 & f_{11} & f_{12} \\ f_2 & f_{21} & f_{22} \end{vmatrix} \leq 0, |Q_3| = \begin{vmatrix} 0 & f_1 & f_2 & f_3 \\ f_1 & f_{11} & f_{12} & f_{13} \\ f_2 & f_{21} & f_{22} & f_{23} \\ f_3 & f_{31} & f_{32} & f_{33} \end{vmatrix} \leq 0, \ldots$$

We present these tests of quasiconcavity and quasiconvexity in Table 3.5.2 for easy reference.

3.5.8 Application examples

Example 1. Assume that the inverse demand function for a good is given by $p = f(q) = 1/(1+q)$, where p denotes the price of the good per unit in dollars and q denotes the quantity of the good demanded at price p. What will be the curvature of the demand function for all $q > 0$?

Solution. To see the curvature of the function, we need to find the first and second derivatives of the function. They are $f'(q) = [\{(1+q) \times 0 - 1 \times 1\}/(1+q)^2] = -1/(1+q)^2$ and $f''(q) = [\{(1+q)^2 \times 0 - (-1 \times 1)\}/(1+q)^3] = 1/(1+q)^3$, respectively. Since the first derivative is negative and the second derivative is positive for all $q > 0$, the demand function is diminishing at an increasing rate.

Example 2. Suppose that an individual invests $100 in her bank account. Also suppose that the bank compounds interest continuously at 8 percent. Then the amount of money (F), after t years, in her account will be given by the function $F = g(t) = 100e^{0.08t}$. Determine whether this function is strictly concave (or concave) or strictly convex (or convex).

Solution. To determine whether the function is strictly concave or strictly convex, we need to find the second derivative of the function. The first and second derivatives are $g'(t) = 100e^{0.08t}(0.08) = 8e^{0.08t}$ and $g''(t) = 8e^{0.08t}(0.08) = (0.64)e^{0.08t}$, respectively. Since the second derivative is positive for all $t > 0$, the function is strictly convex.

Example 3. Suppose that a firm's output y when it employs l units of labor is given by the function $y = f(l) = l^{3/2}$. Determine whether this production function is strictly concave (or concave) or strictly convex (or convex).

Solution. To determine whether the production function is strictly concave or strictly convex, we need to find the second derivative of the function. The first and second derivatives are $f'(l) = (3/2)l^{1/2}$ and $f''(l) = (3/2)(1/2)l^{-1/2}$, respectively. Since the second derivative is positive for all $l > 0$, the function is strictly convex.

Example 4. Suppose that the total monthly consumption expenditure (C) in dollars of a household is given by the function $C = h(Y) = 20Y + 30Y^2 - 0.4Y^3$, where Y denotes the household's monthly *disposable income* in dollars. Does the curvature of the *consumption function* change, and (if it does) at what level of income does it change?

Solution. To determine whether the consumption function changes it curvature, we need to determine the value of the second derivative of the function when it is equal to zero. The first and second derivatives are $h'(Y) = 20 + 60Y - 1.2Y^2$ and $h''(Y) = 60 - 2.4Y$, respectively. Therefore, the value of the second derivative when it is equal to zero is $h''(Y)|_{Y=0} = 60 - 2.4Y|_{Y=0} = 60 - 2.4Y^* = 0$ or $Y^* = 25$. Notice that for all $0 < Y < 25$, the second derivative is positive. Similarly, it is negative for all $25 > Y$; that is, $h''(Y)|_{Y>25} = 60 - 2.4Y|_{Y>25} < 0$. Therefore, the consumption function is strictly convex until $Y = 25$ and is strictly concave after $Y = 25$. This implies that the inflection point occurs or the curvature of the consumption function changes at $Y = 25$.

3.5.9 Exercises

1. Determine the curvature of the following functions:
   (i) $y = f(x) = 1 + 2x$; (ii) $y = g(x) = 2 - 2x$; (iii) $y = h(x) = 1 - (1/x)$;
   (iv) $y = h(x) = 1 + (1/x)$.
2. Determine whether the following functions are strictly concave (or concave) or strictly convex (or convex):
   (i) $y = f(x) = 1 + 2x$; (ii) $y = g(x) = 2 - 2x$; (iii) $y = h(x) = 1 - (1/x)$;
   (iv) $y = h(x) = 1 + (1/x)$.
3. Determine whether the following functions possess an infection point for $x > 0$. If they do, determine the value of x at which it happens.
   (i) $y = f(x) = 25x^2 - 3x^3$; (ii) $y = g(x) = -10x + x^2$; (iii) $y = h(x) = 10x^2 - x^3$.
4. *Application exercise.* Suppose that the total utility (U) an individual obtains from the consumption of different units of a good is given by the function $U = f(x) = \ln x$, where x denotes the quantity of the good that the individual consumes. Determine whether the individual's total utility increases (decreases) at an increasing (decreasing) rate or remains constant.
5. *Application exercise.* Assume that the total profit (Π) of a firm that produces a good is given by the function $\Pi = g(x) = (0.2)e^{\ln x}$, where x denotes the quantity of the good produced. Determine whether the firm's total profit function is strictly concave (or concave) or strictly convex (or convex).
6. *Application exercise.* Assume that the *budget set* of a consumer, who buys two goods x and y, is given by the inequality $4x + 8y \leq 40$, where $x \geq 0$ and $y \geq 0$. Show this budget set in a figure, and determine whether it is a convex set or not.

3.6 Derivatives and transformation of univariate functions: Maclaurin and Taylor series

There may be situations where one has to deal with the transformed forms of some functions. In such situations, one has to transform the given function into another form. The *transformation* of a function is also called the *expansion* of the function or the *approximation* of the function.

Two methods widely used to expand a function are *Maclaurin series* or *Maclaurin expansion* and *Taylor series* or *Taylor expansion*. Let there be a function $y = f(x)$. Then, the Maclaurin series expands this function around the point $x = 0$ in the domain of the function, while the Taylor series expands this function at any point $x = x_0$ in its domain. Both methods use the derivatives of the function $y = f(x)$ evaluated at $x = 0$ (in the former) and at $x = x_0$ (in the latter). The expanded or transformed function will be a new *polynomial function* called a *power series*.

3.6.1 Meaning of linearization of a univariate nonlinear function

Assume that our function $y = f(x)$ takes the form given by the graph in Figure 3.6.1(A). We can give a description of the linear approximation of $y = f(x)$ with this figure. We can see that $f(x_1) = Dx_1$ and $f(x_0) = x_0A = x_1B$. Notice that $f(x_1) = Dx_1 = x_1B + BC + CD$. Since $f(x_0) = x_1B$ and $BC = f'(x_0).AB = f'(x_0)(x_1 - x_0)$, $f(x_1)$ can be written as $f(x_1) = Dx_1 = f(x_0) + f'(x_0)(x_1 - x_0) + CD$. The last equation means that we have transformed the function $y = f(x)$ into the form $f(x_1)$, which is equal to the value of the function and its derivative times $(x_1 - x_0)$ at the point $x = x_0$ plus a value equal to the distance CD. Using the standard terminology, we state that the function $f(x)$ has been expanded about the point $x = x_0$ and the *centre of expansion* is x_0.

The vertical distance CD (or R) in Figure 3.6.1(A) is called the *remainder*. It is easy to see that the quality of approximation improves as the remainder becomes smaller. We will see in the next section that as the degree of the transformed polynomial function increases, the value of R will decrease and, thereby, the quality of approximation will improve.

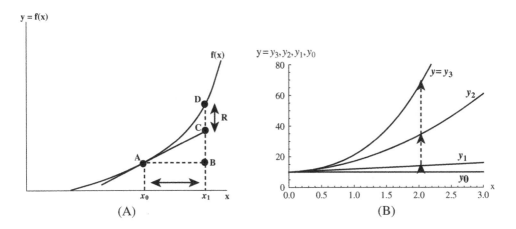

Figure 3.6.1

3.6.2 Maclaurin series

Suppose that we have an nth-degree polynomial function as

$$y = f(x) = a_0 x^0 + a_1 x^1 + a_2 x^2 + a_3 x^3 + \cdots + a_n x^n \tag{3.6.1}$$

We know that $f'(x) = a_1 + 2a_2 x + 3a_3 x^2 + \cdots + na_n x^{n-1}$, $f''(x) = 2a_2 + 6a_3 x + \cdots + n(n-1)a_n x^{n-2}$, $f'''(x) = 6a_3 x + \cdots + n(n-2)a_n x^{n-3}, \ldots$, and $f^n(x) = n(n-1)(n-2)\ldots(2)(1)a_n$. We have seen in Section 1.10.6 that $n(n-1)(n-2)\ldots(2)(1)a_n = n!$. If we evaluate these derivatives at $x = 0$, we obtain $f'(0) = a_1$, $f''(0) = 2a_2, \ldots$ and $f^n(0) = n!a_n$. Therefore, these equations can be written as $[f'(0)/1!] = a_1, [f''(0)/2!] = a_2, \ldots$, and $[f^n(0)/n!] = a_n$. When we substitute these results into equation (3.6.1), we obtain

$$y = f(x) = \left[\frac{f(0)}{0!}\right] + \left[\frac{f'(0)}{1!}\right]x + \left[\frac{f''(0)}{2!}\right]x^2 + \cdots + \left[\frac{f^n(0)}{n!}\right]x^n \tag{3.6.2}$$

which is called the Maclaurin series or expansion of the polynomial function (3.6.1). Notice that the original function has been expanded around $x = 0$. This means that when we expand a function of x around $x = 0$ we must obtain the original function.

As an example, consider the first-degree polynomial function $y = f(x) = 10 + 2x$. Then, $f(0) = 10$ and $f'(x) = 2$; and when the derivative is evaluated at $x = 0$, we have $f'(0) = 2$. We do not need to take higher derivatives as they are all zeros. Then, equation (3.6.2) implies that $y = f(x) = f(0)/0! + [f'(0)/1!]x = 10/0! + (2/1!)x = 10 + 2x$, which is exactly our original function. This implies that the Maclaurin series describes the original function exactly.

As a second example, consider the second-degree polynomial function $y = f(x) = 10 + 2x + 5x^2$. Then, $f'(x) = 2 + 10x$ and $f''(x) = 10$; and when these derivatives are evaluated at $x = 0$, we have $f'(0) = 2$ and $f''(0) = 10$, respectively. As before, we do not take higher derivatives as they are all zeros. Notice that $f(0) = 10$. Then, equation (3.6.2) implies that $y = f(x) = [f(0)/0!] + [f'(0)/1!].x + [f''(0)/2!].x^2 = [10/0!] + [2/1!].x + [[10/2!].x^2 = 10 + 2x + 5x^2$. Again the Maclaurin series describes the original function exactly.

As a third example, consider the third-degree polynomial function $y = f(x) = 10 + 2x + 5x^2 + 4x^3$. Then, $f'(x) = 2 + 10x + 12x^2$, $f''(x) = 10 + 24x$, and $f'''(x) = 24$; and when we evaluate these derivatives at $x = 0$, we have $f'(0) = 2$, $f''(0) = 10$, and $f'''(0) = 24$, respectively. Notice that $f(0) = 10$. Then, equation (3.6.2) implies that $y = f(x) = [f(0)/0!] + [f'(0)/1!].x + [f''(0)/2!].x^2 + [f'''(0)/3!].x^3 = [10/0!] + [2/1!].x + [10/2!].x^2 + [24/3!].x^3 = 10 + 2x + 5x^2 + 4x^3$. Once again the Maclaurin series describes the original function exactly.

We stated at the end of the previous section that as the degree of the transformed polynomial function increases, the value of the remainder R will decrease and, thereby, the quality of approximation will increase. Consider our last example with $y = f(x) = 10 + 2x + 5x^2 + 4x^3$. If we consider only the zero-degree polynomial or $f(x) = 10$, then the Maclaurin series gives $y_0 = f_0(x) = f(0)/! = 10$ (where the subscript shows the series of the new polynomial function (the Maclaurin series) when only the first term is considered). Similarly, if we consider the first-degree polynomial, the series will be $y_1 = f_1(x) = [f(0)/0!] + [f'(0)/1!].x = 10 + 2x$; if we consider the second-degree polynomial, the series will be $y_2 = f_2(x) = [f(0)/0!] + [f'(0)/1!].x + [f''(0)/2!].x^2 = [10/0!] + [2/1!].x + [[10/2!].x^2 = 10 + 2x + 5x^2$; and when we consider the third-degree polynomial, the series will be $y_3 = f_3(x) = [f(0)/0!] + [f'(0)/1!].x +$

$[f''(0)/2!].x^2 + [f'''(0)/3!].x^3 = [10/0!] + [2/1!].x + [10/2!].x^2 + [24/3!].x^3 = 10 + 2x + 5x^2 + 4x^3$. The graphs of all these polynomial functions (from zero degree to third degree) are illustrated in Figure 3.6.1(B). These graphs clearly show that as the degree of the polynomial increases, the Maclaurin series approaches the original function that is expanded. The declining vertical distances represented by arrows between graphs show the reduction in errors or remainders in successive approximations.

3.6.3 *Taylor series*

In Maclaurin series, we expanded the function $y = f(x)$ around $x = 0$. It is not necessary that we always expand the function around $x = 0$. One can use any other value in the domain of the function, say $x = x_0$, to expand the function. This is, in fact, done in Taylor series.

For a discussion of the Taylor series or expansion, consider the first-degree polynomial function we used in the previous section: $y = f(x) = 10 + 2x$. Let us expand this function around $x = x_0$. Assume that, as in Figure 3.6.1(A), $x = x_1 = x_0 + \varepsilon$. Then, the function $y = f(x) = 10 + 2x$ becomes $f(x) = 10 + 2(x_0 + \varepsilon)$. Notice that, in the last function, x_0 is a constant and $\Delta x = \varepsilon$ is the variable. Given these, $f(x)$ can be written as $g(\varepsilon) = 10 + 2(x_0 + \varepsilon) \equiv f(x)$. Then the derivative of $g(\varepsilon)$ is $g'(\varepsilon) = 2$. Notice that the derivatives higher than $g'(\varepsilon)$ are all zeros, and therefore we discard them.

Notice that when we expand $g(\varepsilon)$ around $\varepsilon = 0$ we obtain the Maclaurin series as in equation (3.6.2). That is, we will get the following equation when we expand $g(\varepsilon)$ around $\varepsilon = 0$:

$$g(\varepsilon) = \left[\frac{g(0)}{0!}\right] + \left[\frac{g'(0)}{1!}\right].\varepsilon \qquad (3.6.3)$$

But, $x_1 = x_0 + \varepsilon$ implies that, when $\varepsilon = 0$, $x_1 = x_0$. Then, we obtain $g(0) = f(x_0)$ and $g'(0) = f'(x_0)$. When we substitute the last two results in equation (3.6.3), we get

$$g(\varepsilon) = \left[\frac{f(x_0)}{0!}\right] + \left[\frac{f'(x_0)}{1!}\right].\varepsilon = \left[\frac{f(x_0)}{0!}\right] + \left[\frac{f'(x_0)}{1!}\right].(x_1 - x_0) \qquad (3.6.4)$$

We know that x_1 can be any value of x. Therefore, we replace x_1 by x. This implies that we can rewrite equation (3.6.4) as

$$g(\varepsilon) = \frac{f(x_0)}{0!} + \frac{f'(x_0)}{1!}\varepsilon = \frac{f(x_0)}{0!} + \frac{f'(x_0)}{1!}(x - x_0) \qquad (3.6.5)$$

Equation (3.6.5) is the Taylor series expansion of the linear polynomial function $y = f(x) = 10 + 2x$. In the case of our example of linear polynomial function $y = f(x) = 10 + 2x$, $f(x_0) = 10 + 2x_0$ and $f'(x_0) = 2$. Substituting these in equation (3.6.5) yields $g(\varepsilon) = [f(x_0)/0!] + [f'(x_0)/1!](x - x_0) = 10 + 2x_0 + 2(x - x_0) = 10 + 2x$, which shows that the Taylor series approximates exactly the function in our example.

Now suppose that, instead of a first-degree polynomial function, we have a second-degree polynomial function $y = f(x) = 2 + 3x + 4x^2$. Then, following the above arguments, we can write the Taylor series for the second-degree polynomial function in our example as

$$g(\varepsilon) = [f(x_0)/0!] + [f'(x_0)/1!].\varepsilon + [f''(x_0)/2!].\varepsilon^2$$
$$= [f(x_0)/0!] + [f'(x_0)/1!].(x - x_0) + [f''(x_0)/2!].(x - x_0)^2 \qquad (3.6.6)$$

With respect to our present example, we can obtain $f(x_0) = 2 + 3x_0 + 4x_0^2$, $f'(x_0) = 3 + 8x_0$, and $f''(x_0) = 8$. Substituting these in equation (3.6.6) gives $g(\varepsilon) = [f(x_0)/0!] + [f'(x_0)/1!].(x-x_0) + [f''(x_0)/2!].(x-x_0)^2 = 2 + 3x_0 + 4x_0^2 + (3+8x_0)(x-x_0) + 4(x-x_0)^2 = 2 + 3x + 4x^2 = y = f(x)$. This, once again, shows that the Taylor series correctly approximates the original second-degree polynomial function.

Let us now generalize the above result to an nth-degree polynomial function as the function in equation (3.6.1). The Taylor series of the expansion of this nth-degree polynomial function is given as

$$f(x) = [f(x_0)/0!] + [f'(x_0)/1!].(x-x_0) + [f''(x_0)/2!].(x-x_0)^2$$
$$+ [f'''(x_0)/3!].(x-x_0)^3 + \cdots + [f^n(x_0)/n!].(x-x_0)^n \qquad (3.6.7)$$

Notice that the only difference between Taylor series and Maclaurin series (equations (3.6.7) and (3.6.2), respectively) is that if we use $x_0 = 0$ in the former, we obtain the latter. Therefore, the Maclaurin series is a special case of the Taylor series. What equation (3.6.7) says is that if we pick two numbers in the domain of the function (say, $x = x_1 = 10$ and $x_0 = 5$) and evaluate the RHS of equation (3.6.7), then the RHS will be equal to $f(x = x_1 = 10)$.

3.6.4 Maclaurin and Taylor series with remainders

So far, we have been concerned with the expansion of an nth-degree polynomial function into another nth-degree polynomial function. In fact, we can expand (around, say, x_0) any function, say $h(x)$, not necessarily a polynomial function, into a polynomial form. But, the condition for the expansion of an arbitrary function $h(x)$ is that it must have finite, continuous derivatives up to the required degree at x_0.

Taylor series with remainder is based on the *Taylor theorem*, which states that a function $h(x)$ can be expanded around x_0 as

$$h(x) = [h(x_0)/0!] + [h'(x_0)/1!].(x-x_0) + [h''(x_0)/2!].(x-x_0)^2 + \cdots$$
$$+ [h^n(x_0)/n!].(x-x_0)^n + [h^{n+1}(x_0)/(n+1)!].(x-x_0)^{n+1} \qquad (3.6.8)$$

where the last term $[h^{n+1}(x_0)/(n+1)!].(x-x_0)^{n+1} = R_n$ is called the remainder or error of approximation. It can be shown that $R_n \to 0$ as $n \to \infty$. In this case, we will have

$$h(x) = [h(x_0)/0!] + [h'(x_0)/1!].(x-x_0) + [h''(x_0)/2!].(x-x_0)^2 + \cdots + [h^n(x_0)/n!].(x-x_0)^n$$
$$(3.6.9)$$

Notice that if we expand a polynomial function into another polynomial function of the same degree, then the term R_n will be equal to zero. If we expand $h(x)$ around $x = 0$, then we have

$$h(x) = [h(0)/0!] + [h'(0)/1!].x + [h''(0)/2!].x^2 + \cdots + [h^n(0)/n!].x^n + R_n \qquad (3.6.10)$$

which is called the *Maclaurin series with remainder*. Suppose we assume now that $n = 0$. In this case equation (3.6.8) will reduce to $h(x) = [h(x_0)/0!] + [h'(x_0)/1!].(x-x_0) = h(x_0) + h'(x_0)(x-x_0)$. This result is popularly called the *mean-value theorem*. Notice that we have already used this theorem few times earlier.

3.6.5 Exercises

1. Expand the following functions using Maclaurin series:
   (i) $f(x) = 2x + 3x^2 + 5x^3$; (ii) $g(x) = 3x + 2x^2 + 5x^3$; (iii) $h(x) = 5x + 3x^2 + 2x^3$.
2. Expand the above functions using Taylor series when $x = 5$ and $x_0 = 3$.

 Web supplement: S3.6.6 *Euler relations*

 Web supplement: S3.6.7 Mathematica applications

3.7 Differentiation of multivariate functions: partial derivatives

So far in this chapter we were dealing with differentiation of univariate functions and their applications in specific cases. However, most of the relationships in economics, business, and finance involve more than one independent variable. Functions that involve more than one independent variable are called multivariate functions. We noted in Section 3.2 that the derivative of a univariate function $y = f(x)$ was interpreted as the rate of change of y when x changes by an infinitesimally small amount. One may wonder if a similar concept can be developed in the case of a multivariate function.

As an illustration, consider the example of a firm that produces Q units of output using K units of capital and L units of labor. We can write the output produced as a function of capital and labor as $Q = f(K, L)$. This production function is an example of a multivariate function with two independent variables, which is also called a bivariate function. An important question that arises now is: what impact does a change in capital have on the output produced? Another question is: if there is an impact, how does one find it? These questions are answered below.

3.7.1 Partial differentiation: meaning, notations, and method

Consider the bivariate function discussed above: $Q = f(K, L)$. We know that when K changes by a particular amount, keeping L constant at $\overline{L}$, there will be a corresponding change in Q. Let us denote the change in K by ΔK and the change in Q by ΔQ. As we have seen in the case of univariate differentiation, we divide the latter change by the former change to yield

$$\frac{\Delta Q}{\Delta K} = \frac{f(K + \Delta K, \overline{L}) - f(K, \overline{L})}{\Delta K} \tag{3.7.1}$$

which gives the difference quotient in the case of the bivariate function under consideration. Notice that equation (3.7.1) is based on the assumption that L is held constant at $\overline{L}$. If we take the limit, if it exists, on both sides of equation (3.7.1) when $\Delta K \to 0$, we obtain what is called the *partial derivative* of Q with respect to K. Notice the term "partial derivative." This derivative is called the partial derivative because the change on the RHS is partial or the change on the RHS is taking place in K only as L is held constant. The process of finding the partial derivative is called *partial differentiation*. Therefore, the partial derivative of Q with respect to K is $\lim\limits_{\Delta K \to 0} [\Delta Q / \Delta K] = \lim\limits_{\Delta K \to 0} [\{f(K + \Delta K, \overline{L}) - f(K, \overline{L})\} / \Delta K]$.

Although we can denote the partial derivative of Q with respect to K in many forms, we denote it by either $\partial Q/\partial K$ or f_K. Therefore, we have $\partial Q/\partial K = f_K = \lim_{\Delta K \to 0} [\Delta Q/\Delta K]$. The partial derivative of the production function with respect to capital, $\partial Q/\partial K = f_K$, is called the *marginal product of capital*.

Exactly as we held L constant and allowed K to vary, we now can hold K constant at $\overline{K}$ and allow L to vary (say, by ΔL). This will also produce a change in Q. Then, we obtain a ratio similar to the one in equation (3.7.1): $\Delta Q/\Delta L = [f(\overline{K}, L + \Delta L) - f(\overline{K}, L)]/\Delta L$. If we follow a process similar to the one we used in the derivation of $\partial Q/\partial K$ or of f_K, we obtain the partial derivative of Q with respect to L as $\partial Q/\partial L$ or f_L. This partial derivative with respect to labor is called the *marginal product of labor*. Notice that $\partial Q/\partial L$ or f_L is derived assuming that K is held constant at $\overline{K}$.

We now extend the above results to the case of a multivariate function involving n independent variables. Assume that we have the function $y = f(x_1, x_2, x_3, \ldots, x_i, \ldots, x_n)$. Then the partial derivative of the function with respect to x_i is defined as

$$\lim_{\Delta x_i \to 0} \left[\frac{\Delta y}{\Delta x_i} \right] = \lim_{\Delta x_i \to 0} \left[\frac{f(x_1, x_2, x_3, \ldots, x_i + \Delta x_i, \ldots, x_n) - f(x_1, x_2, x_3, \ldots, x_i, \ldots, x_n)}{\Delta x_i} \right]$$

(3.7.2)

and is denoted by $\partial y/\partial x_i = f_{x_i}$. From now on we will denote $\partial y/\partial x_i = f_{x_i}$, for convenience, by f_i. Notice that the rules of univariate differentiation are equally applicable in the case of multivariate differentiation. Therefore, one does not need to learn a different set of rules for multivariate differentiation.

Let us apply the above definition to specific cases. Suppose that our function is $y = f(x_1, x_2) = x_1^2 + x_2^2$. Then, using the above definition and holding x_2 constant, the partial derivative of the function with respect to x_1 is $\partial y/\partial x_1 = f_{x_1} = f_1 = 2x_1$. Similarly, the partial derivative of the function with respect to x_2, holding x_1 constant, is $\partial y/\partial x_2 = f_{x_2} = f_2 = 2x_2$.

Consider another function: $y = g(x_1, x_2) = x_1^2 + x_2^2 + x_1 x_2$. Then, holding x_2 constant, the partial derivative of the function with respect to x_1 is $\partial y/\partial x_1 = g_{x_1} = g_1 = 2x_1 + x_2$, and the partial derivative of the function with respect to x_2, holding x_1 constant, is $\partial y/\partial x_2 = g_{x_2} = g_2 = 2x_2 + x_1$.

3.7.2 *Partial derivatives: geometric illustrations*

As they will help easier understanding, we present here the geometric illustrations of partial derivatives. Suppose that we have our last bivariate production function: $Q = f(K, L)$. Suppose that we keep L constant at $L = L_0$. Then the production function can be written as $Q = f(K, L_0)$. Also suppose that this production function takes the form of the graph in the three-dimensional space as illustrated in Figure 3.7.1(A). Notice that the graph in this figure is obtained by cutting the space with a vertical plane parallel to the LQ space. The slope of the graph at $K = K_i$ and $L = L_0$ (which is equal to the slope of the tangent AB at $K = K_i$ and $L = L_0$) is equal to the derivative of $Q = f(K, L)$ with respect to K at $K = K_i$ and $L = L_0$ or $\partial Q/\partial K|_{K=K_i, L=L_0} = f_K|_{K=K_i, L=L_0}$.

Similarly, we can present the geometric illustration of the derivative of $Q = f(K, L)$ with respect to L at $L = L_i$ and holding K constant at K_0. This is illustrated in Figure 3.7.1(B). The slope of the graph at $K = K_0$ and $L = L_i$ (which is equal to the slope of the tangent AB at $K = K_0$ and $L = L_i$) is equal to the derivative of $Q = f(K, L)$ with respect to L at $K = K_0$ and $L = L_i$ or $\partial Q/\partial L|_{K=K_0, L=L_i} = f_L|_{K=K_0, L=L_i}$.

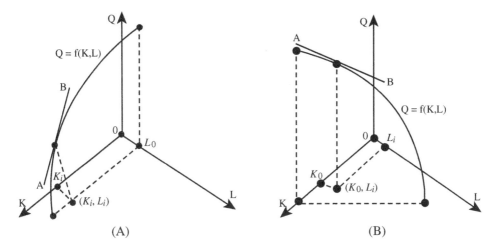

Figure 3.7.1

3.7.3 Higher partial derivatives

In the case of univariate functions, we found higher derivatives by successively differentiating the function until the higher derivative became zero. Similarly, we can obtain *higher partial derivatives* by successively differentiating the multivariate function until the higher partial derivative becomes zero. But, one important point to remember here is that when we find the higher derivative of the multivariate function with respect to one independent variable, the other independent variable(s) is (are) to be held constant.

Suppose that we have a general function $y = f(x_1, x_2, \ldots, x_i, \ldots, x_n)$. Then, the *first partial derivative* of the function with respect to x_i is $\partial y/\partial x_i = f_{x_i} = f_i$. The *second partial derivative* of the function with respect to x_i is found by partially differentiating the first partial derivative with respect to x_i, holding x_j (where $i \neq j$) constant. This is normally denoted by $\partial(\partial y)/\partial x_i(\partial x_i) = \partial^2 y/\partial x_i^2 = f_{x_i x_i} = f_{ii}$. The *third partial derivative* of the function with respect to x_i is $\partial^3 y/\partial x_i^3 = f_{x_i x_i x_i} = f_{iii}, \ldots$, and the *nth partial derivative* is $\partial^n y/\partial x_i^n = f_{x_i x_i x_i \ldots x_i} = f_{iii \ldots i}$.

As an example, consider the function $z = f(x, y) = x^3 y^3$. Then, we can find that $\partial z/\partial x = f_x = 3x^2 y^3$; $\partial(\partial z)/\partial x(\partial x) = \partial^2 z/\partial x^2 = f_{xx} = 6xy^3$; $\partial^3 z/\partial x^3 = f_{xxx} = 6y^3$; and all the higher partial derivatives of the function with respect to x are zeros. Similarly, we can obtain $\partial z/\partial y = f_y = 3x^3 y^2$; $\partial^2 z/\partial y^2 = f_{yy} = 6x^3 y$; $\partial^3 z/\partial y^3 = f_{yyy} = 6x^3$; and all the higher partial derivatives of the function with respect to y are zeros.

Notice that we can also find the *mixed partial derivatives* or *cross partial derivatives* from a multivariate function. As an example, consider the function in the last example: $z = f(x, y) = x^3 y^3$. We found above that $\partial z/\partial x = f_x = 3x^2 y^3$. Now, suppose that we want to find the second partial derivative of the function with respect to y. This can be found as $\partial(\partial z)/\partial y(\partial x) = \partial^2 z/\partial y \partial x = f_{xy} = 9x^2 y^2$. Similarly, we can find $\partial^3 z/\partial x \partial y \partial x = f_{xyx} = 18xy^2$; $\partial z/\partial y = f_y = 3x^3 y^2$; $\partial^2 z/\partial x \partial y = f_{yx} = 9x^2 y^2$; and $\partial^3 z/\partial y \partial x \partial y = f_{yxy} = 18x^2 y$. Notice that, in our example, $f_{xy} = 9x^2 y^2 = f_{yx}$. This result is called *Young's theorem*. All other higher partial derivatives are called mixed partial derivatives or cross partial derivatives. Most of the examples that we consider in this book involve functions whose second mixed partial derivatives have the feature that conforms to Young's theorem.

3.7.4 Application examples

Example 1. Properties of Cobb–Douglas production function. Assume that a manufacturer produces Q units of output using K units of capital and L units of labor. Also assume that the manufacturer's production function takes the form of the Cobb–Douglas production function, with K and L as the independent variables (disregarding technology for convenience), as $Q = f(K, L) = K^\alpha L^{1-\alpha}$, where α represents the *elasticity of output* with respect to capital (K) and $0 < \alpha < 1$. Show that the Cobb–Douglas production function: (i) is a *linear homogeneous production function* and, therefore, exhibits *constant returns to scale*; (ii) yields *diminishing returns* to both K and L; (iii) gives α = *output elasticity of capital* and $(1 - \alpha)$ = *output elasticity of labor*; (iv) is fully correspondent with *Euler's theorem*; (v) α and $1 - \alpha$ represent the *relative share of capital* and the *relative share of labor*, respectively, in *national income*; and (vi) has *elasticity of substitution* equal to 1. Notice that these are the properties of the Cobb–Douglas production function.

Solution. (i) A function is said to be a *homogeneous function* if when each independent variable in the function is multiplied by a positive real constant x, then the constant can be factored out. If the power of this factored constant is 1, then the function is homogeneous of degree one or is a linear homogeneous function; if its power is greater than one, then the function is homogeneous of degree greater than one; and if its power is less than one, then the function is homogeneous of degree less than one.

Now consider the Cobb–Douglas production function: $Q = f(K, L) = K^\alpha L^{1-\alpha}$. Let us now multiply the production function by the positive real constant x to yield $f(xK, xL) = (xK)^\alpha (xL)^{1-\alpha}$. We can now factor out the constant as $xf(K, L) = (x)^{\alpha+1-\alpha} K^\alpha L^{1-\alpha} = (x)^{\alpha+1-\alpha} K^\alpha L^{1-\alpha} = (x)^{\alpha+1-\alpha} Q$. This implies that the Cobb–Douglas production function is a homogeneous production function. Since we assumed that $0 < \alpha < 1$, the last equation can be written as $xf(K, L) = (x)^{\alpha+1-\alpha} K^\alpha L^{1-\alpha} = (x)^{\alpha+1-\alpha} K^\alpha L^{1-\alpha} = xQ$. This means that Cobb–Douglas production function is a linear homogeneous production function.

Since we assumed that $0 < \alpha < 1$ and since the Cobb–Douglas production function is a homogeneous production function, the equation $xf(K, L) = (x)^{\alpha+1-\alpha} K^\alpha L^{1-\alpha} = (x)^{\alpha+1-\alpha} K^\alpha L^{1-\alpha} = xQ$ implies that when we change both factors by the constant x, output Q also changes by the same factor (xQ). This shows that the Cobb–Douglas production function exhibits constant returns to scale. Notice that if the power of x in xQ were greater than one, then the Cobb–Douglas production function would exhibit *increasing returns to scale*; and if the power of x in xQ were less than one, then it would exhibit *diminishing returns to scale*.

(ii) We shall differentiate the production function to answer this question. Differentiating $Q = f(K, L) = K^\alpha L^{1-\alpha}$ partially with respect to K and L, we obtain, respectively

$$\partial Q / \partial K = f_K = \alpha K^{\alpha-1} L^{1-\alpha} = \alpha[\{K^\alpha L^{1-\alpha}\}/K] = \alpha[Q/K] \tag{3.7.3}$$

and

$$\partial Q / \partial L = f_L = (1-\alpha) K^\alpha L^{1-\alpha-1} = (1-\alpha)[\{K^\alpha L^{1-\alpha}\}/L] = (1-\alpha)[Q/L] \tag{3.7.4}$$

Notice that the marginal product of capital (f_K) in equation (3.7.3) is positive for all values of $K > 0$. Differentiating equation (3.7.3) again with respect to K, we obtain $f_{KK} = \alpha(\alpha - 1)K^{\alpha-2}L^{1-\alpha}$. But, since $\alpha < 1$, $(\alpha - 1) < 0$ and, therefore, $f_{KK} < 0$. This implies that the marginal product of capital diminishes as K increases. Similarly, the marginal product of labor (f_L) in equation (3.7.4) is also positive for all values of $L > 0$. Differentiating equation (3.7.4) again with respect to L, we obtain $f_{LL} = (1-\alpha)(-\alpha)K^\alpha L^{-\alpha-1}$. Since $0 < \alpha < 1$ and $-\alpha < 0$, $f_{LL} < 0$. This implies that the marginal product of labor diminishes as L increases. These results show that the Cobb–Douglas production function yields diminishing returns to both K and L.

(iii) The output elasticity of capital (α) and the output elasticity of labor $(1-\alpha)$ are defined, respectively, by $[(\partial Q/\partial K).(K/Q)]$ and $[(\partial Q/\partial L).(L/Q)]$. Substituting equations (3.7.3) and (3.7.4) in these two expressions, we obtain $[(\partial Q/\partial K).(K/Q)] = \alpha[(Q/K).(K/Q)] = \alpha$ and $[(\partial Q/\partial L).(L/Q)] = (1-\alpha)[(Q/L).(L/Q)] = 1 - \alpha$. This shows that the output elasticity of capital is α and the output elasticity of labor is $(1-\alpha)$.

(iv) Before we show that the Cobb–Douglas production function is fully correspondent with Euler's theorem, let us first discuss this theorem. Euler's theorem states that, given a multivariate function $y = f(x_1, x_2, x_3, \ldots, x_n)$, the following result holds:

$$y = x_1[\partial y/\partial x_1] + x_2[\partial y/\partial x_2] + x_3[\partial y/\partial x_3] + \cdots + x_n[\partial y/\partial x_n] \tag{3.7.5}$$

We now apply Euler's theorem in equation (3.7.5) treating $x_1 = K$ and $x_2 = L$. The result is $Q = K[\partial Q/\partial K] + L[\partial Q/\partial L]$. Substituting equations (3.7.3) and (3.7.4) into equation (3.7.5), we obtain $Q = K[\alpha.(Q/K) + L(1-\alpha)(Q/L) = \alpha Q + Q - \alpha Q = Q$. This shows that the Cobb–Douglas production function is fully correspondent with Euler's theorem.

(v) The *share of capital* in national income is defined as $[K(\partial Q/\partial K)]/Q$ and the *share of labor* in national income is defined as $[L(\partial Q/\partial L)]/Q$. Substituting equations (3.7.3) and (3.7.4) into the last two expressions, we obtain $[K(\partial Q/\partial K)]/Q = [K(\alpha Q/K)]/Q = \alpha$ and $[L(\partial Q/\partial L)/Q] = L[(1-\alpha)Q/L]/Q = (1-\alpha)$. This shows that α represents the share of capital in national income and $(1-\alpha)$ represents the share of labor in national income.

(vi) The elasticity of substitution is defined as the percentage change in the optimum (or, the least-cost) K/L ratio due to a small percentage change in the input-price ratio P_L/P_K, and is denoted by σ. Therefore, $\sigma = [d(K^*/L^*)/K^*/L^*]/[d(P_L/P_K)/P_L/P_L] = [d(K^*/L^*)/d(P_L/P_K)]/[(K^*/L^*)/P_L/P_K]$. As shown in example 5(ii) in Section 4.4.9, both the numerator and the denominator of the last equation are equal to $\alpha/(1-\alpha)$. This means that $\sigma = [\alpha/(1-\alpha)]/[\alpha/(1-\alpha)] = 1$.

Example 2. Properties of constant elasticity of substitution (CES) production function. Assume that a manufacturer produces Q units of output using K units of capital and L units of labor. Also assume that the manufacturer's production function takes the form of CES production function with K and L as the independent variables (again disregarding technology for convenience), as $Q = f(K, L) = (\alpha K^{-\beta} + (1-\alpha)L^{-\beta})^{-1/\beta}$, where $\alpha(0 < \alpha < 1)$ represents the *distribution parameter* and β $(\beta > -1)$ represents the *substitution parameter*. Show that the CES production function: (i) is a linear homogeneous production function, and, therefore, exhibits constant returns to scale; (ii) yields diminishing returns to both K and L; (iii) is fully correspondent with Euler's theorem; (iv) exhibits constant elasticity of substitution; and (v) approaches the Cobb–Douglas production function as $\beta \to 0$. Notice that these are the properties of the CES production function.

Solution. (i) We are given $Q = f(K, L) = (\alpha K^{-\beta} + (1-\alpha)L^{-\beta})^{-1/\beta}$. Multiplying both sides of this function by the positive real constant x, we obtain $Qx = [\alpha(Kx)^{-\beta} + (1-\alpha)(Lx)^{-\beta}]^{-1/\beta} = [x^{-\beta}\alpha K^{-\beta} + x^{-\beta}(1-\alpha)L^{-\beta}]^{-1/\beta} = [x^{-\beta}\alpha K^{-\beta} + x^{-\beta}(1-\alpha)L^{-\beta}]^{-1/\beta} = [x^{-\beta}(\alpha K^{-\beta} + (1-\alpha)L^{-\beta})^{-1/\beta}] = x(\alpha K^{-\beta} + (1-\alpha)L^{-\beta})^{-1/\beta} = xQ$. This shows that the

CES production function is a linear homogeneous production function and, therefore, exhibits constant returns to scale if power of x equals one.

(ii) Differentiating $Q = (\alpha K^{-\beta} + (1 - \alpha)L^{-\beta})^{-1/\beta}$ or $Q^{-\beta} = \alpha K^{-\beta} + (1 - \alpha)L^{-\beta}$ with respect to K and L, we obtain the marginal products of capital and labor, respectively, as

$$-\beta Q^{-\beta-1}[\partial Q/\partial K] = (-\beta)\alpha K^{-\beta-1}, \text{ or}$$

$$\partial Q/\partial K = [-\beta\alpha K^{-\beta-1}/-\beta Q^{-\beta-1}] = \alpha K^{-\beta-1}/Q^{-\beta-1}, \text{ and} \qquad (3.7.6)$$

$$-\beta Q^{-\beta-1}[\partial Q/\partial L] = (-\beta)(1-\alpha)L^{-\beta-1}, \text{ or}$$

$$\partial Q/\partial L = [-\beta(1-\alpha)L^{-\beta-1}/-\beta Q^{-\beta-1}] = (1-\alpha)L^{-\beta-1}/Q^{-\beta-1} \qquad (3.7.7)$$

Notice that the marginal product of capital (f_K) in equation (3.7.6) is positive for all values of $K > 0$. Differentiating equation (3.7.6) again with respect to K, we obtain $f_{KK} = [\alpha/Q^{-\beta-1}](-\beta - 1)K^{-\beta-2}$. But, if $\beta > -1$ (as we assumed above), $f_{KK} < 0$. This implies that the marginal product of capital diminishes as K increases. Similarly, the marginal product of labor (f_L) in equation (3.7.7) is also positive for all values of $L > 0$. Differentiating equation (3.7.7) again with respect to L, we obtain $f_{LL} = [(1 - \alpha)/Q^{-\beta-1}](-\beta - 1)L^{-\beta-2}$. But, if $\beta > -1$ (again, as we assumed above), $f_{LL} < 0$. This implies that the marginal product of labor diminishes as L increases. These results show that the CES production function yields diminishing returns to both K and L.

(iii) Applying Euler's theorem given in equation (3.7.5), we obtain $Q = K[\alpha K^{-\beta-1}/Q^{-\beta-1}] + L[(1-\alpha)L^{-\beta-1}/Q^{-\beta-1}] = [\{\alpha K^{-\beta} + (1-\alpha)L^{-\beta}\}/Q^{-\beta-1}]$. The las equation can be simplified to obtain $Q^{-\beta} = (\alpha K^{-\beta} + (1-\alpha)L^{-\beta})$ or $Q = (\alpha K^{-\beta} + (1-\alpha)L^{-\beta})^{-1/\beta} = Q$. This shows that the CES production function is fully correspondent with Euler's theorem.

(iv) As in example 1(vi) above, the elasticity of substitution (σ) is defined as $\sigma = [d(K^*/L^*)/K^*/L^*]/[d(P_L/P_K)/P_L/P_L] = [d(K^*/L^*)/d(P_L/P_K)]/[(K^*/L^*)/P_L/P_K]$. As shown in example 6(ii) in Section 4.4.9, the numerator and the denominator of the last equation can be given by $[\{\alpha/(1-\alpha)\}^{1/(1+\beta)}/(1 + \beta)].[P_L/P_K]^{1/(1+\beta)-1}$ and $[\{\alpha/(1-\alpha)\}^{1/(1+\beta)}.[P_L/P_K]^{1/(1+\beta)-1}$, respectively. Therefore, we can write $\sigma = [[[\{\alpha/(1 - \alpha)\}^{1/(1+\beta)}/(1 + \beta)].[P_L/P_K]^{1/(1+\beta)-1}]/[[\{\alpha/(1 - \alpha)\}^{1/(1+\beta)}.[P_L/P_K]^{1/(1+\beta)-1}]] = 1/(1 + \beta)$. Notice that if $\beta = -1$, then $\sigma = 1/(1 + \beta) = 1/(1 - 1) = \infty$; if $\beta = 0$, then $\sigma = 1$; and if $\beta = \infty$, then $\sigma = 0$. This implies that the elasticity of substitution of the CES production function lies between 0 and ∞; that is, $0 \le \sigma \le \infty$.

(v) We need to show that as $\beta \to 0$, the CES production function approaches the Cobb–Douglas production function; that is, we need to show that the limit of the CES production function as $\beta \to 0$ is the Cobb–Douglas production function. For this, we first take the natural logarithm on both sides of the function to obtain $\ln Q = [-\ln(\alpha K^{-\beta} + (1-\alpha)L^{-\beta})]/\beta$. Our task here is to find the limit of $\ln Q$ when $\beta \to 0$ (that is, $\lim_{\beta \to 0} \ln Q = \lim_{\beta \to 0}[\{-\ln(\alpha K^{-\beta} + (1-\alpha)L^{-\beta})\}/\beta]$. But, when $\beta \to 0$, both the numerator and denominator on the RHS will give us a meaningless expression (0/0) necessitating the use of l'Hôpital's rule. For this, consider the numerator and denominator as functions $f(\beta) = -\ln(\alpha K^{-\beta} + (1 - \alpha)L^{-\beta})$ and $g(\beta) = \beta$, respectively; and differentiate them (as in l'Hôpital's rule) with respect to β. The required derivatives are $f'(\beta) = [\{-1[-\alpha K^{-\beta}\ln K - (1 - \alpha)L^{-\beta}\ln L]\}/\{\alpha K^{-\beta} + (1-\alpha)L^{-\beta}\}] = [\{\alpha K^{-\beta}\ln K + (1-\alpha)L^{-\beta}\ln L\}/\{\alpha K^{-\beta} + (1-\alpha)L^{-\beta}\}]$. Similarly, $g'(\beta) = 1$. We can now write, as stated by l'Hôpital's rule, $\lim_{\beta \to 0} \ln Q = \lim_{\beta \to 0} [\{-\ln(\alpha K^{-\beta} + $

$(1 - \alpha)L^{-\beta})\}/\beta] = \lim_{\beta \to 0}[f'(\beta)/g'(\beta)] = \lim_{\beta \to 0}[\{\alpha K^{-\beta}\ln K + (1 - \alpha)L^{-\beta}\ln L\}/\{\alpha K^{-\beta} + (1 - \alpha)L^{-\beta}\}] = \{\alpha \ln K + (1 - \alpha)\ln L\}/1$, because $K^{-\beta}$ and $L^{-\beta}$ both tend to 1 when $\beta \to 0$. Therefore, we can write the last result as $\lim_{\beta \to 0}\ln Q = \ln K^{\alpha} + \ln L^{1-\alpha} = \ln(K^{\alpha}L^{1-\alpha})$, or $\lim_{\beta \to 0}Q = K^{\alpha}L^{1-\alpha}$. This shows that the CES production function approaches the Cobb–Douglas production function when $\beta \to 0$.

Example 3. Partial elasticities. Suppose that a consumer's demand for good x manufactured by a firm is given by the multivariate function $Q_x = f(P_x, P_r, Y) = 5000 - 2P_x + 3P_r + 0.05Y$, where P_x, P_r, and Y denote the price of the good x, the price of the related (*substitute*) good, and the income of the consumer, respectively. Find the partial elasticities (*own-price elasticity*, *cross-price elasticity*, and *income elasticity*) when P_x, P_r, and Y are $1000, $400, and $10 000, respectively.

Solution. Equation (3.3.21) gives the price elasticity or own-price elasticity. Then, applying equation (3.3.21) in the case of the present multivariate function, we obtain $\Phi_{P_x} = (P_x/Q_x) \div (\partial P_x/\partial Q_x)$, where Φ_{P_x} shows that the elasticity under consideration is own-price elasticity. When P_x, P_r, and Y are $1000, $400, and $10 000, respectively, $Q_x = 5000 - 2 \times 1000 + 3 \times 400 + 0.05 \times 10\,000 = 4700$. The partial derivative of $Q_x = f(P_x, P_r, Y) = 20 - 2P_x + 5P_r + 0.05Y$ with respect to P_x is $\partial Q/\partial P_x = -2$. Substituting $Q_s = 4700$, $P_x = $1000, and $\partial Q/\partial P_x = -2$ into $\Phi_{P_x} = (P_x/Q_x) \div (\partial P_x/\partial Q_x)$ yields $\Phi_{P_x} = (1000/4700)/(-2) = -0.11$.

Cross-price elasticity denoted by Φ_{P_r} is defined as the percentage change in Q_x due to a percentage change in P_r. This is given by the equation

$$\Phi_{P_r} = \frac{P_r}{Q_x} \div \frac{\partial P_r}{\partial Q_x} \tag{3.7.8}$$

We can find Φ_{P_r} by substituting $Q_x = 4700$, $P_r = $400, and $\partial P_r/\partial Q_x = 3$ into equation (3.7.8): $\Phi_{P_r} = (P_r/Q_x) \div (\partial P_r/\partial Q_x) = (400/4700) \div 3 = 0.028$. Notice that the cross-price elasticity (Φ_{P_r}) is positive if the related good is a substitute (as in our present example) and is negative if the good is a *complementary good*.

Income elasticity denoted by Φ_Y is defined as the percentage change in Q_x due to a percentage change in Y. This is given by the equation

$$\Phi_Y = \frac{Y}{Q_x} \div \frac{\partial Y}{\partial Q_x} \tag{3.7.9}$$

We can find Φ_Y by substituting $Q_x = 4700$, $Y = $10 000, and $\partial Y/\partial Q_x = 0.05$ into equation (3.7.9): $\Phi_Y = (Y/Q_x) \div (\partial Y/\partial Q_x) = (10\,000/4700) \div 0.05 = 42.6$. Notice that income elasticity (Φ_Y) is positive in our present example. This implies that as income increases, the quantity demanded of the good (Q_x) also increases. Such a good is called a *normal good*. If $\Phi_Y < 0$, then the good is called an *inferior good*.

3.7.5 Exercises

1. Find the partial derivatives of the following functions with respect to each independent variable:
   (i) $z = x + y$; (ii) $z = x - y$; (iii) $z = x/y$; (iv) $z = y/x$; (v) $z = (x + y)/(y - x)$.

2. Find the partial derivatives of the following functions with respect to each independent variable:

   (i) $z = xy$; (ii) $z = xy - y$; (iii) $z = xy/y^2$; (iv) $z = y^2/xy$; (v) $z = (x^2 + y)/(y^2 + x)$.

3. Find the partial derivatives of the following functions with respect to each independent variable:

   (i) $u = xyz$; (ii) $u = xy - zy$; (iii) $z = x/y$; (iv) $z = y/x$; (v) $z = (x + y)/(x - y)$.

4. Find the second partial derivatives, with respect to each independent variable, of the functions in exercises 1 through 3 above.

5. *Application exercise.* Suppose that the total cost, in dollars, of producing two goods, x and y, by a firm is given by $C = 1000 + 0.1x^2 + 50x + 0.1y^2 + 40y$. Find the marginal costs when $x = 50$ and $y = 100$ units.

6. *Application exercise.* Suppose that the total revenue, in dollars, of producing two goods, x and y, by a firm is given by $R = x^2 - 10x + y^2 - 10y$. Find the marginal revenues when $x = 100$ and $y = 100$ units.

7. *Application exercise.* Suppose that a consumer's demand for good x is given by the function $Q_x = 10 - P_x + 0.5P_r + 0.1Y$, where Q_x, P_x, P_r, and Y denote the quantity demanded of the good, price of the good, price of the related good, and income of the consumer. Find the own-price, cross-price, and income elasticities when $P_x = \$10$, $P_r = \$5$, and $Y = \$100$, and interpret the results.

8. *Application exercise.* Suppose that the quantities demanded, q_x and q_y, of two goods, x and y, are given by the functions $q_x = \sqrt{p_y}/2\sqrt{p_x}$ and $q_y = \sqrt{p_x}/4\sqrt{p_y}$, respectively, where p_x and p_y denote the prices of the goods. Find the *marginal demand functions* ($\partial q_x/\partial p_x$, $\partial q_x/\partial p_y$, $\partial q_y/\partial p_x$, and $\partial q_y/\partial p_y$) and determine whether the two goods are substitutes (or *competitive goods*), or complementary goods, or neither.

9. *Application exercise.* Suppose that the quantities demanded of two goods, q_x and q_y, are given by the functions $q_x = 100/\sqrt{p_x}\sqrt[3]{p_y}$ and $q_y = 200/\sqrt{p_y}\sqrt[3]{p_x}$, respectively. Find the marginal demand functions ($\partial q_x/\partial p_x$, $\partial q_x/\partial p_y$, $\partial q_y/\partial p_x$, and $\partial q_y/\partial p_y$) and determine whether the two goods are substitutes (or competitive goods), or complementary goods, or neither.

10. *Application exercise.* Suppose that the quantities demanded q_x and q_y of two goods, x and y, are given by the functions $q_x = \sqrt{p_y}/2\sqrt{p_x}$ and $q_y = \sqrt{p_x}/4\sqrt{p_y}$, respectively, where p_x and p_y denote the prices of the goods. The total cost of producing these two goods is given by the function $C = 100 + 20q_x + 10q_y + 0.1q_xq_y$. Find the rate of change of total cost with respect to both p_x and p_y when $p_x = 10$ and $p_y = 20$.

 Web supplement: S3.7.6 *Differentiation of matrices*

 Web supplement: S3.7.7 Mathematica applications

3.8 Differentials, total derivatives, and multivariate implicit differentiation

3.8.1 Meaning of differentials of univariate functions

Earlier we defined the derivative of a function as the rate of change of that function when the independent variable changes by an infinitesimally small amount. But, how can we find the total change in the dependent variable when the independent variable changes by

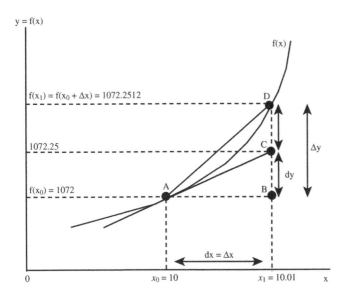

Figure 3.8.1

an infinitesimally small amount? This question leads us to the concept of *differentials*. The idea behind differentials can be best explained with the help of simple example.

Suppose that a firm produces a good y using labor x. Then, the quantity of y produced by the firm can be expressed as $y = f(x)$. Assume, for simplicity, that this function takes the form $y = f(x) = 20 - 15x + 12x^2$, whose graph resembles the one shown in Figure 3.8.1. It can be seen from Figure 3.8.1 that $\Delta y = f(x_0 + \Delta x) - f(x_0)$. Let us now multiply and divide the RHS of the last equation by Δx to obtain $\Delta y = \{[f(x_0 + \Delta x) - f(x_0)]/\Delta x\}\Delta x$. Therefore, $\Delta y = \{[f(x_0 + \Delta x) - f(x_0)]/\Delta x\}\Delta x$ gives the change in y when x changes by Δx. Notice that finding the change in y when x changes by Δx using the difference quotient is a laborious task. Therefore, we attempt to find the approximate change in y when x changes by an infinitesimally small amount. For this, we use the concept of derivative that we have been discussing so far.

Notice that the term inside the brackets of the last equation represents the difference quotient which becomes the slope of the function at point A (that is, the slope of the function at $x = x_0$) when $\Delta x \to 0$, and is given by the equation $dy/dx = f'(x)$. We know that $\{[f(x_0 + \Delta x) - f(x_0)]/\Delta x\}\Delta x$ is $dy/dx = f'(x)$. Therefore, we can write $\Delta y = \{[f(x_0 + \Delta x) - f(x_0)]/\Delta x\}\Delta x$ as $\Delta y = f'(x_0)\Delta x$. However, when the change in x is infinitesimally small, the last equation can be written as

$$dy = f'(x_0)dx \tag{3.8.1}$$

The terms dy and dx in equation (3.8.1) are called the *first differentials* of y and x, respectively, in the univariate function $y = f(x)$, and the process of finding dy is called differentiation. Notice that if we divide both sides of equation (3.8.1) by dx, we will end up with the derivative of the function $y = f(x)$ at $x = x_0$. This shows the close connection between the differentials and the derivatives.

It should be noticed that the result in equation (3.8.1) is valid only when the change in x is infinitesimally small. Instead of such a small change in x, if one uses a large value for Δx, equation (3.8.1) would give only an approximation to the change in y with a large error (or remainder). This can be shown through an example. Consider our function $y = f(x) = 20 - 15x + 12x^2$. Now suppose that the change in x is $\Delta x = x_1 - x_0 = 10.01 - 10 = 0.01$. We know that $f'(x) = -15 + 24x$. Plugging these values in equation (3.8.1), we obtain the change in y equal to $dy = f'(x)dx = (-15 + 24 \times 10) \times 0.01 = 2.25$. But, the actual change in y is $\Delta y = y_1 - y_0 = f(x_1) - f(x_0) = (20 - 15x_1 + 12x_1^2) - (20 - 15x_0 + 12x_0^2) = [20 - 15 \times 10.01 + 12(10.01)^2] - [20 - 15 \times 10 + 12 \times 10^2] = 1072.2512 - 1070 = 2.2512$, where $y = y_1$ and $y = y_0$ represent the values of the function corresponding to $x = x_1$ and $x = x_0$, respectively. The change in y if we use equation (3.8.1) is 2.25. Therefore, the difference is $0.0012 = (2.2512 - 2.25)$, which is the error in approximation. This error will be larger (smaller) if we use $\Delta x > 0.01$ ($\Delta x < 0.01$).

Why did the error noted above occur? Notice in Figure 3.8.1 that the actual change in y is $\Delta y = BD$ and the actual change in x is $\Delta x = AB$. If we used $\Delta y = BD$ and $\Delta x = AB$ in $\Delta y = \{[f(x_0 + \Delta x) - f(x_0)]/\Delta x\}\Delta x$ we would have obtained $\Delta y = (BD/AB)AB = BD$, where BD/AB is the slope of the secant line AD; and this result does not involve any error. But, when we applied equation (3.8.1) we used the slope of the tangent line AC, and not that of the secant line. When we use the slope of the tangent line AC, the part CD is excluded from the calculation and the error (0.0012 in our example) is equal to the vertical distance CD. What all this means is that, when we find the differential Δx must be infinitesimally small so that the error in approximation will be small.

3.8.2 Differentials of multivariate functions: total and partial differentials

In the last section we discussed how we can approximate the change in the dependent variable of a univariate function when its independent variable changes by a very small amount. A pertinent question that arises now is: can one find such a change in the value of the dependent variable when all of the independent variables in a multivariate function change by very small amounts? The answer is yes. The multivariate analogue to the univariate differential is called *total differential*. Total differentials give us the approximate change in the dependent variable when the independent variables change by infinitesimally small amounts.

Let us cite a simple example to drive home the meaning of total differentials. We know from the *principles of microeconomics* that the quantity demanded of good x (denoted by Q_x) by a consumer depends, other things remaining the same, on the price of the good x (denoted by P_x), price of related goods (either substitutes or complements denoted by P_r), and income of the consumer (denoted by Y). Then, the consumer's demand function can be written as

$$Q_x = f(P_x, P_r, Y) \tag{3.8.2}$$

We know, from Section 3.7, that the partial derivative $\partial Q_x / \partial P_x$ measures the rate of change of Q_x when P_x changes by an infinitesimally small amount (or by dP_x), holding P_r and Y constant. Therefore, following equation (3.8.1), the change in Q_x when P_x changes by dP_x is given by the product of $\partial Q_x / \partial P_x$ and dP_x; that is, by $[\partial Q_x / \partial P_x] \times dP_x$. Similarly, the changes in Q_x when P_r and Y change, respectively, by dP_r and dY are given, respectively, by $[\partial Q_x/\partial P_x] \times dP_r$ and $[\partial Q_x/\partial P_x] \times dY$. If we add all these three component changes, we obtain the total change in Q_x as

$$dQ_x = \frac{\partial Q_x}{\partial P_x} \times dP_x + \frac{\partial Q_x}{\partial P_r} \times dP_r + \frac{\partial Q_x}{\partial Y} \times dY \tag{3.8.3}$$

where dQ_x is called the total differential of the demand function in equation (3.8.2). The process of finding this total differential is called *total differentiation*. We now generalize the result in equation (3.8.3) to the case of a multivariate function with n independent variables. Suppose that the multivariate function is $y = f(x_1, x_2, x_3, \ldots, x_n)$. Then, the total differential of this function, evaluated at $(x_1^0, x_2^0, x_3^0, \ldots, x_n^0)$, is

$$dy = f_1(x_1^0, x_2^0, x_3^0, \ldots, x_n^0)dx_1 + f_2(x_1^0, x_2^0, x_3^0, \ldots, x_n^0)dx_2 + f_3(x_1^0, x_2^0, x_3^0, \ldots, x_n^0)dx_3 + \cdots$$

$$+ f_n(x_1^0, x_2^0, x_3^0, \ldots, x_n^0)dx_n = \sum_{i=1}^{n} f_i(x_1^0, x_2^0, x_3^0, \ldots, x_n^0)dx_i \tag{3.8.4}$$

where $f_i(x_1^0, x_2^0, x_3^0, \ldots, x_n^0)$ denotes the partial derivative of the function $y = f(x_1, x_2, x_3, \ldots, x_n)$ with respect to the ith independent variable, each evaluated at $(x_1^0, x_2^0, x_3^0, \ldots, x_n^0)$.

Let us now consider *partial differentials*. Consider the multivariate function in equation (3.8.2). The total differential of this function is given in equation (3.8.3). Assume now that only P_x in the function changes, while P_r and Y remain constant. This implies that $dP_r = dY = 0$. Therefore, equation (3.8.3) reduces to $dQ_x = [\partial Q_x/\partial P_x] \times dP_x$, which is called a partial differential of $Q_x = f(P_x, P_r, Y)$. Similarly, we can get the other partial differentials of the function as $dQ_x = [\partial Q_x/\partial P_r] \times dP_r$ and $dQ_x = [\partial Q_x/\partial Y] \times dY$.

3.8.3 Higher differentials and higher total differentials

We found differentials of univariate functions and total differentials of multivariate functions in the previous two sections. However, students of economics, business, and finance sometimes need to use *higher differentials* and *higher total differentials*, particularly in optimization problems. Therefore, we shall present them here.

Equation (3.8.1) gives the differential dy of the univariate function $y = f(x)$. How do we find the *second differential* of this function? It can be found by differentiating dy again. It should be noted, however, that when we differentiate dy again it is with respect to $f'(x)$. The reason is that $f'(x)$ is the independent variable in equation (3.8.1), and dx is just a constant as it is a real number. Therefore, differentiating dy in equation (3.8.1) again, we obtain

$$d^2y = d(dy) = d[f'(x)dx] = [df'(x)]dx = [f''(x)dx]dx = f''(x)dx^2 \tag{3.8.5}$$

Following the same arguments as above, we can find the *third differential* of the function $y = f(x)$, and it is given by $d^3y = f'''(x)dx^3$. Continuing analogously, we obtain the *nth differential* of $y = f(x)$, if it exists, as

$$d^ny = f^n(x)dx^n \tag{3.8.6}$$

We shall now turn to the higher total differentials (that is, higher differentials of multivariate functions). Let us first consider the case of a multivariate function with two independent variables, $y = f(x_1, x_2)$. Using equation (3.8.4), the *first total differential* of this function can be found as $dy = f_1 dx_1 + f_2 dx_2$, which is identical to equation (2.8.8). The *second total differential* of this equation can be found by differentiating again the first total differential. But, when we find the second total differential, as in the case of the first total differential, dx_1 and dx_2 are considered constants, and f_1 and f_2 are considered to be variables. Therefore, differentiating $dy = f_1 dx_1 + f_2 dx_2$ again totally, we find $d^2y = d(dy) = [\partial(dy)/\partial x_1]dx_1 + [\partial(dy)/\partial x_2]dx_2 = [\partial(f_1 dx_1 + f_2 dx_2)/\partial x_1]dx_1 + [\partial(f_1 dx_1 + f_2 dx_2)/\partial x_2]dx_2 = (f_{11}dx_1 + f_{12}dx_2)dx_1 + (f_{21}dx_1 + f_{22}dx_2)dx_2 = f_{11}dx_1^2 + f_{12}dx_2 dx_1 + f_{21}dx_1 dx_2 + f_{22}dx_2^2$. Since $f_{12} = f_{21}$, the last result can be written as

$$d^2y = f_{11}dx_1^2 + f_{12}dx_2 dx_1 + f_{21}dx_1 dx_2 + f_{22}dx_2^2 = f_{11}dx_1^2 + 2f_{12}dx_1 dx_2 + f_{22}dx_2^2 \quad (3.8.7)$$

Notice that equation (3.8.7), which is identical to equation (2.8.9), gives us the second total differential of the multivariate function with two independent variables, $y = f(x_1, x_2)$. Following the same procedure, one can obtain the second total differentials of multivariate functions with three and n independent variables ($y = f(x_1, x_2, x_3)$ and $y = f(x_1, x_2, x_3, \ldots, x_n)$), respectively, as

$$d^2y = f_{11}dx_1^2 + f_{22}dx_2^2 + f_{33}dx_3^2 + 2f_{12}dx_1 dx_2 + 2f_{13}dx_1 dx_3 + 2f_{23}dx_2 dx_3 \quad (3.8.8)$$

and

$$d^2y = f_{11}dx_1^2 + f_{22}dx_2^2 + f_{33}dx_3^2 + \cdots + f_{nn}dx_n^2$$
$$+ 2f_{12}dx_1 dx_2 + 2f_{13}dx_1 dx_3 + \cdots + 2f_{1n}dx_1 dx_n + 2f_{23}dx_2 dx_3 + 2f_{24}dx_2 dx_4 + \cdots$$
$$+ 2f_{2n}dx_2 dx_n + \cdots + 2f_{n(n-1)}dx_n dx_{n-1} \quad (3.8.9)$$

Notice that the total differentials we used as quadratic forms (along with their matrix representations) in Sections 2.8.2–2.8.4 (with borders in the case of bordered Hessian matrices and bordered Hessians in Section 2.8.3) were special cases (with $n = 2$ and with $n = 3$) of equation (3.8.9).

3.8.4 Rules of differentials

Differentials obey some important rules. Suppose that we have two functions $y = f(x_1, x_2)$ and $z = g(x_1, x_2)$, and a constant s. Then, the following *rules of differentials* are valid. Notice that many of these rules resemble the rules of differentiation.

Rule I.   Constant function: $ds = 0$.
Rule II.   Function-with-constant: $d(sy) = s\,dy$.
Rule III.   Power function: $d(y^n) = ny^{n-1}dy$.
Rule IV.   Sum–difference: $d(y \pm z) = dy \pm dz$.
Rule V.   Product: $d(y \times z) = z \times dy + y \times dz$.
Rule VI.   Quotient: $d(y/z) = (z \times dy - v \times dz)/z^2$.

3.8.5 Derivatives of multivariate composite functions: total derivatives

So far, we have been concerned with multivariate functions whose independent variables were *independent*. This meant that in a multivariate function $y = f(x_1, x_2)$, x_1 and x_2 were

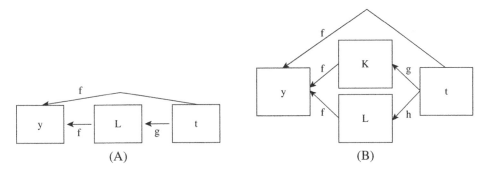

(A)   (B)

Figure 3.8.2

assumed to be independent or unrelated. But, one can cite many examples of multivariate functions in economics, business, and finance with independent variables that are related.

One simple example is the quantity of output (y) produced by a firm. We know that y depends on many factors. Assume for convenience that y depends on time t (because time can affect technology, which, in turn, will affect y) and the quantity of labor (L) used. Therefore, the firm's production function can be written as $y = f(L, t)$. But, we know that the quantity of labor that the firm uses will change over time; that is, L depends on t. This implies that L is a function of t; that is, $L = g(t)$. What all these mean is that a change in t will affect y directly through the function f and indirectly through the function g. This is illustrated in Figure 3.8.2(A). The question now is: how can we find the effect of a change in t on y when t and L are dependent? The answer is that, to find the effect of a change in t on y when t and L are dependent, we have to use the method called *total derivative*, which is discussed below.

As Figure 3.8.2(A) shows, there are two impacts on y of a change in t: (1) the direct impact through f and (2) the indirect impact through f and g. To obtain the total impact on y (or the total derivative), we first find the differential of y: it is given as $dy = f_L dL + f_t dt$. We can now divide both sides of this equation by dt to yield

$$\frac{dy}{dt} = \dot{y} = f_L \frac{dL}{dt} + f_t \frac{dt}{dt} = f_L \dot{L} + f_t \tag{3.8.10}$$

where $dL/dt = \dot{L}$. Equation (3.8.10) gives the total derivative of y with respect to t. The process of finding the total derivative is called *total differentiation*.

Suppose now that the firm employs two factors, capital (K) and labor (L). Also suppose that $K = g(t)$ and $L = h(t)$. Therefore, we can write the firm's production function as $y = f(K, L, t)$. As above, the relationship among y, K, L, and t can be illustrated through Figure 3.8.2(B). Following the same line of arguments as that we used in the derivation of equation (3.8.10), we obtain the total derivative of y with respect to t as

$$\frac{dy}{dt} = \dot{y} = f_K \frac{dK}{dt} + f_L \frac{dL}{dt} + f_t \frac{dt}{dt} = f_K \dot{K} + f_L \dot{L} + f_t \tag{3.8.11}$$

where $dL/dt = \dot{L}$ and $dK/dt = \dot{K}$. Let us now generalize the above results to a general function $y = f(x_1, x_2, x_3, \ldots, x_n, t)$, where $x_1, x_2, x_3, \ldots, x_n$ are different functions of t. Then the total derivative of y with respect to t is given by

$$\frac{dy}{dt} = \dot{y} = f_1 \frac{dx_1}{dt} + f_2 \frac{dx_2}{dt} + f_3 \frac{dx_3}{dt} + \cdots + f_n \frac{dx_n}{dt} = f_1 \dot{x}_1 + f_2 \dot{x}_2 + f_3 \dot{x}_3 + \cdots + f_n \dot{x}_n + f_t$$

(3.8.12)

where $dx_1/dt = \dot{x}_1$, $dx_2/dt = \dot{x}_2$, $dx_3/dt = \dot{x}_3$, $\ldots$, $dx_n/dt = \dot{x}_n$.

Notice that if a variable (t in our present case) does not appear directly in a function (but only indirectly through other independent variable(s)), then that variable does not directly influence the dependent variable. This means that that the panels in Figure 3.8.2 will be devoid of the kinked arrows. This also means that the total derivatives of such functions (such as equations (3.8.10)–(3.8.12)) will be devoid of the term representing the partial derivative of the function with respect to that variable (f_t).

As an example, consider the function $y = f(x, t) = 10xt + 2x^2$, where $x = g(t) = t^2$. Then, applying equation (3.8.12), we can find the total derivative of y with respect to t as $\dot{y} = f_x \dot{x} + f_t(dt/dt) = f_x \dot{x} + f_t = (10t + 4x)(2t) + (10x) = 20t^2 + 8tx + 10x$. Since $x = g(t) = t^2$, substituting this into the last result gives us $\dot{y} = (10t + 4x)(2t) + (10x) = 20t^2 + 8t^3 + 10t^2 = 30t^2 + 8t^3$. This result can be verified by substituting $x = g(t) = t^2$ into $y = f(x, t) = 10xt + 2x^2$ and differentiating the result with respect to t. Substituting $x = g(t) = t^2$ into $y = f(x, t) = 10xt + 2x^2$, we obtain $y = f(x, t) = 10t^2 t + 2(t^2)^2 = 10t^3 + 2t^4$. Then, differentiating this last equation with respect to t yields $\dot{y} = [d(10t^3 + 2t^4)/dt] = 30t^2 + 8t^3$, which is identical to the result we obtained above.

Assume now that our function is $y = f(x_i, t_i)$, where $x_i = g^i(t_i)$, and $i = 1, 2, \ldots, n$. Then the total derivative of $y = f(x_i, t_i)$ with respect to t_i (holding t_j constant, where $i \neq j$) is obtained as

$$\frac{dy}{dt_i} = \frac{\partial y}{\partial x_1} \frac{dx_1}{dt_i} + \frac{\partial y}{\partial x_2} \frac{dx_2}{dt_i} + \cdots + \frac{\partial y}{\partial x_n} \frac{dx_n}{dt_i} + \frac{\partial y}{\partial t_i} \frac{dt_i}{dt_i} + \frac{\partial y}{\partial t_2} \frac{dt_2}{dt_i} + \cdots + \frac{\partial y}{\partial t_n} \frac{dt_n}{dt_i}$$

$$= \frac{\partial y}{\partial x_1} \frac{dx_1}{dt_i} + \frac{\partial y}{\partial x_2} \frac{dx_2}{dt_i} + \cdots + \frac{\partial y}{\partial x_n} \frac{dx_n}{dt_i} + \frac{\partial y}{\partial t_i}$$

(3.8.13)

where $(dt_i/dt_i) = 1$ and $(\partial y/\partial t_2)(dt_2/dt_i) = \cdots = (\partial y/\partial t_n)(dt_n/dt_i) = 0$ (because t_j, where $i \neq j$, is held constant).

So far, we have been concerned with the *first total derivative* of different multivariate functions. We now consider the *second total derivative* of such functions. As an example, assume that our function is $y = f(x_1, x_2)$, where $x_1 = g(t)$ and $x_2 = h(t)$. Notice that the function is not directly influenced by the variable t. Applying equation (3.8.12) we obtain the first total derivative of $y = f(x_1, x_2, t)$ with respect to t as $dy/dt = f_1(dx_1/dt) + f_2(dx_2/dt)$. If we totally differentiate the last equation with respect to t again, we obtain the second total derivative of $y = f(x_1, x_2)$ with respect to t. Therefore, differentiating $dy/dt = f_1(dx_1/dt) + f_2(dx_2/dt) + f_t$ totally with respect to t, we obtain $d^2y/dt^2 = \{\partial[f_1(dx_1/dt) + f_2(dx_2/dt)]/\partial x_1\}(dx_1/dt) + \{\partial[f_1(dx_1/dt) + f_2(dx_2/dt)/\partial x_2\}(dx_2/dt)$. Notice that the first term on the RHS, $\{\partial[f_1(dx_1/dt) + f_2(dx_2/dt)]/\partial x_1\}(dx_1/dt)$, can be written as $\{f_{11}(dx_1/dt) + f_1[\partial(dx_1/dt)/\partial x_1] + f_{21}(dx_2/dt) + f_2[(\partial/\partial x_1)(dx_2/dt)](dx_1/dt)$. This expression can be written as $f_{11}(dx_1/dt)^2 + f_1(d^2x_1/dt^2)(dt/dx_1)(dx_1/dt) + f_{21}(dx_2/dt)(dx_1/dt) = f_{11}(dx_1/dt)^2 + f_1(d^2x_1/dt^2) + f_{21}(dx_2/dt)(dx_1/dt)$, because $f_1 \partial(dx_1/dt)/\partial x_1 = f_1[d(dx_1/dt)/dt](dt/dx_1) = f_1(d^2x_1/dt^2)(dt/dx_1)$ and $f_2[\partial(dx_2/dt)/\partial x_1] = f_2[0]$ since $dx_2 = 0$ when we hold x_2 constant and differentiate with respect to x_1. Similarly, the second term on the RHS, $\{\partial[f_1(dx_1/dt) + f_2(dx_2/dt)]/\partial x_2\}(dx_2/dt)$, can be written as $f_{12}(dx_1/dt)(dx_2/dt) + f_1(0) + f_{22}(dx_2/dt)^2 + f_2(d^2x_2/dt^2)$. Using these two results, we can rewrite the second total derivative as $d^2y/dt^2 = f_{11}(dx_1/dt)^2 + f_1(d^2x_1/dt^2) + f_{21}(dx_2/dt)(dx_1/dt) + f_{12}(dx_1/dt)(dx_2/dt) + f_{22}(dx_2/dt)^2 + f_2(d^2x_2/dt^2) = f_{11}(dx_1/dt)^2 + f_{22}(dx_2/dt)^2 + 2f_{12}(dx_1/dt)(dx_2/dt) + f_1(d^2x_1/dt^2) + f_2(d^2x_2/dt^2)$.

3.8.6 Differentiation of multivariate implicit functions: implicit partial derivatives

We discussed univariate implicit functions in Section 1.8.4. We learned, in Section 3.3.12, how to differentiate univariate implicit functions. Here we extend our exposition to multivariate implicit functions, and attempt to see how one can differentiate such functions and obtain *implicit partial derivatives*. But, a proper understanding of the differentiation of multivariate implicit functions requires an understanding of the *implicit function theorem*. Therefore, we begin this section by discussing the implicit function theorem.

Suppose that we have a multivariate implicit function given by $F(y, x_1, x_2, x_3, \ldots, x_n) = s$, where s is a constant. Also let that this implicit function is defined at $(y^0, x_1^0, x_2^0, x_3^0, \ldots, x_n^0)$ and that this function has continuous partial derivatives at $(y^0, x_1^0, x_2^0, x_3^0, \ldots, x_n^0)$ with $F_y(y^0, x_1^0, x_2^0, x_3^0, \ldots, x_n^0) \neq 0$. Then, the implicit function theorem states that there exists a function $y = f(x_1, x_2, x_3, \ldots, x_n)$, defined in the *neighborhood* of $(x_1^0, x_2^0, x_3^0, \ldots, x_n^0)$ corresponding to $F(y, x_1, x_2, x_3, \ldots, x_n) = s$, such that (1) $F[f(x_1^0, x_2^0, x_3^0, \ldots, x_n^0), x_1^0, x_2^0, x_3^0, \ldots, x_n^0] = s$, (2) $y^0 = f(x_1^0, x_2^0, x_3^0, \ldots, x_n^0)$, and (3) $f_i(x_1^0, x_2^0, x_3^0, \ldots, x_n^0) = F_{x_i}(x_1^0, x_2^0, x_3^0, \ldots, x_n^0)/F_y(x_1^0, x_2^0, x_3^0, \ldots, x_n^0)$, with $F_{x_i}(x_1^0, x_2^0, x_3^0, \ldots, x_n^0) = \partial F(y, x_1, x_2, x_3, \ldots, x_n)/\partial x_i$ and $F_y(x_1^0, x_2^0, x_3^0, \ldots, x_n^0) = \partial F(y, x_1, x_2, x_3, \ldots, x_n)/\partial y$. In simple terms, what the implicit function theorem says is that, among others, given an implicit function $F(y, x_1, x_2, x_3, \ldots, x_n) = s$, one can derive an explicit function of the form $y = f(x_1, x_2, x_3, \ldots, x_n)$. We have already applied the above implicit function theorem in Section 3.3.12 to find the derivatives of univariate implicit functions.

Now consider a multivariate implicit function $F(y, x_1, x_2, x_3, \ldots, x_n) = s$. One can now apply both the implicit function theorem stated above and the implicit function rule of differentiation to obtain the total differential of this function (using equation (3.8.4)) as $F_y dy + F_1 dx_1 + F_2 dx_2 + F_3 dx_3 + \cdots + F_n dx_n = 0$. Assume now that only y and x_1 vary while all other variables $(x_2, x_3, \ldots, x_n)$ are assumed to be constant. This implies that $dy \neq 0$, $dx_1 \neq 0$, and $dx_2 = dx_3 = \cdots = dx_n = 0$. Then the equation $F_y dy + F_1 dx_1 + F_2 dx_2 + F_3 dx_3 + \cdots + F_n dx_n = 0$ becomes $F_y dy + F_1 dx_1 = 0$, and solving for $dy/dx_1 = \partial y/\partial x_1$ yields $dy/dx_1 = \partial y/\partial x_1 = -F_1/F_y$, which is an implicit partial derivative. Following the arguments above, one can derive the other implicit partial derivatives. In general, if there exists a multivariate implicit function $F(y, x_1, x_2, x_3, \ldots, x_n) = s$ with its explicit form $y = f(x_1, x_2, x_3, \ldots, x_n)$, as defined by the implicit function theorem, one can obtain the implicit partial derivative of $y = f(x_1, x_2, x_3, \ldots, x_n)$ with respect to x_i (where $i = 1, 2, 3, \ldots, n$), and is defined by

$$\frac{dy}{dx_i} = \frac{\partial y}{\partial x_i} = -\frac{F_i}{F_y} \tag{3.8.14}$$

Notice that in the case of univariate implicit function $F(y, x) = s$, equation (3.8.14) can be written as $dy/dx = -F_x/F_y$. Notice also that this was the result that we used in the example of Section 3.3.12. Therefore, we do not repeat here the differentiation of univariate implicit functions.

As an example of the application of equation (3.8.14), consider the multivariate implicit function $F(y, x_1, x_2) = y^2 + x_1 y + x_2 y = 0$, where $s = 0$. Therefore, applying equation (3.8.14), we can find the partial derivatives $\partial y/\partial x_1$ and $\partial y/\partial x_2$ as $\partial y/\partial x_1 = -F_1/F_y = [-(2y + y)]/[2y + x_1 + x_2] = -3y/(2y + x_1 + x_2)$, and $\partial y/\partial x_2 = -F_2/F_y = -y/(2y + x_1 + x_2)$, respectively.

3.8.7 Application examples

Example 1. Suppose that the national income in a three-sector economy is given by $Y = C + I_0 + G_0$, where $C = C_0 + bY_d$, $Y_d = Y - T$, $T = T_0 + tY$, $C_0, I_0, G_0, T_0 > 0$, $0 < b < 1$, and $0 < t < 1$; and where Y, C, I_0, G_0, C_0, Y_d, T, T_0, b, and t denote national income, *consumption expenditure, autonomous investment, autonomous government expenditure, autonomous consumption expenditure, disposable income*, tax receipts, *autonomous tax*, MPC, and the fraction of national income collected as taxes respectively. Assume now that $I_0 = \$100$, $G_0 = \$400$, $C_0 = \$50$, $T_0 = \$200$, $b = 0.75$, and $t = 0.25$. Find the change in equilibrium level of income (Y^*) when: (i) government autonomous expenditure increases by \$100; (ii) government autonomous expenditure decreases by \$100; (iii) autonomous tax increases by \$100; (iv) autonomous tax decreases by \$100; (v) autonomous investment increases by \$100; and (vi) autonomous investment decreases by \$100. All the values are in billions of dollars.

Solution. This problem can be solved using partial differentials. But, before this, we need to find the equilibrium level of income in the economy. Substituting $C = C_0 + bY_d$, $Y_d = Y - T$, $T = T_0 + tY$ into $Y = C + I_0 + G_0$ and simplifying the resulting expression yields the equilibrium level of income: $Y^* = [1/(1 - b + bt)][C_0 - bT_0 + I_0 + G_0]$. Substituting $I_0 = 100$, $G_0 = 400$, $C_0 = 50$, $T_0 = 200$, $b = 0.75$, and $t = 0.25$ into the last equation gives $Y^* = [1/(1 - b + bt)][C_0 - bT_0 + I_0 + G_0] = 400/0.4375 = \$915$ billion.

(i) To find the change in equilibrium level of income when the government expenditure increases by \$100 billion, we can use the equation of partial differential given in Section 3.8.2. The partial differential of income with respect to government autonomous expenditure is given by $\Delta Y^* = (\partial Y^*/\partial G_0) \times \Delta G_0$ (which is equal to $dY^* = (\partial Y/\partial G_0) \times dG_0$ when ΔG_0 is infinitesimally small). Since $\partial Y^*/\partial G_0 = 1/(1 - b + bt) = 2.29$ and $\Delta G_0 = +\$100$, the change in equilibrium level of income is $\Delta Y^* = (\partial Y^*/\partial G_0) \times \Delta G_0 = [1/(1 - b + bt)] \times \Delta G_0 = 2.29 \times \$100 = \$229$ billion.

(ii) The only difference between the answer to this question and the answer to the last question is that now $\Delta G_0 = -\$100$ while it was $\Delta G_0 = +\$100$ in the last question. Therefore, we can use all the last results with $\Delta G_0 = -\$100$. Then the change in the equilibrium level of income when autonomous government expenditure decreases by \$100 billion is $\Delta Y^* = (\partial Y^*/\partial G_0) \times -\Delta G_0 = [1/(1 - b + bt)] \times -\Delta G_0 = 2.29 \times -\$100 = -\$229$ billion.

(iii) Following arguments similar to those above, the partial differential of Y^* with respect to autonomous tax is given by $\Delta Y^* = (\partial Y^*/\partial T_0) \times \Delta T_0$. Since $\partial Y^*/\partial T_0 = [-b/(1 - b + bt)] = -1.715$ and $\Delta T_0 = +\$100$, the change in equilibrium level of income when autonomous tax increases by \$100 billion is $\Delta Y^* = (\partial Y^*/\partial T_0) \times \Delta T_0 = [-b/(1-b+bt)] \times \Delta T_0 = -1.715 \times \$100 = -\$171.5$ billion.

(iv) The partial differential of Y^* with respect to autonomous tax is given by $\Delta Y^* = (\partial Y^*/\partial T_0) \times \Delta T_0$. Since $\partial Y^*/\partial T_0 = [-b/(1-b+bt)] = -1.715$ and $\Delta T_0 = -\$100$ billion, the change in equilibrium level of income when autonomous tax decreases by \$100 billion is $\Delta Y^* = (\partial Y^*/\partial T_0) \times -\Delta T_0 = [-b/(1-b+bt)] \times -\Delta T_0 = -1.715 \times -\$100 = \$171.5$ billion.

(v) The partial differential of Y^* with respect to autonomous investment is given by $\Delta Y^* = (\partial Y^*/\partial I_0) \times \Delta I_0$. Since $\partial Y/\partial I_0 = 1/(1-b+bt) = 2.29$ and $\Delta I_0 = \$100$, the change in equilibrium level of income when autonomous investment increases by \$100 billion is $\Delta Y^* = (\partial Y^*/\partial I_0) \times \Delta I_0 = 1/(1-b+bt) \times \Delta I_0 = 2.29 \times \$100 = \$229$ billion.

(vi) The partial differential of Y^* with respect to autonomous investment is given by $\Delta Y^* = (\partial Y^*/\partial I_0) \times \Delta I_0$. Since $\partial Y/\partial I_0 = 1/(1-b+bt) = 2.29$ and $\Delta I_0 = -\$100$, the change in equilibrium level of income when autonomous investment decreases by \$100 billion is $\Delta Y^* = (\partial Y^*/\partial I_0) \times -\Delta I_0 = 1/(1-b+bt) \times -\Delta I_0 = 2.29 \times -\$100 = -\$229$ billion.

Example 2. Suppose that a firm, producing Q units of output using K units of capital and L units of labor, has the production function given by the Cobb–Douglas form $Q = f(K,L) = K^\alpha L^{1-\alpha}$. Find the approximate change in the total output when both K and L change by infinitesimally small amounts.

Solution. To solve this problem, we have to find the total differential of Q using equation (3.8.4). But, to use this equation, we first need to find the partial derivatives of Q with respect to both K and L (marginal products of capital and labor, respectively). These are given, respectively, by $\partial Q/\partial K = \alpha K^{\alpha-1}L^{1-\alpha} = \alpha Q/K$ and $\partial Q/\partial L = (1-\alpha)K^\alpha L^{-\alpha} = [(1-\alpha)Q]/L$. Therefore, the approximate change in the total output Q when both K and L change by infinitesimally small amounts is given, by equation (3.8.4), as $dQ = (\partial Q/\partial K)dK + (\partial Q/\partial L)dL = [\alpha K^{\alpha-1}L^{1-\alpha}]dK + [(1-\alpha)K^\alpha L^{1-\alpha}]dL = (\alpha Q/K)dK + [(1-\alpha)Q/L]dL$.

Example 3. Suppose that a firm, producing Q units of output using K units of capital, L units of labor, and using t level of technology, has the production function given by the form $Q = f(K,L,t) = K^\alpha L^\beta t^{1-\alpha-\beta}$. Also suppose that $K = g(t) = 2t$ and $L = h(t) = 3t$. Find the change in the total output produced when technology changes by an infinitesimally small amount.

Solution. To solve this problem, we need to use the total derivative in equation (3.8.11) as a special case of equation (3.8.12). The change in output due to a very small change in the level of technology is given by the total derivative of Q with respect to t: $dQ/dt = \dot{Q} = f_K(dK/dt) + f_L(dL/dt) + f_t = f_K\dot{K} + f_L\dot{L} + f_t$. From $Q = f(K,L,t) = K^\alpha L^\beta t^{1-\alpha-\beta}$ we can find $\partial Q/\partial K = \alpha K^{\alpha-1}L^\beta t^{1-\alpha-\beta}$; $\partial Q/\partial L = \beta K^\alpha L^{\beta-1}t^{1-\alpha-\beta}$; and $\dot{Q} = (1 - \alpha - \beta)K^\alpha L^\beta t^{-\alpha-\beta}$. Plugging these partial derivatives in $\dot{Q}$ yields $\dot{Q} = f_K\dot{K} + f_L\dot{L} + f_t = \alpha K^{\alpha-1}L^\beta t^{1-\alpha-\beta}\dot{K} + \beta K^\alpha L^{\beta-1}t^{1-\alpha-\beta}\dot{L} + (1 - \alpha - \beta)K^\alpha L^\beta t^{-\alpha-\beta} = (\alpha Q/K)\dot{K} + (\beta Q/L)\dot{L} + [(1-\alpha-\beta)Q]/t$. Since $\dot{K} = dK/dt = 2$ and $\dot{L} = dL/dt = 3$, the last result can be written as $\dot{Q} = (\alpha Q/K)\dot{K} + (\beta Q/L)\dot{L} + [(1-\alpha-\beta)Q/t] = (2\alpha Q/K) + (3\beta Q/L) + [(1-\alpha-\beta)Q/t]$.

Example 4.   *Consumer equilibrium.* Assume that an *indifference curve* representing the utility (U) obtained by a consumer when two goods (y and x) are consumed is defined by the *implicit utility function* $U(y,x) = U_0$, where U_0 is a constant. Also assume that the consumer's budget line is represented by the equation $p_y y + p_x x = I$, where p_y, p_x, and I denote the price of good y, the price of good x, the income of the consumer, respectively. Show that, when the consumer is in equilibrium, the slope of the indifference curve is equal to the slope of the budget line (that is, $U_x/U_y = p_x/p_y$, where U_x and U_y represent the marginal utilities that the consumer obtains from x and y, respectively).

Solution.   Let us write the budget line as $y = (I/p_y) - (p_x/p_y)x$. This shows that the vertical intercept of the budget line is I/p_y and its horizontal intercept is I/p_x. Differentiating y in $y = (I/p_y) - (p_x/p_y)x$ with respect to x gives the slope of the budget line: $dy/dx = -p_x/p_y$.

We now find the slope of the indifference curve. This can be found by applying the implicit function theorem and the technique of total differentiation. Using these, we obtain $(\partial U/\partial y)dy + (\partial U/\partial x)dx = d(U_0) = 0$. Notice that $\partial U/\partial y = U_y$ and $\partial U/\partial x = U_x$. Substituting these two into $(\partial U/\partial y)dy + (\partial U/\partial x)dx = d(U_0) = 0$ yields $U_y dy + U_x dx = 0$. By rearranging this result we get $U_y dy = -U_x dx$ or $dy/dx = -U_x/U_y$. Therefore, the slope of the indifference curve representing utility U_0 is $dy/dx = -U_x/U_y$.

We know from the principles of microeconomics that the slope of the budget line is equal to the slope of the indifference curve at the point of equilibrium (E). This is shown in Figure 3.8.3(A). Therefore, equating the slope of the indifference curve with the slope of the budget line, we get $-U_x/U_y = -p_x/p_y = -MU_x/MU_y$. As shown in example 1 in Section 4.4.11, this is indeed the condition for consumer equilibrium. Notice that the slope of the indifference curve is called *marginal rate of substitution* between x and y (denoted by MRS_{xy}). Therefore, when the consumer is in equilibrium, we have the condition that $MRS_{xy} = -U_x/U_y = -p_x/p_y = -MU_x/MU_y$.

Example 5.   Producer equilibrium. Assume that an *isoquant* representing the output (Q) manufactured by a producer when two factors (capital, K, and labor, L) are employed is defined by an *implicit production function* $Q(K,L) = Q_0$. Also assume that the producer's *isocost line* is represented by the equation $rK + wL = M$, where r, w, and M denote interest

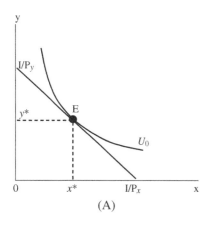

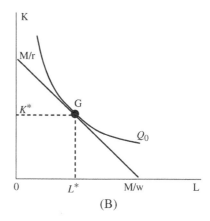

(A)   (B)

Figure 3.8.3

rate (the price of capital), wage rate (the price of labor), and the amount of investment, respectively. Show that, when the producer is in equilibrium (that is, when the cost of production is minimized), the slope of the isocost line is equal to the slope of the isoquant (that is, $MP_L/MP_K = w/r$, where MP_K and MP_L denote the marginal products of capital and labor, respectively).

Solution. Let us write the isocost line as $K = (M/r) - (w/r)L$. This shows that the vertical intercept of the isocost line is M/r and its horizontal intercept is M/w. Differentiating K in $K = (M/r) - (w/r)L$ with respect to L gives the slope of the isocost line: $dK/dL = -w/r$.

Let us now find the slope of the isoquant. This can be found, as before, by applying the implicit function theorem and the technique of total differentiation. Using these, we obtain $(\partial Q/\partial K)dK + (\partial Q/\partial L)dL = d(Q_0) = 0$. Notice that $\partial Q/\partial K = MP_K$ and $\partial Q/\partial L = MP_L$. Substituting these two into $(\partial Q/\partial K)dK + (\partial Q/\partial L)dL = d(Q_0) = 0$ yields $MP_K dK + MP_L dL = 0$. By rearranging this equation we get $MP_K dK = -MP_L dL$ or $dK/dL = -MP_L/MP_K$. Therefore, the slope of the isoquant representing output Q is $dK/dL = -MP_L/MP_K$.

We know that the slope of the isocost line is equal to the slope of the isoquant at the point of equilibrium (G). This is shown in Figure 3.8.3(B). Therefore, equating the slope of the isoquant with the slope of the isocost line, we get $-MP_L/MP_K = -w/r$. As shown in example 2 in Section 4.4.11, this is indeed the condition for producer equilibrium. Notice that the slope of the isoquant is called *marginal rate of technical substitution* between L and K (denoted by $MRTS_{LK}$). Therefore, when the producer is in equilibrium, we have the condition that $MRTS_{LK} = -MP_L/MP_K = -w/r$.

3.8.8 Exercises

1. Find the differentials of the following functions:
   (i) $y = 2x^3 + 4x^2 + 8x + 10$; (ii) $y = (x+2)^2$; (iii) $y = (x+2)^3$; (iv) $y = 2x/(x+1)$;
   (v) $y = x/(x+1)$.

2. Find the total differentials of the following functions:
   (i) $y = x_1 x_2$; (ii) $y = x_1^2 x_2^2$; (iii) $y = x_1/x_2$; (iv) $y = (x_1 + x_2)^2$.

3. Find the second differentials of the following functions:
   (i) $y = 2x^3 + 4x^2 + 8x + 10$; (ii) $y = (x+2)^2$; (iii) $y = x/(x+1)$; (iv) $y = x_1 x_2$;
   (v) $y = x_1^2 x_2^2$; (vi) $y = x_1/x_2$.

4. Find the total derivatives of the following functions:
   (i) $y = x_1 x_2$, where $x_1 = 2x_2^2$; (ii) $y = x_1^2 x_2^2$, where $x_1 = 2x_2^2$; (iii) $y = x_1 + x_2$ where $x_1 = 2x_2^2$.

5. Find dy/dx of the following functions:
   (i) $2x^2 + 3y = 0$; (ii) $2x + 3y = 0$; (iii) $y/(x+1) = 2$; (iv) $(x+1)/y = 2$.

6. *Application exercise.* Suppose that the national income in a three-sector economy is given by $Y = C + I_0 + G_0$, where $C = C_0 + bY_d$, $Y_d = Y - T$, $T = T_0 + tY$, $C_0, I_0, G_0, T_0 > 0$, $0 < b < 1$, and $0 < t < 1$, and where Y, C, I_0, G_0, C_0, Y_d, T, T_0, b, and t denote national income, consumption expenditure, autonomous investment, autonomous government expenditure, autonomous consumption expenditure, disposable income, tax receipts, autonomous tax, MPC, and the fraction of national income collected as taxes, respectively. Assume that $I_0 = \$50$, $G_0 = \$500$, $C_0 = \$100$, $T_0 = \$300$, $b = 0.80$, and $t = 0.20$. (i) Find the equilibrium level of income in the economy. (ii) What should be the change in autonomous investment so that

the equilibrium level of income will be $1000? (iii) What should be the change in autonomous government expenditure so that the equilibrium level of income will be $1100? (iv) What should be the change in autonomous tax so that the equilibrium level of income will be $1200? All values are in billions.

7. *Application exercise.* Suppose that the total utility, U, that a consumer obtains from the consumption two goods, x and y, is given by the function $U = f(x,y) = x^\alpha y^{1-\alpha}$. Find the approximate change in total utility when both x and y change by infinitesimally small amounts.

8. *Application exercise.* Suppose that the total revenue R that a seller receives from the sale of two goods, x and y, along with the *advertisement expenditure*, v, is given by the function $R = f(x,y,v) = x^{0.4}y^{0.4}v^{0.2}$. Also suppose that $x = g(v) = 0.5v$ and $y = h(v) = v$. Find the change in the total revenue when advertisement expenditure changes by an infinitesimally small amount.

9. *Application exercise.* Assume that an indifference curve representing the utility (U) obtained by a consumer when two goods $(y$ and $x)$ are consumed is defined by the *implicit utility function* $U(y,x) = y^{0.4}x^{0.6} = 100$. Also assume that the consumer's budget line is represented by the equation $p_y y + p_x x = I$, where $p_y = \$2$ denotes the price of good y, $p_x = \$4$ denotes the price of good x, and $I = \$10\,000$ denotes the income of the consumer. Show the equilibrium of this consumer (assuming that the equilibrium occurs when the slope of the indifference curve is equal to the slope of the budget line).

10. *Application exercise.* Assume that an isoquant representing the output (Q) manufactured by a producer when two factors (capital, K, and labor, L) are employed is defined by the *implicit production function* $Q(K,L) = K^{0.4}L^{0.6} = 500$. Also assume that the producer's isocost line is represented by the equation $rK + wL = M$, where $r = 0.2$ denotes interest rate, $w = \$10$ denotes wage rate, and $M = \$100\,000$ denotes the amount of investment. Show the equilibrium of this producer (assuming that the equilibrium occurs when the slope of the isoquant is equal to the slope of the isocost line).

 Web supplement: S3.8.9 Mathematica applications

4 Classical optimization

4.1 Introduction

Every student of economics begins the study of the subject with an introduction to the relationship between available resources and human wants. It is a reality that the resources to satisfy unlimited human wants are scarce. This necessitates choice. Every economic agent, whether the agent is a consumer, or a producer, or a government, is compelled to make choices. Since the agent is assumed to be rational, the agent attempts to allocate the scarce resources in such a way that the agent's objective is "optimized." Such an allocation of resources is called the *optimal allocation* of resources.

The science of economics deals with the economic behavior of economic agents. Every consumer is assumed to allocate his or her income to different goods and services that he or she buys so that his or her objective is optimized. Every business firm is assumed to allocate its resources so that its objective is optimized. Similarly, government agents are assumed to allocate resources so that society's benefits or welfare will be optimized. It is needless to say that optimal allocation of resources lies at the heart of the science of economics. The states of affairs are similar in the fields of business and finance. Therefore, a reasonably good understanding of the topic of optimization is indispensable for students of economics, business, and finance.

However, a very important question that arises now is how one can say that a particular economic agent has, in fact, optimized the agent's objective. There exist several mathematical approaches, concerned with differing *states of nature*, to answer this and the related questions. These include *classical approach*, *linear programming approach*, *nonlinear programming approach*, *game theory approach*, and so on. It should be noticed that these approaches are complementary in that they are concerned with problems with *different* states of nature.

This chapter deals with the *classical approach to optimization*. We consider other relatively newer approaches in the following chapters. We have already presented, in the previous three chapters, most of the mathematical prerequisites for a comprehensive but introductory discussion of optimization problems that follow in this chapter. We begin this chapter with an introduction to different concepts that students normally encounter in optimization topics.

4.2 Optima and extrema of univariate objective functions

4.2.1 Objective function and optima

In this section we are concerned with univariate functions. We mentioned above that the aim of every economic agent is to attain the *optimum* of a goal, called the *objective function*.

Consider, for example, the case of a consumer who consumes one good, x. The consumer consumes this good because the consumer obtains utility or satisfaction from its consumption. Therefore, the consumer's aim is to consume that quantity of x so as to obtain the "*maximum*" possible level of utility expressed by a *utility function*. Therefore, the objective function in this example is the consumer's utility function that is to be optimized.

As another example, consider the case of a firm that produces good x. We know from the principles of microeconomics that the aim of the firm, other things remaining the same, is to obtain maximum profits. But, again other things remaining the same, maximization of profits depends on the cost of production. Therefore, the aim of the firm is to choose its output level so that its cost is "*minimum*" or, in other words, its profit is maximum. In short, the objective function in this example is the producer's *cost function* or *profit function* that is to be minimized or maximized, respectively.

As the last example, consider the case of a local government that plans to build a recreation facility for the local people. As above, the aim of the government is to build the facility such that the *social benefit* to the local people is maximum. Therefore, the objective function in this example is the *social benefit function* that is to be maximized.

In all three illustrative examples above, the economic agents are either maximizing or minimizing their respective *objective functions*. This implies that optimization of an objective function is either *maximization* or *minimization* of the objective function. And the optimum value of the function is either the *maximum value* or the *minimum value* depending upon whether the function is maximized or minimized, respectively. These values are also called *optima*.[1]

4.2.2 Optima and extrema

We stated above that optimization is either maximization or minimization of an objective function, and that optima refer to either the maximum value or the minimum value of the function. But, this classification is purely from an application or economic point of view, and it does not have a mathematical connotation.

However, when we refer to the optima of an objective function what we mean is the maximum or minimum possible value of the function; that is, the *extreme values* or the *extrema* of the function. Therefore, the correct mathematical term for maximum and minimum values (or, for optima) is extrema. However, we will use these terms interchangeably in our following discussions.

4.2.3 Extrema: graphical illustrations and definitions

We saw that in Figures 3.5.1(A) when the first derivative of a function (say, $y = f(x)$) is greater than zero or if $f'(x) > 0$ at $x = x_0$, then the graph of the function would increase at $x = x_0$, as represented by point A. We also saw in the same figure that if the first derivative is less than zero or if $f'(x) < 0$ at $x = x_2$, then the graph of the function would decrease at $x = x_2$, as represented by point C. One can give similar interpretations to points F and D in Figure 3.5.1(B), which represents the graph of the function $y = g(x)$. However, what will be the values of $f'(x)$ and $g'(x)$ at $x = x_1$ (with corresponding points B and E)? To answer this question, consider these figures again, which are reproduced in Figure 4.2.1 for convenience.

In equation (3.2.4) we defined the derivative of the function $y = f(x)$ at a particular point on its graph, which is equal to the slope of the function at that point on the graph, as the ratio of the change in y to a very small change in x. Now consider points B and E

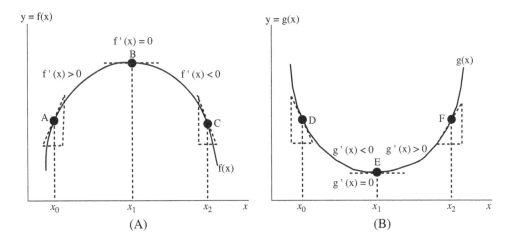

Figure 4.2.1

in Figures 4.2.1(A) and (B), respectively. At both points the change in y is zero while the change in x is different from zero. This implies that the ratio of the former to the latter must be zero. Therefore, the derivatives of $y = f(x)$ and $y = g(x)$ at points B and E, respectively, must be zero: $f'(x) = 0|_{x=x_1}$ and $g'(x) = 0|_{x=x_1}$. This answers our last question.

We are now ready to illustrate extrema or optima, the extreme or optimum values such as maximum or minimum values. Assume that our objective function is $y = f(x)$, the graph of which resembles the one in Figure 4.2.2. Notice that the graph of the function $y = f(x)$ in Figure 4.2.2 is a smooth graph except at points F and G and, therefore, the function is differentiable at every point on its graph except at F and G. We know that $f'(x) < 0$ between points A and B, between D and F, and on the RHS of G; and that $f'(x) > 0$ between points B and D, and between F and G. We also know that $f'(x) = 0$ at points B and D.

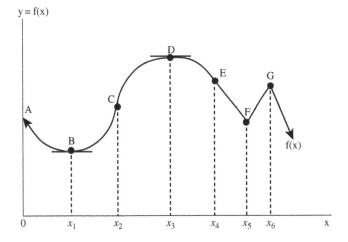

Figure 4.2.2

Now consider the interval $[0, x_3]$ on the real line in Figure 4.2.2. Notice that this is a closed interval. The lowest value of the function in this interval is at point B, corresponding to x_1. This lowest value of the function in the interval $[0, x_3]$ is called a *relative minimum*, or a *local minimum*, of the function, and it occurs when $x = x_1$. Now consider another interval, $[x_2, x_6]$, on the same real line. We know that this is, as before, a closed interval. As can be seen from the figure, the highest value of the function in this interval is at point D, corresponding to x_3. This highest value of the function in the interval $[x_2, x_6]$ is called a *relative maximum*, or a *local maximum*, of the function, and it occurs at $x = x_3$. Although $f'(x)$ is not defined at points F and G, yet they represent local minimum and local maximum values of the function in the intervals $[x_3, x_6]$ and $[x_4, x_6]$, respectively.

What will be the lowest and the highest values of the function if we discard the intervals on the real line we mentioned above? If the lowest value of the function occurs at point B when x takes any value on the real line, then that value of the function is called the *global minimum* or the *absolute minimum* of the function. Similarly, if the highest value of the function occurs at point D when x takes any value on the real line, then that value of the function is called the *global maximum* or the *absolute maximum* of the function. The relative or local minimum (or maximum) is called a *relative extremum* and the global or absolute minimum (or maximum) is called the *global extremum*.

Let us now formally define local minimum (and maximum) and global minimum (and maximum) of a function. A function $f(x)$ has a relative minimum (or maximum) at $x = x_0$ in an interval containing x_0 if $f(x_0) \leq f(x)$ (or $f(x_0) \geq f(x)$) for all x in the interval, and $f(x_0)$ is a local minimum (or maximum) of the function. Similarly, the function has the global minimum (or maximum) if $f(x_0) \leq f(x)$ (or $f(x_0) \geq f(x)$) for all x in the domain of the function, and $f(x_0)$ is the global minimum (or maximum) of the function. It should be noticed that the absolute minimum or maximum of a function, if it exists, is unique; and it may occur at more than one value of x.

4.2.4 Extrema: necessary and sufficient conditions

Let us now discuss the formal tests of relative minimum and relative maximum (in other words, of relative optima or extrema). For this we use Figure 4.2.2. Notice that we discard points such as F and G from our discussion as the function is not differentiable at these points. Our following discussion will pertain only to those points on the graph of the function where the function is differentiable or the derivative of the function exists at those points.

The reader will have noticed an important feature of the graph of the function illustrated in Figure 4.2.2. This feature is that the relative minimum (point B) or the relative maximum (point D) occurs when the first derivative of the function is zero (or $f'(x) = 0$).[2] Another important feature of the graph is that $f'(x) < 0$ and $f'(x) > 0$ to the immediate LHS and RHS, respectively, of point B; and that $f'(x) > 0$ and $f'(x) < 0$ to the immediate LHS and RHS, respectively, of point D. These suggest that at the point of minimum of the function, $f'(x)$ changes sign from negative to positive (and vice versa); and at the point of maximum of the function, $f'(x)$ changes sign from positive to negative (and vice versa). These results lead us to the following condition, which is called the first-order condition (FOC) or the *necessary condition* for extrema at $x = x_0$. The FOC for a relative minimum of the function $f(x)$ are

(i) $f'(x_0) = 0$ or $f'(x_0)$ is not defined

(ii) $f'(x)$ changes sign from negative to positive at small values around $x = x_0$

$$(4.2.1)$$

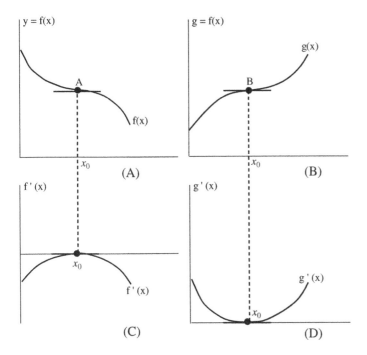

Figure 4.2.3

Similarly, the FOC or necessary condition for a relative maximum of the function $f(x)$:

(i) $f'(x_0) = 0$ or $f'(x_0)$ is not defined
(ii) $f'(x)$ changes sign from positive to negative at small values around $x = x_0$ $\left.\right\}$

$$(4.2.2)$$

What the above conditions imply can be easily seen from Figure 4.2.2. Notice that for both a relative minimum and a relative maximum at $x = x_0$, $f'(x_0)$ must be zero at $x = x_0$. However, the condition $f'(x_0) = 0$ does not guarantee the existence of either a relative maximum or a relative minimum. This can be illustrated through Figure 4.2.3. As can be seen from Figure 4.2.3(A), the first derivative of the function $y = f(x)$ (as shown in Figure 4.2.3(C)) corresponding to point A is zero; that is, $f'(x_0) = 0$. But point A is not a relative minimum (or maximum) point of the function $y = f(x)$. Similarly, we can see in Figure 4.2.3(B) that the first derivative of the function $y = g(x)$ (as shown in Figure 4.2.3(D)) corresponding to point B is zero; that is, $g'(x_0) = 0$. But, again, point B is not a minimum (or maximum) of the function $y = g(x)$.[3] All this shows that $f'(x_0) = 0$ (or $g'(x_0) = 0$) is only a necessary or a FOC for an optimum of $f(x)$ (or $g(x)$).

What, then, is the *sufficient condition* for a function to have relative extrema? The sufficient condition for a function to have relative extrema is also called the second-order condition (SOC). Let us use Figure 4.2.2 again to explain the SOC. We know from the results of our discussions in Sections 3.5.1 and 3.5.2 that a function is a strictly convex function if its second derivative is positive and is a strictly concave function if its second derivative is negative. That is, if $f''(x) > 0$ then $f(x)$ is strictly convex; and if $f''(x) < 0$ then $f(x)$ is strictly concave.

Table 4.2.1

Condition	Minimum	Maximum
FOC or necessary condition	$f'(x) = 0$	$f'(x) = 0$
SOC or sufficient condition	$f''(x) > 0$	$f''(x) < 0$
	Inflection point or inconclusive	
	$f''(x) = 0$	

Following these results, we know that the function $f(x)$ in Figure 4.2.2 is strictly convex for all x in the neighborhood of x_1 and is strictly concave for all x in the neighborhood of x_3. Therefore, we state below the SOC or the sufficient condition for extrema. The SOC or the sufficient condition for a relative minimum of $f(x)$ at $x = x_0$ is

$$f''(x_0) > 0 \tag{4.2.3}$$

Similarly, the SOC or the sufficient condition for a relative maximum of $f(x)$ at $x = x_0$ is

$$f''(x_0) < 0 \tag{4.2.4}$$

Therefore, for a function $y = f(x)$ to have a relative minimum at $x = x_0$, the FOC and the SOC are $f'(x_0) = 0$ and $f''(x_0) > 0$, respectively. Similarly, for the function to have a relative maximum at $x = x_0$, the FOC and the SOC are $f'(x_0) = 0$ and $f''(x_0) < 0$, respectively.[4] We present these conditions in Table 4.2.1 for easy reference.

Let us now explain three important concepts related to optimization problems: *critical value*, *stationary value*, and *critical point* or *stationary point*. In Figure 4.2.2 the value $x = x_1$ corresponds to the relative optimum (minimum) value of the function $y = f(x)$. This value of x (x_1) is called a critical value. It is called a critical value because it is very important, or "critical", in locating the relative optimum. Another critical value in the same figure is x_3. The y-coordinate of a critical value is called a stationary value. And its corresponding coordinate point (such as B or D) is called a critical point or stationary point. Let us now define critical values, stationary values, and critical points formally. Assume that x_0 is in the domain of the function $y = f(x)$ and that $f'(x_0) = 0$ or $f'(x_0)$ is not defined. Then, x_0 is called the critical value of $f(x)$, $f(x_0)$ is called the stationary value of $f(x)$, and the coordinate point $[x_0, f(x_0)]$ is called the critical point or stationary point.

Notice that, sometimes, the second derivative may turn out to be either indeterminate or zero at some points on the graph of the function. In these cases, we will have to put the first derivative to further analysis or determine the sign of the third or higher derivatives to check whether the function has a relative optimum at the specified point on the graph of the function. These cases are rare in the fields under consideration, and most of the functions that we use in this book possess nonzero second derivatives. However, we state below the nth derivative test for relative optima. Suppose that we have a function $y = f(x)$, with $f'(x_0) = 0$ at $x = x_0$ and $f^n(x_0) \neq 0$. Then

$$\left.\begin{array}{l} \text{(i) } f(x_0) \text{ is a relative minimum if } n \text{ is even and } f^n(x_0) > 0 \\ \text{(ii) } f(x_0) \text{ is a relative maximum if } n \text{ is even and } f^n(x_0) < 0 \\ \text{(iii) } f(x_0) \text{ is an inflection point if } n \text{ is odd} \end{array}\right\} \tag{4.2.5}$$

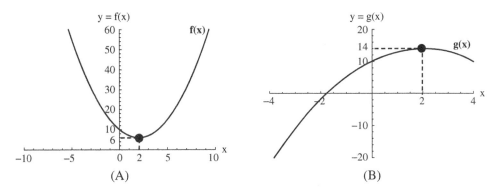

Figure 4.2.4

As an example, consider the function $y = f(x) = 10 - 4x + x^2$ for relative optima. Let us first find the first derivative of the function. The first derivative of the function is $f'(x) = -4 + 2x$. We now apply the FOC given in Table 4.2.1. Applying this, we obtain $x = 2$ when $f'(x) = -4 + 2x = 0$. We denote this value of x by x^*. This suggests that the critical value is $x^* = 2$. Finally, we apply the SOC. Applying this yields $f''(x^*) = 2$. Since $f''(x^*) = 2 > 0$, the function has a relative minimum at $x = 2 = x^*$, and the minimum value of the function, the stationary value, is $y^*|_{x=2} = f(x)|_{x=2} = 10 - 4x + x^2|_{x=2} = 10 - 4 \times 2 + 2^2 = 6$. This is shown in Figure 4.2.4(A).

As another example, consider the function $y = g(x) = 10 + 4x - x^2$. The first derivative of the function is $g'(x) = 4 - 2x$. We can now apply the FOC to obtain $x^* = 2$ when $g'(x) = 4 - 2x = 0$. This suggests that the critical value is, as before, $x^* = 2$. Application of the SOC yields $g''(x^*) = -2$. Since $g''(x^*) = -2 < 0$, the function has a relative maximum at $x^* = 2$, and the maximum value of the function (its stationary value) is $y^*|_{x=2} = g(x)|_{x=2} = 10 + 4x - x^2|_{x=2} = 10 + 4 \times 2 - 2^2 = 14$. This is shown in Figure 4.2.4(B).

4.2.5 Conditions of extrema of univariate functions: differential version

We used only the first and the second derivative of the univariate function $y = f(x)$ in the development of the conditions for the optima of the function presented in Table 4.2.1. However, one can develop a similar set of conditions using differentials, which we discussed in Section 3.8. To develop these conditions, consider the graph of the function in Figure 4.2.5(A).[5]

Let us now develop the FOC for the relative extrema of the function $y = f(x)$ using differentials. For this, we use the differential of a univariate function ($y = f(x)$) given in equation (3.8.1): $dy = f'(x)dx$. Now consider point A in Figure 4.2.5(A). Since $f'(x) = 0$ at point A, point A is a point of relative minimum of the function $f(x)$ as per the FOC we stated in the previous section. Since $f'(x) = 0$ and $dx > 0$ at point A, $dy = f'(x)dx = 0.dx = 0$ at point A. The same is true at point C, which is a relative maximum of the function. Notice that $dy \neq 0$ at points such as B (where $dy = f'(x)dx > 0$ because $f'(x) > 0$ and $dx > 0$) and D (where $dy = f'(x)dx < 0$ because $f'(x) < 0$ and $dx > 0$); therefore, points such as B and D do not qualify for the points of relative optima. In short, for a function $y = f(x)$ to have an optimum point (either minimum or maximum), the FOC given in the following equation

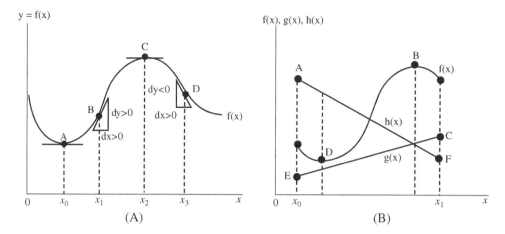

Figure 4.2.5

must be satisfied for $dx > 0$:

$$dy = f'(x)dx = 0.dx = 0 \qquad (4.2.6)$$

Notice that, as we had the FOC when we used only derivatives in the previous section, equation (4.2.6) is only the necessary condition or the FOC for function $y = f(x)$, in terms of differentials, to have a relative optimum point. The SOC or the sufficient condition must also be satisfied, as before, for $y = f(x)$ to have a relative optimum point. This SOC requires that we check the sign of the second differential of the function. The required second differential can be found by differentiating the first differential (equation (4.2.6)) again. Then, applying equation (3.8.6), we obtain the second differential of equation (4.2.6) as $d^2y = f''(x)dx^2$. Since dx^2 is the square of dx, it must always be positive. This implies that the sign of d^2y depends solely on the sign of $f''(x)$: if $f''(x) > 0$, then $d^2y > 0$; if $f''(x) < 0$, then $d^2y < 0$. We know, from the results presented in Table 4.2.1, that the SOC for a minimum of the function is $f''(x) > 0$. Therefore, for the function to have a relative minimum, the second differential of the function must be positive; that is, $d^2y > 0$. Similarly, for the function to have a relative maximum, the second differential of the function must be negative; that is, $d^2y < 0$. Therefore, we have the following inequalities as the SOC:

$$\left. \begin{array}{l} \text{for } y = f(x) \text{ to have a relative minimum, given } dy = 0, d^2y > 0 \\ \text{for } y = f(x) \text{ to have a relative maximum, given } dy = 0, d^2y < 0 \end{array} \right\} \qquad (4.2.7)$$

As before, we present these differential versions of the FOC and SOC for a function $y = f(x)$ to have a relative optimum in Table 4.2.2.

4.2.6 Optima of univariate functions on closed intervals

So far, we have mainly used open intervals on the domains of functions to find their relative extrema. Let us now use closed intervals on the domains of functions to find their optima (to be more precise, absolute optima).

Table 4.2.2

Condition	Minimum	Maximum
FOC or necessary condition	$dy = 0$	$dy = 0$
SOC or sufficient condition	$d^2y > 0$	$d^2y < 0$
	Inflection point or inconclusive	
	$d^2y = 0$	

But, before this, we present an important theorem, called the *extreme-value theorem*, which guarantees the existence of relative optima for a function with a closed interval on its domain. The extreme-value theorem states that if a function $y = f(x)$ is a smooth function on a closed interval (say, $[x_0, x_1]$) on its domain, then the function possesses both a maximum value and a minimum value on that interval. These are the absolute maximum and minimum values of the function on that interval. This can be shown through Figure 4.2.5(B), which illustrates the graphs of three functions: $f(x)$, $g(x)$, and $h(x)$. For convenience we have fixed the same closed interval, $[x_0, x_1]$, on their domains for all these functions. It can be seen that points D, E, and F represent the absolute minimum values of functions $f(x)$, $g(x)$, and $h(x)$, respectively, on the interval $[x_0, x_1]$. Similarly, their respective absolute maximum values are represented by points B, C, and A. What this implies is that every smooth function has both an absolute minimum value and an absolute maximum value on a closed interval on its domain. This is what the extreme-value theorem states.

How does one determine these absolute extrema? The procedure is as follows: (1) Find the stationary value(s); (2) find the values of the function at the lower and upper bounds of the interval; (3) compare these results with the stationary value(s); and (4) the absolute minimum (maximum) value of the function is the smallest (highest) of the values in step (3).

As an example, consider the function $y = f(x) = 10 - 2x + x^2$. Assume that we want to find the absolute extrema of this function in the closed interval $[0, 5]$. We follow the above procedures. First, we obtain the critical value $x^* = 1$ from $f'(x) = -2 + 2x = 0$, and the stationary value $y = f(1) = 10 - 2 \times 1 + 1^2 = 9$. Second, we find the values of the function at the lower and upper bounds of the interval. The value of the function at $x = 0$ (the *lower bound* of the interval) is $y = f(0) = 10 - 2 \times 0 + 0^2 = 10$ and the value of the function at $x = 5$ (the *upper bound* of the interval) is $y = f(5) = 10 - 2 \times 5 + 5^2 = 25$. Third, we can compare these values. Notice that the stationary value is lower than the values of the function at both the lower and the upper bounds of the interval. Fourth, based on these results, we conclude that the function has, on the closed interval $[0, 5]$ on its domain, an absolute minimum (equal to 9) at the critical value $x = 1$ and has an absolute maximum (equal to 25) at the upper bound of the interval. This is shown in Figure 4.2.6(A). Notice that the graph of the function $f(x)$ is similar to the graph of the function $g(x)$ in Figure 4.2.5(B).

As another example, consider the function $y = g(x) = 10 + 2x - x^2$. Assume that we want to find the absolute extrema of this function on the closed interval $[0, 4]$. As in the last example, we follow the above procedures. First, we obtain the critical value $x^* = 1$ from $g'(x) = 2 - 2x = 0$, and the stationary value $y = g(1) = 10 + 2 \times 1 - 1^2 = 11$. Second, we find the values of the function at the lower and upper bounds of the interval. The value of the

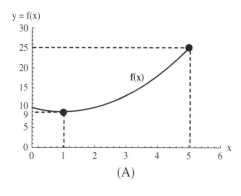
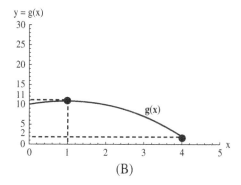

Figure 4.2.6

function at $x = 0$ (the lower bound of the interval) is $y = g(0) = 10 + 2 \times 0 - 0^2 = 10$ and the value of the function at $x = 4$ (the upper bound of the interval) is $y = g(4) = 10 + 2 \times 4 - 4^2 = 2$. Third, we compare these values. Notice that the stationary value is higher than the values of the function at both the lower and upper bounds of the interval. Fourth, based on these results, we can state that the function has, on the closed interval $[0, 4]$ on its domain, an absolute maximum (equal to 11) at the critical value $x = 1$ and has an absolute minimum (equal to 2) at the upper bound of the interval. This is illustrated in Figure 4.2.6(B).

4.2.7 *Convexity and concavity, and extrema of univariate functions*

We discussed convexity and concavity of univariate functions in Section 3.5. In our discussion we found that the graphs in Figures 3.5.1(A) and (B) were strictly concave and strictly convex, respectively. These are due to the fact that the second derivative of the first function is negative throughout ($f''(x) < 0$) and the second derivative of the second function is positive throughout ($g''(x) > 0$) on the closed interval $[x_0, x_2]$ on their respective domains.

The reader would have noticed an important feature of the graph in Figure 3.5.1(A). This feature is that the function $f(x)$ has a maximum at $x = x_1$ and two minimum points at around $x = x_0$ and around $x = x_2$ on the closed interval $[x_0, x_2]$ on its domain. Similarly, the graph of the function $g(x)$ in Figure 3.5.1(B) has a minimum at $x = x_1$ and two maximum points at around $x = x_0$ and around $x = x_2$ on the closed interval $[x_0, x_2]$ on its domain. What this means is that a strictly concave function will always have an absolute maximum and a strictly convex function will always have an absolute minimum on the closed intervals on their respective domains. This implies that a knowledge of the curvature (convexity and concavity) of functions obviates the need to check the SOC in optimization. This again confirms the extreme-value theorem we stated in the previous section. Let us now state these results formally.

Suppose that $f(x)$ is a convex (concave) function on a closed interval $I = [x_i, x_j]$, where $i \neq j$. Also suppose that x_0 in $I = [x_i, x_j]$ is a critical value of $f(x)$, and $f(x_0)$ is a stationary value of $f(x)$. Then, $f(x_0)$ is a local minimum (maximum) of $f(x)$. Suppose, instead, that $f(x)$ is a strictly convex (concave) function on its entire domain; x_0 is the critical value of $f(x)$; and $f(x_0)$ is the stationary value of $f(x)$. Then, $f(x_0)$ is the global or absolute minimum (maximum) of $f(x)$.

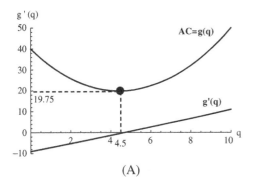
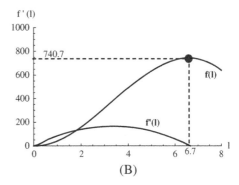

(A)   (B)

Figure 4.2.7

4.2.8 Application examples

Example 1.   Suppose that the total cost (in thousands of dollars) of producing q units of output (assuming that there is no fixed cost) of a firm is given by the function $C = f(q) = 40q - 9q^2 + q^3$. Find the level of output at which the average cost of the firm is a minimum, and the minimum average cost. Show this minimum on the graph of the average cost function.

Solution.   The average cost function ($\overline{C} = g(q)$) is obtained by dividing the total cost function by the output produced, q. Therefore, the average cost function is $\overline{C} = g(q) = 40 - 9q + q^2$. The FOC for an optimum requires $g'(q) = 0$. Differentiating $g(q)$ with respect to q yields $g'(q) = -9 + 2q$. Setting this derivative to zero and solving yields $q^* = 9/2 = 4.5$. Therefore, the critical value is $q^* = 4.5$. The SOC requires that we check the sign of the second derivative of the average cost function. Differentiating $g'(q) = -9 + 2q$ again with respect to q, we obtain $g''(q) = 2 > 0$. This implies that the function has a relative minimum at the critical value; that is, the average cost is a minimum when the quantity of output produced is 4.5 units. The minimum average cost (or the stationary value of the average cost function) is $\overline{C} = g(q) = 40 - 9 \times 4.5 + (4.5)^2 = \$19.75$ (in thousands). This relative minimum of the average cost function is shown in Figure 4.2.7(A). Notice that this relative minimum is the minimum of the function for all $x > 0$. Therefore, this relative minimum is also the absolute or global minimum of the average cost function.

Example 2.   Assume that a firm produces q units of output employing l units of labor. Also assume that the firm's production function is given by $q = f(l) = 50l^2 - 5l^3$. How many units of labor should the firm employ in order to maximize the total output produced? Show this maximum on the graph of the firm's production function.

Solution.   We need to check whether the function $q = f(l) = 50l^2 - 5l^3$ has a maximum. If it has one, then the number of workers corresponding to that maximum output will be the answer to the question. For an optimum, as before, the FOC must be satisfied: it is $f'(l) = 0$. We know that $f'(l) = 100l - 15l^2$. Setting $f'(l) = 100l - 15l^2$ to zero, and solving for l, yields $l^* = (0, 6.7)$. Since $l^* = 0$ is meaningless in our example, we discard it; instead, we choose $l^* = 6.7$. This means that the admissible critical value is $l^* = 6.7$. We can now check the sign of $f''(l)$ to see if this critical value corresponds to an optimum of the function. We obtain

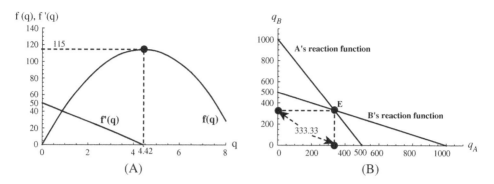

Figure 4.2.8

that $f''(l) = 100 - 30l$. When we substitute the critical value $l^* = 6.7$ into $f''(l^*) = 100 - 30l^*$ we get $f''(6.7) = 100 - 30 \times 6.7 = -101 < 0$. This satisfies the SOC for $f(l)$ for a relative maximum at $l^* = 6.7$. This implies that the firm must employ 6.7 units of labor to maximize its output, and the maximum output will be $q^* = f(6.7) = 50(6.7)^2 - 5(6.7)^3 = 740.7$ units. This relative maximum of the firm's production function is shown in Figure 4.2.7(B). Notice that this relative maximum is the maximum of the function for all $l > 0$. Therefore, this relative maximum is also the absolute or global maximum of the production function.

Example 3.   Assume that the total revenue, R, of a company from the sale of q units of a good is given by the function $R = f(q) = 50q - 5q^2 - 0.1q^3$, where R is in thousands of dollars. How many units of the good should the company sell to maximize its total revenue? Show this maximum on the graph of the company's total revenue function.

Solution.   As before, we need to check whether the function $R = f(q) = 50q - 5q^2 - 0.1q^3$ has a maximum. If it has one, then the units of the good sold corresponding to that maximum revenue will be the solution to the problem. For an optimum, as before, the FOC must be satisfied: it is $f'(q) = 0$. We know that $f'(q) = 50 - 10q - 0.3q^2$. Setting $f'(q) = 50 - 10q - 0.3q^2$ to zero and solving for q yields $q^* = (-37.75, 4.42)$. Since $q^* = -37.75$ is inadmissible in our example, we discard it; instead, we choose $q^* = 4.42$. This means that the admissible critical value is $q^* = 4.42$. We now check the sign of $f''(q)$ to see if this critical value corresponds to an optimum of the function. We can obtain that $f''(q) = -10 - 0.6q$. When we substitute the critical value $q^* = 4.42$ into $f''(q^*) = -10 - 0.6q^*$ we get $f''(4.42) = -10 - 0.6 \times 4.42 = -12.7 < 0$. This satisfies the SOC for $f(q)$ for a relative maximum at $q^* = 4.42$. This implies that the company must sell 4.42 units of the good to maximize its total revenue, and the maximum total revenue will be $R^* = f(4.42) = 50 \times 4.42 - 5(4.42)^2 - 0.1(4.42)^3 = \$115$ (in thousands). The relative maximum of the company's total revenue is shown in Figure 4.2.8(A). Notice that this relative maximum is the maximum of the function for all $q > 0$. Therefore, this relative maximum is also the absolute or global maximum of the total revenue function.

Example 4.   Suppose that the price, P, in dollars, in a *Cournot duopoly market*, which is a special type of *oligopoly market*, is given by $P = f(q) = 1100 - (q_A + q_B) = 1100 - q_A - q_B$, where q_A and q_B denote *duopolist* A's output and duopolist B's output, respectively (and the goods are assumed to be homogeneous). Notice that the total duopoly market output is $q = q_A + q_B$. Also suppose, for convenience, that the duopolists have equal and constant marginal and average costs of $100; that is, $AC_A = MC_A = AC_B = MC_B = \$100$. (i) Draw the *reaction function* of each duopolist. (ii) Find the output that maximizes the total profits of each duopolist. (iii) Find the total profits of each duopolist. (iv) Find the market price. (v) Find the total quantity produced by the duopolists. (vi) How do these results change if we assume that these duopolists collude?

Solution. (i) The reaction functions can be derived by maximizing the total profits of the duopolists. We are given that the market demand function is $P = f(q) = 1100 - (q_A + q_B) = 1100 - q_A - q_B$. Then the total revenue of A is $R_A = P.q_A = (1100 - q_A - q_B)q_A = 1100q_A - q_A^2 - q_Aq_B$ and the total cost of A is $C_A = 100q_A$. Therefore, A's total profit is $\Pi_A = R_A - C_A = Pq_A - 100q_A = 1100q_A - q_A^2 - q_Aq_B - 100q_A = 1000q_A - q_A^2 - q_Aq_B$. The FOC for a maximum of the total profit of A is that $d\Pi_A/dq_A = 0$. We know that $d\Pi_A/dq_A = 1000 - 2q_A - q_B$; setting this to zero and solving for q_A yields $q_A = h(q_B) = 500 - 0.5q_B$. This function is the reaction function of A, which has the horizontal intercept (500, 0) and the vertical intercept (0, 1000). This is shown in Figure 4.2.8(B).

We now derive the reaction function of duopolist B. The total revenue of B is $R_B = P.q_B = (1100 - q_A - q_B)q_B = 1100q_B - q_Aq_B - q_B^2$ and the total cost of B is $C_B = 100q_B$. Therefore, B's total profit is $\Pi_B = R_B - C_B = Pq_B - 100q_B = 1100q_B - q_B^2 - q_Aq_B - 100q_B = 1000q_B - q_B^2 - q_Aq_B$. The FOC for a maximum of the total profits of B is that $d\Pi_B/dq_B = 0$. We know that $d\Pi_B/dq_B = 1000 - 2q_B - q_A$; setting this to zero and solving for q_B yields $q_B = f(q_A) = 500 - 0.5q_A$. This function is the reaction function of B, which has the horizontal intercept (1000, 0) and the vertical intercept (0, 500). This is shown in Figure 4.2.8(B). Since $d^2\Pi_A/dq_A^2 = -2 < 0$ and $d^2\Pi_B/dq_B^2 = -2 < 0$, the SOC is also satisfied in the case of both duopolists.

(ii) The output that maximizes each duopolist's profit can be found by equating the reaction functions of the duopolists and solving for q_A and q_B. Equating the two reaction functions gives us $q_A = q_B$. This implies that we can substitute one reaction function in the other reaction function to solve for either q_A or q_B. Therefore, we substitute q_B into q_A to obtain $q_A = 500 - 0.5q_B = 500 - 0.5(500 - 0.5q_A)$, which gives us $q_A^* = 333.33$. Since $q_A^* = q_B^*$, $q_B^* = 333.33$. Therefore, the output levels that maximize the profits of each duopolist are identical: $q_A = q_B = 333.33$ units.

(iii), (iv), and (v) Since $q_A^* = q_B^* = 333.33$, $\Pi_A^* = \Pi_B^* = 1000q_A^* - q_A^{*2} - q_A^*q_B^* = 1000(333.33) - (333.33)^2 - (333.33)^2 = \$111\ 112$. The combined profit in the duopoly market is $\Pi_A^* + \Pi_B^* = \$111\ 112 + \$111\ 112 = \$222\ 224$. The market price is $P = f(q^*) = 1100 - (q_A^* + q_B^*) = 1100 - q_A^* - q_B^* = 1100 - 333.33 - 333.33 = \$433.34$. The total quantity produced by the duopolists is $q^* = q_A^* + q_B^* = 333.33 + 333.33 = 666.66$ units. Notice that point E in Figure 4.2.8(B) is called the *Cournot equilibrium* point.

(vi) If the two duopolists collude, the price in the market will be $P = f(q) = 1100 - q$. Then, their combined total revenue and total cost will be $R = Pq = 1100q - q^2$ and $C = 100q$, respectively. The profit function (of the industry), when there is collusion of duopolists, is $\Pi = R - C = Pq - C = 1100q - q^2 - 100q = 1000q - q^2$. The FOC requires, for this profit function to have a maximum, that $d\Pi/dq = 0$. We know that $d\Pi/dq = 1000 - 2q$. Setting this to zero yields $q^* = 500$ units. Moreover, we have $d^2\Pi/dq^2 = -2 < 0$, which means that the industry profits will be maximum when output produced is 500 units. This also means that the collusive (or cooperative) output (500 units) is smaller than the noncollusive (or noncooperative) output (666.66 units). The market price when there is *collusion* will be $P = f(q^*) = 1100 - q^* = 1100 - 500 = \$600$, which is higher than the market price (\$433.34) without collusion. The industry profit under collusion will be $\Pi^* = 1000q^* - q^{*2} = 1000(500) - (500)^2 = \$250\,000$, which is higher than the industry profit (\$222\,224) when there is no collusion.

Example 5.   Assume that the demand and the cost functions of the duopolists in a *Stackelberg duopoly market*, which is another special case of oligopoly market, are the same as those in example 4 above: $P = f(q) = 1100 - (q_A + q_B) = 1100 - q_A - q_B$ and $AC_A = MC_A = AC_B = MC_B = \$100$. (i) If A is the leader, find the output and profit of each duopolist. (ii) If B is the leader, find the output and profit of each duopolist.

Solution.   Notice that the demand and the cost conditions here are the same as those in example 4. We know, from the solution for example 4, that the reaction functions of the duopolists A and B are $q_A = 500 - 0.5q_B$ and $q_B = 500 - 0.5q_A$, respectively.

(i) *Stackelberg equilibrium* with A as leader. When A is the leader, it will substitute B's reaction function into its profit function, which is $\Pi_A = 1000q_A - q_A^2 - q_Aq_B$. Substituting $q_B = 500 - 0.5q_A$ into $\Pi_A = 1000q_A - q_A^2 - q_Aq_B$, we obtain $\Pi_A = 1000q_A - q_A^2 - q_A(500 - 0.5q_A) = 500q_A - 0.5q_A^2$. We now need to maximize this profit function of A. Then the FOC is $d\Pi_A/dq_A = 500 - q_A = 0$, which gives us the critical output for the leader, A: $q_A^* = 500$ units. Since $d^2\Pi_A/dq_A^2 = -1 < 0$, this critical value ($q_A^* = 500$) corresponds to the maximum point on the total profit function of A. Hence, the output that maximizes A's total profit is $q_A^* = 500$. Therefore, A's total profit will be $\Pi_A^* = 500q_A - 0.5q_A^{*2} = 500 \times 500 - 0.5(500)^2 = \$125\,000$.

Since B is the follower, it will assume that A will continue to produce $q_A^* = 500$ units. Therefore, B substitutes this quantity into its reaction function. Substituting $q_A^* = 500$ into $q_B = 500 - 0.5q_A$ yields $q_B^* = 500 - 0.5 \times 500 = 250$ units. Therefore, B's total profit will be $\Pi_B^* = 1000q_B^* - q_B^{*2} - q_A^*q_B^* = 1000 \times 250 - (250)^2 - 500 \times 250 = \$62\,500$. This Stackelberg equilibrium is represented by point E_A in Figure 4.2.9(A).

(ii) Stackelberg equilibrium with B as leader. When B is the leader, it will substitute A's reaction function into its profit function, which is $\Pi_B = 1000q_B - q_B^2 - q_Aq_B$. Substituting $q_A = 500 - 0.5q_B$ into $\Pi_B = 1000q_B - q_B^2 - q_Aq_B$, we obtain $\Pi_B = 1000q_B - q_B^2 - q_B(500 - 0.5q_B) = 500q_B - 0.5q_B^2$. We can now maximize this profit function of B. Then the FOC is $d\Pi_B/dq_B = 500 - q_B = 0$, which gives us the critical output for the leader, B: $q_B^* = 500$ units. Since $d^2\Pi_B/dq_B^2 = -1 < 0$, this critical value ($q_B^* = 500$) corresponds to the maximum point

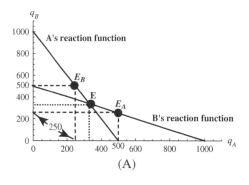
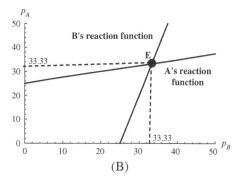

(A)   (B)

Figure 4.2.9

on the total profit function of B. Hence, the output that maximizes B's total profit is $q_B^* = 500$. Therefore, B's total profit will be $\Pi_B^* = 500q_B^* - 0.5q_B^{*2} = 500 \times 500 - 0.5(500)^2 = \$125\,000$.

Since A is the follower now, it will assume that B will continue to produce $q_B^* = 500$ units. Therefore, A substitutes this quantity into its reaction function. Substituting $q_B^* = 500$ in $q_A = 500 - 0.5q_B$ yields $q_A^* = 500 - 0.5 \times 500 = 250$ units. Therefore, A's total profit will be $\Pi_A^* = 1000q_A^* - q_A^{*2} - q_A^* q_B^* = 1000 \times 250 - (250)^2 - 250 \times 500 = \$62\,500$. This Stackelberg equilibrium is represented by point E_B in Figure 4.2.9(A).

Example 6. Assume that the demand functions of the duopolists in a *Bertrand duopoly market*, which is yet another special case of oligopoly market, are given by $q_A = 30 - p_A + 0.5p_B$ and $q_B = 30 - p_B + 0.5p_A$, where q_A, q_B, p_A, and p_B are the quantities produced and prices fixed by duopolists A and B, respectively. Also assume that $AC_A = MC_A = AC_B = MC_B = \$20$. (i) Determine the duopolists' reaction functions and graph them. (ii) Find the *Bertrand equilibrium*.

Solution. Let us first find the total profit functions of each duopolist. The total profit of A is given by $\Pi_A = p_A q_A - AC_A q_A = 50p_A - p_A^2 + 0.5p_A p_B - 10p_B - 600$. Similarly, the total profit of B is given by $\Pi_B = p_B q_B - AC_B q_B = 50p_B - p_B^2 + 0.5p_A p_B - 10p_A - 600$.

(i) Differentiating the above two profit functions with respect to p_A and p_B, respectively, and setting the results to zero yields $p_A = 25 + 0.25p_B$ and $p_B = 25 + 0.25p_A$ ($= p_A = -100 + 4p_B$). The previous two equations give us the reaction functions of A and B, respectively. Their graphs are shown in Figure 4.2.9(B).

(ii) The Bertrand duopoly market equilibrium can be found by equating the reaction functions of the two duopolists. Therefore, equating $p_A = 25 + 0.25p_B$ and $p_B = 25 + 0.25p_A$ ($= p_A = -100 + 4p_B$) gives us $p_A^* = p_B^* = 33.33$. Notice that in the present Bertrand duopoly market equilibrium the prices set by the duopolists are equal.

4.2.9 Exercises

1. Test the following functions for relative extrema:
   (i) $y = f(x) = 10 + 10x - x^2$; (ii) $y = f(x) = 10 + 10x + 10x^2 - x^3$; (iii) $y = f(x) = x^2 - 5x$; (iv) $y = f(x) = 5x - x^2$; (v) $y = f(x) = (x + 1)(x + 1)$; (vi) $y = f(x) = (x - 1)(x - 1)$; (vii) $y = f(x) = (x + 1)(x + 1)(x - 1)$; (viii) $y = f(x) = (x - 1)(x - 1)(x + 1)$.
2. Test the following functions for absolute extrema on the indicated intervals:
   (i) $y = f(x) = 10 - 10x + x^2$, $[0, 10]$; (ii) $y = f(x) = 10 + 10x - x^2$, $[0, 10]$; (iii) $y = f(x) = 10 - 5x + x^3$, $[-10, 10]$; (iv) $y = f(x) = (x + 1)(x - 1)$, $[-5, 5]$; (v) $y = f(x) = (x - 1)(x - 1)$, $[-5, 5]$; (vi) $y = f(x) = (x + 1)(x + 1)$, $[-10, 10]$.
3. *Application exercise.* Suppose that the total utility (U) obtained by a consumer when q units of a good are consumed is given by $U = f(q) = 50q + 10q^2 - q^3$. Find the number of units of the good that must be consumed to maximize the utility.
4. *Application exercise.* Suppose that the total revenue (R) that a company obtains from the sale of q units of a good is given by $R = g(q) = (100q - 2q^2)/50$. How many units of the good must the company sell in order to maximize its total revenue in dollars?
5. *Application exercise.* Assume that the inverse demand function for a company's product is given by $P = f(q) = 200 - 3q$, where P denotes the price per unit of the good in dollars, and q denotes the quantity of the good demanded. Also assume that the total cost of the company is given by $C = g(q) = 200 + 2q + 0.5q^2$. Find the level of output at which the profit of the company is maximum. What will be the profit-maximizing price?

6. *Application exercise.* Suppose that the total cost of producing q units of a good by a firm is given by $C = f(q) = 50q - 10q^2 + q^3$. Find the level of output at which the firm's average cost is minimum. What is the average cost at this level of output?

7. *Application exercise.* Suppose that the demand functions of duopolists in a Cournot duopoly market are the same as those in example 4 in Section 4.2.8. Also suppose that the government gives a subsidy of \$25 per unit of the good produced to duopolist A; and $AC_B = MC_B = \$100$ remains the same as before. How does this subsidy affect the solutions to example 4?

8. *Application exercise.* Suppose that the demand functions of duopolists in a Stackelberg duopoly market are the same as those in example 5 in Section 4.2.8. Also suppose that the government imposes a tax of \$25 per unit of the good on duopolist B; and that $AC_A = MC_A = \$100$ remains the same as before. How does this tax affect the solutions to example 5?

9. *Application exercise.* Assume that the demand functions of the duopolists in a Bertrand duopoly market are the same as those in example 6 in Section 4.2.8. Also assume that $AC_A = MC_A = \$10$ and $AC_B = MC_B = \$20$. How does this difference in costs affect the solutions to example 6?

 Web supplement: S4.2.10 Mathematica applications

4.3 Extrema of unconstrained multivariate objective functions

So far we have been concerned with finding the extrema, if they exist, of objective functions that involve only one independent variable, or of univariate objective functions. But we know that most of the relationships among variables in economics, business, and finance are expressed in the form of multivariate functions. Therefore, it is important that students of these subjects possess a good understanding of the topic of optimization of objective functions that involve more than one independent variable. We provide in this section the techniques of optimization of multivariate functions.

4.3.1 Extrema of unconstrained bivariate functions: FOCs and SOCs

As a simple case, consider the production function of a firm that uses two factors, labor (L) and capital (K), to produce a particular quantity (Q) of a good. We normally express the production function in the form $Q = f(K, L)$. We know from our discussion in Section 3.7 that Q will change when K changes for given L; or when L changes for given K; or when both K and L change.

We stated in Section 4.2 that, in the case of a univariate function $f(x)$ (or $g(x)$), the FOC for the function to attain a relative optima is $f'(x) = 0$ (or $g'(x) = 0$). That is, the function has a relative maximum when the function forms itself a "hill" as in Figure 4.2.1(A) or a relative minimum when it forms itself a "valley" as in Figure 4.2.1(B). Can we think of such hills and valleys in the case of bivariate functions? Yes, we can; they are shown in Figure 4.3.1.

Consider, to begin with, the graph of $Q = f(K, L)$ in Figure 4.3.1(A). Notice that at point A the graph attains the maximum possible value. An important feature of point A is that the partial derivatives of $Q = f(K, L)$ with respect to K holding L constant (represented by the tangent T_K) and the partial derivative of $Q = f(K, L)$ with respect to L holding K constant

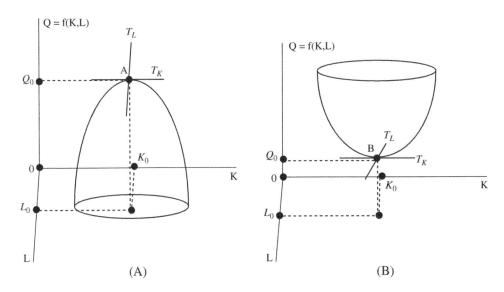

Figure 4.3.1

(represented by the tangent T_L) are both zero. This implies that, at point A, $\partial Q/\partial K = \partial f/\partial K = f_K = 0$ and $\partial Q/\partial L = \partial f/\partial L = f_L = 0$.

Similarly, at point B in Figure 4.3.1(B), the graph attains the minimum possible value. An important feature of point B is that the partial derivatives of $Q = g(K, L)$ with respect to K holding L constant (represented by the tangent T_K) and the partial derivative of $Q = g(K, L)$ with respect to L holding K constant (represented by the tangent T_L) are both zero. This implies that, at point B, $\partial Q/\partial K = \partial g/\partial K = g_K = 0$ and $\partial Q/\partial L = \partial g/\partial L = g_L = 0$. Therefore, the FOC for a relative optimum (either a maximum or a minimum) is that the first partial derivatives of the functions are all zero:

$$\frac{\partial Q}{\partial K} = \frac{\partial f}{\partial K} = f_K = 0 \text{ and } \frac{\partial Q}{\partial L} = \frac{\partial f}{\partial L} = f_L = 0 \tag{4.3.1}$$

However, as we saw in the case of univariate functions, the FOC does not guarantee optima. The reason is that we may find $f_K = f_L = 0$ at a particular point on the graph of $Q = f(K, L)$ (as represented by the middle point B on the graph in Figure 4.3.2) but it does not represent either a maximum or a minimum. Such a point is called a *saddle point*.

The above discussion implies that, as in the case of univariate functions, the FOC is only a necessary condition for an optimum. Again, as in the case of univariate functions, a SOC must be satisfied for an optimum. Therefore, once the FOC is satisfied, the SOC for a minimum of $Q = f(K, L)$ is

$$f_{KK} > 0, f_{LL} > 0 \text{ and } f_{KK} \cdot f_{LL} > f_{KL}^2 \tag{4.3.2}$$

and the SOC for a maximum of $Q = f(K, L)$ is

$$f_{KK} < 0, f_{LL} < 0 \text{ and } f_{KK} \cdot f_{LL} > f_{KL}^2 \tag{4.3.3}$$

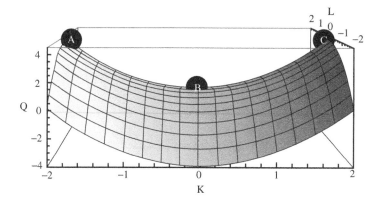

Figure 4.3.2

Table 4.3.1

Condition	Minimum	Maximum
FOCs or necessary conditions	$f_1 = f_2 = 0$	$f_1 = f_2 = 0$
SOCs or sufficient conditions	$f_{11} > 0, f_{22} > 0$, and	$f_{11} < 0, f_{22} < 0$, and
	$f_{11}.f_{22} > (f_{12})^2$	$f_{11}.f_{22} > (f_{12})^2$

Inflection point	
$f_{11} < 0, f_{22} < 0$, and $f_{11}.f_{22} < (f_{12})^2$ or	
$f_{11} < 0, f_{22} < 0$, and $f_{11}.f_{22} < (f_{12})^2$	
Saddle point	
$f_{11} > 0, f_{22} < 0$, and $f_{11}.f_{22} < (f_{12})^2$, or	
$f_{11} < 0, f_{22} > 0$, and $f_{11}.f_{22} < (f_{12})^2$	
Inconclusive	
$f_{11}.f_{22} = (f_{12})^2$	

We present these FOCs and SOCs in the case of a general bivariate function $y = f(x_1, x_2)$ in Table 4.3.1 for easy reference.

As an example, consider the function $U = f(x_1, x_2) = x_1^2 + x_2^2 - x_1 - x_2$. Following the FOCs given in Table 4.3.1, we obtain $f_1 = 2x_1 - 1, f_2 = 2x_2 - 1, f_{11} = 2, f_{22} = 2$, and $f_{12} = f_{21} = 0$. Now setting $f_1 = f_2 = 0$ and solving, as before, we obtain the critical values $(x_1^*, x_2^*) = (0.5, 0.5)$. Since $f_{11} = 2, f_{22} = 2, f_{12} = f_{21} = 0$, the SOC gives us $f_{11}.f_{22} = 4 > 0 \, (= (f_{12})^2)$. Therefore, following Table 4.3.1, both the FOC and the SOC are satisfied for the function to have a relative minimum corresponding to the critical points $(x_1^*, x_2^*) = (0.5, 0.5)$. The minimum or the stationary value of the function at the critical values $(x_1^*, x_2^*) = (0.5, 0.5)$ is $U^* = f(x_1^*, x_2^*) = x_1^{*2} + x_2^{*2} - x_1^* - x_2^* = 0.5^2 + 0.5^2 - 0.5 - 0.5 = -0.5$.

As another example, consider the function $U = f(x_1, x_2) = x_1 + x_2 - x_1^2 - x_2^2$. Following the FOCs given in Table 4.3.1, we obtain $f_1 = 1 - 2x_1, f_2 = 1 - 2x_2, f_{11} = -2, f_{22} = -2$, and $f_{12} = f_{21} = 0$. Now setting $f_1 = f_2 = 0$ and solving, we obtain the critical values $(x_1^*, x_2^*) = (0.5, 0.5)$. Since $f_{11} = -2, f_{22} = -2, f_{12} = f_{21} = 0$, the SOC gives us $f_{11}.f_{22} = 4 > 0 (= (f_{12})^2)$. Therefore, following Table 4.3.1, both the FOC and the SOC are satisfied for the function to have a relative maximum corresponding to the critical point $(x_1^*, x_2^*) = (0.5, 0.5)$. The maximum or the stationary value of the function at the critical values $(x_1^*, x_2^*) = (0.5, 0.5)$ is $U^* = f(x_1^*, x_2^*) = x_1^* + x_2^* - x_1^{*2} - x_2^{*2} = 0.5 + 0.5 - 0.5^2 - 0.5^2 = 0.5$.

4.3.2 Conditions of extrema of bivariate functions: differential version

As we developed the conditions of extrema of univariate functions using differentials in Section 4.2.5, we can develop the conditions of extrema of bivariate functions using total differentials. To develop these conditions, assume that we have a general bivariate function $U = f(x_1, x_2)$. The first total differential of this function, as a special case of equation (3.8.4) with $n = 2$, is dU. But, we know from our discussion in Section 3.8.2 that $dU = (\partial f / \partial x_1)dx_1 + (\partial f / \partial x_2)dx_2 = f_1 dx_1 + f_2 dx_2$.

Notice that the function $U = f(x_1, x_2)$ is stationary only when $dU = 0$, for given dx_1 and dx_2, not both zero. This implies that an extreme point of the function must be a stationary point. Therefore, as in the case of univariate functions, $dU = 0$ is the FOC for a relative extremum of the function $U = f(x_1, x_2)$. But, since $dU = f_1 dx_1 + f_2 dx_2$, $dU = 0$ only when both partial derivatives are zero, that is, only when $f_1 = f_2 = 0$. Therefore, the FOC in terms of total differential that $dU = 0$ is, in fact, identical with the FOC using derivatives ($f_1 = f_2 = 0$).

We now consider the SOC using total differential. The second total differential of the function $U = f(x_1, x_2)$ is given in equation (3.8.7), which is a special case of equation (3.8.9). Therefore, the second total differential of the function is $d^2 U = f_{11} dx_1^2 + 2f_{12} dx_1 dx_2 + f_{22} dx_2^2$. It is clear from this equation that the sign of $d^2 U$ depends on the signs of the partial derivatives. It is easy to see that $d^2 U > 0$ only if $f_{11} > 0, f_{22} > 0$, and $f_{11}.f_{22} > (f_{12})^2$; and $d^2 U < 0$ only if $f_{11} < 0, f_{22} < 0$, and $f_{11}.f_{22} > (f_{12})^2$. Therefore, the SOC for the function $U = f(x_1, x_2)$ to have a minimum is $d^2 U > 0$ (given $dU = 0$) and the SOC for the function to have a maximum is $d^2 U < 0$ (given $dU = 0$). We present the FOCs and the SOCs for the bivariate function $U = f(x_1, x_2)$ to have relative extrema in Table 4.3.2 for easy reference. Notice that these conditions for extrema are identical to the conditions presented in Table 4.3.1 using derivatives.

Notice that the second total differential $d^2 U = f_{11} dx_1^2 + 2f_{12} dx_1 dx_2 + f_{22} dx_2^2$, of the function $U = f(x_1, x_2)$, is a quadratic form that we discussed and utilized in Section 2.8.2. We stated in Section 2.8.2 that $d^2 U$ is positive definite if $d^2 U > 0$ and negative definite if $d^2 U < 0$. Therefore, a minimum of $U = f(x_1, x_2)$ requires (given $dU = 0$) $d^2 U$ to be positive definite (that is, $d^2 U > 0$) and a maximum of $U = f(x_1, x_2)$ requires $d^2 U$ to be negative definite (that is, $d^2 U < 0$).

Table 4.3.2

Condition	Minimum	Maximum
FOC or necessary condition	$dU = 0$	$dU = 0$
SOC or sufficient condition	$d^2 U > 0$	$d^2 U < 0$

Table 4.3.3

Condition	Minimum	Maximum
FOC or necessary condition	$dU = 0$	$dU = 0$
SOC or sufficient condition	$d^2U > 0$, or positive definite, if $\mathbf{D_{R1}} > 0$ and $\mathbf{D_{R2}} = \mathbf{D_R} > 0$	$d^2U < 0$, or negative definite, if $\mathbf{D_{R1}} < 0$ and $\mathbf{D_{R2}} = \mathbf{D_R} > 0$

How do we determine the sign of d^2U?[6] We have already discussed the methods of determining the sign of d^2U in Chapter 2. In Section 2.8.2, we represented the quadratic form $d^2U = f_{11}dx_1^2 + 2f_{12}dx_1dx_2 + f_{22}dx_2^2$ (equation (2.8.9)) in terms of matrices as $d^2U = \mathbf{x^T Dx}$ (equation (2.8.14)), where $\mathbf{x^T} = \begin{bmatrix} dx_1 & dx_2 \end{bmatrix}$ and $\mathbf{D} = \begin{bmatrix} f_{x_1x_1} & f_{x_1x_2} \\ f_{x_2x_1} & f_{x_2x_2} \end{bmatrix} = \begin{bmatrix} f_{11} & f_{12} \\ f_{21} & f_{22} \end{bmatrix}$.

We called the determinant of $\mathbf{D}$ the discriminant of the quadratic form d^2U and denoted it by $\mathbf{D_R}$. Therefore, $|\mathbf{D}| = \mathbf{D_R}$. We found there that the first principal minor of $\mathbf{D_R}$, denoted by $\mathbf{D_{R1}}$, is equal to $|f_{11}| = f_{11}$. The second principal minor is $\mathbf{D_{R2}} = \mathbf{D_R} = |\mathbf{D}|$. We also found there that if d^2U is positive definite (or $d^2U > 0$), then the function $U = f(x_1, x_2)$ has a minimum (given $dU = 0$), if $\mathbf{D_{R1}} > 0$ and $\mathbf{D_{R2}} = \mathbf{D_R} > 0$. Similarly, we found that if d^2U is negative definite (or $d^2U < 0$), then the function $U = f(x_1, x_2)$ has a maximum (given $dU = 0$), if $\mathbf{D_{R1}} < 0$ and $\mathbf{D_{R2}} = \mathbf{D_R} > 0$. If these conditions of extrema are not satisfied, then $U = f(x_1, x_2)$ has a saddle point at the corresponding critical values. We present these FOCs and SOCs, using differentials and quadratic forms, of extrema of the bivariate function $U = f(x_1, x_2)$ in Table 4.3.3 for easy reference.

As an example, consider the function in the first example in the last section: $U = f(x_1, x_2) = x_1^2 + x_2^2 - x_1 - x_2$. The first total differential of this function can be written as $dU = f_1 dx_1 + f_2 dx_2$. The FOC is that $dU = 0$; and it implies, as before, that $f_1 = f_2 = 0$. We obtained in the last section that $f_1 = 2x_1 - 1$ and $f_2 = 2x_2 - 1$; and obtained, by setting them to zero and simplifying, the critical values $(x_1^*, x_2^*) = (0.5, 0.5)$. We also obtained $f_{11} = 2, f_{22} = 2$, and $f_{12} = f_{21} = 0$. The quadratic form of $U = f(x_1, x_2) = x_1^2 + x_2^2 - x_1 - x_2$ with $f_{11} = 2, f_{22} = 2$, and $f_{12} = f_{21} = 0$ is $d^2U = f_{11}dx_1^2 + 2f_{12}dx_1dx_2 + f_{22}dx_2^2 = 2dx_1^2 + 2 \times 0 \times dx_1dx_2 + 2dx_2^2$, which in matrix form is $d^2U = \mathbf{x^T Dx}$, where $\mathbf{x^T}$ and $\mathbf{D}$ are in accordance with our earlier definitions. Since, in this example, $\mathbf{D} = \begin{bmatrix} f_{x_1x_1} & f_{x_1x_2} \\ f_{x_2x_1} & f_{x_2x_2} \end{bmatrix} = \begin{bmatrix} f_{11} & f_{12} \\ f_{21} & f_{22} \end{bmatrix} \begin{bmatrix} 2 & 0 \\ 0 & 2 \end{bmatrix}$, $\mathbf{D_{R1}} = 2 > 0$,

and $\mathbf{D_{R2}} = \mathbf{D_R} = 4 > 0$, d^2U is positive definite; therefore, the function $U = f(x_1, x_2) = x_1^2 + x_2^2 - x_1 - x_2$ qualifies for a minimum at the critical value $(x_1^*, x_2^*) = (0.5, 0.5)$. The stationary value of the function at the critical values $(x_1^*, x_2^*) = (0.5, 0.5)$ is $U^* = f(x_1^*, x_2^*) = x_1^{*2} + x_2^{*2} - x_1^* - x_2^* = 0.5^2 + 0.5^2 - 0.5 - 0.5 = -0.5$. These were precisely the results we obtained in the last section when we used the derivatives to find the optimum of the given function.

As another example, consider the function we used in the second example in the last section: $U = f(x_1, x_2) = x_1 + x_2 - x_1^2 - x_2^2$. The first total differential of this function can be written as $dU = f_1 dx_1 + f_2 dx_2$. The FOC is that $dU = 0$; and it implies, as we stated above, that $f_1 = f_2 = 0$. We obtained in the last section that $f_1 = 1 - 2x_1$ and $f_2 = 1 - 2x_2$; and obtained, by setting them to zero and simplifying, the critical values $(x_1^*, x_2^*) = (0.5, 0.5)$. We also obtained there that $f_{11} = -2, f_{22} = -2$, and $f_{12} = f_{21} = 0$. The quadratic form of

$U = f(x_1, x_2) = x_1 + x_2 - x_1^2 - x_2^2$ with $f_{11} = -2$, $f_{22} = -2$, and $f_{12} = f_{21} = 0$ is $d^2U = f_{11}dx_1^2 + 2f_{12}dx_1dx_2 + f_{22}dx_2^2 = -2dx_1^2 + 2 \times 0 \times dx_1dx_2 - 2dx_2^2$, which in matrix form is $d^2U = \mathbf{x}^T\mathbf{D}\mathbf{x}$, where $\mathbf{x}^T$ and $\mathbf{D}$ are in accordance with our earlier definitions. Since, in this example, $\mathbf{D} = \begin{bmatrix} f_{x_1x_1} & f_{x_1x_2} \\ f_{x_2x_1} & f_{x_2x_2} \end{bmatrix} = \begin{bmatrix} f_{11} & f_{12} \\ f_{21} & f_{22} \end{bmatrix} \begin{bmatrix} -2 & 0 \\ 0 & -2 \end{bmatrix}$, $\mathbf{D_{R1}} = -2 < 0$, and $\mathbf{D_{R2}} = \mathbf{D_R} = 4 > 0$, d^2U is negative definite; therefore, the function $U = f(x_1, x_2) = x_1 + x_2 - x_1^2 - x_2^2$ qualifies for a maximum at the critical value $(x_1^*, x_2^*) = (0.5, 0.5)$. The stationary value of the function at the critical values $(x_1^*, x_2^*) = (0.5, 0.5)$ is $U^* = f(x_1^*, x_2^*) = x_1^* + x_2^* - x_1^{*2} - x_2^{*2} = 0.5 + 0.5 - 0.5^2 - 0.5^2 = 0.5$. Again, these were precisely the results we obtained in the last section when we used the derivatives to find the extremum of the function.

4.3.3 Conditions of extrema of trivariate functions: differential version

Now assume that the unconstrained objective function is a *trivariate function* such as $U = f(x_1, x_2, x_3)$. Since it is difficult to state the conditions of extrema of $U = f(x_1, x_2, x_3)$ using derivatives similar to those presented in Table 4.3.1, we rely here on total differentials. Even in the case of trivariate functions, the FOC is the same as before: the first total differential of $U = f(x_1, x_2, x_2)$ is equal to zero or $dU = f_1dx_1 + f_2dx_2 + f_3dx_3 = 0$ (which is a special case of equation (3.8.4) with $n = 3$). Therefore, this FOC implies that $f_1 = f_2 = f_3 = 0$.

We know from the discussion in Section 3.8.3 that the second total differential of $U = f(x_1, x_2, x_3)$, which is a special case of $U = f(x_1, x_2, x_3, \ldots, x_n)$ with $n = 3$, is given by equation (2.8.12) or (3.8.8): $d^2U = f_{11}dx_1^2 + f_{22}dx_2^2 + f_{33}dx_3^2 + 2f_{12}dx_1dx_2 + 2f_{13}dx_1dx_3 + 2f_{23}dx_2dx_3$. The SOC for extrema of $U = f(x_1, x_2, x_3)$ requires that we check the sign of d^2U. Notice that the matrix representation of d^2U, as shown in Section 2.8.2, is $d^2U = \mathbf{x}^T\mathbf{D}\mathbf{x}$, where $\mathbf{x}^T = \begin{bmatrix} dx_1 & dx_2 & dx_3 \end{bmatrix}$ and $\mathbf{D} = \begin{bmatrix} f_{11} & f_{12} & f_{13} \\ f_{21} & f_{22} & f_{23} \\ f_{31} & f_{32} & f_{33} \end{bmatrix}$. We stated in Section 2.8.2 that d^2U would be positive definite (given $dU = 0$) so that $U = f(x_1, x_2, x_3)$ would have a minimum if $\mathbf{D_{R1}} > 0$, $\mathbf{D_{R2}} > 0$, and $\mathbf{D_{R3}} = \mathbf{D_R} > 0$. And, d^2U would be negative definite (given $dU = 0$) so that $U = f(x_1, x_2, x_3)$ would have a maximum if $\mathbf{D_{R1}} < 0$, $\mathbf{D_{R2}} > 0$, and $\mathbf{D_{R3}} = \mathbf{D_R} < 0$. These conditions are presented in Table 4.3.4.

As an example, consider the function $U = f(x_1, x_2, x_3) = x_1^2 + x_2^2 + x_3^2 - x_1 - x_2 - x_3$. The FOC, as per Table 4.3.4, requires that $dU = 0$ for an optimum of the function. For this to happen, we must have $f_1 = f_2 = f_3 = 0$. We can obtain from the function that $f_1 = 2x_1 - 1$, $f_2 = 2x_2 - 1$, and $f_3 = 2x_3 - 1$. Setting these partial derivatives to zero and simplifying yields the critical values $(x_1^*, x_2^*, x_3^*) = (0.5, 0.5, 0.5)$.

Table 4.3.4

Condition	Minimum	Maximum
FOC or necessary condition	$dU = 0$	$dU = 0$
SOC or sufficient condition	$d^2U > 0$, or positive definite, if $\mathbf{D_{R1}} > 0$, $\mathbf{D_{R2}} > 0$, and $\mathbf{D_{R3}} = \mathbf{D_R} > 0$	$d^2U < 0$, or negative definite, if $\mathbf{D_{R1}} < 0$, $\mathbf{D_{R2}} \geq 0$, and $\mathbf{D_{R3}} = \mathbf{D_R} > 0$

We now need to check the SOC to know whether these critical values correspond to an optimum of the function. The SOC requires that the sign of the associated quadratic form $d^2U = f_{11}dx_1^2 + f_{22}dx_2^2 + f_{33}dx_3^2 + 2f_{12}dx_1dx_2 + 2f_{13}dx_1dx_3 + 2f_{23}dx_2dx_3$ be checked for a relative optimum of the function. We can obtain from the function that $f_{11} = 2$, $f_{22} = 2$, $f_{33} = 2$, $f_{12} = 0$, $f_{13} = 0$, and $f_{23} = 0$. Substituting these values into the quadratic form we obtain $d^2U = 2dx_1^2 + 2dx_2^2 + 2dx_3^2 + 2 \times 0dx_1dx_2 + 2 \times 0dx_1dx_3 + 2 \times 0dx_2dx_3$. The matrix representation of this equation is $d^2U = \mathbf{x}^T\mathbf{Dx}$, where $\mathbf{x}^T$ and $\mathbf{D}$ are as we defined earlier. Then, we have $\mathbf{D} = \begin{bmatrix} f_{11} & f_{12} & f_{13} \\ f_{21} & f_{22} & f_{23} \\ f_{31} & f_{32} & f_{33} \end{bmatrix} = \begin{bmatrix} 2 & 0 & 0 \\ 0 & 2 & 0 \\ 0 & 0 & 2 \end{bmatrix}$, which gives us $\mathbf{D_{R1}} = 2 > 0$,

$\mathbf{D_{R2}} = 4 > 0$, and $\mathbf{D_{R3}} = \mathbf{D_R} = 8 > 0$. Then, following Table 4.3.4, we conclude that d^2U is positive definite; therefore, $U = f(x_1, x_2, x_3) = x_1^2 + x_2^2 + x_3^2 - x_1 - x_2 - x_3$ qualifies for a minimum at the critical values $(x_1^*, x_2^*, x_3^*) = (0.5, 0.5, 0.5)$. Therefore, the minimum value of the function is $U^* = f(x_1^*, x_2^*, x_3^*) = x_1^{*2} + x_2^{*2} + x_3^{*2} - x_1^* - x_2^* - x_3^* = -0.75$.

As another example, consider the function $U = f(x_1, x_2, x_3) = x_1 + x_2 + x_3 - x_1^2 - x_2^2 - x_3^2$. The FOC, as above, requires that $dU = 0$ for an optimum of the function. For this to happen, we must have $f_1 = f_2 = f_3 = 0$. We can obtain from the function that $f_1 = 1 - 2x_1$, $f_2 = 1 - 2x_2$, and $f_3 = 1 - 2x_3$. Setting these partial derivatives to zero and simplifying yields the critical values, as before, $(x_1^*, x_2^*, x_3^*) = (0.5, 0.5, 0.5)$.

We now need to check the SOC to know whether the above critical values correspond to an optimum of the function. The SOC requires that the sign of the associated quadratic form $d^2U = f_{11}dx_1^2 + f_{22}dx_2^2 + f_{33}dx_3^2 + 2f_{12}dx_1dx_2 + 2f_{13}dx_1dx_3 + 2f_{23}dx_2dx_3$ be checked for a relative optimum of the function. We can obtain from the function that $f_{11} = -2$, $f_{22} = -2, f_{33} = -2, f_{12} = 0, f_{13} = 0$, and $f_{23} = 0$. Substituting these values into the quadratic form we obtain $d^2U = -2dx_1^2 - 2dx_2^2 - 2dx_3^2 + 2 \times 0dx_1dx_2 + 2 \times 0dx_1dx_3 + 2 \times 0dx_2dx_3$. The matrix representation of this equation is $d^2U = \mathbf{x}^T\mathbf{Dx}$, where $\mathbf{x}^T$ and $\mathbf{D}$ are as we defined earlier. Then, we have $\mathbf{D} = \begin{bmatrix} f_{11} & f_{12} & f_{13} \\ f_{21} & f_{22} & f_{23} \\ f_{31} & f_{32} & f_{33} \end{bmatrix} = \begin{bmatrix} -2 & 0 & 0 \\ 0 & -2 & 0 \\ 0 & 0 & -2 \end{bmatrix}$, which gives us

$\mathbf{D_{R1}} = -2 < 0$, $\mathbf{D_{R2}} = 4 > 0$, and $\mathbf{D_{R3}} = \mathbf{D_R} = -8 < 0$. Then, following Table 4.3.4, we conclude that d^2U is negative definite; therefore, $U = f(x_1, x_2, x_3) = x_1 + x_2 + x_3 - x_1^2 - x_2^2 - x_3^2$ qualifies for a maximum at the critical values $(x_1^*, x_2^*, x_3^*) = (0.5, 0.5, 0.5)$. Therefore, the maximum value of the function is $U^* = f(x_1^*, x_2^*, x_3^*) = x_1^* + x_2^* + x_3^* - x_1^{*2} - x_2^{*2} - x_3^{*2} = 0.75$.

4.3.4 Conditions of extrema of n-variable functions: differential version

Assume now that the unconstrained objective function is $U = f(x_1, x_2, x_3, \ldots, x_n)$, a function with n independent variables. As before, the FOC for optima in the case of an n-variable function is that the first total differential of the function be equal to zero or $dU = f_1dx_1 + f_2dx_2 + f_3dx_3 + \cdots + f_ndx_n = 0$, which is identical with equation (3.8.4). Therefore, this FOC implies, as earlier, that $f_1 = f_2 = f_3 = \cdots = f_n = 0$.

We know from Section 3.8.3 that the second total differential of $U = f(x_1, x_2, x_3, \ldots, x_n)$ is $d^2U = f_{11}dx_1^2 + f_{22}dx_2^2 + f_{33}dx_3^2 + \cdots + f_{nn}dx_n^2 + 2f_{12}dx_1dx_2 + 2f_{13}dx_1dx_3 + \cdots + 2f_{1n}dx_1dx_n + 2f_{23}dx_2dx_3 + 2f_{24}dx_2dx_4 + \cdots + 2f_{2n}dx_2dx_n + \cdots + 2f_{n-1,n}dx_{n-1}dx_n + \cdots$, which is identical with equation (3.8.9). The SOC for an optimum of $U = f(x_1, x_2, x_3, \ldots, x_n)$ requires that we check the sign of d^2U. Notice that the matrix representation of d^2U, as we saw

Table 4.3.5

Condition	Minimum	Maximum
FOC or necessary condition	$dU = 0$	$dU = 0$
SOC or sufficient condition	$d^2U > 0$, or positive definite, if $\mathbf{D_{R1}} > 0$, $\mathbf{D_{R2}} > 0$, $\mathbf{D_{R3}} \geq 0,\ldots,$ $\mathbf{D_{Rn}} = \mathbf{D_R} \geq 0$	$d^2U < 0$, or negative definite, if $\mathbf{D_{R1}} < 0$, $\mathbf{D_{R2}} > 0$, $\mathbf{D_{R3}} < 0,\ldots,$ $(-1)^n\mathbf{D_{Rn}} = (-1)^n\mathbf{D_R} \geq 0$

at the end of Section 2.8.2, is $d^2U = \mathbf{x^T}\mathbf{Dx}$, where $\mathbf{x^T} = \begin{bmatrix} dx_1 & dx_2 & \ldots & dx_n \end{bmatrix}$ and $\mathbf{D}$

$$= \begin{bmatrix} f_{11} & f_{12} & \cdots & f_{1n} \\ f_{21} & f_{22} & \cdots & f_{2n} \\ \cdots & \cdots & \cdots & \cdots \\ f_{n1} & f_{n2} & \cdots & f_{nn} \end{bmatrix}.$$ We stated at the end of Section 2.8.2 that d^2U would be positive

definite (given $dU = 0$) so that $U = f(x_1, x_2, x_3, \ldots, x_n)$ would have a minimum if $\mathbf{D_{R1}} > 0$, $\mathbf{D_{R2}} > 0$, $\mathbf{D_{R3}} > 0, \ldots$, $\mathbf{D_{Rn}} = \mathbf{D_R} > 0$. And, d^2U would be negative definite (given $dU = 0$) so that $U = f(x_1, x_2, x_3, \ldots, x_n)$ would have a maximum if $\mathbf{D_{R1}} < 0$, $\mathbf{D_{R2}} > 0$, $\mathbf{D_{R3}} < 0, \ldots, (-1)^n \mathbf{D_{Rn}} = (-1)^n \mathbf{D_R} > 0$. These conditions are presented in Table 4.3.5.

4.3.5 Concavity and convexity, and extrema of n-variable functions

In Section 4.2.7 we discussed how a knowledge of the curvature (convexity and concavity) of a univariate function, $y = f(x)$, obviated the need to check the SOCs to determine whether the function has an optimum. We found from our discussion there that if the function was convex (concave) over a closed interval on its domain, then the function had a local or relative minimum (maximum). We also found that if the function was strictly convex (concave) on its entire domain as the graph in Figure 4.2.1(A) (Figure 4.2.1(B)), then the function had a global or absolute minimum (maximum).

So far in the present section we were presenting the FOCs and the SOCs for optima of multivariate functions in terms of both derivatives and differentials. Suppose now that we have a multivariate function such as $U = f(x_1, x_2, x_3, \ldots, x_n)$. A reasonable question that arises now is: can one dispense with the SOCs and only use the information on the curvature of the function to judge whether the function has an optimum (as in the case of univariate functions)? The answer is yes, and the procedure is outlined as follows.

To begin with, assume that we have two bivariate functions $Q = f(K, L)$ and $Q = g(K, L)$. Also assume that the graphs of these two functions resemble those illustrated in Figures 4.3.1(A) and (B), respectively. We know from our discussion in Section 4.3.1 that the graphs in these figures (which are planes in three-dimensional spaces; a three-dimensional extension of the graphs in two-dimensional spaces as those in Figures 4.2.1(A) and (B), respectively) are strictly concave and convex, respectively. The graph in Figure 4.3.1(A) attains a global maximum (at point A) and that in Figure 4.3.1(B) attains a global minimum (at point B). They attain global optima because they are strictly concave or convex. If these functions were defined over closed intervals of their respective domains, then they would have attained local or relative optima. These results suggest that information on the curvature of bivariate functions can be used to determine whether the function attains local or global optima.

Let us now generalize the above results to the optima of an n-variable function, $U = f(x_1, x_2, x_3, \ldots, x_n)$. If the function is convex (concave) over a closed interval on its domain, then the function has a local minimum (maximum) over that interval. If, instead, this function is strictly convex (concave) over its entire domain, then the function has a global minimum (maximum). Therefore, we need only check whether the function under consideration is convex (concave) to determine whether it has a minimum (maximum), and, therefore, we can dispense with the need to check the SOCs (as in the case of univariate functions).

But, the question is how one can check the curvature of functions. This is an easy problem, as we saw in Section 3.5.1, if we have a univariate function such as $y = f(x)$. We learned there that the function is convex (concave) if its second derivative is positive (negative), that is, if $f'' > 0$ ($f'' < 0$). In terms of differentials, as we saw in Section 4.2.5, the function $y = f(x)$ is convex (concave) if its second differential is positive (negative). The problem will be a bit cumbersome if we have a multivariate function such as $U = f(x_1, x_2, x_3, \ldots, x_n)$. We know from our discussions in Section 2.8.2 that if $d^2 U$ is positive (negative) definite, then the function has a global minimum (maximum); if $d^2 U$ is positive (negative) semidefinite, then the function has a local minimum (maximum). It can be shown that the curvature of the function $U = f(x_1, x_2, x_3, \ldots, x_n)$, the nature of the n-dimensional hyperplane, is determined by the sign of the associated quadratic form $d^2 U$ (equation (3.8.9)). If $d^2 U$ is positive (negative) definite, then the function is strictly convex (concave) and the function has a global minimum (maximum); if $d^2 U$ is positive (negative) semidefinite, then the function is convex (concave) and the function has a local minimum (maximum). We know from Sections 2.8.2 and 4.3.2–4.3.4 how one can determine the sign of $d^2 U$.

4.3.6 Application examples

Example 1.   Assume that the inverse demand functions in two markets that a *discriminating monopolist* faces for its product are given by $p_1 = f(q_1) = 100 - 2q_1$ and $p_2 = g(q_2) = 60 - 2q_2$, where $p_1, p_2, q_1,$ and q_2 denote the price charged in market one, the price charged in market two, the quantity demanded in market one, and the quantity demanded in market two, respectively. Also assume that the total cost (C) of the monopolist in supplying the good in the two markets is given by $C = h(q_1, q_2) = 10 + 20q_1 + 20q_2$. Find the levels of q_1 and q_2 that should be supplied to the two markets so that the combined profit of the monopolist will be maximized. Assume that the prices are in dollars.

Solution.   The objective function to be optimized in this example is the monopolist's total profit function and it is the difference between the total revenue and the total cost. The total cost of the monopolist is already given: $C = h(q_1, q_2) = 10 + 20q_1 + 20q_2$. The total revenue ($R$) that the monopolist obtains from both markets together is the sum of the revenue from market one (R_1) and the revenue from market two (R_2), and is given by $R = R_1 + R_2 = p_1 q_1 + p_2 q_2 = (100 - 2q_1)q_1 + (60 - 2q_2)q_2 = 100q_1 - 2q_1^2 + 60q_2 - 2q_2^2$. Therefore, the total profit (Π) of the monopolist is $\Pi = j(q_1, q_2) = R - C = 100q_1 - 2q_1^2 + 60q_2 - 2q_2^2 - 10 - 20q_1 - 20q_2$ or $\Pi = j(q_1, q_2) = -10 + 80q_1 + 40q_2 - 2q_1^2 - 2q_2^2$.

Notice that this is a problem of optimizing an objective function that has two independent variables (q_1 and q_2); that is, optimizing a bivariate function ($\Pi = j(q_1, q_2)$). The conditions for an optimum of $\Pi = j(q_1, q_2)$ are presented in Table 4.3.1 (in terms of derivatives) and in Table 4.3.3 (in terms of differentials and quadratic forms) or in Table 4.3.5, which is the general case of the conditions presented in Table 4.3.3. One can apply any one of these; we shall apply the conditions presented in Table 4.3.3. The FOC for an optimum of $\Pi = j(q_1, q_2)$ requires that $d\Pi = (\partial j/\partial q_1)dq_1 + (\partial j/\partial q_2)dq_2 = j_1 dq_1 + j_2 dq_2 = 0$. This implies that $\partial j/\partial q_1 = j_1 = 0$ and $\partial j/\partial q_2 = j_2 = 0$. Therefore, the partial derivatives of $\Pi = j(q_1, q_2) = -10 + 80q_1 + 40q_2 - 2q_1^2 - 2q_2^2$ with respect to q_1 and q_2 are $j_1 = 80 - 4q_1$ and $j_2 = 40 - 4q_2$, respectively. Setting these partial derivatives to zero and simplifying yields $q_1^* = 20$ and $q_2^* = 10$. Therefore, the critical values are $(q_1^*, q_2^*) = (20, 10)$. These values imply that the price in market one is $p_1^* = f(q_1^*) = 100 - 2 \times 20 = \$60$ and the price in market two is $p_2^* = g(q_2^*) = 60 - 2 \times 10 = \$40$.

The SOC for an optimum of $\Pi = j(q_1, q_2)$ requires that we check the sign of its second total differential. The second total differential of $\Pi = j(q_1, q_2)$ is $d^2\Pi = j_{11}dq_1^2 + 2j_{12}dq_1 dq_2 + j_{22}dq_2^2$, the matrix representation of which is $d^2\Pi = \mathbf{x}^T\mathbf{D}\mathbf{x}$, where $\mathbf{x}^T$ and $\mathbf{D}$ are as we defined earlier. We can obtain, from $\Pi = j(q_1, q_2) = -10 + 80q_1 + 40q_2 - 2q_1^2 - 2q_2^2$, that $j_{11} = -4$, $j_{22} = -4$, and $j_{12} = j_{21} = 0$. Therefore, the quadratic form in the present example can be written as $d^2\Pi = -4dq_1^2 + 2 \times 0dq_1 dq_2 - 4dq_2^2$. Substituting these partial derivatives in $\mathbf{D}$ we obtain $\mathbf{D} = \begin{bmatrix} -4 & 0 \\ 0 & -4 \end{bmatrix}$. We obtain $\mathbf{D_{R1}} = j_{11} = -4 < 0$ and $\mathbf{D_{R1}} = \mathbf{D_R} = |\mathbf{D}| = 16 > 0$. Since $\mathbf{D_{R1}} < 0$ and $\mathbf{D_{R2}} = \mathbf{D_R} > 0$, the quadratic form is negative definite. This implies, following the conditions in Table 4.3.3, that the function $\Pi = j(q_1, q_2) = -10 + 80q_1 + 40q_2 - 2q_1^2 - 2q_2^2$ qualifies for a maximum at the critical values $(q_1^*, q_2^*) = (20, 10)$. Therefore, the stationary value of the function (or the maximum profit) at the critical values $(q_1^*, q_2^*) = (20, 10)$ is $\Pi^* = j(q_1^*, q_2^*) = -10 + 80q_1^* + 40q_2^* - 2q_1^{*2} - 2q_2^{*2} - 10 + 80 \times 20 + 40 \times 10 - 2 \times 20^2 - 2 \times 10^2 = 990$.

Example 2. Suppose that the total cost (C), in dollars, of producing two goods by a *multiproduct firm* is given by $C = f(q_1, q_2) = 100 + 3q_1^2 + 2q_2^2 - 2q_1 q_2 - 4q_1 - 4q_2$, where q_1 and q_2 represent the quantities of good 1 and good 2, respectively. How many units of the two goods must the firm produce in order to minimize the total cost? What will be minimum cost to the firm?

Solution. Notice that this is a problem, as before, of optimizing a bivariate function: $C = f(q_1, q_2)$. The conditions for an optimum, in terms of differentials and quadratic forms, of a bivariate function are given in Table 4.3.3. We know, from this table, that the FOC for an optimum of $C = f(q_1, q_2)$ requires that $dC = (\partial f/\partial q_1)dq_1 + (\partial f/\partial q_2)dq_2 = f_1 dq_1 + f_2 dq_2 = 0$. This implies that $f_1 = f_2 = 0$. The partial derivatives of $C = f(q_1, q_2) = 100 + 3q_1^2 + 2q_2^2 - 2q_1 q_2 - 4q_1 - 4q_2$ with respect to q_1 and q_2 are $f_1 = 6q_1 + 2q_2 - 4$ and $f_2 = 4q_2 + 2q_1 - 4$, respectively. Setting these partial derivatives to zero and simplifying yields $q_1^* = 0.4$ and $q_2^* = 0.8$. Therefore, the critical values are $(q_1^*, q_2^*) = (0.4, 0.8)$. The other critical values $((q_1^*, q_2^*) = (0, 0))$ are discarded as they are meaningless in the present example.

The SOC for an optimum of $C = f(q_1, q_2)$ requires that we check the sign of the second total differential of $C = f(q_1, q_2)$: $d^2 C = f_{11}dq_1^2 + 2f_{12}dq_1 dq_2 + f_{22}dq_2^2$. Its matrix representation is $d^2 C = \mathbf{x}^T\mathbf{D}\mathbf{x}$, where $\mathbf{x}^T$ and $\mathbf{D}$ are as we defined earlier. We can obtain, from $C = f(q_1, q_2) = 100 + 3q_1^2 + 2q_2^2 - 2q_1 q_2 - 4q_1 - 4q_2$, that $f_{11} = 6$, $f_{22} = 4$, and $f_{12} = f_{21} = 2$. Therefore, the quadratic form in the present example can be written as $d^2 C = 6dq_1^2 + 4dq_1 dq_2 + 4dq_2^2$. Substituting these partial derivatives in $\mathbf{D}$, we obtain $\mathbf{D} = \begin{bmatrix} 6 & 2 \\ 2 & 4 \end{bmatrix}$. We know that $\mathbf{D_{R1}} = f_{11} = 6 > 0$ and $\mathbf{D_{R2}} = \mathbf{D_R} = |\mathbf{D}| = 24 - 4 = 20$. Since $\mathbf{D_{R1}} = 6 > 0$ and $\mathbf{D_{R2}} = \mathbf{D_R} = 20 > 0$, the quadratic form is positive definite. This implies that, following the conditions in Table 4.3.3, the function $C = f(q_1, q_2)$

$= 100 + 3q_1^2 + 2q_2^2 - 2q_1q_2 - 4q_1 - 4q_2$ qualifies for a minimum at critical values $(q_1^*, q_2^*) = (0.4, 0.8)$. Therefore, the stationary value of the function (or the minimum cost) at the critical values $(q_1^*, q_2^*) = (0.4, 0.8)$ is $C^* = f(q_1^*, q_2^*) = 100 + 3q_1^{*2} + 2q_2^{*2} - 2q_1^*q_2^* - 4q_1^* - 4q_2^* = 100 + 3(0.4)^2 + 2(0.8)^2 - 2(0.4)(0.8) - 4(0.4) - 4(0.8) = 96.26$.

Example 3.   Assume that a firm uses capital (K) and labor (L) to produce a good. The quantity (q) of the good produced is given by the function $q = f(K, L) = 4K^{1/3}L^{1/2}$. Also assume that the total cost of production (C) is given by the function $C = g(K, L) = 0.1K + 2L$, and the price (p) of the good is \$1 per unit. Find the quantities of capital and labor that the firm must employ to maximize its profits. Also find the profit-maximizing quantity of the good produced and the maximum profit.

Solution.   The total revenue (R) of the firm is $R = h(K, L) = p \times q = 1 \times q = q = 4K^{1/3}L^{1/2}$. Therefore, the profit (Π) of the firm is $\Pi = j(K, L) = R - C = 4K^{1/3}L^{1/2} - 0.1K - 2L$. The FOC for an optimum of $\Pi = j(K, L)$ requires that $d\Pi = (\partial j / \partial K)dK + (\partial j / \partial L)dL = j_K dK + j_L dL = 0$. This implies that $j_K = j_L = 0$. Therefore, the partial derivatives of $\Pi = j(K, L) = 4K^{1/3}L^{1/2} - 0.1K - 2L$ with respect to K and L are $j_K = (4/3)K^{-2/3}L^{1/2} - 0.1$ and $j_L = 2K^{1/3}L^{-1/2} - 2$, respectively. Setting these partial derivatives to zero and simplifying yields $K^* = \$2370.4$ and $L^* = 177.8$ units. Therefore, the critical values are $(K^*, L^*) = (\$2370.4, 177.8)$.

The SOC for an optimum of $\Pi = j(K, L)$ requires that we check the sign of the second total differential of $\Pi = j(K, L)$: $d^2\Pi = j_{KK}dK^2 + 2j_{KL}dKdL + j_{LL}dL^2$, the matrix representation of which is $d^2\Pi = \mathbf{x}^T\mathbf{D}\mathbf{x}$, where $\mathbf{x}^T$ and $\mathbf{D}$ are as we defined earlier. We can obtain, from $\Pi = j(K, L) = 4K^{1/3}L^{1/2} - 0.1K - 2L$, that $j_{KK} = (-8/9)K^{-5/3}L^{1/2}$, $j_{LL} = -K^{1/3}L^{-3/2}$ and $j_{KL} = j_{LK} = (2/3)K^{-2/3}L^{-1/2}$. Substituting the critical values in these second partial derivatives gives $j_{KK} = -28.4$, $j_{LL} = -5.61$ and $j_{KL} = j_{LK} = 0.281$. Therefore, the quadratic form in this example can be written as $d^2\Pi = -28.4dK^2 + 2 \times 0.281dKdL - 5.61dL^2$. Substituting these partial derivatives in $\mathbf{D}$ we obtain $\mathbf{D} = \begin{bmatrix} -28.4 & 0.281 \\ 0.281 & -5.61 \end{bmatrix}$. We obtain $\mathbf{D_{R1}} = j_{KK} = -28.4 < 0$ and $\mathbf{D_{R2}} = \mathbf{D_R} = |\mathbf{D}| = 159.22 > 0$. Since $\mathbf{D_{R1}} < 0$ and $\mathbf{D_{R2}} = \mathbf{D_R} > 0$, the quadratic form is negative definite. This implies that, following the conditions in Table 4.3.3, the function $\Pi = 4K^{1/3}L^{1/2} - 0.1K - 2L$ qualifies for a maximum at the critical values $(K^*, L^*) = (\$2370.4, 177.8)$. The stationary value of the function (or the maximum profit) at the critical values $(K^*, L^*) = (\$2370.4, 177.8)$ is $\Pi^* = j(K^*, L^*) = \$115$. The profit-maximizing output is $q^* = f(K^*, L^*) = 707.6$ units.

Example 4.   Assume that the quantities of two goods (A and B) produced by a multi-product firm are given by the functions $q_A = f(p_A, p_B) = 10p_B - 10p_A$ and $q_B = g(p_A, p_B) = 100 + 10p_A - 20p_B$, where q_A, q_B, p_A, and p_B denote the quantity of A produced, the quantity of B produced, the price per unit (in dollars) of A, and the price per unit (in dollars) of B, respectively. The average cost of producing A is \$1 and the average cost of producing B is \$2. Find p_A, and p_B that will maximize the firm's total profit, Π. Also find the profit-maximizing levels of q_A and q_B and the maximum profit.

Solution.   The profit per unit from either good is equal to the price of the good minus per unit cost of the good. Since the per unit costs of goods A and B are \$1 and \$2, respectively, the per unit profits from these goods are $(p_A - 1)$ and $(p_B - 2)$, respectively. The total profits from goods A and B are given by $(p_A - 1)q_A$ and $(p_B - 2)q_B$, respectively. Therefore, the multiproduct firm's total profit is given by $\Pi = j(p_A, q_A, p_B, q_B) = (p_A - 1)q_A + (p_B - 2)q_B$. Substituting $q_A = 10p_B - 10p_A$ and $q_B = 100 + 10p_A - 20p_B$ into $\Pi = (p_A - 1)q_A + (p_B - 2)q_B$ and simplifying, we obtain $\Pi = h(p_A, p_B) = -200 + 20p_A p_B - 10p_A^2 - 20p_B^2 + 130p_B - 10p_A$.

The FOC for an optimum of $\Pi = h(p_A, p_B)$ requires that $d\Pi = (\partial h / \partial p_A)dp_A + (\partial h / \partial p_B)dp_B = h_{p_A}dp_A + h_{p_B}dp_B = 0$, which implies that $\partial h / \partial p_A = h_{p_A} = 0$ and $\partial h / \partial p_B = h_{p_B} = 0$. Therefore, the partial derivatives of $\Pi = h(p_A, p_B)$ with respect to p_A and p_B are $\Pi_{p_A} = 20p_B - 20p_A - 10$ and $\Pi_{p_B} = 20p_A - 40p_B + 130$, respectively. Setting these partial derivatives to zero and simplifying yields $p_A^* = 5.5$ and $p_B^* = 6$. Therefore, the critical values are $(p_A^*, p_B^*) = (5.5, 6)$.

The SOC for an optimum of $\Pi = h(p_A, p_B)$ requires that we check the sign of its second total differential. The second total differential of $\Pi = h(p_A, p_B)$ is $d^2\Pi = h_{p_{AA}}dp_A^2 + 2h_{p_{A}p_{B}}dp_A dp_B + h_{p_{BB}}dp_B^2$, the matrix representation of which is $d^2\Pi = \mathbf{x}^T \mathbf{D} \mathbf{x}$, where $\mathbf{x}^T$ and $\mathbf{D}$ are as we defined earlier. We can obtain, from $\Pi = h(p_A, p_B) = -200 + 20p_A p_B - 10p_A^2 - 20p_B^2 + 130p_B - 10p_A$, that $h_{p_{AA}} = -20$, $h_{p_{BB}} = -40$, and $h_{p_{AB}} = 20 = h_{p_{BA}}$. Therefore, the quadratic form in this example with these values of the partial derivatives can be written as $d^2\Pi = -20dp_A^2 + 2 \times 20dp_A dp_B - 40dp_B^2$. Substituting these partial derivatives into $\mathbf{D}$, we obtain $\mathbf{D} = \begin{bmatrix} -20 & 20 \\ 20 & -40 \end{bmatrix}$. We know that $\mathbf{D_{R1}} = h_{p_{AA}} = -20 < 0$ and $\mathbf{D_{R2}} = \mathbf{D_R} = |\mathbf{D}| = 400 > 0$. Since $\mathbf{D_{R1}} < 0$ and $\mathbf{D_{R2}} = \mathbf{D_R} = |\mathbf{D}| > 0$, the quadratic form is negative definite. This implies, following the conditions in Table 4.3.3, that the function $\Pi = h(p_A, p_B) = -200 + 20p_A p_B - 10p_A^2 - 20p_B^2 + 130p_B - 10p_A$ qualifies for a maximum at the critical values $(p_A^*, p_B^*) = (5.5, 6)$. The stationary value of the function (or the maximum profit) at the critical values $(p_A^*, p_B^*) = (5.5, 6)$ can be found by substituting the critical values into $\Pi^* = h(p_A^*, p_B^*) = -200 + 20p_A^* p_B^* - 10p_A^{*2} - 20p_B^{*2} + 130p_B^* - 10p_A^*$. Therefore, the maximum profit is $\Pi^* = h(p_A^*, p_B^*) = \$162.50$. The profit-maximizing quantities of the two goods produced are $q_A^* = 5$ units and $q_B^* = 35$ units.

4.3.7 Exercises

1. Find the critical values of the following functions. Also determine the optima, if they exist.
    (i) $U = f(x, y) = 10 + x^2 + y^2 - 4x - 4y$; (ii) $U = f(x, y) = 10 - x^2 - y^2 10x + 10y$; (iii) $U = f(x, y) = x^3 + y^3 - xy$; (iv) $U = f(x, y) = x^3 + y^3 - 9x - 9y$; (v) $U = f(x, y) = x^2 - 4xy - y^2$; (vi) $U = f(x, y) = x^2 + 2xy + y^3$; (vii) $U = f(x, y) = x^3 - xy + y^3$; (viii) $U = f(x, y) = xy - x^3 - y^3$.
2. *Application exercise.* Suppose that the production function of a firm that produces q units of the output of a good using k units of capital and l units of labor is given by $q = f(k, l) = 10k + 10l - k^2 - l^2 - kl$. Find the quantities of capital and labor that the firm must use in order to maximize its output. Also find the maximum output.
3. *Application exercise.* Assume that the total cost (C) of producing two goods (A and B) by a multiproduct firm is given by $C = f(q_A, q_B) = q_A^2 + 4q_B^2 + 2q_A q_B - 2q_A - 4q_B$, where q_A and q_B represent the quantities of goods A and B produced, respectively. Find q_A and q_B that the firm should produce so that its total cost will be minimum. Also find the minimum total cost.
4. *Application exercise.* Suppose that two firms, A and B, are duopolists in a market, and that they produce q_A and q_B units, respectively, of a homogeneous good. Also suppose that the industry demand for the good is given by $P = f(q_A, q_B) = 80 - q_A - q_B$, where P

denotes market price; and that the total cost of producing the good by firm A is $C_A = 5q_A$ and the total cost of producing the good by firm B is $C_B = 0.1q_B^2$. Suppose again that the two firms decided to enter into collusion and form a monopoly. Find the quantities that firms A and B (i.e. q_A and q_B) should produce so that the monopoly profit will be maximum. Also find the maximum monopoly profit and the market price.

5. *Application exercise.* Suppose that the inverse demand functions of two goods that a monopolist sells are given by $q_A = 10 - 2p_A$ and $q_B = 5 - p_B$, where q_A, p_A, q_B, and p_B denote the quantity of good A demanded, price of good A, quantity of good B demanded, and price of good B, respectively. Also suppose that the monopolist's joint cost of producing these two goods is given by $C = q_A + q_B + q_A q_B$. Prices and costs are in dollars. Find p_A and p_B that maximize the monopolist's total profit. Also find the profit-maximizing q_A and q_B, and the maximum total profit.

6. *Application exercise.* Assume that the inverse demand functions that a discriminating monopolist faces in two markets (1 and 2) are given by $p_1 = 90 - q_1$ and $p_2 = 80 - q_2$, where p_1, q_1, p_2, and q_2 denote price in market 1, quantity demanded in market 1, price in market 2, and quantity demanded in market 2, respectively. Also assume that the monopolist's total cost is given by $C = 500 + 5q_1 + 5q_2$. Prices and costs are in dollars. Find q_1 and q_2 that the monopolist should sell to maximize its total profit. Also find the profit-maximizing p_1 and p_2, and the maximum total profit.

7. *Application exercise.* Suppose that the total utility (U) that a consumer obtains from the consumption of two goods, q_1 and q_2, is given by the function $U = f(q_1, q_2) = 2q_1 q_2 - q_1^2 - q_2^3$. Find the quantities of q_1 and q_2 that the consumer should consume to maximize the total utility. Also find the maximum total utility.

 Web supplement: S4.3.8 Mathematica applications

4.4 Extrema of multivariate objective functions with constraints

4.4.1 Introduction

We began this chapter with an introduction to the objectives of economic agents. We stated that every economic agent aimed at optimizing some objective function. A consumer wanted to maximize the total utility from the bundle of goods consumed or minimize the expenditure on the bundle of goods purchased; a producer wanted to maximize the total profit or minimize the total cost; a government agent wanted to maximize social welfare or minimize social costs; and so on.

However, in optimizing all the objective functions that we have considered so far, the agent was assumed to be free to choose the quantities that optimized the agent's objective function. For example, suppose that a consumer aims at maximizing the total utility (U) obtained from the consumption of two goods, x_1 and x_2. For convenience, suppose that the consumer's total utility is given by the function $U = f(x_1, x_2) = x_1^{1/2} + x_2^{1/2}$. Given this utility function, the consumer was assumed to choose any quantities of the goods x_1 and x_2 that would maximize the total utility. Notice that the utility will be a maximum only when infinite quantities of the two goods are consumed, and that the consumer is free to consume these quantities. An optimization problem such as this is called a *free optimization* problem or *unconstrained optimization* problem. It is unnecessary to mention that a free or unconstrained optimization problem has little significance in the subjects of our interest.

We also stated at the beginning of this chapter that the resources to achieve the objectives of the economic agents were limited and that this "limitation" forced the agents to make choices. In the face of this limitation, the consumer in our above example cannot consume infinite quantities of the two goods to maximize total utility. Consuming infinite quantities of the two goods would not have been a problem if the consumer had an infinite amount of money or if the goods were free. But, the truth is that the amount of money (or, in general, resources) of the consumer or of any other economic agent is limited and the goods have to be paid for at their respective prices. These constraints, as mentioned, force one to make choices. Therefore, given the prices of the two goods and the limited amount of money, the consumer can consume more of one good only by reducing the consumption of the other good. This means that there exists a *trade-off* in the consumption of goods. This also means that the consumer has to maximize utility taking into account the limitation imposed by the prices of the goods and the amount of money the consumer has to spend on the two goods. Suppose that one unit of x_1 costs \$2 and one unit of x_2 costs \$4 and the amount of money at the consumer's disposal for spending on these two goods is \$40. Then the resource limitation that the consumer faces can be written as $2x_1 + 4x_2 = 40$. This equation is called the *budget constraint* or, simply, the *constraint* or the *subsidiary condition*. Notice that the problem of scarcity we noted earlier enters into the optimization problem through the budget constraint. What all this means is that, instead of a free optimization problem, the consumer now has a *constrained optimization* problem; specifically, a *constrained maximization* problem. In fact, the optimization problems of all the economic agents we referred to above are constrained optimization problems.

4.4.2 Constrained optimization: concepts and geometric interpretation

We stated above that every economic agent aims at optimizing some objective function. Moreover, we saw above that every economic agent faces at least one constraint. In certain optimization problems, the agent(s) may confront more than one constraint, and such problems are called *multiconstraint optimization* problems.

The constraints that the agent faces may be *equality constraints* (such as the one in the last example) or *inequality constraints* (which we will consider in Chapters 5 and 6). In the present chapter we will consider constrained optimization problems with only equality constraints.

One term that is frequently used in constrained optimization theory is *feasible set*. A feasible set is the set of combinations of the independent or choice variables of the objective function that satisfies all the constraints of the optimization problem simultaneously.

We have also seen above the difference between free or unconstrained and constrained optimization problems. We can now present a geometric illustration of this difference. For this we use the utility maximization example presented in the last section. Notice that we can represent the budget constraint given in that example by the straight line BC in Figure 4.4.1(A). The problem of the consumer now is to maximize the total utility $U = f(x_1, x_2) = x_1^{1/2} + x_2^{1/2}$ subject to the budget constraint $2x_1 + 4x_2 = 40$. If there were no constraint, then the domain of the objective function would be the nonnegative values on the real lines representing x_1 and x_2; that is, the nonnegative (x_1, x_2) plane in a three-dimensional figure. Consequently, the range of the function would be nonnegative values on the real line representing U in Figure 4.4.1(B). However, with the introduction of the constraint, the domain of the objective function is curtailed (to the set of points on the budget line BC in Figure 4.4.1(A)). This curtailment in the domain leads to a corresponding curtailment of the

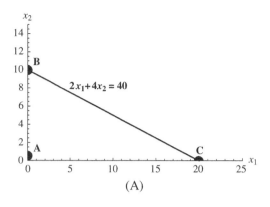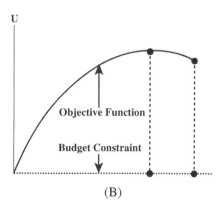

(A)

(B)

Figure 4.4.1

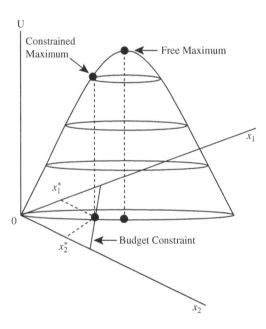

Figure 4.4.2

range of the function (to the set of points of the utility plane lying directly above the budget line) as shown in Figure 4.4.1(B).

The curtailment mentioned above in the objective function due to the introduction of the budget constraint can be better understood with a three-dimensional illustration as presented in Figure 4.4.2. As can be seen, the constrained maximum is, in most cases, lower than the free maximum, and it can be utmost as high as (and not greater than) the free maximum.

Our aim here is to find the optimum (in the present example, maximum) of the objective function. The two methods that are widely used to find the optimum of the objective functions in constrained optimization problems are the *substitution method* and the *Lagrange multiplier method*. We first present the substitution method followed by the Lagrange multiplier method.

4.4.3 Solution of constrained optimization problems: the substitution method

As an illustration of the substitution method of solving constrained optimization problems, consider our previous example. The objective function in our example was to maximize

$$U = f(x_1, x_2) = x_1^{1/2} + x_2^{1/2} \tag{4.4.1}$$

subject to the constraint

$$2x_1 + 4x_2 = 40 \tag{4.4.2}$$

Notice that there are two independent or choice variables in the objective function in equation (4.4.1). Therefore, this is the case of a *bivariate constrained optimization* problem with one constraint. We will later consider optimization problems involving more than two choice variables and one or more constraints. We now use the substitution method to solve this problem.

We begin by solving for x_2 in the budget constraint given in equation (4.4.2). Solving for x_2, we obtain $x_2 = 10 - 0.5x_1$. We can now substitute this value for x_2 in the objective function in equation (4.4.1) to obtain $U = g(x_1) = x_1^{1/2} + (10 - 0.5x_1)^{1/2}$. Notice that the original bivariate objective function is now converted into a univariate objective function. Therefore, we can use the FOC and the SOC in Table 4.2.1 to check whether the last function has an optimum.

The FOC for $g(x_1)$ to have a maximum is $g'(x_1) = 0$. Differentiating $U = g(x_1) = x_1^{1/2} + (10 - 0.5x_1)^{1/2}$ with respect to x_1 we obtain $g'(x_1) = 0.5x_1^{-1/2} + 0.5(10 - 0.5x_1)^{-1/2} \times -0.5$. Setting this first derivative to zero and simplifying yields $x_1^* = 160/12 = 13.33$. Substituting this value into the budget constraint gives $x_2^* = 40/12 = 3.33$. Therefore, the critical values are $(x_1^*, x_2^*) = (13.33, 3.33)$. The SOC for $g(x_1)$ to have a maximum is $g''(x_1) < 0$. The second derivative of $g(x_1)$ is $g''(x_1) = -0.5x_1^{-3/2} + (1/8)(10 - 0.5x_1)^{-3/2} \times -0.5$. The value of this second derivative at $x_1^* = 13.33$ is $g''(x_1^*) = -0.0154 < 0$. Since the second derivative is negative, the function $U = g(x_1) = x_1^{1/2} + (10 - 0.5x_1)^{1/2}$ is strictly concave at $x_1^* = 13.33$ and, therefore, it has a maximum at $x_1^* = 13.33$. Therefore, the maximum or the stationary value of the function at the critical values $(x_1^*, x_2^*) = (13.33, 3.33)$ is $U^* = f(x_1^*, x_2^*) = x_1^{*1/2} + x_2^{*1/2} = g(x_1^*) = x_1^{*1/2} + (10 - 0.5x_1^*)^{1/2} = 5.5$.

The above procedure for solving a constrained optimization problem involving two choice variables and one constraint can be summarized as follows. Firstly, solve the constraint for one of the variables. Secondly, substitute this solution in the original bivariate objective function to convert it into a univariate objective function. Thirdly, find the critical value of the choice variable in the converted objective function from the FOC. Fourthly, substitute this value of the choice variable into the constraint to solve for the other choice variable, and this will give its critical value. Fifthly, check the second derivative of the converted objective function to determine whether the function has a maximum or a minimum corresponding to the first critical value. Lastly, substitute these two critical values into the original bivariate objective function to obtain the optimum or the stationary value of the function.

As another example, assume that a firm received an order for ten units of its product, which is produced at two separate plants. Also assume that the quantities of the goods produced at these two plants are q_1 and q_2, respectively, and that the total cost (C) of producing the product is given by the function $C = f(q_1, q_2) = q_1^2 + q_2^2 - q_1 q_2$. How many units of the good must be produced in each plant such that the firm's total cost will be minimum?

The constraint in this problem can be written as $q_1 + q_2 = 10$. We shall now follow the steps outlined above. Solving $q_1 + q_2 = 10$ for q_2 yields $q_2 = 10 - q_1$. Substituting this value of q_2 into the objective function $C = f(q_1, q_2) = q_1^2 + q_2^2 - q_1 q_2$ and simplifying, we obtain $C = g(q_1) = 100 + 3q_1^2 - 30q_1$. Notice that we converted the original bivariate objective function into a univariate objective function. We know that its critical values can be found by differentiating it with respect to q_1 and setting the derivative to zero. The derivative of $C = g(q_1) = 100 + 3q_1^2 - 30q_1$ with respect to q_1 is $g'(q_1) = 6q_1 - 30$, and setting this to zero yields $q_1^* = 5$. Since $q_2 = 10 - q_1$, we obtain $q_2^* = 10 - q_1^* = 10 - 5 = 5$. Therefore, the critical values are $(q_1^*, q_2^*) = (5, 5)$. Since $g''(q_1) = 6 > 0$, the SOC for a minimum of $C = g(q_1) = 100 + 3q_1^2 - 30q_1$ is also satisfied. This implies that the function $C = g(q_1) = 100 + 3q_1^2 - 30q_1$ has a minimum and it occurs at $q_1^* = q_2^* = 5$. The minimum cost or the stationary value of the function is $C^* = f(q_1^*, q_2^*) = q_1^{*2} + q_2^{*2} - q_1^* q_2^* = 5^2 + 5^2 - 5 \times 5 = g(q_1^*) = 100 + 3q_1^{*2} - 30q_1^* = 100 + 3 \times 5^2 - 30 \times 5 = \$25$.

The reader will have noticed that the substitution method of solving constrained optimization problems is easier when there is only one constraint and the objective function is a simple expression of two variables. But this method will be more cumbersome to use if the optimization problem involves more constraints and more choice variables or when the original objective function is a complicated expression. This necessitates the employment of a more formal method to solve constrained optimization problems. This method is called the Lagrange multiplier method and is presented in the following section.

4.4.4 Solution of constrained optimization problems: the Lagrange multiplier method

As before, we are still dealing with the optimization of bivariate objective functions subject to one constraint. In the last section we presented the substitution method of solving constrained optimization problems. There is an alternative method which, as in the case of the substitution method, also converts the original objective function into another form. This alternative method is called the Lagrange multiplier method. The converted form of the original objective function is called the *Lagrangian function* which comprises the original objective function, the constraint, and an undetermined variable called the *Lagrangian multiplier*. One might wonder why one needs to convert the original objective function into the Lagrangian function. The reason is that (as in the case of the substitution method), once the Lagrangian function is formed, we can still use the FOC of free optimization problems. However, how does one do this? The procedure is outlined below.

We use the first example in the last section. The objective function and the constraint of this example are given in equations (4.4.1) and (4.4.2), respectively. We can now combine the objective function with the constraint and with the Lagrangian multiplier (λ) to form the Lagrangian function[7] as

$$L = h(x_1, x_2, \lambda) = x_1^{1/2} + x_2^{1/2} + \lambda(40 - 2x_1 - 4x_2) \qquad (4.4.3)$$

Notice that the terms inside the brackets on the RHS of equation (4.4.3) are nothing but a restatement of the constraint. It should be noted that most of the constraints that appear in this chapter are *binding constraints*.[8] Since, at maximum, the consumer utilizes the available resources completely or since the constraint is binding at extremum, the terms inside the brackets on the RHS of equation (4.4.3) will be zero. In this case the Lagrangian function in equation (4.4.3) will reduce to the original objective function in equation (4.4.1). Therefore, when the Lagrangian function is optimized, the original function will also be optimized. This suggests that we can still use, as in the case of the substitution method, the simple FOC of free optimization problems.

Therefore, the FOCs for the objective function to have an optimum require that the partial derivatives of the Lagrangian function with respect to each choice variable be equal to zero. That is, the FOCs are

$$L_1 = \partial L/\partial x_1 = 0.5x_1^{-1/2} - 2\lambda = 0; \quad L_2 = \partial L/\partial x_2 = 0.5x_2^{-1/2} - 4\lambda = 0; \text{ and}$$

$$L_\lambda = 40 - 2x_1 - 4x_2 = 0 \tag{4.4.4}$$

Notice that L_λ in equation (4.4.4) is a mere restatement of the budget constraint. Solving the three equations in the system (4.4.4) simultaneously, we obtain the critical values $(x_1^*, x_2^*) =$ (13.33, 3.33). Notice also that these are precisely the critical values we obtained when we used the substitution method. With these critical values we can solve, using any one of the first two equations, for λ: $\lambda = 0.06849$. As before, the stationary value or the maximum utility is $U^* = 5.5$. The next task here is to determine whether the function has a maximum corresponding to these critical values. This requires the satisfaction of the SOC, which is discussed in the following section.

As another example, consider the problem in the second example we solved in the previous section. The objective function and the constraint in this example are $C = f(q_1, q_2) = q_1^2 + q_2^2 - q_1 q_2$ and $q_1 + q_2 = 10$, respectively. We now set up the Lagrangian function as

$$L = h(q_1, q_2, \lambda) = q_1^2 + q_2^2 - q_1 q_2 + \lambda(10 - q_1 - q_2) \tag{4.4.5}$$

Therefore, the FOCs for the objective function to have an optimum require that the partial derivatives of the Lagrangian function with respect to each choice variable be equal to zero. That is, the FOCs are

$$L_1 = \partial L/\partial q_1 = 2q_1 - q_2 - \lambda = 0; \quad L_2 = \partial L/\partial q_2 = 2q_2 - q_1 - \lambda = 0; \text{ and}$$

$$L_\lambda = 10 - q_1 - q_2 = 0 \tag{4.4.6}$$

Solving the first two equations in the system (4.4.6) simultaneously, we obtain $q_1^* = q_2^*$. Substituting this in the last equation and simplifying, we obtain $q_1^* = q_2^* = 5$. Substituting this in the first or the second equation yields $\lambda = 5$. Notice that these are identical to the critical values we obtained when we used the substitution method. The minimum cost or the stationary value of the function is $C^* = f(q_1^*, q_2^*) = q_1^{*2} + q_2^{*2} - q_1^* q_2^* = 5^2 + 5^2 - 5 \times 5 = \$25$. The next task, as in the last example, is to determine whether the function has a minimum corresponding to the above critical values. This requires the satisfaction of the SOC, which is presented in the following section.

We now consider a general bivariate constrained optimization problem. Suppose that the objective function is $U = f(x_1, x_2)$ and the constraint is $g(x_1, x_2) = c$, where c is a constant.

Then, the associated Lagrangian function is

$$L = h(x_1, x_2, \lambda) = f(x_1, x_2) + \lambda[c - g(x_1, x_2)] \tag{4.4.7}$$

Therefore, an optimum of $U = f(x_1, x_2)$ requires that the partial derivatives of $L = h(x_1, x_2, \lambda)$ with respect to x_1, x_2, and λ are all zero. That is, the FOCs are

$$L_1 = \partial L / \partial x_1 = f_1 - \lambda g_{x_1} = 0; \quad L_2 = \partial L / \partial x_2 = f_2 - \lambda g_{x_2} = 0; \text{ and}$$
$$L_\lambda = c - g(x_1, x_2) = 0 \tag{4.4.8}$$

The above system of three equations gives us, when solved simultaneously, the critical values of x_1 and x_2 denoted by x_1^* and x_2^*, respectively. Since the budget constraint is expected to be binding at the optimum, the expression in the square brackets in equation (4.4.7) will be zero. Therefore, the critical values that guarantee an optimum of the Lagrangian function will also guarantee an optimum of the original objective function. That is, the optimum of the Lagrangian function will be identical with the optimum of the original objective function. We will see, as stated before, whether this "optimum" is indeed an optimum using the SOCs presented in the following section. Notice that, from the first two expressions in the system of equations (4.4.8), we can obtain

$$\frac{f_1}{g_1} = \lambda = \frac{f_2}{g_2}, \tag{4.4.9}$$

which we will use later. Notice also that in equation (4.4.9) we used notations $g_1 = g_{x_1}$ and $g_2 = g_{x_2}$ for convenience.

4.4.5 Constrained optimization with Lagrange multiplier method: FOC and SOC using differentials

As in the case of unconstrained optimization problems, we first attempt to express the above (derivative) FOCs in terms of total differentials. In the case of unconstrained, free optimization of $U = f(x_1, x_2)$, the FOC in terms of total differential (following Table 4.3.2) can be written as $dU = 0$. It can be shown that this condition remains valid even if we add the constraint $(g(x_1, x_2) = c)$ to the problem. Therefore, the FOC for an optimum of $U = f(x_1, x_2)$, given the constraint $g(x_1, x_2) = c$, is $dU = f_1 dx_1 + f_2 dx_2 = 0$ or, in other words, $f_1 = f_2 = 0$.

As in the case of the free optimization problem, the SOC in constrained optimization also depends on the sign of the second total differential or the quadratic form of $U = f(x_1, x_2)$. But what will be its second total differential? Differentiating the first total differential $(dU = f_1 dx_1 + f_2 dx_2)$ with respect to dx_1 and dx_2, we can find the second total differential of $U = f(x_1, x_2)$. But, when we do this, we cannot treat dx_1 and dx_2 as independent of each other (as we did in the case of free optimization problems); instead we have to treat dx_2 as a variable and dx_1 as a constant or vice versa. Using this condition, we can write the second total differential of $U = f(x_1, x_2)$ as

$$d(dU) = d^2 U = \frac{\partial(dU)}{\partial x_1} dx_1 + \frac{\partial(dU)}{\partial x_2} dx_2 = \frac{\partial}{\partial x_1}[f_1 dx_1 + f_2 dx_2] dx_1 + \frac{\partial}{\partial x_2}[f_1 dx_1 + f_2 dx_2] dx_2$$

$$= \left[f_{11} dx_1 + f_{12} dx_2 + f_2 \frac{\partial dx_2}{\partial x_1} \right] dx_1 + \left[f_{21} dx_1 + f_{22} dx_2 + f_2 \frac{\partial dx_2}{\partial x_2} \right] dx_2$$

$$= f_{11} dx_1^2 + f_{12} dx_1 dx_2 + f_2 \frac{\partial dx_2}{\partial x_1} dx_1 + f_{21} dx_1 dx_2 + f_{22} dx_2^2 + f_2 \frac{\partial dx_2}{\partial x_2} dx_2$$

$$= f_{11} dx_1^2 + 2 f_{12} dx_1 dx_2 + f_{22} dx_2^2 + f_2 d^2 x_2 \tag{4.4.10}$$

since $f_2[(\partial dx_2/\partial x_1)dx_1 + (\partial dx_2/\partial x_2)dx_2] = f_2 d^2 x_2$. Notice that the second total differential of the objective function in the constrained optimization problem (equation (4.4.10)) differs from the second total differential of the objective function in the free optimization problem (equation (3.8.7)) only with regard to the last term in the former, which does not make $d^2 U$ a quadratic form. How can we convert equation (4.4.10) so that $d^2 U$ will become a quadratic form? The way out of this problem is to find the second total differential of the constraint and substitute the value of $d^2 x_2$ from that into equation (4.4.10). The first total differential of the constraint $g(x_1, x_2) = c$ is

$$dg = g_1 dx_1 + g_2 dx_2 = 0 \tag{4.4.11}$$

and its second total differential is

$$d^2 g = g_{11} dx_1^2 + g_{12} dx_1 dx_2 + g_{21} dx_2 dx_1 + g_{22} dx_2^2 + g_2 d^2 x_2 = 0$$

or

$$d^2 g = g_{11} dx_1^2 + 2 g_{12} dx_1 dx_2 + g_{22} dx_2^2 + g_2 d^2 x_2 = 0 \tag{4.4.12}$$

Solving equation (4.4.12) for $d^2 x_2$, we obtain $d^2 x_2 = [-g_{11} dx_1^2 - 2 g_{12} dx_1 dx_2 - g_{22} dx_2^2]/(g_2)$. If we substitute this last result (with $g_{11} = g_{x_1 x_1}$, $g_{22} = g_{x_2 x_2}$, $g_{12} = g_{x_1 x_2}$, and $g_{21} = g_{x_2 x_1}$) into equation (4.4.10) for $d^2 x_2$, we can obtain the quadratic form of $d^2 U$ as

$$d^2 U = \left[f_{11} - \frac{f_1}{g_2} g_{11} \right] dx_1^2 + 2 \left[f_{12} - \frac{f_2}{g_2} g_{12} \right] dx_1 dx_2 + \left[f_{22} - \frac{f_2}{g_2} g_{22} \right] dx_2^2 \tag{4.4.13}$$

which can be rewritten, using equation (4.4.9), as $d^2 U = [f_{11} - \lambda g_{11}] dx_1^2 + 2[f_{12} - \lambda g_{12}] dx_1 dx_2 + [f_{22} - \lambda g_{22}] dx_2^2$, which can again be rewritten as

$$d^2 U = L_{11} dx_1^2 + 2 L_{12} dx_1 dx_2 + L_{22} dx_2^2 = L_{11} dx_1^2 + 2 L_{12} dx_1 dx_2 + L_{22} dx_2^2 \tag{4.4.14}$$

where $L_{11} = L_{x_1 x_1} = f_{11} - \lambda g_{11}$, $L_{12} = L_{x_1 x_2} = f_{12} - \lambda g_{12}$, and $L_{22} = L_{x_2 x_2} = f_{22} - \lambda g_{22}$. Solving equation (4.4.11) for dx_2 and substituting the result in equation (4.4.14), we obtain

$$d^2 U = [L_{11} g_2^2 - 2 L_{12} g_1 g_2 + L_{22} g_1^2][dx_1^2/g_2^2] \tag{4.4.15}$$

Table 4.4.1

Condition	Minimum	Maximum
FOC or necessary condition	$dU = 0$	$dU = 0$
SOC or sufficient condition	$d^2U > 0$, or positive definite, if $\|\mathbf{H}_2\| = \|\mathbf{H}\| < 0$	$d^2U < 0$, or negative definite, if $\|\mathbf{H}_2\| = \|\mathbf{H}\| > 0$

As we saw in Section 2.8.3, the bordered Hessian $\|\mathbf{H}_2\| = \|\mathbf{H}\| = \begin{vmatrix} 0 & g_1 & g_2 \\ g_1 & L_{11} & L_{12} \\ g_2 & L_{21} & L_{22} \end{vmatrix} = -[L_{11}g_2^2 -$

$2L_{12}g_1g_2 + L_{22}g_1^2]$ of the bordered Hessian matrix $\mathbf{H} = \begin{bmatrix} 0 & g_1 & g_2 \\ g_1 & L_{11} & L_{12} \\ g_2 & L_{21} & L_{22} \end{bmatrix}$ is the negative of

the value in the first brackets of equation (4.4.15). Therefore, we can state that $d^2U > 0$, or positive definite, given $dg = g_1dx_1 + g_2dx_2 = 0$, if and only if $\|\mathbf{H}_2\| = \|\mathbf{H}\| < 0$; that is, if the bordered Hessian is negative. Then, the objective function $U = f(x_1, x_2)$ subject to the constraint $g(x_1, x_2) = c$ will have a minimum. Similarly, $d^2U < 0$, or negative definite, given again $dg = g_1dx_1 + g_2dx_2 = 0$, if and only if $\|\mathbf{H}_2\| = \|\mathbf{H}\| > 0$; that is, if the bordered Hessian is positive. Then, the objective function $U = f(x_1, x_2)$ subject to the constraint $g(x_1, x_2) = c$ will have a maximum. These are the SOCs for $U = f(x_1, x_2)$, subject to the constraint $g(x_1, x_2) = c$, to have a minimum or maximum. We present the above FOCs and SOCs in Table 4.4.1.

Let us now check whether the critical values, $(x_1^*, x_2^*) = (13.33, 3.33)$, we obtained for the first example in the previous section indeed correspond to a maximum. For this we can use the SOC in Table 4.4.1. We can obtain from equation (4.4.4) that $L_{11} = -x_1^{-3/2}/4 = -(13.33)^{-3/2}/4$, $L_{22} = -x_2^{-3/2}/4 = -(3.33)^{-3/2}/4$, and $L_{12} = L_{21} = 0$, and from equation (4.4.2) that $g_1 = 2$ and $g_2 = 4$. The quadratic form (corresponding to equation (4.4.15)) in the case of this example can be written as $d^2U = [\{-(13.33)^{-3/2}/4\}4^2 - 2 \times 0 \times 2 \times 4 + \{-(3.33)^{-3/2}/4\}2^2](dx_1^2/4^2)$. The expression in the square brackets of the last equation is the negative of the bordered Hessian matrix $\mathbf{H}$ and its determinant (the corresponding bordered Hessian) $\|\mathbf{H}_2\| = \|\mathbf{H}\|$, which can,

respectively, be written as $\mathbf{H} = \begin{bmatrix} 0 & 2 & 4 \\ 2 & -(13.33)^{-3/2}/4 & 0 \\ 4 & 0 & -(3.33)^{-3/2}/4 \end{bmatrix}$ and $\|\mathbf{H}_2\| = \|\mathbf{H}\| =$

$\begin{vmatrix} 0 & 2 & 4 \\ 2 & -(13.33)^{-3/2}/4 & 0 \\ 4 & 0 & -(3.33)^{-3/2}/4 \end{vmatrix} = 0.25 > 0$. These results suggest that the quadratic

form is negative definite ($d^2U < 0$). Therefore, following Table 4.4.1, the SOC is satisfied for the function $U = f(x_1, x_2) = x_1^{1/2} + x_2^{1/2}$, subject to the constraint $2x_1 + 4x_2 = 40$, to have a maximum at critical values $(x_1^*, x_2^*) = (13.33, 3.33)$. Notice that this is the same as the conclusion we obtained when we used the substitution method in Section 4.4.3.

We also check the SOC for the second example in the previous section. The critical values we obtained from the FOC were $(q_1^*, q_2^*) = (5, 5)$. We can obtain from equation (4.4.6), that $L_{11} = 2$, $L_{22} = 2$, and $L_{12} = L_{21} = 0$, and from the constraint $g(q_1, q_2) = q_1 + q_2 = 10 = c$ that $g_1 = 1$ and $g_2 = 1$. The quadratic form (corresponding to equation (4.4.15)) in the case of this example can be written as $d^2U = [2 \times 1^2 - 2 \times 0 \times 1 \times 1 + 2 \times 1^2](dx_1^2/1^2)$. The expression in the square brackets of the last equation is the negative of the bordered Hessian matrix $\bar{\mathbf{H}}$ and its determinant (the corresponding bordered Hessian) is $|\bar{\mathbf{H}}_2| = |\bar{\mathbf{H}}|$, which can, respectively, be written as $\bar{\mathbf{H}} = \begin{bmatrix} 0 & 1 & 1 \\ 1 & 2 & 0 \\ 1 & 0 & 2 \end{bmatrix}$ and $|\bar{\mathbf{H}}_2| = |\bar{\mathbf{H}}| = \begin{vmatrix} 0 & 1 & 1 \\ 1 & 2 & 0 \\ 1 & 0 & 2 \end{vmatrix} = -4 < 0$. These results suggest that the quadratic form is positive definite $(d^2C > 0)$. Therefore, following Table 4.4.1, the SOC is satisfied for the function $C = f(q_1, q_2) = q_1^2 + q_2^2 - q_1 q_2$, subject to the constraint $q_1 + q_2 = 10$, to have a minimum at critical values $(q_1^*, q_2^*) = (5, 5)$. Notice that, as before, this is the same as the conclusion we obtained when we used the substitution method in Section 4.4.3.

We now present, for easy reference, the steps involved in optimizing $U = f(x_1, x_2)$, subject to the constraint $g(x_1, x_2) = c$, using the Lagrangian method.

Step 1. Set up the Lagrangian function $L = h(x_1, x_2, \lambda) = f(x_1, x_2) + \lambda[c - g(x_1, x_2)]$.

Step 2. Find the partial derivatives of L with respect to x_1, x_2, and λ, and set them to zero.

Step 3. Solve the simultaneous equations in step 2 for the critical values $(x_1^*$ and $x_2^*)$ and λ. This will satisfy the FOCs.

Step 4. To determine whether the critical values obtained in step 3 correspond to an optimum, check the sign definiteness of the quadratic form d^2U. If $d^2U > 0$ or positive definite (which happens when $|\bar{\mathbf{H}}_2| = |\bar{\mathbf{H}}| < 0$), $U = f(x_1, x_2)$, subject to $g(x_1, x_2) = c$, has a minimum. If $d^2U < 0$ or negative definite (which happens when $|\bar{\mathbf{H}}_2| = |\bar{\mathbf{H}}| > 0$), $U = f(x_1, x_2)$, subject to $g(x_1, x_2) = c$, has a maximum. These results will satisfy the SOC.

4.4.6 Constrained optimization with three variables and one constraint

So far we were dealing with the solution of constrained optimization problems with two independent or choice variables and one constraint. We now consider constrained optimization problems with three independent variables and one constraint.

Suppose that the function we wish to optimize is $U = f(x_1, x_2, x_3)$ and the constraint is $g(x_1, x_2, x_3) = c$. Since most of the ideas involved in optimizing $U = f(x_1, x_2, x_3)$ subject to $g(x_1, x_2, x_3) = c$ are similar to those used in the previous sections, we do not repeat them here and, instead, we will only present the results. Then, following the procedures outlined at the end of the previous section, we set up the Lagrangian function as

$$L = h(x_1, x_2, x_3, \lambda) = f(x_1, x_2, x_3) + \lambda[c - g(x_1, x_2, x_3)] \tag{4.4.16}$$

Notice that equation (4.4.16) is an extension of equation (4.4.7) to the case of three variables. Then, the FOCs require that the first partial derivatives of equation (4.4.16) be zero:

$$L_1 = f_1 - \lambda g_1 = 0; L_2 = f_2 - \lambda g_2 = 0; L_3 = f_3 - \lambda g_3 = 0; \text{ and } L_\lambda = c - g(x_1, x_2, x_3) = 0 \tag{4.4.17}$$

As we stated in the previous section, the solution of the above system of equations will yield the critical values (x_1^*, x_2^*, x_3^*) and λ. These critical values will satisfy the FOCs for an optimum of $U = f(x_1, x_2, x_3)$ subject to $g(x_1, x_2, x_3) = c$.

Table 4.4.2

Condition	Minimum	Maximum
FOC or necessary condition	$dU = 0$	$dU = 0$
SOC or sufficient condition	$d^2U > 0$ or positive definite, or $|\mathbf{H_2}| < 0$ and $|\mathbf{H_3}| = |\mathbf{H}| < 0$	$d^2U < 0$ or negative definite or $|\mathbf{H_2}| > 0$ and $|\mathbf{H_3}| = |\mathbf{H}| < 0$

If we follow the same line of thought as that was adopted in the derivation of the quadratic form of the bivariate case in equation (4.4.15), we can derive the quadratic form in the present trivariate case too. Then the associated bordered Hessian matrix and the bordered

$$\text{Hessian will, respectively, be } \mathbf{H} = \begin{bmatrix} 0 & g_1 & g_2 & g_3 \\ g_1 & L_{11} & L_{12} & L_{13} \\ g_2 & L_{21} & L_{22} & L_{23} \\ g_3 & L_{31} & L_{32} & L_{33} \end{bmatrix} \text{ and } |\mathbf{H}| = \begin{vmatrix} 0 & g_1 & g_2 & g_3 \\ g_1 & L_{11} & L_{12} & L_{13} \\ g_2 & L_{21} & L_{22} & L_{23} \\ g_3 & L_{31} & L_{32} & L_{33} \end{vmatrix}.$$

Following the results presented in Section 2.8.3, d^2U is positive definite (then $U = f(x_1, x_2, x_3)$, subject to $g(x_1, x_2, x_3) = c$, will have a minimum) if $|\mathbf{H_2}| < 0$ and $|\mathbf{H_3}| = |\mathbf{H}| < 0$; and d^2U is negative definite (then $U = f(x_1, x_2, x_3)$, subject to $g(x_1, x_2, x_3) = c$, will have a maximum) if $|\mathbf{H_2}| > 0$ and $|\mathbf{H_3}| = |\mathbf{H}| < 0$. We present these FOCs and SOCs in Table 4.4.2 for easy reference.

As an example, suppose that we want to optimize the function $U = f(x_1, x_2, x_3) = x_1 + x_2 + x_3 - x_1^2 - x_2^2 - x_3^2$ subject to the constraint $x_1 + x_2 + x_3 = 6$. The Lagrangian function associated with this problem can be written as

$$L = h(x_1, x_2, x_3, \lambda) = x_1 + x_2 + x_3 - x_1^2 - x_2^2 - x_3^2 + \lambda(6 - x_1 - x_2 - x_3) \qquad (4.4.18)$$

Differentiating equation (4.4.18) with respect to each choice variable and setting the results to zero, we obtain the FOCs as

$$L_1 = 1 - 2x_1 - \lambda = 0; \quad L_2 = 1 - 2x_2 - \lambda = 0; \quad L_3 = 1 - 2x_3 - \lambda = 0; \text{ and}$$

$$L_\lambda = \partial L / \partial \lambda = 6 - x_1 - x_2 - x_3 = 0 \qquad (4.4.19)$$

Solving the above system of simultaneous equations, we obtain the critical values $(x_1^*, x_2^*, x_3^*) = (2, 2, 2)$. These critical values satisfy the FOCs. We can obtain from equation (4.4.19) that

$$L_{11} = -2, \quad L_{22} = -2, \quad L_{33} = -2, \quad L_{12} = 0, \quad L_{13} = 0, \quad L_{23} = 0, \quad g_1 = 1,$$

$$g_2 = 1, \quad \text{and} \quad g_3 = 1 \qquad (4.4.20)$$

Therefore, with the partial derivatives given in equation (4.4.20), the bordered Hessian matrix and bordered Hessian in the present example can be written, respectively, as

$$\mathbf{H} = \begin{bmatrix} 0 & g_1 & g_2 & g_3 \\ g_1 & L_{11} & L_{12} & L_{13} \\ g_2 & L_{21} & L_{22} & L_{23} \\ g_3 & L_{31} & L_{32} & L_{33} \end{bmatrix} = \begin{bmatrix} 0 & 1 & 1 & 1 \\ 1 & -2 & 0 & 0 \\ 1 & 0 & -2 & 0 \\ 1 & 0 & 0 & -2 \end{bmatrix} \text{ and } |\mathbf{H}| = \begin{vmatrix} 0 & g_1 & g_2 & g_3 \\ g_1 & L_{11} & L_{12} & L_{13} \\ g_2 & L_{21} & L_{22} & L_{23} \\ g_3 & L_{31} & L_{32} & L_{33} \end{vmatrix} =$$

$$\begin{vmatrix} 0 & 1 & 1 & 1 \\ 1 & -2 & 0 & 0 \\ 1 & 0 & -2 & 0 \\ 1 & 0 & 0 & -2 \end{vmatrix}.$$ We now need to check the signs of $|\mathbf{H}_2|$ and $|\mathbf{H}_3| = |\mathbf{H}|$ to determine

whether d^2U is negative definite or positive definite. Since $|\mathbf{H}_2| = 4 > 0$ and $|\mathbf{H}_3| = |\mathbf{H}| = -12 < 0$, we conclude (using the conditions given in Table 4.4.2) that d^2U is negative definite. Therefore, the function $U = f(x_1, x_2, x_3) = x_1 + x_2 + x_3 - x_1^2 - x_2^2 - x_3^2$, subject to the constraint $x_1 + x_2 + x_3 = 6$, qualifies for a maximum at the critical values $(x_1^*, x_2^*, x_3^*) = (2, 2, 2)$. The maximum value of the function is $U = f(x_1^*, x_2^*, x_3^*) = x_1^* + x_1^* + x_3^* - x_1^{*2} - x_2^{*2} - x_3^{*2} = 2 + 2 + 2 - 2^2 - 2^2 - 2^2 = -6$.

As another example, suppose that we want to optimize the function $U = f(x_1, x_2, x_3) = x_1^2 + x_2^2 + x_3^2 - x_1 - x_2 - x_3$ subject to the constraint $x_1 + x_2 + x_3 = 6$. The Lagrangian function associated with this problem can be written as

$$L = h(x_1, x_2, x_3, \lambda) = x_1^2 + x_2^2 + x_3^2 - x_1 - x_2 - x_3 + \lambda(6 - x_1 - x_2 - x_3) \qquad (4.4.21)$$

Differentiating equation (4.4.21) with respect to each choice variable and setting the results to zero, we obtain the FOCs as

$$L_1 = 2x_1 - 1 - \lambda = 0; \quad L_2 = 2x_2 - 1 - \lambda = 0; \quad L_3 = 2x_3 - 1 - \lambda = 0; \quad \text{and}$$

$$L_\lambda = 6 - x_1 - x_2 - x_3 = 0 \qquad (4.4.22)$$

Solving the above system of simultaneous equations, we obtain the critical values $(x_1^*, x_2^*, x_3^*) = (2, 2, 2)$. These critical values satisfy the FOCs. We can obtain from equation (4.4.22) that

$$L_{11} = 2, \quad L_{22} = 2, \quad L_{33} = 2, \quad L_{12} = 0, \quad L_{13} = 0, \quad L_{23} = 0, \quad g_1 = 1,$$

$$g_2 = 1, \quad \text{and} \quad g_3 = 1 \qquad (4.4.23)$$

Therefore, with the partial derivatives in equation (4.4.23), the bordered Hessian matrix and the bordered Hessian in the present example can be written, respectively, as

$$\mathbf{H} = \begin{bmatrix} 0 & g_1 & g_2 & g_3 \\ g_1 & L_{11} & L_{12} & L_{13} \\ g_2 & L_{21} & L_{22} & L_{23} \\ g_3 & L_{31} & L_{32} & L_{33} \end{bmatrix} = \begin{bmatrix} 0 & 1 & 1 & 1 \\ 1 & 2 & 0 & 0 \\ 1 & 0 & 2 & 0 \\ 1 & 0 & 0 & 2 \end{bmatrix} \quad \text{and} \quad |\mathbf{H}| = \begin{vmatrix} 0 & g_1 & g_2 & g_3 \\ g_1 & L_{11} & L_{12} & L_{13} \\ g_2 & L_{21} & L_{22} & L_{23} \\ g_3 & L_{31} & L_{32} & L_{33} \end{vmatrix} = \begin{vmatrix} 0 & 1 & 1 & 1 \\ 1 & 2 & 0 & 0 \\ 1 & 0 & 2 & 0 \\ 1 & 0 & 0 & 2 \end{vmatrix}.$$

We now need to check the signs of $|\mathbf{H}_2|$ and $|\mathbf{H}_3| = |\mathbf{H}|$ to determine whether d^2U is negative definite or positive definite. Since $|\mathbf{H}_2| = -4 < 0$ and $|\mathbf{H}_3| = |\mathbf{H}| = -12 < 0$, we conclude (using the conditions given in Table 4.4.2) that d^2U is negative definite. Therefore, the function $U = f(x_1, x_2, x_3) = x_1^2 + x_2^2 + x_3^2 - x_1 - x_2 - x_3$, subject to the constraint $x_1 + x_2 + x_3 = 6$, qualifies for a minimum at the critical values $(x_1^*, x_2^*, x_3^*) = (2, 2, 2)$. The minimum value of the function is $U = f(x_1^*, x_2^*, x_3^*) = x_1^{*2} + x_2^{*2} + x_3^{*2} - x_1^* - x_1^* - x_3^* = 2^2 + 2^2 + 2^2 - 2 - 2 - 2 = 6$.

4.4.7 Constrained optimization with 'n' choice variables and one constraint

So far in this section we have been concerned with optimization of bivariate or trivariate objective functions subject to one constraint. We now consider the more general case of optimization of objective functions with n choice variables subject to one constraint.

Suppose that we want to optimize $U = f(x_1, x_2, x_3, \ldots, x_n)$ subject to $g(x_1, x_2, x_3, \ldots, x_n) = c$. The Lagrangian function of this problem can be written as

$$L = h(x_1, x_2, x_3, \ldots, x_n, \lambda) = f(x_1, x_2, x_3, \ldots, x_n) + \lambda[c - g(x_1, x_2, x_3, \ldots, x_n)] \quad (4.4.24)$$

Then, the usual FOCs can be stated as

$$L_1 = f_1 - \lambda g_1 = 0; \quad L_2 = f_2 - \lambda g_2 = 0; \quad L_3 = f_3 - \lambda g_3 = 0; \quad \ldots;$$
$$L_n = f_n - \lambda g_n = 0; \quad \text{and} \quad L_\lambda = c - g(x_1, x_2, x_3) = 0 \quad (4.4.25)$$

Solving the above system of simultaneous equations, we can obtain the critical values $(x_1^*, x_2^*, x_3^*, \ldots, x_n^*)$. These critical values satisfy the FOCs. From equation (4.4.25) we can obtain

$$L_{11}, L_{22}, L_{33}, \ldots, L_{nn}, L_{12}, L_{13}, \ldots, L_{1n}, \ldots, L_{21}, L_{23}, \ldots, L_{2n}, \ldots,$$
$$L_{n1}, L_{n2}, L_{n3}, \ldots, L_{(n-1),n}, \text{ and } g_1, g_2, g_3, \ldots, g_n \quad (4.4.26)$$

Therefore, with the partial derivatives in (4.4.26), the bordered Hessian matrix and the bordered Hessian in the present example can be written as

$$\mathbf{H} = \begin{bmatrix} 0 & g_1 & g_2 & \cdots & g_n \\ g_1 & L_{11} & L_{12} & \cdots & L_{1n} \\ g_2 & L_{21} & L_{22} & \cdots & L_{2n} \\ \cdots & \cdots & \cdots & \cdots & \cdots \\ g_n & L_{n1} & L_{n2} & \cdots & L_{nn} \end{bmatrix} \text{ and } |\mathbf{H}| = \begin{vmatrix} 0 & g_1 & g_2 & \cdots & g_n \\ g_1 & L_{11} & L_{12} & \cdots & L_{1n} \\ g_2 & L_{21} & L_{22} & \cdots & L_{2n} \\ \cdots & \cdots & \cdots & \cdots & \cdots \\ g_n & L_{n1} & L_{n2} & \cdots & L_{nn} \end{vmatrix}$$

We now need to check, again as before, the signs of the principal minors of the bordered Hessian to determine whether d^2U is negative definite or positive definite. As we stated in Section 2.8.3, d^2U is positive definite (and, therefore, $U = f(x_1, x_2, x_3, \ldots, x_n)$, subject to the constraint $g(x_1, x_2, x_3, \ldots, x_n) = c$, has a minimum) if

$$|\mathbf{H}_2| = \begin{vmatrix} 0 & g_1 & g_2 \\ g_1 & L_{11} & L_{12} \\ g_2 & L_{21} & L_{22} \end{vmatrix} < 0, \quad |\mathbf{H}_3| = \begin{vmatrix} 0 & g_1 & g_2 & g_3 \\ g_1 & L_{11} & L_{12} & L_{13} \\ g_2 & L_{21} & L_{22} & L_{23} \\ g_3 & L_{31} & L_{32} & L_{33} \end{vmatrix} < 0, \ldots,$$

$$|\mathbf{H}_n| = |\mathbf{H}| = \begin{vmatrix} 0 & g_1 & g_2 & \cdots & g_n \\ g_1 & L_{11} & L_{12} & \cdots & L_{1n} \\ g_2 & L_{21} & L_{22} & \cdots & L_{2n} \\ \cdots & \cdots & \cdots & \cdots & \cdots \\ g_n & L_{n1} & L_{n2} & \cdots & L_{nn} \end{vmatrix} < 0$$

The minimum of $U = f(x_1, x_2, x_3, \ldots, x_n)$ will be given by $U^* = f(x_1^*, x_2^*, x_3^*, \ldots, x_n^*)$. Similarly, d^2U is negative definite (and, therefore, $U = f(x_1, x_2, x_3, \ldots, x_n)$, subject to the

Table 4.4.3

Condition	Minimum	Maximum
FOC or necessary condition	$dU = 0$	$dU = 0$
SOC or sufficient condition	$d^2U > 0$, or positive definite, if $\|\mathbf{H}_2\| < 0, \|\mathbf{H}_3\| < 0, \ldots, \|\mathbf{H}_n\| = \|\mathbf{H}\| < 0$	$d^2U < 0$, or negative definite, if $\|\mathbf{H}_2\| > 0, \|\mathbf{H}_3\| < 0, \ldots,$ $(-1)^n\|\mathbf{H}_n\| = (-1)^n\|\mathbf{H}\| > 0$

constraint $g(x_1, x_2, x_3, \ldots, x_n) = c$, has a maximum) if

$$|\mathbf{H}_2| = \begin{vmatrix} 0 & g_1 & g_2 \\ g_1 & L_{11} & L_{12} \\ g_2 & L_{21} & L_{22} \end{vmatrix} > 0, \quad |\mathbf{H}_3| = \begin{vmatrix} 0 & g_1 & g_2 & g_3 \\ g_1 & L_{11} & L_{12} & L_{13} \\ g_2 & L_{21} & L_{22} & L_{23} \\ g_3 & L_{31} & L_{32} & L_{33} \end{vmatrix} < 0, \ldots,$$

$$(-1)^n|\mathbf{H}_n| = (-1)^n|\mathbf{H}| = \begin{vmatrix} 0 & g_1 & g_2 & \cdots & g_n \\ g_1 & L_{11} & L_{12} & \cdots & L_{1n} \\ g_2 & L_{21} & L_{22} & \cdots & L_{2n} \\ \cdots & \cdots & \cdots & \cdots & \cdots \\ g_n & L_{n1} & L_{n2} & \cdots & L_{nn} \end{vmatrix} > 0$$

These conditions are presented in Table 4.4.3 for easy reference.

4.4.8 Constrained optimization with n choice variables and m constraints (n > m)

Suppose now that we want to optimize a function with n choice variables subject to m constraints such that $n > m$. Also suppose that the objective function that we want to optimize is $U = f(x_1, x_2, x_3, \ldots, x_n)$ and that the constraints are $g^1(x_1, x_2, x_3, \ldots, x_n) = c^1$, $g^2(x_1, x_2, x_3, \ldots, x_n) = c^2, \ldots, g^m(x_1, x_2, x_3, \ldots, x_n) = c^m$. Therefore, the Lagrangian function can be written as

$$L = h(x_1, x_2, \ldots, x_n, \lambda_1, \lambda_2, \ldots, \lambda_m) = f(x_1, x_2, \ldots, x_n) + \sum_{k=1}^{m} \lambda_k [c^k - g^k(x_1, x_2, x_3, \ldots, x_n)]$$

(4.4.27)

where $k = 1, 2, 3, \ldots, m$. Notice that equation (4.4.27) is similar to equation (2.8.24). The FOC for $U = f(x_1, x_2, x_3, \ldots, x_n)$, subject to the constraints, to have an optimum is that the first partial derivatives of equation (4.4.27) are all equal to zero. From these, one can obtain the critical values $(x_1^*, x_2^*, x_3^*, \ldots, x_n^*)$ and λ_k. The SOCs for an optimum of $U = f(x_1, x_2, x_3, \ldots, x_n)$, subject to the constraint $g^k(x_1, x_2, x_3, \ldots, x_n) = c^k$, are given at the end of Section 2.8.3. Therefore, we do not repeat them here. We simply state these conditions in Table 4.4.4 for easy reference. Notice that the SOCs of constrained extrema for the bivariate case with one constraint (Table 4.4.1), the trivariate case with one constraint (Table 4.4.2), and the n-variable case with one constraint (Table 4.4.3) are special cases of the SOCs presented in Table 4.4.4.

Table 4.4.4

Condition	Minimum	Maximum
FOC or necessary condition	$dU = 0$	$dU = 0$
SOC or sufficient condition	$d^2U > 0$, or positive definite, if all the principal minors have the same sign as that of $(-1)^m$	$d^2U < 0$, or negative definite, if the principal minors alternate in sign (starting with the sign of $|\mathbf{H}_{n+1}|$ equals the sign of $(-1)^{m+1}$)

As an example, consider optimizing $U = f(x_1, x_2, x_3) = x_1^2 + x_2^2 + x_3^2$ subject to two constraints $g^1(x_1, x_2, x_3) = 2x_1 + x_2 + x_3 = 10 = c^1$ and $g^2(x_1, x_2, x_3) = x_1 + 2x_2 + x_3 = 20 = c^2$. Following equation (4.4.27), the Lagrangian function for this problem, with $n = 3$ and $m = 2$, can be set up as

$$L = h(x_1, x_2, x_3, \lambda_1, \lambda_2) = x_1^2 + x_2^2 + x_3^2 + \lambda_1[10 - 2x_1 - x_2 - x_3] + \lambda_2[20 - x_1 - 2x_2 - x_3]$$

The FOCs require that $L_1 = 2x_1 - 2\lambda_1 - \lambda_2 = 0$, $L_2 = 2x_2 - \lambda_1 - 2\lambda_2 = 0$, $L_3 = 2x_3 - \lambda_1 - \lambda_2 = 0$, $L_{\lambda_1} = 10 - 2x_1 - x_2 - x_3 = 0$, and $L_{\lambda_2} = 20 - x_1 - 2x_2 - x_3 = 0$. Solving this system of simultaneous equations will yield the critical values $(x_1^*, x_2^*, x_3^*) = (-10/11, 100/11, 30/11)$, and $\lambda_1 = -80/11$ and $\lambda_2 = 140/11$.

The SOC for an optimum of $U = f(x_1, x_2, x_3) = x_1^2 + x_2^2 + x_3^2$ requires, as we presented in Section 2.8.3, that we check the sign of its quadratic form. We can obtain from the above FOCs that $L_{11} = 2$, $L_{22} = 2$, $L_{33} = 2$, $L_{12} = L_{21} = 0$, $L_{13} = L_{31} = 0$, and $L_{23} = L_{32} = 0$, and that $g_1^1 = 2$, $g_2^1 = 1$, $g_3^1 = 1$, $g_1^2 = 1$, $g_2^2 = 2$, and $g_3^2 = 1$. If we follow the procedures outlined in Section 2.8.3 and equation (2.8.26), we can construct the bordered Hessian for the present problem as

$$|\mathbf{H}_n| = |\mathbf{H}_{m+1}| = |\mathbf{H}_3| = \begin{vmatrix} 0 & 0 & g_1^1 & g_2^1 & g_3^1 \\ 0 & 0 & g_1^2 & g_2^2 & g_3^2 \\ g_1^1 & g_1^2 & L_{11} & L_{12} & L_{13} \\ g_2^1 & g_2^2 & L_{21} & L_{22} & L_{23} \\ g_3^1 & g_3^2 & L_{31} & L_{32} & L_{33} \end{vmatrix} = \begin{vmatrix} 0 & 0 & 2 & 1 & 1 \\ 0 & 0 & 1 & 2 & 1 \\ 2 & 1 & 2 & 0 & 0 \\ 1 & 2 & 0 & 2 & 0 \\ 1 & 1 & 0 & 0 & 2 \end{vmatrix}$$

As discussed at the end of Section 2.8.3, we now need to check the sign of $(n - m) = (3 - 2) = 1$ principal minor, which is the Hessian itself. Since $|\mathbf{H}_n| = |\mathbf{H}_{m+1}| = |\mathbf{H}_3| = 22 > 0$, it has the same sign as that of $(-1)^m = (-1)^2 = 1$. This suggests that, following Table 4.4.4, the quadratic form of $U = f(x_1, x_2, x_3) = x_1^2 + x_2^2 + x_3^2$ is positive definite. Therefore, we conclude that the function $U = f(x_1, x_2, x_3) = x_1^2 + x_2^2 + x_3^2$ has a minimum at the critical values $(x_1^*, x_2^*, x_3^*) = (-10/11, 100/11, 30/11)$, and the stationary value or the minimum value of the function is $U^* = f(x_1^*, x_2^*, x_3^*) = x_1^{*2} + x_2^{*2} + x_3^{*2} = 1000/11 = 90.90$.

As another example, consider optimizing $U = f(x_1, x_2, x_3) = -x_1^2 - x_2^2 - x_3^2$ subject to two constraints $g^1(x_1, x_2, x_3) = 2x_1 + x_2 + x_3 = 10 = c^1$ and $g^2(x_1, x_2, x_3) = x_1 + 2x_2 + x_3 = 20 = c^2$. Following equation (4.4.27), the Lagrangian function for this problem, with $n = 3$ and $m = 2$, can be set up as

$$L = h(x_1, x_2, x_3, \lambda_1, \lambda_2) = -x_1^2 - x_2^2 - x_3^2 + \lambda_1[10 - 2x_1 - x_2 - x_3] + \lambda_2[20 - x_1 - 2x_2 - x_3]$$

The FOCs require that $L_1 = -2x_1 - 2\lambda_1 - \lambda_2 = 0$, $L_2 = -2x_2 - \lambda_1 - 2\lambda_2 = 0$, $L_3 = -2x_3 - \lambda_1 - \lambda_2 = 0$, $L_{\lambda_1} = 10 - 2x_1 - x_2 - x_3 = 0$, and $L_{\lambda_2} = 20 - x_1 - 2x_2 - x_3 = 0$. Solving this system of simultaneous equations will yield the critical values $(x_1^*, x_2^*, x_3^*) = (-10/11, 100/11, 30/11)$, and $\lambda_1 = 80/11$ and $\lambda_2 = -140/11$.

The SOC for an optimum of $U = f(x_1, x_2, x_3) = -x_1^2 - x_2^2 - x_3^2$ requires, as we presented in Section 2.8.3, that we check the sign of its quadratic form. We can obtain from the above FOCs that $L_{11} = -2, L_{22} = -2, L_{33} = -2, L_{12} = L_{21} = 0, L_{13} = L_{31} = 0$, and $L_{23} = L_{32} = 0$, and that $g_1^1 = 2, g_2^1 = 1, g_3^1 = 1, g_1^2 = 1, g_2^2 = 2$, and $g_3^2 = 1$. If we follow the procedures outlined in Section 2.8.3 and equation (2.8.26), we can construct the bordered Hessian for the present problem as

$$|\mathbf{H}_n| = |\mathbf{H}_{m+1}| = |\mathbf{H}_3| = \begin{vmatrix} 0 & 0 & g_1^1 & g_2^1 & g_3^1 \\ 0 & 0 & g_1^2 & g_2^2 & g_3^2 \\ g_1^1 & g_1^2 & L_{11} & L_{12} & L_{13} \\ g_2^1 & g_2^2 & L_{21} & L_{22} & L_{23} \\ g_3^1 & g_3^2 & L_{31} & L_{32} & L_{33} \end{vmatrix} = \begin{vmatrix} 0 & 0 & 2 & 1 & 1 \\ 0 & 0 & 1 & 2 & 1 \\ 2 & 1 & -2 & 0 & 0 \\ 1 & 2 & 0 & -2 & 0 \\ 1 & 1 & 0 & 0 & -2 \end{vmatrix}$$

As presented in Section 2.8.3, we now need to check the sign of $(n - m) = (3 - 2) = 1$ principal minor, which is the Hessian itself. Since $|\mathbf{H}_n| = |\mathbf{H}_{m+1}| = |\mathbf{H}_3| = -22 < 0$, it has the same sign as that of $(-1)^{m+1} = (-1)^{2+1} = (-1)^3 = -1$. This suggests that, following Table 4.4.4, the quadratic form of $U = f(x_1, x_2, x_3) = -x_1^2 - x_2^2 - x_3^2$ is negative definite. Therefore, we conclude that the function $U = f(x_1, x_2, x_3) = -x_1^2 - x_2^2 - x_3^2$ has a maximum at the critical values $(x_1^*, x_2^*, x_3^*) = (-10/11, 100/11, 30/11)$, and the stationary value or the maximum value of the function is $U^* = f(x_1^*, x_2^*, x_3^*) = -x_1^{*2} - x_2^{*2} - x_3^{*2} = -79.98$.

4.4.9 Meaning of the Lagrange multiplier, λ

One important difference between the substitution method and the *Lagrangian method* is that the latter makes use of a new variable called the Lagrangian multiplier. One might wonder what significance this variable possesses. This new variable represents the rate of change of the optimal value of the objective function with respect to the constraint. In other words, the Lagrangian multiplier, as we show below, represents the effect of a small change in the constraint on the optimal value of the objective function. Assume that our objective function is $U = f(x_1, x_2)$ and the constraint is $g(x_1, x_2) = c$. Also assume that the optimal value of this objective function is $U^* = f[x_1^*(c), x_2^*(c)]$ and the constraint at the optimum position is $g[x_1^*(c), x_2^*(c)] = c$. The total derivative of $U^* = f[x_1^*(c), x_2^*(c)]$ with respect to c will yield

$$\frac{\partial f[x_1^*(c), x_2^*(c)]}{dc} = \frac{\partial f[x_1^*(c), x_2^*(c)]}{\partial x_1^*}\frac{dx_1^*(c)}{dc} + \frac{\partial f[x_1^*(c), x_2^*(c)]}{\partial x_2^*}\frac{dx_2^*(c)}{dc} \tag{4.4.28}$$

and the total derivative of $g[x_1^*(c), x_2^*(c)] = c$ with respect to c will yield

$$\frac{\partial g[x_1^*(c), x_2^*(c)]}{\partial x_1^*}\frac{dx_1^*(c)}{dc} + \frac{\partial g[x_1^*(c), x_2^*(c)]}{\partial x_2^*}\frac{dx_2^*(c)}{dc} = \frac{dc}{dc} = 1 \tag{4.4.29}$$

Our previous presentations show that the FOCs require that

$$\frac{\partial f[x_1^*(c), x_2^*(c)]}{\partial x_1^*} = \lambda \frac{\partial g[x_1^*(c), x_2^*(c)]}{\partial x_1^*} \tag{4.4.30}$$

and

$$\frac{\partial f[x_1^*(c), x_2^*(c)]}{\partial x_2^*} = \lambda \frac{\partial g[x_1^*(c), x_2^*(c)]}{\partial x_2^*} \tag{4.4.31}$$

Substituting equations (4.4.30) and (4.4.31) into equation (4.4.28) yields

$$\frac{df[x_1^*(c), x_2^*(c)]}{dc} = \frac{dU^*}{dc} = \lambda \frac{\partial g[x_1^*(c), x_2^*(c)]}{\partial x_1^*} \frac{dx_1^*(c)}{dc} + \lambda \frac{\partial g[x_1^*(c), x_2^*(c)]}{\partial x_2^*} \frac{dx_1^*(c)}{dc}$$

$$= \lambda \left[\frac{\partial g[x_1^*(c), x_2^*(c)]}{\partial x_1^*} \frac{dx_1^*(c)}{dc} + \frac{\partial g[x_1^*(c), x_2^*(c)]}{\partial x_2^*} \frac{dx_2^*(c)}{dc} \right] \tag{4.4.32}$$

Notice that the terms inside the bracket in equation (4.4.32) are identical with equation (4.4.29), which, in turn, is equal to 1. Therefore, equation (4.4.32) can be rewritten as

$$\frac{\partial f[x_1^*(c), x_2^*(c)]}{dc} = \frac{dU^*}{dc} = \lambda \times 1 = \lambda \tag{4.4.33}$$

Equation (4.4.33) is the result we required. It states that the rate of change of the optimal value of the objective function with respect to the constraint (or, the effect of a small change in the constraint on the optimal value of the objective function) is equal to the Lagrangian multiplier, λ. However, for this interpretation of the Lagrange multiplier, we need to form the Lagrangian function as given in equation (4.4.3) and not as given in footnote 8 in this chapter. In certain circumstances, the Lagrangian multiplier, λ, is referred to as the *shadow price*.

4.4.10 Quasiconcavity, quasiconvexity, and constrained optima

We know from Section 4.2.7 that the knowledge of the nature of a function (that is, its concavity and convexity) obviates the need to check the SOC in an unconstrained optimization or free optimization problem with a univariate objective function. We found from our discussion there that if the function was convex (concave) over a closed interval on its domain, then the function had a local minimum (maximum). We also found that if the function was strictly convex (concave) on its entire domain, then the function had a global minimum (maximum).

Similarly, we found in Section 4.3.5 that if the quadratic form of a multivariate function such as $U = f(x_1, x_2, x_3, \ldots, x_n)$ was positive (negative) definite, then the function was strictly convex (concave) and that the function had a global minimum (maximum); and if its quadratic form was positive (negative) semidefinite, then the function was convex (concave) and that the function had a local minimum (maximum). A question that arises now is: can we use a similar knowledge that obviates the need to check the SOC in a constrained optimization problem? The answer is yes, although the required knowledge is about *quasiconcavity* and *quasiconvexity* (discussed in Sections 3.5.6 and 3.5.7) and not about concavity and convexity (strict or nonstrict).

Consider, for example, the general constrained optimization problem given in equation (4.4.24). The quadric form of this Lagrangian function and its Hessian ($|\mathbf{H}_n|$) can be derived as presented in Section 4.4.7. We used the determinant of the partial derivatives of the function $f(x_1, x_2, x_3, \ldots, x_n)$, in Section 3.5.7, to determine whether the function is quasiconcave or quasiconvex, and that determinant is $|\mathbf{Q}_n|$. These two, respectively, are

$$|\mathbf{H}| = |\mathbf{H}_n| = \begin{vmatrix} 0 & g_1 & g_2 & \cdots & g_n \\ g_1 & L_{11} & L_{12} & \cdots & L_{1n} \\ g_2 & L_{21} & L_{22} & \cdots & L_{2n} \\ \cdots & \cdots & \cdots & \cdots & \cdots \\ g_n & L_{n1} & L_{n2} & \cdots & L_{nn} \end{vmatrix} \quad \text{and} \quad |\mathbf{Q}| = |\mathbf{Q}_n| = \begin{vmatrix} 0 & f_1 & f_2 & \cdots & f_n \\ f_1 & f_{11} & f_{12} & \cdots & f_{1n} \\ f_2 & f_{21} & f_{22} & \cdots & f_{2n} \\ \cdots & \cdots & \cdots & \cdots & \cdots \\ f_n & f_{n1} & f_{n2} & \cdots & f_{nn} \end{vmatrix}$$

Notice the two differences between the Hessian and the determinant $|\mathbf{Q}|$. The first difference is in the nonborder elements in the two determinants. In the Hessian they are the second partial derivatives of the associated Lagrangian function, while in the determinant $|\mathbf{Q}|$ they are the second partial derivatives of the original function. The second difference is that the borders in the Hessian are the first partial derivatives of the constraint, while the borders in the determinant $|\mathbf{Q}|$ are the first partial derivatives of the original function.

Notice that when the constraint is linear we obtain from the Lagrangian function that $L_j = f_j - \lambda g_j$, where $j = 1, 2, 3, \ldots, n$. Similarly, when the constraint is linear, we obtain from the FOC that $f_j - \lambda g_j = 0$ or $f_j = \lambda g_j$. This implies that the borders in $|\mathbf{Q}|$ are nothing but the borders in the Hessian multiplied by the Lagrangian multiplier, λ. Therefore, we can factor out λ in the Hessian and can write $|\mathbf{Q}| = \lambda^2 |\mathbf{H}|$. This can be done for all the principal minors. Thus, we arrive at an important result that, when the constraint is linear, the sign of $|\mathbf{Q}|$ is the same as the sign of $|\mathbf{H}|$. Therefore, we can conclude that if the objective function is quasiconcave (quasiconvex), given linear constraint and the FOC, the objective function has a maximum (minimum).

4.4.11 Application examples

Example 1. Assume that a consumer's utility, from the consumption of two goods x_1 and x_2, is given by

$$U = f(x_1, x_2) \tag{4.4.34}$$

and that the consumer faces the budget constraint given by

$$p_1 x_1 + p_2 x_2 = c \tag{4.4.35}$$

where p_1, p_2, and c denote the price of good x_1, the price of good x_2, and the income of the consumer, respectively. Maximize the consumer's utility subject to the constraint.

Solution. The Lagrangian function for this example can be set up as

$$L = h(x_1, x_2, p_1, p_2, \lambda, c) = f(x_1, x_2) + \lambda[c - p_1 x_1 - p_2 x_2] \tag{4.4.36}$$

Differentiating the Lagrangian function with respect to x_1, x_2, and λ and setting the results to zero will yield the FOCs for a maximum of $U = f(x_1, x_2)$. These FOCs are

$$L_1 = f_1 - \lambda p_1 = 0; L_2 = f_2 - \lambda p_2 = 0 \text{ and } L_\lambda = c - p_1 x_1 - p_2 x_2 = 0 \tag{4.4.37}$$

Solving the above system of simultaneous equations yields the critical values (x_1^*, x_2^*) and λ. Notice that from equation (4.4.37) we can obtain (because the derivatives of the objective function $U = f(x_1, x_2)$ with respect to x_1 and x_2 represent the marginal utility from x_1 (MU_1) and the marginal utility from x_2 (MU_2), respectively)

$$(f_1/p_1) = (f_2/p_2) = \lambda, \quad \text{or} \quad (MU_1/p_1) = (MU_2/p_2) = \lambda \tag{4.4.38}$$

Equation (4.4.38) shows that, at the utility-maximizing point, each good purchased must yield the same marginal utility (MU) per dollar spent on that good. This result is popularly called *equi-marginal utility* or, in general, the *equi-marginal principle*. Therefore, the Lagrangian multiplier, λ, can be interpreted as the marginal utility of an additional dollar spent, or the *marginal utility of money*. Notice that equation (4.4.38) can written as

$$(f_1/f_2) = (MU_1/MU_2) = (p_1/p_2) = \lambda \tag{4.4.39}$$

The result in equation (4.4.39) is the result we used in application example 4 in Section 3.8.7. Notice that we can write the budget constraint (equation (4.4.35)) as $x_2 = (c/p_2) - (p_1/p_2)x_1$. This means that the vertical intercept, horizontal intercept, and slope of the budget constraint are c/p_2, c/p_1, and $-p_1/p_2$, respectively. As we showed in application example 4 in Section 3.8.7, the slope of the indifference curve is given by $dx_2/dx_1 = -MU_1/MU_2$. We also stated there that the slope of the indifference curve is called the marginal rate of substitution between x_1 and x_2 ($MRS_{x_1 x_2}$); that is, $dx_2/dx_1 = -MU_1/MU_2 = MRS_{x_1 x_2}$. Therefore, equation (4.4.39) can also be expressed as

$$MRS_{x_1 x_2} = -(MU_1/MU_2) = -(p_1/p_2) \tag{4.4.40}$$

which states that when the consumer is in equilibrium or maximizes utility, the slope of the indifference curve is equal to the slope of the budget line. This is shown via point E in Figure 3.8.3(A).

We now consider the SOC to ensure that the critical values (x_1^*, x_2^*) qualify for a maximum of $U = f(x_1, x_2)$. We know that this requires that we check the sign of the quadratic form of $U = f(x_1, x_2)$ with constraint $p_1 x_1 + p_2 x_2 = c$. Assume that we write the constraint as $g(x_1, x_2, p_1, p_2, c) = c - p_1 x_1 - p_2 x_2 = 0$. Then, given the Lagrangian function presented in equation (4.4.36), the concerned quadratic form can be written, as we did before, as

$$d^2 U = [L_{11} g_2^2 - 2L_{12} g_1 g_2 + L_{22} g_1^2][dx_1^2/g_2^2] \tag{4.4.41}$$

Therefore, the associated Hessian can be written as

$$|\mathbf{H}_2| = |\mathbf{H}| = \begin{vmatrix} 0 & g_1 & g_2 \\ g_1 & L_{11} & L_{12} \\ g_2 & L_{21} & L_{22} \end{vmatrix} \tag{4.4.42}$$

We know from Table 4.4.1 that the SOC for a maximum of $U = f(x_1, x_2)$ is that the quadratic form in equation (4.4.41) must be negative definite or $d^2 U < 0$. For this to happen, the Hessian in equation (4.4.42) must be positive; that is,

$$|\mathbf{H}_2| = |\mathbf{H}| = \begin{vmatrix} 0 & p_1 & p_2 \\ p_1 & L_{11} & L_{12} \\ p_2 & L_{21} & L_{22} \end{vmatrix} = -p_1^2 L_{11} - p_2^2 L_{22} + 2p_1 p_2 L_{12} > 0 \qquad (4.4.43)$$

We assume that the indifference curve that we deal with in the present example is strictly convex at the point of equilibrium (or the critical values). This assumption guarantees that $|\mathbf{H}_2| = |\mathbf{H}| > 0$. Therefore, we state that the function $U = f(x_1, x_2)$ qualifies for a maximum corresponding to the critical values (x_1^*, x_2^*).

Example 2. Assume that the total cost, C, to a firm that uses capital, k, and labor, l, to produce output, q, is given by

$$C = f(k, l, r, w) = r.k + w.l \qquad (4.4.44)$$

where r denotes interest rate and w denotes wage rate. Also assume that the firm's total output is given by the constraint

$$j(k, l) = q_0 \qquad (4.4.45)$$

Minimize the firm's total cost subject to the constraint.

Solution. The Lagrangian function for this example can be set up as

$$L = h(k, l, r, w, \lambda, q_0) = f(k, l, r, w) + \lambda[q_0 - j(k, l)] \qquad (4.4.46)$$

Differentiating the Lagrangian function with respect to k, l, and λ, we obtain the FOCs as

$$L_k = r - \lambda.j_k = 0, \quad L_l = w - \lambda.j_l = 0, \quad \text{and} \quad L_\lambda = q_0 - r.k - w.l = 0 \qquad (4.4.47)$$

Solving the above system of simultaneous equations yields the critical values (k^*, l^*) and λ. Notice that from equation (4.4.47) we can obtain (because the derivatives of the production function $j(k, l)$ with respect to k and l represent the marginal product of capital (MP_k) and the marginal product of labor (MP_l), respectively)

$$(j_k/r) = (j_l/w) = \lambda, \quad \text{or} \quad (MP_k/r) = (MP_l/w) = \lambda \qquad (4.4.48)$$

which shows that, at the cost-minimizing point, each factor employed must yield the same marginal product (MP) per dollar spent on that factor. Notice that equation (4.4.48) can be written as

$$(j_l/j_k) = (MP_l/MP_k) = (w/r) = \lambda \qquad (4.4.49)$$

The result in equation (4.4.49) is the result we used in application example 5 in Section 3.8.7. Notice that we can write equation (4.4.44) as $k = C/r - [w/r] l$. This means that the vertical intercept, horizontal intercept, and slope of the isocost line are C/w, C/r, and $-w/r$, respectively. As we showed in application example 5 in Section 3.8.7, the slope of the isoquant was given by $dk/dl = -MP_l/MP_k$. We also stated there that the slope of the isoquant was called the marginal rate of technical substitution between l and k (MRTS$_{lk}$); that is, $dk/dl = -MP_l/MP_k = $ MRTS$_{lk}$. Therefore, equation (4.4.49) can also be expressed as

$$\text{MRTS}_{lk} = -(MP_l/MP_k) = -(w/r) \tag{4.4.50}$$

which states that when the firm is in equilibrium or when the firm minimizes cost, the slope of the isoquant is equal to the slope of the isocost line. This is shown via point G in Figure 3.8.3.(B).

We now consider the SOC to ensure that the critical values (k^*, l^*) qualify for a minimum of $C = f(k, l, r, w)$. We know that this requires that we check the sign of the quadratic form of $C = f(k, l, r, w)$ with constraint $j(k, l) = q_0$. Assume that we write the constraint as $g(k, l, q_0) = j(k, l) - q_0 = 0$. Then, given the Lagrangian function presented in equation (4.4.46), the concerned quadratic form can be written, as we did before, as

$$d^2C = [L_{kk}g_k^2 + L_{ll}g_l^2 + 2L_{kl}g_kg_l][dk^2/g_k^2] \tag{4.4.51}$$

and the associated Hessian as

$$|\mathbf{H_2}| = |\mathbf{H}| = \begin{vmatrix} 0 & g_k & g_l \\ g_k & L_{kk} & L_{kl} \\ g_l & L_{lk} & L_{ll} \end{vmatrix} \tag{4.4.52}$$

We know from Table 4.4.1 that the SOC for a minimum of $C = f(k, l, r, w)$ is that the quadratic form in equation (4.4.51) be positive definite or $d^2U > 0$. For this to happen, the Hessian in equation (4.4.52) must be negative; that is,

$$|\mathbf{H_2}| = |\mathbf{H}| = \begin{vmatrix} 0 & g_k & g_l \\ g_k & L_{kk} & L_{kl} \\ g_l & L_{lk} & L_{ll} \end{vmatrix} = -g_k^2L_{kk} - g_l^2L_{ll} + 2g_kg_lL_{kl} < 0 \tag{4.4.53}$$

We assume that the isoquant that we deal with in the present example is strictly convex at the point of equilibrium (or the critical values). This assumption guarantees that $|\mathbf{H_2}| = |\mathbf{H}| < 0$. Therefore, we state that the function $C = f(k, l, r, w)$ qualifies for a minimum corresponding to the critical values (k^*, l^*).

Example 3.   Assume that the output q produced by a firm using k units of capital and l units of labor is given by the Cobb–Douglas production function $q = f(k, l) = k^\alpha l^{1-\alpha}$. Also assume that the firm's budget constraint is given by $r.k + w.l = s$, where r, w, and s denote interest rate, wage rate, and available fund in dollars, respectively. (i) Find the quantities of k and l that the firm must use to maximize its output assuming $\alpha = 0.5$, $r = 0.1$ (or 10 percent), $w = \$10$, and $s = \$100$. (ii) Find the elasticity of substitution between the two factors.

Solution.   (i) The Lagrangian function for this example can be written as $L = k^\alpha l^{1-\alpha} + \lambda[s - r.k - w.l]$. Then the FOCs are $L_k = \alpha k^{\alpha-1}l^{1-\alpha} - \lambda r = 0$, $L_l = (1-\alpha)k^\alpha l^{-\alpha} - \lambda w = 0$, and $L_\lambda = s - r.k - w.l = 0$. These three equations can be solved to obtain the critical values $(k^*, l^*) = \alpha.s/r, (1 - \alpha).s/w$, and $\lambda = (\alpha/r)^\alpha.[(1 - \alpha 0/w]^{1-\alpha}$. If we substitute $\alpha = 0.5$, $r = 0.1$ (or 10 percent), $w = \$10$, and $s = \$100$ into the last two equations, we find that $(k^*, l^*) = (\$500, 5)$ and $\lambda = 0.5$. Therefore, the maximum output will be $q^* = f(k^*, l^*) = k^{*\alpha} l^{*1-\alpha} = 50$ units.

Let us now check whether the critical points $(k^*, l^*) = (\$500, 5)$ correspond to a maximum of $q = f(k, l) = k^\alpha l^{1-\alpha}$. The quadratic form associated with this problem is $d^2 q = [L_{kk}g_k^2 + L_{ll}g_l^2 + 2L_{kl}g_kg_l](dk^2/g_k^2)$ and the associated Hessian is similar to equation (4.4.43): $|\mathbf{H}_2| = |\mathbf{H}|$. We can find $L_{kk} = \alpha(\alpha - 1)k^{\alpha-2}l^{1-\alpha}$, $L_{ll} = (1 - \alpha)(-\alpha)k^\alpha l^{-\alpha-1}$, and $L_{kl} = \alpha(1 - \alpha)k^{\alpha-1}l^{-\alpha} = L_{lk}$. If we evaluate these second partial derivatives at the critical values (with $\alpha = 0.5$), we obtain $L_{kk} = 0.0001$, $L_{ll} = 1$, and $L_{kl} = L_{lk} = -0.01$. We can find from the constraint that $g_k = r = 0.1$ and $g_l = w = \$10$. Substituting these values into the Hessian

yields $|\mathbf{H}_2| = |\mathbf{H}| = \begin{vmatrix} 0 & 0.1 & 10 \\ 0.1 & -0.00005 & 0.025 \\ 10 & 0.025 & -0.5 \end{vmatrix} = 0.06 > 0$. Therefore, following Table 4.4.1,

the quadratic form is negative definite. This suggests that the function $q = f(k, l) = k^\alpha l^{1-\alpha}$ qualifies for a maximum corresponding to the critical values $(k^*, l^*) = (\$500, 5)$.

(ii) Let us now find the elasticity of substitution. We defined in example 1(vi) in Section 3.7.4 the elasticity of substitution (σ) as the percentage change in the optimum (that is, the least or minimum cost) k/l ratio due to a small percentage change in the input-price ratio, p_l/p_k. Notice that $p_l = w$ and $p_k = r$ in the present example. The least cost *capital-labor ratio* in the present example is k^*/l^* and the input-price ratio is w/r. Therefore, we define elasticity of substitution as $\sigma = [d(k^*/l^*)k^*/l^*]/[d(w/r)/w/r] = [d(k^*/l^*)/d(w/r)]/[k^*/l^*w/r]$. The numerator of the last equation is called the *marginal function* and the denominator is called the *average function*. Therefore, the elasticity of substitution is the ratio of the marginal function to the average function.

We found above the critical (optimal) values of k and l as $k^* = \alpha.s/r$ and $l^* = (1 - \alpha).s/w$, respectively. Taking the ratios of the last two equations and simplifying yields $k^*/l^* = [\alpha/(1 - \alpha)].(w/r)$. Then the derivative of this ratio with respect to w/r (the marginal function) is $d(k^*/l^*)/d(w/r) = \alpha/(1 - \alpha)$. Similarly, the denominator (the average function) is $(k^*/l^*)/(w/r) = \{[\alpha/(1 - \alpha)].(w/r)\}/(w/r) = \alpha/(1 - \alpha)$. Therefore, the ratio of the marginal function to the average function is $\sigma = [d(k^*/l^*)/d(w/r)]/[k^*/l^*/w/r] = [\alpha/(1 - \alpha)]/[\alpha/(1 - \alpha)] = 1$. This was the result we used in example 1(vi) in Section 3.7.4.

Example 4. Assume that the output q produced by a firm using k units of capital and l units of labor is given by the CES production function $q = f(k, l) = [a.k^{-\beta} + b.l^{-\beta}]^{-1/\beta}$. Also assume that the firm's budget constraint is given by $r.k + w.l = s$, where r, w, and s represent interest rate, wage rate, and available fund, respectively. (i) Find the quantities of k and l that the firm must employ to maximize its output assuming $a = 0.5$, $b = 0.5$, $\beta = 0.5$, $r = 0.1$ (or 10 percent), $w = \$10$, and $s = \$100$. (ii) Find the elasticity of substitution between the two factors.

Solution. (i) The Lagrangian function for this example can be written as $L = [a.k^{-\beta} + b.l^{-\beta}]^{-1/\beta} + \lambda[s - r.k - w.l]$. Then the FOCs are $L_k = a.[a.k^{-\beta} + b.l^{-\beta}]^{-(1+\beta)/\beta}.k^{-(1+\beta)} - \lambda r = 0$, $L_l = b.[a.k^{-\beta} + b.l^{-\beta}]^{-(1+\beta)/\beta}.l^{-(1+\beta)} - \lambda w = 0$, and $L_\lambda = s - r.k - w.l = 0$. Dividing the first equation by the second equation and simplifying yields $k = [(r/w).(b/a)]^{-1/(1+\beta)}.l = x.l$, where $x = [(r/w).(b/a)]^{-1/(1+\beta)}$. Substituting this into the budget constraint and solving for l gives the optimal value of l as $l^* = s/(r.x + w)$. Substituting l^* into the budget constraint again yields the optimal value of k as $k^* = s.x/(r.x + w)$. Therefore, the critical values are $(k^*, l^*) = \{[s.x/(r.x + w)], [s/(r.x + w)]\}$. If we substitute $a = 0.5$, $b = 0.5$, $\beta = 0.5$, $r = 0.1$ (or 10 percent), $w = \$10$, and $s = \$100$ into the last equation, we find that $(k^*, l^*) = (\$177, 8.23)$ and $\lambda = 464$. Therefore, the maximum output will be $q^* = f(k^*, l^*) = [a.k^{*-\beta} + b.l^{*-\beta}]^{-1/\beta} = 21.84$ units.

Let us now check whether the critical values $(k^*, l^*) = (177, 8.23)$ correspond to a maximum of $q = f(k, l) = [a.k^{-\beta} + b.l^{-\beta}]^{-1/\beta}$. The quadratic form associated with this problem is $d^2q = [L_{kk}g_k^2 + L_{ll}g_l^2 + 2L_{kl}g_kg_l](dk^2/g_k^2)$, and the associated Hessian can be written as before. We can find the second partial derivatives (L_{kk}, L_{ll}, and L_{kl}) from the Lagrangian. If we evaluate these derivatives at the critical values $(k^*, l^*) = (177, 8.23)$ (and with $a = 0.5$, $b = 0.5$, and $\beta = 0.5$), we can obtain $L_{kk} = -0.034$, $L_{ll} = -0.891$, and $L_{kl} = L_{lk} = 0.018$. We can find from the constraint that $g_k = r = 0.1$ and $g_l = w = \$10$. Substituting these values into the Hessian yields $|\mathbf{H}_2| = |\mathbf{H}| = \begin{vmatrix} 0 & 0.1 & 10 \\ 0.1 & -0.034 & 0.018 \\ 10 & 0.018 & -0.891 \end{vmatrix} = $ $3.8 > 0$. Therefore, following Table 4.4.1, the quadratic form is negative definite. This means that the function $q = f(k, l) = [a.k^{-\beta} + b.l^{-\beta}]^{-1/\beta}$ qualifies for a maximum corresponding to the critical values $(k^*, l^*) = (177, 8.23)$.

(ii) Let us now find the elasticity of substitution. We defined in example 2(iv) in Section 3.7.4 the elasticity of substitution (σ) as the percentage change in the optimum k/l ratio due to a small percentage change in the input-price ratio, p_l/p_k. We will follow our solution to the previous example. We found above that the critical or optimal values of k and l as $k^* = s.x/(r.x + w)$ and $l^* = s/(r.x + w)$, respectively. Taking the ratio of the last two equations and simplifying yields $k^*/l^* = x = [(r/w).(b/a)]^{-1/(1+\beta)} = (a/b)^{1/(1+\beta)}.(w/r)^{1/(1+\beta)}$. Then, the derivative of this ratio with respect to w/r (the marginal function) is $d(k^*/l^*)/d(w/r) = (a/b)^{1/(1+\beta)}.[1/(1+\beta)].(w/r)^{-\beta/(+\beta)}$. Similarly, the denominator (the average function) is $(k^*/l^*)/(w/r) = (a/b)^{1/(1+\beta)}.(w/r)^{-\beta/(1+\beta)}$. Therefore, the ratio of the marginal function to the average function is $\sigma = [d(k^*/l^*)/d(w/r)]/[k^*/l^*/w/r] = \{(a/b)^{1/(1+\beta)}.[1/(1+\beta)].(w/r)^{-\beta/(1+\beta)}\}/\{(a/b)^{1/(1+\beta)}.(w/r)^{-\beta/(1+\beta)}\} = 1/(1+\beta)$. This was the result that we used in example 2(iv) in Section 3.7.4. Notice that we used there $a = \alpha$ and $b = 1 - \alpha$.

4.4.12 Exercises

1. Use the substitution method to solve the following constrained optimization problems, all subject to $x_1 + x_2 = 2$:
    (i) $U = f(x_1, x_2) = x_1^2 + x_2^2$; (ii) $U = f(x_1, x_2) = x_1^2x_2^2$; (iii) $U = f(x_1, x_2) = x_1^2 + 10x_2^2 + 10$.
2. Use the Lagrangian method to solve the following constrained optimization problems, all subject to $x_1 + x_2 + x_3 = 3$:
    (i) $U = f(x_1, x_2, x_3) = x_1^2 + x_2^2 + x_3^2$; (ii) $U = f(x_1, x_2, x_3) = x_1^2x_2^2x_3^2$; (iii) $U = f(x_1, x_2, x_3) = -x_1 - x_2 - x_3$.
3. Use the Lagrangian method to solve the following constrained optimization problems, all subject to $x_1 + x_2 + x_3 = 3$ and $x_1 + x_2 + x_3 = 6$:
    (i) $U = f(x_1, x_2, x_3) = x_1^2 + x_2^2 + x_3^2$; (ii) $U = f(x_1, x_2, x_3) = x_1^2x_2^2x_3^2$; (iii) $U = f(x_1, x_2, x_3) = -x_1^2 - x_2^2 - x_3^2$.

4. *Application exercise.* Assume that a consumer's total utility, U, from the consumption of two goods, x_1 and x_2, is given by the *Cobb–Douglas utility function* $U = f(x_1, x_2) = 10x_1^{0.7} . x_2^{0.3}$. Also assume that the price of one unit of x_1 is \$4, the price of one unit of x_2 is \$5, and the money available for spending is \$20. Find the quantities of x_1 and x_2 that maximize the consumer's total utility.

5. *Application exercise.* Suppose that a firm's total output, q, from the employment of two factors, capital, k, and labor, l, is given by the Cobb–Douglas production function $q = f(k, l) = 10k^{0.6}l^{0.4}$. Also suppose that the price of one unit of k is \$0.05, the price of one unit of l is \$4, and the money available for spending is \$100. Find the quantities of capital and labor that maximize the firm's output.

6. *Application exercise.* Suppose that the total cost, C, of producing $q_0 = 100$ units of output by a firm is given by the function $C = f(k, l) = 0.2k + 5l$, where k and l denote capital and labor, respectively. Also suppose that the firm's production function is given by the Cobb–Douglas form $g(k, l) = 10k^{0.4}l^{0.6} = q_0$. Find the quantities of capital and labor that minimize the total cost to the firm.

7. *Application exercise.* Suppose that the total cost, C, of producing $q_0 = 1000$ units of output by a firm is given by the function $C = f(k, l) = 0.2k + 5l$, where k and l denote capital and labor, respectively. Also suppose that the firm's production function is given by the CES form $g(k, l) = [0.5k^{0.5} + 0.5l^{0.5}]^{-1/0.5} = q_0$. Find the quantities of capital and labor that minimize the total cost to the firm.

 Web supplement: S4.4.12 Mathematica applications

5 Linear programming

5.1 Introduction

Ever since economics became a systematic branch of knowledge, economists have been concerned with, among other issues, devising general rules for optimization of functions representing economic relationships. This quest culminated in the development of so-called *neoclassical marginalism*, which was based mainly on differential calculus. This was the topic that we introduced in the last chapter.

However, the method of neoclassical marginalism based on differential calculus possesses two perceptible limitations. The reader might have noticed that all the constrained optimization problems we dealt with in the last chapter had constraints in the form of linear equalities. This was due to the restriction that, when an economic agent attains an optimum, the agent is assumed to expend the available resources completely. This means that the agent, given the agent's preferences, is assumed to choose a particular point *on* the budget line or the isocost line. This also means that the constraints in the optimization problem are binding. The fact is that, in many real-world cases, an agent may sometimes attempt to optimize the objective function without expending the available resources completely. This is the case of constrained optimization when the constraints are not binding. The second limitation is that the classical approach, as we saw in the last chapter, may yield negative critical or optimal values for the choice variables. However, these negative optimal values are inadmissible in the subjects of our interest.

An important question, therefore, is: can the methods of constrained optimization we introduced in the last chapter be used to find the optima when the constraints are not binding or when the constraints are *linear inequalities* and when the optimal values of the choice variables have to be nonnegative? The answer is, obviously, no. Then the question is: how can one find the optima of problems if the constraints are linear inequalities and if the optimal values of the choice variables are to be nonnegative? Here lies the significance of the *linear programming* (LP) approach to optimization. We have already presented all the prerequisites for a comprehensive discussion of the LP approach. These prerequisites include the solution and geometric representation of linear inequalities in Chapter 1 and linear algebra in Chapter 2.

Even before the development of the LP technique, economists had recognized optimization problems that were similar to LP problems. However, it was the needs of the United States Air Force that stimulated the development of the LP technique. During the Second World War the United States Air Force needed a more effective and efficient way of allocating resources. This led to the development of the LP technique in 1947 by George B. Dantzig who was a member of the Air Force.

The development of the LP technique has been dubbed one the most significant scientific developments of the twentieth century. There have been remarkable theoretical developments related to LP. And there have been extraordinary developments in the applications of the LP technique. One must admit that, in the history of mathematics, only very rarely has a new mathematical technique found such a wide range of practical applications as the LP technique. It has been applied in many branches of knowledge, including the subjects of our interest. Therefore, it is important that students of economics, business, and finance, which make heavy use of LP, have a good knowledge of this highly valuable technique.

We begin this chapter with a definition of LP and an introduction to concepts that are used in the solution of LP problems. We will then explore the graphical method of solving LP problems, and then the more general LP method called the *simplex method*. Subsequently, we will consider duality and sensitivity analyses. We will also explore two of the extensions of LP problems called the *transportation problem* and *assignment problem*.

5.2 LP: introduction

5.2.1 LP: definition, concepts, and assumptions

One can define LP as a technique that, given a linear objective function in n variables and m linear inequality constraints with the same variables, helps find nonnegative values of these variables which will satisfy the constraints and optimize the objective function. This definition suggests that the LP technique employs a mathematical model to represent the optimization problem. It also suggests that the objective function and the constraints are expressed as *linear functions*. It should be noted that the word "programming" in LP does not connote computer programming. Instead, it stands for "planning"; in other words, for "planning of activities."

The above definition of the LP technique implies that a LP problem possesses three important ingredients. The first ingredient is that, as in the classical approach, any LP problem has an objective function to be optimized. One difference between the *classical optimization approach* and the *LP optimization approach* is that in the former the objective function *may or may not* be linear, while in the latter it is strictly linear. The second important ingredient of a LP problem is that, again like a classical constrained optimization problem, it involves one or more constraints. But, the constraints in the former are, like the objective function, also linear. Moreover, these constraints are in the form of inequalities, while they are equalities in classical constrained optimization problems. The third important ingredient of the LP technique is that it, in most cases, gives the optimal solution(s) for the choice variables (also called *decision variables*). But, a condition imposed is that the optimal values of the choice variables be nonnegative. Although classical optimization may also give an optimal solution to a constrained optimization problem, it does not impose the just mentioned *nonnegativity constraint* or *nonnegativity condition*.

Let us now explain some of the concepts that are normally found in the literature on LP. Notice that a clear comprehension of the meanings of some of these concepts often requires specific examples, which will be presented shortly. We begin with the concept of *solution*. Any set of values that solve a SSLEs is called a solution. Notice that a SSLEs may, sometimes, possess *multiple solutions*. The sets of values that solve a SSLEs are called *feasible solutions* or *feasible points*. A plane that represents the set of all feasible solutions or feasible points is called the *feasible region*. A feasible region may be either a *bounded feasible region* or an *unbounded feasible region*. Similarly, a feasible region may be either an *empty feasible region* or a *nonempty feasible region*.

Notice that when *n* (the number of variables) > *m* (the number of constraints) in the system, the set of feasible points may contain infinite points. However, the number of *extreme points* or *corner points* in the set of feasible points will be finite. An extreme point in a convex set is a point in the set which does not lie on a line segment that connects two other interior points of the same set. All these extreme points[1] are called *basic solutions*. Some of these basic solutions may violate one or more of the constraints. A subset of these basic solutions is called *basic feasible solutions* (BFSs) as they do not violate any of the constraints. The aim in solving a LP problem is to choose the best of all the basic feasible solutions; that is, to choose that basic feasible solution which gives the optimum value of the objective function. Such a solution is also called an *optimum feasible solution* or *optimal solution*. We will use the extreme-point theorem introduced in Section 4.2.6, which states that if an optimal solution of the objective function exists, it will be located at one of the extreme points or corner points of a convex set.

Let us now consider the *assumptions of the LP* technique, which are, in fact, implicit in the definition of LP. The first is the *proportionality* assumption. The assumption means that the change in the objective function and in the LHS of the constraints is proportional to the values of the decision variable(s). This assumption rules out the possibility of exponents for the decision variables other than one, and, thus, ensures that both the objective function and the LHS of the constraints are linear. The second assumption is *additivity*. Additivity in LP means that both the objective function and the LHS of the constraints are expressed as the sum of individual contributions of the decision variables. This assumption rules out the possibility of decision variables occurring as cross products. The third, and the last, assumption is called *divisibility*. This assumption implies that the decision variables in a LP problem may take any value, not just integral or discrete values.

5.2.2 Illustrative LP example: maximization problem with two variables

We now present an illustrative example of a LP problem with two variables. The example we present here is that of a maximization problem. We will present the example of a minimization problem in the next section.

For the purpose of exposition and for simplicity, a slightly modified form of example 4 in Section 1.7.7. For convenience we reproduce the example here. Assume that a firm uses two types of inputs *A* and *B* in order to produce two goods *x* and *y* (which are the decision or choice variables in the example). Also assume that to produce 1 unit of *x* the firm has to use 2 units of input *A* and 3 units of input *B*, and to produce 1 unit of *y* it has to use 3 units of input *A* and 2 units of input *B*. Assume again that the maximum amounts of inputs *A* and *B* available are 90 and 120 units, respectively, and that the unit profits from *x* and *y* are $2 and $2, respectively. We now impose the condition that the quantities of the two goods produced must be greater than or equal to zero because negative quantities of the goods are inadmissible. The firm's aim is to maximize its profits from the production of the two goods while still complying with the input or resource constraints. A problem such as this is called a LP maximization problem. How can the firm achieve its objective of maximizing its profits given the constraints? Before we solve this problem, we summarize the above information in Table 5.2.1 for visual clarity.

The next step is to convert the information in Table 5.2.1 into inequalities and equations. The last row of the table shows the unit profits from the production of the two goods.

Table 5.2.1

	Output x requirement	*Output y requirement*	*Input availability*
Input A	2	3	90
Input B	3	2	120
	Profit per unit of x	Profit per unit of y	
	$2	$2	

Therefore, the total profit, Π, can be written as

$$\Pi = 2x + 2y \tag{5.2.1}$$

which is called the objective function of the given LP problem. Notice that the total profit in equation (5.2.1) is proportional to the values of the two goods and is the sum of the individual contributions of the two goods as stated in the proportionality and additivity assumptions. Notice also that equation (5.2.1) can be written as $2y = \Pi - 2x$ or as

$$y = (\Pi/2) - x \tag{5.2.2}$$

which suggests that y is a linear function of x (hence the name "linear" programming) with $(0,\ \Pi/2)$ as the vertical intercept, $(\Pi/2,\ 0)$ as the horizontal intercept, and -1 as the slope.

Let us now convert the information in Table 5.2.1 into constraints. Since there are upper limits for the availability of inputs, the total use of any particular input must be less than or equal to the available quantity of that input. Therefore the constraints can be written as

$$2x + 3y \le 90 \quad \text{and} \quad 3x + 2y \le 120 \tag{5.2.3}$$

which are called *technical constraints* determined by the existing state of technology and the availability of inputs. Notice that the constraint functions also obey, as does the objective function in equation (5.2.1), the proportionality and additivity assumptions. Another constraint we mentioned in the presentation of the problem above is

$$x, y \ge 0 \tag{5.2.4}$$

which is called the nonnegativity constraint. This constraint precludes negative or unacceptable values from the solutions of the choice variables (x and y). It is more convenient to state the objective function and the constraints of the optimization problems together as

$$\text{Maximize } \Pi = 2x + 2y, \text{ subject to } 2x + 3y \le 90, 3x + 2y \le 120, \text{ and } x, y \ge 0 \tag{5.2.5}$$

Assume now that we treat the first two inequalities above as equalities and solve them for y to obtain

$$y = 30 - (2/3)x \quad \text{and} \quad y = 60 - (3/2)x \tag{5.2.6}$$

Notice that the first function in equation (5.2.6) has vertical intercept $(0, 30)$ and horizontal intercept $(45, 0)$. Similarly, the second function in equation (5.2.6) has vertical intercept $(0, 60)$ and horizontal intercept $(40, 0)$.

Table 5.2.2

	Dish x requirements	Dish y requirements	Minimum vitamin requirements
Vitamin A	15	10	90
Vitamin B	10	15	90
	Cost per unit of x	Cost per unit of y	
	$9	$12	

5.2.3 Illustrative LP example: minimization problem with two variables

We now present an illustrative minimization example of a LP problem with two variables. Assume that an order received by a catering firm contains two types of dishes, x and y. Also assume that each unit of dish x must contain at least 15 units of vitamin A and 10 units of vitamin C; and each unit of dish y must contain at least 10 units of vitamin A and 15 units of vitamin C. Again assume that the minimum amount of vitamin A in the dishes must be 90 units and the minimum amount of vitamin C must also be 90 units, and that the cost of preparation of one unit of x is $9 and the cost of preparation of one unit of y is $12. We also impose the condition, as before, that the quantities of the two dishes prepared must be greater than or equal to zero because negative quantities of the dishes are unacceptable. The firm's aim is to minimize its total cost while still complying with the vitamin requirements or constraints. A problem such as this is called a LP minimization problem or a *diet problem*. How can the firm achieve its objective of minimizing its costs given the constraints? We present the above information in Table 5.2.2 for visual clarity.

The next step is to convert the information in Table 5.2.2 into inequalities and equations. The last row of the table shows the unit costs to the firm of preparation of the two dishes. Therefore, the total cost, C, can be written as

$$C = 9x + 12y \tag{5.2.7}$$

which is the objective function of the present LP problem. Notice that the total cost in equation (5.2.7) is proportional to the values of the two dishes and is the sum of the individual contributions of the two dishes as stated in the proportionality and additivity assumptions. Notice also that equation (5.2.7) can be written as $12y = C - 9x$ or as

$$y = (C/12) - (9/12)x \tag{5.2.8}$$

which means that y is a linear function of x (hence the name, as before, "linear" programming) with $(0, C/10)$ as the vertical intercept, $(C/5, 0)$ as the horizontal intercept, and $-3/4$ as the slope.

Let us now convert the constraints implied by Table 5.2.2. Since there are lower limits for the requirement of vitamins, the total content of any particular vitamin must be greater than or equal to the minimum units stipulated of the vitamins. Therefore the constraints can be written as

$$15x + 10y \geq 90 \quad \text{and} \quad 10x + 15y \geq 90 \tag{5.2.9}$$

The constraints presented in inequality (5.2.9) are called technical constraints of the minimization problem. Notice that the constraint functions also obey, as does the objective function in equation (5.2.7), the proportionality and additivity assumptions. Another constraint we mentioned in the presentation of the problem above is

$$x, y \geq 0 \qquad (5.2.10)$$

which is called, as in the case of the maximization problem, the nonnegativity constraint. This constraint precludes negative or unacceptable values from the solutions of the decision variables (x and y). As earlier, it is more convenient to state the objective function and the constraints of the minimization problems together as

$$\text{Minimize } C = 9x + 12y, \text{ subject to } 15x + 10y \geq 90, \, 10x + 15y \geq 90, \text{ and } x, y \geq 0$$
$$(5.2.11)$$

Assume now, as in the maximization problem, that we treat the first two inequalities above as equalities and solve them for y to obtain

$$y = 9 - (3/2)x \quad \text{and} \quad y = 6 - (2/3)x \qquad (5.2.12)$$

Notice that the first function in equation (5.2.12) has vertical intercept (0, 9) and horizontal intercept (6, 0). Similarly, the second function in equation (5.2.12) has vertical intercept (0, 6) and horizontal intercept (9, 0).

5.2.4 Solution of LP problems: maximization problem with graphical method

There are two popular methods to solve LP problems. They are the *graphical method* or *geometric method*, and the *simplex method*. The graphical method is useful when there are only two decision variables in the optimization problem and the simplex method is useful when the problem involves more than two decision variables. Since we are concerned with LP problems that involve only two decision variables in the present section, we use the graphical method here. The simplex method is introduced in Section 5.3.

Consider our maximization problem condensed in equation (5.2.5). We presented the constraints in functional forms in equation (5.2.6). These functions are $y = 30 - (2/3)x$ and $y = 60 - (3/2)x$. The graphs of these functions are illustrated in Figure 5.2.1(A).

Notice that there are six corner points (A, B, C, D, E, and F) in Figure 5.2.1(A). They are all called the basic solutions. But, point E violates the constraint $2x + 3y \leq 90$ and point F violates the constraint $3x + 2y \leq 120$. Corner points A, B, C, and D do not violate any of the constraints and, therefore, they are the feasible solutions in the example. The set of all the points in the shaded region ABCD is the feasible set, and the space represented by the feasible set is the feasible region. According to extreme point theorem, if an optimal solution occurs in the present example it would occur at one of the above feasible solutions or corner points. Therefore, we need only check the coordinates of the corner points to know whether there exists an optimal solution or not. To do this, we illustrate the feasible region along with the graphs of the transformed objective function (equation (5.2.2)), for different levels of profit Π_i, in Figure 5.2.1(B). These levels of profit are represented by the dashed lines.

We now attempt to find the optimal solution. But before this we need to find the coordinates corresponding to corner points A, B, C, and D. We know the coordinates of

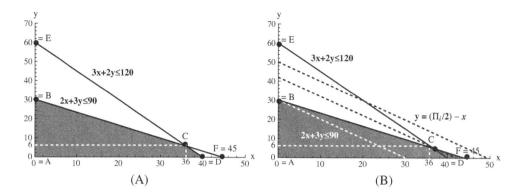

Figure 5.2.1

A, B, and D are (0, 0), (0, 30), and (40, 0), respectively. The coordinates corresponding to point C can be found by solving the two equations $3x + 2y = 120$ and $2x + 3y = 90$ simultaneously. This yields the coordinates corresponding to point C as (36, 6). We can now find the optimal solution. To find this, we can substitute the coordinate values of the two variables into the profit function (equation (5.2.1)). This yields

Corner point A: $\Pi = 2x + 2y = 2 \times 0 + 2 \times 0 = 0$;

corner point B: $\Pi = 2x + 2y = 2 \times 0 + 2 \times 30 = 60$;

corner point C: $\Pi^* = 2x^* + 2y^* = 2 \times 36 + 2 \times 6 = 84$;

corner point D: $\Pi = 2x + 2y = 2 \times 36 + 2 \times 0 = 72$ 　　　　(5.2.13)

which show that profit is the highest at corner point C. This means that the optimal solution is $(x^*, y^*) = (36, 6)$, or the firm must produce 36 units of x and 6 units of y to maximize its profit and the maximum profit will be $\Pi^* = \$84$. It can be verified that this optimal solution satisfies the inequalities in (5.2.3). Since $x^* = 36 > 0$ and $y^* = 6 > 0$, the solution satisfies the nonnegativity constraint given in inequality (5.2.4).

Notice that in Figure 5.2.1(B) we have drawn three *isoprofit lines* (dashed lines) for three different levels of profits ($\Pi = 60$, $\Pi = 84$, and $\Pi = 120$). An isoprofit line represents the loci of different combinations of the two goods that give the same level of profit. One can draw many isoprofit lines that touch one or more of the corner points. Since the present problem is a maximization problem, our aim is to choose the highest isoprofit line that touches any of the corner points of the feasible region. In Figure 5.2.1(B), there is only one such isoprofit line and that passes through point C. This implies that there is only one optimal solution in the present example, and it occurs at the corner point C. This was exactly the result we obtained above.

5.2.5 Solution of LP problems: minimization problem with graphical method

Consider our minimization problem condensed in equation (5.2.11). We presented the constraints of the problem in functional forms in equation (5.2.12). These functions are $y = 9 - (3/2)x$ and $y = 6 - (2/3)x$. The graphs of these functions are illustrated in

Figure 5.2.2(A). Notice that there are six corner points (A, B, G, D, E, and F) in the figure. They are all called the basic solutions. But, point B violates the constraint $15x + 10y \geq 90$, point D violates the constraint $10x + 15y \geq 90$, and point A violates both constraints. Corner points E, F, and G do not violate any of the constraints and, therefore, they are the feasible solutions in the example. The set of all the points in the shaded region on or above the line EGF is the feasible set, and the space represented by the feasible set is the feasible region. According to extreme point theorem, if an optimal solution occurs in the present example it would occur at one of the above feasible solutions or corner points. Therefore, we need only check the coordinates of these corner points to know whether there exists an optimal solution or not. To do this, we illustrate the feasible region along with the graphs of the transformed objective function (equation (5.2.8)), for different levels of cost C_i, in Figure 5.2.2(B). These levels of cost are represented by the dashed lines.

We now attempt to find the optimal solution. But before this, as in the case of the maximization problem, we need to find the coordinates corresponding to corner points E, F, and G. We know that the coordinates of E and F are (0, 9) and (9, 0), respectively. The coordinates corresponding to point G can be found by solving the two equations $15x + 10y = 90$ and $10x + 15y = 90$ simultaneously. This yields the coordinates corresponding to point C as (3.6, 3.6). We can now find the optimal solution. To find this, we can substitute the coordinate values of the two variables into the cost function (equation (5.2.7)). This yields

Corner point E : $C = 9x + 12y = 9 \times 0 + 12 \times 9 = 108$;

corner point F : $C = 9x + 12y = 9 \times 9 + 12 \times 0 = 81$;

corner point G : $C^* = 9x^* + 12y^* = 9 \times 3.6 + 12 \times 3.6 = 75.6$ $\qquad$ (5.2.14)

Equation (5.2.14) shows that the cost is the lowest at corner point G. This means that the optimal solution is $x^* = 3.6$ and $y^* = 3.6$, or the firm must prepare 3.6 units of x and 3.6 units of y to minimize its cost and the minimum cost will be $C^* = \$75.6$. It can be verified that this optimal solution satisfies the inequalities in (5.2.9). Since $x^* = 3.6 > 0$ and $y^* = 3.6 > 0$, the solution satisfies the nonnegativity constraint given in inequality (5.2.10).

Notice that in Figure 5.2.2(B) we have drawn three isocost lines (dashed lines) for three different levels of costs ($C = 50$, $C = 75.6$, and $C = 90$). An isocost line represents the loci of different combinations of the two dishes that cost the firm equally. One can draw

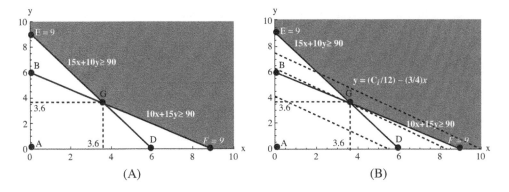

(A)　　　　　　　　　　　　　　(B)

Figure 5.2.2

many isocost lines that touch one or more of the corner points. Since the present problem is a minimization problem, our aim is to choose the lowest isocost line that touches any of the corner points of the feasible region. In Figure 5.2.2(B), there is only one such isocost line and that passes through point G. This implies that there is only one optimal solution in the present example, and it occurs at the corner point G. This was exactly the result we obtained above.

We now summarize the steps in the graphical method of finding the solution to a two-variable LP problem.

Step 1. Set up the equation of the objective function and the inequalities of the constraints based on the verbal or tabular information.

Step 2. Convert the objective function and inequalities into equations and then solve for one of the variables.

Step 3. Graph the equations in step 2 in a figure.

Step 4. Determine the feasible points.

Step 5. Find the coordinates of the feasible points.

Step 6. Substitute these coordinates in the objective function in step 1.

Step 7. The maximum (minimum) of the objective function will be the highest (lowest) value in step 6. This result can also be obtained by drawing the isoprofit (isocost) lines and choosing the highest (lowest) isoprofit (isocost) line that touches a corner point.

5.2.6 Additional examples

In this section we consider an additional example each of a maximization and minimization LP problem. Let us first consider a hypothetical maximization problem. Suppose that a firm wants to produce two goods x and y using three factors 1, 2, and 3 whose maximum available quantities are 30, 20, and 45 units, respectively. Also suppose that the production of one unit of good x requires 4, 2, and 3 units of factors 1, 2, and 3, respectively; and that the production of one of unit of good y requires 2, 2, and 9 units of factors 1, 2, and 3, respectively. Suppose again that the firm can obtain $10 and $20 profit from each unit of x and y, respectively. How many units of the two goods should the firm produce in order to maximize its total profit and what will be its total profit? For convenience, we summarize the above verbal presentation of the problem in Table 5.2.3.

Let us follow the steps outlined at the end of the last section. The first step stipulates that we set up the functions representing the objective function and the inequalities. These can be

Table 5.2.3

	Output x requirement	Output y requirement	Input availability
Factor 1	4	2	30
Factor 2	2	2	20
Factor 3	3	9	45
	Profit per unit of x	Profit per unit of y	
	$10	$20	

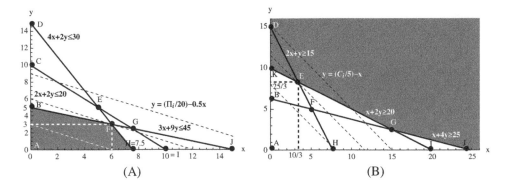

Figure 5.2.3

condensed as

Maximize $\Pi = 10x + 20y$, subject to $4x + 2y \le 30, 2x + 2y \le 20,$

$3x + 9y \le 45$, and $x, y \ge 0$ (5.2.15)

The second step stipulates that we convert the inequalities in the constraints in problem (5.2.15) into equations and then express the equations in terms of one of the variables (y). Then, expressing the equations and the objective function in terms of one of the variables, we obtain

$$y = (\Pi/20) - (1/2)x, \quad y = 15 - 2x, y = 10 - x, \quad \text{and} \quad y = 5 - (1/3)x \quad (5.2.16)$$

As the third step, we now graph the functions in equations (5.2.16), as shown in Figure 5.2.3(A). Notice that unlike Figures 5.2.1 and 5.2.2, we have plotted the graphs of both the objective function and the constraints together in Figure 5.2.3(A). The dashed lines represent isoprofit lines (the graphs of the transformed objective function $y = (\Pi_i/20) - 0.5x$) for different levels of profit ($\Pi = 40$, $\Pi = 58.33$, and $\Pi = 75$).

The next step is to determine the feasible points. Notice that out of the corner points in Figure 5.2.3(A), points C, D, E, G, I, and J violate one or more constraints. Corner points A, B, F, and H do not violate any of the constraints and, therefore, their set constitutes the set of feasible solutions. The set of points in the shaded area ABFH is the set of feasible points, and that area is the feasible region in the present example. Since the corner points B, F, and H constitute the set of feasible solutions, we only need to check the coordinates of these corner points. The coordinates of points B and H are (0, 5) and (7.5, 0), respectively. We can obtain the coordinates of point F (by solving equations $4x + 2y = 30$ and $3x + 9y = 45$ simultaneously) as (6, 3). We can now substitute, as the fourth step, the above coordinates in the objective function to obtain

Point B: $\Pi = 10x + 20y = 10 \times 0 + 20 \times 5 = 100$;

point F: $\Pi^* = 10x^* + 20y^* = 10 \times 6 + 20 \times 3 = 120$;

point H: $\Pi = 10x + 20y = 10 \times 7.5 + 20 \times 0 = 75$ (5.2.17)

Table 5.2.4

	One unit of x requires	One unit of y requires	Minimum vitamin requirements
Vitamin A	2	1	15
Vitamin C	1	4	25
Vitamin D	1	2	20

	Cost per unit of x	Cost per unit of y
	$5	$5

As can be seen from equation (5.2.17), the total profit of the firm will be maximum if the firm produces 3 units of y and 6 units of x, and the maximum total profit is $120. This result can be verified if we use the isoprofit lines as shown in Figure 5.2.3(A). Since the present problem is a maximization problem, we choose the highest possible isoprofit line that touches at least one corner point of the feasible region. This happens only at the corner point F. The coordinates of point F are (6, 3). If we substitute these values in the objective function we obtain total profit equal to $120.

Let us now consider a diet problem as a hypothetical minimization problem. Assume that the diet prescribed to a person by a dietician contains dishes x and y, and these dishes must contain vitamin A, vitamin C, and vitamin D. Also assume that one unit of dish x must contain 2, 1, and 1 units of vitamins A, C, and D, respectively; and one unit of dish y must contain 1, 4, and 2 units of vitamins A, C, and D respectively. The dietician also prescribed that the minimum quantities in the two dishes of vitamins A, C, and D must be 15, 25, and 20 units, respectively. Assume again that the unit cost of dishes x and y containing the prescribed units of vitamins is the same $5. Find the quantities of the two dishes, still complying with the vitamin requirements, that minimize the person's total expenditure.

We first set up Table 5.2.4 based on the above information, and follow the steps outlined at the end of the last section. Based on Table 5.2.4 we can set up the condensed mathematical form of the minimization problem as

$$\text{Minimize } C = 5x + 5y, \text{ subject to } 2x + y \geq 15, \ x + 4y \geq 25, \ x + 2y \geq 20, \text{ and } x, y \geq 0$$
$$(5.2.18)$$

We now convert the inequalities in problem (5.2.18) into equations and solve both the converted inequalities and the objective function to obtain

$$y = (C/5) - x, y = 15 - 2x, y = (25/4) - 0.25x, \text{ and } y = 10 - 0.5x \quad (5.2.19)$$

We can now graph the functions in equation (5.2.19), as shown in Figure 5.2.3(B). As in the last example, we have plotted the graphs of both the objective function and the constraints. The dashed lines represent the isocost lines (the graph of the transformed objective function $y = (C_i/5) - x$) for different levels of expenditures (i.e. $C = 75$, $C = 58.33$, and $C = 40$).

Let us now determine the feasible points. Notice that out of the corner points in Figure 5.2.3(B), points A, B, K, F, H, and I violate one or more constraints. Corner points D, E, G, and J do not violate any of the constraints and, therefore, their set constitutes the set of feasible solutions. The set of points in the area on or above the line DEGJ is the set of feasible points and that shaded area is the feasible region in the present example. Since the corner points D, E, G, and J constitute the set of feasible solutions, we need only check the coordinates of these corner points. The coordinates of points D and J are (0, 15) and (25, 0), respectively. We can obtain the coordinates of points E (by solving equations $2x + y = 15$ and $x + 2y = 20$ simultaneously) and G (by solving equations $x + 4y = 25$ and $x + 2y = 20$ simultaneously) as (10/3, 25/3) and (15, 2.5), respectively.

We can now substitute the above coordinates into the objective function (in equation (5.2.18)) to obtain

Point D: $C = 5x + 5y = 5 \times 0 + 5 \times 15 = 75$;

point E: $C^* = 5x^* + 5y^* = 5 \times (10/3) + 5 \times (25/3) = 58.33$;

point G: $C = 5x + 5y = 5 \times 15 + 5 \times 2.5 = 87.5$;

point J: $C = 5x + 5y = 5 \times 25 + 5 \times 0 = 75$ \hfill (5.2.20)

As can be seen from equation (5.2.20), the total expenditure of the person will be minimum if they buy 10/3 units of x and 25/3 of y, and the minimum expenditure will be \$58.33. This result can be verified if we use the isocost lines as in Figure 5.2.3(B). Since the present problem is a minimization problem, we choose the lowest possible isocost line that touches at least one corner point of the feasible region. This happens only at the corner point E. The coordinates of point E are, as we found above, (10/3, 25/3). If we substitute these values in the objective function we obtain total expenditure equal to \$58.33.

5.2.7 *Slack and surplus variables and LP problems*

Consider, for simplicity, our maximization problem (5.2.5). We reproduce that problem here for convenience. The problem is to maximize $\Pi = 2x + 2y$, subject to the constraints $2x + 3y \leq 90$, $3x + 2y \leq 120$, and $x, y \geq 0$. Notice that the LHSs of the first two of these inequalities represent the amounts of the resources actually used and the RHSs show the available quantities of the two resources. Since the inequalities are "less than or equal to" inequalities, there are chances that the available quantities of the resources may not be used completely. Suppose that the differences, if any, between the two in the cases of the first and second inequalities are denoted by the variables s_1 and s_2, respectively. These variables are called *slack variables*. Since the unused quantity of the resources cannot be negative, slack variables are always greater than or equal to zero (or, $s_1, s_2 \geq 0$).

We know that it is always easier to work with equalities than with inequalities. Since the slack variables represent unused quantities of the two resources in the present example, they should not make any impact on the objective (profit) function, $\Pi = 2x + 2y$. Because of this fact they appear in the objective function with zero coefficients. Therefore we can write the *augmented objective function*, augmented by including the slack variables and their zero coefficients, as

Maximize $\Pi = 2x + 2y + 0s_1 + 0s_2$ \hfill (5.2.21)

We now augment the constraints in the problem by including the slack variables. This will give us *augmented constraints*. But, when we do this, we *add* the slack variables to the LHS of the inequalities so that the inequalities become equalities. Therefore, the augmented constraints can be written as

$$\text{Subject to } 2x + 3y + s_1 = 90 \quad \text{and} \quad 3x + 2y + s_2 = 120 \tag{5.2.22}$$

Notice that the nonnegativity constraint now becomes $x, y, s_1, s_2 \geq 0$. Notice also that the augmented problem represented by equations (5.2.21) and (5.2.22) cannot be solved by the graphical method we introduced earlier, because it now contains more than two variables. Therefore, we defer the solution of the augmented problem to Section 5.3. We will show there that addition of slack variables in the problem will not alter the optimal solution.

Now consider our minimization problem (5.2.11). The objective in this problem was to minimize $C = 5x + 10y$ subject to constraints $15x + 10y \geq 90$, $10x + 15y \geq 90$, and $x, y \geq 0$. Notice that the LHSs of the first two of these inequalities represent the amounts of the vitamins actually contained in the two dishes and the RHSs show the minimum quantities of the vitamins required. There are chances that the quantities of the vitamins contained in the dishes may be more than the minimum requirements. Suppose that the differences, if any, between the two in the cases of the first and second inequalities are denoted by the variables s_3 and s_4, respectively. These variables are called *surplus variables*. Since the excess quantities of the vitamins contained in the dishes cannot be negative, the surplus variables, like the slack variables, are always greater than or equal to zero ($s_3, s_4 \geq 0$).

Since the surplus variables represent the excess quantities of the two vitamins in the present example, they should not make any impact on the objective function, $C = 5x + 10y$. Because of this fact, the surplus variables, like the slack variables, appear in the objective function with zero coefficients. Therefore, we can write the augmented objective function, augmented by including the surplus variables and their zero coefficients, as

$$\text{Minimize } C = 5x + 10y + 0s_3 + 0s_4 \tag{5.2.23}$$

Let us now augment the constraints in the problem by including the surplus variables. But, when we do this, we must *subtract* the surplus variables from the LHSs of the inequalities so that the inequalities become equalities. Therefore, the augmented constraints can be written as

$$\text{Subject to } 15x + 10y - s_3 = 90 \text{ and } 10x + 15y - s_4 = 90 \tag{5.2.24}$$

and the nonnegativity constraint becomes $x, y, s_3, s_4 \geq 0$. Since it involves four variables, like the augmented maximization problem, the augmented minimization problem represented by equations (5.2.23) and (5.2.24) cannot be solved by the graphical method we introduced earlier. Therefore, we defer the solution of the augmented minimization problem to Section 5.3. We will show in Section 5.3 that the subtraction of surplus variables will not alter the optimal solution.

5.2.8 LP problems: some special cases

LP problems, in most cases, yield unique optimal solutions. However, there are few special LP problems that yield *empty feasible region* or no optimal solution, *unbounded feasible region* or *unbounded solution*, *multiple optimal solutions*, and *degeneracy* or *degenerate BFS*.

Let us begin with the case of empty feasible region. As an illustration, consider the problems in the following equation:

Maximize $\Pi = 5x + 6y$, subject to $4x + 2y \leq 10, 4x + 2y \geq 20$, and $x, y \geq 0$

Minimize C $= 4x + 6y$, subject to $4x + 2y = 10, 4x + 2y \geq 20$, and $x, y \geq 0$ (5.2.25)

The graphs of the associated constraints are illustrated in Figures 5.2.4(A) and (B), respectively. Figure 5.2.4(A) shows the graphs of constraints in the maximization problem in (5.2.25). Notice that the feasible solutions that satisfy the constraint $4x + 2y \leq 10$ are given by the set of points lying on or below the lower line. Similarly, the feasible solutions that satisfy the constraint $4x + 2y \geq 20$ are given by the set of points lying on or above the upper line. Notice also that no point in this figure can satisfy both constraints simultaneously. Therefore, there is no optimal solution.

Figure 5.2.4(B) shows the graphs of constraints in the minimization problem in (5.2.25). Notice that the feasible solutions that satisfy the constraint $4x + 2y \geq 20$ are given by the set of points lying on or above the upper line. Similarly, the feasible solutions that satisfy the equality constraint $4x + 2y = 10$ are given by the set of points lying on the lower line. Notice also that, as in the maximization problem, no point in this figure can satisfy both constraints simultaneously. Therefore, again as in the maximization problem, there is no optimal solution in the minimization problem.

The second case is that of unbounded solution or of unbounded feasible region. To illustrate this, consider the example

Maximize $\Pi = x + y$, subject to $y = 4$ and $x, y \geq 0$ (5.2.26)

The graph of the constraint in problem (5.2.26) is illustrated in Figure 5.2.5. Notice that the feasible region in the present problem is the line $y = 4$ in Figure 5.2.5. It is easy to see that this feasible region is unbounded. If we substitute the constraint $y = 4$ in the objective function, we obtain $\Pi = 4 + x$. This equation implies that Π will increase without bounds as x increases. This also implies that the feasible region is unbounded. Therefore, no feasible point maximizes the objective function and no unique optimum solution exists; specifically, the solution is unbounded. Notice also that the present problem would have an optimum solution at $y = 4$ if the problem were a minimization problem. In that case we would have $x = 0$.

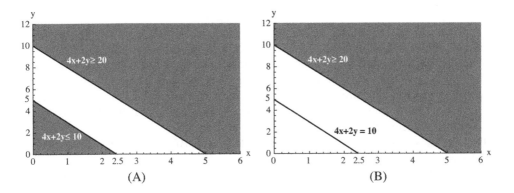

Figure 5.2.4

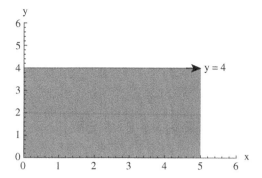

Figure 5.2.5

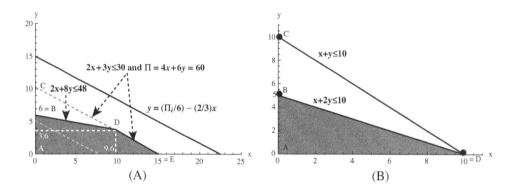

Figure 5.2.6

The third case is that of multiple optimal solutions. This happens when the objective function and the LHS of one (or more) of the constraints are linearly dependent. As an illustration, consider the problem

$$\text{Maximize } \Pi = 4x + 6y, \text{ subject to } 2x + 8y \le 48, 2x + 3y \le 30, \text{ and } x, y \ge 0 \quad (5.2.27)$$

If we plot the transformed forms of the constraints and the objective function, we obtain the graphs illustrated in Figure 5.2.6(A). The feasible region in Figure 5.2.6(A) is the set of points on and inside the area ABDE. Notice that the isoprofit line (for $\Pi = 60$) coincides with the constraint $2x + 3y \le 30$. This implies that there is no unique optimal solution; instead, the set of coordinate points from D to E (including D and E) constitutes the optimal solution. In other words, the problem has multiple solutions and each solution satisfies the constraints as well as the objective function for $\Pi = 60$.

The last case we consider here is degeneracy or degenerate BFS. Degeneracy or degenerate BFS happens when the number of positive choice variables in the optimal solution is less than the number of the constraints. As an illustration, consider the maximization problem

$$\text{Maximize } \Pi = 25x + 15y, \text{ subject to } x + y \le 10, x + 2y \le 10, \text{ and } x, y \ge 0 \quad (5.2.28)$$

Table 5.2.5

Resources	Resource requirement per unit of x_j						Maximum availability of the ith resource
	x_1	x_2	...	x_j	...	x_n	
1	a_{11}	a_{12}	...	a_{1j}	...	a_{1n}	b_1
2	a_{21}	a_{22}	...	a_{2j}	...	a_{2n}	b_2
...	...	...	...	...	...	...	...
i	a_{i1}	a_{i21}	...		...	a_{in}	b_i
...	...	...	...	...	...	...	...
m	a_{m1}	a_{m2}	...	a_{m3}	...	a_{mn}	b_m
Contribution to Π per unit of x_j	π_1	π_2	...	π_j	...	π_n	

The graphs of the constraints in problem (5.2.28) are illustrated in Figure 5.2.6(B). Notice that the feasible region is the triangle ABD. The coordinates of the corner points A, B, and D in the feasible region are (0, 0), (0, 5), and (10, 0), respectively. The values of the objective function corresponding to these corner points can be obtained by substituting their values into the objective function. Therefore, the values of the objective function corresponding to A, B, and D are $\Pi = 25x + 15y = 25 \times 0 + 15 \times 0 = 0$, $\Pi = 25x + 15y = 25 \times 0 + 15 \times 5 = 75$, and $\Pi = 25x + 15y = 25 \times 10 + 15 \times 0 = 250$. This means that the optimal solution in the present example happens at the corner point D, where $x = 10$ and $y = 0$. Since there are two constraints and two choice variables in the problem and since one of the choice variables is equal to zero at the optimal solution ($y = 0$), the number of positive choice variable at the optimal solution is less than the number of constraints. Therefore, the optimal solution in the present problem is an example of a degenerate BFS.

5.2.9 Standard forms of LP problems

We now consider the *standard form* or the *mathematical form* of a LP problem. We first consider the standard form of a maximization problem. Suppose that a firm has to use a_{ij} (where $i = 1, 2, \ldots, m$ and $j = 1, 2, \ldots, n$) units of the ith resource to produce one unit of good x_j, and that the maximum available quantity of the ith resource is b_i. Also suppose that the profit from one unit of good x_j is π_j. We present this information in Table 5.2.5, as we constructed tables earlier in the chapter. We assume here that the number of choice variables (n) is greater than the number of constraints (m).

As we did in the last two sections, we can condense the information in Table 5.2.5 as given in the following equation and inequalities:

$$\text{Maximize } \Pi = \text{ profit } = \pi_1 x_1 + \pi_2 x_2 + \cdots + \pi_j x_j + \cdots + \pi_n x_n \tag{5.2.29}$$

$$\text{Subject to } a_{11}x_1 + a_{12}x_2 + \cdots + a_{1j}x_j + \cdots + a_{1n}x_n \le b_1,$$

$$a_{21}x_1 + a_{22}x_2 + \cdots + a_{2j}x_j + \cdots + a_{2n}x_n \le b_2,$$

$$\ldots, a_{i1}x_1 + a_{i2}x_2 + \cdots + a_{ij}x_j + \cdots + a_{in}x_n \le b_i,$$

$$\ldots, a_{m1}x_1 + a_{m2}x_2 + \cdots + a_{mj}x_j + \cdots + a_{mn}x_n \le b_m,$$

$$\text{and } x_j \ge 0, j = 1, 2, 3, \ldots, n \text{ and } i = 1, 2, \ldots, m \tag{5.2.30}$$

This is called the *longhand form* of a standard LP maximization problem. Notice that the last equation and inequalities can be written in *matrix form* as

Maximize $\Pi = \boldsymbol{\pi}^{\mathrm{T}}\mathbf{x}$ (5.2.31)

Subject to $\mathbf{Ax} \leq \mathbf{b}$ and $\mathbf{x} \geq 0$ (5.2.32)

where $\boldsymbol{\pi}^{\mathrm{T}} = [\ \pi_1 \quad \pi_2 \quad .. \quad \pi_j \quad .. \quad \pi_n\ ]$, $\mathbf{x}^{\mathrm{T}} = [\ x_1 \quad x_2 \quad .. \quad x_j \quad .. \quad x_n\ ]$,

$$\mathbf{A} = \begin{bmatrix} a_{11} & a_{12} & .. & a_{1j} & .. & a_{1n} \\ a_{21} & a_{22} & .. & a_{2j} & .. & a_{2n} \\ .. & .. & .. & .. & .. & .. \\ a_{i1} & a_{i2} & .. & a_{ij} & .. & a_{in} \\ .. & .. & .. & .. & .. & .. \\ a_{m1} & a_{m2} & .. & a_{mj} & .. & a_{mn} \end{bmatrix},$$

and $\mathbf{b}^{\mathrm{T}} = \begin{bmatrix} b_1 & b_2 & .. & b_i & .. & b_m \end{bmatrix}$; or using the sum $(\sum)$ notation as

Maximize $\Pi = \sum\limits_{j=1}^{n} \pi_j x_j$ (5.2.33)

Subject to $\sum\limits_{j=1}^{n} a_{ij}x_j \leq b_i$, and $x_j \geq 0$, where $i = 1, 2, \ldots, m$ and $j = 1, 2, \ldots, n$ (5.2.34)

The above matrix and sigma forms of the problem are called the *shorthand forms* of a standard LP maximization problem. Let us now set up the standard or mathematical form of a minimization problem. Suppose that a person has been prescribed to include e_{ij} (where $i = 1, 2, \ldots, m$ and $j = 1, 2, \ldots, n$) units of the ith nutrient in one unit of dish x_j and that the minimum quantity of the ith nutrient in all the dishes together must be d_i. Also suppose that the cost of one unit of dish x_j is c_j. We present this information in Table 5.2.6, as we constructed Table 5.2.5. We assume here that, as before, the number of choice variables (n) is greater than the number of constraints (m).

Table 5.2.6

Nutrients	Nutrient requirement per unit of x_j					Minimum requirement of ith nutrient
	x_1	x_2	$\ldots$	x_j	$\ldots$	x_n	
1	e_{11}	e_{12}	$\ldots$	e_{1j}	$\ldots$	e_{1n}	d_1
2	e_{21}	e_{22}	$\ldots$	e_{2j}	$\ldots$	e_{2n}	d_2
$\ldots$	$\ldots$	$\ldots$	$\ldots$	$\ldots$	$\ldots$	$\ldots$	$\ldots$
i	e_{i1}	e_{i2}	$\ldots$	e_{ij}	$\ldots$	e_{in}	d_i
$\ldots$	$\ldots$	$\ldots$	$\ldots$	$\ldots$	$\ldots$	$\ldots$	$\ldots$
m	e_{m1}	e_{m2}	$\ldots$	e_{m3}	$\ldots$	e_{mn}	d_m
Contribution to C per unit of x_j	c_1	c_2	$\ldots$	c_j	$\ldots$	c_n	

As we did in the case of the maximization problem, we can condense the information in Table 5.2.6 and write the longhand form of the LP minimization problem as given in the following equation and inequalities:

$$\text{Minimize } C = \text{cost} = c_1 x_1 + c_2 x_2 + \cdots + c_j x_j + \cdots + c_n x_n \tag{5.2.35}$$

$$\text{Subject to } e_{11} x_1 + e_{12} x_2 + \cdots + e_{ij} x_j + \cdots + e_{1n} x_n \geq d_1,$$

$$e_{21} x_1 + e_{22} x_2 + \cdots + e_{2j} x_j + \cdots + e_{2n} x_n \geq d_2, \ldots,$$

$$e_{i1} x_1 + e_{i2} x_2 + \cdots + e_{ij} x_j + \cdots + e_{in} x_n \geq d_i, \ldots,$$

$$e_{m1} x_1 + e_{m2} x_2 + \cdots + e_{mj} x_j + \cdots + e_{mn} x_n \geq d_m,$$

$$\text{and } x_j \geq 0, j = 1, 2, 3, \ldots, n \text{ and } i = 1, 2, \ldots, m \tag{5.2.36}$$

Notice that the above equation and inequalities can be written in matrix form as

$$\text{Minimize } C = \mathbf{c}^T \mathbf{x} \tag{5.2.37}$$

$$\text{Subject to } \mathbf{Ex} \geq \mathbf{d} \text{ and } \mathbf{x} \geq 0 \tag{5.2.38}$$

where $\mathbf{c}^T = [\, c_1 \quad c_2 \quad .. \quad c_j \quad .. \quad c_n \,]$, $\mathbf{x}^T = [\, x_1 \quad x_2 \quad .. \quad x_j \quad .. \quad x_n \,]$,

$$\mathbf{E} = \begin{bmatrix} e_{11} & e_{12} & .. & e_{1j} & .. & e_{1n} \\ e_{21} & e_{22} & .. & e_{2j} & .. & e_{2n} \\ .. & .. & .. & .. & .. & .. \\ e_{i1} & e_{i2} & .. & e_{ij} & .. & e_{in} \\ .. & .. & .. & .. & .. & .. \\ e_{m1} & e_{m2} & .. & e_{mj} & .. & e_{mn} \end{bmatrix}$$

and $\mathbf{d}^T = [\, d_1 \ d_2 \ .. \ d_i \ .. \ d_m \,]$; or using the sum ($\sum$) notation as

$$\text{Minimize } C = \sum_{j=1}^{n} c_j x_j \tag{5.2.39}$$

$$\text{Subject to } \sum_{j=1}^{n} e_{ij} x_j \geq d_i, \text{ and } x_j \geq 0, \text{ where } i = 1, 2, \ldots, m \text{ and } j = 1, 2, \ldots, n \tag{5.2.40}$$

5.2.10 Convex sets and LP problems: existence of optimal solution

We discussed vector spaces, hyperplanes, and convex sets in Sections 2.2.5, 2.2.7, and 3.5.5, respectively. We now extend here our discussions of these topics to some related aspects so that we can apply them to demonstrate the existence of solutions to LP problems.

We know from our discussion in Section 2.2.7 that lines and planes in two- or higher-dimensional spaces are special cases of hyperplanes. We defined a hyperplane by equation (2.2.6). Hyperplanes can be defined in alternative forms. One such form is, in an n-dimensional space, the set of points whose coordinates satisfy a linear equation

$$a_1 x_1 + a_2 x_2 + \cdots + a_j x_j + \cdots + a_n x_n = \mathbf{a}^T \mathbf{x} = b \tag{5.2.41}$$

where $\mathbf{a}^T = [\, a_1 \quad a_2 \quad .. \quad a_j \quad .. \quad a_n \,]$ and $\mathbf{x}^T = [\, x_1 \quad x_2 \quad .. \quad x_j \quad .. \quad x_n \,]$. We can also write equation (5.2.41) in set-builder form as

$$H(\mathbf{a}, b) = \{\mathbf{x} \,|\, \mathbf{a}^T \mathbf{x} = b\} \tag{5.2.42}$$

where H represents the hyperplane. If $n = 2$ (two-dimensional space), equation (5.2.41) gives us a line; in higher dimensions it gives us a plane. If we substitute π_j (or c_j) for a_j and Π (or C) for b into equation (5.2.41), we obtain the objective function – profit (cost) function – in a standard LP problem. Therefore, an objective function in a LP problem is a hyperplane.

Now suppose that we have a linear inequality such as

$$a_{11}x_1 + a_{12}x_2 + \cdots + a_{1j}x_j + \cdots + a_{1n}x_n = \mathbf{a}^\mathbf{T}\mathbf{x} \leq b_1 \qquad (5.2.43)$$

which, in an n-dimensional space, gives a *half-space*. In fact, it is called a *closed half-space*. Consider a special case of inequality (5.2.43): $a_{11}x_1 + a_{12}x_{2n} = \mathbf{a}^\mathbf{T}\mathbf{x} \leq b_1$, where $\mathbf{a}^\mathbf{T} = \begin{bmatrix} a_{11} & a_{12} \end{bmatrix}$ and $\mathbf{x}^\mathbf{T} = \begin{bmatrix} x_1 & x_2 \end{bmatrix}$. Then $a_{11}x_1 + a_{12}x_{2n} = \mathbf{a}^\mathbf{T}\mathbf{x} \leq b_1$ is the set of points defined by one of the constraints in a two-variable general maximization problem of LP. This set divides, with a hyperplane (a line), the total two-dimensional space (in the graphs of maximization problems we have considered so far) into two *sub-spaces*; one above the hyperplane and the other below the hyperplane. This is why the space (or the set of points) represented by the inequality $a_{11}x_1 + a_{12}x_{2n} = \mathbf{a}^\mathbf{T}\mathbf{x} \leq b_1$ is called a half-space. Since the inequality is in the form of "less than or equal to," the lower sub-space includes the points on the hyperplane, and, therefore, it is called a closed half-space. If the inequality were $a_{11}x_1 + a_{12}x_{2n} = \mathbf{a}^\mathbf{T}\mathbf{x} < b_1$, then it would be called an *open half-space* and it would not contain the points on the hyperplane. Similarly, in a two-variable minimization LP problem, the closed half-space is defined by the inequality $a_{11}x_1 + a_{12}x_{2n} = \mathbf{a}^\mathbf{T}\mathbf{x} \geq b_1$. It is needless to say that if one sub-space is a closed half-space, the other sub-space must be an open half-space. Our discussion so far suggests that every constraint in a LP problem represents a half-space.

One can show that hyperplanes, whether they are lines or planes in different dimensions, are convex sets. We defined convex sets as the set of points in different dimensions with the property that a line that joins any two points in the set lies completely inside the set. We now provide here an intuitive explanation for the statement that hyperplanes are closed convex sets. Consider a line in a two-dimensional space, such as the graph of the objective function; or a plane, such as the feasible region, in a standard LP problem. If we draw a line connecting any two points on this line or in this plane, the line will lie completely on the graph of the objective function or inside the feasible region, respectively. Therefore, the *objective hyperplane* (the hyperplane representing the objective function) and the *feasible hyperplane* (the hyperplane representing the feasible region), which we denote by F, of a standard two-variable LP problem are closed convex sets. One can generalize this to dimensions higher than two. This result also applies to closed half-spaces.

One can also show that the intersection of a finite number of convex sets is a convex set; if each of these finite convex sets is a closed convex set, their intersection will also be a closed convex set. This can be shown through Figure 5.2.7 for the case of two closed

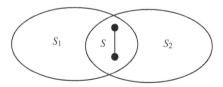

Figure 5.2.7

convex sets. Suppose that we have two closed convex sets, S_1 and S_2. We denote their intersection by S. If we draw a line connecting two points in S, then the line lies completely in S itself, as illustrated in Figure 5.2.7. This suggests that the intersection of two closed convex sets is a closed convex set.

We explained the boundary and interior points of a closed convex set in Section 3.5.5. Another closely related concept is the *extreme point* (or the *corner point* or the *vertex*) of a closed convex set. To elucidate this concept, consider Figure 3.5.7(E). This figure shows the closed convex set of a triangle. There are three corner points or extreme points in this closed convex set. One property of these extreme points is that they do not lie on line segments that connect two other different points in the set. Another property is that the set of the extreme points is the subset of the boundary points of a closed convex set.

We now attempt to apply all the ideas discussed in this section so far. Our aim in solving a LP problem, as we saw in the numerical, illustrative examples in the previous few sections, is to push upward (downward) in maximization (minimization) problems the objective hyperplane until it touches the highest (lowest) possible extreme point of the feasible hyperplane. Such an objective hyperplane is called a *supporting hyperplane* and we denote it by H^*. Notice that H^* will intersect F only at the latter's extreme point or will coincide with the latter's side. If H^* intersects with a side of F, then every point on that line segment will give an optimal solution, which leads to multiple optimal solutions (as illustrated in Figure 5.2.6(A) as the solution to the maximization problem (5.2.27)). Barring this and the other special cases noted in Section 5.2.8, in most practical applications H^* intersects with only one extreme or corner point of F. This is the crux of the extreme point theorem. This is the idea we used when we solved the numerical, illustrative examples of LP problems in earlier sections.

The last important point we note here is that the local maximum (minimum) solution to a LP problem will also be a global maximum (minimum) solution. This result is guaranteed by the *globality theorem*, which states that if F is a closed convex set and if the objective function is a concave (convex) function, then any local maximum (minimum) will also be global maximum (minimum); and if the objective function is a strictly concave (convex) function, then the global maximum (minimum) will be unique.

5.2.11 Exercises

1. Solve the following LP problems:
   (i) maximize $\Pi = 3x + 3y$, subject to $x \le 60$, $y \le 70$, $x + y \le 90$, and $x, y \ge 0$; (ii) maximize $\Pi = x + y$, subject to $x \le 60$, $y \le 50$, and $x, y \ge 0$; (iii) maximize $\Pi = 10x + 5y$, subject to $2x + \le 10$, $x + y \le 20$, $x + y \le 20$, $4x + y \le 30$, and $x, y \ge 0$; (iv) maximize $\Pi = 4x + 3y$, subject to $2x + 4y \le 40$, $4x + 2y \le 50$, $2x + 2y \le 40$, and $x, y \ge 0$.

2. Solve the following LP problems:
   (i) minimize $C = 3x + 2y$, subject to $x + 10y \ge 20$, $10x + y \ge 40$, and $x, y \ge 0$; (ii) minimize $C = 3x + 3y$, subject to $x \ge 60$, $y \ge 70$, $x + y \ge 90$, and $x, y \ge 0$; (iii) minimize $C = x + y$, subject to $2x + y \ge 30$, $x + y \ge 40$, $x + y \ge 50$, and $x, y \ge 0$; (iv) minimize $C = 2x + 3y$, subject to $2x + 3y \ge 30$, $3x + 2y \ge 30$, $x + 2y \ge 20$, and $x, y \ge 0$.

3. *Application exercise.* Suppose that a firm plans to produce two goods, x and y, using three factors: capital, labor, and land. The production of 1 unit of x requires 4, 8, and 1 units of capital, labor, and land, respectively. Similarly, the production of 1 unit of y requires 1 unit each of capital, labor, and land. The maximum quantities of capital, labor, and land available are 40, 60, and 30 units, respectively; and the profit the firm obtains

from 1 unit of both goods is the same $1. Find the quantities of the goods that the firm must produce to maximize its profits satisfying the constraints.

4. *Application exercise.* Suppose that a farmer cultivates two crops, x and y. For this, the farmer has to use both fertilizer and pesticide. One unit of x requires a minimum of 1 unit of both fertilizer and pesticide. Similarly, 1 unit of y requires a minimum of 6 and 1 units of fertilizer and pesticide, respectively. The farmer has to use a minimum of 30 and 10 units of fertilizer and pesticides, respectively, for the two crops. The unit costs of x and y are $2 and $3, respectively. Find the quantities of the two crops that minimize the farmer's total cost while still complying with the minimum requirements of fertilizer and pesticide.

 Web supplement: S5.2.12 Mathematica applications

5.3 The simplex method

5.3.1 Introduction

In Section 5.2 we considered optimization of LP problems involving two choice variables. But, in most practical applications, LP problems involve more than two choice variables. Solving such problems using a graphical method is difficult. Therefore, we need an algebraic method that can handle problems involving any number of choice variables. This is suggested by the *simplex method* or the *simplex algorithm*.

5.3.2 Concepts and definitions

Let us first explain and define here some of the concepts that we will use throughout the rest of this chapter. Most of these concepts will be clearer as we proceed through the next few sections. We begin with the term *algorithm*. An algorithm is a fixed set of computational rules or procedures that are used repetitively to the problem under consideration for finding the solution to it. Each repetition of the algorithm is called an *iteration*. In each iteration we move the solution closer and closer to the optimum.

The simplex method or simplex algorithm is a fixed set of computational rules or procedures for finding the BFSs in a LP problem. For any BFS, some variables are held at 0. Such variables are called *nonbasic variables* (NBVs) and all other variables are called *basic variables* (BVs). The set of BVs is called the *basis*.

5.3.3 Iteration: the basic nature of the simplex method

As an example to illustrate the iterative nature of the simplex method, consider the two-variable maximization LP problem (5.2.5) in Section 5.2.2, which we solved graphically using Figure 5.2.1 and found the optimal solution $x^* = 36$ and $y^* = 6$. We reproduce that problem here for convenience:

$$\text{Maximize } \Pi = 2x + 2y, \text{ subject to } 2x + 3y \leq 90, \, 3x + 2y \leq 120, \text{ and } x, y \geq 0 \quad (5.3.1)$$

The graphs of the objective function and of the functional forms of the constraints of this problem are illustrated in Figure 5.3.1(A), which is similar to Figure 5.2.1. We converted the inequality constraints of this problem to equality constraints using slack variables and

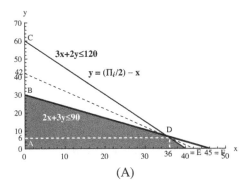
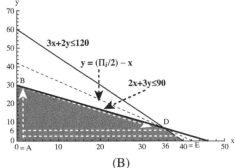

(A)   (B)

Figure 5.3.1

Table 5.3.1

	Variables			
Corner point	x	y	s_1	s_2
A	0	0	90	120
B	0	30	0	60
D	36	6	0	0
E	40	0	10	0

the converted problem is given in equations (5.2.21) and (5.2.22). This is called the augmented form of the original LP maximization problem (5.2.5). Let us now include the slack variables in our discussion and construct Table 5.3.1 that gives the coordinates of the corner points in Figure 5.3.1.

We know from our presentation in Section 5.2.4 that the optimum solution lies at the corner point D where $x^* = 36$ and $y^* = 6$, and then the profit will be a maximum ($\Pi^* = \$84$). Notice an important feature of Table 5.3.1 that the coordinate values of two variables are zero for any corner point. These corner points, as we noted earlier, are the BFSs. Our problem now involves four variables (x, y, s_1, and s_2) and two constraints ($n = 4$ and $m = 2$). Therefore, for every BFS at least $n - m = 4 - 2 = 2$ variables must be zero. We refer to the variables that are zero for a BFS as the NBVs and others as the BVs. Therefore, at corner point D, x and y are BVs and s_1, and s_2 are NBVs. Similarly, at corner point B, y and s_2 are BVs and x and s_1 are NBVs. Notice that $s_1 = s_2 = 0$ at the optimum solution (corner point D), implying that the two inputs are utilized completely.

We are now ready to show the iterative nature of the simplex method. For this we use Table 5.3.1 and the modified form of Figure 5.3.1(A) as illustrated in Figure 5.3.1(B). The augmented objective function is $\Pi = 2x + 2y + 0s_1 + 0s_2$ (equation (5.2.21)). Normally the starting point of iteration is the corner point A, where both x and y are NBVs: $x = y = 0$. At point A, the objective function is zero: $\Pi = 2x + 2y + 0s_1 + 0s_2 = 0 \times x + 0 \times y + 0 \times s_1 + 0 \times s_2 = 0$. With this we have completed the first iteration. We can now try to see whether the objective function attains higher values if we move along the horizontal axis to corner point E or along the vertical axis to corner point B. In which direction we have to move depends on the size of the coefficient of the variable in the objective function. In a maximization problem, we choose the variable whose coefficient has the highest positive value. If all the NBVs have the same coefficients, as in our present problem, then the choice is arbitrary. Therefore, we move to corner point E where, as can be seen in Table 5.3.1, x and s_1 are BVs (because $x = 40 > 0$ and $s_1 = 10 > 0$) and y and s_2 are NBVs. The value of the objective function corresponding to corner point E is $\Pi = 2x + 2y + 0s_1 + 0s_2 = 2 \times 40 + 0 \times y + 0 \times 10 + 0 \times s_2 = 80$, which is higher than that at point A. Thus we have completed the second iteration. We now move to corner point D at which the value of the objective function is $\Pi = 2x + 2y + 0s_1 + 0s_2 = 2 \times 36 + 2 \times 6 + 0 \times 0 + 0 \times 0 = 84$, which is higher than that at point E. With this, we have completed the third iteration. So far our iterations were through the route A → E → D. We can also do similar iterations through the route A → B → D. In our present problem, we know that the optimum occurs at the corner point D. What all this means is that the simplex algorithm involves a number of iterations in which we move from one BFS to another, always improving upon the former, until we reach the optimal solution.

We know that, in Figure 5.3.1, at point A both x and y are NBVs and s_1 and s_2 are BVs. But as we move to point E, x becomes a BV and s_2 becomes a NBV. Therefore, we say that x is the *entering variable* and s_2 is the *departing variable* at point A. Similar interpretations can be given to these or the other variables at points B and E.

However, a question that arises now is: how does one carry out the iterations outlined above and arrive at the optimum solution nongeometrically? The answer can be found in two widely used approaches of the simplex method. They are the *tabular approach* and the *revised simplex method*. Let us first explore the tabular approach of the simplex method. We will discuss the revised simplex method in Section 5.3.6.

5.3.4 The simplex method: the tabular approach

In order to illustrate the tabular approach of the simplex method of solving LP problems, we consider the augmented form of the example we used in Section 5.2.7 (equations (5.2.21) and (5.2.22)). Notice that this problem can be expressed in matrix form as

$$\text{Maximize } \Pi = \pi^T x, \text{ subject to } Ax \leq b \text{ and } x \geq 0 \tag{5.3.2}$$

where $\pi^T = \begin{bmatrix} 2 & 2 & 0 & 0 \end{bmatrix}$, $x^T = \begin{bmatrix} x & y & s_1 & s_2 \end{bmatrix}$, $b^T = \begin{bmatrix} 90 & 120 \end{bmatrix}$, and $A = \begin{bmatrix} 2 & 3 & 1 & 0 \\ 3 & 2 & 0 & 1 \end{bmatrix}$.

Notice that the maximization problem in (5.3.2) can be written alternatively as

$$\text{Maximize } -2x - 2y - 0s_1 - 0s_2 + \Pi = 0, \text{ subject to } 2x + 3y + s_1 + 0s_2 = 90,$$

$$3x + 2y + 0s_1 + s_2 = 120, \text{ and } x, y, s_1, s_2 \geq 0 \tag{5.3.3}$$

which we represent in a table called the *initial simplex tableau*, as shown in Table 5.3.2.[2]

Let us first explain the features of Table 7.3.2. The column "RHS" in the table represents the RHS of the problem in (5.3.3). The values in each column (other than those under RHS) represent the coefficients of the corresponding variables in the problem in (5.3.3). If we begin with $x = y = 0$ (as at point A in Figure 5.3.1(B)), then the slack variables (s_1 and s_2) are the BVs and are represented by the column "BVs." This gives us, as can be seen in the last column, $s_1 = 90$ and $s_2 = 120$. Since all the variables are nonnegative, this represents a BFS. The value of the objective function, as shown at the bottom of the last column, with $x = 0$, $y = 0$, $s_1 = 90$, and $s_2 = 120$, is $\Pi = 0$.

Table 5.3.2

BVs	x	y	s_1	s_2	RHS
s_1	2	3	1	0	90
s_2	3	2	0	1	120
$\Pi_j - \pi_j$	−2	−2	0	0	0

Indicators

Iteration 1

Let us now see if we can find a BFS with a larger value of Π. Since the slack variables do not contribute to the objective function, Π may increase only due to an increase in x or an increase in y. But, which one shall we choose? The answer depends on the coefficients of these decision variables among the indicators given in the last row. Since the coefficient with the *highest absolute value* makes the highest contribution to Π, that variable is chosen as the *entering variable*. If there is a tie among the coefficients of the decision variable, as in our present case with −2 and −2, the choice is arbitrary. Therefore, we choose x as the entering variable and it becomes a BV or enters the basis. The column of the entering variable is called the *pivot column*.

As stated earlier, in any BFS in the present problem two of the variables must be BVs and the other two must be NBVs. This is true in the first BFS presented in the initial simplex tableau. But, when we made x a BV we have in total three BVs and one NBV. Therefore, one of the BVs must become a NBV (so that there will be two NBVs) for a new BFS. How do we determine which variable will become a NBV or leaves the basis? The variable that leaves the basis is the slack variable with the constant 1 in the row with the least *displacement ratio* (DR), ignoring zeroes and negative values. DRs are obtained by dividing the elements in the column of constants (RHS column) by the corresponding elements in the pivot column. Therefore, the DRs in the present example are $90/2 = 45$ and $120/3 = 40$, and the least DR is $120/4 = 40$. This means that the slack variable s_2 must leave the basis and must become a NBV. The row of the leaving or departing variable (s_2 in the present example) is called the *pivot row*. The element, in the simplex tableau, that lies at the intersection of the pivot column and pivot row (3 in the present example) is called the *pivot element* or the *pivot entry*.

Notice that this is true in terms of Table 5.3.1 when $x = 40$. Notice also that we now have (in the new BFS) exactly two BVs (x and s_1) and two NBVs (y and s_2). Let us now update Table 5.3.2 so that it will reflect the above mentioned points, as presented in Table 5.3.3. Since s_2 departed the basis and x entered the basis, we have to update Table 5.3.3 to obtain Table 5.3.4.

Table 5.3.3

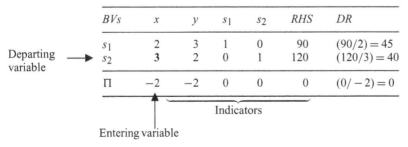

	BVs	x	y	s_1	s_2	RHS	DR
Departing variable →	s_1	2	3	1	0	90	$(90/2) = 45$
	s_2	3	2	0	1	120	$(120/3) = 40$
	Π	−2	−2	0	0	0	$(0/-2) = 0$

Indicators

Entering variable

Table 5.3.4

BVs	x	y	s_1	s_2	RHS
s_1	2	3	1	0	90
x	3	2	0	1	120
Π	−2	−2	0	0	0

Indicators

Table 5.3.5

BVs	x	y	s_1	s_2	RHS
s_1	0	5/3	1	−2/3	10
x	1	2/3	0	1/3	40
Π	0	−2/3	0	2/3	80

Indicators

Let us now carry out elementary row operations, as presented in Section 2.3.7, to convert the pivot element in Table 5.3.4 to 1 and all other elements in the pivot column to zeros. This process is also called *pivoting*. If we do the elementary row operations ($R_2 \rightarrow (1/3)R_2$, $R_1 \rightarrow R_1 - 2R_2$, and $R_3 \rightarrow R_3 + 2R_2$), we obtain Table 5.3.5.

We now interpret Table 5.3.5, which corresponds to the second BFS. Since x and s_1 are the BVs in the second BFS, the values of x, s_1, and Π are (as can be read from the last column) 40, 10, and 80, respectively. This shows that the value of the objective function improved from 0 to 80 when we moved from the first to the second BFS. Notice that this was the value of the objective function we obtained at point E as shown in Table 5.3.1 or in Figure 5.3.1(B).

Iteration 2

We need to continue iteration as above as long as there are negative indicators in the last row of the last simplex tableau (Table 5.3.5). Since the absolute values of the coefficients of y and s_2 in the objective function (the last row) make a tie, they make equal contribution to the objective function. Therefore, we arbitrarily choose y as the entering variable. The new DRs are $10/(5/3) = 6$ and $40/(2/3) = 60$, which show that s_1 is the departing variable. Therefore, y enters the basis and s_1 leaves the basis. We present this information, by updating Table 5.3.5, in Table 5.3.6. Notice that, in the new or third BFS presented in Table 5.3.6, the pivot column is the column of y, the pivot row is the row of s_1, and the pivot element is 5/3. Since s_1 departed the basis and y entered the basis, we have to update Table 5.3.6 to obtain Table 5.3.7. We shall again carry out the elementary row operations on Table 5.3.7 ($R_1 \rightarrow (3/5)R_1$, $R_2 \rightarrow R_2 - (2/3)R_1$, and $R_3 \rightarrow R_3 + (2/3)R_1$) to obtain Table 5.3.8.

As before, we interpret Table 5.3.8. The last row in the table contains no negative value. This implies that the objective function is maximized. Therefore, the optimum solution in the third and the final BFS constitutes $x^* = 36$, $y^* = 6$, and $\Pi^* = 84$. These solutions can be read from the last column. The slack variables have zero values in the optimum solution ($s_1 = s_2 = 0$). This means that the two factors are completely utilized. Notice that these are precisely the results we obtained when we used the graphical method in Section 5.2.4.

Table 5.3.6

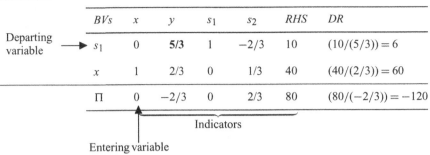

	BVs	x	y	s_1	s_2	RHS	DR
Departing variable →	s_1	0	**5/3**	1	-2/3	10	$(10/(5/3)) = 6$
	x	1	2/3	0	1/3	40	$(40/(2/3)) = 60$
	Π	0	-2/3	0	2/3	80	$(80/(-2/3)) = -120$

Indicators

Entering variable

Table 5.3.7

BVs	x	y	s_1	s_2	RHS
y	0	**5/3**	1	-2/3	10
x	1	2/3	0	1/3	40
Π	0	-2/3	0	2/3	80

Indicators

Table 5.3.8

BVs	x	y	s_1	s_2	RHS
y	0	1	3/5	-2/5	6
x	1	0	-2/5	9/5	36
Π	0	0	2/5	2/5	84

Indicators

Let us now summarize the tabular approach of the simplex method of solving a standard maximization LP problem involving n variables and m constraints. Assume that we want to

$$\text{Maximize } \Pi = \pi_1 x_1 + \pi_2 x_2 + \cdots + \pi_n x_n, \text{ subject to } a_{11}x_1 + a_{12}x_2 + \cdots + a_{1n}x_n \leq b_1,$$

$$a_{21}x_1 + a_{22}x_2 + \cdots + a_{2n}x_n \leq b_2, \ldots, a_{m1}x_1 + a_{m2}x_2 + \cdots + a_{mn}x_n \leq b_m,$$

$$\text{and } x_j \geq 0, \quad j = 1, 2, 3, \ldots, n \tag{5.3.4}$$

Then the steps involved in the solution to the problem are the following.

Step 1. Set up the problem using only slack variable (when the constraint appears with $\leq$ sign), surplus variable and *artificial variable* (when the constraint appears with $\geq$ sign), and only artificial variable (when the constraint appears with = sign). The use of artificial variables and surplus variables are explained in the next section.

Step 2. Convert the objective function, by including the n slack variables, into
$$\Pi = \pi_1 x_1 + \pi_2 x_2 + \cdots + \pi_n x_n + 0s_1 + 0s_2 + \cdots + 0s_n \text{ and then into}$$

$$-\pi_1 x_1 - \pi_2 x_2 - \cdots - \pi_n x_n - 0s_1 - 0s_2 - \cdots . - 0s_2 + \Pi = 0 \tag{5.3.5}$$

Table 5.3.9

BVs	x_1	x_2	...	x_n	s_1	s_2	...	s_n	RHS
s_1	a_{11}	a_{12}	...	a_{1n}	1	0	0	0	b_1
s_2	a_{21}	a_{22}	...	a_{2n}	0	1	0	0	b_2
...	...	...	...	...	...	...	...	...	...
s_n	a_{m1}	a_{m2}	...	a_{mn}	0	0	0	1	b_m
Π	$-\pi_1$	$-\pi_2$	...	$-\pi_n$	0	0	0	0	0

$$\underbrace{\qquad\qquad\qquad\qquad\qquad\qquad\qquad\qquad\qquad}_{\text{Indicators}}$$

Step 3. Convert the constraints (excluding the nonnegative constraints), by including the n slack variables, into

$$a_{11}x_1 + a_{12}x_2 + \cdots + a_{1n}x_n + s_1 + 0s_2 + \cdots + 0s_n = b_1,$$

$$a_{21}x_1 + a_{22}x_2 + \cdots + a_{2n}x_n + 0s_1 + s_2 + \cdots + 0s_n = b_2, \ldots,$$

$$a_{m1}x_1 + a_{m2}x_2 + \cdots + a_{mn}x_n + 0s_1 + 0s_2 + \cdots + s_n = b_m,$$

$$a_{m1}x_1 + a_{m2}x_2 + \cdots + a_{mn}x_n + 0s_1 + 0s_2 + \cdots + s_n = b_m \qquad (5.3.6)$$

Step 4. Construct the initial simplex tableau as in Table 5.3.9.
Step 5. If all the indicators in last row are nonnegative, then Π has a maximum. If some of the indicators are negative, then locate the pivot column (the column with the largest absolute value among the indicators) or the entering variable.
Step 6. Find the DR by dividing the elements in the last column (RHS) by the corresponding elements in the pivot column.
Step 7. Locate the pivot row (the row with the lowest DR found in step 6). The element at the intersection of the pivot row and pivot column will be the pivot element. The slack variable with coefficient 1 in the pivot row will be the departing variable.
Step 8. Carry out elementary row operations to convert the pivot element in step 7 to 1 and all other elements in the pivot column to 0.
Step 9. If all the indicators after step 8 are nonnegative, then the objective function is maximized and the maximum value of the function (Π^*) will be the value in the right bottom cell of the tableau. Then the optimum values of the BVs can be read from the corresponding cells under the column RHS and the values of all the NBVs will be zero. If, instead, there is at least one negative indicator in the last column, repeat steps 5 to 9 until all the indicators appear with nonnegative values.

5.3.5 Tabular approach of the simplex method: solution of minimization problems

In the last section we applied the tabular approach of the simplex method to solve a maximization problem. We now apply the tabular approach with slight modification to solve a minimization problem. There are two widely used methods (under the simplex method) to solve minimization problems. They are the *big-M method* and the *two-phase method*. Let us first present the big-M method followed by the two-phase method.

Big-M method

Notice that in a maximization problem, in which the constraints appear with ≤ sign, we added slack variables to the left of the constraints to make them equalities. These slack variables offered us an initial BFS. But, as we saw in the graphical solution to the minimization problem we considered in Section 5.2.5, the origin in the graph is not a BFS because it violated all the constraints of the problem. Moreover, if we proceed with a BFS containing only slack variables, these variables will produce negative values in the BFS which will violate the nonnegativity condition. Sometimes the problems may originally involve some, or all, constraints with = or ≥ signs, which are called *alternative LP problems*. In such situations, we introduce variables called *artificial variables* to the left of the constraints with = or ≥ signs. Notice also that we still have to use slack variables and surplus variables on the LHS of constraints with ≤ and ≥ signs, respectively. The artificial variables are introduced mainly to generate an initial BFS, and these variables will be, as shown below, forced out of the optimum solution.

As an illustration, consider the minimization problem we introduced in (5.2.11) and solved graphically in Section 5.2.5. For convenience we reproduce the problem here with slight variation, by incorporating the above points, as

Minimize $C = 9x_1 + 12x_2 + 0x_3 + 0x_4 + MA_1 + MA_2$,

subject to $15x_1 + 10x_2 - x_3 - 0x_4 + A_1 + 0A_2 = 90$,

$10x_1 + 15x_2 - 0x_3 - x_4 + 0A_1 + A_2 = 90$, and $x_1, x_2, x_3, x_4, A_1, A_2 \geq 0$ (5.3.7)

where we use x_1 for x and x_2 for y; x_3 and x_4 denote the surplus variables in the first and the second constraints, respectively; and A_1 and A_2 denote the artificial variables in the first and the second constraints, respectively. Notice the corresponding changes in the objective function. Notice also that the coefficient(s) of the artificial variable(s) in the objective function must be a large positive number (M). This is why this method is called the big-M method. We can now construct Table 5.3.10 as the initial simplex tableau on the basis of the above problem.

Notice that Table 5.3.10 contains an inconsistency. If $x_1 = x_2 = x_3 = x_4 = 0$, then $A_1 = A_2 = 90$ and, therefore, $C = 180M$. But, the last cell in the last row shows that $C = 0$. This is the said inconsistency. To eliminate this inconsistency, we can multiply the first and the second rows by M and add them (column-wise) to the last row. The result can be shown, through a modified tableau, as in Table 5.3.11. Notice that an initial BFS can be read now from Table 5.3.11: when $x_1 = x_2 = x_3 = x_4 = 0$, $A_1 = A_2 = 90$, $C = 180M$. Notice also that this initial BFS has eliminated the inconsistency. Because $C = 180M$ is an inconceivably huge amount (since M is a huge value), we need to check for a new BFS that may produce a lower value for C.

Table 5.3.10

BVs	x_1	x_2	x_3	x_4	A_1	A_2	RHS
A_1	15	10	−1	0	1	0	90
A_2	10	15	0	−1	0	1	90
C	−9	−12	0	0	−M	−M	0

Indicators

Table 5.3.11

BVs	x_1	x_2	x_3	x_4	A_1	A_2	RHS
A_1	15	10	−1	0	1	0	90
A_2	10	15	0	−1	0	1	90
C	$25M − 9$	$25M − 12$	$−M$	$−M$	0	0	$180M$

Indicators

Table 5.3.12

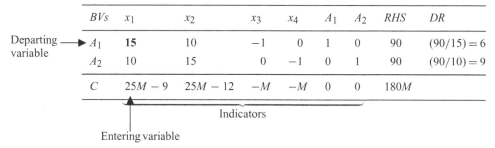

	BVs	x_1	x_2	x_3	x_4	A_1	A_2	RHS	DR
Departing → variable	A_1	**15**	10	−1	0	1	0	90	$(90/15) = 6$
	A_2	10	15	0	−1	0	1	90	$(90/10) = 9$
	C	$25M − 9$	$25M − 12$	$−M$	$−M$	0	0	$180M$	

Indicators

Entering variable

Iteration 1. The NBV that enters the basis in a maximization problem, as we saw in the previous section, is the NBV with the largest negative indicator. In a minimization problem, the entering variable is that with the largest positive indicator; and, as can be seen in Table 5.3.11, it is x_1. The column that represents x_1 is the pivot column. The variable that leaves the basis can be determined, as before, by finding the DRs. The DRs (obtained by dividing each element of the RHS column by the corresponding elements of the pivot column) are 6 and 9. Since the smallest DR is 6, and since 1 is associated with A_1, A_1 leaves the basis. This means that the row that represents A_1 is the pivot row. The element at the intersection of the pivot row and pivot column is the pivot element, and it is 15 in the present example. Therefore, we have Table 5.3.12. Since A_1 departed the basis and x_1 entered the basis, we have to update Table 5.3.12 to obtain Table 5.3.13. Let us now carry out elementary row operations on Table 5.3.13 to convert the pivot element to 1 and all other elements in the pivot column to zeros. If we do the elementary row operations $(R_1 \rightarrow (1/15)R_1, R_2 \rightarrow R_2 − 10R_1$, and $R_3 \rightarrow R_3 − (25M − 9)R_1)$, we obtain Table 5.3.14.

Iteration 2. Since x_2 has the largest positive value in the last row in Table 5.3.14, x_2 enters the basis. And since the DR is the smallest in the case of row A_2, A_2 leaves the basis. We can now modify Table 5.3.14 on the basis of this information to obtain Table 5.3.15. Notice that the pivot element now is 25/3. Since A_2 departed the basis and x_2 entered the basis, we have to

Table 5.3.13

BVs	x_1	x_2	x_3	x_4	A_1	A_2	RHS
x_1	15	10	−1	0	1	0	90
A_2	10	15	0	−1	0	1	90
C	$25M − 9$	$25M − 12$	$−M$	$−M$	0	0	$180M$

Indicators

Table 5.3.14

BVs	x_1	x_2	x_3	x_4	A_1	A_2	RHS
x_1	1	2/3	$-1/15$	0	1/15	0	90
A_2	0	25/3	2/3	-1	$-2/3$	1	90
C	0	$(25M - 18)/3$	$(10M - 9)/15$	$-M$	$(-25M - 9)/15$	0	$30M + 54$

Indicators

Table 5.3.15

	BVs	x_1	x_2	x_3	x_4	A_1	A_2	RHS	DR
Departing	x_1	1	2/3	$-1/15$	0	1/15	0	6	$(6/(2/3)) = 9$
variable ⟶	A_2	0	**25/3**	2/3	-1	$-2/3$	1	30	$(30/(25/3)) = 3.6$
	C	0	$(25M - 18)/3$	$(10M - 9)/15$	$-M$	$(-25M - 9)/15$	0	$30M + 54$	

Indicators

Entering variable

Table 5.3.16

BVs	x_1	x_2	x_3	x_4	A_1	A_2	RHS
x_1	1	2/3	$-1/15$	0	1/15	0	6
x_2	0	**25/3**	2/3	-1	$-2/3$	1	30
C	0	$(25M - 18)/3$	$(10M - 9)/15$	$-M$	$(-25M - 9)/15$	0	$30M + 54$

Indicators

Table 5.3.17

BVs	x_1	x_2	x_3	x_4	A_1	A_2	RHS
x_1	1	0	$-3/25$	2/25	3/25	$-2/25$	3.6
x_2	0	1	2/25	$-3/25$	$-2/25$	3/25	3.6
C	0	0	$-3/25$	$-18/25$	$(-50M - 81)/75$	$-M - 18/25$	75.6

Indicators

update Table 5.3.15 to obtain Table 5.3.16. Let us again carry out elementary row operations on Table 5.3.16 to convert the pivot element to 1 and all other elements in the pivot column to zeros. If we do the elementary row operations ($R_2 \rightarrow (3/25)R_2$, $R_1 \rightarrow R_1 - (2/3)R_2$, and $R_3 \rightarrow R_3 - [(25M - 18)/3]R_2$), we obtain Table 5.3.17.

Notice that iteration will continue as long as a positive indicator remains in the last row and it will stop when all the indicators are *nonpositive*. Since all the indicators are nonpositive in Table 5.3.17, we can stop iteration with it. This means that we have achieved the optimal solution: $C^* = 75.6$, $x_1^* = 3.6$, $x_2^* = 3.6$, $x_3 = 0$, and $x_4 = 0$. Notice also that these are exactly the same results as those we obtained when applying the graphical method in Section 5.2.5.

Two-phase method

The reader would have noticed that solving a minimization problem by hand using the big-M method was a laborious task as it involved a number of calculations. The chance for errors in these calculations is high. Moreover, when one solves minimization problems with a computer using the big-M method, one has to specify a value for M no matter how large it is. Due to these issues with the big-M method, an alternative method that eliminates these issues is generally adopted and it is called the two-phase method.

As the name suggests, the two-phase method involves two phases. In phase I, we specify the objective function in terms of the artificial variables only: minimize $\sum A_i$ or maximize $-\sum A_i$, where A_i represents the ith artificial variable. This objective function will eliminate the artificial variables from the basis if the problem has a feasible solution. Then we have to eliminate the possible discrepancy in the RHS column as we did in the big-M method. After this, we need to apply the simplex method. This is the end of the first phase of the two-phase method; that is, when we have eliminated all the artificial variables from the basis. If we cannot eliminate them, then there is no feasible solution to the problem. Once this is done, the original objective function will be introduced and, then, we apply the simplex method again.

As an illustration, let us use the example we considered with the big-M method:

$$\text{Minimize } C = 9x_1 + 12x_2 + 0x_3 + 0x_4 + MA_1 + MA_2,$$

$$\text{subject to } 15x_1 + 10x_2 - x_3 - 0x_4 + A_1 + 0A_2 = 90,$$

$$10x_1 + 15x_2 - 0x_3 - x_4 + 0A_1 + A_2 = 90, \text{ and } x_1, x_2, x_3, x_4, A_1, A_2 \geq 0$$

which we transform, as stated above, to

$$\text{Minimize } C = 0x_1 + 0x_2 + 0x_3 + 0x_4 + A_1 + A_2,$$

$$\text{subject to } 15x_1 + 10x_2 - x_3 - 0x_4 + A_1 + 0A_2 = 90,$$

$$10x_1 + 15x_2 - 0x_3 - x_4 + 0A_1 + A_2 = 90, \text{ and } x_1, x_2, x_3, x_4, A_1, A_2 \geq 0$$

On the basis of the above problem, we can construct the initial simplex tableau in Table 5.3.18. Notice that, in Table 5.3.18, when $x_1 = x_2 = x_3 = x_4 = 0$, $A_1 = A_2 = 90$. But, as can be seen from the initial tableau, $C = 0$. This is a discrepancy, like the one we saw in the application of the big-M method. To eliminate this discrepancy, we multiply the first two rows by unity (1) and add the first two cells in every column to the last cell of the same column. After doing this, we obtain Table 5.3.19. Notice that Table 5.3.19 is devoid of the said discrepancy. Let us now apply the simplex method to solve this minimization problem. Following the same procedure as that adopted in the big-M method, and skipping some of the routine steps therein, we obtain Table 5.3.20.

Table 5.3.18

BVs	x_1	x_2	x_3	x_4	A_1	A_2	RHS
A_1	15	10	−1	0	1	0	90
A_2	10	15	0	−1	0	1	90
C	0	0	0	0	−1	−1	0

$\underbrace{}$
Indicators

Table 5.3.19

BVs	x_1	x_2	x_3	x_4	A_1	A_2	RHS
A_1	15	10	−1	0	1	0	90
A_2	10	15	0	−1	0	1	90
C	25	25	−1	−1	0	0	180

Indicators

Table 5.3.20

BVs	x_1	x_2	x_3	x_4	A_1	A_2	RHS
x_1	1	0	−3/25	2/25	3/25	−2/25	3.6
x_2	0	1	2/25	−3/25	−2/25	3/25	3.6
C	0	0	0	0	−1/6	−1	0

Indicators

Table 5.3.21

BVs	x_1	x_2	x_3	x_4	RHS
x_1	1	0	−3/25	2/25	3.6
x_2	0	1	2/25	−3/25	3.6
C	−9	−12	0	0	0

Indicators

Table 5.3.22

BVs	x_1	x_2	x_3	x_4	RHS
x_1	1	0	−3/25	2/25	3.6
x_2	0	1	2/25	−3/25	3.6
C	0	0	−3/25	−18/25	75.6

Indicators

With Table 5.3.20, we have eliminated both artificial variables from the basis. Therefore, we have completed the first phase. Let us now begin the second phase. In the second phase we introduce the original objective function. With the original objective function (after removing the artificial variables), Table 5.3.20 can be reformulated as Table 5.3.21. We now carry out elementary row operations ($R_3 \rightarrow R_3 + 9R_1$ and $R_3 \rightarrow R_3 + 12R_2$) to obtain the final tableau as in Table 5.3.22. Notice that in Table 5.3.22, all the values in the last row (except in the column RHS) are nonpositive, the condition for an optimum in the case of minimization. Therefore, we stop with the last tableau and the optimum values are $x_1^* = 3.6$, $x_2^* = 3.6$, and $C^* = 75.6$. These are precisely the optimum values we obtained when we used the graphical method in Section 5.2.5 and the big-M method earlier in the current section.

5.3.6 Revised simplex method

We saw in the last section how the tabular approach of the simplex method can be utilized to solve LP problems. We now consider the *revised simplex method* of solving LP problems. One may wonder why we need this new method. The reason is that the revised simplex method is computationally more efficient than the tabular approach. Instead of the elementary row operations, the revised simplex method utilizes matrices, which reduce computational errors as it uses, as shown below, the original data and only one inverse. Assume that we want to

$$\text{Maximize } \Pi = \pi^{\mathrm{T}} \mathbf{x}, \text{ subject to } \mathbf{Ax} = \mathbf{b} \text{ and } \mathbf{x} \geq \mathbf{0} \tag{5.3.8}$$

where $\pi^{\mathrm{T}} = \begin{bmatrix} \pi_1 & \pi_2 & \dots & \pi_n & 0 & 0 & \dots & 0 \end{bmatrix}$, $\mathbf{x}^{\mathrm{T}} = \begin{bmatrix} x_1 & x_2 & \dots & x_n & s_1 & s_2 & \dots & s_n \end{bmatrix}$,

$$\mathbf{A} = \begin{bmatrix} a_{11} & a_{12} & \dots & a_{1n} & 1 & 0 & \dots & 0 \\ a_{21} & a_{22} & \dots & a_{2n} & 0 & 1 & \dots & 0 \\ \dots & \dots & \dots & \dots & 0 & 0 & \dots & 0 \\ a_{m1} & a_{m2} & \dots & a_{mn} & 0 & 0 & \dots & 1 \end{bmatrix}, \quad \mathbf{b} = \begin{bmatrix} b_1 \\ b_2 \\ \dots \\ b_m \end{bmatrix}, \text{ and } \mathbf{0} = \begin{bmatrix} 0 \\ 0 \\ \dots \\ 0 \end{bmatrix}$$

We will later use the definition that

$$\mathbf{a}_1 = \begin{bmatrix} a_{11} \\ a_{21} \\ \dots \\ a_{m1} \end{bmatrix}, \quad \mathbf{a}_2 = \begin{bmatrix} a_{12} \\ a_{22} \\ \dots \\ a_{m2} \end{bmatrix}, \quad \dots, \quad \mathbf{a}_n = \begin{bmatrix} 0 \\ 0 \\ \dots \\ 1 \end{bmatrix}$$

Now suppose that $\mathbf{B}$ is a *feasible basis* of the problem in (5.3.8). Also suppose that $\mathbf{x_B}$ constitutes, corresponding to $\mathbf{B}$, the set of BVs and that $\pi_{\mathbf{B}}^{\mathrm{T}}$ is the vector of the corresponding coefficients in the objective function. Now eliminating n variables by treating them as zeros gives m equations in m unknown BVs. This system of equations can be written as

$$\mathbf{Bx_B} = \mathbf{b} \tag{5.3.9}$$

which can be pre-multiplied on both sides by the inverse of $\mathbf{B}$, $\mathbf{B}^{-1}$, to obtain

$$\mathbf{x_B} = \mathbf{B}^{-1}\mathbf{b} \tag{5.3.10}$$

Notice that we introduced the inverse of the feasible basis $\mathbf{B}$ in equation (5.3.10). Since the simplex method introduces only the BVs, $\mathbf{B}$ is always nonsingular and, therefore, $\mathbf{B}^{-1}$ will always exist. We know that the objective function is given by $\Pi = \pi^{\mathrm{T}}\mathbf{x}$ and that the current set of BVs is given by $\mathbf{x_B}$. Therefore, the value of the objective function in the current BFS can be written as $\Pi = \pi_{\mathbf{B}}^{\mathrm{T}}\mathbf{x_B}$, which, using equation (5.3.10), can be written as

$$\Pi = \pi_{\mathbf{B}}^{\mathrm{T}}\mathbf{B}^{-1}\mathbf{b} \tag{5.3.11}$$

Equation (5.3.10) gives the first BFS and equation (5.3.11) gives the value of the objective function corresponding to the first BFS. We can now introduce a NBV, x_j, into the basis to see if it increases the value of the objective function. Then equation (5.3.12) will hold:

$$\mathbf{B}\mathbf{x_B} + \mathbf{a}_j x_j = \mathbf{b} \tag{5.3.12}$$

where $\mathbf{a}_j$ represents the jth $(j = 1, 2, \ldots, n)$ column of $\mathbf{A}$. We can now pre-multiply equation (5.3.12) by $\mathbf{B}^{-1}$ and rearrange it to obtain

$$\mathbf{x_B} = \mathbf{B}^{-1}\mathbf{b} - \mathbf{B}^{-1}\mathbf{a}_j x_j \tag{5.3.13}$$

Equation (5.3.11) gives the old value of the objective function (that is, the value of the objective function in the last BFS). Its new value (in the new BFS) is given by

$$\Pi_{\text{new}} = \pi_{\mathbf{B}}^{\mathsf{T}}\mathbf{B}^{-1}\mathbf{b} - (\pi_{\mathbf{B}}^{\mathsf{T}}\mathbf{B}^{-1}\mathbf{a}_j - \pi_j)x_j \tag{5.3.14}$$

where π_j represents the coefficient of x_j in the objective function. Notice that the first part on the RHS of equation (5.3.14) is Π given in equation (5.3.11). Therefore, using equation (5.3.11) and denoting $\Pi_j = \pi_{\mathbf{B}}^{\mathsf{T}}\mathbf{B}^{-1}\mathbf{a}_j$, we can rewrite equation (5.3.14) as

$$\Pi_{\text{new}} = \Pi - (\Pi_j - \pi_j)x_j \tag{5.3.15}$$

Equation (5.3.15) implies that the objective function of the maximization problem will increase only if $\Pi_j - \pi_j$ is negative. Therefore, the NBV is to be introduced and a new iteration is to be carried out in a maximization problem only if $\Pi_j - \pi_j$ is negative. *This condition is to be checked for every NBV.* If $\Pi_j - \pi_j$ is nonnegative for all x_j in the case of a maximization problem, then the current solution is optimal (which is called the *optimality criterion* of the revised simplex method); if $\Pi_j - \pi_j = 0$ for a NBV, then there will be multiple optimal solutions.

Equation (5.3.15) also implies that the objective function of the minimization problem will decrease only if $\pi_j - \Pi_j$ is negative. Therefore, the NBV is to be introduced and a new iteration is to be carried out in a minimization problem only if $\pi_j - \Pi_j$ is negative. As in the case of a maximization problem, *this condition is to be checked for every NBV.* If $\pi_j - \Pi_j$ is nonnegative for all x_j in the case of a minimization problem, then the current solution, as before, is optimal and the optimality criterion of the revised simplex method is met; if $\pi_j - \Pi_j = 0$ for a NBV, then there will be multiple optimal solutions.

So far we have checked for optimality in successive iterations and determed the variable that enters basis. Let us now consider the *feasibility criterion* of the revised simplex method. Using this criterion we can determine which variable leaves the basis. For this we use equation (5.3.13): $\mathbf{x_B} = \mathbf{B}^{-1}\mathbf{b} - \mathbf{B}^{-1}\mathbf{a}_j x_j$. This equation implies that $\mathbf{x_B}$ will become negative (and thus will violate the nonnegative condition $\mathbf{x_B} \geq 0$) if x_j increases for $\mathbf{B}^{-1}\mathbf{a}_j > 0$. Therefore, we impose the condition $(\mathbf{B}^{-1}\mathbf{b})_i - (\mathbf{B}^{-1}\mathbf{a}_j)_i x_j \geq 0$ or $(\mathbf{B}^{-1}\mathbf{b})_i/(\mathbf{B}^{-1}\mathbf{a}_j)_I \geq x_j$, where $i = 1, 2, \ldots, m$. This inequality implies that the maximum value of the entering variable, x_j, is given by

$$x_j = \min\{(\mathbf{B}^{-1}\mathbf{b})_i/(\mathbf{B}^{-1}\mathbf{a}_j)_i | (\mathbf{B}^{-1}\mathbf{a}_j) > 0\} \tag{5.3.16}$$

which implies that the basic variable x_j that gives the minimum DR given in this equation leaves the basis. This is the feasibility criterion. We can now summarize the steps involved in the application of the revised simplex method.

Iteration 1

Step 1. Set up the problem using only slack variable (when the constraint appears with $\leq$ sign), surplus variable and artificial variable (when the constraint appears with $\geq$ sign), and only artificial variable (when the constraint appears with $=$ sign).

Step 2. Set up the first BFS such that $\mathbf{B}$ and $\pi_{\mathbf{B}}^{\mathrm{T}}$ are its respective basis and objective coefficient vectors.

Step 3. Find $\mathbf{B}^{-1}$ using any of the methods we discussed in Chapter 2.

Step 4. Find $\Pi_j - \pi_j = \pi_{\mathbf{B}}^{\mathrm{T}} \mathbf{B}^{-1} \mathbf{a}_j - \pi_j$ for each nonbasic variable x_j. If we find $\Pi_j - \pi_j \geq 0$ ($\Pi_j - \pi_j \leq 0$) in maximization (minimization) problem for NBVS, stop at this point. Then, the optimum solution to the problem is given by $\mathbf{X_B} = \mathbf{B}^{-1}\mathbf{b}$ and $\Pi = \pi_{\mathbf{B}}^{\mathrm{T}} \mathbf{X}_B$.

Step 5. If, instead, in step 4 we find $\Pi_j - \pi_j < 0$ ($\Pi_j - \pi_j > 0$) for NBVs in a maximization (minimization) problem, apply the optimality criterion and determine the variable that enters the basis (the variable with the most negative (positive) $\Pi_j - \pi_j$ value in maximization (minimization) problems).

Step 6. Find $\mathbf{B}^{-1}\mathbf{a}_j$. If we find that all the elements of $\mathbf{B}^{-1}\mathbf{a}_j$ are nonpositive, stop at this point. Then the problem has an unbounded solution. Otherwise, find $\mathbf{B}^{-1}\mathbf{b}$ and find the DRs for all positive elements of $\mathbf{B}^{-1}\mathbf{a}_j$. Then determine the BV that leaves the basis (the variable with the smallest DR).

Step 7. Make a new basis, from the current basis $\mathbf{B}$, by replacing the vector of the leaving BV ($\mathbf{a}_j$) with the vector of the entering NBV.

Iteration 2

Step 8. Redo steps 1 to 7, and continue with fresh iterations and stop further iterations when $\Pi_j - \pi_j \geq 0$ ($\Pi_j - \pi_j \leq 0$) in maximization (minimization) problems for NBVs.

As an example, consider the maximization problem we solved with the graphical and tabular methods (problem (5.3.2)). For convenience we reproduce it here with $x_1 = x$ and $x_2 = y$: maximize $\Pi = 2x_1 + 2x_2 + 0s_1 + 0s_2$, subject to $2x_1 + 3x_2 + s_1 + 0s_2 = 90$, $3x_1 + 2x_2 + 0s_1 + s_2 = 120$, and $x_1, x_2, s_1, s_2 \geq 0$, which can be written in matrix form as

$$\text{Maximize } \Pi = \pi^{\mathrm{T}}\mathbf{x}, \quad \text{subject to } \mathbf{Ax} = \mathbf{b} \text{ and } \mathbf{x} \geq \mathbf{0} \qquad (5.3.17)$$

where $\pi^{\mathrm{T}} = \begin{bmatrix} \pi_1 & \pi_2 & \pi_3 & \pi_4 \end{bmatrix} = \begin{bmatrix} 2 & 2 & 0 & 0 \end{bmatrix}$, $\mathbf{x}^{\mathrm{T}} = \begin{bmatrix} x_1 & x_2 & s_1 & s_2 \end{bmatrix}$, $\mathbf{A} = \begin{bmatrix} 2 & 3 & 1 & 0 \\ 3 & 2 & 0 & 1 \end{bmatrix}$,

$\mathbf{b} = \begin{bmatrix} 90 \\ 120 \end{bmatrix}$, $\mathbf{0} = \begin{bmatrix} 0 \\ 0 \end{bmatrix}$; and $\mathbf{a}_1 = \begin{bmatrix} 2 \\ 3 \end{bmatrix}$, $\mathbf{a}_2 = \begin{bmatrix} 3 \\ 2 \end{bmatrix}$, $\mathbf{a}_3 = \begin{bmatrix} 1 \\ 0 \end{bmatrix}$, and $\mathbf{a}_4 = \begin{bmatrix} 0 \\ 1 \end{bmatrix}$.

Iteration 1

We begin the initial iteration treating the choice variables (x_1 and x_2) as NBVs and slack variables (s_1 and s_2) as BVs. Therefore, $\mathbf{x_{B,1}} = \begin{bmatrix} s_1 \\ s_2 \end{bmatrix}$, $\mathbf{B_1} = \begin{bmatrix} 1 & 0 \\ 0 & 1 \end{bmatrix}$, $\mathbf{B_1^{-1}} = \begin{bmatrix} 1 & 0 \\ 0 & 1 \end{bmatrix}$, and $\boldsymbol{\pi_{B,1}}^T = \begin{bmatrix} 0 & 0 \end{bmatrix}$, where the numerical subscript denotes the number of the iteration. These imply that $\mathbf{x_{B,1}} = \mathbf{B_1^{-1}b} = \begin{bmatrix} 1 & 0 \\ 0 & 1 \end{bmatrix} \begin{bmatrix} 90 \\ 120 \end{bmatrix} = \begin{bmatrix} 90 \\ 120 \end{bmatrix} = \begin{bmatrix} s_1 \\ s_2 \end{bmatrix}$, and $\Pi_1 = \boldsymbol{\pi_{B,1}}^T \mathbf{x_{B,1}} = \begin{bmatrix} 0 & 0 \end{bmatrix} \begin{bmatrix} 90 \\ 120 \end{bmatrix} = 0$.

Therefore, iteration 1 gives the solution, as expected, $s_1 = 90$, $s_2 = 90$, and $\Pi_0 = 0$. Let us now carry out the optimality and feasibility tests. *Optimality test*: $\Pi_{j,1} - \boldsymbol{\pi_{B,1}}^T = \boldsymbol{\pi_{B,1}}^T \mathbf{B_1^{-1}a_j} - \pi_j$. Since the existing NBVs are x_1 and x_2, we have $\mathbf{a_1}$ and $\mathbf{a_2}$. Therefore we obtain $\Pi_{1,1} - \boldsymbol{\pi_{B,1}}^T = \boldsymbol{\pi_{B,1}}^T \mathbf{B_1^{-1}a_1} - \pi_1 = \begin{bmatrix} 0 & 0 \end{bmatrix} \begin{bmatrix} 1 & 0 \\ 0 & 1 \end{bmatrix} \begin{bmatrix} 2 \\ 3 \end{bmatrix} - 2 = -2$, and

$\Pi_{2,1} - \boldsymbol{\pi_{B,1}}^T = \boldsymbol{\pi_{B,1}}^T \mathbf{B_1^{-1}a_2} - \pi_2 = \begin{bmatrix} 0 & 0 \end{bmatrix} \begin{bmatrix} 1 & 0 \\ 0 & 1 \end{bmatrix} \begin{bmatrix} 3 \\ 2 \end{bmatrix} - 2 = -2$, where the first subscript of Π denotes the number of NBV in the current basis. Since both give us the same negative value (and, therefore, the introduction of both x_1 and x_2 can increase the objective function equally), we arbitrarily choose x_1 as the entering variable. *Feasibility test*: Notice that $\mathbf{B_1^{-1}a_1} = \begin{bmatrix} 1 & 0 \\ 0 & 1 \end{bmatrix} \begin{bmatrix} 2 \\ 3 \end{bmatrix} = \begin{bmatrix} 2 \\ 3 \end{bmatrix}$ and $\mathbf{x_{B,1}} = \mathbf{B_1^{-1}b} = \begin{bmatrix} 90 \\ 120 \end{bmatrix}$. Therefore, $x_j = \min\{(\mathbf{x_{B,1}} = \mathbf{B_1^{-1}b})/(\mathbf{B_1^{-1}a_1})\} = \min\{90/2, 120/3\} = \min\{45, 40\}$, which implies that the current BV s_2 must leave the basis (and, thus, become a NBV).

Iteration 2

The current BVs are x_1 and s_1, and the NBVs are x_2 and s_2. Therefore, $\mathbf{x_{B,2}} = \begin{bmatrix} x_1 \\ s_1 \end{bmatrix}$, $\mathbf{B_2} = \begin{bmatrix} 2 & 1 \\ 3 & 0 \end{bmatrix}$, $\mathbf{B_2^{-1}} = \begin{bmatrix} 0 & 1/3 \\ 1 & -2/3 \end{bmatrix}$, and $\boldsymbol{\pi_{B,2}}^T = \begin{bmatrix} 2 & 0 \end{bmatrix}$. These quantities yield $\mathbf{x_{B,2}} = \mathbf{B_2^{-1}b} = \begin{bmatrix} 0 & 1/3 \\ 1 & -2/3 \end{bmatrix} \begin{bmatrix} 90 \\ 120 \end{bmatrix} = \begin{bmatrix} 40 \\ 10 \end{bmatrix} = \begin{bmatrix} x_1 \\ s_1 \end{bmatrix}$ and $\Pi_2 = \boldsymbol{\pi_{B,2}}^T \mathbf{x_{B,2}} = \begin{bmatrix} 2 & 0 \end{bmatrix} \begin{bmatrix} 40 \\ 10 \end{bmatrix} = 80$, which shows that the value of the objective function increased when x_1 entered the basis and s_2 left the basis. We can now carry out, as above, the optimality and feasibility tests. *Optimality test*: $\Pi_{j,2} - \boldsymbol{\pi_{B,2}}^T = \boldsymbol{\pi_{B,2}}^T \mathbf{B_2^{-1}a_j} - \pi_j$. Since the existing NBVs are x_2 and s_2, we are using $\mathbf{a_2} = \begin{bmatrix} 3 \\ 2 \end{bmatrix}$ and $\mathbf{a_4} = \begin{bmatrix} 0 \\ 1 \end{bmatrix}$. Therefore, we obtain $\Pi_{2,2} - \boldsymbol{\pi_{B,2}}^T = \boldsymbol{\pi_{B,2}}^T \mathbf{B_2^{-1}a_2} - \pi_2 = \begin{bmatrix} 2 & 0 \end{bmatrix} \begin{bmatrix} 0 & 1/3 \\ 1 & -2/3 \end{bmatrix} \begin{bmatrix} 3 \\ 2 \end{bmatrix} - 2 = (-2/3) < 0$, and $\Pi_{2,2} - \boldsymbol{\pi_{B,2}}^T = \boldsymbol{\pi_{B,2}}^T \mathbf{B_2^{-1}a_4} - \pi_{s2} = \begin{bmatrix} 2 & 0 \end{bmatrix} \begin{bmatrix} 0 & 1/3 \\ 1 & -2/3 \end{bmatrix} \begin{bmatrix} 0 \\ 1 \end{bmatrix} - 0 = 2/3 > 0$, where the first subscript of Π denotes the number of NBV in the current basis. Since x_2 is the only NBV with a negative value, it will only contribute to the value of the objective function. Therefore, x_2 enters the basis. *Feasibility test*: We have $\mathbf{B_2^{-1}a_2} = \begin{bmatrix} 0 & 1/3 \\ 1 & -2/3 \end{bmatrix} \begin{bmatrix} 3 \\ 2 \end{bmatrix} = \begin{bmatrix} 2/3 \\ 5/3 \end{bmatrix}$ and $\mathbf{x_{B,2}} = \mathbf{B_2^{-1}b} = \begin{bmatrix} 40 \\ 10 \end{bmatrix}$. Therefore, $x_j = \min\{(\mathbf{x_{B,2}} = \mathbf{B_2^{-1}b})/(\mathbf{B_2^{-1}a_2})\} = \min\{40/(2/3), 10/(5/3)\} = \min\{60, 6\}$, which implies that the current BV, s_1, must leave the basis.

Iteration 3

We now have x_1 and x_2 as the BVs and, therefore, we have $\mathbf{x_{B,3}} = \begin{bmatrix} x_1 \\ x_2 \end{bmatrix}$, $\mathbf{B_3} = \begin{bmatrix} 2 & 3 \\ 3 & 2 \end{bmatrix}$, $\mathbf{B_3}^{-1} = \begin{bmatrix} -2/5 & 3/5 \\ 3/5 & -2/5 \end{bmatrix}$, and $\pi_{B,3}^{\mathbf{T}} = \begin{bmatrix} 2 & 2 \end{bmatrix}$. These values yield $\mathbf{x_{B,3}} = \mathbf{B_3}^{-1}\mathbf{b} = \begin{bmatrix} -2/5 & 3/5 \\ 3/5 & -2/5 \end{bmatrix}\begin{bmatrix} 90 \\ 120 \end{bmatrix} = \begin{bmatrix} 36 \\ 6 \end{bmatrix} = \begin{bmatrix} x_1 \\ x_2 \end{bmatrix}$, and $\Pi_3 = \pi_{B,3}^{\mathbf{T}}\mathbf{x_{B,3}} = \begin{bmatrix} 2 & 2 \end{bmatrix}\begin{bmatrix} 36 \\ 6 \end{bmatrix} = 84$.

Notice that the current NBVs (s_1 and s_2) were the BVs when we started with iteration 1. This implies that the entry of any of these NBVs will reduce the value of the objective function. In addition, we know from equation (5.3.15) that the value of the objective function will increase only if $\Pi_j - \pi_j < 0$. And we know that $\Pi_j - \pi_j > 0$ for both NBVs (because $\Pi_j - \pi_j = 84 - 0 = 84 > 0$). Therefore, the current iteration (iteration 3) yields the optimum solution: $x_1 = 36$, $x_2 = 6$, and $\Pi^* = 84$. Notice also that these are exactly the same solutions as those we obtained in Sections 5.2.4 (through the graphical method) and 5.3.4 (through the tabular approach of the simplex method).

Let us now apply the revised simplex method to a minimization problem. For this we consider the minimization problem we solved with the graphical method (problem (5.2.11) in Section 5.2.5) and with the tabular approach of the simplex method (equation (5.3.8) in Section 5.3.5). For convenience we reproduce that problem here with $x_1 = x$ and $x_2 = y$: minimize $C = 9x_1 + 12x_2$, subject to $15x_1 + 10x_2 \geq 90$, $10x_1 + 15x_2 \geq 90$, and $x_1, x_2 \geq 0$. This can again be written with surplus variables (s_1 and s_2) and artificial variables (A_1 and A_2) as: minimize $C = 9x_1 + 12x_2 + 0s_1 + 0s_2 + MA_1 + MA_2$, subject to $-15x_1 - 10x_2 - s_1 - 0s_2 + A_1 + 0A_2 = -90$, $-10x_1 - 15x_2 + 0s_1 - s_2 + 0A_1 + A_2 = -90$, and $x_1, x_2, s_1, s_2, A_1, A_2 \geq 0$. As earlier, we can write this problem in matrix form as

$$\text{Minimize } C = \mathbf{c^T x}, \text{ subject to } \mathbf{Ex = d} \text{ and } \mathbf{x} \geq \mathbf{0} \tag{5.3.18}$$

where $\mathbf{c^T} = \begin{bmatrix} c_1 & c_2 & c_3 & c_4 & c_5 & c_6 \end{bmatrix} = \begin{bmatrix} 9 & 12 & 0 & 0 & M & M \end{bmatrix}$, $\mathbf{x^T} = \begin{bmatrix} x_1 & x_2 & s_1 & s_2 & R_1 & R_2 \end{bmatrix}$, $\mathbf{E} = \begin{bmatrix} 15 & 10 & -1 & 0 & 1 & 0 \\ 10 & 15 & 0 & -1 & 0 & 1 \end{bmatrix}$, $\mathbf{d} = \begin{bmatrix} 90 \\ 90 \end{bmatrix}$, and $\mathbf{0} = \begin{bmatrix} 0 \\ 0 \end{bmatrix}$, $\mathbf{e_1} = \begin{bmatrix} 15 \\ 10 \end{bmatrix}$, $\mathbf{e_2} = \begin{bmatrix} 10 \\ 15 \end{bmatrix}$, $\mathbf{e_3} = \begin{bmatrix} -1 \\ 0 \end{bmatrix}$, $\mathbf{e_4} = \begin{bmatrix} 0 \\ -1 \end{bmatrix}$, $\mathbf{e_5} = \begin{bmatrix} 1 \\ 0 \end{bmatrix}$, and $\mathbf{e_6} = \begin{bmatrix} 0 \\ 1 \end{bmatrix}$.

Iteration 1

We begin the initial iteration (iteration 1) treating the choice variables (x_1 and x_2) and the surplus variables (s_1 and s_2) as NBVs and artificial variables (A_1 and A_2) as BVs. Therefore, we have $\mathbf{x_{B,1}} = \begin{bmatrix} A_1 \\ A_2 \end{bmatrix}$, $\mathbf{B_1} = \begin{bmatrix} 1 & 0 \\ 0 & 1 \end{bmatrix}$, $\mathbf{B_1}^{-1} = \begin{bmatrix} 1 & 0 \\ 0 & 1 \end{bmatrix}$, and $\mathbf{c_{B,1}^T} = \begin{bmatrix} M & M \end{bmatrix}$, where the numerical subscript denotes the number of the iteration. These imply that $\mathbf{x_{B,1}} = \mathbf{B_1}^{-1}\mathbf{d} = \begin{bmatrix} 1 & 0 \\ 0 & 1 \end{bmatrix}\begin{bmatrix} 90 \\ 90 \end{bmatrix} = \begin{bmatrix} 90 \\ 90 \end{bmatrix} = \begin{bmatrix} A_1 \\ A_2 \end{bmatrix}$, and $C_1 = \mathbf{c_{B,1}^T x_{B,1}} = \begin{bmatrix} M & M \end{bmatrix}\begin{bmatrix} 90 \\ 90 \end{bmatrix} = 90M + 90M = 180M$.

Therefore, iteration 1 gives the solution, as expected, $x_1 = 0$, $x_2 = 0$, $s_1 = 0$, $s_2 = 0$, $A_1 = 90$, $A_2 = 90$, and $C = 180M$. Since $C_1 = 180M$ is a huge number, this solution to the minimization problem is unacceptable. We may now check whether we can reduce the value of C through further iterations. For this, let us now carry out the optimality and feasibility tests to determine the variables that enter and leave the basis. *Optimality test*: $\mathbf{c_{B,1}} - C_{j,1} = c_j - \mathbf{c_{B,1}^T B_1}^{-1}\mathbf{e_j}$.

Since the existing NBVs are x_1, x_2, s_1 and s_2, we have $\mathbf{e}_1 = \begin{bmatrix} 15 \\ 10 \end{bmatrix}, \mathbf{e}_2 = \begin{bmatrix} 10 \\ 15 \end{bmatrix}, \mathbf{e}_3 = \begin{bmatrix} -1 \\ 0 \end{bmatrix}$, and

$\mathbf{e}_4 = \begin{bmatrix} 0 \\ -1 \end{bmatrix}$. Therefore $\mathbf{c}_{B,1}{}^T - C_{1,1} = c_1 - \mathbf{c}_{B,1}{}^T \mathbf{B}_1{}^{-1}\mathbf{e}_1 = 9 - [M \ \ M]\begin{bmatrix} 1 & 0 \\ 0 & 1 \end{bmatrix}\begin{bmatrix} 15 \\ 10 \end{bmatrix} =$

$9 - 25M; \ \mathbf{c}_{B,2}{}^T - C_{2,1} = c_2 - \mathbf{c}_{B,2}{}^T\mathbf{B}_1{}^{-1}\mathbf{e}_2 = 12 - [M \ \ M]\begin{bmatrix} 1 & 0 \\ 0 & 1 \end{bmatrix}\begin{bmatrix} 10 \\ 15 \end{bmatrix} = 12 - 25M;$

$\mathbf{c}_{B,s1}{}^T - C_{s1,1} = c_{s1} - \mathbf{c}_{B,s1}{}^T\mathbf{B}_1{}^{-1}\mathbf{e}_3 = 0 - [M \ \ M]\begin{bmatrix} 1 & 0 \\ 0 & 1 \end{bmatrix}\begin{bmatrix} -1 \\ 0 \end{bmatrix} = M;$ and $\mathbf{c}_{B,s2}{}^T - C_{s2,1}$

$= c_{s2} - \mathbf{c}_{B,s2}{}^T\mathbf{B}_1{}^{-1}\mathbf{e}_4 = 0 - [M \ \ M]\begin{bmatrix} 1 & 0 \\ 0 & 1 \end{bmatrix}\begin{bmatrix} 0 \\ -1 \end{bmatrix} = M.$ Since $\mathbf{c}_{B,1} - C_{1,1} = 9 - 25M$ is

the most negative of all the above four equations, x_1 is the entering variable. *Feasibility test*:

We know that $\mathbf{B}_1{}^{-1}\mathbf{e}_1 = \begin{bmatrix} 1 & 0 \\ 0 & 1 \end{bmatrix}\begin{bmatrix} 15 \\ 10 \end{bmatrix} = \begin{bmatrix} 15 \\ 10 \end{bmatrix}$ and $\mathbf{x}_{B,1} = \mathbf{B}_1{}^{-1}\mathbf{d} = \begin{bmatrix} 90 \\ 90 \end{bmatrix}$. Therefore, $x_j =$

$\min\{(\mathbf{x}_{B,1} = \mathbf{B}_1{}^{-1}\mathbf{d})/(\mathbf{B}_1{}^{-1}\mathbf{e}_1)\} = \min\{90/15, 90/10\} = \min\{6, 9\}$, which implies that the current BV R_1 must leave the basis (and, thus, become a NBV).

Iteration 2

The current BVs are x_1 and A_2 and x_2, s_1, s_2, and A_1 are NBVs. Therefore, we have $\mathbf{x}_{B,2} =$

$\begin{bmatrix} x_1 \\ A_2 \end{bmatrix}, \mathbf{B}_2 = \begin{bmatrix} 15 & 0 \\ 10 & 1 \end{bmatrix}, \mathbf{B}_1{}^{-1} = \begin{bmatrix} 1/15 & 0 \\ -2/3 & 1 \end{bmatrix}$, and $\pi_{B,1}{}^T = [9 \ \ M]$. These quantities yield $\mathbf{x}_{B,2}$

$= \mathbf{B}_2{}^{-1}\mathbf{d} = \begin{bmatrix} 1/15 & 0 \\ -2/3 & 1 \end{bmatrix}\begin{bmatrix} 90 \\ 90 \end{bmatrix} = \begin{bmatrix} 6 \\ 30 \end{bmatrix} = \begin{bmatrix} x_1 \\ A_2 \end{bmatrix}$, and $C_2 = \mathbf{c}_{B,2}{}^T\mathbf{x}_{B,2} = [9 \ \ M]\begin{bmatrix} 6 \\ 30 \end{bmatrix} =$

$54 + 30M$, which shows that the value of the objective function decreased when x_1 entered the basis and A_1 left the basis. We can now carry out, as above, the optimality and feasibility tests. *Optimality test*: $\mathbf{c}_{B,2}{}^T C_{j,2} = c_j - \mathbf{c}_{B,2}{}^T\mathbf{B}_2{}^{-1}\mathbf{e}_j$. Since the existing NBVs are x_2, s_1, s_2,

and A_1, we obtain $\mathbf{c}_{B,2}{}^T - C_{2,2} = c_2 - \mathbf{c}_{B,2}{}^T\mathbf{B}_2{}^{-1}\mathbf{e}_2 = 12 - [9 \ \ M]\begin{bmatrix} 1/15 & 0 \\ -2/3 & 1 \end{bmatrix}\begin{bmatrix} 10 \\ 15 \end{bmatrix} =$

$-42 - 15M; \ \mathbf{c}_{B,s1}{}^T - C_{s1,2} = c_{s1} - \mathbf{c}_{Bs1}{}^T\mathbf{B}_2{}^{-1}\mathbf{e}_3 = 12 - [9 \ \ M]\begin{bmatrix} 1/15 & 0 \\ -2/3 & 1 \end{bmatrix}\begin{bmatrix} -1 \\ 0 \end{bmatrix} = 17.4;$

$\mathbf{c}_{B,s2}{}^T - C_{s2,2} = c_{s2} - \mathbf{c}_{B,s2}{}^T\mathbf{B}_2{}^{-1}\mathbf{e}_4 = 0 - [9 \ \ M]\begin{bmatrix} 1/15 & 0 \\ -2/3 & 1 \end{bmatrix}\begin{bmatrix} 0 \\ -1 \end{bmatrix} = M;$ and $\mathbf{c}_{B,A1}{}^T -$

$C_{A1,2} = c_{A1} - \mathbf{c}_{B,A1}{}^T\mathbf{B}_2{}^{-1}\mathbf{e}_5 = M - [9 \ \ M]\begin{bmatrix} 1/15 & 0 \\ -2/3 & 1 \end{bmatrix}\begin{bmatrix} 1 \\ 0 \end{bmatrix} = M - 5.4.$ Since $\mathbf{c}_{B,2}{}^T -$

$C_{2,2} = -42 - 15M$ is the most negative of all the above four values, x_2 is the entering

variable. *Feasibility test*: We know that $\mathbf{B}_2{}^{-1}\mathbf{e}_2 = \begin{bmatrix} 1/15 & 0 \\ -2/3 & 1 \end{bmatrix}\begin{bmatrix} 10 \\ 15 \end{bmatrix} = \begin{bmatrix} 2/3 \\ 25/3 \end{bmatrix}$ and $\mathbf{x}_{B,2} =$

$\mathbf{B}_2{}^{-1}\mathbf{d} = \begin{bmatrix} 6 \\ 30 \end{bmatrix}$. Therefore, $x_j = \min\{(\mathbf{x}_{B,2} = \mathbf{B}_2{}^{-1}\mathbf{d})/(\mathbf{B}_2{}^{-1}\mathbf{e}_1)\} = \min\{6/(2/3), 30/(25/3)\}$

$= \min\{9, 3.6\}$, which implies that the current BV A_2 must leave the basis (and, thus, become a NBV).

Iteration 3

We now have x_1 and x_2 as the basic variables and, therefore, we obtain $\mathbf{x}_{B,3} = \begin{bmatrix} x_1 \\ x_2 \end{bmatrix}$,

$\mathbf{B}_3 = \begin{bmatrix} 15 & 10 \\ 10 & 15 \end{bmatrix}, \ \mathbf{B}_3{}^{-1} = \begin{bmatrix} 3/25 & -2/25 \\ -2/25 & 3/25 \end{bmatrix}$ and $\mathbf{c}_{B,3}{}^T = [9 \ \ 12]$, which yield

$$\mathbf{x}_{B,3} = \mathbf{B}_3^{-1}\mathbf{d} = \begin{bmatrix} 3/25 & -2/25 \\ -2/25 & 3/25 \end{bmatrix}\begin{bmatrix} 90 \\ 90 \end{bmatrix} = \begin{bmatrix} 3.6 \\ 3.6 \end{bmatrix} = \begin{bmatrix} x_1 \\ x_2 \end{bmatrix}, \text{ and } C_3 = \mathbf{c}_{B,3}{}^T\mathbf{x}_{B,3} =$$

$$\begin{bmatrix} 9 & 12 \end{bmatrix}\begin{bmatrix} 3.6 \\ 3.6 \end{bmatrix} = 75.6.$$

Notice that two of the current NBVs (A_1 and A_2) were the BVs when we started with iteration 1. This implies that the entry of any of these NBVs into the basis will increase the value of the objective function. Since the surplus variables do not contribute to the objective function, entry of these into the basis will leave the value of the objective function unchanged. Moreover, we know that $c_j - C_j < 0$ for all NBVs (because $c_j - C_j = 0 - 75.6 = -75.6 < 0$). Therefore, the current iteration (iteration 3) yields the optimum solution: $x_1^* = 3.6$, $x_2^* = 3.6$, and $C^* = 75.6$. Notice also that these are exactly the same solutions as those we obtained in Sections 5.2.5 (through the graphical method) and 5.3.5 (through the tabular approach (the big-M and the two-phase methods) of the simplex method).

5.3.7 Application examples

Example 1. Suppose that a small pharmaceutical firm produces three types of drugs, x_1, x_2, and x_3, using three types of chemicals, A, B, and C, the maximum available quantities of which are 45, 40, and 25 units, respectively. The firm needs to use 1, 0, and 2 units, respectively, of A, B, and C to produce one unit of x_1; 0, 2, and 1 units of A, B, and C, respectively, to produce one unit of x_2; and 3, 2, and 0 units of A, B, and C, respectively, to produce one unit of x_3. The revenue that the firm obtains from one unit of x_1, x_2, and x_3 is $6, $5, and $3, respectively. Find the quantities of the drugs that the firm must produce to maximize its revenue.

Solution. Let us first set up a table that summarizes the above information, as presented in Table 5.3.23. On the basis of the information presented in Table 5.3.23, we can set up the LP problem as

$$\text{Maximize } \Pi = 6x_1 + 5x_2 + 3x_3, \text{ subject to } x_1 + 0x_2 + 3x_3 \le 45, x_1 + 2x_2 + 2x_3 \le 40,$$

$$2x_1 + x_2 + 0x_3 \le 25, \text{ and } x_1, x_2, x_3 \ge 0 \tag{5.3.19}$$

Let us first convert, by including slack variables, the above problem to

$$\text{Maximize } \Pi = 6x_1 + 5x_2 + 3x_3 + 0s_1 + 0s_2 + 0s_3,$$

$$\text{subject to } x_1 + 0x_2 + 3x_3 + s_1 + 0s_2 + 0s_3 = 45,$$

Table 5.3.23

	Output x_1 requirement	Output x_2 requirement	Output x_3 requirement	Input availability
Input A	1	0	3	45
Input B	1	2	2	40
Input C	2	1	0	25
	Profit per unit of x_1	Profit per unit of x_2	Profit per unit of x_3	
	$6	$5	$3	

Table 5.3.24

BVs	x_1	x_2	x_3	s_1	s_2	s_3	RHS
s_1	1	0	3	1	0	0	45
s_2	1	2	2	0	1	0	40
s_3	2	1	0	0	0	1	25
Π	-6	-5	-3	0	0	0	0

Indicators

Table 5.3.25

BVs	x_1	x_2	x_3	s_1	s_2	s_3	RHS
x_3	0	1	0	$-4/11$	6/11	$-5/11$	35/11
x_2	0	0	1	3/11	1/11	1/11	125/11
x_1	1	0	0	2/11	$-3/11$	8/11	120/11
Π	0	0	0	1/11	15/11	15/11	1270/11

Indicators

$$x_1 + 2x_2 + 2x_3 + 0s_1 + s_2 + 0s_3 = 40,$$

$$2x_1 + x_2 + 3x_3 + 0s_2 + s_3 = 25, \text{ and } x_1, x_2, x_3, s_1, s_2, s_3 \geq 0 \tag{5.3.20}$$

which is represented in the initial simplex tableau in Table 5.3.24.

Let us now carry out the operations exactly as we did earlier. Then, after three iterations, the final tableau will be as in Table 5.3.25. Notice that in Table 5.3.25, all the values in the last row are nonnegative (the condition for optimum in the case of maximization). Therefore, we attained optimum. The optimum solutions are $x_1^* = 120/11$, $x_2^* = 35/11$, $x_3^* = 125/11$, $s_1^* = 0$, $s_1^* = 0$, $s_1^* = 0$, and $\Pi^* = \$1270/11$.

Example 2. Assume that a company employs three salespeople, A, B, and C, temporarily to sell its three products, x_1, x_2, and x_3. It was directed that 1, 0, and 2 hours of A, B, and C to be spent for the sale of 1 unit of x_1; 0, 2, and 1 hours of A, B, and C to be spent for the sale of 1 unit of x_2; and 4, 2, and 0 hours of A, B, and C to be spent for the sale of 1 unit of x_3. The minimum number of hours to be spent by A, B, and C must be 40, 45, and 30 hours, respectively. The cost of sale of one unit of x_1, x_2, and x_3 is \$4, \$3, and \$2, respectively. How many units of the three goods should be sold so that the company's total cost will be minimum?

Solution. Let us first summarize the above information, as presented in Table 5.3.26. On the basis of this information, we can set up the LP problem as

$$\text{Minimize } C = 4x_1 + 3x_2 + 2x_3, \text{ subject to } x_1 + 0x_2 + 4x_3 \geq 40, 0x_1 + 2x_2 + 2x_3 \geq 45,$$

$$2x_1 + x_2 + 0x_3 \geq 30, \text{ and } x_1, x_2, x_3 \geq 0 \tag{5.3.21}$$

Table 5.3.26

Salesperson	One unit of x_1 requires	One unit of x_2 requires	One unit of x_3 requires	Minimum hours
A	1	0	4	40
B	0	2	2	45
C	2	1	0	30
	Cost per unit of x_1	Cost per unit of x_2	Cost per unit of x_3	
	\$4	\$3	\$2	

which we write, using surplus and artificial variables, as

Minimize $C = 4x_1 + 3x_2 + 2x_3 + 0s_1 + 0s_2 + 0s_3 + MA_1 + MA_2 + MA_3$,

subject to $x_1 + 0x_2 + 4x_3 - s_1 - 0s_2 - 0s_3 + A_1 + 0A_2 + 0A_3 = 40$,

$0x_1 + 2x_2 + 2x_3 - 0s_1 - s_2 - 0s_3 + 0A_1 + A_2 + 0A_3 = 45$,

$2x_1 + x_2 + 0x_3 - 0s_1 - 0s_2 - s_3 + 0A_1 + 0A_2 + A_3 = 30$,

and $x_1, x_2, x_3, s_1, s_2, s_3, A_1, A_2, A_3 \geq 0$ \hfill (5.3.22)

We can use either the big-M method or the two-phase method to solve this problem. Since the latter is computationally more efficient, we use that method. As required in the two-phase method, let us write the problem by including only slack and artificial variables in the objective function as

Minimize $C = 0s_1 + 0s_2 + 0s_3 + MA_1 + MA_2 + MA_3$,

subject to $x_1 + 0x_2 + 4x_3 - s_1 - 0s_2 - 0s_3 + A_1 + 0A_2 + 0A_3 = 40$,

$0x_1 + 2x_2 + 2x_3 - 0s_1 - s_2 - 0s_3 + 0A_1 + A_2 + 0A_3 = 45$,

$2x_1 + x_2 + 0x_3 - 0s_1 - 0s_2 - s_3 + 0A_1 + 0A_2 + A_3 = 30$,

and $x_1, x_2, x_3, s_1, s_2, s_3, A_1, A_2, A_3 \geq 0$ \hfill (5.3.23)

On the basis of equation (5.3.23), let us construct the initial simplex tableau as in Table 5.3.27. Now following exactly as we did in the application of the two-phase method we can obtain, after three iterations, the final tableau of the first phase as presented in Table 5.3.28. With this table, we have eliminated the artificial variables from the basis. Therefore, we have completed the first phase. Let us now begin the second phase. In the second phase we introduce the original objective function. With the original objective function (after removing the artificial variables), Table 5.3.28 can be reformulated to obtain Table 5.3.29. Let us now carry out elementary row operations ($R_4 \rightarrow R_4 + 4R_3$, $R_4 \rightarrow R_4 + 3R_2$, and $R_4 \rightarrow R_4 + 2R_1$) to obtain the final tableau as presented in Table 5.3.30.

Notice that, in Table 5.3.30, all the values in the last row (except those in the column RHS) are nonpositive, the condition for optimum in the case of minimization. Therefore, we stop the iteration with the last tableau and the optimum solution constitutes $x_1^* = 70/9$, $x_2^* = 520/36$, $x_3^* = 260/36$, $s_1^* = 0$, $s_2^* = 0$, $s_3^* = 0$, and $C^* = \$815/9$.

Table 5.3.27

BVs	x_1	x_2	x_3	s_1	s_2	s_3	A_1	A_2	A_3	RHS
A_1	1	0	4	−1	0	0	1	0	0	40
A_2	0	2	2	0	−1	0	0	1	0	45
A_3	2	1	0	0	0	−1	0	0	1	30
C	0	0	0	0	0	0	−1	−1	−1	0

Indicators

Table 5.3.28

BVs	x_1	x_2	x_3	s_1	s_2	s_3	A_1	A_2	A_3	RHS
x_3	0	0	1	−8/36	−2/36	1/9	8/36	2/36	−1/9	260/36
x_2	0	1	0	8/36	−16/36	−1/9	−8/36	16/36	1/9	520/36
x_1	1	0	0	−1/9	2/9	−4/9	1/9	−2/9	4/9	70/9
C	0	0	0	0	0	0	−1	−1	−1	0

Indicators

Table 5.3.29

BVs	x_1	x_2	x_3	s_1	s_2	s_3	RHS
x_3	0	0	1	−8/36	−2/36	1/9	260/36
x_2	0	1	0	8/36	−16/36	−1/9	520/36
x_1	1	0	0	−1/9	2/9	−4/9	70/9
C	−4	−3	−2	0	0	0	0

Indicators

Table 5.3.30

BVs	x_1	x_2	x_3	s_1	s_2	s_3	RHS
x_3	0	0	1	−8/36	−2/36	1/9	260/36
x_2	0	1	0	8/36	−16/36	−1/9	520/36
x_1	1	0	0	−1/9	2/9	−4/9	70/9
C	0	0	0	−16/36	−20/36	−17/9	815/9

Indicators

5.3.8 Exercises

1. Solve the following LP problems using the tabular approach of the simplex method:
   (i) maximize $\Pi = 30x_1 + 25x_2$, subject to $25x_1 + 20x_2 \leq 80$, $20x_1 + 25x_2 \leq 90$, and $x_1, x_2 \geq 0$; (ii) maximize $\Pi = 10x_1 + 5x_2$, subject to $5x + 2x_2 \leq 34$, $4x_1 + 4x_2 \leq 40$, and $x_1, x_2 \geq 0$; (iii) maximize $\Pi = 30x_1 + 25x_2 + 20x_3$, subject to $25x_1 + 10x_2 + 15x_3 \leq 80$, $20x_1 + 25x_2 + 10x_3 \leq 90$, and $x_1, x_2, x_3 \geq 0$; (iv) maximize $\Pi = 3x_1 + 2x_2 + x_3$, subject to

$2x_1 + x_2 + 0x_3 \leq 60$, $0x_1 + 2x_2 + x_3 \leq 80$, $10x_1 + 15x_2 + 20x_3 \leq 100$, $x_1 + x_2 + 2x_3 \leq 90$, and $x_1, x_2, x_3 \geq 0$.

2. Solve the LP problems in exercise 1 above using the revised simplex method.
3. Solve the following LP problems using the tabular approaches (the big-M method and the two-phase methods) of the simplex method:
    (i) minimize $C = 8x_1 + 10x_2$, subject to $x_1 + 3x_2 \geq 30$, $4x_1 + x_2 \geq 20$, and $x_1, x_2 \geq 0$;
    (ii) minimize $C = 5x_1 + 3x_2$, subject to $6x + 3x_2 \geq 30$, $x_1 + 4x_2 \geq 20$, and $x_1, x_2 \geq 0$;
    (iii) minimize $C = 4x_1 + 3x_2 + 2x_3$, subject to $x_1 + 4x_2 + x_3 \geq 40$, $5x_1 + 4x_2 + x_3 \geq 50$, $2x_1 + x_2 + x_3 \geq 60$, and $x_1, x_2, x_3 \geq 0$; (iv) minimize $C = 8x_1 + 5x_2 + 3x_3$, subject to $0x_1 + 4x_2 + x_3 \geq 20$, $5x_1 + 0x_2 + 4x_3 \geq 40$, $4x_1 + 4x_2 + 0x_3 \geq 40$, and $x_1, x_2, x_3 \geq 0$.
4. Solve the LP problems in exercise 3 using the revised simplex method.
5. Solve the following LP problems using the revised simplex method:
    (i) maximize $\Pi = 3x_1 + 2x_2 + x_3$, subject to $x_1 + x_2 5x_3 \geq 10$, $5x_1 + 2x_2 + 3x_3 \leq 15$, $x_1 + 2x_2 + x_3 = 10$, and $x_1, x_2, x_3 \geq 0$; (ii) minimize $C = x_1 + 2x_2 + 3x_3$, subject to $x_1 + x_2 5x_3 \geq 10$, $5x_1 + 2x_2 + 3x_3 \leq 15$, $x_1 + 2x_2 + x_3 = 10$, and $x_1, x_2, x_3 \geq 0$.
6. *Application exercise.* Suppose that a firm produces three goods (x_1, x_2, and x_3) using three factors (capital, labor, and land). In order to produce 1 unit of x_1, the firm has to use 2, 1, and 1 units of capital, labor, and land, respectively. Similarly, to produce 1 unit of x_2, the firm has to use 1, 2, and 1 units of capital, labor, and land, respectively. To produce 1 unit of x_3, the firm has to use 1, 1, and 2 units of capital, labor, and land, respectively. The maximum quantities of capital, labor, and land available for use are 100, 120, and 80, respectively. The unit profits that the firm obtains from x_1, x_2, and x_3 are \$20, \$18, and \$16, respectively. Find the quantities of the three goods that the firm must produce to maximize its total profits using both the tabular approach of the simplex method and the revised simplex method.
7. *Application exercise.* Assume that a catering firm received an order for supplying three dishes (x_1, x_2, and x_3) from a hostel. The dishes must contain three nutrients, A, B, and C. Each unit of dish x_1 must contain 4, 1, and 1 units of nutrients A, B, and C, respectively. Similarly, each unit of dish x_2 must contain 1, 4, and 1 units of nutrients A, B, and C, respectively. Each unit of dish x_3 must contain 1, 1, and 4 units of nutrients A, B, and C, respectively. The minimum quantities of these nutrients in all dishes must be 60, 30, and 60 units, respectively. The cost of one unit of x_1, x_2, and x_3 is \$3, \$2, and \$1, respectively. Find the quantities of the three dishes that the firm must supply to minimize its total cost using both the tabular approach of the simplex method and the revised simplex method.

 Web supplement: S5.3.9 Mathematica applications

5.4 LP: duality and sensitivity

5.4.1 Introduction

So far in this chapter, we have attempted to solve LP maximization and minimization problems independently of one another. One might wonder if there exists any relationship between these problems. Yes, there exists; because every LP problem is related to another LP problem. An example can make clear this relationship between two LP problems. Assume that a producer wants to maximize the total profits obtained from the goods produced. We can approach the producer's optimization problem in two ways. One approach is to maximize the total output produced, given the contributions of these outputs to total profit and the resource constraints. The second approach is to minimize the total *opportunity cost* of producing these goods, given the contributions to this cost of these goods and the profit constraints. The former (called the *primal*) gives the maximum profit that can be achieved, and the latter (called the *dual*) gives the minimum opportunity cost needed to obtain a certain level of profit. What all this implies is that if the primal is a maximization problem, then the dual is a minimization problem, and vice versa. However, it must be noted that both approaches give identical optimal values of the choice variables. We will consider duality in the context of LP problems from Section 5.4.2 onwards. The study of the relationship between the primal and dual problems will take us, as we will see shortly, to important theorems and interesting results. We will also consider what is popularly called the *sensitivity analysis* or *postoptimality analysis* in Section 5.4.9.

5.4.2 The duals of LP problems

Assume that our primal problem is to maximize the objective function in equation (5.2.29) subject to constraints (5.2.30), the matrix representations of which were given in equation (5.2.31) and inequalities (5.2.32), respectively. The dual of this maximization problem is a minimization problem and can be written as

$$\text{Minimize } \Pi_d = b_1 y_1 + b_2 y_2 + \cdots + b_i y_i + \cdots + b_m y_m \tag{5.4.1}$$

$$\text{Subject to } a_{11} y_1 + a_{21} y_2 + \cdots + a_{i1} y_i + \cdots + a_{m1} y_m \geq \pi_1,$$

$$a_{12} y_1 + a_{22} y_2 + \cdots + a_{i2} y_i + \cdots + a_{m2} y_m \geq \pi_2, \ldots,$$

$$a_{1j} y_1 + a_{2j} x_2 + \cdots + a_{ij} y_i + \cdots + a_{mj} y_m \geq \pi_j, \ldots,$$

$$a_{1n} y_1 + a_{2n} y_2 + \cdots + a_{in} y_i + \cdots + a_{mn} y_m \geq \pi_n,$$

$$\text{and } y_i \geq 0, i = 1, 2, 3, \ldots, m, \ j = 1, 2, \ldots, n \tag{5.4.2}$$

which can be written in matrix notations as

$$\text{Minimize } \Pi_d = \mathbf{b}^T \mathbf{y} \tag{5.4.3}$$

$$\text{Subject to } \mathbf{A}^T \mathbf{y} \geq \boldsymbol{\pi}, \text{ and } \mathbf{y} \geq 0 \tag{5.4.4}$$

where $\mathbf{b}^T = \begin{bmatrix} b_1 & b_2 & .. & b_i & .. & b_m \end{bmatrix}$,

$$\mathbf{y} = \begin{bmatrix} y_1 \\ y_2 \\ .. \\ y_i \\ .. \\ y_m \end{bmatrix}, \mathbf{A}^T = \begin{bmatrix} a_{11} & a_{21} & .. & a_{i1} & .. & a_{m1} \\ a_{12} & a_{22} & .. & a_{i2} & .. & a_{m2} \\ .. & .. & .. & .. & .. & .. \\ a_{1j} & a_{2j} & .. & a_{ij} & .. & a_{mj} \\ .. & .. & .. & .. & .. & .. \\ a_{1n} & a_{2n} & .. & a_{in} & .. & a_{mn} \end{bmatrix}, \text{ and } \boldsymbol{\pi} = \begin{bmatrix} \pi_1 \\ \pi_2 \\ .. \\ \pi_j \\ .. \\ \pi_n \end{bmatrix}$$

Assume now that our primal problem is to minimize the objective function in equation (5.2.35) subject to constraints (5.2.36), the matrix forms of which were given in equation (5.2.37) and inequalities (5.2.38), respectively. The dual of this minimization problem is a maximization problem and can be written as

$$\text{Maximize } C_d = d_1 y_1 + d_2 y_2 + \cdots + d_i y_i + \cdots + d_m y_m \tag{5.4.5}$$

$$\text{Subject to } e_{11} y_1 + e_{21} y_2 + \cdots + e_{i1} y_i + \cdots + e_{m1} y_m \leq c_1,$$

$$e_{12} y_1 + e_{22} y_2 + \cdots + e_{i2} y_i + \cdots + e_{m2} y_m \leq c_2, \ldots,$$

$$e_{1j} y_1 + e_{2j} x_2 + \cdots + e_{ij} y_i + \cdots + e_{mj} y_m \leq c_j, \ldots,$$

$$e_{1n} y_1 + e_{2n} y_2 + \cdots + e_{in} y_i + \cdots + e_{mn} y_m \leq c_n,$$

$$\text{and } y_i \geq 0, i = 1, 2, 3, \ldots, m, \ j = 1, 2, \ldots, n \tag{5.4.6}$$

which can be written in matrix notations as

$$\text{Maximize } C_d = \mathbf{d}^{\mathrm{T}} \mathbf{y} \tag{5.4.7}$$

$$\text{Subject to } \mathbf{E}^{\mathrm{T}} \mathbf{y} \leq \mathbf{c}, \text{ and } \mathbf{y} \geq 0 \tag{5.4.8}$$

where $\mathbf{d}^{\mathrm{T}} = \begin{bmatrix} d_1 & d_2 & .. & d_i & .. & d_m \end{bmatrix}$,

$$\mathbf{y} = \begin{bmatrix} y_1 \\ y_2 \\ .. \\ y_i \\ .. \\ y_m \end{bmatrix}, \mathbf{E}^{\mathrm{T}} = \begin{bmatrix} e_{11} & e_{21} & .. & e_{i1} & .. & e_{m1} \\ e_{12} & e_{22} & .. & e_{i2} & .. & e_{m2} \\ .. & .. & .. & .. & .. & .. \\ e_{1j} & e_{2j} & .. & e_{ij} & .. & e_{mj} \\ .. & .. & .. & .. & .. & .. \\ e_{1n} & e_{2n} & .. & e_{in} & .. & e_{mn} \end{bmatrix}, \text{ and } \mathbf{c} = \begin{bmatrix} c_1 \\ c_2 \\ .. \\ c_j \\ .. \\ c_n \end{bmatrix}$$

As an example of the dual of a maximization LP problem, consider the maximization problem (5.2.5), which we reproduce below with $x_1 = x$ and $x_2 = y$:

$$\text{Maximize } \Pi = 2x_1 + 2x_2, \text{ subject to } 2x_1 + 3x_2 \leq 90, 3x_1 + 2x_2 \leq 120, \text{ and } x_1, x_2 \geq 0 \tag{5.4.9}$$

The dual of the maximization problem (5.4.9) can be written as

$$\text{Minimize } \Pi_d = 90y_1 + 120y_2, \text{ subject to } 2y_1 + 3y_2 \geq 2, 3y_1 + 2y_2 \geq 2, \text{ and } y_1, y_2 \geq 0 \tag{5.4.10}$$

Similarly, as an example of a minimization LP problem, consider the minimization problem (5.2.11), which we reproduce below with $x_1 = x$ and $x_2 = y$:

$$\text{Minimize } C = 9x_1 + 12x_2, \text{ subject to } 15x_1 + 10x_2 \geq 90,$$

$$10x_1 + 15x_2 \geq 90, \text{ and } x_1, x_2 \geq 0 \tag{5.4.11}$$

The dual of the minimization problem (5.4.11) can be written as

$$\text{Maximize } C_d = 90y_1 + 90y_2, \text{ subject to } 15y_1 + 10y_2 \leq 9,$$

$$10y_1 + 15y_2 \leq 12, \text{ and } y_1, y_2 \geq 0 \tag{5.4.12}$$

As another example of the dual of a minimization problem, consider the problem

Minimize $C = 25x_1 + 20x_2 + 15x_3$, subject to $9x_1 + 6x_2 + 3x_3 \geq 30$,

$2x_1 + 4x_2 + 8x_3 \geq 20$, and $x_1, x_2, x_3 \geq 0$ (5.4.13)

The dual of the minimization problem (5.4.13) can be written as

Maximize $C_d = 30y_1 + 20y_2$, subject to $9y_1 + 2y_2 \leq 25, 6y_1 + 4y_2 \leq 20$,

$3y_1 + 8y_2 \leq 15$, and $y_1, y_2 \geq 0$ (5.4.14)

We will solve the dual problems in (5.4.10), (5.4.12), and (5.4.14) in Section 5.4.5. We will explore there the relationships between the solutions to the dual problems in (5.4.10) and (5.4.12) and the solutions to their respective primal problems (5.4.9) and (5.4.11).

5.4.3 Relationships between primal and dual LP problems

Let us list below some of the important relationships that exist between the primal LP problem and its dual.

1. If the primal LP problem is a maximization (minimization) problem, then its dual is a minimization (maximization) problem. This means that we have to change the $\leq (\geq)$ sign to $\geq (\leq)$ in the constraints when we set up the dual of a primal maximization (minimization) problem.
2. The matrix of the coefficients of the constraints of the dual problem is the transpose of the matrix of the coefficients of the constraints of the primal problem. Therefore, when we set up the dual of a LP problem we need to transpose the matrix of the coefficients of the constraints.
3. The column vector of the constants (the RHS) in the dual problem is the transpose of the row vector of the coefficients of the objective function in the primal problem. This implies that when we set up the dual of a LP problem we need to transpose the coefficients of the objective function in the primal problem to form the column vector of the constants in the dual problem.
4. The row vector of the coefficients of the objective function in the dual problem is the transpose of the column vector of constants (the RHS) in the primal problem. Therefore, when we set up the dual of a LP problem we need to transpose the column vector of constants in the primal problem to obtain the row vector of the coefficients of the objective function in the dual problem.
5. The dual of a dual LP problem is the primal LP problem.
6. The variables in both the primal problem and the dual problem are nonnegative.
7. If a LP problem involving three choice variables and two constraints (such as problem (5.4.13)) is difficult to solve by the graphical method; its dual with two choice variables and three constraints (given in problem (5.4.14)) can be solved by the graphical method.

5.4.4 Duality: two important theorems

We found above the important relationships that exist between the primal and the dual LP problems. We now turn our attention to two important theorems, called *duality theorems*, which are highly useful, as we will see shortly, in deriving inferences about the solution to the primal problem from the dual problem. These theorems are as follows.

Theorem 1.   If an optimal feasible solution exists, the optimal values of the primal problem and the optimal values of the dual problem are identical. This means that, as per our previous notations, $\Pi^* = \Pi_d^*$ and $C^* = C_d^*$.

Theorem 2.   Suppose that the slack or surplus variable (also called *dummy variable*) in the ith constraint of the primal problem is denoted by s_i and the jth dummy variable in the dual problem is denoted by t_j. If, in the optimum, $x_j^* > 0$, then $t_j^* = 0$; if $y_i^* > 0$, then $s_i^* = 0$. This means that if a choice variable in the primal problem is nonzero in the optimum, its corresponding dummy variable in the dual problem must be zero in the optimum; and if a choice variable in the dual problem is nonzero in the optimum, its corresponding dummy variable in the primal problem must be zero in the optimum. This theorem is called *complementary slackness property* or *complementary slackness condition* or *complementary slackness principle*.

5.4.5 Solution of dual problems

We attempt in this section to solve the dual problems presented in Section 5.4.2. Let us first consider the problem in (5.4.10), which is a minimization problem. Using the two-phase method discussed in Section 5.3.5, we can obtain the final tableau that presents the optimum solution to this minimization problem as in Table 5.4.1.

Table 5.4.1 shows that $y_1^* = 2/5$, $y_2^* = 2/5$, $t_1^* = 0$, $t_2^* = 0$, and $C_d^* = 84$. Notice the few remarkable outcomes we obtain when we solve a dual problem using the tabular approach of the simplex method. The first is that the table gives the optimal solution, as just stated, to the dual problem. The second outcome is the sufficient information it gives about the solution to the primal problem even without directly solving it. This outcome comprises three parts. Firstly, the optimal value of the dual objective function ($C_d^* = 84$) is the same as the optimal value of the primal objective function we obtained when we solved the same primal problem in Section 5.3.4. This is exactly what duality theorem 1 states. Secondly, the absolute values among the indicators (36 and 6) under the columns of the dummy variables (t_1 and t_2 in the present example) are the same as the optimal values of the choice variables we obtained when we solved the primal problem in Sections 5.2.4 and 5.3.4. This can be taken as a general rule. Since $x_1^* = 36 > 0$ and $x_2^* = 6 > 0$ in the primal solution, $t_1^* = 0$ and $t_2^* = 0$ in the dual solution. This result is in accordance with duality theorem 2. Thirdly, the values (zeros in the present example) among the indicators under the columns of the choice variables (y_1 and y_2 in the present example) are the same as the optimal values of the dummy variables ($s_1^* = 0$

Table 5.4.1

BVs	y_1	y_2	t_1	t_2	RHS
y_1	0	1	$-3/5$	$2/5$	$2/5$
y_2	1	0	$2/5$	$-3/5$	$2/5$
C_d	0	0	-36	-6	84

Indicators

Table 5.4.2

BVs	y_1	y_2	t_1	t_2	RHS
y_1	1	0	3/25	$-2/25$	3/25
y_2	0	1	$-2/25$	3/25	18/25
C_d	0	0	3.6	3.6	75.6

Indicators

and $s_2^* = 0$) we obtained in the primal solution in Section 5.3.4. This can also be taken as a general rule. Since $t_1^* = 0$ and $t_2^* = 0$, $x_1^* = 36 > 0$ and $x_2^* = 6 > 0$; and since $y_1^* = 2/5 > 0$, $y_2^* = 2/5 > 0$, $s_1^* = 0$, and $s_2^* = 0$. This result is also in accordance with duality theorem 2.

Let us now consider the problem in (5.4.12), which is a maximization problem (the primal of which we solved in Sections 5.2.5 and 5.3.5). Following the tabular approach of the simplex method discussed in Section 5.3.4, we can obtain the final tableau that presents the solution to this maximization problem as presented in Table 5.4.2. Table 5.4.2 shows that the optimal solution comprises $y_1^* = 3/25$, $y_2^* = 18/25$, $t_1^* = 0$, $t_2^* = 0$, and $C_d^* = 75.6$. We can read, as stated above, the solution to the primal problem from the table as $x_1^* = 3.6$, $x_2^* = 3.6$, $s_1^* = 0$, $s_2^* = 0$, and $C^* = 75.6$. Notice that $C_d^* = 75.6 = C^*$, as stated by duality theorem 1. Since $x_1^* = 3.6 > 0$ and $x_2^* = 3.6 > 0$, $t_1^* = 0$ and $t_2^* = 0$; and since $s_1^* = 0$ and $s_2^* = 0$, $y_1^* = 3/25 > 0$ and $y_2^* = 18/25 > 0$, as stated by duality theorem 2.

Lastly, we shall consider the solution of the minimization problem in (5.4.13). We know that we have to use artificial variables to solve this problem and that the use of these variables is more cumbersome than to solve maximization problems. Another issue with problem (5.4.13) is that it involves three choice variables and, therefore, we cannot use the graphical method. However, finding the solution to this minimization problem is a simple matter if we solve its dual because the solution to this dual, as stated above, will yield the solution to its primal. Therefore, we attempt to solve the maximization problem (5.4.14), which is the dual of the minimization problem (5.4.13). Following the tabular approach of the simplex method discussed in Section 5.3.4, we can obtain the final tableau that presents the solution to this maximization problem as in Table 5.4.3. Table 5.4.3 shows that the optimal solution comprises $y_1^* = 255/99$, $y_2^* = 10/11$, $t_1^* = 0$, $t_2^* = 10/11$, $t_3^* = 0$, and $C_d^* = 95.4$. We can read from Table 5.4.3, as in the previous examples, the solution to the primal problem: $x_1^* = 2.73$, $x_2^* = 0$, $x_3^* = 1.81$, $s_1^* = 0$, $s_2^* = 0$, $s_2^* = 0$, and $C^* = 95.4$. This example illustrates how we can find comparatively easily the solution to a complicated primal problem using the solution to its dual.

Table 5.4.3

BVs	y_1	y_2	t_1	t_2	t_3	RHS
y_1	1	0	12/99	0	$-1/33$	255/99
t_2	0	0	$-6/11$	1	$-24/66$	10/11
y_2	0	1	$-1/22$	0	9/66	10/11
C_d	0	0	2.73	0	1.81	95.4

Indicators

5.4.6 Duality: economic interpretations

We saw above that the solution to the dual of a primal LP problem could be used to read the solution to the primal too. However, the dual problems also possess some economic meanings. We now attempt to see these economic meanings, popularly called *economic interpretations*, of dual LP problems.

Let us begin with the general maximization problem in matrix forms in (5.2.31) and (5.2.32). For convenience we reproduce the problem here: Maximize $\Pi = \pi^T x$ subject to $Ax \leq b$ and $x \geq 0$, where π^T is the row vector of the coefficients of the objective function, x is the column vector of the choice variables, A is the matrix of the coefficients of the LHS of the constraints, and b is the column vector of the RHS of the constraints. The dual of this maximization problem is a minimization problem and is given in (5.4.3) and (5.4.4) in matrix forms. We reproduce it also here minimize $\Pi_d = b^T y$ subject to $A^T y \geq \pi$ and $y \geq 0$. The only new factor in this dual is the column vector y.

As we saw, every LP problem has three elements: the objective function, the technical constraints, and the nonnegativity constraints. In all the maximization problems our objective was to maximize total profit, Π, which was expressed in monetary units (dollars). For Π to be a monetary unit, the contribution of the jth good (x_j) must be in monetary units; that is, π_j must be in monetary unit. Therefore, we considered π_j in terms of dollars. This meant that our aim was to maximize the dollar value of the objective function. However, the elements of the matrix A are the technical coefficients and the elements of b are the maximum availabilities of the resources to be used in the production of the goods given by the elements of x, which is a sensible restriction. This means that the total quantities of the resources used up in the production process must be less than or equal to the total availabilities of the resources, which is a meaningful restriction. The nonnegativity condition means that the quantities of the goods (choice variables) produced must be nonnegative, which, again, is a meaningful restriction.

We can now consider the dual of the primal maximization problem. The dual is a minimization problem. In the dual objective function, $\Pi_d = b^T y$, b^T is the transpose of the column vector of resources. We know that $\Pi_d^* = \Pi^*$ in optimum. This implies that Π_d must also be in monetary or dollar terms. For Π_d to be in dollar terms, y must be expressed in dollar terms. In fact, the vector y is the vector of *imputed values*, also called *accounting price*, or *shadow price*, or *opportunity cost*, of the resources (in the vector b^T) used in the production process. Therefore, in the dual problem our objective is to minimize the total opportunity cost because Π_d represents the total opportunity cost. Now consider the constraints in the dual problem. A particular element, say a_{ji}, of A^T of a particular constraint shows the quantity of the ith resource used in the production of different goods. If we multiply this element by y_i we obtain the opportunity cost of producing one unit of good x_j. Therefore, the LHS of the ith constraint gives the total opportunity cost of using one unit of a particular resource. The RHS of the corresponding constraint shows the profit from one unit of x_j. What this means is that the total opportunity cost of producing one unit of a good must be equal to (or greater than, if the good is not produced in the optimum) the profit from that unit. As before, this is a sensible restriction, for otherwise profit can be increased through a reallocation of the resources and the current position would not be optimum. Lastly, the nonnegativity condition states that the opportunity costs of each resource must be nonnegative, which, again, is sensible.

Let us now give an economic interpretation to the dual of the primal minimization problem given in (5.2.37) and (5.2.38), which we reproduce here for convenience: minimize $C = c^T x$ subject to $Ex \geq d$, and $x \geq 0$, where c^T is the row vector of the coefficients of the objective function, $\mathbf{x}$ is the column vector of the choice variables, $\mathbf{E}$ is the matrix of the coefficients of the LHS of the constraints, and $\mathbf{d}$ is the column vector of the RHS of the constraints. The dual of this minimization problem is a maximization problem and is given in (5.4.7) and (5.4.8) in matrix forms. We reproduce it here: maximize $C_d = \mathbf{d}^T\mathbf{y}$, subject to $\mathbf{E}^T\mathbf{y} \leq \mathbf{c}$, and $\mathbf{y} \geq 0$. The only new notation in this dual, as before, is the column vector $\mathbf{y}$.

As in the case of a primal maximization problem, $\mathbf{c}^T$ is expressed in dollar terms (i.e. the imputed value per unit of the nutrient). This implies that C also is expressed in terms of dollars. Since $C^* = C_d^*$ at the optimum, C_d^* must also be in dollars. For C_d^* to be in dollars, $\mathbf{y}$, as before, must be in dollars because $\mathbf{d}^T$ is in physical units. This means that, as in the dual of a maximization problem, $\mathbf{y}$ represents the imputed values of the nutrients in the dual of a minimization problem. Therefore, the constraint $\mathbf{E}^T\mathbf{y} \leq \mathbf{c}$ implies that the total imputed value of a particular nutrient in one unit of the corresponding food item must be equal to (or less than, when the corresponding food item is not produced at the optimum) the cost per unit of the food item. It is needless to state that $\mathbf{y} \geq 0$, as imputed values cannot be negative.

5.4.7 Revised simplex method for dual problems

In Section 5.3.6 we presented the revised simplex method. In the revised simplex method we employed two criteria: the optimality criterion and the feasibility criterion. In Section 5.4.5, we solved dual LP problems using the tabular approach of the simplex method. As we have seen that the revised simplex method is computationally more efficient than the tabular approach, we explore here the closely related revised simplex method for solving dual problems, which is more efficient as it is useful in *sensitivity analysis*, a topic to be dealt with in Section 5.4.9. This method also involves the application of two criteria with the same names as those referred to above. Since we have already learned how to solve dual LP problems using the tabular approach of the simplex method, our aim here is not to apply this new method but to use it later in sensitivity analysis. Therefore, we will not present the applications of the revised simplex method for dual problems here.

Assume that our primal problem is to maximize $\Pi = \pi^T\mathbf{x}$, subject to $\mathbf{Ax} \leq \mathbf{b}$ and $\mathbf{x} \geq 0$. The dual of this maximization problem is to minimize $\Pi_d = \mathbf{b}^T\mathbf{y}$, subject to $\mathbf{A}^T\mathbf{y} \geq \pi$ and $\mathbf{y} \geq 0$. Notice that π, π^T, $\mathbf{b}$, $\mathbf{b}^T$, $\mathbf{x}$, $\mathbf{y}$, $\mathbf{A}$, and $\mathbf{A}^T$ are all defined the same as before. Before we proceed further, we need to augment $\mathbf{A}^T$ by the coefficients of the surplus and artificial variables and define $\mathbf{a}_1^T, \mathbf{a}_2^T, \ldots, \mathbf{a}_m^T$ so that they will respectively be

$$\mathbf{A^TaugI} = \begin{bmatrix} a_{11} & a_{21} & .. & a_{m1} & 1 & 0 & .. & 0 & -1 & 0 & .. & 0 \\ a_{12} & a_{22} & .. & a_{m2} & 0 & 1 & .. & 0 & 0 & -1 & .. & 0 \\ .. & .. & .. & .. & .. & .. & .. & .. & .. & .. & .. & .. \\ a_{1n} & a_{2n} & .. & a_{mn} & 0 & 0 & .. & 1 & 0 & 0 & .. & -1 \end{bmatrix},$$

$$\text{and } \mathbf{a_1}^T = \begin{bmatrix} a_{11} \\ a_{12} \\ .. \\ a_{1n} \end{bmatrix}, \mathbf{a_2}^T = \begin{bmatrix} a_{21} \\ a_{22} \\ .. \\ a_{2n} \end{bmatrix}, \ldots, \mathbf{a_m}^T = \begin{bmatrix} 0 \\ 0 \\ .. \\ -1 \end{bmatrix}$$

Now suppose that $\mathbf{F}$ is a feasible basis of the above dual. Also suppose that $\mathbf{y_F}$ constitutes, corresponding to $\mathbf{F}$, the set of BVs and that $\mathbf{b_F}^T$ is the vector of the corresponding coefficients in the dual objective function. Now eliminating n variables by treating them as zeros gives

m equations in m unknown BVs. This system of equations can be written as

$$\mathbf{F}\mathbf{y_F} = \pi \tag{5.4.15}$$

which can be pre-multiplied on both sides by the inverse of $\mathbf{F}$, $\mathbf{F}^{-1}$, to obtain

$$\mathbf{y_F} = \mathbf{F}^{-1}\pi \tag{5.4.16}$$

Notice that we introduced the inverse of the feasible basis $\mathbf{F}$ in equation (5.4.16). Since the simplex method introduces only the BVs, $\mathbf{F}$ is always nonsingular and, therefore, $\mathbf{F}^{-1}$ will always exist. We know that the objective function is given by $\Pi_d = \mathbf{b}^\mathrm{T}\mathbf{y}$ and that the current set of BVs is given by $\mathbf{y_F}$. Therefore, the value of the objective function in the current BFS can be written as $\Pi_d = \mathbf{b_F}^\mathrm{T}\mathbf{y_F}$, which, using equation (5.4.16), can be written as

$$\Pi_d = \mathbf{b_F}^\mathrm{T}\mathbf{F}^{-1}\pi \tag{5.4.17}$$

Equation (5.4.16) gives the first BFS and equation (5.4.17) gives the value of the objective function corresponding to the first BFS. We can now introduce a NBV, y_i, into the basis to see if it decreases the value of the objective function. Then the following equation will hold:

$$\mathbf{F}\mathbf{y_F} + \mathbf{a}_i^\mathrm{T} y_i = \pi \tag{5.4.18}$$

where $\mathbf{a}_i^\mathrm{T}$ represents the ith $(i = 1, 2, \ldots, m)$ column of $\mathbf{A}^\mathrm{T}\mathbf{augI}$. We can now pre-multiply equation (5.4.18) by $\mathbf{F}^{-1}$ and rearrange it to obtain

$$\mathbf{y_F} = \mathbf{F}^{-1}\pi - \mathbf{F}^{-1}\mathbf{a}_i^\mathrm{T} y_i \tag{5.4.19}$$

Equation (5.4.17) gives the old value of the objective function (that is, the value of the objective function in the last BFS). Its new value (in the new BFS) is given by

$$\Pi_{d,new} = \mathbf{b_F^T}\mathbf{F}^{-1}\pi - (\mathbf{b_F^T}\mathbf{F}^{-1}\mathbf{a}_i^T - b_i)y_i \tag{5.4.20}$$

Notice that the first part on the RHS of equation (5.4.20) is Π_d in equation (5.4.17). Therefore, using equation (5.4.17) and denoting $\Pi_{d,i} = \mathbf{b_F^T}\mathbf{F}^{-1}\mathbf{a}_i^T$, we can rewrite equation (5.4.20) as

$$\Pi_{d,new} = \Pi_d - (\Pi_{d,j} - b_i)y_i \tag{5.4.21}$$

Equation (5.4.21) implies that the objective function of the dual minimization problem will decrease only if $\Pi_{d,i} - b_i$ is positive. Therefore, the NBV is to be introduced and a new iteration is to be carried out in a dual minimization problem only if $\Pi_{d,i} - b_i$ is positive. *This condition is to be checked for every NBV.* If $\Pi_{d,i} - b_i$ is nonpositive for all y_i in the case of a dual minimization problem, then the current solution is optimal, which is called the optimality criterion in the case of the revised simplex method for dual minimization problems.

Equation (5.4.21) also implies that the objective function of the dual maximization problem will increase only if $\Pi_{d,i} - b_i$ is negative. Therefore, the NBV is to be introduced and a new iteration is to be carried out in a dual maximization problem only if $\Pi_{d,i} - b_i$ is negative. As in the case of a dual minimization problem, *this condition is to be checked for every NBV.* If $\Pi_{d,i} - b_i$ is nonnegative for all y_i in the case of a dual maximization problem, then the current solution, as before, is optimal.

So far, we were checking for optimality in successive iterations and determining the variable that enters basis. Let us now consider the feasibility criterion in the case of the revised simplex method for dual minimization problems. This is, as in the case of the revised simplex method, determined by the DRs. The leaving variable is the one with the smallest DR. This condition is valid for both dual minimization and dual maximization cases.

5.4.8 Dual simplex method

In Section 5.4.5 we applied the tabular approach of the simplex method along with artificial variables to solve dual LP problems. There is a closely related method, called the *dual simplex method*, which can be used to solve dual minimization LP problems without the use of artificial variables. We demonstrate this method here.

Recall that in the normal simplex method, also called the *primal simplex method*, we continued iterations, holding feasibility, until we reached the optimum solution. That is, in the primal simplex method we were searching for optimality holding feasibility in successive iterations. In the dual simplex method we do the opposite; that is, we search for feasibility holding optimality. What this means is that while in the primal simplex method we start with a basic and feasible solution and continue iterations until the optimum solution is reached, in the dual simplex method we start with a basic optimum but infeasible solution and continue iterations until feasibility is attained. As in the case of the primal simplex method, we have two criteria in the dual simplex method also. They are the *dual feasibility criterion* and the *dual optimality criterion*.

The dual feasibility criterion determines the variable that leaves the current basis. Therefore, the dual feasibility criterion states that the variable with the most negative value (ties are broken arbitrarily) leaves the current basis. The row of this variable is named as the pivot row. Notice that the iteration is terminated if all the variables in the current basis are nonnegative.

The dual optimality criterion is used to determine the current NBV that enters the basis. This criterion states that the NBV with the lowest absolute ratio, obtained by dividing the coefficients of the objective function in each iteration by the corresponding elements in the pivot row (except those under the RHS column), which must necessarily be negative, enters the basis. The column of this variable is named as the pivot column. The element at the intersection of the pivot row and pivot column is the pivot element.

As an application of the dual simplex method, consider the minimization problem (5.4.10), which is the dual of the maximization problem (5.4.9). Let us first multiply both sides of the constraints of this dual problem by -1 to convert $\geq$ inequalities into $\leq$ inequalities, and then use the slack dummy variables t_1 and t_2 to convert the $\leq$ inequalities into equalities ($=$). After doing this the problem becomes

$$\text{Minimize } \Pi_d = 90y_1 + 120y_2 + 0t_1 + 0t_2, \text{ subject to } -2y_1 - 3y_2 + t_1 + 0t_2 = -2,$$

$$-3y_1 - 2y_2 + 0t_1 + t_2 = -2, \text{ and } y_1, y_2, t_1, t_2 \geq 0 \tag{5.4.22}$$

On the basis of problem (5.4.22), we can set up the initial dual simplex tableau as given in Table 5.4.4. Notice that no indicator in Table 5.4.4 has a negative value indicating that the current basic solution is optimal. Since the current *dummy variables* take negative values

Table 5.4.4

BVs	y_1	y_2	t_1	t_2	RHS
t_1	−2	−3	1	0	−2
t_2	−3	−2	0	1	−2
Π_d	90	120	0	0	0

Indicators

Table 5.4.5

BVs	y_1	y_2	t_1	t_2	RHS
y_2	2/3	1	−1/3	0	2/3
t_2	**−5/3**	0	−2/3	1	−2/3
Π_d	10	0	40	0	−80

Indicators

($t_1 = -2$ and $t_2 = -2$), the current basic solution is not feasible. Therefore, we continue to search for optimal but feasible solution.

Iteration 1

Let us first invoke the two criteria of the dual simplex method to determine the leaving and entering variables. *Feasibility criterion*: Since both current BVs (t_1 and t_2) have the same negative values, we choose t_1 as the leaving variable and therefore the row of t_1 is the pivot row. *Optimality criterion*: We divide each element of the last row (except under the RHS column) of Table 5.4.4 by the corresponding elements of the pivot row to obtain the lowest absolute ratio 40 (discarding 0 and infinite values). Therefore, y_2 is the current NBV that enters the basis, and the column of y_2 is the pivot column. This implies that −3 is the pivot element. We can now carry out the elementary row operations ($R_1 \to (-1/3)R_1$, $R_2 \to R_2 + 2R_1$, and $R_3 \to R_3 - 120R_1$) to obtain Table 5.4.5. Notice that, as before, no indicator in Table 5.4.5 has negative value indicating that the current basic solution is optimal. Since the current dummy variable t_2 takes a negative value ($t_2 = -2/3$), the current basic solution is not feasible. Therefore, we continue to search for optimal but feasible solution.

Iteration 2

Let us again invoke the two criteria of the dual simplex method to determine the leaving and entering variables. *Feasibility criterion*: Since t_2 is the only current basic variable with a negative value, t_2 is the leaving variable and therefore the row of t_2 is the pivot row. *Optimality criterion*: Since y_1 has the least absolute ratio (6), y_1 is the current NBV that enters the basis. This implies that −5/3 is the pivot element. We can now carry out the elementary row operations ($R_2 \to (-3/5)R_1$, $R_1 \to R_1 - (2/3)R_2$, and $R_3 \to R_3 - 10R_2$) to obtain Table 5.4.6.

Notice that no indicator in Table 5.4.6 has a negative value indicating that the current basic solution is optimal. Moreover, no variable in the current basis has a negative value

Table 5.4.6

BVs	y_1	y_2	t_1	t_2	RHS
y_2	0	1	$-3/5$	$2/5$	$2/5$
y_1	1	0	$2/5$	$-3/5$	$2/5$
Π_d	0	0	36	6	-84

$\underbrace{\hspace{6cm}}_{\text{Indicators}}$

indicating that the current basis is feasible. Since we have achieved optimality and feasibility simultaneously, the last dual simplex tableau gives the solution to the dual minimization problem and it is $y_1^* = 2/5$, $y_2^* = 2/5$, and $\Pi_d^* = 84$. This is exactly the same solution as that we obtained for problem (5.4.10) in Section 5.4.5. Table 5.4.6 also gives us the solution to the primal problem in (5.4.9), $x_1^* = 36$, $x_2^* = 6$, and $\Pi^* = 84$, again as in the solution to problem (5.4.10) in Section 5.4.5. The reader will have noticed the advantage of the dual simplex method over the primal simplex method in solving dual LP problems: the former does not require the use of the artificial variables.

Let us now outline the steps involved in the application of the dual simplex method in solving dual minimization LP problems. These steps are the following.

Step 1. Transform those constraints in the problem with $\geq$ sign to $\leq$ sign by multiplying by -1.

Step 2. After step 1, introduce dummy variables.

Step 3. Set up the initial dual simplex tableau. Notice that if the optimality condition is satisfied and if at least one of the current BVs has a negative value, the dual simplex method must be applied.

Step 4. If the dual simplex method is applicable, carry out the feasibility criterion test. The current BV with the most negative value will leave the basis. The row of this variable is the pivot row.

Step 5. Carry out the optimality criterion test. The current NBV with the least absolute value (obtained by dividing the coefficients of the objective function by the corresponding elements of the pivot row, barring the element under the RHS column, and discarding zero and infinite values) enters the basis. The column of the entering variable is the pivot column. The element at the intersection of the pivot row and pivot column is the pivot element.

Step 6. Carry out elementary row operations to convert the pivot element to 1 and all other elements in the pivot column to 0.

Step 7. Repeat steps 4 to 6 above until all basic variables have nonnegative values.

5.4.9 Sensitivity analysis

We live in a dynamic world and, therefore, most economic activities change over time. Since LP is an approach to solving some of the problems related to economic activities, one may visualize corresponding changes in the structure of a LP problem. It is possible to cite many forms of changes in the structure of a LP problem. Since a meaningful analysis of all these changes is beyond the scope of this book, we confine our analysis to the effects of only the following three changes:

1. Changes in the coefficients of the objective function; that is, changes in the vector π (in the maximization problem).
2. Changes in the vector $\mathbf{a}_j$ of a NBV x_j in the maximization problem, or changes in the vector $\mathbf{a}_i$ of a NBV y_i in the dual minimization problem.
3. Changes in the constants of the constraints or changes in the RHS of constraints; that is, changes in the vector $\mathbf{b}$.

These changes in a particular LP problem may produce corresponding changes in the optimal solution as well as its feasibility. Sensitivity analysis is defined as the analysis or study of how sensitive the optimal solution and its feasibility are to the above listed changes in a LP problem. This analysis is also called post-optimality analysis because it is carried out after the optimum solution is achieved.

Sensitivity analysis can be carried out using the matrix approach (presented in the revised primal simplex method and revised dual simple method presented in Sections 5.3.6 and 5.4.7, respectively) or the tabular approach of the simplex method with specific, numerical examples. Since the former is more general than the latter, we employ the matrix approach to carry out the sensitivity analyses of the above mentioned changes.

Our main aim here is to see how the above listed three changes affect the optimal solution, the optimality criterion, and the feasibility criterion (presented in Sections 5.3.6 and 5.4.7) of both the primal and the dual. In the case of the primal problem, the optimal solution (equation (5.3.10)), the optimality criterion (which can be derived from equation (5.3.14)), and the feasibility criterion (equation (5.3.13)), respectively, are

$$\mathbf{x_B} = \mathbf{B}^{-1}\mathbf{b} \tag{5.4.23}$$

$$\pi_\mathbf{B}{}^\mathbf{T}\mathbf{B}^{-1}\mathbf{a}_j - \pi_j \geq 0 \tag{5.4.24}$$

and

$$\mathbf{x_B} = \mathbf{B}^{-1}\mathbf{b} - \mathbf{B}^{-1}\mathbf{a}_j\, x_j \tag{5.4.25}$$

Similarly, in the case of the dual problem, the optimal solution (equation (5.4.16)), the optimality criterion (which can be derived from equation (5.4.20)), and the feasibility criterion (equation (5.4.19)), respectively, are

$$\mathbf{y_F} = \mathbf{F}^{-1}\pi \tag{5.4.26}$$

$$\mathbf{b_F}{}^\mathbf{T}\mathbf{F}^{-1}\mathbf{a}_i{}^\mathbf{T} - b_i \leq 0 \tag{5.4.27}$$

and

$$\mathbf{y_F} = \mathbf{F}^{-1}\pi - \mathbf{F}^{-1}\mathbf{a}_i{}^\mathbf{T} y_i \tag{5.4.28}$$

Let us first consider the effects of changes in the vector π on the optimal solutions, and the optimality and feasibility criteria, of both primal and dual problems. The vector π does not appear in equations (5.4.23) and (5.4.25) and inequality (5.4.27). Therefore, changes in the vector π have no effect on the optimal solution to the primal problem, its feasibility criterion, and the optimality criterion of the dual problem. But the changes in the vector π do affect the optimality criterion of the primal problem, the optimal solution to the dual problem, and the feasibility criterion of the dual problem because π appears in equations (5.4.24) and (5.4.26) and in inequality (5.4.28), respectively.

We now consider the effect of a change in the vector $\mathbf{a}_j$ ($\mathbf{a}_i$) of a NBV x_j (y_i) in a maximization (dual minimization) problem. Notice that this change may affect equations (5.4.23)–(5.4.28). Therefore, a change in the vector $\mathbf{a}_j$ ($\mathbf{a}_i$) may affect the optimal solution, the optimality criterion, and feasibility criterion of both the primal and the dual problems.

Let us now consider the third case: the effect of changes in the vector $\mathbf{b}$. The vector $\mathbf{b}$ appears in equations (5.4.23), (5.4.25), and (5.4.27). Therefore, changes in the vector $\mathbf{b}$ have effects on the optimal solution to the primal problem and its feasibility criterion, and on the optimality criterion of the dual problem. These changes do not create impacts on the optimality criterion of the primal problem, and on the optimal solution to the dual problem and on its feasibility criterion.

The sensitivity analyses we have carried out so far in this section are based on the assumption that the changes mentioned in π, $\mathbf{a}_i$ (or $\mathbf{a}_j$), and $\mathbf{b}$ are discrete. But, we are aware that this assumption is restrictive because these changes could be continuous. An analysis of the impacts of continuous changes in π, $\mathbf{a}_i$ (or $\mathbf{a}_j$), and $\mathbf{b}$ is called *parametric programming*. Since this topic is beyond the scope of this book, we do not attempt to present it here.

5.4.10 Application examples

Example 1. Solve the dual of problem (5.3.19) in example 1 in Section 5.3.7 and interpret the solution to the dual.

Solution. The primal problem in (5.3.19) is a maximization problem with solution $x_1^* = 120/11$, $x_2^* = 35/11$, $x_3^* = 125/11$, $s_1^* = 0$, $s_2^* = 0$, $s_3^* = 0$, and $\Pi^* = 1270/11$. The dual of this problem can be set up as

$$\text{Minimize } \Pi_d = 45y_1 + 40y_2 + 25y_3, \text{ subject to } y_1 + y_2 + 2y_3 \geq 6, 0y_1 + 2y_2 + y_3 \geq 5,$$

$$3y_1 + 2y_2 + 0y_3 \geq 3, \text{ and } y_1, y_2, y_3 \geq 0 \qquad (5.4.29)$$

Using the two-phase method discussed in Section 5.3.5, we can obtain the final tableau that presents the solution to this minimization problem as presented in Table 5.4.7. This shows that the optimal solution comprises $y_1^* = 1/11$, $y_2^* = 15/11$, and $y_3^* = 25/11$; $t_1^* = 0$, $t_2^* = 0$, and $t_3^* = 0$; and $\Pi_d^* = 1270/11$. We can read from the last row of the table, as we obtained and presented above, the solution to the primal problem: $x_1^* = 120/11$, $x_2^* = 35/11$, and $x_3^* = 125/11$; $s_1^* = 0$, $s_2^* = 0$, and $s_3^* = 0$; and $\Pi^* = 1270/11$. Notice that $\Pi_d^* = 1270/11 = \Pi^*$, as stated by duality theorem 1. Since $x_1^* = 120/11 > 0$, $x_2^* = 35/11 > 0$, and $x_3^* = 125/11 > 0$, $t_1^* = 0$, $t_2^* = 0$, and $t_3^* = 0$; and since $s_1^* = 0$, $s_2^* = 0$, and $s_3^* = 0$, $y_1^* = 1/11 > 0$, $y_2^* = 15/11 > 0$, and $y_3^* = 25/11 > 0$, as stated by duality theorem 2.

The economic interpretation of the solution to the dual problem presented above is as follows. Since the optimal value of the primal objective function is equal to the optimal value of the dual objective function, the total profit to the firm is equal to the total opportunity cost of using the resources. If the optimal values of the choice variables in the dual problem are substituted into its constraints, we will find that the LHSs of the constraints equal their RHSs. This means that the total opportunity cost of producing one unit of each good is equal

Table 5.4.7

BVs	y_1	y_2	y_3	t_1	t_2	t_3	RHS
y_1	1	0	0	$-2/11$	$4/11$	$-3/11$	$1/11$
y_3	0	0	1	$-6/11$	$1/11$	$2/11$	$25/11$
y_2	0	1	0	$3/11$	$-6/11$	$-1/11$	$15/11$
Π_d	0	0	0	$-120/11$	$-35/11$	$-125/11$	$1270/11$

Indicators

Table 5.4.8

BVs	y_1	y_2	y_3	t_1	t_2	t_3	RHS
y_1	1	0	0	$1/9$	$-2/9$	$2/9$	$2/9$
y_3	0	0	1	$4/9$	$1/9$	$-1/9$	$17/9$
y_2	0	1	0	$-2/9$	$4/9$	$1/18$	$5/9$
C_d	0	0	0	$70/9$	$520/36$	$260/36$	$815/9$

Indicators

to the profit per unit from that good. Since the optimal values of the dual choice variables are positive (that is, $y_i^* > 0$), the nonnegativity condition is also satisfied.

Example 2.   Solve the dual of problem (5.3.21) in example 2 in Section 5.3.7 and interpret the solution to dual.

Solution.   The primal problem in (5.3.21) is a minimization problem with solution $x_1^* = 70/9$, $x_2^* = 520/36$, $x_3^* = 260/36$, $s_1^* = 0, s_2^* = 0, s_3^* = 0$, and $C^* = 815/9$. The dual of this problem can be set up as

$$\text{Maximize } C_d = 40y_1 + 45y_2 + 30y_3, \text{ subject to } y_1 + 0y_2 + 2y_3 \leq 4,$$

$$0y_1 + 2y_2 + y_3 \leq 3, 4y_1 + 2y_2 + 0y_3 \leq 2, \text{ and } y_1, y_2, y_3 \geq 0 \tag{5.4.30}$$

Following the tabular approach of the simplex method discussed in Section 5.3.4, we can obtain the final tableau that presents the solution to this maximization problem as in Table 5.4.8. This shows that the optimal solution comprises $y_1^* = 2/9$, $y_2^* = 5/9$, and $y_3^* = 17/9$; $t_1^* = 0, t_2^* = 0, t_3^* = 0$; and $C_d^* = 815/9$. We can read from the table, as before, the solution to the primal problem: $x_1^* = 70/9$, $x_2^* = 520/36$, and $x_3^* = 260/36$; $s_1^* = 0, s_2^* = 0, s_3^* = 0$; and $C^* = 815/9$. Notice that $C_d^* = 815/9 = C^*$, as stated by duality theorem 1. Since $x_1^* = 70/9 > 0$, $x_2^* = 520/36 > 0$, and $x_3^* = 260/36 > 0 t_1^* = 0, t_2^* = 0$, and $t_3^* = 0$; and since $s_1^* = 0, s_2^* = 0$, and $s_3^* = 0, y_1^* = 2/9 > 0, y_2^* = 5/9 > 0$, and $y_3^* = 17/9 > 0$, as stated by duality theorem 2.

The economic interpretation of the solution to the dual problem presented above is as follows. Since the optimal value of the primal objective function is equal to the optimal value of the dual objective function, the total cost to the firm is equal to the total imputed value. If the optimal values, in the dual problem, of the choice variables are substituted into its constraints, we will find that the LHSs of the constraints equal their RHSs. This means that the total imputed value of producing one unit of each food item is equal to the price per unit of that item. Since the optimal values of the dual choice variables are positive (that is, $y_i^* > 0$), the nonnegativity condition is also satisfied.

5.4.11 Exercises

1. Solve the duals of the LP problems in exercise 1 of Section 5.3.8 using both the tabular approach of the simplex method and the dual simplex method.
2. Solve the duals of the LP problems in exercise 3 of Section 5.3.8 using both the tabular approach of the simplex method and the dual simplex method.
3. Solve the duals of the LP problems in exercise 5 of Section 5.3.8 using both the tabular approach of the simplex method and the dual simplex method.
4. *Application exercise.* Solve the dual of the LP problem in exercise 6 of Section 5.3.8 using both the tabular approach of the simplex method and the dual simplex method.
5. *Application exercise.* Solve the dual of the LP problem in exercise 7 of Section 5.3.8 using both the tabular approach of the simplex method and the dual simplex method.

5.5 Transportation and assignment problems

5.5.1 Introduction

So far in this chapter we have been concerned with primal and dual LP problems, their solutions, and the related topics. In this section we shall present and solve a particular type of LP problem called the *transportation problem*. We shall also present and solve a problem, which is closely related to transportation problem, called the *assignment problem*.

Let us first consider the transportation problem. The assignment problem will be dealt with in Section 5.5.6. We bring home the meaning of the transportation problem with a simple example. Suppose that a company has two production plants, one each in Mumbai and New Delhi in India, which produce an identical good. This company also has two distribution centers for the good, one each in Kolkata and Chennai, also in India. The places where the plants are situated are called the *sources* (m) and the places where the distribution centers are situated are called the warehouses, or the shops, or the *destinations* (n). Therefore, we have $m = 2$ and $n = 2$ in our present example. The company transports all the good it produces at the sources (Mumbai and New Delhi) to the destinations (Kolkata and Chennai). The total amount of the good available at the ith source and the total quantity of the good transported to the jth destination (where $i = 1, 2$ and $j = 1, 2$ in the present example) are called the supply and the demand for the good, from the ith source and at the jth destination, respectively. For convenience, we represent Mumbai and New Delhi by source 1 or supply 1 (denoted by s_1) and source 2 or supply 2 (denoted by s_2), respectively; and Kolkata and Chennai by destination 1 or demand 1 (denoted by d_1) and destination 2 or demand 2 (denoted by d_2), respectively.

To transport the good from a particular source to a particular destination, there is a particular cost per unit of the good. Thus, the total transportation cost of the company is the total cost of transporting goods from m sources to n destinations. Therefore, the objective of the company is to devise a transportation plan that minimizes its total transportation cost. This is the essence of the transportation problem.

5.5.2 Structure of the transportation problem

Like a LP problem, a transportation problem also has three elements: the objective function, the constraints, and the nonnegativity condition. We assume that the cost of transporting one unit of the good from the ith source to the jth destination is constant, and we denote it by c_{ij}. We denote the quantity of the good transported from the ith source to the jth destination by x_{ij}. This means that the total cost of transporting x_{ij} units of the good from the ith source to the jth destination is given by $c_{ij}x_{ij}$. Therefore, the total transportation cost to the company in our present two-source–two-destination example is $C = c_{11}x_{11} + c_{12}x_{12} + c_{21}x_{21} + c_{22}x_{22}$. This means that the company's objective is to minimize this total cost. Notice that, as in the LP problems, the objective function in the transportation problem also is linear.

Another important ingredient of the transportation problem is the set of constraints. For the time being, we assume that the total quantity of the good supplied from m sources is equal to the total quantity of the good demanded at n destinations. This type of transportation problem where the total supply is equal to the total demand is called a *balanced transportation problem*. We will consider *unbalanced transportation problems* in examples 2 and 3 in Section 5.5.5. Thus the constraints of the problem, in the case of our present example, can be written as $x_{11} + x_{12} = s_1, x_{21} + x_{22} = s_2, x_{11} + x_{21} = d_1, x_{12} + x_{22} = d_2$. The last important element of the transportation problem is the nonnegativity constraint, $x_{ij} > 0$. This implies that the quantity of the good transported from the ith source to the jth destination must be nonnegative.

Therefore, the transportation problem with our present example can be written as

$$\text{Minimize } C = c_{11}x_{11} + c_{12}x_{12} + c_{21}x_{21} + c_{22}x_{22},$$

$$\text{subject to } x_{11} + x_{12} + 0x_{21} + 0x_{22} = s_1, 0x_{11} + 0x_{12} + x_{21} + x_{22} = s_2,$$

$$x_{11} + 0x_{12} + x_{21} + 0x_{22} = d_1, 0x_{11} + x_{12} + 0x_{21} + x_{22} = d_2,$$

$$\text{and } x_{ij} \geq 0, i = 1, 2, j = 1, 2 \tag{5.5.1}$$

The problem in (5.5.1) can be written using sigma notations as

$$\text{Minimize } \sum_{i=1}^{m=2} \sum_{j=1}^{n=2} c_{ij}x_{ij}, \text{ subject to } \sum_{j=1}^{n=2} x_{ij} = s_i, \sum_{i=1}^{m=2} x_{ij} = d_j, \text{ and } x_{ij} \geq 0 \tag{5.5.2}$$

or using matrices as

$$\text{Minimize } C = \mathbf{c}^T\mathbf{x}, \text{ subject to } \mathbf{A}\mathbf{x} = \mathbf{s} \text{ and } \mathbf{x} \geq 0 \tag{5.5.3}$$

where $\mathbf{c}^T = \begin{bmatrix} c_{11} & c_{12} & c_{21} & c_{22} \end{bmatrix}$, $\mathbf{x}^T = [x_{11} \ \ x_{12} \ \ x_{21} \ \ x_{22}]$, $\mathbf{A} = \begin{bmatrix} 1 & 1 & 0 & 0 \\ 0 & 0 & 1 & 1 \\ 1 & 0 & 1 & 0 \\ 0 & 1 & 0 & 1 \end{bmatrix}$,

and $\mathbf{s}^T = \begin{bmatrix} s_1 & s_2 & d_1 & d_2 \end{bmatrix}$.

5.5.3 General transportation problem

Instead of two sources and two destinations as in the example considered in the last section, assume now that the company has i sources and j destinations, where $i = 1, 2, \ldots, m$ and

$j = 1, 2, \ldots, n$. Then the company's problem is to

$$\text{Minimize } C = \sum_{i=1}^{m} \sum_{j=1}^{n} c_{ij} x_{ij}, \text{ subject to } \sum_{j=1}^{n} x_{ij} = s_i$$

$$\text{and } \sum_{i=1}^{m} x_{ij} = d_j, i = 1, 2, \ldots, m, j = 1, 2, \ldots, n, \text{ and } x_{ij} \geq 0 \tag{5.5.4}$$

or in matrix notation as

$$\text{Minimize } C = \mathbf{c}^{\mathbf{T}} \mathbf{x}, \text{ subject to } \mathbf{A}\mathbf{x} = \mathbf{s} \text{ and } \mathbf{x} \geq 0 \tag{5.5.5}$$

where $\mathbf{c}^{\mathbf{T}} = \begin{bmatrix} c_{11} & c_{12} & .. & c_{1n} & c_{21} & c_{22} & .. & c_{2n} & .. & c_{m1} & c_{m2} & .. & c_{mn} \end{bmatrix}$,
$\mathbf{x}^{\mathbf{T}} = \begin{bmatrix} x_{11} & x_{12} & .. & x_{1n} & x_{21} & x_{22} & .. & x_{2n} & .. & x_{m1} & x_{m2} & .. & x_{mn} \end{bmatrix}$,
$\mathbf{s}^{\mathbf{T}} = \begin{bmatrix} s_1 & s_2 & .. & s_m & d_1 & d_2 & .. & d_n \end{bmatrix}$, and

$$\mathbf{A} = \begin{bmatrix}
1 & 1 & .. & 1 & 0 & 0 & .. & 0 & .. & 0 & 0 & .. & 0 \\
0 & 0 & 0 & 0 & 1 & 1 & .. & 1 & .. & 0 & 0 & 0 & 0 \\
.. & .. & .. & .. & .. & .. & .. & .. & .. & .. & .. & .. \\
0 & 0 & 0 & 0 & 0 & 0 & 0 & 0 & 0 & 1 & 1 & .. & 1 \\
.. & .. & .. & .. & .. & .. & .. & .. & .. & .. & .. & .. \\
1 & 0 & 0 & 0 & 1 & 0 & 0 & 0 & .. & 1 & 0 & 0 & 0 \\
0 & 1 & 0 & 0 & 0 & 1 & 0 & 0 & .. & 0 & 1 & 0 & 0 \\
0 & 0 & 1 & 0 & 0 & 0 & 1 & 0 & .. & 0 & 0 & 1 & 0 \\
0 & 0 & 0 & 1 & 0 & 0 & 0 & 1 & .. & 0 & 0 & 0 & 1
\end{bmatrix}$$

The problem in (5.5.4) or (5.5.5) is the general transportation problem.

5.5.4 General transportation algorithm: methods of finding the initial BFS and the optimality test

Methods of finding the initial BFS

Before we begin the general transportation algorithm, we need to be clear about the nature of the solution to a transportation problem. Consider our m-source–n-destination transportation problem presented in problem (5.5.4). Notice that there are $m + n$ constraints in the problem. But there are only $m + n - 1$ independent constraints because any BFS that satisfies the $m + n - 1$ constraints of the problem will satisfy the last constraint and, therefore, the last constraint is redundant. This means that only $m + n - 1$ routes will be used in the cost-minimizing transportation plan. However, the total number of available routes is equal to $m \times n$.

Let us now state the general transportation algorithm. It involves four steps: (1) find an initial BFS; (2) test for optimality; (3) improve the BFS if it is not optimal; and (4) repeat the last two steps until optimal solution is achieved. Let us first consider the case of the initial BFS. In the case of a standard LP problem, we begin iterations with an initial BFS,

Table 5.5.1

		Destinations					Supply	w_i
		1	2	3	...	n		
Sources	1	c_{11} x_{11}	c_{12} x_{12}	c_{13} x_{13}	... ...	c_{1n} x_{1n}	s_1	w_1
	2	c_{21} x_{21}	c_{22} x_{22}	c_{23} x_{23}	... ...	c_{2n} x_{2n}	s_2	w_2
	3	c_{31} x_{31}	c_{32} x_{32}	c_{33} x_{33}	... ...	c_{3n} x_{3n}	s_3	w_3
	...	... ...	... ...	... ...	... ...	... ...	...	...
	m	c_{m1} x_{m1}	c_{m2} x_{m2}	c_{m3} x_{m3}	... ...	c_{mn} x_{mn}	s_m	w_m
	Demand	d_1	d_2	d_3	...	d_n		
	v_j	v_1	v_2	v_3	...	v_n		

and then move to BFSs through successive iterations. Exactly like this, we need to find an initial BFS to move forward in the transportation problem also. The question now is how we can find an initial BFS in the transportation problem. There are several methods in the literature for this. We demonstrate three of the most frequently used methods: the *north-west corner method*, the *least-cost method*, and *Vogel's approximation method*. But, before we explore these methods, we shall construct Table 5.5.1, corresponding to problem (5.5.4), which can be used in the demonstrations of all three methods mentioned above. Notice that all the elements in this table are familiar to us except those in the row v_j and in the column w_i, which will be explained shortly.

(1) **The north-west corner method.** In this method we begin with cell (1, 1), the north-west corner cell. Hence the name the north-west corner method. Allocate the maximum possible amount to this cell (that is, to x_{11}) such that the constraints are not violated (that is, the minimum of s_1 and d_1). After this and if supply still remains, we move one cell to the right. Notice that at each step we allocate the maximum possible amount to the cell under consideration without violating the constraints. In any allocation, the sum of a particular row should not exceed the supply indicated in that row and the sum of a particular column should not exceed the demand indicated in that column. Moreover, any allocation should be nonnegative.

(2) **The least-cost method.** This method uses the cheapest destination first; that is, we allocate the maximum possible quantity to the cell which has the lowest per unit cost (break ties arbitrarily). Then, allocate the maximum possible quantity to the cell which has the second lowest per unit cost; and continue like this until we achieve a feasible plan. It must be emphasized that the constraints must be satisfied in these processes.

(3) **Vogel's approximation method.** This method is an improvement over the least-cost method. It takes into account both the row and column cost differences (i.e. the difference in costs in supplying from a particular source to different destinations and the

difference in costs in meeting the demand of a particular destination from different sources, respectively). In this method, for each row (or column) we calculate the difference between the second smallest and the smallest unit costs of the unassigned cells (variables). Find the row (column) with the largest above difference, and allocate the maximum possible quantity of the good to the cell with the smallest unit cost (break ties arbitrarily) in that row (column). We have to recalculate the differences and repeat the last procedure until all demands are met.

Notice that in all the above three methods of finding the initial BFS, the variables (cells) that are assigned values are called BVs and the variables that are not assigned values (or are assigned zero) are called NBVs in the initial BFS. In our examples that follow, we denote the values that are assigned to BVs in boldfaces.

Optimality test

Now consider the column w_i and the row v_j. Choose any one w_i or v_j and assign it zero value, and find the remaining w_i and v_j such that, for each basic variable, the condition $w_i + v_j = c_{ij}$ is satisfied. After this, find the value $c_{ij} - w_i - v_j$ for each NBV. If the value $c_{ij} - w_i - v_j$ is nonnegative for each NBV, then the solution is optimal. Otherwise, the current solution is not optimal. This method of testing optimality using the multipliers w_i and v_j (with the condition $w_i + v_j = c_{ij}$) is called the *method of multipliers*.

Improving the initial BFS

If the optimality test gives us a solution that is not optimal, how do we improve the initial BFS? The procedure is as follows. The cell (variable) with the most negative value in $c_{ij} - w_i - v_j$ (in the optimality test above) is the entering variable. Now establish a loop that connects the entering variable with the current nonbasic cells, and allocate the maximum possible quantity (without violating the constraints) to this entering variable.

5.5.5 Application examples of solutions to transportation problems

Example 1. Consider our example presented in problem (5.5.1). Assume that the supply from the first source, Mumbai, is 100 units ($s_1 = 100$) and the supply from the second source, New Delhi, is 50 units ($s_2 = 50$); and that the demand at the first destination, Kolkata, is 80 units ($d_1 = 80$) and the demand at the second destination, Chennai, is 70 units ($d_2 = 70$). Also assume that the unit transportation costs (in dollars) between the sources and the destinations are as given in Table 5.5.2. Find the transportation plan that minimizes the company's total transportation cost.

Table 5.5.2

		Destinations	
		d_1	d_2
Sources	s_1	5	2
	s_2	3	4

Table 5.5.3

		Destinations			
		1	2	Supply	w_i
Sources	1	5 x_{11}	2 x_{12}	100	w_1
	2	3 x_{21}	4 x_{22}	50	w_2
	Demand	80	70		
	v_j	v_1	v_2		

Table 5.5.4

		Destinations			
		1	2	Supply	w_i
Sources	1	5 x_{11} **80**--→	2 x_{12} **20**	100	w_1
	2	3 x_{21}	4 x_{22} **50**	50	w_2
	Demand	80	70		
	v_j	v_1	v_2		

Solution.   Notice that the total supply ($s_1 + s_2 = 100 + 50 = 150$) is equal to the total demand ($d_1 + d_2 = 80 + 70 = 150$). Therefore, this is an example of a balanced transportation problem. Let us first set up Table 5.5.3, which can be used to find the initial BFS. Although we can use any of the three methods presented in the previous section to find an initial BFS, the easiest one is the north-west corner method and we will use only that method in this book. Following this method, we can allocate the maximum possible quantity of the good from source 1 to destination 1. This quantity is 80 and it is shown by bold type in cell (1, 1) in Table 5.5.4. Since the demand at destination 1 is met by this allocation, we move to cell (1, 2) and allocate to it the remaining quantity (20) in the supply from source 1. These two allocations exhaust s_1 and, therefore, we move down to cell (2, 2). Since we have already allocated 20 units to destination 2, we need only allocate the remaining requirement at destination 2, which is 50 units. These 50 units can be allocated from s_2 which exhausts the supply at s_2. These allocations give us the initial BFS without violating the constraints (that is, the total quantity supplied is equal to the total quantity demanded and there is no negative allocation). These allocations are shown in Table 5.5.4.

It can be seen from Table 5.5.4 that we have, in the initial BFS, $x_{11} = 80$, $x_{12} = 20$, and $x_{22} = 50$. Therefore, the total transportation cost in the initial BFS is $C = c_{11}x_{11} + c_{12}x_{12} + c_{21}x_{21} + c_{22}x_{22} = 5 \times 80 + 2 \times 20 + 3 \times 0 + 4 \times 50 = \$640$. Let us now determine whether

Table 5.5.5

		Destinations			
		1	2	Supply	w_i
Sources	1	5 x_{11} **30**	2 x_{12} **70**	100	w_1
	2	3 x_{21} **50**	4 x_{22}	50	w_2
	Demand	80	70		
	v_j	v_1	v_2		

the initial BFS found above is optimal or not. For this we use the optimality test described in the last section. But, to carry out the optimality test, we need to find w_i and v_j with respect to the BVs (those variables that are assigned values in the initial BFS). Since the first row contains more BVs, we choose $w_1 = 0$. We know from the last section that $w_i + v_j = c_{ij}$. Therefore, we have the following result:

Cell $(1, 1) \Rightarrow w_1 + v_1 = c_{11} \Rightarrow 0 + v_1 = 5 \Rightarrow v_1 = 5$

Cell $(1, 2) \Rightarrow w_1 + v_2 = c_{12} \Rightarrow 0 + v_2 = 2 \Rightarrow v_2 = 2$

Cell $(2, 2) \Rightarrow w_2 + v_2 = c_{22} \Rightarrow w_2 + 2 = 4 \Rightarrow w_2 = 2$

The next step in this method is to find the value $c_{ij} - w_i - v_j$ in the case of each NBV (cell). There is only one NBV in the last tableau and its cell is cell $(2, 1)$. Therefore, $c_{21} - w_2 - v_1 = 3 - 2 - 5 = -4$. Since this value is negative, the current (or the initial) BFS is not optimal. This implies that we have to carry out the above procedure again. Notice that the above optimality test showed that cell $(2, 1)$ has the most negative (the only negative) value and, therefore, we have to allocate the maximum possible quantity to that cell without violating the constraints. This gives us Table 5.5.5. It can be seen from Table 5.5.5 that we have, in the current BFS, $x_{11} = 30$, $x_{12} = 70$, and $x_{21} = 50$. Therefore, the total transportation cost in the current BFS is $C = c_{11}x_{11} + c_{12}x_{12} + c_{21}x_{21} + c_{22}x_{22} = 5 \times 30 + 2 \times 70 + 3 \times 50 + 4 \times 0 = \$440$. Notice that the total transportation cost (C) in the current BFS is smaller (\$440) than that (\$640) in the initial BFS.

Let us now determine, as before, whether the current BFS found above is optimal or not. For this we use the optimality test with $w_1 = 0$. We know that $w_i + v_j = c_{ij}$. Therefore, we have the following result:

Cell $(1, 1) \Rightarrow w_1 + v_1 = c_{11} \Rightarrow 0 + v_1 = 5 \Rightarrow v_1 = 5$

Cell $(1, 2) \Rightarrow w_1 + v_2 = c_{12} \Rightarrow 0 + v_2 = 2 \Rightarrow v_2 = 2$

Cell $(2, 1) \Rightarrow w_2 + v_1 = c_{21} \Rightarrow w_2 + 5 = 3 \Rightarrow w_2 = -2$

The next step is to find the value $c_{ij} - w_i - v_j$ in the case of each NBV (cell). There is only one NBV in the last tableau and its cell is cell $(2, 2)$, which gives us $c_{22} - w_2 - v_2 =$

Table 5.5.6

		Destinations		
		d_1	d_2	d_3
Sources	s_1	5	2	2
	s_2	3	6	4
	s_3 (dummy)	0	0	0

$4 - (-2) - 2 = 4$. Since this value is nonnegative, the current BFS is optimal. Therefore, the optimum solution is $x_{11}^* = 30, x_{12}^* = 70, x_{21}^* = 50, x_{22}^* = 0$, and $C^* = \$440$. This means that the company must ship 30 units from Mumbai to Kolkata, 70 units from Mumbai to Chennai, 50 units from New Delhi to Kolkata, and zero units from New Delhi to Chennai to minimize its total transportation cost. The minimum total transportation cost will be \$440.

Example 2.   We continue with example 1 above. Assume now that the company has a third destination in the city of Kochi (denoted by d_3) in India to which (along with transportation to Kolkata and Chennai) it has to transport the good. The total quantity of the good supplied is 350 units (200 units from the plant in Mumbai and 150 from the plant in New Delhi) and the total quantity demanded is 375 units (150, 75, and 150 units demanded at Kolkata, Chennai, and Kochi, respectively). The unit costs of transportation to these three destinations are given in Table 5.5.6. Find the quantities of the good that the company has to transport from the sources to the destinations so that the total cost of transportation is a minimum.

Solution.   Notice that the total quantity demanded (375 units) is larger than the total quantity supplied (350 units). Therefore, this is a case of an *unbalanced transportation problem*. To make the problem balanced, we add a *dummy source* (denoted by s_3) which accounts for the difference in the demand and supply (25 units $=$ 375 units $-$ 350 units). Since the dummy source never supplies any quantity, its coefficients in the objective function are zero (as can be seen in Table 5.5.6). Therefore, we can construct Table 5.5.7, which gives the complete information on the problem. We can solve this problem following the same procedures as those we applied in the solution to the problem in example 1 above. Rather than repeating this, we present the optimum solution in Table 5.5.8. This table shows that the optimal solution to the problem in the present example is $x_{11}^* = 0, x_{12}^* = 50, x_{13}^* = 150, x_{21}^* = 150, x_{22}^* = 0, x_{23}^* = 0, x_{31}^* = 0, x_{32}^* = 25, x_{33}^* = 0$, and $C^* = \$850$. Notice that in the optimum allocation, destination 2 obtains 25 units fewer than its demand.

Example 3.   We continue with example 1 above. Assume now that the company has a third source in the city of Kochi (denoted by s_3) in India from which (along with sources in Mumbai and New Delhi) it has to transport the good. The total quantity of the good supplied now is 400 units (200, 150, and 50 units from the plants in Mumbai, New Delhi, and Kochi, respectively) and the total quantity demanded is 350 units (175 units each at Kolkata and Chennai). The unit costs of transportation to these three destinations are given in Table 5.5.9. Find the quantities of the good that the company has to transport from the sources to the destinations so that the total cost of transportation is minimum.

Table 5.5.7

		Destinations				
		1	2	3	Supply	w_i
Sources	1	5 x_{11}	2 x_{12}	2 x_{13}	200	w_1
	2	3 x_{21}	6 x_{22}	4 x_{23}	150	w_2
	3 (dummy)	0 x_{31}	0 x_{32}	0 x_{33}	25	w_3
	Demand	150	75	150		
	v_j	v_1	v_2	v_3		

Table 5.5.8

		Destinations			
		1	2	3	Supply
Sources	1	5 x_{11}	2 x_{12} **50**	2 x_{13} **150**	200
	2	3 x_{21} **150**	6 x_{22}	4 x_{23}	150
	3 (dummy)	0 x_{31}	0 x_{32} **25**	0 x_{33}	25
	Demand	150	75	150	

Table 5.5.9

		Destinations		
		d_1	d_2	d_3
Sources	s_1	5	2	0
	s_2	3	6	0
	s_3	4	2	0

Solution.   Notice that the total quantity supplied (400 units) is larger than the total quantity demanded (350 units). This is another case of an unbalanced transportation problem. To make the problem balanced, we add a *dummy destination* (denoted by d_3) which accounts for the difference in the supplied and demanded (50 units $=$ 400 units $-$ 350 units) quantities. Since the dummy destination never accepts any quantity, its coefficients in the

Table 5.5.10

		Destinations			Supply	w_i
		1	2	3 (dummy)		
Sources	1	5 x_{11}	2 x_{12}	0 x_{13}	200	w_1
	2	3 x_{21}	6 x_{22}	0 x_{23}	150	w_2
	3	4 x_{31}	2 x_{32}	0 x_{33}	50	w_3
	Demand	175	175	50		
	v_j	v_1	v_2	v_3		

Table 5.5.11

		Destinations			Supply
		1	2	3 (dummy)	
Sources	1	5 x_{11}	2 x_{12} **150**	0 x_{13} **50**	200
	2	3 x_{21} **150**	6 x_{22}	0 x_{23}	150
	3	4 x_{31} **25**	2 x_{32} **25**	0 x_{33}	50
	Demand	175	175	50	

objective function are zero (as can be seen in Table 5.5.9). Therefore, we can construct Table 5.5.10, which gives the complete information on the problem. As earlier, we present the optimum solution to the problem in Table 5.5.11. It shows that the optimal solution to the problem in the present example is $x_{11}^* = 0, x_{12}^* = 150, x_{13}^* = 50, x_{21}^* = 150, x_{22}^* = 0, x_{23}^* = 0, x_{31}^* = 25, x_{32}^* = 25, x_{33}^* = 0$, and $C^* = \$900$. Notice that in the optimum allocation source 1 supplies 50 units more than what is demanded.

5.5.6 *Assignment problem*

Assignment problem is an optimization problem in which a firm or a company assigns a particular job or task to a particular worker, and still attempts to minimize the total assignment cost. This is a special kind of transportation problem in which workers are sources and jobs or tasks are destinations. As in the case of transportation problems, the total number of workers

Table 5.5.12

		Jobs								
		1	2	3	...	n	...	n	
Workers	1	c_{11}	c_{12}	c_{13}	...	c_{1j}	...	c_{1n}	1
	2	c_{21}	c_{22}	c_{23}	...	c_{2j}	...	c_{2n}	1
	3	c_{31}	c_{32}	c_{33}	...	c_{3j}	...	c_{3n}	1
	...	...	...	...	...	...	...	...	...
	in	c_{i1}	c_{i2}	c_{i3}	...	c_{ij}	...	c_{in}	1
	...	...	...	...	...	...	...	...	...
	n	c_{n1}	c_{n2}	c_{n3}	...	c_{nj}	...	c_{nn}	1
		1	1	1	...	1	...	1	

(the supply) may be equal to the total number of jobs (the demand) giving us the *balanced assignment problem*; or *unbalanced assignment problem* if the number of workers is not equal to the number of jobs. We can always balance an unbalanced assignment problem, as we did in the case of the transportation problem, by introducing dummy workers or dummy jobs.

The general assignment problem with $i(i = 1, 2, \ldots, n)$ workers and j $(j = 1, 2, \ldots, n)$ jobs can be constructed as presented in Table 5.5.12, where c_{ij} is the cost of assigning the ith worker to the jth job. The values in the last column and the last row show that the problem is a balanced assignment problem.

We can now state that the company's problem is to

$$\text{Minimize } C = \sum_{i=1}^{n}\sum_{j=1}^{n} c_{ij}x_{ij}, i = 1, 2, \ldots, n, j = 1, 2, \ldots, n,$$

$$\text{subject to } \sum_{i=1}^{n} x_{ij} = 1, \sum_{j=1}^{n} x_{ij} = 1, \text{ and } x_{ij} = 0 \text{ or } 1 \tag{5.5.6}$$

Notice that x_{ij} in problem (5.5.6) represents worker i assigned to job j. The value $C = \sum_{i=1}^{n}\sum_{j=1}^{n} c_{ij}x_{ij}$ is the total cost to the company of assigning i workers to j jobs, which the company wants to minimize. Since the problem is a balanced problem, the first constraint states that a particular job is assigned to only one worker and the second constraint states that a worker is assigned to only one job. And the last constraint states that if the ith worker is assigned the jth job, then $x_{ij} = 1$; otherwise $x_{ij} = 0$. Since the assignment problem presented in (5.5.6) obeys the structure of a transportation problem, it can be solved with the methods we used to solve the transportation problem. However, there is a more efficient method, called the *Hungarian method*, to solve assignment problems. The Hungarian method involves the following steps.

Step 1. Find the smallest unit cost in each row of Table 5.5.12. Subtract this from the every other element in that row.

Step 2. After step 1, find the smallest unit cost in each column of Table 5.5.12. Subtract this from the every other element in that column. These two steps give us the table of revised cost matrix.

Table 5.5.13

		Jobs			
		1	2	3	
Workers	1	15	12	11	1
	2	13	10	14	1
	3	12	16	13	1
		1	1	1	

Table 5.5.14

		Jobs			
		1	2	3	
Workers	1	3	2	0	1
	2	1	0	3	1
	3	0	6	2	1
		1	1	1	

Step 3. Determine whether there exists a feasible assignment with only zero costs (and without two zero costs in any row or column) in the revised cost matrix obtained after step 2. If it exists, we have attained the optimal assignment.

Step 4. If there does not exist a feasible assignment, draw the fewest number of straight lines passing through zeros in the revised cost matrix. Find the lowest number not touched by a straight line in the revised cost matrix. Then subtract from this number the every other number not touched by a straight line and add it to the every other number touched by two straight lines. Repeat from step 3.

5.5.7 Application examples of solutions to assignment problems

Example 1. Suppose that a company has three jobs to be assigned to three workers. The costs, in dollars, of assigning these jobs to these workers are given in Table 5.5.13. Determine the assignment plan that minimizes the company's total cost.

Solution. If we carry out the operations in steps 1 and 2 of the Hungarian method detailed in the last section, we obtain the revised cost matrix presented in Table 5.5.14. The result in Table 5.5.14 shows that there exists a feasible assignment with only zero costs (and without two zero costs in any row or column) in the revised cost matrix. This means that we have achieved the optimum assignment plan. In this optimum plan, the company must assign the first, the second, and the third worker the third, the second and the first job respectively. Then the minimum total cost to the company will be $C^* = \$11 + \$10 + \$12 = \$33$.

Example 2. Assume that a company has three jobs to be assigned to four workers. The costs, in dollars, of assigning these jobs to these workers are given in Table 5.5.15. Determine the assignment plan that minimizes the company's total cost.

Table 5.5.15

		Jobs				
		1	2	3	4 *(dummy)*	
Workers	1	6	8	4	0	1
	2	5	3	1	0	1
	3	4	2	1	0	1
	4	3	4	1	0	1
		1	**1**	**1**	**1**	

Table 5.5.16

		Jobs				
		1	2	3	4 *(dummy)*	
Workers	1	3	6	3	0	1
	2	2	1	0	0	1
	3	1	0	0	0	1
	4	0	2	0	0	1
		1	**1**	**1**	**1**	

Solution.   If we carry out the operations in steps 1 and 2 of the Hungarian method detailed in the last section, we obtain the revised cost matrix presented in Table 5.5.16. The result in Table 5.5.16 shows that there is a feasible assignment with only zero costs in the revised cost matrix. This means that we have achieved the optimum assignment plan. In this optimum plan, the company must assign the first, the second, the third, and the fourth worker, the fourth (dummy), the third, the second, and the first job, respectively. Then the minimum total cost to the company will be $C^* = \$0 + \$1 + \$2 + \$3 = \$6$.

5.5.8 Exercises

1. Suppose that a firm has three plants in three cities and three warehouses in three other cities. The firm wants to transport the good produced in these plants to the warehouses. The total supplies from three plants and the total demands at the warehouses along with the associated transportation costs (in dollars) per unit of the good are given in Table 5.5.17. Find the transportation plan that minimizes the firm's total cost.
2. Suppose that a firm has three plants in three cities and two warehouses in three other cities. The firm wants to transport the good produced in these plants to the warehouses. The total supplies from the plants and the total demands at the warehouses along with the associated transportation costs (in dollars) per unit of the good are given in Table 5.5.18. Find the transportation plan that minimizes the firm's total cost.
3. Suppose that a firm has three plants in three cities and three warehouses in three other cities. The firm wants to transport the good produced in these plants to the warehouses. The total supplies from the plants and the total demands at the warehouses along with the

Table 5.5.17

		Warehouses			
		1	2	3	Supplies
Plants	1	5	8	4	**50**
	2	2	3	5	**200**
	3	8	8	9	**50**
	Demands	**75**	**100**	**125**	

Table 5.5.18

		Warehouses		
		1	2	Supplies
Plants	1	10	10	**100**
	2	15	15	**50**
	3	10	15	**50**
	Demands	**75**	**100**	

Table 5.5.19

		Warehouses			
		1	2	3	Supplies
Plants	1	2	3	5	**100**
	2	3	6	5	**125**
	3	5	3	2	**75**
	Demands	**100**	**200**	**100**	

Table 5.5.20

		Jobs				
		1	2	3	4	
Workers	1	8	12	9	10	**1**
	2	15	10	11	12	**1**
	3	16	18	20	22	**1**
	4	19	14	12	10	**1**
		1	**1**	**1**	**1**	

associated transportation costs (in dollars) per unit of the good are given in Table 5.5.19. Find the transportation plan that minimizes the firm's total cost.

4.   Suppose that a company has four jobs to be assigned to four workers. The costs, in dollars, of assigning these jobs to these workers are given in Table 5.5.20. Determine the assignment plan that minimizes the company's total cost.

Table 5.5.21

		Jobs				
		1	2	3	4	
Workers	1	6	4	8	3	**1**
	2	1	5	3	7	**1**
	3	3	1	2	4	**1**
	4	7	6	5	8	**1**
		1	**1**	**1**	**1**	

5. Suppose that a company has four jobs to be assigned to four workers. The costs, in dollars, of assigning these jobs to these workers are given in Table 5.5.21. Determine the assignment plan that minimizes the company's total cost.

 Web supplement: S5.5.9 Mathematica applications

6 Nonlinear programming

6.1 Introduction

We introduced the classical approach to optimization in Chapter 4. In that chapter we considered both unconstrained and constrained optimization problems. These problems possessed two salient features. The first is that the objective functions and the constraint functions in these problems can take *linear* or *nonlinear* forms. The second is that the constraints of these problems are *equality* constraints. We then explored the methods of solving these optimization problems. Notice that the optimal values of the choice variables in classical optimization problems may take any sign, positive or negative.

In Chapter 5 we extended our analysis of optimization to LP problems. The reader would have noticed that we imposed three conditions in the LP approach to optimization. The first is that in optimization involving LP problems both the objective functions and the constraints are *linear*, in contrast to the classical approach where both the objective function and constraints can be linear or nonlinear in forms. The second condition is that although the constraints are linear, they are, again in contrast to the classical approach, *inequality* constraints. The third is the nonnegativity condition – that is, the choice variables should have nonnegative optimum solutions. We then explored several methods of solving LP problems.

We now turn our attention in the present chapter to more general optimization problems – optimization problems that involve *linear* or *nonlinear* objective functions and constraints. Notice that we continue to impose the nonnegativity constraint here too. An optimization problem that involves a linear or nonlinear objective function with linear or nonlinear constraints and the nonnegativity condition is called a *nonlinear programming* (NLP) problem. We shall see shortly that the classical and LP approaches to optimization address problems that are special cases of NLP problems.

6.2 NLP: general ideas

6.2.1 Nonlinear relationships

As we saw in Chapter 4, many of the relationships in economics, business, and finance tend to be nonlinear in nature. For example, consider the graph in Figure 4.2.7(B). This graph shows the case of a short-run production function of a firm with labor as the variable input. It also shows that as the quantity of labor employed increases, total output increases at an increasing rate in the beginning and then increases at a diminishing rate. If the price and cost functions are assumed to be constant, then the firm's total profit function will follow a form similar to that of the shape of its total output function. These imply that the output or profit function (that is, the objective function) of the firm takes a nonlinear form. Similarly, a consumer's utility function rarely takes the form of a linear line. Instead, it usually follows a nonlinear form such as the one exemplified by indifference curves. Moreover, the constraints that a firm or a consumer faces are also normally nonlinear in form. One can cite many examples such as these from economics.

In real-world business and financial problems, the relationships are also normally expressed in the form of nonlinear functions. It is difficult for a manager of a company to assume that the function that represents the objective of the company or the functions that represent its constraints are linear functions of the choice variables. In fact, in most cases, they are nonlinear functions of the choice variables. One can find many examples of these nonlinear relationships in product mix problems, portfolio selection problems, etc.

6.2.2 NLP: a general maximization problem

In this section we set up a general NLP maximization problem involving n choice variables and m constraints. As in LP problems, the number of choice variables may be greater than, less than, or equal to the number of constraints. Suppose that a firm produces x_j ($j = 1, 2, 3, \ldots, n$) goods using b_i ($i = 1, 2, 3, \ldots, m$) resources, and that the firm's aim is to maximize its total profit Π (which is a function, linear or nonlinear, of the quantity of the goods produced) subject to the m constraints as given by

$$\text{Maximize } \Pi = f(x_1, x_2, x_3, \ldots, x_j, \ldots, x_n), \text{ subject to } g^1(x_1, x_2, x_3, \ldots, x_j, \ldots, x_n) \le b_1,$$

$$g^2(x_1, x_2, x_3, \ldots, x_j, \ldots, x_n) \le b_2, g^3(x_1, x_2, x_3, \ldots, x_j, \ldots, x_n) \le b_3, \ldots,$$

$$g^i(x_1, x_2, x_3, \ldots, x_j, \ldots, x_n) \le b_i, \ldots, g^m(x_1, x_2, x_3, \ldots, x_j, \ldots, x_n) \le b_m, \text{ and } x_j \ge 0$$
$$(6.2.1)$$

Notice that some or all of the constraints in the NLP maximization problem (6.2.1) may appear with equality signs. All the functions in problem (6.2.1) are assumed to be differentiable. If all of the constraints appear with equality signs, then we have a classical maximization problem, as those presented in Section 4.4. If both the objective function and the constraints are linear, then we have a LP problem, as those presented throughout in Chapter 5. This is the reason why we stated at the end of Section 6.1 that a NLP problem is the most general optimization problem we found so far in this book and that the classical and LP optimization problems are special cases of the general NLP problem.

Notice that the set of critical values $(x_1^*, x_2^*, x_3^*, \ldots, x_j^*, \ldots, x_n^*)$ that satisfies all the constraints in problem (6.2.1) is called the *feasible set*. The reader would have noticed that we can convert a maximization problem into a minimization problem by multiplying the objective function and the constraints of the former by -1. Similarly, a particular $\le$ constraint can be converted into a $\ge$ constraint (and vice versa) by multiplying the former by -1. The maximization problem (6.2.1) can also be written succinctly as

$$\text{Maximize } \Pi = f(x_j), j = 1, 2, 3, \ldots, n, \text{ subject to } g^i(x_j) \le b_i,$$

$$i = 1, 2, 3, \ldots, m, \text{ and } x_j \ge 0 \tag{6.2.2}$$

6.2.3 NLP: a general minimization problem

Let us now set up a general NLP minimization problem involving n choice variables and m constraints. Again the number of choice variables may be greater than, less than, or equal to the number of constraints. Suppose that a firm employs c_i $(i = 1, 2, \dots, m)$ factors to produce x_j $(j = 1, 2, \dots, n)$ goods and that the firm's aim is to minimize its total cost C (which is a function, linear or nonlinear, of the quantity of the output produced) subject to the m constraints as

$$\text{Maximize } C = h(x_1, x_2, x_3, \dots, x_j, \dots, x_n), \text{ subject to } g^1(x_1, x_2, x_3, \dots, x_j, \dots, x_n) \geq c_1,$$

$$g^2(x_1, x_2, x_3, \dots, x_j, \dots, x_n) \geq c_2, g^3(x_1, x_2, x_3, \dots, x_j, \dots, x_n) \geq c_3, \dots,$$

$$g^i(x_1, x_2, x_3, \dots, x_j, \dots, x_n) \geq c_i, \dots, g^m(x_1, x_2, x_3, \dots, x_j, \dots, x_n) \geq c_m, \text{ and } x_j \geq 0$$
(6.2.3)

Notice that some or all of the constraints in the NLP minimization problem (6.2.3) may appear with equality signs. All the functions in problem (6.2.3) are assumed to be differentiable. If all of the constraints appear with equality signs, then we have a classical minimization problem as stated previously. If both the objective function and the constraints are linear, then we have a LP problem. Notice that the set of values $(x_1^*, x_2^*, x_3^*, \dots, x_j^*, \dots, x_n^*)$ that satisfies all the constraints in problem (6.2.3) is called, as in a NLP maximization problem, the feasible set. The minimization problem (6.2.3) can also be written succinctly as

$$\text{Minimize } C = h(x_j), j = 1, 2, 3, \dots, n, \text{ subject to } g^i(x_j) \geq c_i,$$

$$i = 1, 2, 3, \dots, m, \text{ and } x_j \geq 0$$
(6.2.4)

6.2.4 NLP: geometric forms of objective functions and constraints

Let us first consider the geometric forms that are normally assumed by objective functions in a NLP problem in the subjects of our interest. An objective function in a NLP problem may assume either a nonlinear form or a linear form. One linear form that is frequently encountered in the literature is similar to the the function

$$\text{Maximize } \Pi = f(x_1, x_2) = 2x_1 + 4x_2$$
(6.2.5)

the graphs of which for different values of Π are illustrated in Figure 6.2.1(A). Notice that higher lines in this figure represent higher levels of Π and, therefore, the aim in a NLP maximization problem that involves the objective function in equation (6.2.5) is to attain the highest possible line.

Another form of the objective function that we will use later is a nonlinear form as

$$\text{Maximize } \Pi = h(x_1, x_2) = 50x_1 + 50x_2 - 5x_1^2 - 5x_2^2$$
(6.2.6)

the graphs of which, for different values of Π, are illustrated in Figure 6.2.1(B). Notice that the smaller circles in this figure represent higher levels of Π, and the maximum of Π (that is, $\Pi = 250$) is when $x_1 = 5$ and $x_2 = 5$. Therefore, the aim in a NLP maximization problem with the objective function in equation (6.2.6) is to attain the smallest possible circle or move as close as possible to point B from points such as A or C. Notice also that we

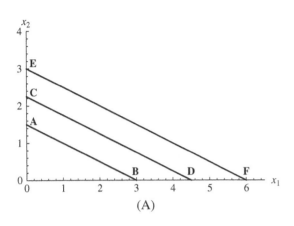

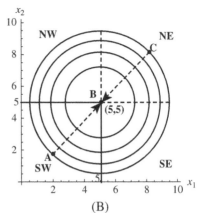

(A)   (B)

Figure 6.2.1

have divided the x–y space into four subspaces, southwest (SW), northwest (NW), northeast (NE), and southeast (SE). The portions of the graphs that lie in the SW space are similar to indifference curves or isoquants.

The third form of the objective function that we will use later is of the form

$$\text{Minimize } C = j(x_1, x_2) = (x_1 - 5)^2 + (x_2 - 5)^2 = x_1^2 - 10x_1 + x_2^2 - 10x_2 + 50 \quad (6.2.7)$$

the graph of which, for different values of C, is illustrated in Figure 6.2.2(A). The reader would have noticed that the smaller circles in this figure represent smaller levels of C, and the minimum of C (that is, $C = 0$) is when $x_1 = 5$ and $x_2 = 5$. Therefore, the aim in a NLP minimization problem with the objective function in equation (6.2.7) is to attain the smallest possible circle or move as close as possible to point B from points such as A or C. Notice that, as before, we have divided the x–y space into four subspaces, southwest (SW), northwest (NW), northeast (NE), and southeast (SE). Notice also that the portions of the graphs that lie in the SW space are similar to indifference curves or isoquants.

The last form of the objective function that we will consider in this section is the linear form

$$\text{Minimize } C = k(x_1, x_2) = x_1 + x_2 \quad (6.2.8)$$

the graph of which is illustrated, for different values of C, in Figure 6.2.2(B). Since the lower lines in this figure represent lower C, the aim in a NLP minimization problem with the objective function in equation (6.2.8) is to attain the lowest possible line.

Let us now illustrate the generally used geometric forms of constraints in NLP problems. These forms, as in the case of the objective functions illustrated in Figures 6.2.1 and 6.2.2, may be either nonlinear or linear. Let us first consider the nonlinear form of a constraint in a NLP problem. Suppose that our constraint is of the form

$$g^1(x_1, x_2) = x_1^2 + x_2^2 \leq 4 \quad (6.2.9)$$

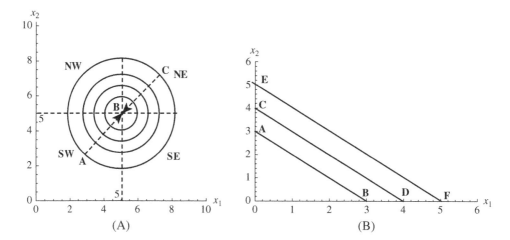

Figure 6.2.2

which can be plotted to yield the curve GH in Figure 6.2.3(A). As can be seen, the set of points in the area OGH (including those on the boundary) is the set of feasible points (the set of points that satisfies the constraint (6.2.9)). Assume now that we have another constraint as

$$g^2(x_1, x_2) = x_1^2 + x_2 \leq 3 \qquad (6.2.10)$$

which is plotted to obtain the curve IJ in Figure 6.2.3(A). The set of points in the area OIJ (including those on the boundary) is the set of feasible points (the set of points that satisfies the constraint (6.2.10)). Therefore, the set of feasible points that satisfies both constraints (inequalities (6.2.9) and (6.2.10)) simultaneously is given by the set of points in the area OGKJ (including the points on the boundary).

Another set of constraints that we will use later consists of constraints that are linear in form and are given by the inequalities

$$g^3(x_1, x_2) = x_1 \leq 10, g^4(x_1, x_2) = x_2 \leq 4, \quad \text{and} \quad g^5(x_1, x_2) = x_1 + 3x_2 \leq 15 \quad (6.2.11)$$

the graphs of which are illustrated in Figure 6.2.3(B). The set of feasible points that satisfies all three constraints simultaneously is given by the set of points in the area DEFGH (including the points on the boundary) in this figure.

The third set of constraints we will use later consists of constraints that are also linear and are given by

$$g^6(x_1, x_2) = x_1 + x_2 \leq 6 \quad \text{and} \quad g^7(x_1, x_2) = x_1 + 2x_2 \geq 8 \qquad (6.2.12)$$

the graphs of which are illustrated in Figure 6.2.4(A). Notice that the set of feasible points that satisfies both constraints simultaneously is given by the set of points in the triangle DEF (including the points on the boundary) in this figure.

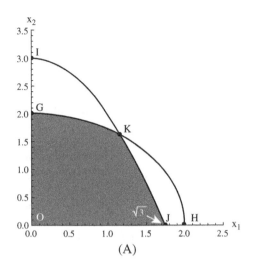

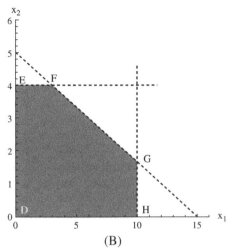

Figure 6.2.3

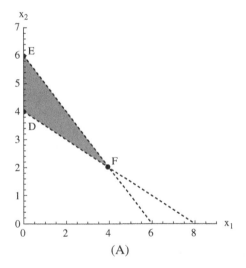

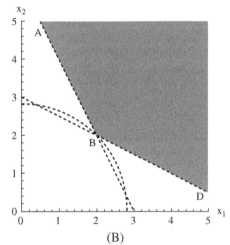

Figure 6.2.4

The last set of constraints we will use later comprises linear and nonlinear inequality constraints as follows:

$$g^8(x_1,x_2)=0.5x_1+x_2\geq3, \quad g^9(x_1,x_2)=x_1+0.5x_2\geq3, \quad \text{and} \quad g^{10}(x_1,x_2)=x_1^2+x_2^2\geq8$$
$$(6.2.13)$$

the graphs of which are illustrated in Figure 6.2.4(B). Notice that the set of feasible points that satisfies all these constraints simultaneously is given by the set of points in the area above the line ABD (including the points on the boundary) in this figure.

6.2.5 *Geometric solution of NLP problems: maximization cases*

Let us now attempt to solve few NLP problems. Although there are many methods to solve NLP problems, we will discuss only three methods in this book. The first of these three methods, the *geometric method*, will be explored in the current section. The second method, popularly called the *Kuhn–Tucker conditions*, will be detailed in Section 6.3. The third method, called the *trial-and-error method*, will be discussed in Section 6.3.10. Although the Kuhn–Tucker conditions are applicable in the case of NLP problems involving any number of choice variables, the geometric method is applicable in the case of NLP problems involving a maximum of two choice variables, and the trial-and-error method can be applied when the number of the choice variables is very few.

Suppose that we want to maximize the objective function in equation (6.2.5), the graph of which is illustrated in Figure 6.2.1(A), subject to the constraint given in inequalities (6.2.9) and (6.2.10), the graphs of which are illustrated in Figure 6.2.3(A). For convenience we reproduce these two sets of graphs in a single figure as illustrated in Figure 6.2.5(A).

The set of feasible points in Figure 6.2.5(A) is the set of points in the region OGNKJ (including the points on the boundary). Our aim in this problem, as in a LP maximization problem, is to push the objective function in the northeast direction. We know that we cannot attain, given the constraints, those lines of the objective function that lie outside the area OGNKJ. The only graph of the objective function that is acceptable is the one that is tangent to one of the constraints $(g^1(x_1,x_2) = x_1^2 + x_2^2 \leq 4)$ at point N. Therefore, the optimum of the objective function occurs at this point of tangency. This implies that the optimum solution satisfies only one constraint $(g^1(x_1,x_2) = x_1^2 + x_2^2 \leq 4)$ exactly.

Notice that at the optimum point N, the slope of the objective function $(\Pi = f(x_1,x_2) = 2x_1 + 4x_2)$ is equal to the slope of the constraint $(g^1(x_1,x_2) = x_1^2 + x_2^2 \leq 4)$. The slope of the objective function is $-1/2$. The slope of the constraint can be found as follows. Let us first write the inequality constraint $g^1(x_1,x_2) = x_1^2 + x_2^2 \leq 4$ as an equality constraint $g^1(x_1,x_2) = x_1^2 + x_2^2 = 4$. We then write it as a new function $G^1(x_1,x_2) = x_1^2 + x_2^2 - 4 = 0$, which is now in the form of an implicit function. We can now apply the technique of differentiation

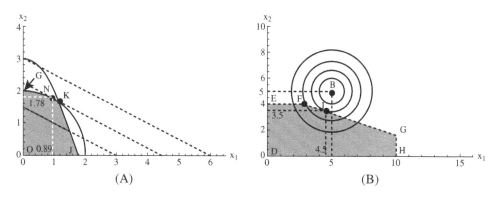

Figure 6.2.5

of implicit functions (discussed in Section 3.3.12) to obtain the slope of the constraint as $dx_2/dx_1 = -[\partial G^1/\partial x_1]/[\partial G^1/\partial x_2] = -2x_1/2x_2$. Since the slope of the objective function is equal to the slope of the constraint, we can write $-2x_1/2x_2 = -1/2$, or $2x_1 = x_2$. If we substitute this equation into $g^1(x_1, x_2) = x_1^2 + x_2^2 = 4$ and simplify, we obtain $x_1^* = 0.894427$. If we substitute this value again into the constraint $g^1(x_1, x_2) = x_1^2 + x_2^2 = 4$ and simplify, we obtain $x_2^* = 1.78885$. Then the optimal (maximum) value of the objective function can be found by substituting these optimal values of the choice variables into the objective function. This yields $\Pi^* = f(x_1^*, x_2^*) = 2x_1^* + 4x_2^* = 2 \times 0.894427 + 4 \times 1.78885 = 8.94427$.

The reader would have noticed an important feature of the solution of the maximization problem illustrated in Figure 6.2.5(A): point N is not a corner point. We know from Chapter 5 that the optimum solution to a LP problem always occurs at a corner point of the feasible region. In contrast to this, the optimum solution to a NLP problem may occur at any point (not just at a corner point), as at point N, of the set of feasible points. Another important feature of the solution illustrated in this figure is that, as mentioned above, it satisfies only one constraint exactly. Therefore, the optimal solution to a NLP problem may not always satisfy all the constraints of the problem exactly, as in classical optimization problems.

So far we have been concerned with maximizing a linear objective function subject to nonlinear constraints. Let us now attempt to maximize a nonlinear objective function subject to linear constraints. For this consider the nonlinear objective function in equation (6.2.6), the graph of which is illustrated in Figure 6.2.1(B), and the constraints in inequality (6.2.11), the graphs of which are illustrated in Figure 6.2.3(B). For convenience we reproduce these two sets of graphs in a single figure as illustrated in Figure 6.2.5(B).

If the present problem were a free optimization problem, then the optimum solution would be $x_1^* = x_2^* = 5$, and the maximum value of the objective function would be $\Pi = h(x_1^*, x_2^*) = 50x_1^* + 50x_2^* - 5x_1^{2*} - 5x_2^{2*} = 50 \times 5 + 50 \times 5 - 5 \times 5^2 - 5 \times 5^2 = 250$. This happens, as can be seen in Figure 6.2.5(B), at the centre of the circle (point B). Notice that the coordinate point (5, 5) corresponding to point B lies outside the set of feasible points (represented by the set of points in the area DEFIGH, including the points on the boundary). This implies that we cannot achieve the level of Π represented by that point ($\Pi = 250$). But, notice that the graph of the constraint $x_1 + 3x_2 \leq 15$ is tangent to one of the graphs of the objective function $\Pi = h(x_1, x_2) = 50x_1 + 50x_2 - 5x_1^2 - 5x_2^2$ at point I. Any movement, from point I, in the southwest direction reduces the value of the objective function; and any movement in the northeast direction will increase the value of the objective function but will violate at least one of the constraints. Hence, this point of tangency, given the constraints, yields the maximum value of the objective function.

How do we find the optimal solution that occurs at point I in Figure 6.2.5(B)? Notice that at point I the slope of the objective function is equal to the slope of the constraint $x_1 + 3x_2 \leq 15$. We know that the slope of this constraint, if we treat it as a function, is $-1/3$. Notice also that we can convert the objective function $\Pi = h(x_1, x_2) = 50x_1 + 50x_2 - 5x_1^2 - 5x_2^2$ into an implicit function as $H(x_1, x_2) = 50x_1 + 50x_2 - 5x_1^2 - 5x_2^2 - \Pi = 0$. We can now use the technique of implicit differentiation to find the slope of the implicit function $H(x_1, x_2) = 50x_1 + 50x_2 - 5x_1^2 - 5x_2^2 - \Pi$. This slope is $dx_2/dx_1 = -[\partial H/\partial x_1]/[\partial H/\partial x_2] = -[(50 - 10x_1)]/[(50 - 10x_2)]$. Equating this to the slope of the constraint $(-1/3)$, we obtain $x_1 = (10 + x_2)/3$. Substituting this result into the equation $x_1 + 3x_2 = 15$, we obtain $x_2^* = 3.5$. Substituting the last result into $x_1 + 3x_2 = 15$ yields $x_1^* = 4.5$. Therefore, the maximum value of the objective function, given the constraints, is $\Pi = h(x_1^*, x_2^*) = 50x_1^* + 50x_2^* - 5x_1^{2*} - 5x_2^{2*} = 50 \times 4.5 + 50 \times 3.5 - 5 \times 4.5^2 - 5 \times 3.5^2 = 237.5$.

6.2.6 Geometric solution of NLP problems: minimization cases

In the previous section, we considered the topic of NLP maximization problems. We now attempt to solve NLP minimization problems geometrically. Suppose that our aim is to minimize the objective function in equation (6.2.7), the graph of which is illustrated in Figure 6.2.2(A), subject to constraints in inequality (6.2.12), the graphs of which are presented in Figure 6.2.4(A). For convenience, we reproduce these two sets of graphs in a single figure as illustrated in Figure 6.2.6(A).

If our problem were a free optimization problem, then the optimum solution would be $x_1^* = x_2^* = 5$, and the minimum value of the objective function would be $C = h(x_1^*, x_2^*) = (x_1^* - 5)^2 + (x_2^* - 5)^2 = 0$. This occurs, as can be seen in Figure 6.2.6(A), at the centre of the circle (point B). But, the present problem is a NLP problem. Therefore, the set of feasible points is the set of points in or on the triangle DEF. The graph of the objective function is tangent to the higher constraint ($x_1 + x_2 \leq 6$) at point G. Any movement, from point G, in the southwest direction increases the value of the objective function, and any movement in the northeast direction will reduce the value of the objective function but will violate at least one of the constraints (specifically, the constraint $x_1 + x_2 \leq 6$). Hence, this point of tangency between the objective function and the constraint $x_1 + x_2 \leq 6$ gives, given the two constraints, the minimum value of the objective function.

How do we find the optimal solution that occurs at point G in Figure 6.2.6(A)? Notice that at point G the slope of the objective function is equal to the slope of the constraint $x_1 + x_2 \leq 6$. We know that the slope of this constraint, if we treat it as a function, is -1. Notice also that we can convert the objective function $C = h(x_1, x_2) = (x_1 - 5)^2 + (x_2 - 5)^2$ into an implicit function as $H(x_1, x_2) = (x_1 - 5)^2 + (x_2 - 5)^2 - C = 0$. We can now use the technique of implicit differentiation to find the slope of the implicit function $H(x_1, x_2) = (x_1 - 5)^2 + (x_2 - 5)^2 - C$ as $dx_2/dx_1 = -[\partial H/\partial x_1]/[\partial H/\partial x_2] = -[2(x_1 - 5)]/[2(x_2 - 5)] = -(x_1 - 5)/(x_2 - 5)$. Equating this to the slope of the constraint (-1), we obtain $x_1 = x_2$. Substituting this last result into the equation $x_1 + x_2 = 6$, we obtain $2x_1 = 6$ or $x^* = 3$.

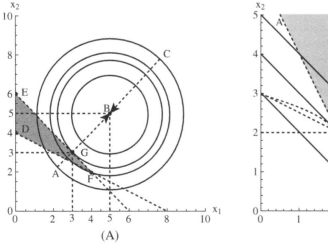

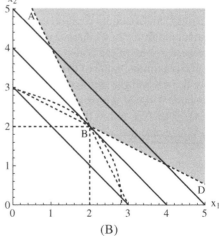

(A) (B)

Figure 6.2.6

Since $x_1 = x_2$, we have $x_1^* = x_2^* = 3$. Therefore, the minimum value of the objective function is $C^* = h(x_1^*, x_2^*) = (x_1^* - 5)^2 + (x_2^* - 5)^2 = (3 - 5)^2 + (3 - 5)^2 = 8$.

Let us now geometrically solve a NLP minimization problem involving a linear objective function and a set of linear and nonlinear constraints. For this we use the objective function in equation (6.2.8), the graph of which is illustrated in Figure 6.2.2(B) for different levels of C, subject to constraints in (6.2.13), the graphs of which are illustrated in Figure 6.2.4(B). As earlier, we reproduce these two sets of graphs in a single figure as illustrated in Figure 6.2.6(B).

We know from Figure 6.2.6(B) that the set of feasible points is the set of points on or above the line ABD. The graph of the objective function is tangent to the constraint $x_1^2 + x_2^2 \geq 8$ at point B. Any movement, from point B, in the southwest direction will violate at least one of the constraints and any movement in the northeast direction will increase the value of the objective function. Hence, this point of tangency between the objective function and the constraint $x_1^2 + x_2^2 \geq 8$ gives, given the constraints, the minimum value of the objective function.

We can find the optimal solution that occurs at point B following the same procedure as the one we adopted earlier. Notice that at point B the slope of the objective function is equal to the slope of the constraint $x_1^2 + x_2^2 \geq 8$. We know that the slope of the objective function is -1. The slope of the constraint $g^{10}(x_1, x_2) = x_1^2 + x_2^2 \geq 8$ can be found by treating it as an equation and differentiating the result implicitly. Converting it into an equation we obtain $G^{10}(x_1, x_2) = x_1^2 + x_2^2 - 8 = 0$. The slope of this implicit function is $dx_2/dx_1 = -[\partial G^{10}/\partial x_1]/[\partial G^{10}/\partial x_2] = -2x_1/2x_2 = -x_1/x_2$. Equating this to the slope of the objective function (-1), we obtain $x_1 = x_2$. Substituting this result into the equation $x_1^2 + x_2^2 = 8$, we obtain $x_2^2 + x_2^2 = 8$ or $x_2^* = 2$. Since $x_1 = x_2$, we have $x_1^* = x_2^* = 2$. Therefore, the minimum value of the objective function is $C^* = k(x_1^*, x_2^*) = x_1^* + x_2^* = 2 + 2 = 4$.

6.2.7 LP and NLP: important differences

We presented the main features of the objective function and the constraints of NLP problems in Section 6.2.4. Let us now identify those important features that distinguish a NLP problem from a LP problem.

1 In a LP problem that has an optimal solution, the feasible region always constitutes a convex set. In a NLP problem that has an optimal solution, the feasible region may not always constitute a convex set.

2 In a minimization (maximization) LP problem our aim was to push down (up) the objective function and, thereby, to decrease (increase) the value of the objective function, as low (high) as possible without violating the constraints. In a minimization (maximization) NLP problem, such a push may increase (decrease) the value of the objective function (as can be seen in Figures 6.2.2(A) and 6.2.1(B)).

3 In a LP problem that has an optimal solution, the optimal solution occurs at one of the corner points of the convex set. In a NLP problem that has an optimal solution, the optimal solution may occur at any point, not just at the corner point, of the feasible region(s) (as can be seen in Figures 6.2.5 and 6.2.6).

4 In a LP problem that has an optimal solution, the number of variables (including dummy variables) with positive optimal values will be equal to the number of constraints in the problem. In a NLP problem that has an optimal solution, the number of variables with

positive optimal values may or may not be equal to the number of constraints in the problem.

5   In a LP problem that has an optimal solution, a local optimum is also a global optimum. In a NLP problem, a local optimum may not be the same as the global optimum.

Because of these differences between LP and NLP problems, the methods applied in Chapter 5 to solve LP problems are not suitable to solve NLP problems. Therefore, new methods are needed. There are a number of methods in the literature that are employed to find quantitative optimal solutions to NLP problems. However, we do not explore these methods in this book; instead we present in Section 6.3 few results that offer qualitative characterization of optimal solutions. These results are called Kuhn–Tucker conditions.

6.2.8 Application examples

Example 1.   Assume that a consumer's total utility, U, from the consumption of two goods, x_1 and x_2, is given by the Cobb–Douglas utility function $U = f(x_1, x_2) = x_1^{1/2} x_2^{1/2}$. Also assume that the consumer has a maximum of \$7 to spend on these two goods, and that the unit price of each good is \$1. Find the quantities of the two goods that maximize the consumer's total utility.

Solution.   Notice that the objective function in this problem is nonlinear in form and, when plotted for different values of U, the utility function will give us an *indifference map*. The constraint the consumer faces (the budget constraint) is linear in form, and can be constructed as $x_1 + x_2 \leq 7$. We can now plot the nonlinear utility function and the linear budget constraint to obtain the graphs in Figure 6.2.7(A). Notice that at point C the indifference curve that represents total utility $U = 3.5$ is tangent to the budget line. This means that the highest attainable indifference curve (the indifference curve that represents total utility $U = 3.5$) has the same slope as that of the budget line. Therefore, we can obtain the solution to the utility maximization problem if we find the values of x_1 and x_2 corresponding to point C. We know that the slope of the budget line is -1. Let us write the utility function as $F(x_1, x_2) = f(x_1, x_2) - U = x_1^{1/2} x_2^{1/2} - U = 0$. Notice that this function is now an implicit function. The slope of the indifference curve (corresponding to $U = 3.5$) can be found as $dx_2/dx_1 = -[\partial F/\partial x_1]/[\partial F/\partial x_2] = -x_2/x_1$. Since the slope of the budget line and the slope of the indifference curve are equal at point B, we have $dx_2/dx_1 = -x_2/x_1 = -1$, which simplifies to $x_1 = x_2$. Substituting this result into the budget constraint and simplifying (after treating it as an equation) we obtain $x_1^* = x_2^* = 3.5$. If we substitute this solution into the objective function we obtain $U^* = f(x_1^*, x_2^*) = x_1^{*1/2} x_2^{*1/2} = (3.5)^{1/2}(3.5)^{1/2} = 3.5$. Therefore, given the constraint, the consumer must consume 3.5 units of both goods to maximize the total utility and the maximum total utility will be 3.5. The reader may want to compare the geometric solution to the present problem illustrated in Figure 6.2.7(A) with the solution presented in Figure 3.8.3(A).

Example 2.   Suppose that the total output produced by a firm using two factors, capital (K) and labor (L), is given by the CES production function $Q = f(K, L) = [0.5K^{-0.5} + 0.5L^{-0.5}]^{-2}$. Also, suppose that the wage rate is \$10, the interest rate is 10 percent, and the maximum fund for investment available with the firm is \$100. How many units of the two factors should the firm employ to maximize its total output?

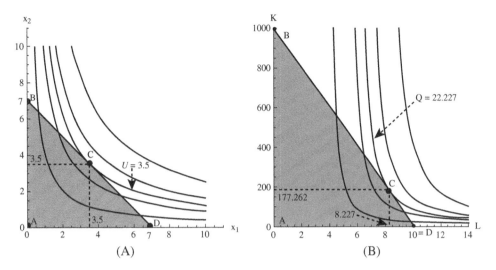

Figure 6.2.7

Solution. Notice that the objective function in this problem is, as in the last example, nonlinear in form and, when plotted for different values of Q, the production function will give us an *isoquant map*. The constraint that the firm faces (the isocost line) is linear in form, and can be written as $0.1K + 10L \leq 100$. We can now plot the nonlinear production function and the linear isocost line to obtain the graphs in Figure 6.2.7(B). Notice that at point C the isoquant that represents total output $Q = 22.227$ is tangent to the isocost line. This means that the highest attainable isoquant (the isoquant that represents total output $Q = 22.227$) has the same slope as that of the isocost line. Therefore, we can obtain the solution to the present output maximization problem if we find the values of K and L corresponding to point C. We know that the slope of the isocost line is -100. Let us write the production function as $F(K, L) = f(K, L) - Q = [0.5K^{-0.5} + 0.5L^{-0.5}]^{-2} - Q = 0$, which is now an implicit function. The slope of the isoquant (treating Q as a constant) can be found as $dK/dL = -[\partial F/\partial L]/[\partial F/\partial K] = -K^{3/2}/L^{3/2}$. Since the slope of the isocost line and the slope of the isoquant are equal at point C, we have $dK/dL = -K^{3/2}/L^{3/2} = -100$, which simplifies to $K = 100^{2/3}L$. Substituting this result into the equation for the isocost line and simplifying, we obtain $K^* = 177.262$ and $L^* = 8.227$. If we substitute this solution into the objective function we obtain $Q^* = f(K^*, L^*) = [0.5K^{*-0.5} + 0.5L^{*-0.5}]^{-2} = [0.5(177.3)^{-0.5} + 0.5(8.2)^{-0.5}]^{-2} = 22.277$. Therefore, given the constraint, the firm must employ 177.262 units of capital and 8.227 units of labor to maximize the total output and the maximum total output will be 22.277 units. As in the last example, compare the geometric solution to the present problem illustrated in Figure 6.2.7(B) with the solution presented in Figure 3.8.3(B).

Example 3. Assume that an individual lives for two periods, period 1 and period 2. Her consumption in period 1 is denoted by C_1 and that in period 2 is denoted by C_2. Her income in period 1 is denoted by Y_1 and that in period 2 is denoted by Y_2. She can either consume or save Y_1 completely, or can consume a part of Y_1 and save the remaining part. Also assume that the interest rate is constant over the periods and is given by r. If she saves a part of Y_1, the total money at her disposal in period 2 will be $Y_2 + (Y_1 - C_1)(1 + r)$. Therefore, her

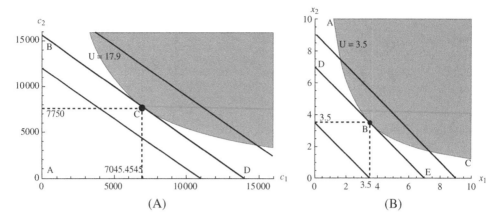

Figure 6.2.8

consumption in period 2 will be $C_2 = Y_2 + (Y_1 - C_1)(1 + r)$, which is called *intertemporal budget constraint*. Also assume that the total utility that she receives from her consumption in the two periods is given by the function $U = g(C_1, C_2) = \ln C_1 + \ln C_2$, which is called an *intertemporal utility function*. Find the minimum amount of money she has to spend in the two periods, given $Y_1 = \$5000$, $Y_2 = \$10\,000$, and $r = 0.1$, such that she attains *intertemporal utility* of at least 17.9 units.

Solution. Given $Y_1 = \$5000$, $Y_2 = \$10\,000$, and $r = 0.1$, we can write the intertemporal budget constraint $C_2 = Y_2 + (Y_1 - C_1)(1 + r)$ as $C_2 = 10\,000 + (5000 - C_1)(1 + 0.1)$ or as $1.1C_1 + C_2 = 15\,500$. In this problem, this budget constraint is the objective function and it can be written as $E = h(C_1, C_2) = 1.1C_1 + C_2 = 15\,500$. Notice that this objective function is linear in form. Moreover, the intertemporal utility function is the constraint in the present problem, and it is nonlinear in form. Let us plot the linear objective function for different values of E and the nonlinear constraint as presented in Figure 6.2.8(A). As can be seen, the constraint (intertemporal utility at least equal to 17.9 units) is given by the nonlinear curve. The linear objective functions (for different levels of E) are represented by the straight lines. The minimum E satisfying the constraint is given by the line BD. This line is tangent to the constraint at point C. This implies that the coordinates of point C will optimize (minimize) the objective function. Notice that both the constraint and the objective function have the same slope at point C. The slope of the objective function is -1.1. To find the slope of the constraint $U = g(C_1, C_2) = \ln C_1 + \ln C_2$, as we did earlier, we first convert it into an implicit function and then differentiate it with respect to C_1. The resulting implicit form is $G(C_1, C_2) = \ln C_1 + \ln C_2 - U = \ln C_1 + \ln C_2 - 17.9 = 0$. Differentiating this implicit function with respect to C_1, we obtain $dC_2/dC_1 = -[\partial G/\partial C_2]/[\partial G/\partial C_1] = -C_2/C_1$. Setting this result to the slope of the objective function (-1.1), we obtain $-C_2/C_1 = -1.1$ or $C_2 = 1.1C_1$. Substituting this result into the objective function $1.1C_1 + C_2 = 15\,500$ and simplifying yields $C_1^* = \$7045.4545$ and $C_2^* = \$7750$. Therefore, the individual must spend $\$7045.4545$ and $\$7750$ in the first and second periods, respectively, to obtain an intertemporal utility of at least 17.9 units.

The reader would have noticed that the individual spends more than her income in the first period. The difference is $Y_1 - C_1 = \$5000 - \$7045.4545 = -\$2045.4545$, which is achieved through borrowing. But, the individual's consumption in the second period is less than her income in that period. This means that she saves a part of the second period's income, and the amount saved is $Y_2 - C_2 = \$10\,000 - \$7750 = \$2250$. The individual uses this amount ($\$2250$) to pay back the principal borrowed in the first period ($\$2045.4545$) plus the interest on that ($\$204.5454$).

Example 4. Assume that a consumer buys two goods x_1 and x_2 and that the unit price of each good is $\$1$. What should be the minimum expenditure on the two goods if the consumer wants to obtain at least 3.5 units of total utility from the two goods if the utility function is of the Cobb–Douglas form: $U = f(x_1, x_2) = x_1^{1/2} x_2^{1/2}$?

Solution. Notice that the objective function in this problem is linear and is given by $E = g = (K, L) = 1x_1 + 1x_2 = x_1 + x_2$. Moreover, the constraint is nonlinear and is given by $U = f(x_1, x_2) = x_1^{1/2} x_2^{1/2} = 3.5$. The graphs of the objective function for different values of E and the constraint are illustrated in Figure 6.2.8(B). As can be seen, the lowest possible graph of the objective function, given the constraint, is represented by the line DE, and this line is tangent to the constraint that represents $U = 3.5$ at point B. Therefore, the coordinates of point B will give the solution to the problem. If we follow the same procedure as that adopted in the solutions to the earlier examples, we find that $x_1^* = 3.5$, $x_2^* = 3.5$, and $E^* = \$7$. This means the consumer has to purchase 3.5 units of both goods such that the minimum total expenditure is $E^* = \$7$ and that the total utility is at least 3.5 units. Notice the similarity between the solution to the present problem and the solution to the problem in example 1 above.

6.2.9 Exercises

1. Solve the following NLP problems geometrically:
   (i) maximize $\Pi = f(x_1, x_2) = \sqrt{x_1 + x_2}$, subject to $6x_1 + x_2 \leq 10$ and $x_1 + 6x_2 \leq 10$;
   (ii) maximize $\Pi = f(x_1, x_2) = \sqrt{x_1} + \sqrt{x_2}$, subject to $x_1 \leq 10$, $x_2 \leq 10$, and $x_1 + x_2 \geq 5$;
   (iii) maximize $\Pi = f(x_1, x_2) = \sqrt{x_1} \times \sqrt{x_2}$, subject to $x_1^2 + x_2^2 \leq 9$.
2. Solve the following NLP problems geometrically:
   (i) minimize $C = h(x_1, x_2) = \sqrt{x_1} + \sqrt{x_2}$, subject to $\sqrt{x_1}\sqrt{x_2} \geq 10$; (ii) minimize $C = h(x_1, x_2) = 6x_1 + 3x_2$, subject to $\sqrt{x_1}\sqrt{x_2} \geq 50$; (iii) minimize $\sqrt{x_1}\sqrt{x_2}$, subject to $x_1 \leq 4$, $x_2 \leq 3$, and $x_1 + x_2 \geq 6$.

 Web supplement: S6.2.10 Mathematica applications

6.3 Algebraic solution of NLP problems: Kuhn–Tucker conditions

6.3.1 Introduction

In Chapter 4 we discussed how to optimize a function, whether linear or nonlinear, without constraints and with linear or nonlinear equality constraints using the methods of the classical approach. We found there that the FOC for a relative optimum of the function was that the first partial derivative(s) of the function with respect to each choice variable(s) was (were) zero in the case of unconstrained problems and the first partial derivative(s) of the Lagrangian function with respect to each choice variable(s) was (were) zero in the case of constrained

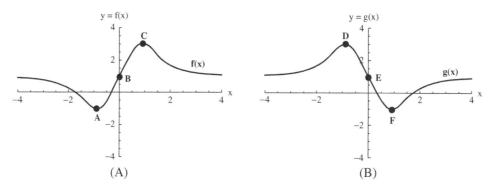

Figure 6.3.1

problems. This FOC in some problems could give us solutions which involved negative optimal values for the choice variables. Although we did not invoke there the *nonnegativity condition*, all our application examples in Chapter 4 and in Section 6.2 in the present chapter were chosen such that the optimal solution involved only nonnegative values for the choice variables. The reason was that it would be nonsensical for a producer to produce or for a consumer to consume negative quantities of some good.

One might wonder what impact it would make on the optimal solution if we invoked the nonnegativity condition explicitly. As an example, consider the function $y = f(x) = 1 + [6x/(x^4 + 2)]$ whose graph on the closed interval $[-4, 4]$ on its domain is illustrated in Figure 6.3.1(A). As can be seen, the minimum of the function $y = f(x) = 1 + [6x/(x^4 + 2)]$ on the closed interval $[-4, 4]$ on its domain occurs at point A (when $x < 0$). If our problem were to minimize the present function without the nonnegativity restriction, then we would choose this point. But, if we restrict the domain of the function to the closed interval $[0, 4]$, the minimum of the function is certainly different: it is at point B ($=1$ when $x = 0$).

As another example, consider the function $y = g(x) = 1 - [6x/(x^4 + 2)]$ whose graph on the closed interval $[-4, 4]$ on its domain is illustrated in Figure 6.3.1(B). This graph shows that the maximum of the function on the closed interval $[-4, 4]$ on its domain occurs at point D (when $x < 0$). If our problem were to maximize the present problem without the nonnegativity restriction, then we would choose this maximum point. But, if we restrict the domain of the function to the closed interval $[0, 4]$, the maximum of the function is different: it is at point E ($=1$ when $x = 0$).

The above two examples show the effect of nonnegative restrictions on the values that the choice variable(s) can take at the optimum. A curious reader might wonder whether one can develop a FOC (similar to those in the case of free or equality-constrained optimization problems) given nonnegativity and inequality constraint(s). The answer is yes; and this FOC was developed by H. W. Kuhn and A. W. Tucker and, hence, is known as the Kuhn–Tucker conditions. Before we begin the presentation and applications of these conditions, let us consider the general, not the specific as given above, consequence of the nonnegativity restrictions.

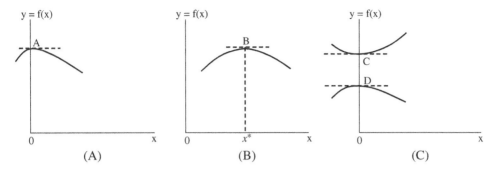

Figure 6.3.2

6.3.2 Nonnegativity condition and Kuhn–Tucker FOCs for optima of univariate functions

Suppose that we wish to maximize a univariate function subject only to the nonnegative restriction. This problem can be written as

$$\text{Maximize } y = f(x), \text{ subject to } x \geq 0 \tag{6.3.1}$$

The problem in (6.3.1) may lead to any one of the three types of solutions illustrated in Figure 6.3.2. The graph in Figure 6.3.2(A) shows that the function attains a relative maximum when $x = 0$ (at point A). Notice that the derivative of the function at $x = 0$ is zero. Since this maximum occurs at the boundary of the feasible region (on the vertical axis), it is called a *boundary solution*. The graph in Figure 6.3.2(B) shows that a relative maximum occurs at point B where the derivative of the function is zero; that is, $f'(x) = 0$ at $x = 0$. Since this relative maximum occurs at an interior point of the feasible region, it is called an *interior solution*. The graphs in Figure 6.3.2(C) show, as in Figure 6.3.2(A), boundary solutions at points C and D. At these two points we have $f'(x) < 0$ at $x = 0$. What these results imply is that, if we invoke the nonnegativity condition, a local maximum of the function is not restricted to the point where $f'(x) = 0$ (as in the FOC for a free maximization problem without nonnegativity restriction), but it may occur even at boundary points where $f'(x) < 0$. These results are presented in Table 6.3.1. We can combine the three results in Table 6.3.1 to obtain

$$f'(x) \leq 0, x > 0, \quad \text{and} \quad xf'(x) = 0 \tag{6.3.2}$$

Let us now interpret the results in (6.3.2). The first one, $f'(x) \leq 0$, states that at the local maximum the first derivative must be either negative or zero. The second one, $x > 0$, is just

Table 6.3.1

	Point	$f'(x)$	x
Figure 6.3.2(A)	A	0	0
Figure 6.3.2(B)	B	0	> 0
Figure 6.3.2(C)	C, D	< 0	0

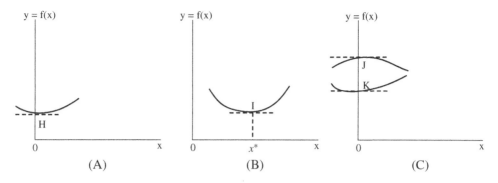

Figure 6.3.3

a restatement of the nonnegativity condition. The third one, $xf'(x) = 0$, says that if $x > 0$ then the first derivative must be zero; or if $x = 0$ then the first derivative must be either negative or zero. Therefore, as observed in Figure 6.3.2, the three results in (6.3.2) constitute the Kuhn–Tucker FOCs (or necessary conditions) for a local maximum with nonnegativity restriction. In fact, as we will see in Section 6.4.2, they are also the sufficient conditions (and the conditions for a global maximum) if certain conditions are satisfied.

Now suppose that our aim is to minimize a univariate function subject only to the nonnegative restriction. This problem can be written as

$$\text{Minimize } y = f(x), \text{ subject to } x \geq 0 \tag{6.3.3}$$

The problem in (6.3.3) may lead to, as in the case of the maximization problem discussed above, any one of the three types of solutions illustrated in Figure 6.3.3. The graph in Figure 6.3.3(A) shows that the function attains a relative minimum when $x = 0$ (at point H). But the derivative of the function at $x = 0$ is zero; that is, $f'(x) = 0$ at $x = 0$. Since this minimum occurs at the boundary of the feasible region (on the vertical axis), it is a boundary solution. Figure 6.3.3(B) shows that a relative minimum occurs at point I where the derivative of the function is zero; that is, $f'(x) = 0$ at $x > 0$. Since this relative minimum occurs at an interior point of the feasible region, it is an interior solution. Figure 6.3.3(C) illustrates, as in Figure 6.3.3(A), boundary solutions at points J and K. At these two points we have $f'(x) > 0$ and $x = 0$. What these results imply is that, if we invoke the nonnegativity condition, a local minimum of the function is not restricted to the point where $f'(x) = 0$ (as in the FOC for a free minimization problem without nonnegativity restriction), but it may occur even at boundary points where $f'(x) > 0$. These results are presented in Table 6.3.2. We can combine the three results in Table 6.3.2 to obtain

$$f'(x) \geq 0, x \geq 0, \quad \text{and} \quad xf'(x) = 0 \tag{6.3.4}$$

Let us now interpret the results in (6.3.4). The first one, $f'(x) \geq 0$, states that at the local minimum the first derivative must be either positive or zero. The second one, $x \geq 0$, is a restatement of the nonnegativity condition. The third one, $xf'(x) = 0$, says that if $x > 0$ then the first derivative must be zero; or if $x = 0$ then the first derivative must be either positive or zero. Therefore, as observed in Figure 6.3.3, the three results in (6.3.4)

Table 6.3.2

	Point	$f'(x)$	x
Figure 6.3.3(A)	H	0	0
Figure 6.3.3(B)	I	0	> 0
Figure 6.3.3(C)	J, K	> 0	0

constitute the Kuhn–Tucker FOCs (or necessary conditions) for a local minimum with nonnegativity restriction. As before, as we will see in Section 6.4.2, they are also the sufficient conditions (and the conditions for a global minimum) if certain conditions are satisfied.

6.3.3 Nonnegativity condition and Kuhn–Tucker FOCs for optima of multivariate functions

So far we have been concerned with the problem of optimizing a simple univariate function with the nonnegativity restriction. Let us now consider the case of optimizing a more general problem involving n variables but with, again, the nonnegativity restriction. Let us begin with the problem

$$\text{Maximize } y = f(x_1, x_2, \ldots, x_n), \text{ subject to } x_j \geq 0, \text{ where } j = 1, 2, \ldots, n \qquad (6.3.5)$$

Generalizing the statements in (6.3.2), we can write the Kuhn–Tucker FOCs for a local maximum of the function in problem (6.3.5) as

$$f_j \leq 0, \ x_j \geq 0, \quad \text{and} \quad x_j f_j = 0 \qquad (6.3.6)$$

Similarly, the general minimization problem is

$$\text{Minimize } y = f(x_1, x_2, \ldots, x_n), \text{ subject to } x_j \geq 0, \text{ where } j = 1, 2, \ldots, n \qquad (6.3.7)$$

As in the case of the above maximization problem, generalizing the statements in (6.3.4), we can write the Kuhn–Tucker FOCs for a local minimum of the function in problem (6.3.7) as

$$f_j \geq 0, \ x_j \geq 0, \quad \text{and} \quad x_j f_j = 0 \qquad (6.3.8)$$

6.3.4 Alternative Kuhn–Tucker FOCs for optima of univariate functions with one constraint: Lagrangian forms

We can obtain the Kuhn–Tucker FOCs (or the necessary conditions) alternatively using Lagrangian functions. Assume that we want to maximize the problem (6.3.1) with one inequality constraint $g(x) \leq c$. Suppose that we treat the inequality constraint as an equality constraint and disregard the nonnegativity restriction. Then we can set up the Lagrangian function (exactly as a classical constrained optimization problem we dealt with in Chapter 4) as

$$L = f(x) + \lambda[c - g(x)] \qquad (6.3.9)$$

We can now differentiate partially the Lagrangian function in equation (6.3.9) with respect to x and λ (the Lagrangian multiplier) to yield

$$L_x = f'(x) - \lambda g'(x) \leq 0, x \geq 0, x L_x = 0, L_\lambda = c - g(x) \geq 0, \lambda \geq 0, \quad \text{and} \quad \lambda L_\lambda = 0 \tag{6.3.10}$$

which constitute the Kuhn–Tucker FOCs (or necessary conditions) for a global maximum of a univariate function with nonnegativity restriction and with one inequality (linear or nonlinear) constraint.

We can obtain similar Kuhn–Tucker FOCs (or necessary conditions) for the univariate minimization problem in (6.3.3) with one inequality constraint $g(x) \geq c$. Assume that we want to minimize the problem (6.3.3) subject to the inequality (linear or nonlinear) constraint $g(x) \geq c$ and the nonnegativity restriction. Then, as we set up equation (6.3.9), we can set up the Lagrangian function as

$$L = f(x) + \lambda[c - g(x)] \tag{6.3.11}$$

As above, we can differentiate partially the Lagrangian function in equation (6.3.11) with respect to x and λ to yield

$$L_x = f'(x) - \lambda g'(x) \geq 0, x \geq 0, x L_x = 0, L_\lambda = c - g(x) \leq 0, \lambda \geq 0, \quad \text{and} \quad \lambda L_\lambda = 0 \tag{6.3.12}$$

which constitute the Kuhn–Tucker FOCs (or necessary conditions) for a global minimum of a univariate function with nonnegativity restriction and with one inequality (linear or nonlinear) constraints.

6.3.5 Kuhn–Tucker FOCs for optima of multivariate functions with many constraints: Lagrangian forms

Assume that our problem now is as the one given in (6.3.5). Assume also that, in addition to the nonnegativity restriction, we also face i, $i = 1, 2, \ldots, m$, inequality constraints, $g^i(x_1, x_2, \ldots, x_n) \leq c_i$. Then, exactly as we set up the Lagrangian function in equation (6.3.9), we can set up the Lagrangian function of the present problem as

$$L = f(x_1, x_2, \ldots, x_n) + \sum_{i=1}^{m} \lambda_i [c_i - g^i(x_1, x_2, \ldots, x_n)] \tag{6.3.13}$$

The Kuhn-Tucker FOCs for a local maximum in the case of the present multivariable, multiconstrained problem (the multivariable, multiconstrained analogue of the conditions given in (6.3.10), which pertains to the maximum of a univariate, single-constrained problem) can be written as

$$L_j \leq 0, x_j \geq 0, x_j L_j = 0, L_{\lambda_i} \geq 0, \lambda_i \geq 0, \quad \text{and} \quad \lambda_i L_{\lambda_i} = 0 \tag{6.3.14}$$

Assume now that our problem is the one given in problem (6.3.7). As above, assume also that, in addition to the nonnegativity restriction, we also face i, $i = 1, 2, \ldots, m$, inequality constraints, $g^i(x_1, x_2, \ldots, x_n) \geq c_i$. Then we can set up the Lagrangian function of the present problem as

$$L = f(x_1, x_2, \ldots, x_n) + \sum_{i=1}^{m} \lambda_i [c_i - g^i(x_1, x_2, \ldots, x_n)] \tag{6.3.15}$$

The Kuhn–Tucker FOCs for a local minimum in the case of the present multivariable, multiconstrained problem (the analogue of the conditions given in (6.3.12) pertaining to the minimum of a univariate, single-constrained problem) can be written as

$$L_j \geq 0, x_j \geq 0, x_j L_j = 0, L_{\lambda_i} \leq 0, \lambda_i \geq 0, \quad \text{and} \quad \lambda_i L_{\lambda_i} = 0 \tag{6.3.16}$$

6.3.6 Economic interpretations of Kuhn–Tucker maximum conditions

Let us first present a general interpretation of the Kuhn–Tucker conditions for a NLP maximum problem, presented in (6.3.14). The first term, $L_j \leq 0$, called the *marginal condition*, is to conform to the first inequality ($f'(x) \leq 0$) in (6.3.2). The second term, $x_j \geq 0$, is to conform to the nonnegativity restriction. The third term, $x_j L_j = 0$, represents the *complementary slackness condition*. This term shows that if $x_j > 0$, then $L_j = 0$; or if $x_j = 0$, then $L_j < 0$. Notice that they both could be zero simultaneously (that is, $x_j = 0$ and $L_j = 0$). The fourth term, $L_{\lambda_i} \geq 0$, is a mere restatement of the inequality constraint. Since the shadow price of a particular resource cannot be negative, the fifth term, the ith Lagrange multiplier (λ_i), must be nonnegative ($\lambda_i \geq 0$). The sixth term, $\lambda_i L_{\lambda_i} = 0$, represents the complementary slackness in the case of the Lagrange multiplier. It shows that if the ith Lagrange multiplier is positive ($\lambda_i > 0$) then $L_{\lambda_i} = 0$; or if $L_{\lambda_i} \geq 0$, then $\lambda_i = 0$; or if $\lambda_i = 0$, then $L_{\lambda_i} = 0$.

We now provide an economic interpretation of the Kuhn–Tucker conditions for a maximum. Assume that the problem that gave rise to the Lagrangian function in equation (6.3.13) is a firm's problem of maximizing its total profit from the production of x_j (where $j = 1, 2, \ldots, n$) goods using c_i (where $i = 1, 2, \ldots, m$) factors. Then the following statements are valid: f_j represents the marginal profit of the jth good, λ_i stands for the shadow price or the imputed cost of the ith factor, c_i shows the total available quantity of the ith factor, g_j^i denotes the quantity of the ith factor employed in producing the marginal unit of the jth good, $\lambda_i g_j^i$ represents the marginal imputed cost of the ith factor employed in producing the marginal unit of the jth good, and $\sum_{i=1, j=1}^{m,n} \lambda_i g_j^i$ denotes the total marginal imputed cost of the jth good. Given these, the first marginal condition ($L_j = f_j - \sum_{i=1, j=1}^{m,n} \lambda_i g_j^i \leq 0$) in (6.3.14) implies that the marginal profit of the jth good must be less than or equal to its total marginal imputed cost. If $x_j^* > 0$ in the optimum, then the complementary slackness condition implies that the marginal profit of the jth good must be equal to its total marginal imputed cost ($L_j = f_j - \sum_{i=1, j=1}^{m,n} \lambda_i g_j^i = 0$). If the marginal profit of the jth good is less than its total marginal imputed cost ($L_j = f_j - \sum_{i=1, j=1}^{m,n} \lambda_i g_j^i < 0$), then the complementary slackness condition implies that the jth good will not be produced in the optimum ($x_j^* = 0$). Another noteworthy term in condition (6.3.14) is $L_{\lambda_i} \geq 0$, which states that the firm must operate within its resource constraint. We know that the shadow price or the imputed cost of the ith factor cannot be negative (that is, $\lambda_i \geq 0$). Then the complementary slackness condition ($\lambda_i L_{\lambda_i} = 0$) in (6.3.14) requires that if the ith factor is not fully used in the production process (if $L_{\lambda_i} = c_i - g^i(x_1, x_2, \ldots, x_n) > 0$), then its imputed cost must be zero (or $\lambda_i = 0$). If the ith factor is completely used in the production process (if $L_{\lambda_i} = c_i - g^i(x_1, x_2, \ldots, x_n) = 0$), then its imputed cost must be positive (that is, $\lambda_i > 0$). In addition to the interpretation we just gave, we can also give another interpretation to the Lagrange multiplier, λ_i. We know from Section 4.4.9 that λ_i is a measure of the effect, in the optimum, on the objective function of a change in the ith constraint. If the ith constraint is binding in the optimum (if $L_{\lambda_i} = c_i - g^i(x_1, x_2, \ldots, x_n) = 0$), then (as required by the complementary slackness condition) profit will increase (decrease) if the ith constraint is increased (decreased). On the other hand, if the ith constraint is not binding in the optimum (if $L_{\lambda_i} = c_i - g^i(x_1, x_2, \ldots, x_n) > 0$), then (again as required by the complementary slackness condition) a change in the ith constraint will leave profit unchanged. This completes the economic interpretation of the Kuhn–Tucker condition for a maximum presented in (6.3.14).

6.3.7 Economic interpretations of Kuhn–Tucker minimum conditions

We begin with a general interpretation of the Kuhn–Tucker conditions for a minimum of NLP problems, presented in (6.3.16). The first marginal condition, $L_j \geq 0$, is to conform to the first inequality $(f'(x) \geq 0)$ in (6.3.4). The second term, $x_j \geq 0$, is to conform to the nonnegativity restriction. The third term, $x_j L_j = 0$, presents the complementary slackness condition. This term shows that if $x_j > 0$, then $L_j = 0$; or if $x_j = 0$, then $L_j \geq 0$; or if $x_j = 0$, then $L_j = 0$. The fourth term, $L_{\lambda_i} \leq 0$, is only a restatement of the inequality constraint. Since the imputed value of a particular product cannot be negative, the ith Lagrange multiplier (λ_i) in the fifth term must be nonnegative $(\lambda_i \geq 0)$. The sixth term, $\lambda_i L_{\lambda_i} = 0$, represents the complementary slackness condition in the case of the Lagrange multiplier. It shows that if the ith Lagrange multiplier is positive $(\lambda_i > 0)$, then $L_{\lambda_i} = 0$; or if $L_{\lambda_i} \leq 0$, then $\lambda_i = 0$.

We now provide an economic interpretation of the Kuhn–Tucker conditions for a minimum. Assume that the problem that gave rise to the Lagrangian function in equation (6.3.15) is a firm's problem of minimizing its total cost of using x_j $(j = 1, 2, \ldots, n)$ factors in producing given quantities of i $(i = 1, 2, \ldots, m)$ goods. Then the following statements are valid: f_j represents the marginal cost of the jth factor, λ_i stands for the imputed value of the ith good, c_i shows the quantity of the ith good, g^i_j denotes the quantity of the ith good produced using the marginal unit of the jth factor, $\lambda_i g^i_j$ represents the total marginal imputed value of the ith good produced using the marginal unit of the jth factor, and $\sum_{i=1, j=1}^{m,n} \lambda_i g^i_j$ denotes the total marginal imputed value of the ith good produced using the marginal unit of the jth factor. Given these, the first marginal condition $\left(L_j = f_j - \sum_{i=1, i=1}^{m,n} \lambda_i g^i_j \geq 0\right)$ in (6.3.16) implies that the marginal cost of the jth factor must be greater than or equal to the total marginal imputed value of the ith good produced using the marginal unit of the jth factor. If $x_j^* > 0$ in the optimum, then the complementary slackness condition in (6.3.16) implies that the marginal cost of the jth good must be equal to the marginal imputed value $\left(L_j = f_j - \sum_{i=1, j=1}^{m,n} \lambda_i g^i_j = 0\right)$. If the marginal cost of the jth factor is greater than the marginal imputed value, then the complementary slackness condition implies that the jth factor will not be used in the optimum (or $x_j^* = 0$). Another noteworthy term in condition (6.3.16) is $L_{\lambda_i} \leq 0$, which states (since $L_{\lambda_i} \leq 0$ is a mere restatement of the constraint $g^i(x_1, x_2, \ldots, x_n) \geq c_i$) that the firm must operate within its output constraints. We know that the imputed value of the ith good cannot be negative $(\lambda_i \geq 0)$. Then the complementary slackness condition $(\lambda_i L_{\lambda_i} = 0)$ in (6.3.16) requires that if the constraint of the ith good is not binding, then its imputed value must be zero $(\lambda_i = 0)$. If that constraint is binding, then its imputed value must be positive $(\lambda_i > 0)$. In addition to these interpretations, we can also give another interpretation (as in the last section) to the Lagrange multiplier, λ_i.

We know that λ_i is a measure of the effect, in the optimum, on the objective function of a change in the ith constraint. If the ith constraint is binding in the optimum, then (as required by the complementary slackness condition) the cost will increase (decrease) if the ith constraint is increased (decreased). On the other hand, if the ith constraint is not binding in the optimum, then (again as required by the complementary slackness condition) a change in the ith constraint will leave cost unchanged. This completes the economic interpretation of the Kuhn–Tucker condition for a minimum presented in (6.3.16).

6.3.8 Examples of Kuhn–Tucker maximum conditions

As an example of the application of the Kuhn–Tucker maximum conditions in (6.3.14), consider the problem in the first example in Section 6.2.5, which we solved geometrically. This problem was to maximize $\Pi = f(x_1, x_2) = 2x_1 + 4x_2$ subject to constraints $g^1(x_1, x_2) = x_1^2 + x_2^2 \leq 4$ and $g^2(x_1, x_2) = x_1^2 + x_2 \leq 3$. Therefore, the Lagrangian function corresponding to equation (6.3.13) in the present example can be set up, with $i = 1, 2$ and $j = 1, 2$, as

$$L = f(x_1, x_2, \ldots, x_n) + \sum_{i=1}^{m} \lambda_i [c_i - g^i(x_1, x_2, \ldots, x_n)]$$

$$= 2x_1 + 4x_2 + \lambda_1[4 - x_1^2 - x_2^2] + \lambda_2[3 - x_1^2 - x_2] \tag{6.3.17}$$

The Kuhn–Tucker conditions corresponding to (6.3.14) in this example are

$$L_1 = 2 - 2\lambda_1 x_1 - 2\lambda_2 x_1 \leq 0, x_1 \geq 0, x_1 L_1 = 0, L_2 = 4 - 2\lambda_1 x_2 - \lambda_2 \leq 0, x_2 \geq 0,$$

$$x_2 L_2 = 0, L_{\lambda_1} = 4 - x_1^2 - x_2^2 \geq 0, \lambda_1 \geq 0, \lambda_1 L_{\lambda_1} = 0, L_{\lambda_2} = 3 - x_1^2 - x_2 \geq 0,$$

$$\lambda_2 \geq 0, \text{ and } \lambda_2 L_{\lambda_2} = 0 \tag{6.3.18}$$

Our problem now is to find the values of x_1, x_2, λ_1, and λ_2 that will satisfy all the conditions in (6.3.18). We have already obtained, from the geometric solution to the present problem in Section 6.2.5, the optimal values of the choice variables: $x_1^* = 0.894427$ and $x_2^* = 1.78885$. Substituting these values in L_1 and L_2 and simplifying yields $\lambda_1^* = 1.11704$ and $\lambda_2^* = 0$.

Let us now check whether the above optimal values satisfy the Kuhn–Tucker conditions in (6.3.18). If we substitute the optimal values into L_1, L_2, L_{λ_1}, and L_{λ_2} in (6.3.18), we can see that they indeed satisfy them. Moreover, these optimal values are such that $x_1 \geq 0$, $x_2 \geq 0$, $\lambda_1 \geq 0$, and $\lambda_2 \geq 0$. Since $x_1^* > 0$, $x_2^* > 0$, and $\lambda_1^* > 0$, one can verify that $L_1 = 0$, $L_2 = 0$, and $L_{\lambda_1} = 0$, as required by the complementary slackness condition. And, since $\lambda_2 = 0$, it can be verified that $L_{\lambda_2} > 0$ (again, as required by the complementary slackness condition). Therefore, the optimal values $x_1^* = 0.894427$, $x_2^* = 1.78885$, $\lambda_1^* = 1.11704$, and $\lambda_2^* = 0$ satisfy all the Kuhn–Tucker conditions for a local maximum of the objective function subject to the constraints in the present NLP problem.

As another example, consider the problem in the second example in Section 6.2.5, which we solved geometrically. This problem was to maximize $\Pi = h(x_1, x_2) = 50x_1 + 50x_2 - 5x_1^2 - 5x_2^2$ subject to constraints $g^3(x_1, x_2) = x_1 \leq 10$, $g^4(x_1, x_2) = x_2 \leq 4$, and

$g^5(x_1, x_2) = x_1 + 3x_2 \leq 15$. Therefore, the Lagrangian function corresponding to equation (6.3.13) in the present example can be set up, with $i = 3, 4, 5$ and $j = 1, 2$, as

$$L = f(x_1, x_2, \ldots, x_n) + \sum_{i=1}^{m} \lambda_i [c_i - g^i(x_1, x_2, \ldots, x_n)]$$

$$= 50x_1 + 50x_2 - 5x_1^2 - 5x_2^2 + \lambda_3[10 - x_1] + \lambda_4[4 - x_2] + \lambda_5[15 - x_1 - 3x_2] \quad (6.3.19)$$

The Kuhn–Tucker conditions corresponding to (6.3.14) in this example are

$$L_1 = 50 - 10x_1 - \lambda_3 - \lambda_5 \leq 0, x_1 \geq 0, x_1 L_1 = 0, L_2 = 50 - 10x_2 - \lambda_4 - 3\lambda_5 \leq 0,$$

$$x_2 \geq 0, x_2 L_2 = 0, L_{\lambda_3} = 10 - x_1 \geq 0, \lambda_3 \geq 0, \lambda_3 L_{\lambda_3} = 0, L_{\lambda_4} = 4 - x_2 \geq 0, \lambda_4 \geq 0,$$

$$\lambda_4 L_{\lambda_4} = 0, L_{\lambda_5} = 15 - x_1 - 3x_2 \geq 0, \lambda_5 \geq 0, \text{ and } \lambda_5 L_{\lambda_5} = 0 \quad (6.3.20)$$

Our problem now is to find the values of x_1, x_2, λ_3, λ_4, and λ_5 that will satisfy all the conditions in (6.3.20). We have already obtained, from the geometric solution to the present problem in Section 6.2.5, the optimal values of the choice variables: $x_1^* = 4.5$ and $x_2^* = 3.5$. Substituting these values into L_1 and L_2 and simplifying yields $5 - \lambda_3 - \lambda_5 = 0$ and $15 - \lambda_4 + 3\lambda_5 = 0$. One set of solutions for the last SLSEs is $\lambda_3^* = 0$, $\lambda_4^* = 0$, and $\lambda_5^* = 5$.

Let us now check whether the above optimal values satisfy the Kuhn–Tucker conditions in (6.3.20). If we substitute the optimal values into L_1 and L_2 in (6.3.20), we can see that they indeed satisfy these inequalities. Moreover, these optimal values are such that $x_1 \geq 0, x_2 \geq 0$, $\lambda_3 \geq 0, \lambda_4 \geq 0$, and $\lambda_5 \geq 0$. Since $x_1^* > 0$, $x_2^* > 0$, $L_1 = 0$, and $L_2 = 0$, the complementary conditions $x_1 L_1 = 0$ and $x_2 L_2 = 0$ are also satisfied. Besides, we obtain $L_{\lambda_3} = 10 - x_1^* = 10 - 4.5 = 5.5 \geq 0$, $L_{\lambda_4} = 4 - x_2^* = 4 - 3.5 = 0.5 \geq 0$, and $L_{\lambda_5} = 15 - x_1^* - 3x_2^* = 15 - 4.5 - 3 \times 3.5 = 0$. Since $\lambda_3^* = 0$, $\lambda_4^* = 0$, and $\lambda_5^* = 5$ and since $L_{\lambda_3} > 0$, $L_{\lambda_4} > 0$, and $L_{\lambda_5} = 0$, all the complementary slackness conditions in (6.3.20) are also satisfied. Therefore, the optimal values $x_1^* = 4.5$ and $x_2^* = 3.5$, and $\lambda_3^* = 0$, $\lambda_4^* = 0$, and $\lambda_5^* = 5$ satisfy all the Kuhn–Tucker conditions for a local maximum of the objective function subject to the constraints in the present NLP problem.

6.3.9 Examples of Kuhn–Tucker minimum conditions

As an example of the application of the Kuhn–Tucker minimum conditions in (6.3.16), consider the problem in the first example in Section 6.2.6, which we solved geometrically. This problem was to minimize $C = j(x_1, x_2) = (x_1 - 5)^2 + (x_2 - 5)^2 = x_1^2 - 10x_1 + x_2^2 - 10x_2 + 50$ subject to the constraints $g^6(x_1, x_2) = x_1 + x_2 \leq 6$ or $g^6(x_1, x_2) = -x_1 + -x_2 \geq -6$ and $g^7(x_1, x_2) = x_1 + 2x_2 \geq 8$. Therefore, the Lagrangian function corresponding to equation (6.3.15) in the present example can be set up, with $i = 6, 7$ and $j = 1, 2$, as

$$L = j(x_1, x_2, \ldots, x_n) + \sum_{i=1}^{m} \lambda_i [c_i - g^i(x_1, x_2, \ldots, x_n)]$$

$$= 50 + x_1^2 - 10x_1 + x_2^2 - 10x_2 + \lambda_6[-6 + x_1 + x_2] + \lambda_7[8 - x_1 - 2x_2] \quad (6.3.21)$$

The Kuhn–Tucker conditions corresponding to (6.3.16) in this example are

$$L_1 = 2x_1 - 10 + \lambda_6 - \lambda_7 \geq 0, x_1 \geq 0, x_1 L_1 = 0, L_2 = 2x_2 - 10 + \lambda_6 - 2\lambda_7 \geq 0, x_2 \geq 0,$$

$$x_2 L_2 = 0, L_{\lambda_6} = 6 - x_1 - x_2 \leq 0, \lambda_6 \geq 0, \lambda_6 L_{\lambda_6} = 0, L_{\lambda_7} = 8 - x_1 - 2x_2 \leq 0,$$

$$\lambda_7 \geq 0, \text{ and } \lambda_7 L_{\lambda_7} = 0 \tag{6.3.22}$$

As earlier, our problem now is to find the values of x_1, x_2, λ_6, and λ_7 that will satisfy all the conditions in (6.3.22). We have already obtained, from the geometric solution to the present problem in Section 6.2.6, the optimal values of the choice variables: $x_1^* = x_2^* = 3$. Substituting these values into L_1 and L_2 in (6.3.22) and simplifying yields $\lambda_6^* = 4$ and $\lambda_7^* = 0$.

We can now check whether the above optimal values satisfy the Kuhn–Tucker conditions in (6.3.22). If we substitute the optimal values into L_1 and L_2 in (6.3.22), we can see that they indeed satisfy them. Moreover, these optimal values satisfy $x_1 \geq 0$, $x_2 \geq 0$, $\lambda_6 \geq 0$, and $\lambda_7 \geq 0$. Since $x_1^* = x_2^* = 3 > 0$, $L_1 = 0$, and $L_2 = 0$, the complementary slackness conditions $x_1 L_1 = 0$ and $x_2 L_2 = 0$ are also satisfied. From the inequalities L_{λ_6} and L_{λ_7} in (6.3.22) we obtain $L_{\lambda_6} = 0$ and $L_{\lambda_7} = -1 < 0$. Since $\lambda_6^* = 4$ and $\lambda_7^* = 0$ and since $L_{\lambda_6} = 0$ and $L_{\lambda_7} = -1 < 0$, all the complementary slackness conditions are satisfied. Therefore, the optimal values $x_1^* = x_2^* = 3$, $\lambda_6^* = 4$, and $\lambda_7^* = 0$ satisfy all the Kuhn–Tucker conditions for a local minimum of the objective function subject to the constraints in the present NLP problem.

As another example, consider the problem in the second example in Section 6.2.6, which we solved geometrically. This problem was to minimize $C = k(x_1, x_2) = x_1 + x_2$ subject to constraints $g^8(x_1, x_2) = 0.5x_1 + x_2 \geq 3$, $g^9(x_1, x_2) = x_1 + 0.5x_2 \geq 3$, and $g^{10}(x_1, x_2) = x_1^2 + x_2^2 \geq 8$. Therefore, the Lagrangian function corresponding to equation (6.3.16) in the present example can be set up, with $i = 8, 9, 10$ and $j = 1, 2$, as

$$L = k(x_1, x_2, \ldots, x_n) + \sum_{i=1}^{m} \lambda_i [c_i - g^i(x_1, x_2, \ldots, x_n)]$$

$$= x_1 + x_2 + \lambda_8 [3 - 0.5x_1 - x_2] + \lambda_9 [3 - x_1 - 0.5x_2] + \lambda_{10} [8 - x_1^2 - x_2^2] \tag{6.3.23}$$

The Kuhn–Tucker conditions corresponding to (6.3.16) in this example are

$$L_1 = 1 - 0.5\lambda_8 - \lambda_9 - 2\lambda_{10}x_1 \geq 0, x_1 \geq 0, x_1 L_1 = 0, L_2 = 1 - \lambda_8 - 0.5\lambda_9 - 2\lambda_{10}x_2 \geq 0,$$

$$x_2 \geq 0, x_2 L_2 = 0, L_{\lambda_8} = 3 - 0.5x_1 - x_2 \leq 0, \lambda_8 \geq 0, \lambda_8 L_{\lambda_8} = 0,$$

$$L_{\lambda_9} = 3 - x_1 - 0.5x_2 \leq 0, \lambda_9 \geq 0, \lambda_9 L_{\lambda_9} = 0, L_{\lambda_{10}} = 8 - x_1^2 - x_2^2 \leq 0,$$

$$\lambda_{10} \geq 0, \text{ and } \lambda_{10} L_{\lambda_{10}} = 0 \tag{6.3.24}$$

As in the previous example, our problem here is to find the values of x_1, x_2, λ_8, λ_9, and λ_{10} that will satisfy all the conditions in (6.3.24). We have already obtained, from the geometric solution to the present problem presented in Section 6.2.6, the optimal values of the choice variables: $x_1^* = x_2^* = 2$. Substituting these values into L_1 and L_2 and simplifying will yield $\lambda_8^* = \lambda_9^* = \lambda_{10}^* = 0.1818$.

We can now check whether the above optimal values satisfy the Kuhn–Tucker conditions in (6.3.24). If we substitute the above optimal values into L_1 and L_2 in (6.3.24), we can see that they indeed satisfy these inequalities. Moreover, these optimal values are such that

$x_1^* \geq 0$, $x_2^* \geq 0$, $\lambda_8 \geq 0$, $\lambda_9 \geq 0$, and $\lambda_{10} \geq 0$. Since $x_1^* = x_2^* = 10 > 0$, $L_1 = 0$, and $L_2 = 0$, the complementary slackness conditions $x_1 \partial L_1 = 0$ and $x_2 L_2 = 0$ are also satisfied. We can also obtain $L_{\lambda 8} = 0$, $L_{\lambda 9} = 0$, and $L_{\lambda 10} = 0$. Since $\lambda_8^* = \lambda_9^* = \lambda_{10}^* = 0.1818$ and since $L_{\lambda 8} = 0$, $L_{\lambda 9} = 0$, and $L_{\lambda 10} = 0$, all the complementary slackness conditions in (6.3.24) are satisfied. Therefore, the optimal values $x_1^* = x_2^* = 2$ and $\lambda_8^* = \lambda_9^* = \lambda_{10}^* = 0.1818$ satisfy all the Kuhn–Tucker conditions for a local minimum of the objective function subject to the constraints in the present NLP problem.

6.3.10 Trial-and-error method of solving NLP problems

So far we have been using the geometric forms of the objective function and the constraints of NLP problems to find the values of the choice variables in the problem that optimize the objective function subject to the constraints and satisfy the associated Kuhn–Tucker conditions. Let us now attempt to see whether we can find these optimal values of the choice variables from the corresponding Kuhn–Tucker conditions. This is basically, as we outline below, a trial-and-error method.

Suppose that we wish to maximize $\Pi = f(x_1, x_2) = x_1 x_2$ subject to the constraint $g(x_1, x_2) = x_1 + x_2 \leq 10 = c$. Then, following our discussion in Section 6.3.5, the Lagrangian function associated with this problem can be set up as

$$L = f(x_1, x_2) + \lambda[c - g(x_1, x_2) = x_1 x_2 + \lambda[10 - x_1 - x_2] \tag{6.3.25}$$

We can derive from function (6.3.25) the Kuhn–Tucker maximum conditions

$$L_1 = x_2 - \lambda \leq 0, x_1 \geq 0, x_1 L_1 = 0, L_2 = x_1 - \lambda \leq 0, x_2 \geq 0,$$

$$x_2 L_2 = 0, L_\lambda = 10 - x_1 - x_2 \geq 0, \lambda \geq 0, \text{ and } \lambda L_\lambda = 0 \tag{6.3.26}$$

Let us now attempt to find the values of x_1, x_2, and λ that satisfy all the Kuhn–Tucker maximum conditions in (6.3.26). Since x_1 and x_2 must be nonnegative, suppose that we begin by choosing the lowest possible nonnegative values: $x_1 = x_2 = 0$. When $x_1 = x_2 = 0$, the first two marginal conditions (L_1 and L_2) imply that $-\lambda \leq 0$ or $\lambda \geq 0$. When $\lambda \geq 0$, the last complementary slackness condition ($\lambda L_\lambda = 0$) implies that the partial derivative of the Lagrangian function with respect to the Lagrangian multiplier (L_λ) must be zero. But, in fact, it is 10 ($L_\lambda = 10 - x_1 - x_2 = 10 - 0 - 0 = 10 \geq 0$), which is contradictory. Therefore, $x_1 = x_2 = 0$ does not satisfy the associated Kuhn–Tucker conditions.

The above trial suggests that we need to use positive values for x_1 and x_2. If we continue to carry out trials similar to the one above with $x_1 > 0$ and $x_2 > 0$, we can see that any value below or above $x_1 = x_2 = 5$ does not satisfy the Kuhn–Tucker maximum conditions. With $x_1 = x_2 = 5$ the first two complementary slackness conditions imply that $L_1 = L_2 = 0$ and, for this to happen, λ must be 5. When $\lambda = 5$, the last complementary slackness condition implies that $L_\lambda = 0$ and this is true only when $x_1 = x_2 = 5$. These results imply that $x_1^* = x_2^* = 5$ and $\lambda = 5$ satisfy all the Kuhn–Tucker maximum conditions and, therefore, the function $\Pi = f(x_1, x_2) = x_1 x_2$ attains a maximum subject to the constraint $g(x_1, x_2) = x_1 + x_2 \leq 10$; and the maximum of the function is $\Pi^* = f(x_1^*, x_2^*) = x_1^* x_2^* = 5 \times 5 = 25$.

As another example, suppose that we wish to minimize the function $C = h(x_1, x_2) = x_1^2 + x_2^2$ subject to the constraint $g(x_1, x_2) = x_1 + x_2 \geq 50$. Then, following our discussion in

Section 6.3.5, the Lagrangian function associated with this problem can be set up as

$$L = f(x_1, x_2) + \lambda[c - g(x_1, x_2)] = x_1^2 + x_2^2 + \lambda[50 - x_1 - x_2] \tag{6.3.27}$$

We can derive from function (6.3.27) the Kuhn–Tucker minimum conditions

$$L_1 = 2x_1 - \lambda \geq 0, x_1 \geq 0, x_1 L_1 = 0, L_2 = 2x_2 - \lambda \geq 0, x_2 \geq 0,$$

$$x_2 L_2 = 0, L_\lambda = 50 - x_1 - x_2 \leq 0, \lambda \geq 0, \text{ and } \lambda L_\lambda = 0 \tag{6.3.28}$$

As before, let us attempt to find the values of x_1, x_2, and λ that satisfy all the Kuhn–Tucker minimum conditions in (6.3.28). Since x_1 and x_2 must be nonnegative, suppose that we begin by choosing the lowest possible nonnegative value: $x_1 = x_2 = 0$. When $x_1 = x_2 = 0$, the first two marginal conditions (L_1 and L_2) imply that $-\lambda \geq 0$ or $\lambda \leq 0$. We know from the second condition in (6.3.28) that λ cannot be negative. When $\lambda = 0$, the last complementary slackness ($\lambda L_\lambda = 0$) condition implies that the partial derivative of the Lagrangian function with respect to the Lagrangian multiplier (L_λ) must be nonpositive. But, in fact, it is 50 ($L_\lambda = 50 - x_1 - x_2 = 50 - 0 - 0 = 50 \geq 0$), which is contradictory. Therefore, $x_1 = x_2 = 0$ does not satisfy the associated Kuhn–Tucker conditions.

As in the last example, the above trial for the present example suggests that we need to use positive values for x_1 and x_2. If we continue to carry out trials similar to the one above with $x_1 > 0$ and $x_2 > 0$, we can see that any value below or above $x_1 = x_2 = 25$ does not satisfy the Kuhn–Tucker minimum conditions. With $x_1 = x_2 = 25$ the first two complementary slackness conditions imply that $L_1 = L_2 = 0$, and for this to happen λ must be 50. When $\lambda = 50$, the last complementary slackness condition implies that $L_\lambda = 0$ and this is true only when $x_1 = x_2 = 25$. These results imply that $x_1^* = x_2^* = 25$ and $\lambda = 50$ satisfy all the Kuhn–Tucker minimum conditions and, therefore, the function $C = h(x_1, x_2) = x_1^2 + x_2^2$ subject to the constraint $g(x_1, x_2) = x_1 + x_2 \geq 50$ attains a minimum; and the minimum of the function is $C^* = h(x_1^*, x_2^*) = x_1^{*2} + x_2^{*2} = 25^2 + 25^2 = 1250$.

6.3.11 Application examples

Example 1. Consider the problem in example 1 in Section 6.2.8. Does the solution we obtained to this problem ($x_1^* = x_2^* = 3.5$) satisfy the Kuhn–Tucker maximum conditions (6.3.14)?

Solution. The problem was to maximize $U = f(x_1, x_2) = x_1^{1/2} x_2^{1/2}$ subject to $x_1 + x_2 \leq 7$. Then, Lagrangian function for this problem is

$$L = f(x_1, x_2) + \lambda[c - g(x_1, x_2)] = x_1^{1/2} x_2^{1/2} + \lambda[7 - x_1 - x_2] \tag{6.3.29}$$

The Kuhn–Tucker conditions associated with this problem are

$$L_1 = (1/2)x_1^{-1/2} x_2^{1/2} - \lambda \leq 0, x_1 \geq 0, x_1 L_1 = 0, L_2 = (1/2)x_1^{1/2} x_2^{-1/2} - \lambda \leq 0, x_2 \geq 0,$$

$$x_2 L_2 = 0, L_\lambda = 7 - x_1 - x_2 \geq 0, \lambda \geq 0, \text{ and } \lambda L_\lambda = 0 \tag{6.3.30}$$

We know from Section 6.2.8 that the optimal solution to the present problem is $x_1^* = x_2^* = 3.5$. If we substitute these values into L_1 and L_2 in (6.3.30), we obtain $\lambda = 0.5$.

Since $L_1 = 0$ and $L_2 = 0$ for $x_1^* = x_2^* = 3.5$ and $\lambda = 0.5$, and since $x_1^* = x_2^* = 3.5 > 0$, the complementary slackness conditions $x_1 L_1 = 0$ and $x_2 L_2 = 0$ are satisfied. Since $L_\lambda = 0$ for $x_1^* = x_2^* = 3.5$, and since $\lambda = 0.5 > 0$, the last complementary slackness condition $(\lambda L_\lambda = 0)$ is also satisfied. Therefore, the optimal solution $(x_1^* = x_2^* = 3.5$ and $\lambda = 0.5)$ to the problem satisfies all the Kuhn–Tucker maximum conditions.

Example 2. Check whether the solution to the problem in example 2 in Section 6.2.8 satisfies the Kuhn–Tucker maximum conditions in (6.3.14).

Solution. The problem in this example was to maximize $Q = f(K, L) = [0.5K^{-0.5} + 0.5L^{-0.5}]^{-2}$ subject to $0.1K + 10L \leq 100$. Then, the Lagrangian function (La) for this problem is

$$La = f(K, L) + \lambda[c - g(K, L)] = [0.5K^{-0.5} + 0.5L^{-0.5}]^{-2} + \lambda[100 - 0.1K - 10L] \tag{6.3.31}$$

The Kuhn–Tucker conditions associated with this problem are

$$La_K = -2[Z]^{-3}(-0.5)K^{-3/2} - 0.1\lambda \leq 0, K \geq 0,$$

$$K La_K = 0, La_L = -2[Z]^{-3}(-0.5)L^{-3/2} - 10\lambda \leq 0, L \geq 0, L La_L = 0,$$

$$La_\lambda = 100 - 0.1K - 10L \geq 0, \lambda \geq 0, \text{ and } \lambda La_\lambda = 0, \text{ where } Z = [0.5K^{-0.5} + 0.5L^{-0.5}] \tag{6.3.32}$$

We know from Section 6.2.8 that the optimal solution to the problem is $K^* = 177.262$ and $L^* = 8.227$. If we substitute these values into La_K and La_L in (6.3.32), we obtain $\lambda = 0.4435$. Since $La_K = 0$ and $La_L = 0$ for $K^* = 177.262$, $L^* = 8.227$, and $\lambda = 0.4435$ and since $K^* = 177.262 > 0$ and $L^* = 8.227 > 0$, the complementary slackness conditions $K La_K = 0$ and $L La_L = 0$ are satisfied. Since $La_\lambda = 0$ for $K^* = 177.262$ and $L^* = 8.227$, and since $\lambda = 0.4435 > 0$, the last complementary slackness condition $(\lambda La_\lambda = 0)$ is also satisfied. Therefore, the optimal solution $(K^* = 177.262, L^* = 8.227,$ and $\lambda = 0.4435)$ to the problem satisfies all the Kuhn–Tucker maximum conditions.

Example 3. Check whether the solution to the problem in example 3 in Section 6.2.8 satisfies the Kuhn–Tucker minimum conditions in (6.3.16).

Solution. The problem in this example was to minimize $E = h(C_1, C_2) = 1.1C_1 + C_2$ subject to $U = g(C_1, C_2) = \ln C_1 + \ln C_2 \geq 17.9$. Then, the Lagrangian function for this problem is

$$L = h(C_1, C_2) + \lambda[c - g(C_1, C_2)] = 1.1C_1 + C_2 + \lambda[17.9 - \ln C_1 - \ln C_2] \tag{6.3.33}$$

The Kuhn–Tucker conditions associated with this problem are

$$L_1 = 1.1 - (\lambda/C_1) \geq 0, C_1 \geq 0, C_1 L_1 = 0, L_2 = 1 - (\lambda/C_2) \geq 0, C_2 \geq 0, C_2 L_2 = 0,$$

$$L_\lambda = 17.9 - \ln C_1 - \ln C_2 \leq 0, \lambda \geq 0, \text{ and } \lambda L_\lambda = 0 \tag{6.3.34}$$

We know from Section 6.2.8 that the optimal solution to the problem is $C_1^* = \$7045.4545$ and $C_2^* = \$7750$. If we substitute these values into L_1 and L_2 in (6.3.34) we obtain $\lambda = 7750$.

Since $L_1 = 0$ and $L_2 = 0$ for $C_1^* = 7045.4545$, $C_2^* = 7750$, and $\lambda = 7750$, and since $C_1^* = \$7045.4545$ and $C_2^* = \$7750$, the complementary slackness conditions $C_1 L_1 = 0$ and $C_2 L_2 = 0$ are satisfied. Since $L_\lambda = 0$ for $C_1^* = \$7045.4545$ and $C_2^* = \$7750$, and since $\lambda = 7750$, the last complementary slackness condition $(\lambda L_\lambda = 0)$ is also satisfied. Therefore, the optimal solution $C_1^* = 7045.4545$, $C_2^* = 7750$, and $\lambda = 7750$ to the problem satisfies all the Kuhn–Tucker minimum conditions.

Example 4.   Does the solution $(x_1^* = 3.5, x_2^* = 3.5,$ and $E^* = \$7)$ obtained for the problem in example 4 in Section 6.2.8 satisfy the Kuhn–Tucker minimum conditions (6.3.16)?

Solution.   The problem in this example was to minimize $E = g = (K, L) = 1x_1 + 1x_2 = x_1 + x_2$ subject to $U = f(x_1, x_2) = x_1^{1/2} x_2^{1/2} = 3.5$. Then, the Lagrangian function for this problem is

$$L = g(x_1, x_2) + \lambda[c - g(x_1, x_2)] = x_1 + x_2 + \lambda[3.5 - x_1^{1/2} x_2^{1/2}] \tag{6.3.35}$$

The Kuhn–Tucker conditions associated with this problem are

$$L_1 = 1 - 0.5\lambda x_1^{-1/2} x_2^{1/2} \geq 0, x_1 \geq 0, x_1 L_1 = 0, L_2 = 1 - 0.5\lambda x_1^{1/2} x_2^{-1/2} \geq 0, x_2 \geq 0,$$

$$x_2 L_2 = 0, L_\lambda = 3.5 - x_1^{1/2} x_2^{1/2} \leq 0, \lambda \geq 0, \text{ and } \lambda L_\lambda = 0 \tag{6.3.36}$$

We found in Section 6.2.8 that the optimal solution to the problem was $x_1^* = 3.5$ and $x_2^* = 3.5$. If we substitute these values into L_1 and L_2 in (6.3.36), we obtain $\lambda = 2$. Since $L_1 = 0$ and $L_1 = 0$ for $x_1^* = 3.5$, $x_2^* = 3.5$, and $\lambda = 2$, and since $x_1^* = x_2^* = 3.5 > 0$, the complementary slackness conditions $x_1 L_1 = 0$ and $x_2 L_2 = 0$ are satisfied. Since $L_\lambda = 0$ for $x_1^* = 3.5$ and $x_2^* = 3.5$, and since $\lambda = 2 > 0$, the last complementary slackness condition $(\lambda L_\lambda = 0)$ is also satisfied. Therefore, the optimal solution $(x_1^* = 3.5, x_2^* = 3.5,$ and $\lambda = 2)$ to the problem satisfies all the Kuhn–Tucker maximum conditions.

6.3.12 Exercises

1. Determine whether the solutions to the problems in exercise 1 in Section 6.2.9 satisfy the associated Kuhn–Tucker conditions.
2. Determine whether the solutions to the problems in exercise 2 in Section 6.2.9 satisfy the associated Kuhn–Tucker conditions.
3. *Application exercise.* Suppose that the total utility that a consumer obtains from the consumption of two goods, x_1 and x_2, is given by the Cobb–Douglas utility function $U = f(x_1, x_2) = x_1^{0.6} x_2^{0.4}$. Also suppose that the consumer's budget constraint is given by $0.5x_1 + 0.5x_2 \leq 10$. Find the quantities of the two goods that maximize the consumer's total utility such that they satisfy the associated Kuhn–Tucker conditions.
4. *Application exercise.* Suppose that the total output that a producer obtains from the employment of two factors, K and L, is given by the Cobb–Douglas production function $Q = f(K, L) = K^{0.6} L^{0.4}$. Also suppose that the producer's budget constraint is given by $2K + 3L \leq 9$. Find the quantities of the two factors that maximize the producer's total output such that they satisfy the associated Kuhn–Tucker conditions.

6.4 NLP: Extensions

6.4.1 Boundary irregularities and constraint qualification

In Section 6.2 we presented, among others, the general nature of solutions of NLP problems. We then employed the geometric method to solve a number of NLP problems. After this, in Section 6.3, we made use of the Kuhn–Tucker conditions, and stated that these conditions were the necessary conditions for an optimal solution. However, this assertion was based on the assumption that the feasible region of the problem under consideration was a *well-behaved* feasible region; that is, the feasible region did not involve *boundary irregularities*.

We found that the optima in all the examples we considered in Section 6.3 satisfied the Kuhn–Tucker conditions. This was so because the feasible regions of the problems were all well-behaved or the feasible regions did not involve boundary irregularities. It could happen that an optimum solution to a NLP problem may not always satisfy the Kuhn–Tucker conditions simply due to boundary irregularities caused by some constraint(s). This is the topic to which we turn our attention in the present section.

A boundary irregularity is often, but not solely, caused by the presence of one or more *cusps* on the boundary of the feasible region of the problem. A cusp is a sharp turning in the graph, such as the one at point C on the graph of the function $x_2 = f(x_1) = (2 - x_1)^3$ shown in Figure 6.4.1(A). Notice that the function has negative slopes on either side of point C. But the slope at point C is zero.

Let us now see how the presence of cusps causes an optimum solution not to satisfy the Kuhn–Tucker optimum conditions. Assume that our objective is to maximize $\Pi = f(x_1, x_2) = x_1$. It is unnecessary to state that we have nonnegativity condition, that is, x_1^*, x_2^*, and $\lambda^* \geq 0$. Let us now plot the graphs of the objective function for different values of Π, together with the constraint illustrated in Figure 6.4.1(A), as shown in Figure 6.4.1(B).

Notice that the feasible region in Figure 6.4.1(B) is represented by the set of points in the area ABC. Since the objective function involves only one choice variable (x_1), our aim is to push the graph of this function forward until it coincides with $x_1 = 2$. Therefore, the objective function will be maximized at $x_1 = 2$. We can now attempt to see whether this optimum solution (point $(x_1^*, x_2^*) = (2, 0)$) satisfies the Kuhn–Tucker condition for a maximum in (6.3.14). For this, we first construct the associated Lagrangian function as

$$L = f(x_1, x_2, \lambda) = f(x_1, x_2) + \lambda[g(x_1, x_2) - c] = x_1 + \lambda[(2 - x_1)^3 - x_2] \qquad (6.4.1)$$

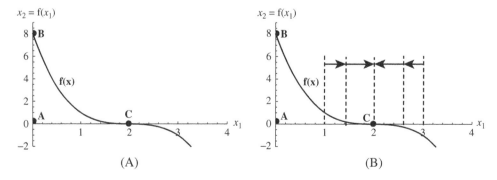

Figure 6.4.1

Then the associated Kuhn–Tucker conditions can be written as

$$L_1 = 1 + 3\lambda(2 - x_1)^2 \le 0, x_1 \ge 0, x_1 L_1 = 0, L_2 = -\lambda \le 0, x_2 \ge 0, x_2 L_2 = 0,$$

$$L_\lambda = (2 - x_1)^3 - x_2 \ge 0, \lambda \ge 0, \text{ and } \lambda L_\lambda = 0 \qquad (6.4.2)$$

If we substitute the optimal solution $(x_1^*, x_2^*) = (2, 0)$ into L_1 in (6.4.2) we obtain the contradictory result $(1 \le 0)$, which violates that inequality. Therefore, the optimal solution we obtained above for the present problem invalidates the Kuhn–Tucker maximum conditions. This is due to the presence of a cusp on the boundary of the feasible region.

So far we have been concerned with the nature of an irregularity of the boundary of the feasible region and how it affects the Kuhn–Tucker conditions. The reader might wonder why the presence of such irregularities invalidates the Kuhn–Tucker conditions. A careful inspection of Figure 6.4.1 would suggest that at point C the graph of the constraint $x_2 = g(x_1) = (2 - x_1)^3$ is tangent to the graph of the nonnegativity condition for the second choice variable (x_2). In other words, the two constraints are indistinguishable at that point. This is the reason why boundary irregularities such as cusps invalidate Kuhn–Tucker conditions.

Therefore, we impose a condition on the constraints of the NLP problems that they do not have cusps on their graphs. This condition is called, in the literature, *constraint qualification*. Since such boundary irregularities as cusps are not normally encountered in the subjects of our interest and since the cost of mathematics involved is larger than the associated benefit, we do not present here a mathematical treatment of constraint qualification. But it can be shown that if this constraint-qualification condition is met, that is, if the constraints of the problem do not involve cusps or if the feasible region of the problem does not involve boundary irregularities, then an optimal solution to a NLP problem will invariably satisfy the associated Kuhn–Tucker conditions. In other words, if the constraint qualification is satisfied (which is implied when the constraints are linear), then the Kuhn–Tucker conditions are necessary conditions for an optimum of the concerned NLP problem.

6.4.2 Concave programming

We stated above that when we imposed the constraint qualification on a NLP problem, then an optimal solution to the problem would satisfy the Kuhn–Tucker conditions. In this event, the Kuhn–Tucker conditions can be considered as the necessary conditions for an optimum. However, if we impose few conditions on the nature of not just the constraints but also of the objective functions, we can be sure that if there exists an optimum point then that point will satisfy the Kuhn–Tucker conditions such that the Kuhn–Tucker conditions are considered as the sufficient (or even necessary and sufficient) conditions for an optimum. Moreover, if there is such an optimum point, then that optimum point will be the global optimum point.

Notice that, in Sections 4.3.5 and 4.4.10, we mentioned that a knowledge of the nature of the objective functions (concavity and convexity in the case of unconstrained problems, and quasiconcavity and quasiconvexity in the case of constrained problems) obviated the need to check the SOCs to determine the optima. In other words, we stated the SOCs or the sufficient condition for an optimum in terms of the nature of the objective functions. In a similar fashion, as we mentioned above and as we state below, we can establish the sufficient condition for optima in the case of NLP problems in terms of concavity and convexity of the objective functions and the constraints. We know from our presentation in Section 4.3.5 that, for the function $U = f(x_1, x_2, x_3, \ldots, x_n)$, if $d^2 U$ is positive (negative) definite, then the function has a global minimum (maximum); and if $d^2 U$ is positive (negative) semidefinite, then the function has a local minimum (maximum). We also know from our presentation there that the curvature (the nature of the n-dimensional hyperplane) of the function $U = f(x_1, x_2, x_3, \ldots, x_n)$ is determined by the sign of the associated quadratic form (equation (3.8.9)); that is, the sign of $d^2 U$. If $d^2 U$ is positive (negative) definite, then the function is strictly convex (concave) and the function has a global minimum (maximum); and if $d^2 U$ is positive (negative) semidefinite, then the function is convex (concave) and the function has a local minimum (maximum). We know from Sections 2.8.2 and 2.8.3 and from Sections 4.3.2–4.3.4 how to determine the sign of $d^2 U$.

Suppose now that we want to maximize $\Pi = f(x_j), j = 1, 2, \ldots, n$, subject to constraints $g^i(x_j) \leq c_i$, $i = 1, 2, \ldots, m$, and to the nonnegativity restriction $(x_j \geq 0)$. Also suppose that $f(x_j)$ and $g^i(x_j)$ are concave differentiable and convex differentiable, respectively, and that $f(x_j^*)$ is a point on $f(x_j)$ that satisfies the Kuhn–Tucker conditions for a maximum. Then, $f(x_j^*)$ is a global maximum of $f(x_j)$ subject to the constraints and the nonnegativity restrictions. This statement is called Kuhn–Tucker necessary and sufficient conditions for a maximum in NLP problems. It is also called the *Kuhn–Tucker sufficiency theorem* for a maximization NLP problem.

Now suppose that we want to minimize $C = h(x_j), j = 1, 2, \ldots, n$, subject to constraints $k^i(x_j) \geq c_i$, $i = 1, 2, \ldots, m$, and to the nonnegativity restriction $(x_j \geq 0)$. Also suppose that $h(x_j)$ and $k^i(x_j)$ are convex differentiable and concave differentiable, respectively, and that $h(x_j^*)$ is a point on $h(x_j)$ that satisfies the Kuhn–Tucker conditions for a minimum. Then, $h(x_j^*)$ is a global minimum of $h(x_j)$ subject to the constraints and the nonnegativity restrictions. This statement is called Kuhn–Tucker necessary and sufficient conditions for a minimum in NLP problems. It is also called the Kuhn–Tucker sufficiency theorem for a NLP minimization problem.

The process of the application of Kuhn–Tucker sufficiency theorems for NLP maximization and minimization problems is called *concave programming*. This process is so named due to the fact that Kuhn and Tucker used the inequality $\geq$ in all the constraints and used the negative of the convex objective function $(h(x_j))$ in minimization problem. This implies that each constraint even in a maximization problem is considered to be concave $(g^i(x_j) \geq c_i)$. Notice that our formulation above can be transformed into the formulation used by Kuhn and Tucker if we multiply the objective function $h(x_j)$ in the minimization problem and the constraints $g^i(x_j) \leq c_i$ in maximization problem by -1. This transformation will give us both the objective function and the constraints in both maximization and minimization problems in concave form. Hence, the name concave programming.

The issue now is to determine whether a function is convex or concave. In the case of a univariate function, we know that the function will be convex (concave) around the point x^* on its domain if its second derivative, evaluated around x^*, is positive (negative), and it is strictly convex (concave) if its second derivative is positive (negative) for every value on its domain. We also know that a straight line is convex (as well as concave) and not strictly convex (concave). We know now, given our recap presented earlier in this section, how to determine the curvature of multivariate functions.

Notice that the application examples we considered in Section 6.2.8 yielded optimal solutions that were consistent with the Kuhn–Tucker conditions. Let us now check whether the objective functions and constraints are concave (convex) and convex (concave), respectively, for maximization (minimization) problems in these examples.

Consider first example 1 in Section 6.2.8. The objective function in this example was $U = f(x_1, x_2) = x_1^{1/2} x_2^{1/2}$ and the constraint was $x_1 + x_2 \leq 7$. The second partial derivatives

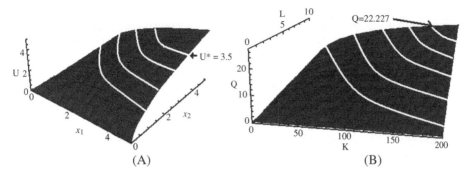

Figure 6.4.2

of the objective function are $f_{11} = (-1/4)x_1^{-3/2}x_2^{1/2}$, $f_{22} = (-1/4)x_1^{1/2}x_2^{-3/2}$, and $f_{12} = (1/4)x_1^{-1/2}x_2^{-1/2} = f_{21}$. If these partial derivatives are evaluated at the optimum values of $x_1^* = x_2^* = 3.5$, we obtain $f_{11} = -3.06$, $f_{22} = -3.06$, and $f_{12} = 0.07 = f_{21}$. This implies that the first and the second principal minors of the associated discriminant are $-3.06 < 0$ and $9.359 > 0$, respectively. Since these principal minors alter in sign with the first being negative, the quadratic form associated with the objective function is negative definite. Therefore, the function is concave around $x_1^* = x_2^* = 3.5$. Since the constraint in the problem forms a straight line when it is considered as an equation, it is convex. Therefore, the Kuhn–Tucker sufficiency conditions for a maximum are satisfied in the case of example 1 in Section 6.2.8.

We just found that the function $U = f(x_1, x_2) = x_1^{1/2}x_2^{1/2}$ in the first example in Section 6.2.8 is concave around $x_1^* = x_2^* = 3.5$. We can deduce this conclusion if we draw the three-dimensional graph of the function as illustrated in Figure 6.4.2(A). The graph shows that the function increases as x_1 and x_2 increase, at a decreasing rate. This suggests that the function is concave. Notice that the curve $U^* = 3.5 = c$ in this figure, which is also called an indifference curve, is the two-dimensional graph of the points x_1 and x_2 for the given value of $U^* = 3.5 = c$. This curve is also called the *level set* of the function $U = f(x_1, x_2) = x^{1/2}x^{1/2}$ for $U^* = 3.5 = c$. If we use different values for $U = c_i$ (such as $c_1, c_2, \ldots, c_n$) we obtain different level sets (such as $U_1 = x_1^{1/2}x_2^{1/2} = c_1$, $U_2 = x_1^{1/2}x_2^{1/2} = c_2, \ldots, U_n = x_1^{1/2}x_2^{1/2} = c_n$) and the two-dimensional graphs of these levels sets constitute an indifference map such as the one illustrated in Figure 6.2.7(A).

Now consider the second example in Section 6.2.8. The problem in this example was to maximize $Q = f(K, L) = [0.5K^{-0.5} + 0.5L^{-0.5}]^{-2}$ subject to $0.1K + 10L \leq 100$. Taking the partial derivatives of the objective function and evaluating them at the optimum values $K^* = 177.262$ and $L^* = 8.227$, we obtain $f_{KK} = -0.000154$, $f_{LL} = -0.07219$, and $f_{KL} = 0.00334 = f_{LK}$. This implies that the first and the second principal minors of the associated discriminant are $-0.000154 < 0$ and $0.00000032 > 0$, respectively. Since these principal minors alter in sign with the first being negative, the quadratic form associated with the function $Q = f(K, L) = [0.5K^{-0.5} + 0.5L^{-0.5}]^{-2}$ is negative definite and, therefore, the function is concave around $K^* = 177.262$ and $L^* = 8.227$. Since the constraint in the problem forms a straight line when it is considered as an equation, it is convex. Therefore, the Kuhn–Tucker sufficiency conditions for a maximum are satisfied in the case of example 2 in Section 6.2.8. We can deduce this conclusion if we draw the three-dimensional graph of the function as illustrated Figure 6.4.2(B). The figure shows that the function increases as K and L increase, at a decreasing rate. This suggests that the function is concave. Notice that the curve $Q^* = 22.227 = q$ in this figure, which is also called an isoquant, is the two-dimensional graph of the points of K and L for the given value of $Q^* = 22.227 = q$. This curve is also called the *level set* of the function $Q = f(K, L) = [0.5K^{-0.5} + 0.5L^{-0.5}]^{-2}$ for $Q^* = 3.5 = q$. If we use different values for $Q = q_i$ (such as $q_1, q_2, \ldots, q_n$), we obtain different level sets (such as $Q_1 = q_1, Q_2 = q_2, \ldots, Q_n = q_n$) and the two-dimensional graphs of these levels sets constitute an isoquant map such as the one illustrated in Figure 6.2.7(B).

The third example in Section 6.2.8 was a minimization problem. The objective function of this problem was $E = h(C_1, C_2) = 1.1C_1 + C_2$ and the constraint was $U = g(C_1, C_2) = \ln C_1 + \ln C_2$. Notice that the objective function in this example is a straight line when it is considered as an equation and, therefore, is convex. The second partial derivatives of the constraint function are $g_{11} = -1/C_1^2$, $g_{22} = -1/C_2^2$ and $g_{12} = 0 = g_{21}$. If these partial derivatives are evaluated at the optimum values of $C_1^* = \$7045.4545$ and $C_2^* = \$7750$, we obtain $g_{11} = -0.00000002$ and $g_{22} = -0.000000017$. This implies that the first and the second principal minors of the associated discriminant are negative and positive, respectively. Since the principal minors alter in sign with the first being negative, the quadratic form associated with the function $U = g(C_1, C_2) = \ln C_1 + \ln C_2$ is negative definite and, therefore, the function is concave around $C_1^* = \$7045.4545$ and $C_2^* = \$7750$, which can be seen from the plot of the function in Figure 6.4.3(A). Therefore, the Kuhn–Tucker sufficiency conditions for a minimum are satisfied in the case of example 3 in Section 6.2.8.

The last example in Section 6.2.8 was also a minimization problem. The objective function of this problem was $E = g = (K, L) = 1x_1 + 1x_2 = x_1 + x_2$ and the constraint function was $U = f(x_1, x_2) = x_1^{1/2}x_2^{1/2} = 3.5$. Notice that the objective function in this example, as in the previous case, is a straight line when it is considered as an equation and, therefore, is convex. The second partial derivatives of the constraint function when evaluated at the optimum values $x_1^* = x_2^* = 3.5$ are $f_{11} = -0.0714$, $f_{22} = -0.0714$, and $f_{12} = -0.0714 = f_{21}$. This implies that the first and the second principal minors of the associated quadratic form are -0.0714 and 0.0051, respectively. Since the principal minors alter in sign with the first being negative, the quadratic form of the function $U = f(x_1, x_2) = x_1^{1/2}x_2^{1/2}$ is negative definite and, therefore, the function is concave, which is evident from the plot of the function in Figure 6.4.3(B). Therefore, the Kuhn–Tucker sufficiency conditions for a minimum are satisfied in the case of example 4 in Section 6.2.8.

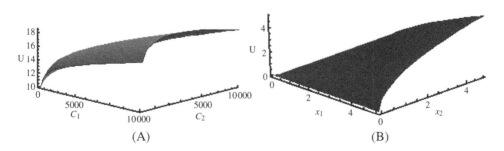

(A)   (B)

Figure 6.4.3

6.4.3 Quasiconcave programming

In the last section, we noticed that if the objective function and constraints are concave (convex) and convex (concave), respectively, in a NLP maximization (minimization) problem and if there exists a point on the function of the objective function that satisfies the associated Kuhn–Tucker conditions, then that point would be the global maximum (minimum) of the objective function. However, these concavity and convexity (of the objective function and the constraints) seem to be rather strong. It seems that the situation would have been much simpler if we had somewhat weaker conditions. In fact, there exists a theorem, called the *Arrow-Enthoven theorem*, which suggests a set of weaker conditions for global optima in NLP problems.

The Arrow–Enthoven sufficiency theorem states that if the objective function, $f(x_j)$, $j = 1, 2, \ldots, n$, and the constraints, $g^i(x_j)$, $i = 1, 2, \ldots, m$, in a NLP maximization problem are quasiconcave and quasiconvex, respectively, and if there exists a point on the objective function $f(x^*)$ that satisfies the Kuhn–Tucker maximum conditions, then that point will be a global maximum of the objective function if $f(x_j)$ is concave, or if $f_j(x^*) > 0$ for some j, or if $f_j(x^*) < 0$ for at least one j.

Similarly, the Arrow–Enthoven sufficiency theorem states that if the objective function, $h(x_j)$, $j = 1, 2, \ldots, n$, and the constraints, $k^i(x_j)$, $i = 1, 2, \ldots, m$, in a NLP minimization problem are quasiconvex and quasiconcave, respectively, and if there exists a point on the objective function $h(x^*)$ that satisfies the Kuhn–Tucker minimum conditions, then that point will be a global minimum of the objective function if $h(x_j)$ is convex, or if $h_j(x^*) < 0$ for some j, or if $h_j(x^*) > 0$ for at least one j.

Arrow and Enthoven used the $\geq$ sign in the constraints of the maximization problem in their original work. Given this, their theorem for a maximization NLP problem stated suggests that both the objective function and the constraints must be quasiconcave. Therefore, the application of the Arrow–Enthoven theorem is also called *quasiconcave programming*. The reader will have noticed that the Arrow–Enthoven theorem is weaker than the Kuhn–Tucker sufficiency theorem.

6.4.4 Quadratic programming

So far in this chapter, we have been concerned with NLP problems in which the objective functions and the constraints are of different functional forms. Suppose now that the objective function is quadratic and the constraint(s) is (are) linear. Such a NLP problem is called a *quadratic NLP problem* and its solution algorithm is called *quadratic programming* (QP).

It can be shown that if a feasible solution to a NLP problem involving a quadratic objective function and linear constraint(s) exists, then that will be a unique global optimal solution. Since the mathematical cost of proving this exceeds the resulting benefit, we do not attempt it here. Instead, we state here that the solution to a QP problem can be obtained by applying the first phase of the two-phase method demonstrated in Section 5.3.5.

As an example, suppose that our problem is the same as that in the first example in Section 6.3.10. The problem is to maximize $\Pi = f(x_1, x_2) = x_1 x_2$ subject to the constraints $g(x_1, x_2) = x_1 + x_2 \leq 10 = c$ and $x_j \geq 0$, where $j = 1, 2$. We can now construct the Lagrangian function as

$$L = f(x_1, x_2) + \lambda[c - g(x_1, x_2)] + \sum_{j=1}^{2} \alpha_j x_j = x_1 x_2 + \lambda[10 - x_1 - x_2] + \alpha_1 x_1 + \alpha_2 x_2$$

$$(6.4.3)$$

Notice that we have introduced nonnegativity condition in the present problem through the last two terms in equation (6.4.3). We can derive from equation (6.4.3) the Kuhn–Tucker maximum conditions as

$$L_1 = x_2 - \lambda + \alpha_1 \leq 0, x_1 \geq 0, x_1 L_1 = 0, L_2 = x_1 - \lambda + \alpha_2 \leq 0, x_2 \geq 0, x_2 L_2 = 0,$$

$$L_{\alpha_1} = x_1 \geq 0, \alpha_1 \geq 0, \alpha_1 L_{\alpha_1} = 0, L_{\alpha_2} = x_2 \geq 0, \alpha_2 \geq 0, \alpha_2 L_{\alpha_2} = 0,$$

$$L_\lambda = 10 - x_1 - x_2 \geq 0, \lambda \geq 0, \text{ and } \lambda L_\lambda = 0 \tag{6.4.4}$$

Based on the Kuhn–Tucker maximum conditions in (6.4.4), we can reformulate the problem using dummy variables (s_1, s_2, and s_3) and artificial variables (A_1 and A_2) as

$$\text{Maximize } -A_1 - A_2, \text{ subject to } 0x_1 + x_2 - \lambda + \alpha_1 + 0\alpha_2 + A_1 + 0A_2 + s_1 + 0s_2 + 0s_3 = 0,$$

$$x_1 + 0x_2 - \lambda + 0\alpha_1 + \alpha_2 + 0A_1 + A_2 + 0s_1 + s_2 + 0s_3 = 0,$$

$$x_1 + x_2 + 0\lambda + 0\alpha_1 + 0\alpha_2 + 0A_1 + 0A_2 + 0s_1 + 0s_2 - s_3 = 10,$$

$$x_1, x_2, \alpha_1, \alpha_2, s_1, s_2, s_3, \lambda \geq 0, \alpha_1 x_1 = 0, \alpha_2 x_2 = 0, \text{ and } \lambda s_1 = 0 \tag{6.4.5}$$

We can now set up the initial tableau based on problem (6.4.5) as in Table 6.4.1. If we now follow the same procedure as the one we used in the first phase of the two-phase method discussed in Section 5.3.5, we can find the optimal solution to the present problem as $x_1^* = x_2^* = \lambda = 5$. We can obtain the optimal value of the objective function if we substitute $x_1^* = x_2^* = \lambda = 5$ into it: $\Pi^* = f(x_1^*, x_2^*) = x_1^* x_2^* = 5 \times 5 = 25$. Notice that this is the same optimal solution as that obtained in Section 6.3.10.

As another example, suppose that our problem is the same as the problem in the second example in Section 6.3.10. This problem is to minimize $C = h(x_1, x_2) = x_1^2 + x_2^2$ subject to the constraint $g(x_1, x_2) = x_1 + x_2 \geq 50 = c$ and $x_j \geq 0$, where $j = 1, 2$. We can now construct the Lagrangian function as

$$L = h(x_1, x_2) + \lambda[c - g(x_1, x_2) - \sum_{j=1}^{2} \alpha_j x_j = x_1^2 + x_2^2 + \lambda[50 - x_1 - x_2] - \alpha_1 x_1 - \alpha_2 x_2$$

$$\tag{6.4.6}$$

Table 6.4.1

BVs	x_1	x_2	λ	α_1	α_2	A_1	A_2	s_1	s_2	s_3	RHS
A_1	0	1	−1	1	0	1	0	1	0	0	0
A_2	1	0	−1	0	1	0	1	0	1	0	0
s_1	1	1	0	0	0	0	0	0	0	−1	10
Π_j	0	0	0	0	0	−1	−1	0	0	0	0

Indicators

Table 6.4.2

BVs	x_1	x_2	λ	α_1	α_2	A_1	A_2	A_3	s_1	s_2	s_3	RHS
A_1	2	0	-1	-1	0	1	0	0	-1	0	0	0
A_2	0	2	-1	0	-1	0	1	0	0	-1	0	0
A_3	1	1	0	0	0	0	0	1	0	0	1	50
Π_j	0	0	0	0	0	-1	-1	-1	0	0	0	0

Indicators

Notice that, as in the last problem, we have introduced the nonnegativity restriction into equation (6.4.6). We can derive from equation (6.4.6) the Kuhn–Tucker minimum conditions

$$L_1 = 2x_1 - \lambda - \alpha_1 \geq 0, x_1 \geq 0, x_1 L_1 = 0, L_2 = 2x_2 - \lambda - \alpha_2 \geq 0, x_2 \geq 0, x_2 L_2 = 0,$$

$$L_{\alpha_1} = -x_1 \leq 0, \alpha_1 \geq 0, \alpha_1 L_{\alpha_1} = 0, L_{\alpha_2} = -x_2 \leq 0, \alpha_2 \geq 0, \alpha_2 L_{\alpha_2} = 0,$$

$$L_\lambda = 50 - x_1 - x_2 \leq 0, \lambda \geq 0, \text{ and } \lambda L_\lambda = 0 \tag{6.4.7}$$

Based on the Kuhn–Tucker maximum conditions in (6.4.7), we can reformulate the problem using dummy variables (s_1, s_2, and s_3) and artificial variables (A_1, A_2, and A_3) as

Maximize $A_1 + A_2 + A_3$, subject to $2x_1 + 0x_2 - \lambda - \alpha_1 - 0\alpha_2 + A_1 + 0A_2 + 0A_3 - s_1$

$$+ 0s_2 + 0s_3 = 0, 0x_1 + 2x_2 - \lambda - 0\alpha_1 - \alpha_2 + 0A_1 + A_2 + 0A_3 + 0s_1 - s_2 + 0s_3 = 0,$$

$$x_1 + x_2 + 0\lambda - 0\alpha_1 - 0\alpha_2 + 0A_1 + 0A_2 + A_3 + 0s_1 + 0s_2 + s_3 = 50,$$

$$x_1, x_2, \alpha_1, \alpha_2, s_1, s_2, s_3, \lambda \geq 0, \alpha_1 x_1 = 0, \alpha_2 x_2 = 0, \text{ and } \lambda s_1 = 0 \tag{6.4.8}$$

We can now set up the initial tableau based on problem (6.4.8) as presented in Table 6.4.2. As in the last example, if we now follow the same procedure as the one we used in the first phase of the two-phase method discussed in Section 5.3.5, we can find the optimal solution as $x_1^* = x_2^* = 25$. We can obtain the optimal value of the objective function if we substitute $x_1^* = x_2^* = 25$ into it: $C^* = h(x_1^*, x_2^*) = x_1^{*2} + x_2^{*2} = 25^2 + 25^2 = 1250$. Notice that this is the same optimal solution as that we obtained in Section 6.3.10.

6.4.5 Exercises

1. Use the trial-and-error method to solve the following NLP problems:
   (i) maximize $\Pi = f(x_1, x_2) = x_1 + x_2 - x_1^2 - x_2^2$, subject to $x_1 + x_2 \leq 10$, and $x_1, x_2 \geq 0$;
   (ii) minimize $C = h(x_1, x_2) = x_1^2 + x_2^2$, subject to $x_1 + x_2 \geq 10$, and $x_1, x_2 \geq 0$.
2. Solve (i) and (ii) in exercise 1 above using the QP algorithm.
3. *Application exercise.* Suppose that the total utility that a consumer obtains from the consumption of two goods, x_1 and x_2, is given by the function $U = f(x_1, x_2) = x_1^{1/2} + x_2^{1/2}$. Also suppose that the consumer's budget constraint is $3x_1 + 2x_2 \leq 30$. Find the quantities of the two goods that maximize the consumer's utility function subject to the budget constraint using the trial-and-error method.

4. *Application exercise.* Suppose that the total expenditure that a firm has to incur by employing two factors, x_1 and x_2, is given by the function $C = g(x_1, x_2) = x_1^2 + x_2^2$. Also suppose that the firm's output constraint is given by $x_1 + x_2 \geq 10$. Find the quantities of the two factors that the firm has to employ such that the firm's total expenditure is minimum subject to the output constraint using the QP algorithm.

7 Game theory

7.1 Introduction

We found in the last three chapters that every economic agent was aiming at optimizing own objective function. Although we did not state it explicitly, in all our analyses so far in this book, except examples 4 through 6 in Section 4.2.8, we maintained an important assumption that the agent's action of optimizing own objective function was independent of other agents' actions of optimizing their objective functions. So far in this book we have developed and applied the mathematical tools that are required to find the answers to questions such as under what conditions a rational economic agent optimizes its independent objective function subject to (or not subject to) constraints.

The above mentioned assumption is a strong restriction given the facts we observe around us. Many of the real-world actions of economic agents are not independent, but interdependent. The interdependencies may be either competitive or cooperative in nature. One good example is the decision of firms in oligopoly markets regarding quantities of output they must produce, the prices they have to fix, advertisement expenditures they must incur, etc. Another example is the decision of countries regarding the duties they have to impose on imports, subsidies to be provided to exports, etc. Examples like these are numerous in the subjects of our interest.

The question that arises, therefore, is how one can incorporate the above mentioned interdependencies into an agent's objective functions and optimize them. An important branch of mathematics called *game theory* deals with optimization of *interdependent objective functions*. The first systematic work on game theory and its applications to the subjects of our interest was the book entitled *Theory of Games and Economic Behavior* authored by J. von Neumann and O. Morgenstern and published in 1944. Since then there have been considerable developments in the theory of games and their applications in various fields including economics, business, and finance. It was in recognition of the importance of game theory and its applications in these branches of knowledge that the 1994 Nobel Prize in Economics was awarded to game theorists John Harsanyi, John Nash, and Reinhard Selten. It seems that it is unnecessary to state that a reasonable understanding of game theory and its applications in economics, business, and finance is indispensable for a proper comprehension of some of the models that undergraduates have to learn in these disciplines. Therefore, we turn our attention to elementary game theory and its applications in this chapter.

7.2 Static games of complete and perfect information

7.2.1 Concepts and definitions

Before we present the basics of game theory, we need to introduce some concepts and define them. The full import of many of these concepts and definitions will become clearer as we proceed through the rest of this chapter. Let us begin with the term *game*. A game is a set of rules in which two or more players interact in a setting of *strategic interdependence*. Strategic interdependency is a situation where the actions of a player or a group of players affect, and are affected by, the actions of other players or other groups of players. We can now define game theory as a mathematical toolkit that helps make optimal decisions under conditions of strategic interdependence among the players who engage in a game.

A game may involve two or more *players*. Players of a game may be individuals, firms, organizations, political parties, governments, etc. They make the *strategic decisions* in the context of games and, therefore, they are also called *strategic decision-makers*. A strategic decision or, in short, a *strategy*, constitutes a set of decisions, plans, or actions to be followed by each player during the course of the game. A strategy may be either a *pure strategy* (also called a *deterministic strategy*) or a *mixed strategy* (also called a *randomized strategy*). A pure strategy is a strategy in which a player makes a deterministic (or nonrandom or specific) action or decision. A mixed strategy is a strategy in which a player makes random choice (that is, choice based on probabilities) among possible actions or decisions (or pure strategies).

In addition to players and strategic decisions, a game involves *rules*, *outcomes*, and *payoffs*. Rules of a game are predetermined and contain a set of information including who starts the game, what information the players possess, how the players start the game, etc. The result of the game represents the outcome of the game. The results may be expressed in cardinal units such as dollars or in ordinal units such as utility obtained. The payoff of a game is the net gain or value of the objective function associated with a possible outcome of a game. A player's utility function or objective function is generally called that player's *payoff function*. An *optimal strategy* in a game is the strategy that maximizes a player's *expected payoffs*.

7.2.2 Classification of games

One can classify games in different ways. One way to distinguish games is based on the timings of decisions taken in games. If decisions are taken by all the players simultaneously, then such games are called *static games* or *simultaneous games*. Static games are also called *one-shot games* or *one-time games* because in these games each player chooses a single action and then the game ends. This implies that static games are *single-period games*. If a single-period game is played over and over again, it becomes a *multiperiod game* and is called a *repeated game*. If in a game a player takes a decision and makes a move only after another player has already done that, such games are called *sequential games*. Sequential games are also called *dynamic games* or *multistage games*. We are primarily concerned with static games in the present section. Dynamic games are discussed in Sections 7.4 and 7.5.

Another way to classify games is based on the nature of information that the players possess. If every player of a game knows every aspect of the game, then the game is called a *game of complete information*. This suggests that the strategies and payoffs are *common knowledge* in games of complete information. If one or more players of a game possess information that is unknown to the other players, or if in a game the strategies and payoffs are not common knowledge, then that game is called a *game of incomplete information*.

Another classification, which is similar to but not identical to the last classification, is based on whether the player has information on the previous decisions or moves in the game. If every player has information on all the previous moves or decisions taken in a game, then that game is called a *game of perfect information*. In contrast, if players do not possess information on some of the previous decisions or moves taken in a game, then that game is called a *game of imperfect information*. Since a meaningful exposition and solution of games of incomplete and imperfect information are beyond the scope of this book, we will not attempt them and, instead, we will concentrate only on games of complete and perfect information.

Games may also be classified on the basis of the value of the payoffs. If in a two-player game the gain of a player is exactly equal to the loss of the other player, then that game is called a *zero-sum game*. Therefore, in a zero-sum game the payoff of one player is just the negative of the payoff of the other player. Otherwise it is called a *nonzero-sum game*. If the sum of payoffs in a game is a nonzero constant, then such a game is called a *constant-sum game*. If the sum of payoffs is not constant but a variable, then such a game is called a *variable-sum game*.

The last classification we introduce here is based on whether the players can enter into binding contracts or not. If in a game the players can make binding contracts and enforce them, then that game is called a *cooperative game*. But, if the players cannot make such contracts and enforce them in a game, then that game is called a *non-cooperative game*.

7.2.3 Representation and solution of games

As in the case of the classification of games, one can represent games in different ways. One way to represent a game is to write the strategies and payoffs of players in a table with columns and rows, which is called a *payoff matrix*. This method is called the *normal-form representation* or the *strategic-form representation* of a game.

Let us demonstrate the normal-form representation of a two-player, two-strategy game. Suppose that we represent the ith player's jth strategy by s_j^i, where $i = 1, 2$ and $j = 1, 2$. Therefore, the complete set of strategies for the ith player is given by $S^i = \{s_1^i, s_2^i\}$. Notice that $s_j^i \in S^i$. As an example, in a two-player game, the set of player A's strategies is given by $S^A = \{s_1^A, s_2^A\}$. Now suppose that we represent the ith player's payoff function is given by $\pi^i : \pi^i = \pi^i(s_j^i)$. This implies that the ith player's payoff function depends on the strategies chosen by all the players (s_j^i) in the game. We denote a two-player game by

Table 7.2.1

		Player B's strategies	
		s_1^B	s_2^B
Player A's strategies	s_1^A	$\pi^A(s_1^A, s_1^B), \pi^B(s_1^A, s_1^B)$	$\pi^A(s_1^A, s_2^B), \pi^B(s_1^A, s_2^B)$
	s_2^A	$\pi^A(s_2^A, s_1^B), \pi^B(s_2^A, s_1^B)$	$\pi^A(s_2^A, s_2^B), \pi^B(s_2^A, s_2^B)$

$G = \{S^i, \pi^i\} = \{S^A, S^B; \pi^A, \pi^B\}$. The payoff matrix of this game is presented in Table 7.2.1. This tabular form of the representation of a game, as mentioned earlier, is called the normal-form representation of the game.

Let us now extend the normal-form representation of a two-person, two-strategy game to the case of an *n*-person, *n*-strategy game. In this type of game the complete set of strategies, or the *strategy profile*, for the *i*th player is given by $S^i = \{s_1^i, s_2^i, \ldots, s_n^i\}$ and the *i*th player's payoff function is given by π^i: $\pi^i = \pi^i(s_j^i)$, where j $(j = 1, 2, \ldots, n)$ represents the *j*th strategy and i $(i = 1, 2, \ldots, n)$ represents the *i*th player. We denote this *n*-person, *n*-strategy game by $G = \{S^i, \pi^i\} = \{S^A, S^B, \ldots, S^n; \pi^A, \pi^B, \ldots, \pi^n\}$. We know that the *i*th player's opponent will choose some strategy when the *i*th player chooses the strategy s_j^i. For convenience we denote this response strategy of the *i*th player's opponent by s_j^{-i}.

Another form of representation of a game is called the *extensive form*, which is usually used in the representation of dynamic or sequential games. We will discuss the extensive-form representation of dynamic games in detail in Sections 7.4 and 7.5. For the purpose of exposition, we continue with the above two-player, two-strategy game. In dynamic games one player moves (i.e. takes a decision) first. Then the other player takes a decision in response to the decision of the first player.

The extensive-form representation of a game is based on *game trees*. A game tree shows the players of a game, the order of their moves, the information available to them, and their payoffs. Notice that a game tree is similar to a *decision tree*. Both are used to determine an optimal course of action. But, they differ in the sense that the former is used to arrive at an optimal course of action in situations that involve strategic decisions and the latter is used to arrive at an optimal course of action in situations that do not involve strategic decisions.

A game tree contains *decision nodes*, or simply nodes, line segments or branches, *root of the game*, *subroots*, and *terminal nodes*. Decision nodes normally represent the location where the designated player's moves are shown and branches represent decisions or actions taken by these players. In our example, assume that player A moves first. Then, this game can be represented in extensive form as shown by the game tree in Figure 7.2.1. In this figure, the ball or the node at point P_A is called the root of the game and the nodes at points P_B are called the subroots of the game. Finally, the balls adjacent to the payoffs are called the terminal nodes of the game.

Figure 7.2.1 shows that, since player A (denoted by P_A) is the first mover, P_A can choose either strategy s_1^A or strategy s_2^A. But player B (denoted by P_B) decides its strategy on the basis of the choice already made by P_A. If P_A decides to choose strategy s_1^A, then P_B decides to

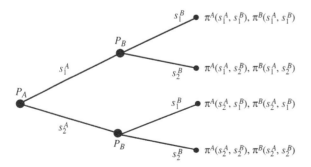

Figure 7.2.1

choose either strategy s_1^B or strategy s_2^B. If P_B decides to choose strategy s_1^B, then P_A's payoff will be $\pi^A(s_1^A, s_1^B)$ and P_B's payoff will be $\pi^B(s_1^A, s_1^B)$. If P_A decides to choose strategy s_2^B, then P_A's payoff will be $\pi^A(s_1^A, s_2^B)$ and P_B's payoff will be $\pi^B(s_1^A, s_2^B)$. Similarly, if P_B's response to P_A's strategy s_2^A is s_1^B, then P_A's and P_B's payoffs will be $\pi^A(s_2^A, s_1^B)$ and $\pi^B(s_2^A, s_1^B)$, respectively. Likewise, if P_B's response to P_A's strategy s_2^A is s_2^B, then the payoffs of P_A and P_B will be $\pi^A(s_2^A, s_2^B)$ and $\pi^B(s_2^A, s_2^B)$, respectively.

Let us now consider the issue of solving a game. Solving a game is the same as finding the equilibrium or equilibria in a game. However, the solution to a game depends on the type of the game; that is, whether the game is static or dynamic. In a static or single-period game, the solution can be achieved by finding the *Nash equilibrium* (or *Nash equilibria*) of the game. But, in a dynamic or multiperiod game the solution can be achieved by finding the *subgame-perfect Nash equilibrium*. We discuss Nash equilibrium and subgame-perfect Nash equilibrium in Sections 7.3 and 7.5, respectively.

7.2.4 Games of conflict or non-cooperation: zero-sum games

We stated in Section 7.2.2 that if the gain of a player in a two-player game is exactly equal to the loss of the other player, then the game was called a zero-sum game. That is, in a zero-sum game the payoff of one player is just the negative of the payoff of the other player. Let us first consider a general zero-sum, two-player, two-strategy game. The normal-form representation of this type of game is given in Table 7.2.2.

Let us now present an example of a zero-sum game called the game of *Matching Pennies*. There are two players, players A and B, in this game. Each player is assumed to put down a penny simultaneously with either heads up or tails up. Therefore, we state that each player has two strategies: heads or tails. The outcome of the game is assumed to be like this: if the players' strategies match or if the pennies match (i.e. if both pennies land heads or tails up), then player B pays \$10 to player A; otherwise (i.e. if the pennies do not match), player A pays \$10 to player B. The normal-form representation of this game is given in Table 7.2.3. We will attempt to see whether this game has an equilibrium or equilibria in Sections 7.2.7 and 7.3.10.

Table 7.2.2

		Player B's strategies	
		s_1^B	s_2^B
Player A's strategies	s_1^A	$\pi^A(s_1^A, s_1^B),\ -\pi^B(s_1^A, s_1^B)$	$-\pi^A(s_1^A, s_2^B),\ \pi^B(s_1^A, s_2^B)$
	s_2^A	$-\pi^A(s_2^A, s_1^B),\ \pi^B(s_2^A, s_1^B)$	$\pi^A(s_2^A, s_2^B),\ -\pi^B(s_2^A, s_2^B)$

Table 7.2.3

		Player B's strategies	
		Heads	Tails
Player A's strategies	Heads	10, −10	−10, 10
	Tails	−10, 10	10, −10

Notice that each player's payoffs, in Table 7.2.3, are equal to the negative of the other player's payoffs. Therefore, the sum of the payoffs is always equal to zero. This is the reason why these types of games are called a zero-sum games. Since the interests of the players in zero-sum games are opposed, these games are also called *games of conflict* or *games of non-cooperation*.

7.2.5 Games of conflict or non-cooperation: nonzero-sum games

In the last section we presented zero-sum games with the example of Matching Pennies. However, most of the games that the students of the subjects of our interest encounter are nonzero-sum (either constant-sum or variable-sum) games. As we stated earlier, in nonzero-sum games, the payoffs do not add up to zero. In most nonzero-sum games the interest of the players are opposed and, therefore, they are also called games of conflict.

For the purpose of illustration, let us consider the famous example of *Prisoner's Dilemma*. The story behind this game is as follows. Two persons are arrested by police for being allegedly involved in a crime. These two persons are brought to a prison and are held in two separate cells such that they cannot interact in any form. Then the police start to extract a confession from them. Each person is secretly offered the deal: "If you confess while your accomplice does not, then you will get a light punishment of 1 year in jail while your recalcitrant accomplice will get a punishment of 8 years in jail. If you are the only one who does not confess, then you will spend 8 years in jail. If you both confess, you both will get some mercy and spend 4 years in jail. If you both do not confess, you both will be jailed for 2 years." The normal-form representation of this game is given in Table 7.2.4. We assume here that both players wish to minimize the time they spend in jail. Notice the negative signs in the payoffs in the table. We also assume that the time spent in jail is a loss and, therefore, negative values are assigned to the payoffs. Notice also that the payoffs in the table do not add up to zero. This is the reason why these types of games are called nonzero-sum games. We will attempt to see whether this game has an equilibrium or equilibria in Sections 7.2.7 and 7.2.8.

7.2.6 Games of cooperation or coordination: nonzero-sum games

So far we have been concerned with games involving conflicts of interests. But, game theory is not just concerned with conflicts of interests; it also deals with cooperation or coordination among the players in games. As an illustration, consider the famous example of the *Battle of the Sexes*. In this game, a man and a woman are planning to spend an evening together. But they have different tastes in entertainment. The man prefers wrestling while the woman prefers to go to the opera. As the payoff matrix of this game presented in Table 7.2.5 shows, the man would most prefer to go to wrestling with the woman, but prefers going to the opera

Table 7.2.4

| | | Prisoner 2's strategies | |
		Don't confess	Confess
Prisoner 1's	Don't confess	−2, −2	−8, −1
strategies	Confess	−1, −8	−4, −4

Table 7.2.5

		Woman's strategies	
		Wrestling	*Opera*
Man's	Wrestling	2, 1	0, 0
strategies	Opera	0, 0	1, 2

with the woman to watching wrestling alone. Similarly, the woman would most prefer to go to the opera, but prefers watching wrestling with the man to going to the opera alone. Notice the special feature of the payoffs in this game. If the two players stick to their preferences, their payoffs will be zero. But, if they cooperate (that is, understand each other's preferences and act accordingly) their payoffs are positive. We will attempt to see whether this game has an equilibrium or equilibria in the next section.

7.2.7 Dominant and dominated strategies

We presented both zero-sum and nonzero-sum games, whether conflict or cooperation in nature, in the last few sections. The reader might wonder what outcome these games will generate. Therefore, our aim in the present section is to find the outcomes of the games we presented above. Consider first the Prisoner's Dilemma game introduced in Section 7.2.5, the payoffs matrix of which is presented in Table 7.2.4.

A careful look at the payoffs in Table 7.2.4 suggests that the prisoners face a dilemma. If the prisoners can communicate, they can cooperate by agreeing not to confess and receive a lighter punishment of 2 years each in jail. But, the assumption is that they cannot communicate. Even if they can, can they trust each other and cooperate? The answer is, probably, not. But then what deters them from cooperating. A careful inspection of the payoffs shows that if they decide to cooperate and, later, if one prisoner decides to cheat the other, the prisoner who cheats will receive a lighter punishment of 1 year in jail (and the other will receive a longer punishment), which is better than the 2-year jail period when they cooperate. This will be the feeling of either prisoner. This suggests that cooperation is highly unlikely in these types of games. Then the best each prisoner can do, *regardless of the decision of the other prisoner*, is to confess and receive a jail term of 4 years. Such a strategy is called a *strictly dominant strategy*. Let us now formally define a strictly dominant strategy. Suppose that in a normal-form game $G = \{S^i, \pi^i\} = \{S^A, S^B, \ldots, S^n; \pi^A, \pi^B, \ldots, \pi^n\}$, strategy s^{i*}, where $s^{i*} \in S^i$, is a possible strategy for player i. Then s^{i*} is a strictly dominant strategy for player i, if for each possible combination of the other players' strategies, player i's payoff from playing s^{i*} is strictly higher than playing any other strategy $s^{i*} \neq s^i$, if for all $s^i \in S^i$:

$$\pi^i(s^{A=1}, s^{B=2}, \ldots, s^{i-1}, s^{i*}, s^{i+1}, \ldots, s^n) > \pi^i(s^{A=1}, s^{B=2}, \ldots, s^{i-1}, s^i, s^{i+1}, \ldots, s^n)$$
(7.2.1)

What inequality (7.2.1) implies is that s^i is a strictly dominant strategy for player i if it maximizes the ith player's payoff regardless of the strategy chosen by the rival player(s). This is exactly the idea we applied in the last paragraph. Notice that if a player has a strictly dominant strategy in a game, then that player is expected to play that strategy no matter what the opponents do. In our present example, each player (prisoner) has a strictly dominant strategy, "Confess." Notice also that the optimal outcome when they can cooperate and when the cooperation is binding is $(-2, -2)$; that is, when they choose "Don't Confess, Don't Confess." This is called a *social optimum*. But, the self-interests of the prisoners result in a *social suboptimum* outcome $(-4, -4)$. If each player has a dominant strategy, then the outcome of the game is called *equilibrium in dominant strategies*. Therefore, "Confess, Confess" is the equilibrium in dominant strategies in the Prisoner's Dilemma game.

So far in this section we were considering strictly dominant strategies. But, the existence of strictly dominant strategies is not common in many games. Notice that the dominance of a particular strategy may depend on the strategy chosen by the rival(s). However, one can still use the idea of dominant strategies to eliminate some other strategies as possible choices. We know that a *rational* player will not play some strategies that are dominated by some other strategies. Strategies that are dominated by some other strategies are called *dominated strategies*. We now formally define a dominated strategy as follows. Suppose that in a normal-form game $G = \{S^i, \pi^i\} = \{S^A, S^B, \ldots, S^n; \pi^A, \pi^B, \ldots, \pi^n\}$, strategies $s^{i\bullet}$ and $s^{i\bullet\bullet}$, where $s^{i\bullet}, s^{i\bullet\bullet} \in S^i$, are two possible strategies for player i. Then $s^{i\bullet\bullet}$ is a strictly dominated strategy for player i, if for each possible combination of the other players' strategies, player i's payoff from playing $s^{i\bullet}$ is strictly higher than that from playing $s^{i\bullet\bullet}$ if

$$\pi^i(s^{A=1}, s^{B=2}, \ldots, s^{i-1}, s^{i\bullet}, s^{i+1}, \ldots, s^n) > \pi^i(s^{A=1}, s^{B=2}, \ldots, s^{i-1}, s^{i\bullet\bullet}, s^{i+1}, \ldots, s^n)$$
$$(7.2.2)$$

Let us now consider the example of the Battle of the Sexes presented in Section 7.2.6. The payoff matrix of this game is presented in Table 7.2.5. Our aim here is to see if there are any strictly dominant strategies in this game. We know from inequality (7.2.1) that a dominant strategy for player i is a strategy that maximizes the payoff of the ith player irrespective of the strategy chosen by the other player(s). Applying this definition, we see from Table 7.2.5 that the man's strictly dominant strategy is to watch wrestling while the woman's strictly dominant strategy is to go to the opera. Therefore, there are two strictly dominant strategies in this game.

Lastly, consider the example of Matching Pennies presented in Section 7.2.4. The payoff matrix of this game is presented in Table 7.2.3. Is there any strictly dominant strategy in this game? A careful inspection of the table reveals that the optimal decision of player 2 depends on what player 1 does, and vice versa. Therefore, there is no strictly dominant strategy in this game.

7.2.8 Iterated elimination of strictly dominated strategies

Having discussed the meaning of strictly dominant and dominated strategies, let us now attempt to see how we can use these concepts to find the equilibrium, if it exists, in a game. One can find the equilibrium or equilibria in some games through successive elimination, or *iterated elimination*, of strictly dominated strategies. This process of elimination will continue until all the strictly dominated strategies are eliminated and will stop when only the strictly dominant strategies are left or when the equilibrium set of strategies remain.

We know from our demonstration in the last section that a rational player will never play a strictly dominated strategy. Since the player is assumed to be rational, elimination of strictly dominated strategies will not affect the outcome of the game. In fact, strictly dominated strategies are redundant. Therefore, a new game can be constructed from the original game if we eliminate a strictly dominated strategy for a player. If this new game still has a strictly dominated strategy for any one of the players, that too can be eliminated. One can continue this process until all the strictly dominated strategies for all the players are eliminated. As an illustration, consider the simple example of the Prisoner's Dilemma game discussed earlier, the payoff matrix of which is presented in Table 7.2.4.

Prisoner 1 conjectures that prisoner 2 is rational and vice versa. Given this rationality assumption, prisoner 1 knows that prisoner 2 will never choose "Don't Confess." The reason for this belief on the part of player 1 is that the number of years in jail for prisoner 2 will be either 2 or 8 if prisoner 2 chooses "Don't Confess." But the jail terms will be either 1 or 4 years if prisoner 2 chooses "Confess." The jail terms in the latter event are shorter than the jail terms in the former event. Since prisoner 2 is rational, prisoner 2 will not choose "Don't Confess." This implies that "Don't Confess" is the strictly dominated strategy for prisoner 2. Therefore, we can eliminate the strictly dominated strategy "Don't Confess" from prisoner 2's strategies. This gives a new game with the payoff matrix given in Table 7.2.6(A). Notice that this table shows that prisoner 2 does not have any choice now but to confess. But, prisoner 1 has two choices: either confess or not. Since prisoner 1 will be much better off with the choice "Confess" and since prisoner 1 is rational, prisoner 1's strictly dominated strategy is "Don't Confess," which can be eliminated. This leads us to a new and final payoff table with only one set of strategies (strictly dominant strategies or *equilibrium strategies*) as presented in Table 7.2.6(B). This process of successive deletion of strictly dominated strategies is called iterated elimination of strictly dominated strategies.

As another illustration, consider an example of a game involving two players and three strategies. The strategies and the associated payoffs (benefits) in the original game are presented in Table 7.2.7. Assume that higher payoffs are preferred.

Let us first consider player 2's strategies. For this player, "Right" is a strictly dominated strategy because the payoffs from the strategies "Left" and "Centre" are strictly higher than the payoffs from the strategy "Right." Since player 2 is assumed to be rational, player 2 will not choose "Right" and, therefore, the strategy "Right" can be eliminated, which generates a new game with its associated payoffs as shown in Table 7.2.8(A). As can be seen, neither of the remaining two strategies of player 2 is strictly dominated. But, for player 1 the strategy

Table 7.2.6(A)

		Prisoner 2's strategy
		Confess
Prisoner 1's strategies	Don't confess	$-8, -1$
	Confess	$-4, -4$

Table 7.2.6(B)

		Prisoner 2's strategy
		Confess
Prisoner 1's strategy	Confess	$-4, -4$

Table 7.2.7

		Player 2's strategies		
		Left	*Center*	*Right*
Player 1's strategies	Top	4, 5	5, 3	3, 0
	Middle	5, 4	4, 3	2, 0
	Bottom	3, 2	2, 3	5, 1

Table 7.2.8(A)

		Player 2's strategies	
		Left	*Center*
Player 1's strategies	Top	4, 5	5, 3
	Middle	5, 4	4, 3
	Bottom	3, 2	2, 3

Table 7.2.8(B)

		Player 2's strategies	
		Left	*Center*
Player 1's strategies	Top	4, 5	5, 3
	Middle	5, 4	4, 3

"Bottom" is strictly dominated by both "Top" and "Middle" strategies. Since player 1 is rational, player 1 will not play the strategy "Bottom" and, therefore, that strategy can be eliminated to obtain the game presented in Table 7.2.8(B). Notice that, as can be seen in Table 7.2.8(B), neither of player 1's strategies is now strictly dominated. But, for player 2 the strategy "Centre" is strictly dominated by the strategy "Left." Therefore, we eliminate the strategy "Centre" to obtain the fourth game presented in Table 7.2.9(A). As can be seen, player 2 is left with only one strategy ("Left") and player 1 is left with two strategies ("Top" and "Middle"). It is now easy to infer that the strategy "Top" for player 1 is a strictly dominated strategy and, therefore, that strategy can be eliminated. After doing this, we arrive at the final payoff matrix presented in Table 7.2.9(B). Notice that, in Table 7.2.9(B), we have arrived

Table 7.2.9(A)

		Player 2's strategy
		Left
Player 1's strategies	Top	4, 5
	Middle	5, 4

Table 7.2.9(B)

		Player 2's strategy
		Left
Player 1's strategies	Middle	5, 4

Table 7.2.10

		Pepsi's strategies	
		Advertise	Don't advertise
Coke's strategies	Advertise	30, 20	40, 0
	Don't advertise	10, 15	30, 5

at the equilibrium of the game with player 1 choosing the strategy "Middle" and player 2 choosing the strategy "Left." Then their respective payoffs will be 5 and 4.

7.2.9 Application examples

Example 1. Suppose that two soft-drinks companies, Coca-Cola (denoted by Coke) and Pepsi, are competing in a small duopoly market. We know that their products are cola drinks. Suppose that their research departments found a positive relationship between advertisement and sales (and the resulting profits). They need to decide whether they have to undertake advertisements. Coke expects immediate response decisions, for any of its decisions, from Pepsi, and vice versa. The strategies and the payoff (in millions of dollars) matrix of this *advertisement game* are presented in Table 7.2.10. Determine whether there is a strictly dominated strategy for each company. Also determine the equilibrium of the game, if any, through the process of iterated elimination of strictly dominated strategies.

Solution. This is an example of a nonzero-sum, non-cooperative, static game. We know from our discussion in Section 7.2.7 that a strategy is a strictly dominated strategy for a particular player if it gives the player a strictly lower payoff than the player's other strategies. We also know from there that a rational player will not play a strictly dominated strategy and, therefore, we can eliminate that strategy from the set of strategies of the player. Following these ideas, we can infer from Table 7.2.10 that "Don't Advertise" is a strictly dominated strategy for Pepsi because the payoffs from "Don't Advertise" are lower than the payoffs from the strategy "Advertise" ($0 < 20$ and $5 < 15$). This implies that "Advertise" is the strictly dominant strategy for Pepsi. Therefore, we can eliminate the strategy "Don't Advertise" from the set of strategies for Pepsi. Similarly, "Don't Advertise" is the strictly dominated strategy for Coke (because $10 < 30$ and $30 < 40$) and its strictly dominant strategy is "Advertise" and, therefore, the strategy "Don't Advertise" can be eliminated from the set of strategies for Coke. After these eliminations, the resulting payoff matrix will comprise only one strategy: "Advertise" for both Pepsi and Coke. Therefore, the equilibrium of the

Table 7.2.11

		AMD's strategies		
		Low price	Medium price	High price
Intel's strategies	Low price	40, 40	50, 40	60, 30
	Medium price	40, 50	50, 50	50, 30
	High price	30, 60	30, 50	30, 30

game is "Advertise, Advertise" with payoffs $30 million and $20 million for Coke and Pepsi, respectively.

Example 2.   Suppose that two of the important computer chip manufacturers in the world, Intel Corporation and AMD Corporation, both based in the USA, need to set the prices of their newly introduced processors. Any price set by Intel is expected to affect the sales and the profits of AMD, and vice versa. Moreover, any price decision by Intel is expected to bring an immediate response from AMD, and vice versa. Suppose that their strategies and the associated payoffs (in millions of dollars) of the two companies are as represented by Table 7.2.11. Determine whether there is a strictly dominated strategy for each company. Also determine the equilibrium of the game, if any, through the process of iterated elimination of strictly dominated strategies.

Solution.   This is an example of a nonzero-sum, cooperative, static game. Following our discussion of strictly dominant and strictly dominated strategies in Section 7.2.7, "High Price" is the strictly dominated strategy (by both the other two strategies) for Intel. Therefore, the strategy "High Price" can be eliminated from the strategies of Intel. After this, neither of Intel's remaining strategies is strictly dominated. But, in the resulting game, AMD's strategy "High Price" is now strictly dominated by its other two strategies and, therefore, it can be eliminated to obtain the second game. Notice that although we started with a three-strategy game, in the second game both Intel and AMD have only two strategies: "Low Price" and "Medium Price." Notice also that in the second game neither of Intel's nor AMD's remaining strategies is strictly dominated.

The absence of strictly dominated strategies suggests us to think differently. Notice that if both companies choose "Low Price" their payoffs will be smaller ($40 million each). Since both are rational, they both will not choose that strategy. If Intel chooses "Low Price" and AMD chooses "Medium Price," then the payoff of the former will be larger than that of the latter suggesting that the latter will move to "Low Price" resulting in the former receiving smaller payoffs. The same result can be inferred if AMD chooses "Low Price" and Intel chooses "Medium Price." If they cooperate, instead of competing, and both choose "Medium Price," their payoffs will be higher. Since the companies are assumed to be rational, they both will choose "Medium Price" and each of their payoffs will be $50 million. Therefore, the only equilibrium in the present game is the cooperative strategy "Medium Price" for both Intel and AMD.

Example 3.   Assume that two of the world's leading passenger aircraft manufacturers, Boeing based in the USA and Airbus based in France, want to frame sales strategies for the coming financial year. Assume also that their strategies will be successful only if

Table 7.2.12

		Airbus's strategies	
		Existing markets	Emerging markets
Boeing's strategies	Existing markets	−2, −2	5, 5
	Emerging markets	5, 5	−2, −2

one of them concentrates on the emerging markets (of Asia, Eastern Europe, Africa, and South America) and the other concentrates on the existing markets (of Western Europe, Oceania and North America). A single market cannot accommodate the aircrafts of both manufacturers and each manufacturer can concentrate only on one market. Their strategies and associated payoffs (in billions of dollars) are presented in Table 7.2.12. Determine the strictly dominant or dominated strategies, if any, in this game. What will be the equilibrium of the game?

Solution. This is another example of a nonzero-sum, cooperative, static game. As the payoffs in Table 7.2.12 suggests, it is not in the interests of both Boeing and Airbus to concentrate in the same market; if they do, they both will make a loss of $2 billion. Notice that there is no strictly dominant or strictly dominated strategy for either manufacturer in this game. The best they can do is to choose separate markets. But, then, how do they choose these markets? Suppose now that Airbus reveals its intention of concentrating on "Emerging Markets" through a press release. Then the best Boeing can do is to concentrate on "Existing Markets." Therefore, the strategy profile "Emerging Markets" and "Existing Markets" constitutes an equilibrium of the game, and the resulting payoffs will be $5 billion each. Once the strategies are decided, no manufacturer will unilaterally deviate from it; if it does, the resulting payoffs will not be in the interests of both manufacturers. Notice also that the strategy profile "Emerging Markets" for Boeing and "Existing Markets" for Airbus is also equally good. Therefore, the present game involves two equilibria or has *multiple equilibria*.

Example 4. Suppose that two of the world's leading breakfast cereal producers, Kellogg and General Mills (GM), both based in the USA, face a duopoly market. Also suppose that they both plan to introduce two new types of cereal (diet and sweet), one each by Kellogg and GM, which will be successful if only one of them produce a particular type. Suppose again that a previous variant of diet cereals produced by GM was not as popular among customers as that produced by Kellogg. If both produce the same type, the market will be glutted with this type of cereal and the payoffs they gain are assumed to be considerably small. The strategies and the associated payoffs (in millions of dollars) of these two companies are given in Table 7.2.13. Determine the strictly dominant or dominated strategies, if any, in this game. What will be the equilibrium of the game?

Solution. This is another example of a nonzero-sum, non-cooperative, static game. It can be inferred from Table 7.2.13 that "Diet Cereal" is a strictly dominated strategy for GM because the payoffs from its "Sweet Cereal" strategy are strictly larger than those from "Diet Cereal" strategy ($100 > 30$ and $40 > 30$). Notice that Kellogg does not have a strictly dominant

Table 7.2.13

		GM's strategies	
		Diet cereal	*Sweet cereal*
Kellogg's	Diet cereal	30, 30	80, 100
strategies	Sweet cereal	80, 30	30, 40

(or dominated) strategy in this game. Its choice depends on the choice of GM. Since GM is a rational player, Kellogg knows for sure that GM will choose "Sweet Cereal." Then, the best Kellogg can do is to produce "Diet Cereal." Therefore, the equilibrium of the game is the strategy profile "Diet Cereal" for Kellogg and "Sweet Cereal" for GM and receive payoffs of $80 million and $100 million, respectively.

7.2.10 Exercises

1. Find the equilibrium, if any, for the games whose payoff matrices are as presented in Tables 7.2.14(A)–(D) using iterated elimination of strictly dominated strategies.

Table 7.2.14(A)

		Player 2's strategies	
		s_1^2	s_2^2
Player 1's	s_1^1	5, 3	4, 6
strategies	s_2^1	1, 2	0, 2

Table 7.2.14(B)

		Player 2's strategies	
		s_1^2	s_2^2
Player 1's	s_1^1	5, 3	4, 6
strategies	s_2^1	1, 2	0, 2

Table 7.2.14(C)

		Player 2's strategies	
		s_1^2	s_2^2
Player 1's	s_1^1	0, 0	5, 8
strategies	s_2^1	8, 5	−1, −1

Table 7.2.14(D)

		Player 2's strategies	
		s_1^2	s_2^2
Player 1's strategies	s_1^1	4, 4	1, 1
	s_2^1	0, 0	5, 5

2. Find the equilibrium, if any, for the games whose payoff matrices are presented in Tables 7.2.15(A) and (B) using iterated elimination of strictly dominated strategies.

Table 7.2.15(A)

		Player 2's strategies		
		s_1^2	s_2^2	s_3^2
	s_1^1	10, 9	9, 8	2, 2
Player 1's strategies	s_2^1	5, 7	4, 4	1, 4
	s_3^1	6, 8	7, 7	3, 5

Table 7.2.15(B)

		Player 2's strategies		
		s_1^2	s_2^2	s_3^2
	s_1^1	9, 10	9, 8	2, 2
Player 1's strategies	s_2^1	5, 7	4, 4	1, 4
	s_3^1	8, 6	7, 7	5, 3

3. *Application exercise.* Assume that two of the world's leading word processor suppliers, Microsoft and Corel, plan to decide the prices of their word processors (Microsoft's Word and Corel's WordPerfect) for the next financial year. They both have three strategies: decrease the price, keep the price constant, or increase the price. They are assumed to decide the strategies simultaneously. Their strategies and the associated payoffs (in billions of dollars) are presented in Table 7.2.16(A). Determine the equilibrium, if any, in the game.

4. *Application exercise.* Assume that two of the leading mobile phone service providers in Europe, Vodafone and Orange, plan to chalk out strategies to attract customers. They both have the same strategies: either charge lower call rates or increase talk-time. Assume that their payoffs (in billions of dollars) from these strategies are as given in Table 7.2.16(B). Determine the equilibrium, if any, in this game.

Table 7.2.16(A)

		Corel's strategies		
		Decrease	*Constant*	*Increase*
Microsoft's strategies	Decrease	3, 2	3, 1	3, 0
	Constant	1, 2	2, 1	3, 0
	Increase	2, 3	1, 1	1, 0

Table 7.2.16(B)

		Vodafone's strategies	
		Low call rate	*Increased talk-time*
Orange's strategies	Low call rate	−1, −1	1, 2
	Increased talk-time	2, 1	0, 0

5. *Application exercise.* Suppose that two of the world's leading personal computer manufacturers, Dell and Compaq, need to decide the quality of the new models they are planning to introduce. Also suppose that the available quality levels are high quality, medium quality, and low quality. The payoff (in millions of dollars) matrix of the two companies is given in Table 7.2.17. Determine the equilibrium, if any, in this game.

Table 7.2.17

		Dell's strategies		
		High quality	*Medium quality*	*Low quality*
Compaq's strategies	High quality	50, 50	40, 75	20, 100
	Medium quality	75, 40	40, 40	30, 80
	Low quality	100, 20	80, 30	100, 100

7.3 Static games of complete and perfect information: Nash equilibrium

7.3.1 Introduction

In Section 7.2 we presented the preliminaries of game theory. We also presented, in Sections 7.2.7–7.2.9, the meaning of strictly dominant and dominated strategies and found how one could determine the equilibrium in a game, if it existed, using the method of the iterated elimination of strictly dominated strategies. Although this method is based on the idea that a rational player will not play strictly dominated strategies, the method is based on two restrictive assumptions.

Notice that the whole idea of iterated elimination of strictly dominated strategies is based on the assumption of *rationality* of the players. Rationality in the present context means that each player makes use of all the available information to choose among strategies so as to maximize the player's payoffs. This assumption is required in each iterative step in the elimination of strictly dominated strategies. Moreover, we need to assume that players know about each other's rationality, or that rationality is a *common knowledge*.

Another assumption that is required for the application of the method of iterated elimination of strictly dominated strategies is that the game has strictly dominated strategies. In some games, all the strategies may survive iterated elimination; that is, there may not always be strictly dominated strategies in some games. In such games it is difficult to find the equilibrium of the game through the application of the method of elimination of strictly dominated strategies.

Therefore, we require a concept that does not rely on the aforesaid assumptions. This is given by *Nash equilibrium*, one of the most widely used concepts in game theory, due to John Nash. We will see shortly that optimal strategies in a Nash equilibrium always survive iterated elimination of strictly dominated strategies. Since the concept of Nash equilibrium is vital in game theory, we devote this section to its definition, discussion, and applications.

7.3.2 Pure-strategy Nash equilibrium: definition

Let us now define a pure-strategy Nash equilibrium in a normal-form game $G = \{S^i, \pi^i\} = \{S^A, S^B, \ldots, S^n; \pi^A, \pi^B, \ldots, \pi^n\}$. The set of strategies $(s^{A^*}, s^{B^*}, \ldots, s^{n^*})$, in a normal-form, pure-strategy game $G = \{S^i, \pi^i\} = \{S^A, S^B, \ldots, S^n; \pi^A, \pi^B, \ldots, \pi^n\}$, is a Nash equilibrium if the payoff for player i from playing s^{i^*} is higher than or equal to the payoff for player i from playing s^i, where $s^{i^*}, s^i \in S^i$, and all other $n-1$ players playing $(s^{A=1^*}, s^{B=2^*}, \ldots, s^{(i-1)^*}, s^{(i+1)^*}, \ldots, s^{n^*})$, or if

$$\pi^i(s^{A=1^*}, s^{B=2^*}, \ldots, s^{(i-1)^*}, s^{i^*}, s^{(i+1)^*}, \ldots, s^{n^*})$$
$$\geq \pi^i(s^{A=1^*}, s^{B=2^*}, \ldots, s^{(i-1)^*}, s^i, s^{(i+1)^*}, \ldots, s^{n^*}) \tag{7.3.1}$$

The definition of the pure-strategy Nash equilibrium given in inequality (7.3.1) implies that each player's strategy is that player's *best response* to the strategies chosen by the other players in the game. Notice that in a Nash equilibrium no player has the incentive to switch strategies because s^{i^*} is as good as or better than any other strategy for player i. This implies that a Nash equilibrium is a stable equilibrium.

The reader would have noticed two important features of the definition of pure-strategy Nash equilibrium given in inequality (7.3.1). The first relates to the weak inequality ($\geq$) in (7.3.1). It is possible that there may exist other strategies for the ith player that are at least as good as the strategy s^{i^*}. But, Nash equilibrium only requires that the ith player does not gain anything additional from switching strategies. The second feature is that more than one set of strategies may satisfy the condition for Nash equilibrium; that is, there may be more than one Nash equilibrium in a single game.

7.3.3 Pure-strategy Nash equilibrium: alternative representation

In the last section we defined a pure-strategy Nash equilibrium assuming that the strategy sets are discontinuous or discrete. One can alternatively define a pure-strategy Nash equilibrium that is especially useful when the strategy sets are continuous. This alternative representation is based on the concept of the *best-response functions*. One can find the *i*th player's best response for each possible strategy chosen by the rivals of the *i*th player. This can be found by solving the problem

$$\text{Maximize } \pi^i(s^{i^*}, s^{A=1^*}, s^{B=2^*}, \ldots, s^{(i-1)^*}, s^{(i+1)^*}, \ldots, s^{n^*}) \qquad (7.3.2)$$

Assume now that the payoff in problem (7.3.2) is continuous and differentiable. We can then solve this problem using the classical approach to unconstrained optimization presented in Chapter 4. From the FOC of this approach, we can obtain the *i*th player's best response (s^{i^*}) as a function of $(s^{A=1^*}, s^{B=2^*}, \ldots, s^{(i-1)^*}, s^{(i+1)^*}, \ldots, s^{n^*})$. Let us denote this function by $s^{i^*} = f^i(s^{A=1^*}, s^{B=2^*}, \ldots, s^{(i-1)^*}, s^{(i+1)^*}, \ldots, s^{n^*})$. Similarly, we can find the best-response functions of the other players in the game. Notice that for a Nash equilibrium to exist, the players' strategies must be best responses to one another. In other words, the Nash equilibrium strategies $(s^{A=1^*}, s^{B=2^*}, \ldots, s^{(i-1)^*}, s^{i^*}, s^{(i+1)^*}, \ldots, s^{n^*})$ will be solution to the system of simultaneous equations

$$s^{A=1^*} = f^1(s^{B=2^*}, \ldots, s^{(i-1)^*}, s^{i^*}, s^{(i+1)^*}, \ldots, s^{n^*}),$$

$$s^{B=2^*} = f^2(s^{A=1^*}, \ldots, s^{(i-1)^*}, s^{i^*}, s^{(i+1)^*}, \ldots, s^{n^*}), \ldots,$$

$$s^{(i-1)^*} = f^{(i-1)}(s^{A=1^*}, s^{B=2^*}, \ldots, s^{i^*}, s^{(i+1)^*}, \ldots, s^{n^*}),$$

$$s^{i^*} = f^i(s^{A=1^*}, s^{B=2^*}, \ldots, s^{(i-1)^*}, s^{(i+1)^*}, \ldots, s^{n^*}),$$

$$s^{(i+1)^*} = f^{(i+1)}(s^{A=1^*}, s^{B=2^*}, \ldots, s^{(n-1)^*}, s^{i^*}, \ldots, s^{n^*}), \ldots,$$

$$\text{and } s^{n^*} = f^n(s^{A=1^*}, s^{B=2^*}, \ldots, s^{(i-1)^*}, s^{i^*}, s^{(i+1)^*}, \ldots, s^{(n-1)^*}) \qquad (7.3.3)$$

Suppose that our game is a normal-form game with two players and two strategies. Also suppose that player A's best response to s^{B^*} by player B is s^{A^*} and that player B's best response to s^{A^*} by player A is s^{B^*}. Then player A's and player B's best-response functions, following the general case presented in system (7.3.3), can be written as

$$s^{A^*} = f^A(s^{B^*}) \quad \text{and} \quad s^{B^*} = f^B(s^{A^*}) \qquad (7.3.4)$$

which were the ideas we used in the name of reaction functions in the solutions to application examples 4 through 6 in Section 4.2.8.

7.3.4 Relationship between dominant and Nash equilibrium strategies

We learned from our earlier discussions that if, in a normal-form game, there exists a strictly dominant strategy for a player, then all other strategies for the player must be strictly dominated strategies. We then used the method of iterated elimination of strictly dominated strategies to find the equilibrium of the game.

Notice that a Nash equilibrium is based on two assumptions. The first assumption relates to the rationality of players as in the case of iterated elimination. Each player is assumed to act rationally. The second assumption is that each player's beliefs or conjectures about the strategies that the other player(s) will choose are correct. Therefore, in a Nash equilibrium the beliefs of the two (or all) players are mutually consistent.

Let us now explore the distinction between the dominant strategy equilibrium, the equilibrium obtained through iterated elimination of strictly dominated strategies, and Nash equilibrium in a normal-form game with two players and two strategies. An equilibrium in dominant strategies implies that player 1 is doing the best it can *no matter what player 2 is doing* and player 2 is doing the best it can *no matter what player 1 is doing*. But, in a Nash equilibrium, player 1 is doing the best it can *given what player 2 is doing* and player 2 is doing the best it can *given what player 1 is doing*.

Let us now see the relationships between strictly dominant strategies and Nash equilibrium strategies. Suppose that we have an *n*-person, *n*-strategy normal-form game. Suppose also that iterated elimination of strictly dominated strategies in this game yields an equilibrium. Then, one can show that this equilibrium obtained through iterated elimination of strictly dominated strategies is also a unique Nash equilibrium in the game. This means that dominant strategy equilibria are always Nash equilibria. But, all Nash equilibria are not dominant strategy equilibria. It can also be shown that Nash equilibrium strategies will always survive iterated elimination of strictly dominated strategies.

7.3.5 Pure-strategy Nash equilibrium: examples

Example 1. Let us now apply the definition of Nash equilibrium presented in inequality (7.3.1) to some concrete cases. For this, consider first the Prisoner's Dilemma example we exposed in Section 7.2.5 the payoff matrix of which is presented in Table 7.2.4. Find the Nash equilibrium, if any, in this game.

Solution. We know that a Nash equilibrium is in which prisoner 1 is doing the best it can *given what prisoner 2 is doing* and prisoner 2 is doing the best it can *given what prisoner 1 is doing*. First of all, let us attempt to find prisoner 1's best responses in the game. For this we need to find the best row response, the best response of prisoner 1, to each column choices by prisoner 2. These best responses of prisoner 1 are shown in boldface in the payoff matrix in Table 7.3.1(A). It can be seen from this table that if prisoner 2 chooses "Don't Confess," then the best response of prisoner 1 is to choose "Confess" because it maximizes prisoner 1's payoff (-1), shown in boldface in the table. Similarly, if prisoner 2 chooses "Confess," then prisoner 1's best response is to choose "Confess," again because it maximizes prisoner 1's payoff (-4), shown in bold type in the table.

Let us now see the best responses of prisoner 2 given the strategies chosen by prisoner 1. For this we need to find the best column response, the best response of prisoner 2, to each row choices by prisoner 1. These best responses of prisoner 2 are shown in boldface in the payoff matrix in Table 7.3.1(B). This shows that if prisoner 1 chooses "Don't Confess," then the best response of prisoner 2 is to choose "Confess" because it maximizes prisoner 2's payoff (-1), shown in boldface in the table. Similarly, if prisoner 1 chooses "Confess," then

Table 7.3.1(A)

		Prisoner 2's strategies	
		Don't confess	Confess
Prisoner 1's	Don't confess	−2, −2	−8, −1
strategies	Confess	−1, −8	**−4**, −4

Table 7.3.1(B)

		Prisoner 2's strategies	
		Don't confess	Confess
Prisoner 1's	Don't confess	−2, −2	−8, **−1**
strategies	Confess	−**1**, −8	−**4**, −**4**

Table 7.3.2(A)

		Prisoner 2's strategies	
		Don't confess	Confess
Prisoner 1's	Don't confess	−2, −2	−8, **−1**
strategies	Confess	−**1**, −8	−**4**, −**4**

Table 7.3.2(B)

		Woman's strategies	
		Wrestling	Opera
Man's	Wrestling	**2, 1**	0, 0
strategies	Opera	0, 0	**1, 2**

prisoner 2's best response is to choose "Confess" because it maximizes prisoner 2's payoff (−4), shown in boldface in the table. We can now present the best responses of both prisoners in a single payoff matrix as given in Table 7.3.2(A).

Notice that the only set of strategies (strategy profile) that is mutually consistent is the set comprising strategies "Confess, Confess." This was exactly the result we obtained in Section 7.2.8 when we applied the method of iterated elimination of strictly dominated strategies. This result corroborates our statement in the last section that if iterated elimination of strictly dominated strategies yields an equilibrium, that equilibrium must be a Nash equilibrium.

Example 2.   Consider the example of the Battle of the Sexes we exposed in Section 7.2.6. The payoff matrix of this game is presented in Table 7.2.5. Find the Nash equilibrium, if any, in this game.

Solution.   Notice that there is no strictly dominated strategy for either the man or the woman in the payoff matrix of this game. We can now attempt to see if there is any Nash equilibrium in this game. For this we apply the definition of Nash equilibrium given in inequality (7.3.1).

First of all, let us attempt to find the man's best responses in the game. The payoff matrix is given in Table 7.2.5. For this we need to find the best row response, the best response of the man, to each column choices by the woman. These best responses of the man are shown in boldface in the payoff matrix in Table 7.3.2(B). Le us now see the best responses of the woman given the strategies chosen by the man. For this we need to find the best column response, the best response of the woman, to each row choices by the man. These best-responses of the woman also are shown in boldface in the payoff matrix in Table 7.3.2(B). As this table shows, there are two mutually consistent sets of strategies or strategy profile in this game: "Wrestling, Wrestling" and "Opera, Opera." Therefore, there are two Nash equilibria in the game of the Battle of the Sexes.

Example 3. Consider the second example of the two-person, three-strategy game we discussed in Section 7.2.8. The payoff matrix of this game is presented in Table 7.2.7. Find the Nash equilibrium, if any, in this game.

Solution. To find if there is any Nash equilibrium in this game, we apply the definition of Nash equilibrium given in inequality (7.3.1). First of all, let us attempt to find player 1's best responses in the game. For this, as earlier, we need to find the best row response, the best response of player 1, to each column choices by player 2. These are shown in boldface in the pay off matrix in Table 7.3.3(A). Let us now see the best responses of player 2 given the strategies chosen by player 1. For this we need to find the best column response, the best response of player 2, to each row choices by player 1. These are also shown in boldface in Table 7.3.3(A). As the payoff matrix in Table 7.3.3(A) shows, there is only one mutually consistent set of strategies or strategy profile in this game: "Middle, Left." Therefore, there is only one Nash equilibrium in this game. Notice that we identified this set of strategies as the set of dominant strategies in the solution to the example at the end of Section 7.2.8.

Example 4. Consider the example of the Matching Pennies game we discussed in Section 7.2.4. The payoff matrix of this game is presented in Table 7.2.3. Find the Nash equilibrium, if any, of this game.

Table 7.3.3(A)

		Player 2's strategies		
		Left	*Center*	*Right*
Player 1's strategies	Top	**4, 5**	**5**, 3	3, 0
	Middle	**5, 4**	4, 3	2, 0
	Bottom	3, 2	2, **3**	**5**, 1

Table 7.3.3(B)

		Player 2's strategies	
		Heads	*Tails*
Player 1's strategies	Heads	**10**, −10	−10, **10**
	Tails	−10, **10**	**10**, −10

Solution.   As the payoff matrix in Table 7.2.3 shows, no player has a strictly dominant or dominated strategy in this game. If we apply the definition of Nash equilibrium given in inequality (7.3.1), we find that the best response for player 1 is to choose "Heads" and "Tails" if player 2 chooses "Heads" and "Tails," respectively. Similarly, the best response for player 2 is to choose "Tails" and "Heads" if player 1 chooses "Heads" and "Tails," respectively. We represent these best responses of both players in bold type in the payoff matrix given in Table 7.3.3(B).

As can be seen from Table 7.3.3(B), no set of strategies or strategy profile is mutually consistent. Therefore, the game of Matching Pennies does not have a Nash equilibrium in pure strategies. One might wonder why there is no Nash equilibrium in this game. The reason is that one player will always try to outguess the other player at every possible set of strategies. What this implies is that one player will always try to switch strategy at every possible set of strategies. Therefore, one cannot predict a stable, consistent set of strategies in this game.

7.3.6 *Best-response functions and Nash equilibrium*

In Section 7.3.3 we observed that the best-response functions of players in a normal-form game can be used to find the Nash equilibrium solution(s) of the game. In this section we attempt to see how one does this in concrete cases.

Suppose that we have a simple normal-form game with two players (A and B) and one strategy for each player (s^A and s^B). Also suppose that the payoff functions of these players are $\pi^A = \pi^A(s^A, s^B) = 1000s^A - (s^A)^2 - s^A s^B$ and $\pi^B = \pi^B(s^A, s^B) = 1000s^B - (s^B)^2 - s^A s^B$. Differentiating the first equation with respect to s^A and the second equation with respect to s^B, setting the results to zero, and solving for s^A and s^B yield $s^A = f(s^B) = 500 - 0.5s^B$ and $s^B = f(s^A) = 500 - 0.5s^A$, respectively. Notice that $s^A = f(s^B) = 500 - 0.5s^B$ is the best-response function of player A and $s^B = f(s^A) = 500 - 0.5s^A$ is the best-response function of player B. As we stated at the end of Section 7.3.3, the solution of these two SLSEs gives us the Nash equilibrium in the present example. Therefore, the Nash equilibrium solutions are $s^A = 333.333$ and $s^B = 333.333$. We can also find this solution if we graph the two best-response functions in a two-dimensional figure.

Notice an interesting relationship between the above game and its solution and the solution to the *Cournot duopoly* problem in example 4 of Section 4.2.8. If we treat $s^A = q_A$, $s^B = q_B$, $\pi^A = \Pi_A$, and $\pi^B = \Pi_B$, then the above game theory problem is identical with the Cournot duopoly problem. This is an example of the direct application of game theory to the subjects of our interest. Notice that the solutions we obtained for the other two duopoly problems in examples 5 (*Stackelberg duopoly* problem) and 6 (*Bertrand duopoly* problem) in Section 4.2.8 can be recast in terms of game theory.

7.3.7 *Mixed strategies*

So far in this chapter we have been concerned exclusively with games of pure strategies: games that involve specific choices or games in which players do not randomly mix pure strategies. In the present section we explore the meaning of *mixed strategy*. The definition of *mixed-strategy Nash equilibrium* is presented in Section 7.3.8. It is assumed that the reader is familiar with the basics of probability for the purposes of this discussion.

Some games, like the Matching Pennies game, may not have a Nash equilibrium in pure strategies. Moreover, in many of the games in the subjects of our interest players randomize their choices of strategies by choosing the probabilities with which they play their strategies. We stated earlier that a mixed strategy is a strategy in which a player makes a random choice (that is, a choice based on probabilities) among possible actions or decisions. The notion of mixed strategy is constructed on the assumption that in a game without strictly dominant or dominated strategies a player has no way of knowing which strategy his opponent(s) will choose. This uncertainty about the opponent's action forces the player to randomize strategies such that the player will be indifferent to the strategies chosen by the opponent.

Let us first define a mixed strategy in an N-person game in M pure strategies. Suppose that player i has $S^i = \{s_1^i, s_2^i, \ldots, s_M^i\}$ pure strategies in a normal-form game $G = \{S^A, S^B, \ldots, S^i, \ldots, S^N; \pi^A, \pi^B, \ldots, \pi^i, \ldots, \pi^N\}$. Then a mixed strategy for player i is a probability distribution $p^i = (p_1^i, p_2^i, \ldots, p_M^i)$, where $0 \leq p_m^i \leq 1$ and $\sum_{m=1}^M p_m^i = p_1^i + p_2^i + \cdots + p_M^i = 1$. In the probability distribution $p^i = (p_1^i, p_2^i, \ldots, p_M^i)$, p_m^i stands for the probability with which the ith player plays the mth pure strategy.

7.3.8 Mixed-strategy Nash equilibrium

We know that in a Nash equilibrium every player of the game has the best response (payoff) given the decision(s) of his opponent(s). But, when players randomize their strategies, what a particular player attempts to optimize, given the choice of his opponent(s), is not the actual associated payoff but the *expected value* of all the player's payoffs in the game. Therefore, we need to calculate first the *expected value* of the payoffs, or the *expected payoffs*, in a game before we define mixed-strategy Nash equilibrium of the game.

Suppose that player i has $S^i = \{s_1^i, s_2^i, \ldots, s_M^i\}$ pure strategies in a normal-form game $G = \{S^A, S^B, \ldots, S^i, \ldots, S^N; \pi^A, \pi^B, \ldots, \pi^i, \ldots, \pi^N\}$. Then a mixed strategy for player i is a probability distribution $p^i = (p_1^i, p_2^i, \ldots, p_M^i)$, where $0 \leq p_m^i \leq 1$ and $\sum_{m=1}^M p_m^i = p_1^i + p_2^i + \cdots + p_M^i = 1$. Then the expected value of the payoffs, or simply the expected payoff, to the ith player can be written as

$$E(\pi^i) = p_1^i p_1^j \pi_{11}^i + p_1^i p_2^j \pi_{12}^i + \cdots + p_1^i p_M^j \pi_{1M}^i + p_2^i p_1^j \pi_{21}^i + p_2^i p_2^j \pi_{22}^i$$

$$+ \cdots + p_2^i p_M^j \pi_{2M}^i + \cdots + p_M^i p_1^j \pi_{M1}^i + p_M^i p_2^j \pi_{M2}^i + \cdots + p_M^i p_M^j \pi_{MM}^i \quad (7.3.5)$$

where p_m^i ($m = 1, 2, \ldots, M$) is the probability with which the ith player, $i \neq j$, chooses the mth strategy. Notice that we can condense equation (7.3.5) as

$$E(\pi^i) = \sum_{m=1}^M p_m^i p_m^j \pi_{mm}^i \quad (7.3.6)$$

We are now ready to define a mixed-strategy Nash equilibrium. Suppose that we have four probability vectors $\mathbf{p_m^{i*}}$, $\mathbf{p_m^{j*}}$, $\mathbf{p_m^{i\bullet}}$, and $\mathbf{p_m^{j\bullet}}$. Then the probabilities in vectors $\mathbf{p_m^{i*}}$ and $\mathbf{p_m^{j*}}$ constitute a Nash equilibrium if, for all other probabilities in the vectors $\mathbf{p_m^{i\bullet}}$ and $\mathbf{p_m^{j\bullet}}$, the following inequalities are satisfied:

$$\sum_{m=1}^M p_m^{i*} p_m^{j*} \pi_{mm}^i \geq \sum_{m=1}^M p_m^{i\bullet} p_m^{j*} \pi_{mm}^i \quad \text{and} \quad \sum_{m=1}^M p_m^{i*} p_m^{j*} \pi_{mm}^j \geq \sum_{m=1}^M p_m^{i*} p_m^{j\bullet} \pi_{mm}^j \quad (7.3.7)$$

where $p_m^{i*} \in \mathbf{p_m^{i*}}$ and $p_m^{j*} \in \mathbf{p_m^{j*}}$, $p_m^{i\bullet} \in \mathbf{p_m^{i\bullet}}$ and $p_m^{j\bullet} \in \mathbf{p_m^{j\bullet}}$. The inequalities in (7.3.7) can also be written alternatively using equation (7.3.6). Suppose that the expected payoffs of the *i*th and the *j*th players when players *i* and *j* choose probabilities in vectors $\mathbf{p_m^{i^*}}$ and $\mathbf{p_m^{j^*}}$, are given, respectively, by $E(\pi^{i^*}) = \sum_{m=1}^{M} p_m^{i^*} p_m^{j^*} \pi_{mm}^{i}$ and $E(\pi^{j^*}) = \sum_{m=1}^{M} p_m^{i^*} p_m^{j^*} \pi_{mm}^{j}$. Also suppose that the expected payoffs of the *i*th and the *j*th players, when players *i* and *j*, respectively, choose probabilities in vectors $\mathbf{p_m^{i\bullet}}$ and $\mathbf{p_m^{j\bullet}}$, are given by $E(\pi^{i\bullet}) = \sum_{m=1}^{M} p_m^{i\bullet} p_m^{j^*} \pi_{mm}^{i}$ and $E(\pi^{j\bullet}) = \sum_{m=1}^{M} p_m^{i^*} p_m^{j\bullet} \pi_{mm}^{j}$. Then, the probabilities in vectors $\mathbf{p_m^{i^*}}$ and $\mathbf{p_m^{j^*}}$ constitute a Nash equilibrium if, for all other probabilities in the vectors $\mathbf{p_m^{i\bullet}}$ and $\mathbf{p_m^{j\bullet}}$, the inequalities in (7.3.8) are satisfied:

$$E(\pi^{i^*}) \geq E(\pi^{i\bullet}) \quad \text{and} \quad E(\pi^{j^*}) \geq E(\pi^{j\bullet}) \tag{7.3.8}$$

Inequalities (7.3.7) or (7.3.8) imply that the probabilities in vectors $\mathbf{p_m^{i^*}}$ and $\mathbf{p_m^{j^*}}$ constitute a Nash equilibrium if no other probabilities (for example, those in vectors $\mathbf{p_m^{i\bullet}}$ and $\mathbf{p_m^{j\bullet}}$) yield a higher payoff for either player (*i* and *j*).

7.3.9 Computation of mixed-strategy Nash equilibrium

Let us now consider the question of how one can find the mixed-strategy Nash equilibrium in practice. We consider one of the simplest possible games: a normal-form game with two players (A and B) and two pure strategies which is a special case of the mixed-strategy Nash equilibrium with N players and M pure strategies given in inequality (7.3.7). In this case, the expected payoffs, similar to equation (7.3.5), of player A and player B can be written, respectively, as

$$\left. \begin{aligned} E(\pi^A) &= p_1^A p_1^B \pi_{11}^A + p_1^A p_2^B \pi_{12}^A + p_2^A p_1^B \pi_{21}^A + p_2^A p_2^B \pi_{22}^A \\ \text{and} & \\ E(\pi^B) &= p_1^A p_1^B \pi_{11}^B + p_1^A p_2^B \pi_{12}^B + p_2^A p_1^B \pi_{21}^B + p_2^A p_2^B \pi_{22}^B \end{aligned} \right\} \tag{7.3.9}$$

We know that $p_1^A + p_2^A = 1$ and $p_1^B + p_2^B = 1$, which imply that $p_2^A = 1 - p_1^A$ and $p_2^B = 1 - p_2^B$, respectively. If we define $p_1^A = p$ and $p_1^B = q$, we obtain $p_2^A = 1 - p$ and $p_2^B = 1 - q$. Using these notations, we can rewrite equation (7.3.9) as

$$\left. \begin{aligned} E(\pi^A) &= pq\pi_{11}^A + p(1-q)\pi_{12}^A + (1-p)q\pi_{21}^A + (1-p)(1-q)\pi_{22}^A \\ \text{and} & \\ E(\pi^B) &= pq\pi_{11}^B + p(1-q)\pi_{12}^B + (1-p)q\pi_{21}^B + (1-p)(1-q)\pi_{22}^B \end{aligned} \right\} \tag{7.3.10}$$

Notice an important feature of the two equations in (7.3.10). The expected payoff of player A ($E(\pi^A)$) is a linear function of p for given q and the expected payoff of player B ($E(\pi^B)$) is a linear function of q for given p. Also notice that player A's problem is to choose the probabilities (p and $(1-p)$) with which to play the strategies such that the expected payoff ($E(\pi^A)$ in (7.3.10)) is maximum. Similarly, player B's problem is to choose the probabilities (q and $(1-q)$) with which to play the strategies such that the expected payoff ($E(\pi^B)$ in (7.3.10)) is maximum. Three of the generally adopted methods to solve this problem are the application of the classical optimization approach (presented in Chapter 4), the NLP approach (presented in Chapter 6), and the geometric or graphical method (presented in Section 7.6.4).

We first apply the classical approach to this optimization problem. Let us now partially differentiate the first equation with respect to p and the second equation with respect to q. These partial derivatives will be

$$\left.\begin{aligned}
\partial E(\pi^A)/\partial p &= q\pi_{11}^A + (1-q)\pi_{12}^A - q\pi_{21}^A - (1-q)\pi_{22}^A \\
\text{and} \\
\partial E(\pi^B)/\partial q &= p\pi_{11}^B - p\pi_{12}^B + (1-p)\pi_{21}^B - (1-p)\pi_{22}^B
\end{aligned}\right\} \tag{7.3.11}$$

Notice that $\partial E(\pi^A)/\partial p$ (or $\partial E(\pi^B)/\partial q$) in (7.3.11) is not a function of p (or q), but of q (or p). Also notice that $\partial E(\pi^A)/\partial p$ (or $\partial E(\pi^B)/\partial q$) may be positive, negative, or zero. If $\partial E(\pi^A)/\partial p$ (or $\partial E(\pi^B)/\partial q$) is positive, player A (or B) will choose the maximum possible value of 1 for p (or q), which will imply that player A (or B) will choose the pure strategy s_1^A (or s_1^B). If $\partial E(\pi^A)/\partial p$ (or $\partial E(\pi^B)/\partial q$) is negative, player A (or B) will choose the minimum possible value of 0 for p (or q), which will imply that player A (or B) will choose the pure strategy s_2^A (or s_2^B). But, when $\partial E(\pi^A)/\partial p$ (or $\partial E(\pi^B)/\partial q$) is zero, $E(\pi^A)$ (or $E(\pi^B)$) is the same for all $0 \le p$ (or q) ≤ 1. Since the payoff of player A (or B) is the same for any mixed strategy (that is, for all possible levels of p or q), player A (or B) is indifferent between any mixed-strategy choices. Therefore, one can state that an equilibrium for player A (or B) to play mixed strategies is valid only if $\partial E(\pi^A)/\partial p$ (or $\partial E(\pi^B)/\partial q$) is zero. In other words, a mixed-strategy Nash equilibrium will exist only if $\partial E(\pi^A)/\partial p = 0$ (or $\partial E(\pi^B)/\partial q = 0$). Notice that this is the FOC in the classical approach to optimization without constraints. Therefore, setting $\partial E(\pi^A)/\partial p$ (and $\partial E(\pi^B)/\partial q$) in equation (7.3.11) to zero and simplifying yields

$$p^* = \frac{\pi_{22}^B - \pi_{21}^B}{\pi_{11}^B - \pi_{12}^B - \pi_{21}^B + \pi_{22}^B} \quad \text{and} \quad q^* = \frac{\pi_{22}^A - \pi_{12}^A}{\pi_{11}^A - \pi_{12}^A - \pi_{21}^A + \pi_{22}^A} \tag{7.3.12}$$

Equations (7.3.12) give us the Nash equilibrium in our simple two-player mixed-strategy game. But how does one make sure that the equilibrium probabilities in equation (7.3.12) are strictly nonnegative? We know that in a classical optimization problem there is no way for one to guarantee that the optimal values of the choice variables (probabilities in the present case) are always nonnegative. Although the above procedure of finding the mixed-strategy Nash equilibrium based on the classical method of optimization may yield nonnegative equilibrium probabilities in most mixed-strategy games, the inherent problem (the possibility of negative optimal probabilities) just noted forces us to use some more robust methods in those cases where we end up with negative optimal probabilities.

We know that one such method is given by the Kuhn–Tucker conditions in the NLP approach we explored in Chapter 6. But, to apply the Kuhn–Tucker conditions, we need to specify the objective function and the constraints. The objective function of player A allowing for mixed strategies, in a two-player, two-pure-strategy normal-form game, is to maximize the expected payoff; that is, to maximize $E(\pi^A) = pq\pi_{11}^A + p(1-q)\pi_{12}^A + (1-p)q\pi_{21}^A + (1-p)(1-q)\pi_{22}^A$ (the first equation in (7.3.10)). The constraints of player A are $p + (1-p) = 1$, $p \ge 0$, and $(1-p) \ge 0$. Therefore, player A's problem is to

Maximize $E(\pi^A) = pq\pi_{11}^A + p(1-q)\pi_{12}^A + (1-p)q\pi_{21}^A + (1-p)(1-q)\pi_{22}^A$,

   subject to $p + (1-p) = 1$, $p \ge 0$, and $(1-p) \ge 0$ $\hspace{2cm}$ (7.3.13)

Assume now that the objective function and the constraints in problem (7.3.13) are as those required for the fulfillment of the Kuhn–Tucker sufficiency theorem presented in Section 6.4.2. Then we can set up the Lagrangian function as

$$
L = pq\pi_{11}^A + p(1-q)\pi_{12}^A + (1-p)q\pi_{21}^A + (1-p)(1-q)\pi_{22}^A
$$
$$
+ \lambda_1[1 - p - (1-p)] + \lambda_2 p + \lambda_3(1-p) \tag{7.3.14}
$$

from which we can derive the Kuhn–Tucker maximum conditions. Solution of these conditions will yield the nonnegative optimal probability p^*.

Similarly, the objective function of player B allowing for mixed strategies, in a two-player, two-pure-strategy normal-form game, is to maximize the expected payoff; that is, to maximize $E(\pi^B) = pq\pi_{11}^B + p(1-q)\pi_{12}^B + (1-p)q\pi_{21}^B + (1-p)(1-q)\pi_{22}^B$ (the second equation in (7.3.10)). The constraints of player B are $q + (1-q) = 1$, $q \geq 0$, and $(1-q) \geq 0$. Therefore, player B's problem is to

$$
\text{Maximize } E(\pi^B) = pq\pi_{11}^B + p(1-q)\pi_{12}^B + (1-p)q\pi_{21}^B + (1-p)(1-q)\pi_{22}^B,
$$
$$
\text{subject to } q + (1-q) = 1, q \geq 0, \text{ and } (1-q) \geq 0 \tag{7.3.15}
$$

Assume now, as above, that the objective function and the constraints in problem (7.3.15) are as those required for the fulfillment of the Kuhn–Tucker sufficiency theorem discussed in Section 6.4.2. Then we can set up the Lagrangian function as

$$
L = pq\pi_{11}^B + p(1-q)\pi_{12}^B + (1-p)q\pi_{21}^B + (1-p)(1-q)\pi_{22}^B
$$
$$
+ \lambda_1[1 - q - (1-q)] + \lambda_2 q + \lambda_3(1-q) \tag{7.3.16}
$$

from which we can derive the Kuhn–Tucker maximum conditions. Solution of these conditions will yield the nonnegative optimal probability q^*.

We now state two important theorems. The first theorem is that *every finite-player, finite-mixed-strategy game has a Nash equilibrium*. The second theorem is that *every finite-player, finite-pure-strategy game has at least one Nash equilibrium in pure or mixed strategies*. The proofs of these theorems require the application of advanced topics such as *fixed-point theorem* and, therefore, we do not present them here.

7.3.10 Computation of mixed-strategy Nash equilibrium: examples

Example 1.   Consider the game of Matching Pennies introduced in Section 7.2.4. We found in example 4 in Section 7.3.5 that this game did not have a Nash equilibrium in pure strategies. Determine the Nash equilibrium of this game in mixed strategies.

Solution.   The payoff matrix of this is presented in Table 7.2.3. Based on this table and following equations (7.3.10), we can write the expected payoffs of players A and B, respectively, as

$$
E(\pi^A) = pq\pi_{11}^A + p(1-q)\pi_{12}^A + (1-p)q\pi_{21}^A + (1-p)(1-q)\pi_{22}^A
$$
$$
= pq10 - p(1-q)10 - (1-p)q10 + (1-p)(1-q)10 \tag{7.3.17}
$$

and

$$E(\pi^B) = pq\pi_{11}^B + p(1-q)\pi_{12}^B + (1-p)q\pi_{21}^B + (1-p)(1-q)\pi_{22}^B$$
$$= -pq10 + p(1-q)10 + (1-p)q10 - (1-p)(1-q)10 \qquad (7.3.18)$$

Differentiating equations (7.3.17) and (7.3.18) partially with respect to p and q, respectively, and setting the results to zero, we obtain equations (7.3.12). We can now plug the payoffs from Table 7.2.3 into equations (7.3.12) to obtain

$$\left.\begin{array}{l} p^* = \dfrac{\pi_{22}^B - \pi_{21}^B}{\pi_{11}^B - \pi_{12}^B - \pi_{21}^B + \pi_{22}^B} = \dfrac{-10-10}{-10-10-10+(-10)} = \dfrac{1}{2} \\[2mm] \text{and} \\[2mm] q^* = \dfrac{\pi_{22}^A - \pi_{12}^A}{\pi_{11}^A - \pi_{12}^A - \pi_{21}^A + \pi_{22}^A} = \dfrac{10-(-10)}{10-(-10)-(-10)+10} = \dfrac{1}{2} \end{array}\right\} \qquad (7.3.19)$$

which can also be obtained through direct simplification of the results after the said differentiation.

Let us now interpret the results in equations (7.3.19). Although there is no pure-strategy Nash equilibrium for the Matching Pennies game, it has a mixed-strategy Nash equilibrium with optimal probabilities $p^* = 1/2$ and $q^* = 1/2$. Each player is doing the best they can given the strategies of the other player. If we substitute the optimal probabilities into equations (7.3.17) and (7.3.18) we find that the expected payoff of each player is zero; that is, $E(\pi^{A^*}) = E(\pi^{B^*}) = 0$. Notice an important feature of the mixed-strategy Nash equilibrium given in equations (7.3.19): neither player has any incentive to play this equilibrium (mixed) strategy. What this means is that if player A chooses $p^* = 1/2$, any probability (including $q^* = 1/2$) will give the same expected payoff to player B. And, if player B chooses $q^* = 1/2$, any probability (including $p^* = 1/2$) will give the same expected payoff to player A. Therefore, there is no guarantee that a player will choose the mixed-strategy equilibrium given that the other player chooses it. This is an issue with most mixed-strategy Nash equilibria.

Example 2.   Consider the game of the Battle of the Sexes introduced in Section 7.2.6. We found in example 2 in Section 7.3.5 that this game has two Nash equilibria in pure strategies ("Wrestling, Wrestling" and "Opera, Opera"). Determine the Nash equilibrium of this game in mixed strategies.

Solution.   The payoff matrix of this game is presented in Table 7.2.5. Based on this table and following equations (7.3.10), we can write the expected payoff of the man and the woman, respectively, as

$$E(\pi^A) = pq\pi_{11}^A + p(1-q)\pi_{12}^A + (1-p)q\pi_{21}^A + (1-p)(1-q)\pi_{22}^A$$
$$= pq2 + p(1-q) \times 0 + (1-p)q \times 0 + (1-p)(1-q) \times 1 \qquad (7.3.20)$$

and

$$E(\pi^B) = pq\pi_{11}^B + p(1-q)\pi_{12}^B + (1-p)q\pi_{21}^B + (1-p)(1-q)\pi_{22}^B$$
$$= pq \times 1 + p(1-q) \times 0 + (1-p)q \times 0 + (1-p)(1-q) \times 2 \qquad (7.3.21)$$

where the superscripts A and B represent the man and the woman, respectively. Differentiating equation (7.3.20) partially with respect to p and equation (7.3.21) partially with respect to q and setting the results to zero, we obtain equations (7.3.12). We can now plug the payoffs from Table 7.2.5 into equations in (7.3.21) to obtain

$$\left. \begin{aligned} p^* &= \frac{\pi_{22}^B - \pi_{21}^B}{\pi_{11}^B - \pi_{12}^B - \pi_{21}^B + \pi_{22}^B} = \frac{2-0}{1-0-0+2} = \frac{2}{3} \\ \text{and} & \\ q^* &= \frac{\pi_{22}^A - \pi_{12}^A}{\pi_{11}^A - \pi_{12}^A - \pi_{21}^A + \pi_{22}^A} = \frac{1-0}{2-0-0+1} = \frac{1}{3} \end{aligned} \right\} \qquad (7.3.22)$$

which can also be obtained through direct simplification of the results after the said differentiation.

As earlier, we shall analyze the results in equations (7.3.22). The mixed-strategy Nash equilibrium in the present example is that the probability of the man going to watch wresting is $p^* = 2/3$ and that of the woman going to watch wrestling is $q^* = 1/3$. Therefore, their probabilities of going to the opera will be $1 - p^* = 1/3$ and $1 - q^* = 2/3$, respectively. Each player is doing the best they can, given the strategies of the other player. If we substitute the optimal probabilities into equations (7.3.20) and (7.3.21) we find that the expected payoff of each player is 2/3; that is, $E(\pi^{A^*}) = E(\pi^{B^*}) = 2/3$. Notice an important feature of the mixed-strategy Nash equilibrium given in equations (7.3.22): both the man and the woman would prefer their pure-strategy Nash equilibria to the mixed-strategy Nash equilibrium because the former gives better payoffs than does the latter. This is an example that in some games mixed strategies may yield undesirable solutions.

7.3.11 Maximin and minimax regret strategies

We stated in Section 7.3.1 that the dominant strategy equilibrium or the equilibrium attained through the iterated elimination of strictly dominated strategies is based on the *cyclic assumption* of the common knowledge of the rationality of each player in a normal-form game in pure strategies. Moreover, as we saw until now in its exposition and applications, the concept of Nash equilibrium also relies heavily on the assumption of players' rationality, though not cyclic in nature.

However, in the real world one or more players in a game may behave irrationally either by mistake or intentionally. This type of behavior on the part of some players may be disastrous to those players who behave with the false assumption that their rivals behave rationally. We can cite an example of a two-player, two-strategy normal-form game to drive home this point. Assume that the two players are the two fast-food giants McDonald's and Kentucky Fried Chicken (KFC) and that their strategies are price decisions. Also assume again that McDonald's is in a better position in terms of market reputations. Let their payoffs (in millions of dollars) be those presented in the payoff matrix in Table 7.3.4(A).

Table 7.3.4(A)

		KFC's strategies	
		Low price	*High price*
McDonald's	Low price	0, 0	5, −50
strategies	High price	10, 5	**10, 10**

Table 7.3.4(B)

		KFC's strategies		Row min.
		Low price	*High price*	
McDonald's strategies	Low price	0, 0	5, −50	0
	High price	10, 5	**10, 10**	⑩
	Column min.	⓪	−50	

Let us first analyze the payoff matrix in Table 7.3.4(A). The payoff matrix shows that "High Price" is the strictly dominant strategy for McDonald's. But, KFC does not have a strictly dominant strategy and its strategy depends on the strategy chosen by McDonald's. Since McDonald's strictly dominant strategy is "High Price," it will choose that strategy and, therefore, the best that KFC can do is to choose "High Price." Therefore, "High Price, High Price" (boldface in the table) is the only Nash equilibrium strategy profile in the game, which yields a payoff of $10 million to both firms.

However, a moment's thought could suggest some other outcome to the game. If KFC is absolutely certain about the assumed rationality of McDonald's, then the above outcome is certain to occur. But, if KFC suspects that McDonald's may choose "Low Price" instead (either by mistake or intentionally to teach KFC a lesson), then it would be disastrous for KFC to choose "High Price" because it would incur a loss of $50 million should McDonald's choose "Low Price." The worst payoffs that KFC can expect when it chooses "Low Price" and "High Price" are $0 and a loss of $50 million, respectively. These column minimum payoffs are given in the last row of Table 7.3.4(B). Similarly, the worst payoffs for McDonald's if it chooses "Low Price" and "High Price" are $0 and $10 million, respectively. These row minimum payoffs are given in the last column of Table 7.3.4(B). KFC will now attempt to maximize the minimum gains, or choose the best of the worst payoffs, that can be obtained from the choice of its strategies, which is the strategy with $0 when it chooses "Low Price" and shown in a circle in the last row. Such a strategy is called a *maximin strategy* or *secure strategy*. A maximin strategy is a strategy that maximizes the minimum gain that can be earned. Notice that McDonald's can also do the same, which is identical to its dominant strategy "High Price" shown in a circle in the last column. Therefore, the equilibrium of the game when both companies choose their maximin strategies is not the Nash equilibrium strategies ("High Price, High Price"), but the "High Price, Low Price" strategies; that is, McDonald's chooses "High Price" strategy and KFC chooses "Low Price" strategy (boldface) yielding $10 million and $5 million, respectively. The reader would have noticed that a maximin strategy is a way of *risk-averse approach* to a game.

Table 7.3.5(A)

		KFC's strategies		Row min.
		Low price	*High price*	
McDonald's	Low price	40, 40	20, 60	⑳
strategies	High price	60, 20	18, 18	18
	Column min.	⑳	18	

Table 7.3.5(B)

		KFC's strategies	
		Low price	*High price*
McDonald's	Low price	$60 - 40 = 20$	$20 - 20 = 0$
strategies	High price	$60 - 60 = 0$	$20 - 18 = 2$

Suppose now that we modify that previous game and the payoff matrix of this modified game is as presented in Table 7.3.5(A). Notice that neither company has a strictly dominant strategy in this game. But, there are two Nash equilibrium strategies, shown in boldface: "Low Price, High Price" and "High Price, Low Price." Notice also that the strategies "Low Price, Low Price," which are boxed, constitute maximin strategies for the companies. This means that, as in the last game, the maximin and Nash equilibrium strategies are different in the modified game.

However, the modified game above leads us to another possibility. Since the players of a game are assumed to be rational, every player of the game believes that the other players take rational decisions. If this does not happen or if some player behaves irrationally, then the player who chooses their best strategy believing that the other players would choose their best strategies may end up with lower payoff and regret later. In this event, every player of the game would attempt to minimize the *opportunity loss* they might incur. Opportunity loss is the difference between a player's best payoff and the lowest payoff when the opponent chooses a particular strategy. These opportunity losses of McDonald's are presented in Table 7.3.5(B). Therefore, in the present modified game, the opportunity losses for McDonald's in deciding "Low Price" and "High Price" when KFC chooses "Low Price" are $60 - 40 = 20$ and $60 - 60 = 0$, respectively. Similarly, the opportunity losses for McDonald's in deciding "Low Price" and "High Price" when KFC chooses "High Price" are $20 - 20 = 0$ and $20 - 18 = 2$, respectively. Since the aim of McDonald's is to choose the strategy that minimizes the possible positive opportunity loss, its strategy is "High Price" because it incurs the minimum of the maximum possible positive opportunity losses. Such a strategy is also called a *minimax regret strategy* or a *savage strategy*. Since the payoff matrix is symmetric, KFC's minimax regret strategy also is "High Price." Therefore, the equilibrium in the modified game, if both companies follow minimax regret strategies, is "High Price, High Price," shown in the dashed box in Table 7.3.5(B). The reader would have noticed that the minimax regret strategy is different from the minimax and Nash equilibrium strategies of the game.

7.3.12 *Application examples of pure- and mixed-strategy Nash equilibria and maximin-strategy equilibria*

Example 1. We know from microeconomics that a firm's objective in a *Cournot oligopoly* market is to choose the quantity of output that it produces, given the quantity of output chosen by its rivals, such that it maximizes its profit. Firms are assumed to produce identical products. Since the quantity of output produced by a firm must be nonnegative and is assumed to be a continuous variable, the set of strategies that the ith firm faces can be written as

$$S^i = \{q_i | q_i \geq 0\} \tag{7.3.23}$$

Assume that the inverse demand function that firms in the market face is linear and is given, with $b > 0$, by

$$P = f(q) = a - bq = a - b\sum_{i=1}^{n} q_i, \text{ where } q = \sum_{i=1}^{n} q_i \tag{7.3.24}$$

and that the average cost denoted by AC and marginal cost denoted by MC of every firm are equal and constant at c

$$AC_i = MC_i = c \tag{7.3.25}$$

which implies that the total cost (denoted by C) of the ith firm is given as

$$C_i = cq_i \tag{7.3.26}$$

Given equations (7.3.23)–(7.3.26), find the output of the ith firm that maximizes its profit; that is, find the *Cournot–Nash equilibrium* output for the ith firm. Also find the Cournot–Nash equilibrium market output, the price, and the payoff (profit) of each firm.

Solution. Given equations (7.3.23)–(7.3.26), we can write the ith firm's total revenue function as $R_i = Pq_i = [a - bq]q_i = \left[a - b\sum_{k=1}^{n-i} q_k - bq_i\right]q_i$, which can be written as $R_i = aq_i - bq_1^2 = aq_i - b\sum_{k=1}^{n-i} q_k q_i - bq_i^2$. Therefore, using equation (7.3.26), we can write the profit function of the ith firm as

$$\Pi_i = aq_i - b\sum_{k=1}^{n-i} q_k q_i - bq_i^2 - cq_i \tag{7.3.27}$$

Let us now partially differentiate equation (7.3.27) with respect to q_i and set the result to zero as the FOC for an optimum of the ith firm's profit:

$$\frac{\partial \Pi_i}{\partial q_i} = a - b\sum_{k=1}^{n-i} q_k - 2bq_i - c = 0 \tag{7.3.28}$$

Notice that the SOC for an optimum of the ith firm's profit is satisfied as the second partial derivative of equation (7.3.28) is negative (since $b > 0$, as assumed). Solving equation (7.3.28) for q_i yields

$$q_i^* = \frac{a-c}{2b} - \frac{1}{2}\sum_{k=1}^{n-i} q_k \tag{7.3.29}$$

The reader would have noticed the two important features of equation (7.3.29). If firm i is the only firm in the market (i.e. if firm i is a *monopoly*), then the last term in the equation will be zero. This implies that the optimum output of the monopoly will be $(a - c)/2b$. Moreover, since costs and demand functions are exactly equal for each firm, equation (7.3.29) also implies that each firm produces the same optimal level of output (i.e. $q_k^* = q_i^*$). This helps us rewrite equation (7.3.28), the FOC for an optimum of the ith firm's profit, as $\partial \Pi_k / \partial q_k = a - b \sum_{k=1}^{n-i} q_k^* - 2bq_k^* - c = 0$. Since firm i is one of the firms in the market, we can write the last equation as $\partial \Pi_k / \partial q_k = a - (n-1)bq_k^* - 2bq_k^* - c = 0$. Solving this equation for q^* yields

$$q_k^* = \frac{a - c}{b(n+1)} \qquad (7.3.30)$$

Therefore, the Cournot–Nash equilibrium output of each firm is $q_k^* = [a - c]/[b(n + 1)]$. Notice that the Cournot–Nash equilibrium market output will be the sum of the Cournot–Nash equilibrium output of each firm:

$$q^* = \sum_{k=1}^{n} q_k^* = nq_k^* = n\left[\frac{a - c}{b(n+1)}\right] = \left[\frac{n}{n+1}\right]\left[\frac{a - c}{b}\right] \qquad (7.3.31)$$

The inverse demand function is already given as $P = f(q) = a - bq = a - \sum_{i=1}^{n} q_i$. This equation becomes, in Cournot–Nash equilibrium, $P^* = f(q^*) = a - bq^* = a - \sum_{k=1}^{n} q_k^*$. Using equation (7.3.30), the last equation can be written as $P^* = a - q^* = a - nq_k^*$ or as

$$P^* = \frac{a + nc}{n+1} \qquad (7.3.32)$$

Lastly, the Cournot–Nash equilibrium payoff or profit of each firm can be found by substituting the cost function (equation (7.3.26)) and the Cournot–Nash equilibrium output of each firm (equation (7.3.30)) into the payoff function of each firm (equation (7.3.27)). Once we simplify the resulting expression we obtain the Cournot–Nash equilibrium payoff of each firm as

$$\Pi_i^* = \frac{(a - c)^2}{b(n+1)^2} \qquad (7.3.33)$$

Now suppose that there are only two firms ($n = 2$) in the oligopoly market; that is, the market is a duopoly market. We denote the output of these two firms by q_A and q_B. Then equations (7.3.30)–(7.3.32) imply that $q_A^* = q_B^* = (a - c)/3b$, $q^* = (2/3)[(a - c)/b]$, and $P^* = (a + 2c)/3$. Also suppose that the inverse demand function (equation (7.3.23)) is given as $P = f(q) = a - bq = a - \sum_{i=1}^{2} q_i = a - q_A - q_B$, where $a = 1100$, and the total cost functions (equation (7.3.26)) of the two firms are $C_A = cq_A = 100q_A$ and $C_B = cq_B = 100q_B$. We know from the inverse demand function that $b = 1$. Substituting the values $a = 1100$, $C_A = cq_A = 100q_A$, $C_B = cq_B = 100q_B$, $n = 2$, and $b = 1$ into equations (7.3.30)–(7.3.33) we obtain $q_A^* = q_B^* = (a - c)/3b = (1100 - 100)/3 \times 1 = 333.333$, $q^* = (2/3)[(a - c)/b] = (2/3)[(1100 - 100)/1] = (2/3) \times 1000 = 666.666$, $P^* = (a + 2c)/3 = (1100 + 2 \times 100)/3 = 1300/3 = 433.333$, and $\Pi_i^* = [(a - c)^2]/[b(n + 1)^2] = [(1100 - 100)^2]/(2 + 1)^2 = 111\,112$. The total industry profit can be found by multiplying the profit of an individual firm by the number of firms. Therefore, in the present case, the industry profit is $n\Pi_i^* = n[(a - c)^2/b(n + 1)^2] = 2 \times [(1100 - 100)^2/(2 + 1)^2] = 2 \times 111\,112 = 222\,224$. Notice that these are exactly the same results as those we obtained when we solved the problem in example 4 in Section 4.2.8.

Example 2.   We know from microeconomics that a firm's objective in a *Bertrand oligopoly market* with differentiated products is to choose the price of the product that the firm produces, given the price chosen by its rivals, such that it maximizes its profit. Since the price of a product must be nonnegative and is assumed be a continuous variable, the set of strategies that the firm faces can be written as

$$S^i = \{p_i | p_i \geq 0\} \tag{7.3.34}$$

Since the products are differentiated, the quantity demanded of the *i*th firm's output can be written as

$$q_i = a - p_i + bp_j \tag{7.3.35}$$

where $0 < b < 1$ and $i \neq j$. The average cost (AC) and the marginal cost (MC) of every firm are equal and constant, and are given by

$$AC_i = MC_i = c = AC_j = MC_j \tag{7.3.36}$$

which implies that the total cost (denoted by C) of the *i*th firm is given by

$$C_i = cq_i \tag{7.3.37}$$

Given equations (7.3.34)–(7.3.37), find the output (i.e. the *Bertrand–Nash equilibrium* output) of the *i*th firm that maximizes its profit. Also find the Bertrand–Nash equilibrium market price and payoff (profit) of each firm.

Solution.   Given equations (7.3.34)–(7.3.37), we can write the *i*th firm's total revenue function as $R_i = p_i q_i = ap_i - p_i^2 + bp_i p_j$. Therefore, using equation (7.3.37), we can write the profit function of the *i*th firm as $\Pi_i = R_i - C_i = ap_i - p_i^2 + bp_i p_j - cq_i$. Then using equation (7.3.35) we can rewrite profit function of the *i*th firm as

$$\Pi_i = R_i - C_i = ap_i - p_i^2 + bp_i p_j - c[a - p_i + bp_j] = ap_i - ac - p_i^2 + cp_i + bp_i p_j - cbp_j \tag{7.3.38}$$

Partially differentiating equation (7.3.38) with respect to p_i we obtain

$$\partial \Pi_i / \partial p_i = a - 2p_i + c + bp_j \tag{7.3.39}$$

and setting it to zero (the FOC for an optimum of the *i*th firm's profit) and solving for p_i, we obtain the *i*th firm's best-response (or reaction) function as

$$p_i = [a + c + bp_j]/2 \tag{7.3.40}$$

Since the cost and demand functions are similar for all other firms in the market or following the same procedure as above with the demand function $q_j = a - p_j + bp_i$, we can write the best-response function of the *j*th firm as

$$p_j = [a + c + bp_i]/2 \tag{7.3.41}$$

Solving equations (7.3.40) and (7.3.41) simultaneously gives us the Bertrand–Nash equilibrium prices chosen by every firm in the market as

$$p_i^* = [a+c]/[2-b] = p_j^* \qquad (7.3.42)$$

The Bertrand–Nash equilibrium output of the ith firm can be found by substituting equation (7.3.42) into equation (7.3.35):

$$q_i^* = a - p_i^* + bp_j^* = a - \frac{a+c}{2-b} + b\frac{a+c}{2-b} = a + (b-1)\frac{a+c}{2-b} = q_j^* \qquad (7.3.43)$$

Lastly, the Bertrand–Nash equilibrium profit of the ith firm can be found by substituting equation (7.3.42) into equation (7.3.38):

$$\Pi_i^* = ap_i^* - ac - p_i^{*2} + cp_i^* + bp_i^{*2} - cbp_i^* = \Pi_j^* \qquad (7.3.44)$$

As in the last example, suppose now that we have two firms in the market, i and j, with demand functions as in equation (7.3.35). Also suppose that $a = 30$, $b = 0.5$, and $c = 20$. Plugging these values in equations (7.3.40) and (7.3.41), we obtain the best-response functions of firms i and j as $p_i = 25 + 0.25p_j$ and $p_j = 25 + 0.25p_i$, respectively. Solving these two equations simultaneously we obtain the Bertrand–Nash equilibrium price (equation (7.3.42)): $p_i^* = (a+c)/(2-b) = (30+20)/(2-0.5) = 33.33 = p_j^*$. This implies that the Bertrand–Nash equilibrium output of the two firms, as per equation (7.3.43), will be $q_i^* = a + (b-1)[(a+c)/(2-b)] = 30 + (0.5-1)[(30+20)/(2-0.5)] = 13.33 = q_j^*$. The Bertrand–Nash equilibrium payoff of the firms can be found by substituting equation (7.3.42) into equation (7.3.44). This yields $\Pi_i^* = \Pi_j^* = 177.75$. Notice that these are exactly the same results as those we obtained when we solved the same problem in example 6 in Section 4.2.8.

Example 3. Suppose that the ith herder ($i = 1, 2, \ldots, n$) in a village plans to graze s_i number of sheep on a green with limited area, which is a common property of the village community. The total number of sheep in the village is $S = \sum_{i=1}^{n} s_i = s_1 + s_2 + \cdots + s_n$. The cost of buying and grazing a sheep is constant, c, and equal for all herders and the payoff (in the form of milk, wool, meat, etc.) that the ith herder obtains from grazing the s_i number of sheep on the green is $s_i V(S) = V(s_1 + s_2 + \cdots + s_{i-1} + s_i + s_{i+1} + \cdots + s_n)$. Since the green is limited in area, there is a maximum number of sheep (S_M) that can be grazed on the green. This implies that if $S < S_M$, $V(S) > 0$; and, if $S \geq S_M$, $V(S) = 0$, which means there are diminishing returns to grazing sheep on the green (that is, $V'(S) > 0$ and $V''(S) < 0$). The ith herder's problem is to decide the number of sheep to graze on the green (or determine the value of s_i, which is assumed to be continuous) such that it maximizes the herder's payoff, given the decisions of other herders. Find the Nash equilibrium payoff of each herder and compare this to the social optimum.

Solution. This problem is called the *problem of the commons*. The strategy space for each herder is any number of sheep in between zero and infinity. Notice that when the ith herder's strategy is to graze s_i number of sheep, the other $n - i$ herders' strategies are to graze

$s_1, s_2, \ldots, s_{i-1}, s_{i+1}, \ldots, s_n$ number of sheep. Then the ith herder's net payoff from the s_i number of sheep can be written as

$$s_i V(s_1 + s_2 + \cdots + s_{i-1} + s_i + s_{i+1} + \cdots + s_n) - c s_i \tag{7.3.45}$$

Notice that for $S^* = (s_i^* + s_2^* + \cdots + s_{i-1}^* + s_i^* + s_{i+1}^* + \cdots + s_n^*)$ to be a Nash equilibrium, then s_i^* (for every i) must maximize the ith herder's payoff function in equation (7.3.45), given that the other $n - i$ herders choose $(s_i^* + s_2^* + \cdots + s_{i-1}^* + s_{i+1}^* + \cdots + s_n^*)$. As the FOC for the maximum of the ith herder's payoff function, we partially differentiate equation (7.3.45) with respect to s_i and set the result to zero to obtain

$$V(S^*) + s_i^* V'(S^*) = c \tag{7.3.46}$$

Since the cost and payoff functions are identical for each herder, $S^* = (s_i^* + s_2^* + \cdots + s_{i-1}^* + s_i^* + s_{i+1}^* + \cdots + s_n^*) = n s_i^*$ or $s_i^* = S^*/n$. Substituting this result into equation (7.3.46), we obtain

$$V(S^*) + [S^*/n] V'(S^*) = c \tag{7.3.47}$$

which gives the Nash equilibrium number of sheep every herder plans to graze on the green.

But, one might wonder what the socially optimum number of sheep is that could be grazed on the grass. To answer this, we can set up the society's payoff function, following the above procedure, as $S^s V(S^s) - S^s c$, where the superscript 's' denotes that the variable represents the society. Then, the society's problem, for $0 \le S < \infty$, is to

$$\text{Maximize } S^s V(S^s) - S^s c \tag{7.3.48}$$

Then, partially differentiating equation (7.3.48) with respect to S^s and setting the result to zero, we obtain the FOC for a maximum of this payoff function as

$$V(S^{s^*}) + S^{s^*} V'(S^{s^*}) = c \tag{7.3.49}$$

Notice that equations (7.3.47) and (7.3.49) both equal the constant c. For this to happen we must have the inequality

$$S^* > S^{s^*} \tag{7.3.50}$$

This implies the sum of the Nash equilibrium number of sheep chosen by each herder exceeds the socially optimal number of sheep. This means that the green would be overutilized. This overutilization of a *public good* (the green in our present example) and the problems associated with it are popularly known as the *tragedy of the commons*.

Let us now consider a special case of the above problem of commons with $n = 2$. Therefore, we have $S = \sum_{i=1}^{2} s_i = s_1 + s_2$. Suppose that the payoff and cost per sheep to both herders when herder 1 chooses to graze s_1 number of sheep and herder 2 chooses to graze s_2 number of sheep are given by \$800 and $(s_1 + s_2)^2$, respectively. Therefore, the total payoffs (corresponding to equation (7.3.45)) to herder 1 and herder 2 (when herder 1 chooses to graze s_1 number of sheep and herder 2 chooses to graze s_2 number of sheep) are given by $s_1 V(s_1, s_2) = 800 s_1 - s_1(s_1 + s_2)^2$ and $s_2 V(s_1, s_2) = 800 s_2 - s_2(s_1 + s_2)^2$, respectively. Then the problem of herder 1 is to find s_1 such that the total payoff $s_1 V(s_1, s_2) = 800s_1 - s_1(s_1 + s_2)^2 = 800s_1 - s_1^3 - 2s_1^2 s_2 - s_1 s_2^2$ is maximum and the problem of herder 2 is to find s_2 such that the total payoff $s_2 V(s_1, s_2) = 800s_2 - s_2(s_1 + s_2)^2 = 800s_2 - s_2 s_1^2 - 2s_1 s_2^2 - s_2^3$ is maximum. Partially differentiating these two functions with respect to s_1 and s_2, respectively, we obtain $\partial s_1 V(s_1, s_2)/\partial s_1 = 800 - 3s_1^2 - 4s_1 s_2 - s_2^2$ and $\partial s_2 V(s_1, s_2)/\partial s_2 = 800 - 3s_2^2 - 4s_1 s_2 - s_1^2$, respectively. The FOCs for optima of the payoff functions $s_1 V(s_1, s_2)$ and $s_2 V(s_1, s_2)$ are $\partial s_1 V(s_1, s_2)/\partial s_1 = 0$ and $\partial s_2 V(s_1, s_2)/\partial s_2 = 0$, respectively. Therefore, setting these partial derivatives to zero we obtain $800 - 3s_1^{*2} - 4s_1^* s_2^* - s_2^{*2} = 0$ and $800 - 3s_2^{*2} - 4s_1^* s_2^* - s_1^{*2} = 0$, respectively. Since the payoff and cost conditions of both herders are identical, we can use $s_1^* = s_2^*$. Then, we can write the FOCs as $800 - 3s_1^{*2} - 4s_1^{*2} - s_1^{*2} = 0$ and $800 - 3s_2^{*2} - 4s_2^{*2} - s_2^{*2} = 0$. The only admissible solution from these two equations is $s_1^* = s_2^* = 10$. Notice that the SOCs are also satisfied: since $s_1^* = s_2^* = 10 > 0$, $\partial^2 V(s_1, s_2)/\partial s_1^2 < 0$ and $\partial^2 V(s_1, s_2)/\partial s_2^2 < 0$. This means that the Nash equilibrium number of sheep each herder must graze on the green such that each herder's payoff will be maximum is 10. In this equilibrium each herder obtains a payoff equal to $s_1 V(s_1, s_2) = s_1 V(s_1, s_2) = 800 \times 10 - 10(10 + 10)^2 = \$4000$ and the two herders' total payoff will be equal to \$8000.

Now consider the society's problem of optimizing its payoff function corresponding to equation (7.3.48): maximize $S^s V(S^s) - S^s c = S^s(800 - S^{s2}) = S^s 800 - S^{s3}$. Then the FOC for an optimum of this payoff function is $\partial(S^s 800 - S^{s3})/\partial S^s = 800 - 3S^{*s2} = 0$, with the meaningful value for $S^{*s} \cong 16$. The SOC is also satisfied: $\partial(S^s 800 - S^{s3})/\partial S^{s2} = -6S^{*s} < 0$. Therefore, the socially optimal number of sheep to be grazed on the green is approximately 16. Then the society's optimal payoff will be $S^{*s}(800 - S^{*s2}) = 16(800 - 16^2) = \$8704$. The implications of these results are that the socially optimal payoff exceeds the sum of the Nash equilibrium payoffs and that the Nash equilibrium use of the green exceeds its socially optimal use. The reader would have noticed the similarity between the socially optimal and Nash equilibrium solutions to the Prisoner's Dilemma game encountered earlier and the socially optimal and Nash equilibrium solutions to the present example of the problem of the commons.

Example 4.   Suppose that two of the leading British banks, NatWest and Barclays, need to decide whether they have to increase or decrease the interest rates on commercial loans. Their payoffs (in billions of dollars) from these strategies are presented in Table 7.3.6(A). Determine the pure-strategy Nash equilibrium, if any, of this game. What will be the mixed-strategy Nash equilibrium of this game? Do you expect the banks to play mixed strategies instead of pure strategies? What will be the outcome of the game if both banks follow maximin strategies?

Table 7.3.6(A)

		Barclay's strategies	
		Decrease	*Increase*
NatWest's	Decrease	−1, −1	4, 0
strategies	Increase	0, 4	2, 2

Table 7.3.6(B)

		Barclay's strategies		Row min.
		Decrease	*Increase*	
NatWest's strategies	Decrease	−1, −1	**4, 0**	−1
	Increase	**0, 4**	2, 2	⓪
	Column min.	−1	⓪	

Solution. The payoff matrix presented in Table 7.3.6(A) shows that neither bank has a strictly dominant strategy. However, if NatWest decides to decrease (increase) its rate, then the best response of Barclays is to increase (decrease) its rate. Similarly, if Barclays decides to decrease (increase) its rate, then the best response of NatWest is to increase (decrease) its rate. Therefore, there are two pure-strategy Nash equilibria in this game, "Decrease, Increase" and "Increase, Decrease," with associated payoffs (4, 0) and (0, 4), respectively. These are shown in boldface in the payoff matrix in the table.

Let us now find the mixed-strategy Nash equilibrium of this game. Following equations (7.3.10), we can write the expected payoffs of NatWest (N) and Barclays (B), respectively, as

$$E(\pi^N) = pq\pi_{11}^N + p(1-q)\pi_{12}^N + (1-p)q\pi_{21}^N + (1-p)(1-q)\pi_{22}^N$$
$$= -pq + p(1-q)4 + (1-p)q \times 0 + (1-p)(1-q)2 \tag{7.3.51}$$

$$E(\pi^B) = pq\pi_{11}^B + p(1-q)\pi_{12}^B + (1-p)q\pi_{21}^B + (1-p)(1-q)\pi_{22}^B$$
$$= -pq + p(1-q) \times 0 + (1-p)q \times 4 + (1-p)(1-q)2 \tag{7.3.52}$$

Differentiating equations (7.3.51) and (7.3.52) partially with respect to p and q, respectively, and setting the results to zero, we obtain (corresponding to equations (7.3.12))

$$p^* = \frac{2-4}{-1-0-4+2} = \frac{2}{3} \quad \text{and} \quad q^* = \frac{2-4}{-1-4-0+2} = \frac{2}{3} \tag{7.3.53}$$

Therefore, equations (7.3.53) imply that NatWest must choose the strategy "Decrease" with probability 2/3 and the strategy "Increase" with probability 1/3; and Barclays must choose the strategy "Decrease" with probability 2/3 and "Increase" with probability 1/3. Then their expected payoffs will be $E(\pi^{*N}) = -p^*q^* + p^*(1-q^*)4 + (1-p^*)q^* \times 0 + (1-p^*)(1-q^*)2 = 2/3$ and $E(\pi^{B^*}) = -p^*q^* + p^*(1-q^*) \times 0 + (1-p^*)q^* \times 4 + (1-p^*)(1-q^*)2 = 2/3$. These results imply that one bank may not be interested in choosing mixed strategy as the expected payoffs of that bank will always be considerably lower than the pure-strategy Nash equilibrium payoff of that bank.

Lastly, we consider the case of maximin strategies by both banks. We know that there is no dominant strategy for either bank in this example. If NatWest decides to choose "Decrease" and "Increase" its worst payoffs are −1 and 0, respectively. The maximum of these minima (or maximin) is 0 and, therefore, NatWest must choose "Increase" as its maximin strategy. Likewise, if Barclays chooses "Decrease" and "Increase" its worst payoffs are −1 and 0, respectively. The maximum of these minima (or maximin) is 0 and, therefore,

Barclays must choose "Increase" as its maximin strategy. These maximin strategies are illustrated in circles in Table 7.3.6(B). These observations imply that both banks choose "Increase" as their maximin strategies, which give them equal payoff (2, 2), which is shown in a box in Table 7.3.6(B). Notice that the maximin payoffs are much better than the mixed-strategy payoffs (2/3, 2/3) for both banks and better than the Nash equilibrium payoff to one bank.

7.3.13 Exercises

1.   Find the pure-strategy Nash equilibrium (equilibria), if any, in the two hypothetical games whose payoff matrices are presented in Tables 7.3.7(A) and (B). Payoffs are in millions of dollars.

Table 7.3.7(A)

		Player 2's strategies	
		s_1^2	s_2^2
Player 1's strategies	s_1^1	1, 1	0, 0
	s_2^1	0, 0	1, 1

Table 7.3.7(B)

		Player 2's strategies		
		s_1^2	s_2^2	s_3^2
	s_1^1	2, 3	2, 3	3, 5
Player 1's strategies	s_2^1	2, 2	4, 4	2, 0
	s_3^1	5, 4	3, 2	2, 3

2.   Find the pure-strategy Nash equilibrium, if any, in the two hypothetical games whose payoff matrices are presented in Table 7.3.8(A) and (B). Find the mixed-strategy Nash equilibrium of these games. What strategy do you expect the players to play? Payoffs are in millions of dollars.

Table 7.3.8(A)

		Player 2's strategies	
		s_1^2	s_2^2
Player 1's strategies	s_1^1	50, 50	100, 0
	s_2^1	0, 100	25, 25

Table 7.3.8(B)

		Player 2's strategies	
		s_1^2	s_2^2
Player 1's strategies	s_1^1	50, 50	0, 0
	s_2^1	0, 0	25, 25

3. Find the pure-strategy Nash equilibrium, if any, in the hypothetical game whose payoff matrix is presented in Table 7.3.9(A). What will be the outcomes of the game if the players are risk averse or if they play maximin strategies? What will be the socially optimal (collusive or cooperative) outcome of the game? Payoffs are in millions of dollars.

Table 7.3.9(A)

		Player 2's strategies	
		s_1^2	s_2^2
Player 1's strategies	s_1^1	25, 25	500, 500
	s_2^1	200, 200	−25, −25

Table 7.3.9(B)

		India's strategies	
		Open	Close
USA's strategies	Open	50, 40	15, 30
	Close	30, 15	10, 10

4. *Application exercise.* Suppose that the USA and India are considering whether to open or close their respective import markets for certain products. The associated payoffs (in billions of dollars) are presented in Table 7.3.9(B). Determine the Nash equilibrium, if any, of this game. What will be the outcome of this game if both countries followed maximin strategies?

5. *Application exercise.* Suppose that Dell (based in the USA) and Sony (based in Japan) operate in a Cournot duopoly market for personal computers. The inverse demand function they face is identical and is given by $P = (q_D + q_S)^{-0.5}$, where P denotes price per unit of personal computer in dollars, and q_D and q_S denote the number of personal computers manufactured by Dell and Sony, respectively. The unit cost is equal to $500 and is the same for both Dell and Sony. Find the Nash equilibrium of this game.

6. *Application exercise.* Suppose that the car manufactures Ford and General Motors (both based in the USA) compete in a Bertrand duopoly market for differentiated models of

two of their cars. The demand functions for Ford's and General Motors' cars are given by $q_F = 10\,000 - p_F + 0.5 p_G$ and $q_G = 10\,000 - p_G + 0.5 p_F$, where subscripts F and G represent Ford and General Motors, respectively. The unit cost of Ford's car is \$8000 and that of General Motors' car is \$9000. Find the Nash equilibrium quantities, prices, and profits of the two manufacturers.

7.4 Dynamic games of complete and perfect information

7.4.1 Introduction

So far in this chapter we are dealing with static or simultaneous games. We found that in these games players take decisions or choose their strategies simultaneously. We represented these games in normal form. In normal-form representation, we used the payoff matrix or the best-response functions for discrete and continuous strategy spaces, respectively. We then attempted to find the solutions to these games using the concept of Nash equilibrium. We stated that every finite game has a Nash equilibrium in either pure or mixed strategies. And we applied the concept of Nash equilibrium to games involving many real-world problems. We also demonstrated minimax and minimax regret strategies and applied them to specific cases.

However, in the real world most of the strategic interactions and interdependencies are not static but dynamic in nature. In oligopoly markets, a firm normally takes its decisions regarding choice variables in response to the decisions already taken by its rival(s). Similarly, a government usually takes its decisions regarding import duties or export subsidies in response to the decisions already taken by other countries with which it trades. One can cite many examples like these. Analyses of these *chronological interactions* cannot be carried out in the framework of static games. This necessitates that we consider another area of game theory called dynamic games.

7.4.2 Differences between static and dynamic games

Let us now turn our attention to the differences between static and dynamic games. We stated in Section 7.2.2 that dynamic games were multistage games of which an important class was sequential games. We know from Section 7.2.2 that if, in a game, a player takes a decision and makes a move only after another player has already done that, such games are called sequential games. This implies that the order with which players play is important in sequential games and it is immaterial in static games.

Another important difference between the two classes of games lies in the form of their representations. As stated above, we use the normal form with payoff matrix and the best-response functions for discrete and continuous strategy spaces, respectively, to represent static games. But, dynamic games with discrete strategy spaces are normally represented in extensive form as illustrated in Figure 7.2.1. Notice that the extensive-form representation of dynamic games is based on the device called game trees. And dynamic games with continuous strategy spaces are represented by best-response functions.

In addition to the differences in the order of play and the methods of representation, static and dynamic games also differ in solution methods. We know that static games rely heavily on the concept of Nash equilibrium to find the solutions. This concept of Nash equilibrium is largely inapplicable in the case of dynamic games as it could suggest unreasonable outcomes. Therefore, we will make use of a richer concept called *subgame-perfect Nash equilibrium* to determine the solutions to dynamic games.

We know from Section 7.2.2 that if every player of a game knows every aspect of the game, then the game is called a game of complete information. If some player(s) of a game possesses information that is unknown to the other player(s), then that game is called a game of incomplete information. We also know that if every player has information on all the previous moves or decisions taken in a game, then that game is called a game of perfect information; and if some players do not possess some information on some of the previous decisions taken in a game, then that game is called a game of imperfect information. As stated there, we will only consider dynamic games of complete and perfect information.

7.4.3 Extensive-form representation of sequential games

We have already given an example of a two-player sequential game in Figure 7.2.1. We will see shortly that it is much easier to represent sequential games in extensive form. Before we do this, we need to stipulate what an extensive form is. An extensive form stipulates the following elements:

1   the set of players;
2   the order of actions by players;
3   the set of actions by players;
4   the set of information that players possess at each action; and
5   the payoffs to players corresponding to combinations of actions.

Let us now consider the representation of a one-player game with two possible actions. As an example, assume that the player is a firm (denoted by F_A) and that the actions are whether the firm wants to enter (denoted by E) a new market or stay out (denoted by O). The nodes, the branches, and the associated payoffs of this simple hypothetical *entry game* are represented by the extensive form shown in Figure 7.4.1(A). Notice that the solution to this game is that F_A chooses the action that yields it the highest payoff.

Now suppose that F_A has already entered the market. Also suppose that another firm, F_B, is considering entering the same market. Suppose also that F_B also has to decide whether to enter the market or stay out of it. If F_B enters the market, the market will be shared by both firms. Let us first represent this hypothetical entry game in normal form as shown in Figure 7.4.1(B). The extensive form of this hypothetical entry game with two firms is illustrated in Figure 7.4.2(A). We will see shortly how we can solve games like these.

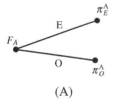

		Firm B's Strategies	
		Enter	Stay Out
Firm A's Strategies	Enter	$\pi_{E,E}^A = \pi_{E,E}^B$	$\pi_{E,O}^A = \pi_{E,O}^B$
	Stay Out	$\pi_{O,E}^A = \pi_{O,E}^B$	$\pi_{O,O}^A = \pi_{O,O}^B$

(A)   (B)

Figure 7.4.1

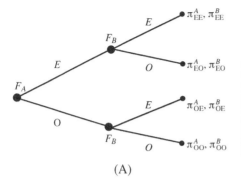

$$\pi_{EE}^A, \pi_{EE}^B$$
$$\pi_{EO}^A, \pi_{EO}^B$$
$$\pi_{OE}^A, \pi_{OE}^B$$
$$\pi_{OO}^A, \pi_{OO}^B$$

(A)

		Firm A's Strategies	
		Fight if Firm B Enters	Accommodate if Firm B Enters
Firm B's Strategies	Stay Out	0, 3	0, 3
	Enter	−2, −1	2, 1

(B)

Figure 7.4.2

7.4.4 Weakness of Nash equilibrium in dynamic games

We mentioned in Section 7.4.2 that the application of the concept of Nash equilibrium to determine the solution to dynamic games might yield unreasonable results. Let us cite an example to confirm this statement. For this purpose we revise the above hypothetical entry game slightly. Suppose that firm A has already entered the market and firm B is planning to enter the market. Then firm A has two strategies: either to fight firm B or accommodate firm B. Suppose also that the payoffs (in millions of dollars) in the revised hypothetical entry game are as those presented in Figure 7.4.2(B).

As the payoff matrix in Figure 7.4.2(B) shows, no firm has a strictly dominant strategy. Notice that there are two pure-strategy Nash equilibria in this game: "Stay Out, Fight if firm B Enters" and "Enter, Accommodate if firm B Enters." But, a moment's thought would suggest that the first Nash equilibrium is not a reasonable one. Notice that "Accommodate if firm B Enters" is the optimal choice for firm A if firm B enters the market. Given this, firm B can foresee that firm A will accommodate if firm B enters the market. In other words, firm A's threat of fighting if firm B enters the market is a *noncredible threat* or an *empty threat*. Then it is certain that firm B will enter the market and firm B and firm A will obtain payoffs $2 million and $1 million, respectively.

7.4.5 Solution to dynamic games: sequential rationality and backward induction

In the last section we found that the application of the concept of Nash equilibrium to dynamic games could yield unreasonable results. Let us now introduce the concept of *sequential rationality* to rule out these types of unreasonable results. The concept of sequential rationality stipulates that each player's strategy should identify optimal actions at every decision node on the game tree. What this implies is that a player's strategy at any node on the game tree must suggest moves that are optimal for that player given the opponents' strategies. We now illustrate the sequential form of the game (whose normal form is presented in Figure 7.4.2(B)) in Figure 7.4.3(A) to explain the meaning of sequential rationality. It can be inferred from this figure that "Fight" if firm B enters the market is not an optimal strategy for firm A. This is what sequential rationality suggests.

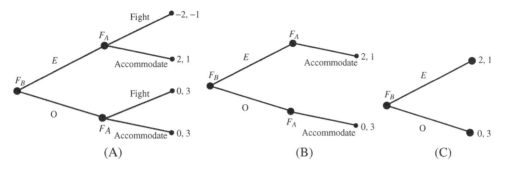

Figure 7.4.3

Let us now attempt to solve the present entry game. For this we use a procedure called *backward induction*, which can eliminate unreasonable Nash equilibria and yield only reasonable Nash equilibria. This procedure involves four steps.

Step 1. Observe the payoffs at the last set of nodes of the game tree.
Step 2. Determine the optimal behavior of the player at the nodes immediately before the last set of nodes on the basis of the observation in step 1.
Step 3. Eliminate strategies that are not optimal as in step 2 and erase (truncate) those branches in the game tree.
Step 4. Redraw the game tree and continue as above.

We can now apply the above steps to our example of the entry game. Since this is a game of complete and perfect information, firm A knows that firm B will enter the market and the best, after firm B entered, for firm A is to choose "Accommodate." Therefore, we can eliminate the branches of the strategy "Fight" from the node F_A, and then redraw the game tree as shown in Figure 7.4.3(B). Given this, firm A now has no choice; and firm B now has two choices: either "Stay Out" or "Enter" with associated payoffs $0 and $2 million, respectively. It is now clear that firm B will enter and earn the payoff equal to $2 million. This is illustrated in Figure 7.4.3(C). Therefore, the solution to the present entry game is that firm B enters the market and firm A accommodates firm B; and firm B and firm A earn payoffs equal to $2 million and $1 million, respectively.

We now state an important theorem, called *Zermelo's theorem*, without proof. This theorem states that every finite, extensive-form game has a pure-strategy Nash equilibrium that can be attained through the procedure of backward induction. Besides, if no player has the same payoffs at any two terminal nodes, then the game has a unique Nash equilibrium that can also be attained through the procedure of backward induction.

7.4.6 Dynamic games of complete and perfect information: application examples

Example 1. Suppose that two of the major Indian automobile manufacturers, Tata (T) and Mahindra (M), plan to decide whether they concentrate on the manufacture of low-end (LE) or high-end (HE) models of cars for the next few years. The payoffs (in billions of dollars) of the two manufacturers are as presented in Figure 7.4.4(A). Assume that T announces first its

		M's Strategies	
		LE	HE
T's Strategies	LE	−1, −1	3, 2
	HE	2, 3	−1, −1

(A)

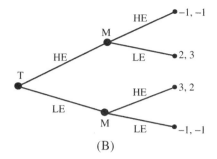

(B)

Figure 7.4.4

decision to manufacture the LE model car called Nano. Present this game in extensive form and solve the game using backward induction.

Solution.   Notice that the payoff matrix in Figure 7.4.4(A) shows that no manufacturer has a strictly dominant strategy. However, there are two pure-strategy Nash equilibria in this game: the strategy profiles "LE, HE" and "HE, LE." The extensive form of this game is illustrated in Figure 7.4.4(B). Let us now apply the procedure of backward induction and the principle of sequential rationality to solve this game. For this we move backwards from the last set of nodes, the nodes at the payoffs. Examining the payoffs at the last set of nodes, we can infer that sequential rationality dictates that M will not choose HE and LE models if T chooses HE and LE models, respectively. The optimal choices for M are to choose LE and HE models if T chooses HE and LE models, respectively. Therefore, we can truncate the game tree in Figure 7.4.4(B) by eliminating nonoptimal choices by M to obtain the game tree illustrated in Figure 7.4.5(A). Since T is supposed to announce the model first, it will clearly announce the LE model because it will give T a higher payoff of $3 billion than if it announces the HE model which yields a payoff of $2 billion. Then the optimal choice for M is to manufacture the HE model cars. We can once again truncate the game in Figure 7.4.5(A) to obtain the game tree in Figure 7.4.5(B). This figure yields the solution that T concentrates on manufacturing LE model cars and M concentrates on manufacturing HE model cars and their payoffs will be $3 billion and $2 billion, respectively.

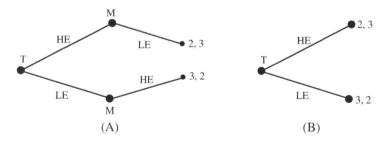

(A)   (B)

Figure 7.4.5

Example 2. Suppose that the inverse demand function that two firms, A and B, face in a Stackelberg duopoly market with identical products is given by $P = a - bq$, where a and b are constants and $q = q_A + q_B$. Suppose also that the average and marginal costs of both duopolists are constant and equal (c), that A announces its quantity first, and that the strategy space for both firms is continuous. Find the profit-maximizing quantities of products, price, total market output, and the maximum profit for both firms if $a = 1100$, $b = 1$, and $c = 100$.

Solution. This is an example of a *leader–follower model* in oligopoly. The price that prevails in the market is $P = a - bq = a - bq_A - bq_B$. The total revenue of A is $R_A = Pq_A = q_A(a - bq) = q_A(a - bq_A - bq_B) = q_A a - bq_A^2 - bq_A q_B$ and its total cost is $C_A = cq_A$. Therefore, A's total profit or payoff is $\pi_A = R_A - C_A = aq_A - bq_A^2 - bq_A q_B - cq_A$. Similarly, the total revenue of B is $R_B = Pq_B = q_B(a - bq) = q_B(a - bq_A - bq_B) = aq_B - bq_B^2 - bq_A q_B$ and its total cost is $C_B = cq_B$. Therefore, B's total profit or payoff is $\pi_B = R_B - C_B = aq_B - bq_B^2 - bq_A q_B - cq_B$.

Notice that A announces its quantity first and, therefore, it is the leader. This implies that B is the follower; that is, it announces its quantity as a response or reaction to the announcement by A. We can now find the reaction function of B to a given quantity announced by A. This can be found by partially differentiating B's payoff function with respect to its quantity, setting the result to zero (the FOC) and solving for q_B. The result will be $q_B = (a - c - bq_A)/2b$. We can now substitute the last equation into the payoff function of A to obtain $\pi_A = R_A - C_A = (1/2)[aq_A - bq_A^2 - cq_A]$. We now partially differentiate the last equation with respect to q_A, set the result to zero, and simplify to yield $q_A^* = (a - c)/2b$. Substituting the last equation into $q_B = (a - c - bq_A)/2b$ and simplifying yields $q_B^* = (a - c)/4b$. Therefore, the optimal total output in the market will be $q^* = q_A^* + q_B^* = [3(a-c)/4b]$ and the optimal market price will be $P^* = a - bq^* = a - bq_A^* - bq_B^* = (a + 3c)/4$. A's optimal payoff will be $\pi_A^* = aq_A^* - bq_A^{2*} - bq_A^* q_B^* - cq_A^* = (a - c)^2/8b$ and B's optimal payoff will be $\pi_B^* = aq_B^* - bq_B^{2*} - bq_A^* q_B^* - cq_B^* = (a - c)^2/16b$.

Let us now plug the values $a = 1100$, $b = 1$, and $c = 100$ into the above results to find the profit-maximizing quantities of output, price, total market output, and the maximum profit of the two firms. The profit-maximizing output of A (the leader) is $q_A^* = (a - c)/2b = (1100 - 100)/2 \times 1 = 500$ and the profit-maximizing output of B (the follower) is $q_B^* = (a - c)/4b = (1100 - 100)/4 \times 1 = 250$. The profit-maximizing market output is $q^* = q_A^* + q_B^* = 3(a - c)/4b = 3(1100 - 100)/4 \times 1 = 750$. The profit-maximizing market price will be $P^* = (a + 3c)/4 = (1100 + 3 \times 100)/4 = \$350$. The maximum profit for A will be $\pi_A^* = (a - c)^2/8b = (1100 - 100)^2/8 \times 1 = \$125\,000$ and that for B will be $\pi_B^* = (a - c)^2/16b = (1100 - 100)^2/16 \times 1 = \$62\,500$. The maximum industry profit will be $\pi_A^* + \pi_B^* = \$187\,500$. Notice that these are exactly the results as those we obtained as the solution to the first problem in example 5 in Section 4.2.8.

Example 3. Suppose that the demand functions that two firms, A and B, face in a Bertrand duopoly market with differentiated products are given by $q_A = a - bp_A + dp_B$ and $q_B = a - bp_B + dp_A$, respectively, where a, b, and d are constants. Suppose also that the average and marginal costs of both duopolists are constant and equal to c, that A announces its price first, and that the strategy space for both firms is continuous. Find the profit-maximizing prices for both firms, quantities, total market output, and the maximum profit if $a = 30$, $b = 1$, $c = 20$, and $d = 0.5$.

Solution.   The total revenue of A is $R_A = p_A q_A = p_A(a - b p_A + d p_B) = a p_A - b p_A^2 + d p_A p_B$ and its total cost is $C_A = c q_A = c(a - b p_A + d p_B) = ac - bc p_A + cd p_B$. Therefore, A's total profit or payoff is $\pi_A = R_A - C_A = a p_A - b p_A^2 + d p_A p_B - ac + bc p_A - cd p_B$. Similarly, the total revenue of B is $R_B = p_B q_B = p_B(a - b p_B + d p_A) = a p_B - b p_B^2 + d p_A p_B$ and its total cost is $C_B = c q_B = c(a - b p_B + d p_A) = ac - bc p_B + cd p_A$. Therefore, B's total profit or payoff is $\pi_B = R_B - C_B = a p_B - b p_B^2 + d p_A p_B - ac + bc p_B - cd p_A$.

Notice that A announces its price first and, therefore, it is the leader. This implies that B, the follower, announces its price as a response or reaction to the announcement by A. We can now find the reaction function of B to a given price announced by A. This can be found by partially differentiating B's payoff function with respect to its price, setting the result to zero (the FOC), and solving for p_B, which will yield $p_B = (a + bc + b p_A)/2b$. We can substitute the last equation into the payoff function of A to obtain $\pi_A = R_A - C_A = a p_A - b p_A^2 + d p_A[(a + bc + d p_A)/2b] - ac + bc p_A - cd[(a + bc + d p_A)/2b]$. We now partially differentiate the last equation with respect to p_A, set the result to zero, and simplify to obtain $p_A^* = (bcd - cd^2 + 2b^2 c + ad + 2ab)/(4b^2 - 2d^2)$. Substitution of the last equation into $p_B = (a + bc + b p_A)/2b$ yields $p_B^* = \{a + bc + [(2ab + ad + bcd + 2b^2 c - ad^2)/(4b^2 - 2d^2)]\}/2b$.

Let us now substitute the values $a = 30$, $b = 1$, $c = 20$, and $d = 0.5$ into p_A^* and p_B^*. This yields the optimal prices $p_A^* = \$34.3$ and $p_B^* = \$42$. The maximum profits of the two firms can be obtained by substituting these optimal prices into their respective profit functions. This will give us $\pi_A^* = a p_A^* - b p_A^{*2} + d p_A^* p_B^* - ac + bc p_A^* - cd p_B^* = \$239$ and $\pi_B^* = a p_B^* - b p_B^{*2} + d p_A^* p_B^* - ac + bc p_B^* - cd p_A^* = \$113.3$. The optimal quantities of outputs produced by the firms will be $q_A^* = a - b p_A^* + d p_B^* = 16.7$ and $q_B^* = a - b p_B^* + d p_A^* = 5.15$. Therefore, the optimal market output will be $q_A^* + q_B^* = 16.7 + 5.15 = 21.85$.

Example 4.   Suppose that the demand function that two firms, A and B, face in a Bertrand duopoly market with identical products is given by $q = f(p)$, where q represents the total quantity demanded in the market and p denotes the price that prevails in the market such that $p = \min(p_A, p_B)$, where p_A and p_B denote the prices set by A and B, respectively. Suppose also that the average and marginal costs of both duopolists are constant and equal to c and that the strategy space for both firms is continuous. Determine the outcome of this game if the duopolists choose their prices simultaneously as in static games, and if A announces its price (p_A) first and B reacts to this by choosing its price (p_B) as in dynamic (sequential) games.

Solution.   Since the products of the duopolists are identical, the duopolist with lower price will serve the entire market. If the duopolists' prices are equal, they will share the market equally. Therefore, the demand function for A's product can be given as $q_A = q = f(p_A)$ if $p_A < p_B$, $q_A = q/2 = f(p_A)/2$ if $p_A = p_B$, and $q_A = 0$ if $p_A > p_B$. Similarly, the demand function for B's product can be given as $q_B = q = f(p_B)$ if $p_B < p_A$, $q_B = q/2 = f(p_B)/2$ if $p_B = p_A$, and $q_B = 0$ if $p_B > p_A$.

If the duopolists choose their prices simultaneously as in static games, A's profit or payoff function can be written as $\pi_A = p_A q_A - c q_A$ if $p_A < p_B$, $\pi_A = (p_A q_A - c q_A)/2$ if $p_A = p_B$, and $\pi_A = p_A q_A - c q_A = 0$ if $p_A > p_B$. Similarly, B's profit or payoff function can be written as $\pi_B = p_B q_B - c q_B$ if $p_B < p_A$, $\pi_B = (p_B q_B - c q_B)/2$ if $p_B = p_A$, and $\pi_B = p_B q_B - c q_B = 0$ if $p_B > p_A$. What these functions imply is that as long as one firm chooses a price slightly lower than that of the other firm but slightly above the average cost, it can capture the entire market and make a positive profit. Therefore, both firms will continue to undercut the price of the other firm until their prices are equal to the constant and identical average cost. Therefore, equilibrium of this simultaneous move game will be when both firms choose prices equal to the identical average cost and their profits are zero; that is, $p_A^* = p_B^* = c$ and $\pi_A^* = \pi_B^* = 0$.

If the duopolists choose their prices sequentially with firm A choosing its price first, the outcome of the game could be different from that outlined above. If the game is a dynamic (sequential) game, then the game could exhibit multiple outcomes or equilibria; the above static outcome (equilibrium) would be just one of them. This can be shown as follows. Since A chooses its price first, B will always react to A's choice. This implies that we can write B's reaction function as $p_B^* = R_B(p_A)$. But, we know from above that B will attempt to undercut the price of A by a very small amount (ε) if A's price is above c or choose a price at least equal to c if A's price is equal to c. Therefore, we have the condition $p_B^* = p_A - \varepsilon$ if $p_A > c$ or $p_B^* \geq c$ if $p_A = c$. The interesting result here is that both firms could end up earning zero profits irrespective of their choices of prices. This implies that all the combinations of prices satisfying $p_A \geq c$ and $p_B^* = R_B(p_A)$ are equilibrium combinations of prices in the present sequential game.

7.4.7 Exercises

1. Suppose that two players (A and B) engage in a hypothetical game of complete and perfect information with two strategies (1 and 2). The payoff matrix of this game is presented in Table 7.4.1(A). Is (are) there any pure-strategy Nash equilibrium (equilibria) in the static form of the game? If yes, what is (are) it (they)? If A moves first in the dynamic (sequential) form, what will be the outcome of the game?

Table 7.4.1(A)

		Player B's strategies	
		Strategy 1	Strategy 2
Player A's	Strategy 1	−10, −5	10, 15
strategies	Strategy 2	15, 10	−5, −10

Table 7.4.1(B)

		Player B's strategies	
		Strategy 1	Strategy 2
Player A's	Strategy 1	15, 10	30, 15
strategies	Strategy 2	20, 20	5, 10

2. Suppose that two players (A and B) engage in a hypothetical game of complete and perfect information with two strategies (1 and 2). The payoff matrix of this game is presented in Table 7.4.1(B). Is (are) there any pure-strategy Nash equilibrium (equilibria) in the static form of the game? If yes, what is (are) it (they)? If A moves first in the dynamic

(sequential) form, what will be the outcome of the game? If B moves first, what will be the outcome of the game?

3. Suppose that two players (A and B) engage in a hypothetical game of complete and perfect information with two strategies (1 and 2). The payoff matrix of this game is presented in Table 7.4.2(A). Is (are) there any pure-strategy Nash equilibrium (equilibria) in the static form of the game? If yes, what is (are) it (they)? If A moves first in the dynamic (sequential) form, what will be the outcome of the game? Is there any possibility for a *credible threat* from one player in this game? Is there any possibility for collusion and better payoffs for both players?

Table 7.4.2(A)

		Player B's strategies	
		Strategy 1	Strategy 2
Player A's	Strategy 1	2, 6	2, 6
strategies	Strategy 2	0, 0	3, 2

Table 7.4.2(B)

		Player B's strategies	
		Strategy 1	Strategy 2
Player A's	Strategy 1	10, 10	−10, 15
strategies	Strategy 2	15, −10	5, 5

4. Suppose that two players (A and B) engage in a hypothetical game of complete and perfect information with two strategies (1 and 2). The payoff matrix of this game is presented in Table 7.4.2(B). Is (are) there any pure-strategy Nash equilibrium (equilibria) in the static form of the game? If yes, what is (are) it (they)? If A moves first in the dynamic (sequential) form, what will be the outcome of the game?

5. *Application exercise.* Assume that we have a Stackelberg oligopoly market with inverse demand function $P = a - Q$ (where Q denotes the market output and p denotes the market price) and that each oligopolist's cost is zero. Also assume that one of the oligopolists, the leader, moves first and announces its quantity of output and all other n oligopolists, the followers, decide their quantities of output in response to the announcement by the leader. Determine the profit-maximizing output of the leader. Do the leader's profit-maximizing quantity of output and total profit depend on the number of the followers?

6. *Application exercise.* Suppose that the demand functions that two firms, A and B, face in a Bertrand duopoly market with identical products are given by $q_A = a - bp_A + dp_B$ and $q_B = a - bp_B + dp_A$, respectively, where a, b, and d are constants. Suppose also that the average and marginal costs of both duopolists are constant and equal to zero, that A announces its price first, and that the strategy space for both firms is continuous.

Find the profit-maximizing prices, quantities, total market output, and the maximum profit of both firms if $a = 30$, $b = 1$, and $d = 0.5$.

7.5 Dynamic games of complete and imperfect information: subgame perfect Nash equilibrium

7.5.1 Introduction

In Sections 7.2 and 7.3 we demonstrated how we could find the equilibrium or equilibria in the case of static games of complete and perfect information using strictly dominant or dominated strategies, Nash's equilibrium concept, maximin strategies, and mixed strategies. In Section 7.4 we demonstrated how one could find the outcomes of dynamic (sequential) games using the ideas of sequential rationality and backward induction.

The reader would have noticed the discussion of the shortcoming of the concept of Nash equilibrium when it is applied to find the equilibrium in the case of dynamic games. We observed that the application of the concept of Nash equilibrium to the determination of the outcomes or equilibrium in the case of dynamic games might yield unreasonable results. We also observed that the reason for this anomaly is the fact that the concept of Nash equilibrium, when it is applied as such and without modification or rectification, may not always satisfy the principle of sequential rationality. Therefore, our attempt in the present section is directed at deriving the concept of Nash equilibrium that satisfies the principle of sequential rationality. In other words, our aim is to present a refined form of the concept of Nash equilibrium that can be applied to find the equilibrium of dynamic games such that it will not yield unreasonable predictions or results.

7.5.2 Subgames

Let us begin this section with the definition of a *subgame*. A subgame can be informally defined as the part of a game that remains to be played beginning at any node of the game. However, we present here a formal definition of a subgame. A subgame in an extensive-form game is a portion of the game that

1   begins at a decision node, other than the terminal node;
2   contains all the decision and terminal nodes that follow the beginning node in 1; and
3   does not contain the nodes that do not follow the beginning node in 1.

As an example of exposition, consider the extensive form of the game of complete and perfect information illustrated in Figure 7.2.1, which we modify slightly as illustrated in Figure 7.5.1(A). The game comprises three subgames. The first subgame is the game starting at the top decision node P_B and the second subgame is the game starting at the bottom decision node P_B. Notice that these two subgames are represented by innermost, dashed rectangles and these rectangles contain the respective decision nodes, strategies, and associated payoffs. The top decision node P_B of the top subgame is called the *subroot* of that subgame, as is the bottom decision node P_B. These two subgames are called *proper subgames*. Therefore, there are two subroots in the present game. Notice also that the original game itself is a subgame and is contained in the outermost, dotted-dashed rectangle and it is called the *trivial subgame*. This suggests that every trivial subgame is the original game itself or every game is a subgame of itself. The root of a trivial subgame, such as point P_1 in the figure, is called

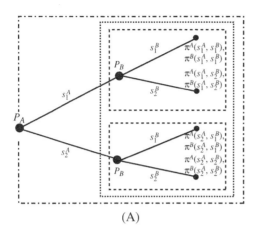
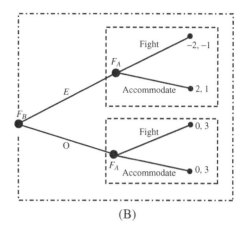

(A)   (B)

Figure 7.5.1

the *trivial subroot*. Therefore, there are three subgames in total in the game presented in Figure 7.5.1(A).

A few more points are worth mentioning in the context of subgames. In a subgame once a player starts to play from the subroot, the player cannot discontinue the play or return and look for alternatives, and has to continue to play the game until its ends. Notice that each subgame in Figure 7.5.1(A) is like a fresh game. It can be shown that each subgame will have a unique subgame Nash equilibrium (SNE) and the total number of SNEs will at least be equal to the number of subgames. It can also be shown that a unique SNE, referred to as *subgame perfect Nash equilibrium* (SPNE) can be identified from the set of all SNEs. These are the topics to which we turn below.

7.5.3 SNEs and SPNE

Our main aim in this section is to state a Nash equilibrium definition that can be applied to dynamic games and that precludes unreasonable predictions. We have already developed and described all the prerequisites for such a definition. We are now ready to define a SPNE. A Nash equilibrium is subgame perfect if *the players' strategies induce a Nash equilibrium in every subgame*. It must be noted that every SPNE is a SNE, but the converse is not true.

It must be emphasized that one can use the idea of SPNE in the case of dynamic games just as one uses the concept of Nash equilibrium in the case of static games. The pertinent question at this juncture is how one finds a SPNE. For this one can move in the spirit of backward induction. The required steps are as follows. Firstly, identify all the smallest subgames that involve terminal nodes in the original game. Secondly, replace each subgame with the payoffs from one of its Nash equilibria. Thirdly, consider the initial nodes in these subgames as the terminal nodes in a truncated version of the original game. Fourthly, find the smallest subgames in this truncated game that involve terminal nodes and replace each of these subgames with the payoffs from one of its Nash equilibria. Lastly, continue backwards as above until a SPNE is reached. This procedure is called *generalized backward induction*.

One might wonder what relationship exists between the idea of backward induction and the idea of SPNE. These two ideas are the same in the case of any game of perfect information.

What this means is that any equilibrium found, in any game of perfect information, through backward induction is a SPNE and any SPNE satisfies backward induction. The main difference between these two ideas is that while the idea of backward induction is useful in the case of games of perfect information, the idea of SPNE can be used in the case of games of perfect information and *games of imperfect information* (which we do not consider in this book) alike. In other words, the latter is more general than the former.

In Section 7.4.5 we stated Zermelo's theorem that every finite, extensive-form game has a pure-strategy Nash equilibrium that can be attained through the procedure of backward induction and that if no player has the same payoffs at any two terminal nodes, then the game has a unique Nash equilibrium that can also be attained through the procedure of backward induction. A similar statement can be made in the case of subgames. Every finite game of perfect information has a pure-strategy SPNE and if no player has the same payoffs at any two terminal nodes, then there is a unique SPNE.

It must be emphasized that any SPNE is a Nash equilibrium because the game itself is a subgame. It must also be emphasized that every Nash equilibrium is not a SPNE. The last statement shows the main purpose behind the development of the idea of SPNE; that is, it eliminates unreasonable Nash equilibrium or equilibria.

7.5.4 SPNE: examples

Example 1. Consider the dynamic game we presented in Section 7.4.3, which is reproduced in Figure 7.5.1(B) with slight modification. Find the SPNE of this game.

Solution. We found in Section 7.4.5 that no firm has a strictly dominant strategy in this game. We also found there that there were two Nash equilibria in this game: "Stay Out, Fight if firm B Enters" and "Enter, Accommodate if firm B Enters." The extensive form of this game presented in Section 7.4.5 yields, through the application of sequential rationality and backward induction and as shown in Figure 7.4.3(C), the solution to the game: firm B enters the market and firm A accommodates firm B.

Notice that the dynamic form of the same game presented in Figure 7.5.1(B) has three subgames: the game itself and the subgames that begin with A's decision nodes, F_A. This implies that there must be at least three SNEs for the game and one of these three SNEs must be the SPNE. Then, following the steps outlined in the last section, we find that the optimal strategy for F_A if F_B decides to enter is to accommodate, and the optimal strategy for F_A if F_B decides to stay out is either fight or accommodate. Therefore, again following the steps mentioned, we can redraw the game tree as illustrated in Figure 7.5.2(A). Notice that the optimal strategy for F_B, given the choices of strategies by F_A after F_B decided to enter or stay out, is to enter. Therefore, the SPNE in the present game is {Enter, (Accommodate, Accommodate)} and the payoffs for F_A and F_B will be $1 million and $2 million, respectively. This result is identical to the result we obtained in Section 7.4.5. The reader would have noticed that SPNE precludes the unreasonable outcome mentioned earlier. This suggests that any equilibrium found, in any game of complete and perfect information, through backward induction is a SPNE and any SPNE satisfies backward induction.

Example 2. Consider the game of the car model decisions of two Indian automobile manufacturers Tata (T) and Mahindra (M) we solved in example 1 in Section 7.4.6. The extensive form of this game illustrated in Figure 7.4.4(B) is reproduced in Figure 7.5.2(B) for convenience. Find the SPNE of this game.

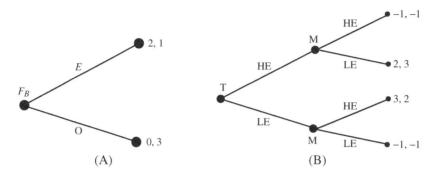

Figure 7.5.2

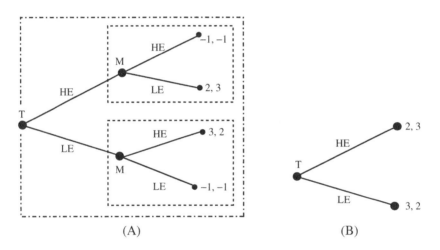

Figure 7.5.3

Solution. Let us first identify the subgames of this game. These are illustrated in Figure 7.5.3(A). As can be seen, there are three subgames in this game: one is the original game enclosed in the outer dotted-dashed rectangle, and the other two are those shown with dashed rectangles, whose subroots start at the two decision nodes of M. This implies that there should be at least three SNEs in this game. We can now use the steps outlined in Section 7.5.3. Following these steps and using the notion of backward induction we see that the optimal strategy for M, if T chooses HE, is LE. Similarly, the optimal strategy for M, if T chooses LE, is HE. On the basis of these results we can redraw the game tree as illustrated in Figure 7.5.3(B). As can be seen, the optimal strategy for T, given the choices of strategies by M, is to manufacture the LE model. Therefore, the SPNE in the present game is {LE, (HE, LE)} and the payoffs for T and M will be $3 million and $2 million, respectively. This result is identical to the result we obtained in Section 7.4.6. This again suggests that any equilibrium found, in any game of complete and perfect information, through backward induction is a SPNE and any SPNE satisfies backward induction.

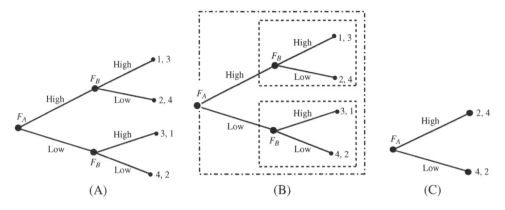

Figure 7.5.4

Example 3. Suppose that F_A and F_B are firms that produce identical products in a duopoly market, and that F_A is the leader and F_B is the follower. Also suppose that F_A has to decide whether it has to set a high price or low price for its product. No matter what price F_A decides, F_B will immediately react by setting either a high price or a low price. Suppose again that the payoffs, in millions of dollars, from these decisions are as those illustrated in the game tree in Figure 7.5.4(A). Find the SPNE of this game.

Solution. Let us first identify the subgames of this game. These are illustrated in Figure 7.5.4(B). As can be seen, there are three subgames in this game: one is the original game enclosed in the outer dotted-dashed rectangle, and the other two are those shown with dashed rectangles, whose subroots start at the two decision nodes of F_B. This implies that there should be at least three SNEs in this game. We can now use the steps outlined in Section 7.5.3. Following these steps and using the notion of backward induction we see that optimal strategy for F_B, if F_A chooses "High," is "Low." Similarly, the optimal strategy for F_B, if F_A chooses "Low," is "Low." On the basis of these results we can redraw the game tree as illustrated in Figure 7.5.4(C). As can be seen, the optimal strategy for F_A, given the choices of strategies by F_B, is to choose the strategy "Low." Therefore, the SPNE in the present game is {Low, (Low, Low)} and the payoffs for F_A and F_B will be $4 million and $2 million, respectively. Notice the advantage for F_A being the first mover, which is called the *first mover advantage* in a game. The first mover advantage in a game is the additional payoff that a player obtains, over and above the payoff that the player would have obtained when all the players of the game moved simultaneously, just because the player happens to be at the trivial subroot of the game.

Example 4. Except for the bilateral air travel services, Air India (AI) was the monopoly of air travel services from India to Western Europe and North America for about half a century. As a consequence of the liberalization policy of the Government of India, private carriers such as Jet Airways (JA) entered the market recently. The entry of JA into the market led to strategic decisions by both companies. AI being the incumbent and the leader of the market, the follower JA often has to take strategic decisions regarding air fares in order to attract passengers. Suppose that both carriers have three choices: increase the fare, keep the fare constant, and decrease the fare. Suppose also that this dynamic game with the payoffs of both

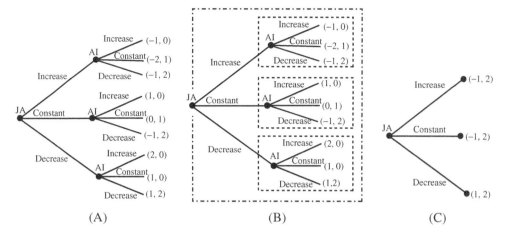

Figure 7.5.5

carriers is as represented by the game tree illustrated in Figure 7.5.5(A). Determine SNEs and SPNE of this game.

Solution.   As before, let us first identify the subgames of this game. These are illustrated in Figure 7.5.5(B). As can be seen, there are four subgames in this game: one is the original game enclosed in the outer dotted-dashed rectangle, and the other three are those, shown with dashed rectangles, whose subroots start at the two decision nodes of AI. This implies that there should be at least four SNEs in this game. We can now use the steps outlined in Section 7.5.3. Following these steps and using the notion of backward induction we see that optimal strategy for AI, if JW chooses "Increase," is "Decrease." Similarly, the optimal strategies for AI, if JW chooses "Constant" and "Decrease," are "Decrease" and "Decrease," respectively. On the basis of these results we can redraw the game tree as illustrated in Figure 7.5.5(C). As can be seen in this panel, the optimal strategy for JA, given the choices of strategies by AI, is to choose the strategy "Decrease." Therefore, the SPNE in the present game is {Decrease, (Decrease, Decrease, Decrease)} and the payoffs for JW and AI will be $1 billion and $2 billion, respectively. Notice the advantage for AI because of being the leader.

7.5.5 Exercises

1. For the extensive-form games involving two players illustrated in Figures 7.5.6(A) and (B), determine the SNEs and SPNE. The payoffs are in dollars.
2. For the extensive-form games involving two players illustrated in Figure 7.5.7(A) and (B), determine the SNEs and SPNE. The payoffs are in dollars.
3. Consider a two-stage game in which player 1 chooses x_1 in the first stage and players 2 and 3 simultaneously choose x_2 and x_3, respectively, in the second stage of the game. Assume that their payoff functions are $\pi_1 = x_2 + x_3 + x_1x_2 + x_1x_3$, $\pi_2 = (20 - x_2 - x_3)x_2 - x_1x_2$, and $\pi_3 = (20 - x_2 - x_3)x_3 - x_1x_3$, respectively. Find the SPNE of this game.
4. Consider a two-stage game in which players 1 and 2 choose x_1 and x_2, respectively, in the first stage, and players 3 and 4 simultaneously choose x_3 and x_4, respectively, in the

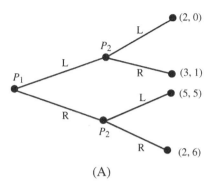
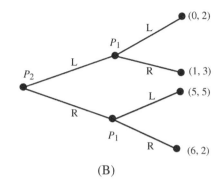

Figure 7.5.6

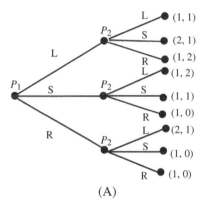
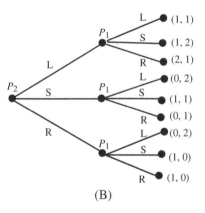

Figure 7.5.7

second stage of the game. Assume that their payoff functions are $\pi_1 = x_1(x_3 + x_4)$, $\pi_2 = x_2(x_3 + x_4)$, $\pi_3 = (20 - x_3 - x_4)x_3 - x_1x_3 - x_2x_3$, and $\pi_4 = (20 - x_3 - x_4)x_4 - x_1x_4 - x_2x_4$, respectively. Find the SPNE of this game.

5. *Application exercise.* Suppose that the demand for the ith firm's product in a Bertrand duopoly market with differentiated products is given by $q_i = 20 - p_i + 0.5p_j$, where $i \neq j$, and that the cost of production is zero. Find the SPNE of this game if firm 1 moves first by choosing its price p_1 and firm 2 follows by choosing its price p_2.

6. *Application exercise.* Suppose that the inverse market demand function in a Stackelberg duopoly market with identical products is given by $P = 100 - 4(q_1 + q_2)$, where P denotes the market price, and that the cost of production is zero for both firms. Find the SPNE of this game if firm 1 moves first by choosing its quantity q_1 and firm 2 follows by choosing its quantity q_2.

7.6 Game theory: extensions

7.6.1 Introduction

Our main aim in this section is the exposition of few aspects of game theory that we have not explored so far. The first aspect we discuss here is a class of games called *repeated games*. The second aspect is an important set of strategies referred to as *minimax strategies*. The third aspect is concerned with the exposition of the geometric solution of games. The last aspect is a discussion of the connection between game theory and LP.

7.6.2 Repeated games: meaning and Nash equilibrium

The reader would have noticed that all the static games of complete and perfect information we considered in Sections 7.2 and 7.3 were played only once. But, in the real world, many of the static games of complete and perfect information are played repeatedly. Such static games are called repeated games of complete and perfect information. In other words, a repeated game is a static, simultaneous move game that is played over and over again, where all players possess complete knowledge of the strategies chosen by all the players in all the previous stages of the game.

If a static game of complete and perfect information is repeated a finite number of times, then it is called a *finitely repeated game* of complete and perfect information. If a static game of complete and perfect information is repeated an infinite number of times, then it is called an *infinitely repeated game* of complete and perfect information. It is unnecessary to state that a finitely repeated game has a final or ending stage while an infinitely repeated game does not have a final or ending stage. A finitely repeated game may have a *certain end* or an *uncertain end*. In the former the players know when the game will end, while in the latter the players know that the game will end but not when the game will end or not know the stage in which the game will end.

One might wonder why we are concerned with repeated games. Consider a static simultaneous move game of complete and perfect information. If this game is played over many stages, each player in the game will be able to base their own actions on the other players' past actions. This will open up the possibilities of threats and promises. A player may threaten to take action that may affect the other players unfavorably; or may promise to cooperate, which may affect the other players favorably. This suggests that the idea of subgame perfection, which is based on the credibility of threats and promises, may have an important role to play in repeated games.

Let us first consider a finitely repeated game of complete and perfect information with a certain end. As an example, consider a variant of the Prisoner's Dilemma game presented in Table 7.6.1(A). Suppose that this game is played in two stages, that the payoff for the two-stage game is the sum of the payoffs from the two stages, and that there is no *discounting*. Assume that higher values in the table are preferred. Find the Nash equilibrium of this two-stage game. What difference will it make if the game is continued for a finite n stages?

Notice that if the game in Table 7.6.1(A) were played just once, there would be two non-cooperative Nash equilibria: (A_2, B_2) and (A_3, B_3) with payoffs $(10, 10)$ and $(2, 2)$, respectively. Notice also that of these two, Nash equilibrium (A_3, B_3) is unreasonable and the only reasonable Nash equilibrium, or the SPNE, is (A_2, B_2). Had the players cooperated, they would have chosen the strategies (A_1, B_1) and their payoffs would have been much

Table 7.6.1(A)

		Player B's strategies		
		B_1	B_2	B_3
Player A's strategies	A_1	20, 20	5, 25	0, 0
	A_2	25, 5	**10, 10**	0, 0
	A_3	0, 0	0, 0	**2, 2**

Table 7.6.1(B)

		Player B's strategies		
		B_1	B_2	B_3
Player A's strategies	A_1	30, 30	15, 35	10, 10
	A_2	35, 15	**20, 20**	10, 10
	A_3	10, 10	10, 10	12, 12

higher (20, 20). The reader would have noticed that this mutually advantageous cooperative outcome is not possible in the first stage of the game as each player has an incentive to defect and switch to strategies (A_2, B_2).

Let us now attempt to see the outcome of the game in its second stage. It is interesting to notice that the first-stage Nash equilibrium will also be the second-stage Nash equilibrium. The reason is that the second stage of the game is a "new game" as the first stage game and no player can improve payoff by switching strategies. Therefore, if both players follow the strategies (A_2, B_2) then their payoff will be $10 + 10 = 20$ as shown in Table 7.6.1(B). The second stage payoffs presented in this table are obtained by adding the Nash equilibrium payoff of each player in the first stage of the game to the payoffs in the first stage (presented in Table 7.6.1(A)). The reader can verify that no other strategy profile can constitute a Nash equilibrium in the second stage of the game. Moreover, one can also verify that a threat of playing A_3 (B_3) by A (B), in the event that both A and B decide to cooperate and choose A_1 and B_1, respectively, but one of them defects, is not a credible threat.

Suppose now that, instead of two stages, the game is repeated for a finite n times. What will be the outcome of the game in its final, nth stage? The reader would have guessed that a retaliatory move by choosing A_3 (B_3) by A (B) in the nth stage is not possible because the nth stage is the final stage and there is no stage after the final stage to carry out retaliation or threat. This suggests that the first-stage Nash equilibrium (A_2, B_2) will also be the nth-stage Nash equilibrium. This also suggests that, if retaliation is possible, it must be in the $(n-1)$th stage, the *effective* final stage. But, the same nth-stage result also occurs in the $(n-1)$th stage with the same Nash equilibrium. Thus, the search for a credible threat moves all the way back from the nth stage to the first stage where we found that cooperation or retaliation is not viable. This suggests that in a finitely repeated game of complete and perfect information with pure strategies and with certain end the Nash equilibrium strategies of the first stage will be played in every subsequent stage of the game.

We now consider a finitely repeated game of complete and perfect information with pure strategies and with uncertain end. Notice that the only difference between this game and the game we considered above is that in the former the players do know only that the game will end in some stage, but do not know the stage in which the game will end; in the latter the players do know both. But, what difference will this make to the outcome of the game?

To answer the above question, we begin by assuming that α, where $0 \leq \alpha \leq 1$, denotes the probability that the game will end in the second stage. This suggests that $(1-\alpha), (1-\alpha)^2, \ldots$, and $(1-\alpha)^{n-1}$ denote the probabilities that the game will be played in the second stage, third stage,..., and nth stage of the game, respectively. Now consider the game presented in Table 7.6.1(A). We know from above (A_2, B_2) is the only reasonable Nash equilibrium strategy profile in this game and the cooperative strategy profile (A_1, B_1) is mutually more beneficial than (A_2, B_2). Then, if both players choose the agreed and cooperative strategy (A_1, B_1), the expected payoff to each player will be $20[1 + (1-\alpha) + (1-\alpha)^2 + \cdots + (1-\alpha)^{n-1}]$. But if one player defects and the other does not in the first stage (and chooses the Nash equilibrium strategy in the remaining stages of the game), the expected payoff to the player will be $25(1) + 10[1 + (1-\alpha) + (1-\alpha)^2 + \cdots + (1-\alpha)^{n-1}]$. It is easy to see that defection is profitable only if $25(1) + 10[(1-\alpha) + (1-\alpha)^2 + \cdots + (1-\alpha)^{n-1}] > 20 + 20[(1-\alpha) + (1-\alpha)^2 + \cdots + (1-\alpha)^{n-1}]$, which can be simplified to obtain $(5/10) > [(1-\alpha) + (1-\alpha)^2 + \ldots + (1-\alpha)^{n-1}]$. Notice that the RHS of the last inequality represents a geometric series with the initial term $s_1 = 1 - \alpha$ and the common ratio $k = 1 - \alpha$. Then applying equation (1.10.10) yields $(5/10) > s_1/(1-k) = (1-\alpha)/\alpha$, or $0.5 > (1-\alpha)/\alpha$, or $\alpha > (1/3) = 0.33$. These results suggest that defection is profitable if the probability that the game will end in the second stage is greater than 0.33. Notice that if $\alpha = 1$ and $\alpha = 0$, a finitely repeated game becomes a single-stage game with certain end and an infinitely repeated game, respectively.

We saw above that cooperation among players in a repeated game is highly unlikely. However, there are a few types of agreements which bind the players to cooperation. One such agreement or *enforcement mechanism* is a *cartel*, which is a formal agreement among firms or countries to allocate market shares and to increase group profits. There are evidences that even these types of formal agreements of cooperation may break down at times. Under these circumstances players of the game often resort to other enforcement mechanisms called *trigger strategies* such as *tit-for-tat strategy, preemptive strategy*, etc. A trigger strategy is a strategy in which a player cooperates with other players if they cooperate and defects if they do not cooperate. It is, in fact, a *retaliatory strategy* adopted by a player when other players adopt unanticipated strategies. A tit-for-tat strategy is a strategy in which a player does exactly the same as other players of the game do; that is, it is a strategy in which the player cooperates if the opponents cooperate and defects if the opponents defect. A preemptive strategy is a strategy in which a player moves before other players of the game so that the player can preempt or forestall some of the strategy options of the other players. A preemptive strategy, if available and if chosen, manifests in the form of first-mover advantage.

Our remaining task in this section is a presentation of infinitely repeated games. We know that an infinitely repeated game does not have a final stage. For the purpose of illustration consider the modified Prisoner's Dilemma game illustrated in Table 7.6.1(A). We found earlier in this section that if this game is played only once or repeated for a finite number of times, the only reasonable Nash equilibrium is the strategy profile (A_2, B_2). We also found that the mutually beneficial, cooperative outcome (A_1, B_1) is not a possible Nash equilibrium strategy profile because of the incentive for each player to switch strategies. What will be the outcome of this game if it is played for an infinite number of times?

We approach the present problem slightly differently. Given the practices of many of the big business corporations around the globe, we assume that player A publicly announces that it will play A_1 in the first stage; that it will play A_1 in the second stage if B cooperates and plays B_1 in the first stage; and that it will play the retaliatory or trigger strategy A_2 in the second stage and thereafter if B does not play B_1 in the first stage. If these happen, then the outcome of the game will be considerably different.

Notice that when A announced its plan to play A_1 and if B plays B_1 in every stage of this infinitely repeated game, the payoffs to each player will be $20 + 20 + 20 + \cdots$. These payoffs are better than that obtained when both players play the Nash equilibrium strategy profile (A_2, B_2): $10 + 10 + 10 + \cdots < 20 + 20 + 20 + \cdots$. Moreover, if B violates the agreement and plays B_2 in the first stage and A sticks to A_1, then the payoff to B will be $25 + 10 + 10 + 10 + \cdots$, which is less than what B obtains when B plays B_1. Thus, it is in the interest of B not to violate the agreement. Since the payoff matrix in Table 7.6.1(A) is symmetric, the same conclusion is true for player A. Therefore, the strategy profile (A_1, B_1) is a Nash equilibrium strategy profile if the game is repeated infinitely. In fact, it is the only reasonable Nash equilibrium or is the SPNE. The reader would have noticed that the strategy profile (A_1, B_1), which is not a Nash equilibrium strategy profile in a single-stage or finitely repeated game with a particular payoff matrix, becomes the SPNE in an infinitely repeated game with the same payoff matrix and that cooperation is now possible. But, it should be emphasized that the strategy profile (A_1, B_1) could cease to be a Nash equilibrium strategy profile and cooperation may again be not possible if there is a slight change in the payoff matrix.

7.6.3 Maximin and minimax strategies

In Section 7.3.11 we defined a maximin strategy for a player as a strategy that maximizes the minimum gain that can be obtained from the choice of different strategies of the player. Here we discuss a closely related concept called *minimax strategy*. But, before this, consider the hypothetical zero-sum, two-player, two-strategy game presented in Table 7.6.2(A).

Since the game presented in Table 7.6.2(A) is a zero-sum game, the payoffs in each cell of the table are such that the gain of player A is equal to the loss of player B. The payoffs of only player A are given in Table 7.6.2(B). This table shows that there exists a clear distinction between the objectives of the two players. The minimum gains that A can obtain from B, no matter what strategy B chooses, are the minimum values in each row against A's strategies. These minimum values are written under the column "Row min." Similarly, the maximum payoffs that B has to concede to A, no matter what strategy A chooses, are the maximum values in each column under B's strategies. Clearly, A wants to choose the strategy that yields the maximum from the minimum gains; that is, A follows the *maximin principle* or *maximin strategy*. Similarly, B wants to choose the strategy that concedes the minimum from

Table 7.6.2(A)

		Player B's strategies		
		B_1	B_2	B_3
Player A's strategies	A_1	2, −2	4, −4	3, −3
	A_2	5, −5	1, −1	2, −2
	A_3	6, −6	5, −5	3, −3

Table 7.6.2(B)

| | | Player B's strategies | | | Row min. |
		B_1	B_2	B_3	
	A_1	2	4	3	2
Player A's	A_2	5	1	2	1
strategies	A_3	6	5	3	③
	Column max.	6	5	③	

the maximum losses; that is, B follows the *minimax principle* or *minimax strategy*. These suggest that A is the *maximizer* and B is the *minimizer* in this game.

Defining the payoffs in of Table 7.6.2(B) by π_{ij}^A, where π_{ij}^A denotes the gain of A (or the loss of B) when A chooses the ith strategy and B chooses the jth strategy, and $i = 1, 2, \ldots, m$ and $j = 1, 2, \ldots, n$, we obtain from the maximin and minimax strategies that maximin $\pi_{ij} = \max\{2, 1, 3\} = 3$ and minimax $\pi_{ij} = \min\{6, 5, 3\} = 3$, respectively. These values are circled in the table. Notice that the strategies that correspond to maximin $\pi_{ij} = 3$ and minimax $\pi_{ij} = 3$ are A_3 and B_3, respectively. The reader would have noticed that maximin $\pi_{ij} \leq$ minimax π_{ij}.

In a two-player, zero-sum, pure strategy, static game with payoff matrix π_{ij}, the payoff determined by the maximin principle is called the *lower value of the game* and we denote it by V_-; and the payoff determined by the minimax principle is called the *upper value of the game* and we denote it by V^-. Notice that $V_- = V^- = 3$ in our present example, and this common value is called the *saddle point* or *equilibrium point* of the game. The payoff determined by maximin and minimax principles is called the *value of the game*, denoted by V. This means that $V = 3$ in our example. The strategies that correspond to maximin and minimax principles or to the value of the game are called *optimal strategies*. Therefore, the optimal strategies in the present example are A_3 and B_3 for A and B, respectively. Since both players do not have an incentive to move away from the optimal strategies, the optimal strategies yield a *stable solution* to the game. A two-player, zero-sum, pure strategy, static game with payoff matrix π_{ij}^A is said to be a *fair game* if $V_- = V^- = V = 0$. A game is a *strictly determinable game* if $V_- = V^- = V \neq 0$ and, therefore, the present game is a strictly determined game. The reader would have guessed that if $V > 0$ ($V < 0$), then the game is in favor of the maximizer (minimizer). It must be emphasized that in most two-player, zero-sum, pure strategy, static games with payoff matrix π_{ij} we have the inequality $V_- \leq V \leq V^-$ or maximin $\pi_{ij} \leq V \geq$ minimax π_{ij}, which is called the *fundamental theorem of game theory*.

7.6.4 Geometric solution of games with mixed strategies

We now attempt to see how zero-sum games with mixed strategies can be solved geometrically. For this consider our Matching Pennies game introduced in Section 7.2.4. The payoff matrix of this game is reproduced in Table 7.6.3(A). The payoff matrix of player A is given in Table 7.6.3(B). It is clear from Table 7.6.3(B) that maximin $\pi_{ij} = -10 \neq$ minimax $\pi_{ij} = 10$. Since the payoff matrix in Table 7.6.3(A) is symmetric, the same is also true for player B. Therefore, the game does not have a stable solution and a saddle point in pure strategies. This suggests that we apply mixed strategies to solve this game.

Table 7.6.3(A)

		Player B's strategies	
		Heads	Tails
Player A's	Heads	10, −10	−10, 10
strategies	Tails	−10, 10	10, −10

Table 7.6.3(B)

		Player B's strategies		Row min.
		Heads	Tails	
Player A's	Heads	10	−10	−10
strategies	Tails	−10	10	−10
	Column max.	10	10	

Therefore, we need to use mixed strategies, introduced in Section 7.3.7, to determine the equilibrium of this game. Suppose that A plays "Heads" with probability p and "Tails" with probability $1 - p$. On the basis of these probabilities, we can write the expected payoffs of A when B plays "Heads" and "Tails," respectively, as

$$E_H^A(p) = 10p + [-10(1-p)] = -10 + 20p \qquad (7.6.1)$$

and

$$E_T^A(p) = -10p + 10(1-p) = 10 - 20p \qquad (7.6.2)$$

Notice that A receives the maximum of these two expected payoffs if B responds correctly to any value that A chooses for p. Therefore, A's expected payoff can be written as

$$E^A(p) = \max(E_H^A, E_T^A) \qquad (7.6.3)$$

One might wonder how the two equations (7.6.1) and (7.6.2) differ from the expected payoff of A given in the first of equations (7.3.10). Equations in (7.6.1) and (7.6.2) can be derived from the first of equations (7.3.10) if we let $q = 1$ and $q = 0$ (and using the numerical payoffs), respectively. We know from Section 7.3.9 that the payoff of A is the same for any mixed strategy (that is, for all possible levels of p) and, therefore, A will be indifferent between any mixed-strategy choices. Using this result, we can equate equations (7.6.1) and (7.6.2) to obtain $p = 1/2$.

We can now plot the expected payoffs of player A given in equations (7.6.1) and (7.6.2) to obtain the graphs in Figure 7.6.1(A). Notice that the lines representing the two equations cross each other at point E. The thick line with a twist at point E represents the part of the line E_H^A to the left of point E and the part of the line E_T^A to the right of point E. This thick line represents $E^A(p)$ in equation (7.6.3). Notice also that point E defines the value of p. It can be seen from the horizontal axis that $p = 1/2$. The expected payoff of player A, as can be

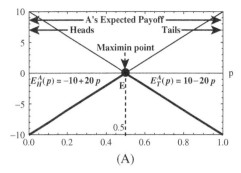
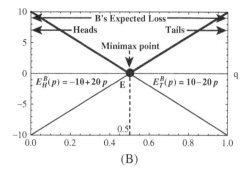

Figure 7.6.1

seen from the vertical axis, is 0. These results are identical to the results we obtained when we solved the same problem using the classical approach to optimization in example 1 in Section 7.3.10.

Similarly, we can find the minimum expected loss of B. Suppose that B plays "Heads" with probability q and "Tails" with probability $1 - q$. Then the expected losses of B when A plays "Heads" and "Tails," respectively, are

$$E_H^B(q) = 10q + [-10(1-q)] = -10 + 20q \tag{7.6.4}$$

and

$$E_T^B(q) = -10q + 10(1-q) = 10 - 20q \tag{7.6.5}$$

Notice that B concedes the minimum of these two expected payoffs. Therefore, B's expected loss can be written as

$$E^B(q) = \min(E_H^B, E_T^B) \tag{7.6.6}$$

We can now plot the expected loss of player B given in equations (7.6.4) and (7.6.5) to obtain the graphs in Figure 7.6.1(B). Notice that the lines representing the two equations cross each other at point E. The thick line with a twist at point E represents the part of the line E_T^B to the left of point E and the part of the line E_H^B to the right of point E. This thick line represents $E^B(q)$ in equation (7.6.6). Notice also that point E defines the value of q. It can be seen from the horizontal axis that $q = 1/2$. The expected loss of B, as can be seen from the vertical axis, is 0. As above, these results are identical to the results we obtained when we solved the same problem using the classical approach to optimization in example 1 in Section 7.3.10.

As another example, consider the payoffs of player A in a two-person, zero-sum game presented in Table 7.6.4(A). It is clear from Table 7.6.4(B) that maximin $\pi_{ij} = 2 \neq$ minimax $\pi_{ij} = 3$. Since the payoff matrix in Table 7.6.4(A) is symmetric, the same is also true for player B. Therefore, the game does not have a stable solution and a saddle point in pure strategies. This suggests that we apply mixed strategies to solve this game.

Therefore, we need to use mixed strategies, as earlier, to determine the equilibrium of this game. Suppose that A plays A_1 with probability p and A_2 with probability $1 - p$. On the

Table 7.6.4(A)

		Player B's strategies	
		B_1	B_2
Player A's	A_1	1	4
strategies	A_2	3	2

Table 7.6.4(B)

		Player B's strategies		Row min.
		B_1	B_2	
Player A's	A_1	1	4	1
strategies	A_2	3	2	2
	Column max.	3	4	

basis of these probabilities, we can write the expected payoff of A when B plays B_1 and B_2, respectively, as

$$E^A_{B_1}(p) = 1p + [3(1-p)] = 3 - 2p \qquad (7.6.7)$$

and

$$E^A_{B_2}(p) = 4p + 2(1-p) = 2 + 2p \qquad (7.6.8)$$

Notice that A receives the maximum of these two expected payoffs if B responds correctly to any value that A chooses for p. Therefore, A's expected payoff can be written as

$$E^A(p) = \max(E^A_{B_1}, E^A_{B_2}) \qquad (7.6.9)$$

We can now plot the expected payoffs of A given in equations (7.6.7) and (7.6.8) to obtain the graphs in Figure 7.6.2(A). Notice that the lines representing the two equation cross each other at point E. The thick line with a twist at point E represents the part of the line $E^A_{B_2}$ to the left of point E and the part of the line $E^A_{B_1}$ to the right of point E. This thick line represents $E^A(p)$ in equation (7.6.9). Notice also that point E defines the value of p. It can be seen from the horizontal axis that $p = 1/4$, which implies that $1 - p = 3/4$. Therefore, the expected value of the payoffs to A is 2.5.

Let us now consider B's problem, which is to minimize the expected loss. Assume that B plays B_1 with probability q and B_2 with probability $1 - q$. Then we can write the expected loss to B when A plays A_1 and A_2, respectively, as

$$E^B_{A_1}(q) = 1q + [4(1-q)] = 4 - 3q \qquad (7.6.10)$$

and

$$E^B_{A_2}(q) = 3q + 2(1-q) = 2 + q \qquad (7.6.11)$$

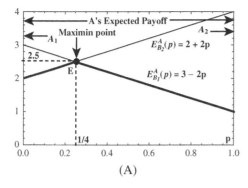
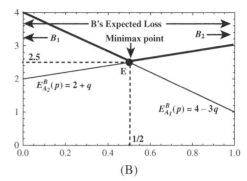

(A)   (B)

Figure 7.6.2

Notice that B concedes the minimum of these two expected payoffs. Therefore, B's expected loss can be written as

$$E^B(q) = \min(E^B_{A1}, E^B_{A2}) \tag{7.6.12}$$

We can now plot the expected loss of B given in equations (7.6.10) and (7.6.11) to obtain the graphs in Figure 7.6.2(B). Notice that the lines representing the two equations cross each other at point E. The thick line with a twist at point E represents the part of the line E^B_{A1} to the left of point E and the part of the line E^B_{A2} to the right of point E. This thick line represents $E^B(q)$ in equation (7.6.12). Notice also that point E defines the value of q. It can be seen from the horizontal axis that $q = 1/2$, which implies that $1 - q = 1/2$. The expected value of loss to B, as can be seen from the vertical axis, is 2.5.

The reader would have noticed that the optimum probabilities in both the examples we have solved in this section are such that the maximum of the minimum expected gains to one player is exactly equal to the minimum of the maximum expected loss to the other player. This is true for any two-person, zero-sum game. Therefore, we have the theorem, called the *minimax theorem*, that states for any two-person, zero-sum game there exist optimal strategies p^* and q^* such that the maximum of the minimum expected gains to one player is exactly equal to the minimum of the maximum expected loss to the other player.

7.6.5 *Connection between game theory and LP*

Since they are both concerned with optimization, one might wonder if there is any connection between game theory and LP. Yes, there is a close connection between game theory and LP. Specifically, every zero-sum, two-person, static game can be solved by converting it into an associated LP problem and every LP problem can be converted into a zero-sum, two-person game. We are now ready to explore this connection. For this we make use of a general payoff matrix (the generalized form of the payoff matrix presented in Table 7.2.1) of the payoffs to player A when player A chooses the strategy s^A_i and player B chooses the strategy s^B_j, where $i = 1, 2, \ldots, m$ and $j = 1, 2, \ldots, n$ as presented in Table 7.6.5.

The reader would have noticed that, since this is a zero-sum game, π_{ij} may be either a positive value or a negative value. We know that A is interested in finding a best mixed

Table 7.6.5

				Player B's strategies				
			s_1^B	s_2^B	...	s_j^B	...	s_n^B
	Prob. $\longrightarrow$		q_1	q_2	...	q_j	...	q_n
	s_1^A	p_1	π_{11}	π_{12}	...	π_{1j}	...	π_{1n}
	s_2^A	p_2	π_{21}	π_{22}	...	π_{2j}	...	π_{2n}
Player A's strategies	...	...	...	...	...	...	...	...
	s_i^A	p_i	π_{i1}	π_{i2}	...	π_{ij}	...	π_{in}
	...	...	...	...	...	...	...	...
	s_m^A	p_m	π_{m1}	π_{m2}	...	π_{mj}	...	π_{mn}

strategy, that is, a best set of probabilities. Notice that $\sum_{i=1}^m p_i = 1$ and $p_i \geq 1$; and $\sum_{j=1}^n q_j = 1$ and $q_j \geq 1$. Player A wants to find the optimal probabilities (or mixed strategies) such that the expected payoff to A is maximum. A's minimum expected payoff when B chooses the strategy s_1^B is given by $E_{s_1^B}^A = p_1\pi_{11} + p_2\pi_{21} + \cdots + p_i\pi_{i1} + \cdots + p_m\pi_{m1}$; and when B plays s_j^B, A's minimum expected gain will be $E_{s_j^B}^A = p_1\pi_{1j} + p_2\pi_{2j} + \cdots + p_i\pi_{ij} + \cdots + p_m\pi_{mj}$. We denote the maximum of these minimum expected gains of A by Z^-. Then we have inequalities corresponding to the last two equations as $p_1\pi_{11} + p_2\pi_{21} + \cdots + p_i\pi_{i1} + \cdots + p_m\pi_{m1} \geq Z^-$ and $p_1\pi_{1j} + p_2\pi_{2j} + \cdots + p_i\pi_{ij} + \cdots + p_m\pi_{mj} \geq Z^-$. Dividing both sides of the last two inequalities by Z^- we obtain $(p_1/Z^-)\pi_{11} + (p_2/Z^-)\pi_{21} + \cdots + (p_i/Z^-)\pi_{i1} + \cdots + (p_m/Z^-)\pi_{m1} \geq 1$ and $(p_1/Z^-)\pi_{1j} + (p_2/Z^-)\pi_{2j} + \cdots + (p_i/Z^-)\pi_{ij} + \cdots + (p_m/Z^-)\pi_{mj} \geq 1$. If $Z^- < 0$, then the direction of the last two inequalities must be reversed by multiplying them by -1. If $Z^- = 0$, then a constant must be added to every element of the payoff matrix in Table 7.6.5 so that the revised value of the game (with the revised payoff matrix) becomes positive. Similarly, if any $\pi_{ij} < 0$ in the payoff matrix, then a constant must be added to every element in the matrix so that the least element will be zero. But, once the optimal solution is obtained (in both cases above), the correct value of the game can be found by subtracting the constant from the revised value of the game. Let us now define $x_i = p_i/Z^-$. Then the last two inequalities can be rewritten as $\pi_{11}x_1 + \pi_{21}x_2 + \cdots + \pi_{i1}x_i + \cdots + \pi_{m1}x_m \geq 1$ and $\pi_{1j}x_1 + \pi_{2j}x_2 + \cdots + \pi_{ij}x_i + \cdots + \pi_{mj}x_m \geq 1$. We know that maximizing Z^- is equivalent to minimizing $1/Z^-$. We can now write the LP version of player A's game problem presented in Table 7.6.5 by expanding i and j in the last inequality as

Minimize $1/Z^- = x_1 + x_2 + \cdots + x_i + \cdots + x_m$,

subject to $\pi_{11}x_1 + \pi_{21}x_2 + \cdots + \pi_{i1}x_i + \cdots + \pi_{m1}x_m \geq 1$,

$\pi_{12}x_1 + \pi_{22}x_2 + \cdots + \pi_{i2}x_i + \cdots + \pi_{m2}x_m \geq 1, \ldots,$

$\pi_{1j}x_1 + \pi_{2j}x_2 + \cdots + \pi_{ij}x_i + \cdots + \pi_{mj}x_m \geq 1, \ldots,$

$\pi_{1n}x_1 + \pi_{2n}x_2 + \cdots + \pi_{in}x_i + \cdots + \pi_{mn}x_m \geq 1,$ and $x_1, x_2, \ldots, x_m \geq 0$ \hfill (7.6.13)

Let us now approach B's problem. Player B wants to find the optimal probabilities (or mixed strategies) such that the expected loss to B is minimum. B's minimum expected loss when A chooses the strategy s_1^A is given by $E_{s_1^A}^B = q_1\pi_{11} + q_2\pi_{12} + \cdots + q_j\pi_{1j} + \cdots + q_n\pi_{1n}$; and when A plays s_i^A, B's minimum expected loss will be $E_{s_i^A}^B = q_1\pi_{i1} + q_2\pi_{i2} + \cdots + q_j\pi_{ij} + \cdots + q_n\pi_{in}$. We denote the minimum of these maximum expected losses of player B by Z_-. Then we have inequalities corresponding to the last two equations as $q_1\pi_{11} + q_2\pi_{12} + \cdots + q_j\pi_{1j} + \cdots + q_n\pi_{1n} \le Z_-$ and $q_1\pi_{i1} + q_2\pi_{i2} + \cdots + q_j\pi_{ij} + \cdots + q_n\pi_{in} \le Z_-$. Dividing both sides of the last two inequalities by Z_- we obtain $(q_1/Z_-)\pi_{11} + (q_2/Z_-)\pi_{12} + \cdots + (q_j/Z_-)\pi_{1j} + \cdots + (q_n/Z_-)\pi_{1n} \le 1$ and $(q_1/Z_-)\pi_{i1} + (q_2/Z_-)\pi_{i2} + \cdots + (q_j/Z_-)\pi_{ij} + \cdots + (q_n/Z_-)\pi_{in} \le 1$. Let us now define $y_j = q_j/Z_-$. Then the last two inequalities can be rewritten as $\pi_{11}y_1 + \pi_{21}y_2 + \cdots + \pi_{i1}y_j + \cdots + \pi_{m1}y_n \le 1$ and $\pi_{i1}y_1 + \pi_{i2}y_2 + \cdots + \pi_{ij}y_j + \cdots + \pi_{in}y_n \le 1$. We know that minimizing Z_- is equivalent to maximizing $1/Z_-$. We can now write the LP version of B's game problem presented in Table 7.6.5 by expanding i and j in the last inequality as

$$\text{Maximize } 1/Z_- = y_1 + y_2 + \cdots + y_j + \cdots + y_n,$$

$$\text{subject to } \pi_{11}y_1 + \pi_{12}y_2 + \cdots + \pi_{1j}y_j + \cdots + \pi_{1n}y_n \le 1,$$

$$\pi_{21}y_1 + \pi_{22}y_2 + \cdots + \pi_{2j}y_j + \cdots + \pi_{2n}y_n \le 1, \ldots,$$

$$\pi_{i1}y_1 + \pi_{i2}y_2 + \cdots + \pi_{ij}y_j + \cdots + \pi_{in}y_n \le 1, \ldots,$$

$$\pi_{m1}y_1 + \pi_{m2}y_2 + \cdots + \pi_{mj}y_j + \cdots + \pi_{mn}y_n \le 1, \text{ and } y_1, y_2, \ldots, y_n \ge 0 \qquad (7.6.14)$$

The reader would have noticed that A's LP problem in (7.6.13) is the dual of B's LP problem in (7.6.14) and vice versa. This implies that $Z^{-*} = Z_-^*$; that is, the expected gain to player A is equal to the expected loss to player B in the optimum. Since the problems in (7.6.13) and (7.6.14) are the duals of each other, the solution to one can be obtained from the optimal simplex tableau of the other.

So far we have converted a two-person, zero-sum game into an equivalent LP problem. One can do the opposite also; that is, convert a LP problem into an equivalent two-person, zero-sum game. Although one can carry out the latter conversion, the benefits from that outweigh the associated costs. Moreover, we know from Chapter 5 that it is much easier to solve LP problems using the methods presented therein. Therefore, we do not present here the conversion of a LP problem into a game. Instead, the interested reader can consult the bibliography at the end of the book.

As an example, consider the two-person, zero-sum Matching Pennies game introduced in Section 7.2.4 and solved using the classical approach to optimization in Section 7.3.10 and geometrically in Section 7.6.4. The matrix showing the payoffs of both players in this game are presented in Table 7.6.3(A). The matrix of the payoffs of A is presented in Table 7.6.3(B). Notice that in Table 7.6.3(B), $\pi_{11} = 10$, $\pi_{12} = -10$, $\pi_{21} = -10$, and $\pi_{22} = 10$; and that $s_1^A = $ Heads, $s_2^A = $ Tails, $s_1^B = $ Heads, and $s_2^B = $ Tails. Our problem is to present the LP versions of the game problems that both players face and solve them using the LP approach to optimization.

Using the values in Table 7.6.3(B) and following the derivation of problems in equations (7.6.13) and (7.6.14), and using constants (10) in the case of negative π_{ij} in the payoff table, we can write the LP versions of the game problems of players A and B, respectively, as

$$\text{Minimize } 1/Z^- = x_1 + x_2, \text{ subject to } 20x_1 - 0x_2 \geq 1, 0x_1 + 20x_2 \geq 1, \text{ and } x_1, x_2 \geq 0$$

$$\text{Maximize } 1/Z_- = y_1 + y_2, \text{ subject to } 20y_1 - 0y_2 \leq 1, 0y_1 + 20y_2 \leq 1, \text{ and } x_1, x_2 \geq 0$$
$$(7.6.15)$$

Notice that in the last two problems $x_i = p_i/Z^-$ and $y_j = q_j/Z_-$. Solving these two problems using the methods presented in Chapter 5, we obtain that $1/Z^{-*\text{revised}} = 1/Z^*_{-\text{revised}} = 1/10$ and $x^* = y^* = 1/20$. These imply that $Z^{-*\text{revised}} = Z^*_{-\text{revised}} = 10$. We now need to subtract the constant (10) added to the elements of the payoff matrix from $Z^{-*\text{revised}} = Z^*_{-\text{revised}} = 10$ to obtain the old but correct Z^{-*} and Z^*_-. Therefore, we have $Z^{-*} = Z^*_- = Z^{-*\text{revised}} - 10 = Z^*_{-\text{revised}} - 10 = 0$. Since $x_1^* = p_1^*/Z^{-*\text{revised}} = 1/20$, $x_2^* = p_2^*/Z^{-*\text{revised}} = 1/20$, $y_1^* = q_1^*/Z^*_{-\text{revised}} = 1/20$, and $y_2^* = q_2^*/Z^*_{-\text{revised}} = 1/20$; we have $p_1^* = x_1^* \times Z^{-*\text{revised}} = (1/20) \times 10 = 1/2$, $p_2^* = x_2^* \times Z^{-*\text{revised}} = (1/20) \times 10 = 1/2$, $q_1^* = y_1^* \times Z^*_{-\text{revised}} = (1/20) \times 10 = 1/2$, and $q_2^* = y_2^* \times Z^*_{-\text{revised}} = (1/20) \times 10 = 1/2$. In short, we obtain $Z^{-*} = Z^*_- = 0$ and $p_1^* = p_2^* = q_1^* = q_2^* = 1/2$. Notice that these results are identical to the results we obtained in Sections 7.3.10 and 7.6.4 when we solved the same problem using the classical approach to optimization and geometrically, respectively.

As another example, consider the second two-person, zero-sum game we solved geometrically in the last section. The matrix showing the payoffs of player A in this game is presented in Table 7.6.4(A). Notice that in this table, $\pi_{11} = 1$, $\pi_{12} = 4$, $\pi_{21} = 3$, and $\pi_{22} = 2$; and that $s_1^A = A_1$, $s_2^A = A_2$, $s_1^B = B_1$, and $s_2^B = B_2$. Our problem is to present the LP versions of the game problems that both players face and solve them using the LP approach to optimization.

Using the values in Table 7.6.4 and following the derivation of problems (7.6.13) and (7.6.14), we can write the LP versions of the game problems of players A and B, respectively, as

$$\text{Minimize } 1/Z^- = x_1 + x_2, \text{ subject to } x_1 + 3x_2 \geq 1, 4x_1 + 2x_2 \geq 1, \text{ and } x_1, x_2 \geq 0$$

$$\text{Maximize } 1/Z_- = y_1 + y_2, \text{ subject to } y_1 + 4y_2 \leq 1, 3y_1 + 2y_2 \leq 1, \text{ and } y_1, y_2 \geq 0$$
$$(7.6.16)$$

Notice that, as before, in the last two problems $x_i = p_i/Z^-$ and $y_j = q_j/Z_-$. Solving these two problems using the methods presented in Chapter 5, we obtain that $1/Z^{-*} = 1/Z^*_- = 2/5$, $x_1^* = 1/10$, $x_2^* = 3/10$, and $y_1^* = y_2^* = 1/5$. These imply that $Z^{-*} = Z^*_- = 5/2 = 2.5$. Since $x_1^* = p_1^*/Z^{-*} = 1/10$, $x_2^* = p_2^*/Z^{-*} = 3/10$, $y_1^* = q_1^*/Z^*_- = 1/5$, and $y_2^* = q_2^*/Z^*_- = 1/5$; we have $p_1^* = x_1^* \times Z^{-*} = (1/10) \times (5/2) = 1/4$, $p_2^* = x_2^* \times Z^{-*} = (3/10) \times (5/2) = 3/4$, $q_1^* = y_1^* \times Z^*_- = (1/5) \times (5/2) = 1/2$, and $q_2^* = y_2^* \times Z^*_{-d} = (1/5) \times (5/2) = 1/2$. In short, we obtain $Z^{-*} = Z^*_- = 5/2 = 2.5$, $p_1^* = 1/4$, $p_2^* = 3/4$, and $q_1^* = q_2^* = 1/2$. Notice that these results are identical to the results we obtained in the last section.

8   Integral calculus

8.1 Introduction

In the first three chapters of this book we developed the necessary mathematical tools required for the analyses of many of the relationships one finds in the fields of economics, business, and finance. We then applied these tools to analyze some of these relationships in Chapters 4–7. These relationships involved variables and *parameters* or constants. In all our analyses so far we have been mainly interested in the properties of the variables when they attain some equilibrium position(s). We have also been interested in the comparison of the properties of the variables in different equilibrium positions attained after changes in the parameters. The former analysis is called *statics* or *static analysis* and the latter analysis is called *comparative statics* or *comparative static analysis*.

We considered numerous examples of static and comparative static analyses in previous chapters. As a recap, consider the problem in example 6 in Section 1.6.7. In this problem we first wanted to find the pre-tax equilibrium (point E in Figure 1.6.6(A)) and then wanted to find the post-tax equilibrium (point E^t). We then compared these two equilibrium positions. Finding the pre-tax equilibrium price and quantity demanded and supplied is called static analysis. Similarly, finding the post-tax equilibrium price and quantity, and its comparison with the pre-tax equilibrium price and quantity, is called comparative static analysis. The reader would have noticed that we have carried out earlier several such static and comparative static analyses.

But, an astute reader would have noticed an important deficiency in static and comparative static analyses. Consider, again for simplicity, the pre-tax equilibrium point E and the post-tax equilibrium point E^t in Figure 1.6.6(A). The reader might ask what happens to the variables, price and quantity, when we move from the pre-tax equilibrium position to the post-tax equilibrium position. In other words, the evolution of the variable(s) is also important, just as their equilibrium values or properties. We could not study or analyze the evolution of variables in different relationships we have considered so far because we did not develop the required mathematical tools for such an analysis. Therefore, our major aim in the present chapter is to develop one of those mathematical tools that can be used to analyze the evolution of variables. This type of analysis is called in the literature *dynamics* or *dynamic analysis*.

An astute reader would have also guessed that dynamic analyses involve time. Once again, consider the example of the impact of a tax on price and quantity we mentioned above. It is difficult for one to think that the pre-tax equilibrium values of price and quantity would change and attain new, post-tax equilibrium values at the same time as the tax is imposed. It is unnecessary to state that it will take *some* time before the post-tax equilibrium values are reached. The justification for this delay in attaining new equilibrium position is that both consumers and producers or suppliers require some time to adjust their demand and supply. What all this implies is that time plays a crucial role in dynamic analysis. Therefore, we need to introduce time explicitly into dynamic analyses.

The introduction of time is sometimes necessary even if we are not interested in the comparison of equilibrium positions of variables. This is particularly so if we wish to explore the independent evolution of some variables. Such variables are numerous in the subjects of our interest. Examples of such variables include the movements of gross domestic product, per capita income, consumption expenditure, national debt, investment, exports and imports, corporate profits, stock or share prices, and so on. In many cases, we may be interested in the evolution, or the chronological movements, of these and other variables. It is again unnecessary to state that the movements of these variables are time-conditioned.

However, a natural question that arises here is how one introduces time. Time can be introduced in two ways: either in discrete form or in continuous form. In the former case, the variable under consideration is expected to change only once in a period of time; for example, in a week, in a month, in a year, etc. But, in the latter case, the variable is expected to change at every possible point in time. An example of the former is the *annual compounding* of interest on a bank deposit and an example of the latter is the *continuous compounding* of interest on another deposit. If a variable is evolving continuously over time, then the study of its evolution can be carried out using *integral calculus* or *differential equations*. And if the variable is evolving discretely over time, its evolution can be studied using *difference equations*. We will study the evolution of variables only using integral calculus in this book. The other two, with some other supplements, are presented on the website of the book.

8.2 Indefinite integrals

8.2.1 Introduction

One of the types of integrals we present in this book is called the *indefinite integral*. For the purpose of exposition of indefinite integrals, we use a simple example. Assume that the total profit, Π, of a company varies as time, t, varies. Also assume that the rate of change of the total profit of the company is given by the first derivative

$$d\Pi/dt = (1/3)t^{-2/3} \tag{8.2.1}$$

which implies that, since $d\Pi/dt = (1/3)t^{-2/3} > 0$ for any $t > 0$, the total profit of the company is an increasing function of time. Moreover, since $d^2\Pi/dt^2 = (-2/9)t^{-5/3} < 0$ for any $t > 0$, the total profit of the company increases at a diminishing rate. The problem now is how one can obtain the time path of the company's total profit function that will yield the rate of change given in equation (8.2.1). In other words, we need to find the original total profit function, which is a function of time, on the basis of the rate of change of that function given in equation (8.2.1). This original function is also called in the literature the *primitive function*. The reader would have noticed that the rate of change in equation (8.2.1) is a *derived function*. Therefore, the question is how one can find the primitive function from the derived function.

The method of finding the primitive function from the derived function is called *integration*. The branch of mathematics concerned with this method is called *integral calculus*. The reader would have noticed that in differential calculus we were interested in finding the derived function from the primitive function. Here in integral calculus we do the opposite: find the original, primitive function from the derived function. For this reason, some authors state that integration is the opposite of differentiation. We explore this idea in detail in the following sections.

8.2.2 Constants of integration and multiple primitive functions

Consider again the rate of change of the total profit given in equation (8.2.1). A visual inspection of that derived function would suggest that *a* form of the primitive total profit function is $\Pi = f(t) = t^{1/3}$. An astute reader might wonder why we said "a" form, instead of "the" form. The reason is that a multitude of primitive functions would give us the derived function in equation (8.2.1).

As an example, consider the primitive function $\Pi = f(t) = t^{1/3}$. The derivative of this function with respect to t would yield the derived function in equation (8.2.1). Now consider the primitive function $\Pi = f(t) = t^{1/3} + C$, where C is a constant that may take any value, which gives the same derived function. The only difference between the two primitive functions $\Pi = f(t) = t^{1/3}$ and $\Pi = f(t) = t^{1/3} + C$ is the presence of the constant C in the latter. Notice that we can generate an infinite number of primitive functions using infinite number of values for the constant. Therefore, the most general primitive function of the derived function in equation (8.2.1) is $\Pi = f(t) = t^{1/3} + C$. We will see in Section 8.3.2 that the last equation is the *integral* of the derived function. The above discussion can be illustrated geometrically, as presented in Figure 8.2.1, with the graph of the derived function $d\Pi/dt = (1/3)t^{-2/3}$ and the graphs of the primitive function $\Pi = f(t) = t^{1/3} + C$ with differing values for the constant C (0, 1, and 2). Since the graphs of the primitive functions are simply vertical translations, they all have the same slope.

Since all the above mentioned primitive functions would yield the same derived function in equation (8.2.1), we cannot derive a unique primitive function unless we are given additional information about the value that the constant takes. What is this additional information?

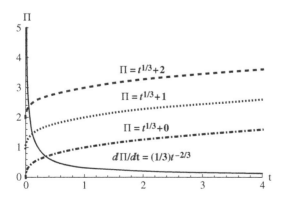

Figure 8.2.1

This additional information is called in the literature *initial condition* or *boundary condition*, which we will discuss in Section 8.3.11.

8.2.3 Integration as anti-differentiation

In the last two sections we mentioned that integration was the process of deriving the primitive function or the *parental function* on the basis of the derived function or the *progeny function*. In this section we use an even simpler example to elucidate the real meaning of integration.

Suppose that we have a univariate function $F(x)$. Also suppose that the first derivative of this function with respect to x is $F'(x) = f(x)$. Then integration is the process of finding $F(x)$ using $F'(x)$ or $f(x)$. What this means is that in integration we are going back from the progeny function to the parental function.

As an example, suppose that $F'(x) = f(x) = x^2$. How do we find $F(x)$ using $F'(x) = f(x) = x^2$? As mentioned in the last section, a large number of parental functions would give us the derivative $F'(x) = f(x) = x^2$. Therefore, following our discussion in the last section and as we shall see in Section 8.3.2, the most general parental function for this progeny function can be written as $F(x) = (1/3)x^3 + C$. Notice that if we differentiate this parental function with respect to x we will obtain the progeny function: $F'(x) = f(x) = d[(1/3)x^3 + C]/dx = x^2$. This result shows that integration of the derivative or the progeny function yields the primitive or parental function. In other words, integration is the opposite of differentiation or is *anti-differentiation*.

8.2.4 Notations and concepts

For our further exposition of integral calculus and its applications in the fields of our interest we need to specify some notations and define some concepts.

We denoted the derived or the derivative function by $F'(x) = f(x)$. And we know that integration is the procedure of finding a parent function or a primitive function $F(x)$ using $F'(x) = f(x)$. This procedure is written in standard symbol as $\int F'(x)\,dx = \int f(x)\,dx$. This symbolic form is called the *indefinite integral* of $F'(x)$, or of $f(x)$, with respect to the variable x. The reason why it is called the indefinite integral is due to both the presence of the general constant C in the parental function and the nonspecification of the interval of values that the variable x can take. We will see in Section 8.4 that integration of a function over a certain interval will remove the said constant.

In the above indefinite integral $\int F'(x)\,dx = \int f(x)\,dx$, the elongated "S," that is, the sign $\int$, is called the *integral sign*. The function $F'(x) = f(x)$, which we intend to integrate, is called the derived or derivative function or the *integrand*. We found in the last section that $\int F'(x)\,dx = \int f(x)\,dx = F(x) + C$. The constant C in the result of integration is called the *constant of integration*. The part dx in $\int F'(x)\,dx = \int f(x)\,dx$ shows the *variable of integration*; in our present case, x is the variable of integration.

8.3 Rules of integration

We presented the rules of differentiation of various types of functions in Section 3.3. Similarly, in the current section, we present some of the important rules of integration. Unlike in the case of the rules of differentiation, we will not attempt to validate these rules of integration. However, we will present numerical examples of these rules.

8.3.1 Constant function rule

Assume that the function we want to integrate is a constant function given by $f(x) = k$, where k is a constant. Then the integral of $f(x)$ with respect to x is

$$\int f(x)dx = \int k\,dx = F(x) + C = kx + C \tag{8.3.1}$$

The reader would have noticed that the basic idea here is that we need to obtain the derivative function if we differentiate the original function with respect to the independent variable. This is easy to check in the present case: $d[\int f(x)dx]/dx = d[kx + C]/dx = k$. The derivative of the primitive function is equal to the integrand, which checks with the result in equation (8.3.1). As a numerical example, consider the constant function $f(x) = 2$. The integral of $f(x) = 2$ with respect to x is $\int f(x)dx = \int 2\,dx = F(x) + C = 2x + C$.

8.3.2 Power function rule

Suppose that the function we want to integrate is a power function given by $f(x) = x^n$, where $n \neq -1$ is a constant. Then the integral of $f(x)$ with respect to x is

$$\int f(x)dx = \int x^n dx = F(x) + C = \frac{x^{n+1}}{n+1} + C \tag{8.3.2}$$

We can check whether the integral is correct or not by differentiating the integral with respect to the independent variable (x). Thus, we obtain $d[\int f(x)dx]/dx = d[\{x^{n+1}/(n+1)\} + C]/dx = x^n$, which checks with equation (8.3.2).

As an example, consider the derived function we used in Section 8.2.2. Equation (8.2.1) gives the function as $F'(t) = f(t) = (1/3)t^{-2/3}$. The integral of this function is $\int f(t)dt = \int (1/3)t^{-2/3}dt = (1/3)\int t^{-2/3}dt = F(t) + C = (1/3)\{1/[(-2/3) + 1]\}t^{(-2/3)+1} + C = t^{1/3} + C$. As before, we can check whether the result we obtained is correct or not. Differentiating the integral with respect to t, we obtain $d[F(t) + C]/dt = d[t^{1/3} + C]/dt = (1/3)t^{(1/3)-1} = (1/3)t^{-2/3}$. This shows that the result we obtained above is correct. As another example, consider the function $f(x) = x^2$. The integral of this function is $\int f(x)dx = \int x^2 dx = F(x) + C = (1/3)x^3 + C$. Notice that this is the same result as that we obtained in Section 8.2.3. If we differentiate the integral with respect to x we obtain $d[F(x) + C]/dx = d[(1/3)x^3 + C]/dx = 3x^2/3 = x^2$, which indeed is equal to the integrand.

8.3.3 Logarithmic rule

Suppose that we a have function of the form $f(x) = 1/x = x^{-1}$, where $x > 0$. Then the integral of $f(x)$ with respect to x is

$$\int f(x)dx = \int \frac{1}{x}dx = F(x) + C = \ln x + C, \quad x > 0 \tag{8.3.3}$$

Notice that we used the restriction $x > 0$ in the integral in equation (8.3.3). The reason is that negative values do not have logarithms. If we use the restriction that $x \neq 0$, then the integral of $f(x) = 1/x = x^{-1}$ is

$$\int f(x)dx = \int \frac{1}{x}dx = F(x)+C = \ln|x|+C, \quad x \neq 0 \tag{8.3.4}$$

As earlier, we can check whether the integral in equation (8.3.3) or (8.3.4) is correct or not by differentiating it with respect to x. Differentiating the integral in equation (8.3.3) we obtain $d[F(x)+C]/dx = d[\ln x + C]/dx = 1/x = x^{-1}$. This result is equal to the integrand and, therefore, the integral we obtained is correct.

8.3.4 Exponential function rule

Assume that we have an exponential function of the form $f(x) = e^x$. Then the integral of $f(x)$ with respect to x is

$$\int f(x)dx = \int e^x dx = F(x)+C = e^x +C \tag{8.3.5}$$

The reader would have noticed the resemblance between the integral and derivative of an exponential function. These two differ only in respect of the constant C. If $C = 0$, then the two are identical. The reader would have also noticed that $d[F(x)+C]/dx = d[e^x +C]/dx = e^x$; that is, the derivative of the integral is the same as the integrand.

A variant of the rule in equation (8.3.5) is the integral of the function $f(x) = e^{kx}$, where k is a constant. Then the integral of $f(x) = e^{kx}$ is

$$\int f(x)dx = \int e^{kx}dx = F(x)+C = [e^{kx}/k]+C \tag{8.3.6}$$

It is unnecessary to say that the derivative of the integral in equation (8.3.6) is the integrand because $d[F(x)+C]/dx = d[(e^{kx}/k)+C]/dx = e^{kx}k/k = e^{kx}$. Instead of an exponential function with e as its base, suppose that we have an exponential function of the form $f(x) = a^{kx}$. Then the integral of $f(x)$ with respect to x is

$$\int f(x)dx = \int a^x dx = F(x)+C = [a^{kx}/k\ln a]+C \tag{8.3.7}$$

Notice that the derivative of the integral in equation (8.3.7) with respect to x is $d[F(x)+C]/dx = d[(a^{kx}/k\ln a)+C] = (1/k\ln a).a^{kx}.\ln a.k = a^{kx}$, which is identical to the integrand. As a numerical example, consider the function $f(x) = e^{2x}$. The integral of this function with respect to x is $\int f(x)dx = \int e^{2x}dx = F(x)+C = (e^{2x}/2)+C$. One can check that this result is correct as the derivative of the integral $(e^{2x}/2)+C$ with respect to x is identical to the integrand $f(x) = e^{2x}$. Similarly, consider the function $f(x) = 5^{2x}$. The integral of this function with respect to x is $\int f(x)dx = \int 5^{2x}dx = F(x)+C = (5^{2x}/2\ln 5)+C$. If we differentiate the integral with respect to x we obtain $d[F(x)+C]/dx = d[(5^{2x}/2\ln 5)+C]/dx = (5^{2x} \times 2\ln 5/2\ln 5) = 5^{2x}$, which is identical to the integrand and, therefore, the integral obtained is correct.

8.3.5 Constant multiple of a function rule

Suppose that we have a function multiplied by a constant, such as $kf(x)$, where k is a constant. The integral of this function is

$$\int kf(x)dx = F(x) + C = k \int f(x)dx \tag{8.3.8}$$

which implies that the integral of a function multiplied by a constant is equal to the constant times the integral of the function. This rule helps us place the constant term on the LHS of the integral sign. As an example, consider the function $f(x) = 6x^5$. The integral of this function can be written as $\int 6x^5 dx = 6 \int x^5 dx = F(x) + C = 6 \times [1/(5+1)]x^{5+1} + C = 6(1/6)x^6 + C = x^6 + C$. Notice that we obtain the integrand when we differentiate the integral with respect to x: $d[x^6 + C]/dx = 6x^5$.

8.3.6 Sum–difference rule

Assume that we have a function generated by combining two functions $f(x)$ and $g(x)$. Then the integral of the sum of these two functions is equal to the sum of the integrals of the two individual functions. Similarly, the integral of the difference of the two functions is equal to the difference of the integrals of the two individual functions. In other words, we have

$$\int [f(x) + g(x)]dx = \int f(x)dx + \int g(x)dx \tag{8.3.9}$$

and

$$\int [f(x) - g(x)]dx = \int f(x)dx - \int g(x)dx \tag{8.3.10}$$

As an example, consider the sum of the functions $f(x) = e^{2x}$ and $g(x) = 5^{2x}$. Then, using equation (8.3.9), we can write the integral of the sum of the two functions as $\int [f(x) + g(x)]dx = \int [e^{2x} + 5^{2x}]dx = \int e^{2x}dx + \int 5^{2x}dx$. We know from Section 8.3.4 that $\int e^{2x}dx = (1/2)e^{2x} + C_1$ and $\int 5^{2x}dx = (1/2\ln 5)5^{2x} + C_2$, where C_1 represents the constant C in $\int e^{2x}dx = (1/2)e^{2x} + C$ and C_2 represents the constant C in $\int 5^{2x}dx = (1/2\ln 5)5^{2x} + C$. Therefore, by combining these two results we obtain $\int [f(x) + g(x)]dx = \int [e^{2x} + 5^{2x}]dx = \int e^{2x}dx + \int 5^{2x}dx = (1/2)e^{2x} + C_1 + (1/2\ln 5)5^{2x} + C_2$. Now letting $C_1 + C_2 = C$, we can write the integral as $\int [f(x) + g(x)]dx = \int [e^{2x} + 5^{2x}]dx = \int e^{2x}dx + \int 5^{2x}dx = (1/2)e^{2x} + (1/2\ln 5)5^{2x} + C$.

As another example, consider the difference of the functions $f(x) = 2x$ and $g(x) = x$. We know by now that the integral of $f(x) = 2x$ is $\int f(x)dx = \int 2xdx = 2 \int xdx = 2(1/2)x^2 + C_1$ and the integral of $g(x) = x$ is $\int g(x)dx = \int xdx = (1/2)x^2 + C_2$. Therefore, the integral of the difference of the two functions can be written, using equation (8.3.10), as $\int [f(x) - g(x)]dx = \int [2x - x]dx = \int 2xdx - \int xdx = (2/2)x^2 + C_1 - (1/2)x^2 - C_2 = x^2 - (1/2)x^2 + C_1 - C_2$. Letting $C_1 - C_2 = C$ we can write the last equation as $\int [f(x) - g(x)]dx = (1/2)x^2 + C$.

8.3.7 Substitution rule

Sometimes we may come across functions that cannot be integrated using any of the rules presented so far. As an example, consider the function $f(x) = 3x^2(x^3 + 5)^{99}$. How do we find the integral of this function? Before attempting to find the integral, we can apply equation (8.3.8) to place the constant 3 to the LHS of the integral sign. But we cannot do this for x^2 as x is the variable of integration. One might think that one can use the binomial theorem to expand the $(x^3 + 5)^{99}$ term of the integrand and then integrate. But this will be a Herculean task and, therefore, is not advisable. Then, what is the way ahead? Here, the chain rule of differentiation (equation (3.3.8)) comes to our help. We first develop the method, and then apply the method to solve the problem.

Suppose that we have two functions of the form $F(u)$ and $u = U(x)$ and that we write $F(u) = F[U(x)] = F(x)$. Differentiating this equation with respect to x using the chain rule we obtain $F'(u) = F'[U(x)]U'(x) = F'(x)$. We now denote $F'(u)$ by $f(u)$, $U'(x)$ by $u(x)$, and $F'(x)$ by $f(x)$. Therefore, the last equation can be written as $f(u) = f(u)u(x) = f(x)$. Let us now take the first differential of the last equation to obtain $f(u)du = f(u).u(x)dx = f(x)dx$. Since $u(x) = du/dx$, the last equation can be rewritten as $f(u)du = f(u)(du/dx)dx = f(x)dx$. If we now integrate the last equation, we obtain $\int f(u)du = \int f(u)(du/dx)dx = \int f(x)dx = F(u) + C = F(x) + C$. Therefore, we can write

$$\int f(x)dx = \int f(u)[du/dx]dx = \int f(u)du \qquad (8.3.11)$$

which is called the *substitution rule* of integration.

As an example, consider the function $f(x) = 3x^2(x^3 + 5)^{99}$. First, we treat $u = U(x) = x^3 + 5$. Notice that the first term in the integrand, $3x^2$, is the first derivative of $u = U(x) = x^3 + 5$ with respect to x; that is, $du/dx = 3x^2$. From this equation we can obtain $dx = du/3x^2$. Therefore, the integral can be written as $\int f(x)dx = \int 3x^2 u^{99} du/3x^2 = \int u^{99}du = (u^{100}/100) + C$. We can now substitute $u = U(x) = x^3 + 5$ back into this integral to obtain $\int f(x)dx = (u^{100}/100) + C = [(x^3 + 5)^{100}/100] + C$. We can check to see if the result we obtained is correct. Differentiating the integral with respect to x yields $d[\int f(x)dx]/dx = d[\{(x^3 + 5)^{100}/100\} + C]dx = [100(x^3 + 5)^{99}/100] \times 3x^2 = 3x^2(x^3 + 5)^{99}$, which is precisely the integrand with which we started.

As another example, consider the function $f(x) = 3e^{3x-5}dx$. Assume now that $u = U(x) = 3x - 5$. This gives us $du/dx = 3$ or $dx = du/3$. Therefore, the integral can be written as $\int f(x)dx = \int 3e^{3x-5}dx = \int 3e^u dx = \int 3e^u du/3 = \int e^u du = e^u + C$. We can now substitute $u = U(x) = 3x - 5$ back into this integral to obtain $\int f(x)dx = e^{3x-5} + C$. Verifying this result confirms that $d[\int f(x)dx]/dx = d[e^{3x-5} + C]/dx = 3e^{3x-5}$, which checks with the integrand.

Consider the function $f(x) = (2x + 1)/(x^2 + x)$. Now let $u = U(x) = x^2 + x$. This gives us $du/dx = 2x + 1$ or $dx = du/(2x + 1)$. Therefore, the integral can be written as $\int f(x)dx = \int (2x + 1)/(x^2 + x)dx = \int [(2x + 1)/u]dx = \int [(2x + 1)/u].[du/(2x + 1)] = \int (1/u)du = \ln|u| + C$. We can now substitute $u = U(x) = x^2 + x$ back into this integral to obtain $\int f(x)dx = \ln|x^2 + x| + C$. We can verify the result we obtained by differentiating the integral with respect to x. Differentiation of the integral with respect to x gives $d[\int f(x)dx]/dx = d[\ln(x^2 + x)]/dx = [1/(x^2 + x)](2x + 1) = (2x + 1)/(x^2 + x)$, which confirms that the result obtained is correct.

We have seen so far in this section that if the integral is a constant multiple of another function and its derivative (that is, $u = U(x)$ and du/dx), then integration by substitution is possible. Notice that this substitution rule reverses the operation of the chain and the power function rules of differentiation. However, a careful reader would have noticed that the substitution method can only be used in the special case where one term of the integrand is the first derivative of the other term of the integrand. But, we rarely encounter these types of integrands. However, in many cases, particularly when the integrand appears as a constant multiple of another function and its derivative, the substitution rule is helpful. The steps of applying the substitution rule are as follows.

Step 1. Find a substitution that simplifies the integrand; that is, try to select $u = U(x)$ so that du becomes a part of the integrand.

Step 2. Try to express the integrand in terms of u and du only (and eliminate the original variable x and its differential dx).

Step 3. Evaluate the new integral and express the result in terms of the original variable x.

8.3.8 *Integration by parts*

We found in the previous section that if the integrand can be expressed as a constant multiple of a function and its first derivative (that is, $u = U(x)$ and du/dx), then we could apply the substitution rule. But, many of the integrands we find in the subjects of our interest cannot be expressed like this. Examples include integrands such as $\int xe^x dx$, $\int x^2 e^{2x} dx$, $\int x \ln x dx$, etc. In such situations we have to apply another rule of integration called *integration by parts*. While the substitution rule of integration relies on the chain rule of differentiation, integration by parts relies on the product rule of differentiation (equation (3.3.6)). In fact, integration by parts reverses the process of differentiating a product of two functions.

Suppose that we have three functions: $u = f(x)$, $v = g(x)$, and $z(x) = f(x)(gx)$. Then, from the product rule of differentiation, we know that $d[z(x)]/dx = d[f(x)g(x)]/dx = f(x)g'(x) + g(x)f'(x)$. Let us now integrate both sides of the last equation to obtain $\int [d[z(x)]/dx] dx = \int [d[f(x)g(x)]/dx] dx = \int f(x)g'(x)dx + \int g(x)f'(x)dx$. This can be written as $\int dz = \int d[f(x)g(x)] = \int f(x)g'(x)dx + \int g(x)f'(x)dx$, or as $z = f(x)g(x) = \int f(x)g'(x)dx + \int g(x)f'(x)dx$, or as $\int f(x)g'(x)dx = f(x)g(x) - \int g(x)f'(x)dx$, or as

$$\int g(x)f'(x)dx = f(x)g(x) - \int f(x)g'(x)dx \qquad (8.3.12)$$

Using the definitions $u = f(x)$ and $v = g(x)$, we can write equation (8.3.12) in a more convenient form as

$$\int udv = uv - \int vdu \quad \text{or} \quad \int vdu = uv - \int udv \qquad (8.3.13)$$

The result in equation (8.3.12) or (8.3.13) is called the *rule of integration by parts*.

As an example, consider the problem $\int xe^x \, dx$. Assume that $f(x) = x$ and that $g'(x) = e^x$. The last equation implies that $g(x) = e^x$. Notice that $f(x)g'(x) = xe^x$ is now equivalent to the integrand on the LHS of equation (8.3.12). Therefore, we can write the integral as $\int f(x)g'(x)dx = \int xe^x dx = f(x)g(x) - \int g(x)f'(x)dx = xe^x - \int e^x \times 1dx = xe^x - \int e^x dx = xe^x - e^x + C$, where we applied the exponential rule of integration presented in Section 8.3.4. We can now check whether the result we obtained is correct or not. Differentiating the integral with respect to x we obtain $d[xe^x - e^x + C]/dx = xe^x + e^x - e^x = xe^x$, which is precisely our integrand and, thus, confirms the result.

As another example, consider the problem $\int x^2 e^{2x} dx$. Assume that $f(x) = x^2$ and that $g'(x) = e^{2x}$. The last equation implies that $g(x) = e^{2x}/2$. Notice that $f(x)g'(x) = x^2 e^{2x}$ is now equivalent to the integrand on the LHS of equation (8.3.12). Therefore, we can write the integral as $\int f(x)g'(x)dx = \int x^2 e^{2x}dx = f(x)g(x) - \int g(x)f'(x)dx = (x^2 e^{2x}/2) - \int (e^{2x}/2) \times 2xdx = (x^2 e^{2x}/2) - \int xe^{2x}dx$. In order to find the term $\int xe^{2x}dx$ in the last result we can again use the rule of integration by parts. Assume that $f(x) = x$ and that $g'(x) = e^{2x}$. The last equation implies, as before, that $g(x) = e^{2x}/2$. Therefore, we can write $\int f(x)g'(x)dx = \int xe^{2x}dx = f(x)g(x) - \int g(x)f'(x)dx = (xe^{2x}/2) - \int (e^{2x}/2) \times 1dx$. Since $\int (e^{2x}/2) \times 1dx = (e^{2x}/2 \times 2) = (e^{2x}/4) + C$, the last equation can be written as $\int f(x)g'(x)dx = \int xe^{2x}dx = f(x)g(x) - \int g(x)f'(x)dx = (xe^{2x}/2) - (e^{2x}/4) + C$. Therefore, combining this result with the result in the last paragraph we can write $\int x^2 e^{2x}dx = f(x)g(x) - \int g(x)f'(x)dx = (x^2 e^{2x}/2) - \int xe^{2x}dx = (x^2 e^{2x}/2) - (xe^{2x}/2) + (e^{2x}/4) + C$. Let us now check the result by differentiating $\int x^2 e^{2x}dx = (x^2 e^{2x}/2) - (xe^{2x}/2) + (e^{2x}/4) + C$ with respect to x to obtain $d[(x^2 e^{2x})/2 - (xe^{2x}/2) + (e^{2x}/4) + C]/dx = (1/2)[2x^2 e^{2x} + 2xe^{2x}] - (1/2)[2xe^{2x} + e^{2x}] + (1/4)2e^{2x} = x^2 e^{2x} + xe^{2x} - xe^{2x} - (e^{2x}/2) + (e^{2x}/2) = x^2 e^{2x}$, which is the integrand with which we started.

As the last example, consider the problem $\int x\ln xdx$. Assume that $f(x) = \ln x$ and that $g'(x) = x$. The last equation implies that $g(x) = x^2/2$. Notice that $f(x)g'(x) = g'(x)f(x) = x\ln x$ is now equivalent to the integrand on the LHS of equation (8.3.12). Therefore, we can write the integral as $\int f(x)g'(x)dx = \int g'(x)f(x)dx = \int x\ln xdx = f(x)g(x) - \int g(x)f'(x)dx = (x^2/2)\ln x - \int (x^2/2)(1/x)dx = (x^2/2)\ln x - (1/2)\int xdx = (x^2/2)\ln x - (1/2)(x^2/2) + C = (x^2/2)\ln x - (x^2/4) + C$. We now check if this result is correct. Differentiation of the integral with respect to x yields $d[(x^2 \ln x/2) - (x^2/4) + C]/dx = (1/2)[x^2(1/x) + 2x\ln x] - (2x/4) = (x/2) + x\ln x - (x/2) = x\ln x$, which is the integrand of the problem.

8.3.9 Integration by partial fractions

Sometimes, one may still come across some functions that are difficult to be integrated using the rules we have presented so far. It is often possible to express these functions as *partial fractions* and then integrate them. This method is called *integration by partial fractions*.

As an example, consider the problem $\int (1/x^2 + x - 2)dx$. How do we find this integral? Notice that the denominator of the integrand can be expressed as $x^2 + x - 2 = (x - 1)(x + 2)$. Let us now write $A/(x - 1) + B/(x + 2) = 1/(x - 1)(x + 2) = 1/(x^2 + x - 2)$. Therefore, the integrand can be written as $\int 1/(x^2 + x - 2)dx = \int [A/(x - 1) + B/(x + 2)]dx$. Our problem now is to find the values of A and B such that they satisfy the last equation. By combining the like terms in the equation $A/(x - 1) + B/(x + 2) = 1/(x - 1)(x + 2) = 1/(x^2 + x - 2)$ we can obtain $1 = A(x + 2) + B(x - 1) = Ax + 2A - B + Bx = (A + B)x + (2A - B)$. Notice that for this equality to hold, the coefficients of the x terms on the LHS of the equation must equal the coefficients of the corresponding x terms on the RHS of the equation. Therefore, we have $(A + B) = 0$ and $(2A - B) = 1$. The last two equations can be solved simultaneously to obtain $A = 1/3$ and $B = -1/3$. This method of finding the coefficients (A and B in the present case) is called the *method of undetermined coefficients*.

The above values of A and B help us rewrite the integral as $\int 1/(x^2 + x - 2)dx = \int [A/(x - 1) + B/(x + 2)]dx = \int [\{(1/3)/(x - 1)\} - \{(1/3)/(x + 2)\}]dx = \int [\{1/3(x - 1)\} - \{1/3(x + 2)\}]dx$. The last integral can now be written using the sum–difference rule of integration as $\int 1/3(x - 1)dx - \int 1/3(x + 2)dx$, or as $(1/3)\int 1/(x - 1)dx - (1/3)\int 1/(x + 2)dx$. We can now use the logarithmic rule of integration to find $(1/3)\int 1/(x - 1)dx - (1/3)\int 1/(x + 2)dx$. The result is $(1/3)\ln |x - 1| - (1/3)\ln |x + 2| + C = (1/3)[\ln |(x - 1)/(x + 2)|] + C$. We now check whether the result obtained is correct or not. Differentiating the integral with respect to x we find that $d[(1/3)\ln |x - 1| - (1/3)\ln |x + 2| + C]dx = (1/3)[1/(x - 1)] - (1/3)[1/(x + 2)] = [(1/3)/(x - 1)] - [(1/3)/(x + 2)]$. Since $A = 1/3$ and $B = -1/3$, and

$d[(1/3)\ln|x-1| - (1/3)\ln|x+2| + C] = [A/(x-1)] + [B/(x+2)] = 1/(x^2 + x - 2)$, the result obtained is correct.

As seen above, the essential problem in integrating functions such as the one we dealt with in the last example or related ones is to convert the integrand into partial fractions and determine the values of the constants. After this, one can apply the rules of integration presented in previous sections. Notice that the integrand may not always be as easy as the one in the last problem. Sometimes the integrand may take forms, among others, such as $5x/(x^2 + 7x + 10)$; $(x^2 - 10)/(x^2 + 7x + 10)$; $(x+2)/(x^2 + 2x + 1)$; $(x+1)/x^2(x-1)$; $(x+1)/(x^3 + x)$; and $1/(x+1)(x^2 + x + 1)$. The first problem, as said above, is to express these integrands as partial fractions: $5x/(x^2 + 7x + 10) = 5x/(x+5)(x+2) = [A/(x+5)] + [B/(x+2)]$, or $5x = A(x+2) + B(x+5))$; $(x^2 - 10)/(x^2 + 7x + 10) = 1 - (7x - 20)/(x^2 + 7x + 10) = 1 - (7x - 20)/(x+5)(x+2)$, which can be fractioned into parts as $(7x - 20)/(x+5)(x+2) = [A/(x+5)] - [B/(x+2)]$ implying $7x - 20 = A(x+2) + B(x+5)$; $(x+2)/(x+1)(x+1) = [A/(x+1)] + [B/(x+1)]$, or $x + 2 = A(x+1) + B$; $(x+1)/x^2(x-1) = [A/x] + [B/x^2] + [C/(x-1)]$, or $x + 1 = Ax(x-1) + B(x-1) + Cx^2$; $(x+1)/(x^3 + x) = (x+1)/x(x^2 + 1) = [A/x] + (Bx + C)/(x^2 + 1)$, which implies that $x + 1 = A(x^2 + 1) + (Bx + C)x$; and $1/(x+1)(x^2 + x + 1) = [A/(x+1)] + (Bx + C)/(x^2 + x + 1)$, or $1 = A(x^2 + x + 1) + (Bx + C)(x+1)$, respectively. We can now determine the values of the constants as presented in the solution to the last example. Once these are obtained, it is straightforward to apply the rules of integration presented so far and find each of the integrals.

8.3.10 Integration by tables

Many of the integration problems we encounter in the subjects of our interest can be solved by applying one or more of the rules or methods of integration presented so far. But, the solution to some of the more complicated problems requires additional rules or formulas. People who evaluate complicated integrals often make use of a table called a *table of integrals*. A table of integrals is a list of integration formulas applied to evaluate integrals. This list contains many formulas applicable to integrals involving different forms of integrands and can be found in most mathematical formula books or handbooks.

The formulas in the table of integrals are normally categorized as "forms involving $(a + bu)$," "forms involving $(\sqrt{a + bu})$," etc. In these categorizations a and b denote constants and u denotes the variable of integration. When one carries out integration by tables, one first finds the category that is closest to the integrand. After this, one finds the formula under the closest category that matches the integrand exactly after the values of the constants are assigned.

One may sometimes find in the table a formula under the closest category that matches the given integrand exactly. But, if one does not find in the table a formula that matches the integrand exactly, the method of substitution may convert the given integrand into a form that matches a particular formula in the table exactly. Notice that, as we have seen before, one may have to apply a particular formula more than once to evaluate a given integral.

As an example, consider the function $f(x) = 1/(x^2 - 4)$. How can we integrate this function? Notice that we cannot apply directly any of the rules we have presented so far to find the integral. Here, the table of integrals comes to our help. Using any standard table of integrals we can obtain the result as $\int f(x)dx = \int [1/(x^2 - 2^2)]dx = (1/4)\ln[(x-2)/(x+2)] + C$.

8.3.11 Integration with initial and boundary conditions

We know from our discussion in Section 8.2.2 that an indefinite integral always contains a constant, C. Our exposition of integral calculus so far has shown that we can derive primitive functions by integrating the derived function. However, we know that we cannot derive a *unique* primitive function unless we are given additional information regarding the value that the constant takes. This additional information is what we referred to earlier as initial conditions or boundary conditions, to which we turn our attention in this section.

It is possible, in many problems, to uniquely determine the constant of integration C, and thereby uniquely specify *the* primitive function. This determination of the constant of integration is done by using initial conditions or boundary conditions. An initial condition specifies a value $F(x) = F(0)$ for the integral function $F(x)$, when $x = 0$. A boundary condition specifies a value $F(x) = F(x_0)$ for the integral function $F(x)$, when $x = x_0$. Determining the unique primitive function using the initial condition that $F(x) = F(0)$ when $x = 0$ and using the boundary condition that $F(x) = F(x_0)$ when $x = x_0$ are called in the literature the *initial-value problem* and the *boundary-value problem*, respectively.

As an example, consider the function $F'(x) = f(x) = x^2$. Then, applying the power function rule of integration, we obtain $\int F'(x)dx = \int f(x)dx = \int x^2 dx = F(x) + C = (1/3)x^3 + C$. We now use the initial condition that $F(x) = F(0) = 1$ when $x = 0$. This gives us the unique value for the constant C as $F(0) = (1/3)x^3 + C = (1/3).0^3 + C = 1$, which implies that $C = 1$. Using this value of the constant we can write the integral as $\int F'(x)dx = \int f(x)dx = \int x^2 dx = F(x) = (1/3).x^3 + 1 = 1 + (1/3).x^3$. Notice that we have now obtained a unique primitive function.

We can also find the unique primitive function using a boundary condition $F(x) = F(x_0)$ when $x = x_0 = 2$. Notice that in the boundary condition, $x = x_0$ is a nonzero value. If we substitute $F(x) = F(x_0) = 3$ when $x = x_0 = 2$ in the integral, we obtain $F(2) = (1/3).x^3 + C = (1/3).2^3 + C = (8/3) + C = 3$. This implies that $C = 1/3$. Therefore, the primitive function with the boundary condition $F(x) = F(x_0) = 3$ when $x = x_0 = 2$ can be written as $\int F'(x)dx = \int f(x)dx = \int x^2 dx = F(x) = (1/3).x^3 + (1/3) = (1/3) + (1/3).x^3$. The reader would have noticed that the primitive function is still a function of x irrespective of the fact that we used a boundary condition or an initial condition. This suggests that the indefinite integral can take any value depending upon the value assumed by x. This also suggests that the indefinite integral is still an indefinite integral, and the use of the above conditions does not change its nature.

As another example, consider the derived function we used in Section 8.2.2. Equation (8.2.1) gives this derived function: $d\Pi/dt = F'(t) = f(t) = (1/3)t^{-2/3}$. We found in Section 8.3.2 that the primitive function that can be obtained by integrating this derived function with respect to t is $\Pi = F(t) = t^{1/3} + C$. Suppose that we use the initial condition that $F(t) = F(0) = 1$ when $t = 0$. Therefore, substituting $t = 0$ in $\Pi = F(t) = t^{1/3} + C$ and setting the result to 1, we obtain $F(0) = 0^{1/3} + C = 1$, which implies that $C = 1$. Therefore, we can write the unique primitive function as $\Pi = F(t) = t^{1/3} + 1 = 1 + t^{1/3}$. If we plot this function in a figure, we obtain a graph identical to that of $\Pi = 1 + t^{1/3}$ in Figure 8.2.1. Notice that we can carry out making the constant definite using boundary or initial conditions in almost all the integrands we have considered so far.

8.3.12 Partial integrals

So far in this chapter we have been concerned with integration of univariate functions. We are now ready to consider integration of multivariate functions. But, before this let us summarize the topic of differentiation of multivariate functions or partial differentiation presented in Section 3.7 as it has a similarity with integration of multivariate functions. Assume that we have a multivariate function $y = F(x_1, x_2, \ldots, x_i, \ldots, x_n)$. Then we found the rate of change of the dependent variable y when one of the independent variables, say x_i, changes by an infinitesimally small amount by differentiating y with respect to x_i by treating all other variables as *constants* and called this rate of change as the partial derivative of y with respect to x_i and denoted it by $\partial y / \partial x_i = \partial F / \partial x_i = F_{xi}$.

One might ask whether one can do a similar operation in integral calculus. The answer is yes; one can integrate a derived multivariate function with respect to one of the independent variables of the function by treating all other independent variables as constants, which will yield an anti-derivative or a primitive function. The process of finding an anti-derivative or a primitive function from the derived multivariate function is called *partial anti-differentiation* or *partial integration*. The integral obtained through partial integration is called the *partial integral*. Notice that partial anti-differentiation is the reverse operation of partial differentiation. Notice also that, as in the case of integration of univariate functions, we can uniquely determine the primitive function by using initial or boundary conditions.

As an example, consider the derived function of two independent variables $F_{x1} = f(x_1, x_2) = 2x_1x_2 + x_2$. Suppose that we want to integrate this function with respect to x_i. Therefore, we can write the integrand as $\int F_{x1} dx_1 = \int f(x_1, x_2) dx_1 = \int (2x_1x_2 + x_2) dx_1$. Applying the rules of integration presented earlier and treating x_2 as a constant, we can rewrite the integrand as $\int F_{x1} dx_1 = 2x_2 \int x_1 dx_1 + x_2 \int 1 dx_1$. This integral can be evaluated, again using the rules presented earlier, to obtain $\int F_{x1} dx_1 = 2x_2.(1/2).x_1^2 + C_1(x_2) + x_2x_1 + C_2(x_2)$. Letting $C = C(x_2) = C_1(x_2) + C_2(x_2)$, we can rewrite the last equation as $\int F_{x1} dx_1 = x_2x_1^2 + x_2x_1 + C(x_2)$. We can now verify the result by partially differentiating the integral with respect to x_1. This yields $\partial [\int F_{x1} dx_1] / \partial x_1 = \partial [x_2x_1^2 + x_2x_1 + C(x_2)] / \partial x_1 = f(x_1, x_2) = 2x_1x_2 + x_2 = F_{x1}$, which is the integrand in the present example. Notice that the constant is a function of x_2, which implies that $\partial [C(x_2)] / \partial x_1 = 0$.

Suppose that, instead of integrating the function $F_{x1} = f(x_1, x_2) = 2x_1x_2 + x_2$ with respect to x_1, we want to integrate it with respect to x_2. Therefore, we can write the integrand as $\int F_{x1} dx_2 = \int f(x_1, x_2) dx_2 = \int (2x_1x_2 + x_2) dx_2$. Following the same procedure as above, we can rewrite the integrand as $\int F_{x_1} dx_2 = 2x_1 \int x_2 dx_2 + \int x_2 dx_2$. This integral can be evaluated to obtain $\int F_{x_1} dx_2 = 2x_1.(1/2).x_2^2 + C_2(x_1) + (1/2).x_2^2 + C_2(x_1)$. Letting $C = C(x_1) = C_2(x_1) + C_2(x_1)$, we may rewrite the last equation as $\int F_{x_1} dx_2 = x_1x_2^2 + (1/2).x_2^2 + C(x_1)$. We can now verify the result by partially differentiating the integral with respect to x_2. This yields $\partial [\int F_{x_1} dx_2] / \partial x_2 = \partial [x_1x_2^2 + (x_2^2/2) + C(x_1)] / \partial x_2 = f(x_1, x_2) = 2x_1x_2 + x_2 = F_{x_1}$, which is the integrand in the present example. Notice that the constant is a function of x_2, which implies that $\partial [C(x_1)] / \partial x_2 = 0$.

As another example, consider the derived function of three independent variables $F_{x1} = f(x_1, x_2, x_3) = x_1x_2 + x_2x_3 + x_1x_3$. Suppose that we want to integrate this function with respect to x_1. Therefore, we can write the integrand as $\int F_{x1} dx_1 = \int f(x_1, x_2, x_3) dx_1 = \int (x_1x_2 + x_2x_3 + x_1x_3) dx_1$. Applying the rules of integration presented earlier and treating x_2 and x_3 as constants, we can rewrite the integrand as $\int F_{x1} dx_1 = x_2 \int x_1 dx_1 + x_2x_3 \int 1 dx_1 + x_3 \int x_1 dx_1$. This integral can be evaluated to obtain $\int F_{x1} dx_1 = x_2(x_1^2/2) + C_1(x_2, x_3) + x_2x_3x_1 + C_2(x_2, x_3) + x_3(x_1^2/2) + C_3(x_2, x_3)$. Letting $C = C(x_2, x_3) = C_1(x_2, x_3) + C_2(x_2, x_3)$

$+C_3(x_2,x_3)$, we may rewrite the last equation as $\int F_{x1}dx_1 = x_2(x_1^2/2) + x_2x_3x_1 + x_3(x_1^2/2) + C(x_2,x_3)$. Notice that this result can be verified by partially differentiating the integral with respect to x_1. Notice also that the constant is a function of x_2 and x_3, which implies that $\partial[C(x_2,x_3)]/\partial x_1 = 0$. One can also integrate the function with respect to x_2 and x_3. Moreover, one can also carry out integration of multivariate functions involving any number of independent variables.

8.3.13 *Multiple integrals*

In the last section we explored partial integrals using a few examples. In the present section we explain the meaning of *multiple integral*. Notice that when we integrate a multivariate derived function with respect to one of its independent variables the result (the primitive function) will still be a function of the same independent variables. We can now use this primitive function again to integrate with respect to the remaining independent variables. In other words, we can continue to integrate the multivariate derived function iteratively. This procedure is called *multiple integration* and the result is called a multiple integral.

The reader would have noticed a connection between partial integration and multiple integration. For example, consider the derived function $F_{x1} = f(x_1,x_2)$. In the last section we found that the integral with respect to one of the variables of the integrand, say x_1, is called the partial integral of the function with respect to x_1 and is denoted by $\int F_{x1}dx_1$. Suppose now that we integrate this result with respect to the other variable of the integrand, x_2. Then we denote it by $\int[\int F_{x1}dx_1]dx_2$ or by $\int\int F_{x1}dx_1\,dx_2$, which is called the *double integral* or, more precisely, the *double-indefinite integral* of the function $F_{x1} = f(x_1,x_2)$. When the derived function is a function of three variables such as $F_{x1} = f(x_1,x_2,x_3)$, then we denote the multiple integral by $\int[\int[\int F_{x1}dx_1]dx_2]dx_3$ or by $\int\int\int F_{x1}\,dx_1dx_2dx_3$, which is called the *triple integral* or *triple-indefinite integral* of the function $F_{x1} = f(x_1,x_2,x_3)$. This suggests that multiple integration is nothing but partial integration carried out iteratively.

As an example, consider the integrand in the first example we considered in the last section: $F_{x1} = f(x_1,x_2) = 2x_1x_2 + x_2$. Suppose that we want to find out $\int[\int F_{x1}dx_1]dx_2 = \int\int F_{x1}dx_1dx_2$. For this, we first integrate the integrand with respect to x_1, and then integrate the result with respect to x_2. We obtained the first result in the last section and, therefore, we simply write it here, $\int F_{x1}dx_1 = x_2x_1^2 + x_2x_1$, where the constant is omitted for convenience. Let us now integrate this result with respect to x_2 to yield $\int[\int F_{x1}dx_1]dx_2 = \int\int[x_2x_1^2 + x_2x_1]dx_2$. This integral can be evaluated to obtain $x_1^2\int x_2dx_2 + x_1\int x_2dx_2 = x_1^2(x_2^2/2) + x_1(x_2^2/2)$, where we have again omitted the constant. Therefore, we may write $\int[\int F_{x1}dx_1]dx_2 = x_1^2(x_2^2/2) + x_1(x_2^2/2)$. Notice that we can now verify this result by partially differentiating the double integral with respect to x_1 and x_2.

As another example, consider the integrand in the third example in the last section: $F_{x1} = f(x_1,x_2,x_3) = x_1x_2 + x_2x_3 + x_1x_3$. Suppose that we want to find out $\int[\int[\int F_{x1}dx_1]dx_2]dx_3 = \int\int\int F_{x1}\,dx_1dx_2dx_3$. For this, we first integrate the integrand with respect to x_1, then integrate the result with respect to x_2, and, lastly, integrate the result with respect to x_3. We obtained at the end of the last section the result of integrating the function with respect to x_1 and, therefore, we simply write it here: $\int F_{x1}dx_1 = x_2(x_1^2/2) + x_2x_3x_1 + x_3(x_1^2/2)$. We can now integrate the last result with respect to x_2 to yield $\int[\int F_{x1}dx_1]dx_2 = \int[x_2(x_1^2/2) + x_2x_3x_1 + x_3(x_1^2/2)]dx_2 = (x_2^2/2)(x_1^2/2) + (x_2^2/2)x_3x_1 + x_3(x_1^2/2)x_2$. Lastly, we can integrate the last result with respect to x_3 to yield $\int[\int[\int F_{x1}dx_1]dx_2]dx_3 = \int[(x_2^2/2)(x_1^2/2) + (x_2^2/2)x_3x_1 + x_3(x_1^2/2)x_2]dx_3 = (x_2^2/2)(x_1^2/2)x_3 + x_1(x_2^2/2)(x_3^2/2) + (x_3^2/2)(x_1^2/2)x_2$. Notice that in the last three results we have omitted the constants for convenience. We can verify that this result is correct by partially (and successively) differentiating the integral with respect to its arguments.

8.3.14 Application examples

Example 1.   Suppose that the rate of change of the total revenue of a firm that sells a good is given by $R' = F'(x) = 50 - x - x^2$, where x is the quantity of the good sold by the firm. Determine the inverse demand function for the firm's good assuming that the firm's total revenue is zero when the firm sells no unit of the good.

Solution.   Total revenue $R = F(x)$ is obtained by integrating the marginal revenue function $R' = F'(x) = 50 - x - x^2$. Once we obtain the total revenue function we can divide it by the total output sold to obtain the demand function. Applying the rules of integration presented earlier, we can find the total revenue function as $R = F(x) = \int F'(x)dx = \int (50 - x - x^2)dx = \int 50dx - \int xdx - \int x^2 dx = 50x + C_1 - (x^2/2) - C_2 - (x^3/3) - C_3$. Let us now define $C = C_1 - C_2 - C_3$. Therefore, the firm's total revenue can be written as $R = F(x) = 50x - (x^2/2) - (x^3/3) + C$. Notice the initial condition: when the firm sells no unit of the good (i.e. when $x = 0$) the total revenue of the firm is zero (that is, $R = F(0) = 0$). This implies that $C = 0$. We can now make the firm's total revenue function definite as $R = F(x) = 50x - (x^2/2) - (x^3/3)$. But, we know that total revenue is the product of price (p) per unit of the good sold and the quantity of the good sold (x). Therefore, we can write the total revenue function as $R = p.x = F(x) = 50x - (x^2/2) - (x^3/3)$. Dividing the total revenue by the number of the goods sold, we obtain the inverse demand function for the good sold by the firm as $R/x = p.x/x = F(x)/x = 50 - (x/2) - (x^2/3)$.

Example 2.   Assume that the marginal cost to a firm of producing x units of a good is given by $F'(x) = e^{1+0.5x}/2$. Assume also that the firm's fixed cost is \$100. Find the firm's total cost function.

Solution.   Total cost function can be found by integrating the marginal cost function with respect to x. Integrating the marginal cost function, using the rules in equations (8.3.1) and (8.3.6), with respect to x yields $F(x) = \int F'(x)dx = (1/2). \int e^{1+0.5x}dx = e^{1+0.5x} + C$. Notice that the firm's total cost is \$100 when the firm does not produce any unit of the good; that is, the initial condition is $F(0) = \$100$ when $x = 0$. Therefore, we can write $F(0) = e^{1+0.5(0)} + C = 100$, which implies that $e + C = 100$ or $C = 100 - e = 100 - 2.178 = 97.282$. Therefore, the firm's total cost function can be written as $F(x) = 97.282 + e^{1.5x}$.

Example 3.   Suppose that a consumer's marginal utility when x units of a good are consumed is given by $F'(x) = 1/x$. Find the consumer's total utility function assuming that the consumer's total utility is zero when the consumer consumes one unit of the good.

Solution.   Total utility function can be found by integrating the marginal utility function with respect to x. Integrating the marginal utility function using the logarithmic rule given in equation (8.3.3) with respect to x yields $F(x) = \int F'(x)dx = \int (1/x)dx = \ln x + C$. Notice that the consumer's total utility is zero when the consumer consumes one unit of the good; that is, the boundary condition is $F(0) = 0$ when $x = 1$. This implies that $C = 0$ (i.e. $F(1) = \ln 1 + C = 0 + C = 0$). Therefore, the consumer's total utility function can be written as $F(x) = \ln x$.

Example 4. Assume that the marginal revenue from the sale of a good by a firm is given by the function $R' = F'(x) = (1/2)x^{-1/2}(1+x^{1/2})$. Find the point elasticity of demand when $x = 4$ and determine the nature of demand.

Solution. The formula for point elasticity of demand is given by equation (3.3.21): $\Phi = (p/x) \div (dp/dx)$. To apply this equation we need to find the inverse demand function. For this we need to integrate the marginal revenue function and obtain the total revenue function. If we divide the total revenue function by the quantity of the good sold, we will obtain the inverse demand function. Therefore, we first integrate the marginal revenue function with respect to x: $\int R'(x)dx = \int F'(x)dx = \int (1/2).x^{-1/2}(1+x^{1/2})dx$. Let us now use the substitution rule in equation (8.3.11). For this assume that $u = (1+x^{1/2})$, which implies that $du/dx = (1/2)x^{-1/2}$ or $dx = 2(du/x^{-1/2})$. Therefore, $R = F(x) = \int R'(x)dx = \int F'(x)dx = \int (1/2)x^{-1/2}u2(du/x^{-1/2}) = \int udu = (u^2/2) + C$. Substituting $u = (1+x^{1/2})$, we obtain $R = F(x) = \int R'(x)dx = \int F'(x)dx = [(1+x^{1/2})^2/2] + C$. Assume for convenience that $C = 0$. This helps us write the total revenue function as $R = (1+x^{1/2})^2/2$. But, total revenue is equal to the product of price (p) and the quantity of the good sold (x). Therefore, we can write $R = (1+x^{1/2})^2/2 = p.x$, or the inverse demand function as $p = (1+x^{1/2})^2/2x$, which implies that when $x = 4$, $p = (1+4^{1/2})^2/2 \times 4 = 9/8$. Having found the inverse demand function, we can apply the formula for point elasticity. But, for this, we need to find the derivative of the inverse demand function with respect to x. This is given by $dp/dx = [x^{1/2}(1+x^{1/2}) - (1+x^{1/2})^2]/2x^2$; and, when $x = 4$, $dp/dx = -3/32$. Therefore, the point elasticity of demand for the firm's good when price $p = 9/8$ and quantity $x = 4$ is $\Phi = (p/x)/(dp/dx) = [(9/8)/4]/(-3/32) = -2.66$. Since $|-2.66| = 2.66 > 1$, the demand for the firm's good is elastic.

Example 5. Suppose that the rate of change of the total profit (Π) of a company with respect to time (t) is given by $\Pi'(t) = \ln(t+1)$. Suppose also that the total profit of the company when $t = 0$ is zero. Find the time path of the company's total profit.

Solution. Total profit function can be found by integrating the marginal profit function $\Pi'(t) = \ln(t+1)$ with respect to t. To integrate this marginal profit function we can use the technique of integration by parts given in equation (8.3.13). For this assume that $u = \ln(t+1)$ and $dv = dt$. These equations imply that $du = [1/(t+1)]dt$ and $v = t$. Therefore, applying equation (8.3.13), we can write the integral as $\Pi(t) = \int \Pi'(t)dt = \int \ln(t+1)dt$ or as $\int udv = uv - \int vdu = t\ln(t+1) - \int [t/(t+1)]dt$. We can now use the method of integration by partial fractions to evaluate the last term in the previous equation. After doing this, we obtain $\Pi(t) = \int \Pi'(t)dt = \int \ln(t+1)dt = t\ln(t+1) - \int [1 - 1/(t+1)]dt = t\ln(t+1) - t - C_1 + \ln(t+1) + C_2$. Denoting $C = C_2 - C_1$, the last result can be written as $\Pi(t) = \int \Pi'(t)dt = \int \ln(t+1)dt = t\ln(t+1) - \int [1 - 1/(t+1)]dt = t\ln(t+1) - t + \ln(t+1) + C$. Since $\Pi(0) = 0$ when $t = 0$, the unique time path of the company's total profit can be written as $\Pi(t) = \ln(t+1)[t+1] - t$.

Example 6. Assume that a firm's marginal cost of producing x units of a good is given by $F'(x) = (25x^2 - 248x + 15)/(x^2 - 10x + 1)$. Find the firm's total cost function if the fixed cost is $1000.

Solution. The firm's total cost can be found by integrating the marginal cost function with respect to x. For this we can apply the technique of integration by partial fractions presented in Section 8.3.9. Notice that the marginal cost function $F'(x) = (25x^2 - 248x + 15)/(x^2 - 10x + 1)$ can be fractioned into $F'(x) = 25 + (2x - 10)/(x^2 - 10x + 1)$. We can now integrate the marginal cost function to obtain $F(x) = \int F'(x)dx = \int [25 + (2x - 10)/(x^2 - 10x + 1)] dx = 25x + C_1 + \ln(x^2 - 10x + 1) + C_2$. Treating $C = C_1 + C_2$, we can write the total cost function as $F(x) = 25x + \ln(x^2 - 10x + 1) + C$. The initial condition is that the fixed cost is \$1000 when the firm does not produce any output. This means that $F(0) = 25 \times 0 + \ln(0^2 - 10 \times 0 + 1) + C = C = \$1000$. Therefore, the total cost function can be written as $F(x) = 1000 + 25x + \ln(x^2 - 10x + 1)$.

8.3.15 Exercises

1. Find the following indefinite integrals:
   (i) $\int 1 dx$; (ii) $\int (1/2x^2) dx$; (iii) $\int [(x^2/2) - (2/x^2)] dx$; (iv) $\int (x^{100}/2) dx$;
   (v) $\int (x^2 + 2x) dx$; (vi) $\int (\sqrt{x} - \sqrt{4x^3}) dx$; (vii) $\int 0 dx$; (viii) $\int (e + e) dx$.

2. Find the following indefinite integrals:
   (i) $\int [e^x + (1/e^{-x})] dx$; (ii) $\int [(e^x + e^{3x})/e^x] dx$; (iii) $\int [e^{5x^2}] dx$;
   (iv) $\int [xe^{5x^2}] dx$; (v) $\int [1/(1+x)] dx$; (vi) $\int [1/(x^2 + x)] dx$; (vii) $\int [1/(1+2x)] dx$;
   (viii) $\int [(e^x - e^{-x})/(e^x + e^{-x})] dx$.

3. Find the following indefinite integrals:
   (i) $\int 2x(1 + x^2) dx$; (ii) $\int 3x^2(1 + x^3)^{49} dx$; (iii) $\int 2^{3x} dx$; (iv) $\int (2x + 1)(x^2 + x + 1)^5 dx$;
   (v) $\int 5x(2 + x^2)^5 dx$; (vi) $\int (1+x)^{9/10} dx$; (vii) $\int [3x^2/(x^3 + 1)] dx$; (viii) $\int [1/(2x+1)] dx$;
   (ix) $\int x^2 e^{x^3} dx$.

4. Find the following indefinite integrals:
   (i) $\int x^2(1 + x)^2 dx$; (ii) $\int x^2 e^x dx$; (iii) $\int [2x/(1 + x)^2] dx$; (iv) $\int [(1-x)/(1+x)^2] dx$;
   (v) $\int [\ln x/x^2] dx$; (vi) $\int xe^{x^2} dx$; (vii) $\int xe^{x+1} dx$; (viii) $\int [x/(1+x)^2] dx$.

5. Find the following indefinite integrals:
   (i) $\int [4x/(x^2 + 3x + 2)] dx$; (ii) $\int [4/(x^2 - 1)] dx$; (iii) $\int [x^2/(x^2 + 3x + 2)] dx$;
   (iv) $\int [(x^2 + 2x)/(x^2 + 3x + 2)] dx$; (v) $\int [(x + 2)/(x^2 + 2x + 1)] dx$;
   (vi) $\int [(x + 2)/(x^2 - 2x + 1)] dx$; (vii) $\int [(x + 2)/(x^3 - 2x^2 + x)] dx$.

6. Find the following partial integrals:
   (i) $\int (y + x + xy) dx$; (ii) $\int [(y/x) + (x/y) + ye^x] dx$; (iii) $\int y dx$; (iv) $\int yz dx$;
   (v) $\int (yz/x) dx$; (vi) $\int (y + x + xy) dy$; (vii) $\int [(y/x) + (x/y) + ye^x] dy$; (viii) $\int y dy$;
   (ix) $\int xz dy$; (x) $\int [xz/y] dy$.

7. Find the following multiple integrals:
   (i) $\int \int dx dy$; (ii) $\int \int \int [(y/x) + (x/y)] dx dy$; (iii) $\int \int xy dx dy$; (iv) $\int \int \int dx dy dz$;
   (v) $\int \int \int [(1/x) + (1/y) + (1/z)] dx dy dz$; (vi) $\int \int \int (xyz) dx dy dz$;
   (vii) $\int \int \int [(x/y) + (y/z) + (z/x)] dx dy dz$.

8. *Application exercise.* Assume that an investor purchased a company's share for \$150. The rate of change of the price ($P'(t)$) of the share with respect to time (t) is given by the function $P'(t) = 7e^{0.07t}$. Find the time path of the price of the share.

9. *Application exercise.* Assume that the marginal propensity to consume in an economy is given by the function $dC/dY = (4/5) - 3/4\sqrt{4Y}$, where C and Y denote the aggregate consumption and national income (in billions of dollars) of the economy, respectively. Find the aggregate consumption function assuming that aggregate consumption is \$80 billion when income is \$100 billion. Also find the economy's aggregate saving function.

10. *Application exercise.* Suppose that a company's marginal cost function and marginal revenue function are given by $dC/dq = 5 + 5q$ and $dR/dq = 200 - 2q$, respectively, where q, C, and R denote the quantity of output produced, the total cost, and the total revenue, respectively. Determine, ignoring the fixed cost, the level of output at which total cost is equal to total revenue.

11. *Application exercise.* Assume that the rate of change of the capital stock of an economy with respect to time is given by $dK/dt = 10t^{1/3}$, where K and t denote the capital stock and time, respectively. Find the function that represents the capital stock of the economy assuming that the economy's capital stock is zero at time $t = 0$.

 Web supplement: S8.3.16 *Integration of trigonometric functions*

 Web supplement: S8.3.17 Mathematica applications

8.4 Definite or Riemann integrals

8.4.1 Introduction

So far in this chapter we were exploring the meaning of indefinite integrals. These integrals are called indefinite integrals because we do not specify the values that the variable(s) of the integrand could take. Moreover, every integration process we have carried out so far had yielded a constant, the value of which had to be specified through initial or boundary conditions. We also explored the important rules and techniques of evaluating indefinite integrals and applied them to solve a number of problems that frequently arise in the fields of our interest.

However, in many problems one may need to evaluate integrals such that the variable of integration assumes values within some specified intervals on the domain of the variable. Such an integral is called the *definite integral* or the *Riemann integral*. We first attempt to explain the meaning of definite integrals before moving on to some of the other aspects and applications of definite integrals.

8.4.2 Method of exhaustion and the meaning of definite integral

Let us begin with a simple function: $y = f(x) = 2x$. The graph of this function is illustrated in Figure 8.4.1(A). Notice that when $x = 1$, $y = f(1) = 2 \times 1 = 2$. This value of y is shown alongside the vertical line starting from $x = 1$. The reader would have noticed that the shaded area in the figure, denoted by A, is a triangle labeled by CBD.

What is the area of this triangle? We know that the area of a triangle is equal to half of the length of the adjacent side (l) times the length of the opposite side (h) of the triangle; that is, $(1/2) \times (l \times h)$ units. Since $l = 1$ and $h = 2$ in our present example, the area of the triangle is $(1/2) \times (l \times h) = (1/2) \times (1 \times 2) = 1$ unit. Let us use another method, called the *method of exhaustion*, to find this area. The method of exhaustion is more general and is based on the summation of areas of *rectangles* or other suitable *polygons*. Notice that a rectangle is an element of the set of general polygons.

Let us now divide the closed interval [0, 1] on the domain of the function into two closed subintervals with equal length: $[x_0, x_1] = [0, 1/2]$ and $[x_1, x_2] = [1/2, 1]$. This is illustrated in Figure 8.4.1(B). Notice that $\Delta x = x_1 - x_0 = x_2 - x_1 = 1/2$; that is, the subintervals have

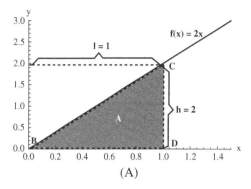
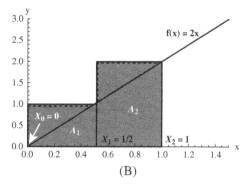

(A)   (B)

Figure 8.4.1

the same length of 1/2 units. As can be seen from this figure, we now have two sub-areas, as represented by A_1 and A_2. What we do here is to use these sub-areas to approximate the area A in Figure 8.4.1(A).

Notice that the bases of the *circumscribed rectangles* in Figure 8.4.1(B) are the associated subintervals and the heights of these rectangles are the associated maximum values of the function on those subintervals. Since the function is an increasing function on the subintervals, the maximum values of the function occur when x takes the right-end values of the subintervals. Therefore, the area represented by A_1, using the right-end value x_1 of the subinterval $[x_0, x_1]$ is given by $\Delta x f(x_1) = \Delta x f(1/2) = (1/2)f(1/2)$; and the area represented by A_2, using the right-end value x_2 of the subinterval $[x_1, x_2]$, is given by $\Delta x f(x_2) = \Delta x f[(2/2) = 1] = (1/2)f(2/2)$. Thus sum of the areas of the two rectangles in this figure can be written as $\Delta x f(x_1) + \Delta x f(x_2) = \Delta x f(1/2) + \Delta x f[(2/2) = 1] = (1/2)f(1/2) + (1/2)f(2/2)$, which we denote by $S_{R,2}$, where the subscript R,2 shows that we used the right-end values of the two subintervals; or (using the function) as

$$S_{R,2} = \sum_{i=1}^{2} \Delta x f(x_i) = \Delta x f(x_1) + \Delta x f(x_2) = \Delta x[f(x_1) + f(x_2)]$$

$$= \Delta x[f(1/2) + f(2/2)] = (1/2)[f(1/2) + f(2/2)]$$

$$= (1/2)[2 \times (1/2) + 2 \times (2/2)] = 1.5 \text{ units} \tag{8.4.1}$$

Above, we approximated the area of the triangle in Figure 8.4.1(A) using the right-end values of the two subintervals in Figure 8.4.1(B). Similarly, we can also approximate the area of the triangle in Figure 8.4.1(A) using the left-end values of the two subintervals (rectangles) in Figure 8.4.1(B), as illustrated in Figure 8.4.2(A). Notice that of the two rectangles in Figure 8.4.1(B), only one rectangle is existent and the other rectangle is nonexistent if we use the left-end values of the subintervals and for all values of $x \geq 0$.

Notice that the bases of the *inscribed rectangles* in Figure 8.4.2(A) are the associated subintervals and the heights of these rectangles are the associated minimum values of the function on these subintervals. Since the function is an increasing function on the subintervals, the minimum values of the function occur when x takes the left-end values on the subintervals. Therefore, the area when $x_0 = 0$ in this figure (the area of the nonexistent rectangle) using the left-end value x_0 of the subinterval $[x_0, x_1]$ is given by $\Delta x f(x_0) = \Delta x f(0) = (1/2)f(0)$;

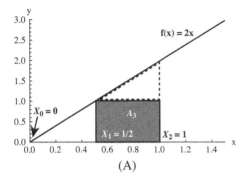

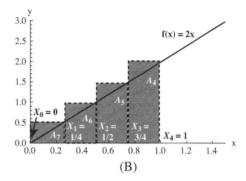

Figure 8.4.2

and the area represented by A_3, using the left-end value x_1 of the subinterval $[x_1, x_2]$, is given by $\Delta x f(x_1) = \Delta x f(1/2) = (1/2) f(1/2)$. Thus the sum of the areas of the two inscribed rectangles in this figure can be written as $\Delta x f(x_0) + \Delta x f(x_1) = (1/2) \times f(0) + (1/2) f(1/2)$, which we denote $S_{L,2}$, where the subscript L,2 shows that we used the left-end values of the two subintervals; or (using the function) as

$$S_{L,2} = \sum_{i=0}^{1} \Delta x f(x_i) = \Delta x f(x_0) + \Delta x f(x_1)$$

$$= \Delta x [f(0) + f(1/2)] = (1/2)[f(0) + f(1/2)] = (1/2)[2 \times 0 + 2 \times (1/2)] = 0.5 \text{ units}$$
(8.4.2)

Let us now compare the results in equations (8.4.1) and (8.4.2) with the result we obtained at the beginning of this section. We found at the beginning of the section that the area of the triangle in Figure 8.4.1(A) was 1 unit. But, through approximations, we obtained the area as 1.5 units when approximated it from above using the circumscribed rectangles (Figure 8.4.1(B) or equation (8.4.1)) and 0.5 units when we approximated the area from below using the inscribed rectangles (Figure 8.4.2(A) or equation (8.4.2)). Clearly, equation (8.4.1) overestimates the area by $1.5 - 1 = 0.5$ units and equation (8.4.2) underestimates it by $1 - 0.5 = 0.5$ units. Notice that when we approximated the area from above using Figure 8.4.1(B), we included the shaded areas of triangles above the graph of the function leading to overestimation and when we approximated the area from below using Figure 8.4.2(A), we excluded the unshaded areas of triangles below the graph of the function leading to underestimation.

A pertinent question is: can we eliminate or reduce the errors of approximations (overestimation and underestimation)? Yes; we can, if we increase the number of subintervals and, thereby, the number of rectangles. Suppose that we divide the interval [0, 1] on the domain of the function into four subintervals of equal length: $[x_0, x_1] = [0, 1/4]$, $[x_1, x_2] = [1/4, 1/2]$, $[x_2, x_3] = [1/2, 3/4]$, and $[x_3, x_4] = [3/4, 1]$. This is illustrated in Figure 8.4.2(B). Notice that $\Delta x = x_1 - x_0 = x_2 - x_1 = x_3 - x_2 = x_4 - x_3 = 1/4$; that is, all the subintervals have the same length of 1/4 units. As can be seen, we now have four sub-areas, as represented by A_7, A_6, A_5, and A_4. We again attempt to use these sub-areas to approximate the area A in Figure 8.4.1(A).

Notice that, as before, the bases of the circumscribed rectangles in Figure 8.4.2(B) are the associated subintervals and the heights of the circumscribed rectangles are the associated maximum values of the function on those subintervals. Since the function is an increasing function on the subintervals, the maximum values of the function occur when x takes the right-end values on the subintervals. Therefore, the area represented by A_7 using the right-end value x_1 of the subinterval $[x_0, x_1]$ is given by $\Delta xf(x_1) = \Delta xf(1/4) = (1/4)f(1/4)$; the area represented by A_6 using the right-end value x_2 of the subinterval $[x_1, x_2]$ is given by $\Delta xf(x_2) = (1/4)f(2/4)$; the area represented by A_5 using the right-end value x_3 of the subinterval $[x_2, x_3]$ is given by $\Delta xf(x_3) = (1/4)f(3/4)$; and the area represented by A_4 using the right-end value x_4 of the subinterval $[x_3, x_4]$ is given by $\Delta xf(x_4) = (1/4)f[(4/4) = 1]$. Thus the sum of the areas of the four rectangles in Figure 8.4.2(B) can be written as $\Delta xf(x_1) + \Delta xf(x_2) + \Delta xf(x_3) + \Delta xf(x_4) = (1/4)[f(1/4) + f(2/4) + f(3/4) + f(4/4) = 1]$. We denote this sum by $S_{R,4}$, where the subscript R,4 shows that we used the right-end values of four subintervals. Now using the function, we can write

$$S_{R,4} = \sum_{i=1}^{4} \Delta xf(x_i) = \Delta xf(x_1) + \Delta xf(x_3) + \Delta xf(x_2) + \Delta xf(x_4)$$

$$= \Delta x \left[f\left(\frac{1}{4}\right) + f\left(\frac{2}{4}\right) + f\left(\frac{3}{4}\right) + f\left(\frac{4}{4}\right) \right] = \frac{1}{4}\left[2\left(\frac{1}{4}\right) + 2\left(\frac{2}{4}\right) + 2\left(\frac{3}{4}\right) + 2(1) \right]$$

$$= \frac{1}{4}[0.5 + 1 + 1.5 + 2] = \frac{5}{4} = 1.25 \text{ units} \tag{8.4.3}$$

Similarly, we can now approximate the area of the triangle in Figure 8.4.1(A) using the left-end values of the four subintervals as illustrated in Figure 8.4.3(A). Following the same procedure as before, the sum of the areas of the four inscribed rectangles, including the rectangle at $x_0 = 0$ that does not exist, when we use the left-end values of the four subintervals can be written as

$$S_{L,4} = \sum_{i=1}^{4} \Delta xf(x_i) = \Delta xf(x_0) + \Delta xf(x_1) + \Delta xf(x_2) + \Delta xf(x_3)$$

$$= \Delta x \left[f(0) + f\left(\frac{1}{4}\right) + f\left(\frac{2}{4}\right) + f\left(\frac{3}{4}\right) \right] = \frac{1}{4}\left[2(0) + 2\left(\frac{1}{4}\right) + 2\left(\frac{2}{4}\right) + 2\left(\frac{3}{4}\right) \right]$$

$$= \frac{1}{4}[0 + 0.5 + 1 + 1.5] = \frac{3}{4} = 0.75 \text{ units} \tag{8.4.4}$$

How do the results in equations (8.4.3) and (8.4.4) compare with those in equations (8.4.1) and (8.4.2)? Compared with the results in equations (8.4.1) and (8.4.2), the results in equations (8.4.3) and (8.4.4) are closer to the true area of the triangle in Figure 8.4.1(A), which is 1 unit. However, the errors in approximations (overestimation and underestimation) still persist. But, the magnitudes of these errors are less now than they were before. One might wonder why the errors in approximations decreased. The reason is that we increased the number of subintervals from two to four. If we increase the number of subintervals more and more, the errors in the approximations of the sum of the areas below and above the graph of the function and above the respective subintervals will decrease.

But, what is meant by increasing the number of subintervals? We first split the closed interval [0, 1] into two closed subintervals [0, 1/2] and [1/2, 1]. Then we found the areas of the rectangles corresponding to these subintervals. After this, we split the same closed interval into four closed subintervals [0, 1/4], [1/4, 1/2], [1/2, 3/4], and [3/4, 4/4 = 1]. And, again, we found the areas of the rectangles corresponding to these subintervals. When we increase the number of the splits, we are in fact reducing the magnitude by which x changes; that is,

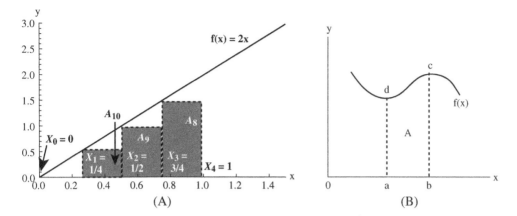

Figure 8.4.3

we are reducing the size of Δx. If we continue to reduce the size of Δx (that is, if we increase the number of subintervals), then Δx will tend to but not be equal to zero. As a result, the sizes of the triangles above (below) the graph of the function that we incorrectly included (excluded) in the calculations of the areas of the circumscribed (inscribed) rectangles will reduce and will tend to zero. This is the reason why we get closer approximations of the area of the triangle in Figure 8.4.1(A) when we increase the subintervals of the rectangles. However, the results in equations (8.4.1)–(8.4.4) suggest that the area of the triangle in Figure 8.4.1(A) lies in between the area of the inscribed rectangles and the area of the circumscribed rectangles. Therefore, we have the inequality

$$S_L \leq A \leq S_R \tag{8.4.5}$$

where we have omitted the subscripts that represent the number of subintervals. Let us now generalize the above results. Suppose that we divide the closed interval [0, 1] into n equal subintervals. Then the length of any particular subinterval, Δx, will be $\Delta x = 1/n$. Moreover, the right-end values of the subintervals will be $x = 1/n, 2/n, \ldots, (n-1)/n$, and $n/n = 1$. Then, following the above procedure, the sum of the areas of the n circumscribed rectangles can be written as

$$S_{R,n} = \sum_{i=1}^{n} \Delta x f(x_i) = \Delta x f(x_1) + \Delta x f(x_2) + \cdots + \Delta x f(x_{n-1}) + \Delta x f(x_n)$$

$$= \Delta x \left[f\left(\frac{1}{n}\right) + f\left(\frac{2}{n}\right) + \cdots + f\left(\frac{n-1}{n}\right) + f\left(\frac{n}{n} = 1\right) \right]$$

$$= \frac{1}{n} \left[2\left(\frac{1}{n}\right) + 2\left(\frac{2}{n}\right) + \cdots + 2\left(\frac{n-1}{n}\right) + 2\left(\frac{n}{n} = 1\right) \right]$$

$$= \frac{2}{n^2}[1 + 2 + \cdots + (n-1) + n]$$

$$= \frac{2}{n^2}[1 + 2 + \cdots + (n-1) + n] \tag{8.4.6}$$

The sum presented in equation (8.4.6) is called a *Riemann sum*. Notice that the terms inside the square brackets of the last line, $[1 + 2 + \cdots + (n-1) + n]$, constitute an arithmetic series with the first term being $s_1 = 1$, the last term being $s_n = n$, and the common difference being $a = 1$. Therefore, we can use equation (1.10.4) to find the sum of those terms. Thus we get the sum $n(1+n)/2$. Therefore, we can rewrite equation (8.4.6) as

$$S_{R,n} = \frac{2}{n^2}\left[\frac{n(1+n)}{2}\right] = \frac{n+1}{n} \tag{8.4.7}$$

Similarly, the left-end values of the subintervals will be $x = 0, 1/n, 2/n, \ldots,$ and $(n-1)/n$. Then, again following the above procedure, the sum of the areas of the n inscribed rectangles can be written as

$$S_{L,n} = \sum_{i=0}^{n} \Delta x f(x_i) = \Delta x f(x_0) + \Delta x f(x_1) + \cdots + \Delta x f(x_{n-1})$$

$$= \Delta x \left[f(0) + f\left(\frac{1}{n}\right) + f\left(\frac{2}{n}\right) + \cdots + f\left(\frac{n-1}{n}\right) \right]$$

$$= \frac{1}{n}\left[2 \times 0 + 2\left(\frac{1}{n}\right) + 2\left(\frac{2}{n}\right) + \cdots + 2\left(\frac{n-1}{n}\right)\right] = \frac{2}{n^2}[0 + 1 + 2 + \cdots + (n-1)]$$

$$= \frac{2}{n^2}[0 + 1 + 2 + \cdots + (n-1)] \tag{8.4.8}$$

The sum presented in equation (8.4.8) is also called a Riemann sum. Notice that, as before, the terms inside the square brackets of the last line, $[0 + 1 + 2 + \cdots + (n-1)]$, constitute an arithmetic series with the first term being $s_1 = 0$, the last term being $s_n = (n-1)$, and the common difference being $a = 1$. Therefore, we can again use equation (1.10.4) to find the sum of those terms. Thus we get the sum $n(n-1)/2$. Therefore, we can rewrite equation (8.4.8) as

$$S_{L,n} = \frac{2}{n^2}\left[\frac{n(n-1)}{2}\right] = \frac{n-1}{n} \tag{8.4.9}$$

An astute reader would have noticed a similarity between equations (8.4.7) and (8.4.9). This similarity lies in the identical values of $S_{R,n}$ and $S_{L,n}$ when $n \to \infty$. If we take limits of both sides of equations (8.4.7) and (8.4.9) we obtain

$$\left. \begin{array}{l} \lim_{n\to\infty} S_{R,n} = \lim_{n\to\infty}\left(\frac{n+1}{n}\right) = \lim_{n\to\infty}\left(\frac{n}{n} + \frac{1}{n}\right) = \lim_{n\to\infty}\left(1 + \frac{1}{n}\right) = 1 \\[2mm] \text{and} \\[2mm] \lim_{n\to\infty} S_{L,n} = \lim_{n\to\infty}\left(\frac{n-1}{n}\right) = \lim_{n\to\infty}\left(\frac{n}{n} - \frac{1}{n}\right) = \lim_{n\to\infty}\left(1 - \frac{1}{n}\right) = 1 \end{array} \right\} \tag{8.4.10}$$

The results in equation (8.4.10) show that, as $n \to \infty$ (that is, as the number of subintervals becomes infinitely large or as $\Delta x \to 0$), the sum of the circumscribed rectangles has the same value of 1 unit as has the sum of the inscribed rectangles. Notice that this is the true area of the triangle. In equation (8.4.10) what we do is find the limiting values of the Riemann sums presented in equations (8.4.6) and (8.4.8). Therefore, as the number of subintervals increases, the Riemann sums approach a common limiting value and that value is the *area under the curve*.

Equation (8.4.10) shows that the limit is common for both sums, $S_{R,n}$ and $S_{L,n}$. This is not always true for any arbitrary function. But, this is true for all the functions we consider in this book. This implies that we can use, for the functions we use in this book, either $S_{R,n}$ or $S_{L,n}$; therefore, we will use $S_{R,n}$ from now onwards.

The common limiting value of 1 of the Riemann sums in equations (8.4.10) is popularly called the *definite integral* or the *Riemann integral* of the function $y = f(x) = 2x$ when x varies from $x = 0$ to $x = 1$ and is generally written as

$$\int_0^1 f(x)dx = \int_0^1 2xdx = 1 \qquad (8.4.11)$$

So far in this section we were attempting to find the area of a given region by circumscribing or inscribing n rectangles. If the sum of the areas of the n circumscribed rectangles and the sum of the areas of the n inscribed rectangles tend to the common limit, as n becomes infinitely large, this limit is defined as the area of the given region. This method of determining the area of a region is called the method of exhaustion. This was the method that we have applied above in the present section.

8.4.3 Notations and concepts

We introduced some notations and concepts of indefinite integral in Section 8.2.4. Similarly, we introduce some notations and concepts of definite integral in the present section. Notice that in equation (8.4.11) we integrated the function $y = f(x) = 2x$ from $x = 0$ to $x = 1$.

Instead of a specific function, suppose that we use a general function $y = f(x)$, the RHS of which may be of any form, defined on the closed interval $[x = a, \ x = b]$. Suppose also that we want to integrate this general univariate function from $x = a$ to $x = b$ (instead of from $x = 0$ to $x = 1$) such that $a < b$. Notice that we can now find the sums (as we did in the last section) $S_{R,n}$ and $S_{L,n}$. Suppose again that the common limit of $S_{R,n}$ and $S_{L,n}$, as $n \to \infty$, exists. Then we have the statement that the common limit of $S_{R,n}$ and $S_{L,n}$ is called the definite integral of $y = f(x)$ over the closed interval $[x = a, \ x = b]$ and is written as

$$\lim_{n \to \infty} \sum_{i=1}^n \Delta x f(x_i) \to \int_a^b f(x)dx \qquad (8.4.12)$$

The statement (8.4.12) gives the definition of a general definite integral. In this definition, the numbers a and b are called the *limits of integration*: a is the *lower limit* and b is the *upper limit*. As before, dx shows the variable with respect to which integration is performed or the variable of integration, and $f(x)$ is the integrand.

One important feature of the definition (8.4.12) is worth mentioning. We know from the last section that the integral of a function over a closed interval on its domain represents the area of the function under the graph of the function and above the horizontal axis corresponding to that interval. If $f(x)$ is negative on that interval, then the integral will also be negative. This does not mean that the area represented by that integral is negative implying that the common limit may not represent an area. Since area can never be negative, one should interpret it as the negative of the area. In short, the definite integral is just a real number and it may or may not represent an area.

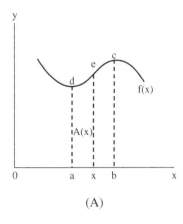
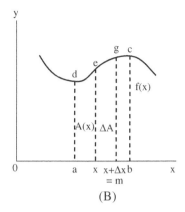

Figure 8.4.4

8.4.4 Fundamental theorem of integral calculus

In Chapter 3 we found that the limit of the difference quotient of a function when the independent variable of the function changed by an infinitesimally small amount was called the derivative of the function. In the last section we found that the limit of the Riemann sum was called the definite integral. In the present section we attempt to explore an important relationship that exists between the derivative of a function and the integral of that function, which will help us evaluate definite integrals more efficiently. It will also help us determine the change in the value of a function if we know the rate of change of that function.

Let us begin with a function that is continuous on a closed interval $[a, b]$ on its domain, such as the function $y = f(x)$ illustrated in Figure 8.4.3(B). How do we find the area A or $abcd$, the area bounded by the function and the interval $[a, b]$? We know from the last section that this area is given by the integral $\int_a^b f(x)dx$, which can be found by applying the method of exhaustion. But, we know by now that the application of the method of exhaustion is a tedious job. Therefore, the question is: is there a more efficient alternative method to find this area? The answer is yes and that is the topic to which we now turn our attention.

Now consider area A in Figure 8.4.3(B). Suppose that we pick a value of x, such that $a \leq x \leq b$, in the closed in interval $[a, b]$, such as x in Figure 8.4.4(A). Notice that area A increases as we push x ahead along the x-axis in the interval. This means that area A or *axed* is a function of x; that is, $A(x)$. Therefore, we have denoted the area by $A(x)$ in the figure and this function is called an *area function*. Notice also that $A(x = a) = A(a) = 0$; that is, there is no area below the graph of the function from one point on the closed interval to that point itself. Similarly, $A(b)$ gives the entire area from point $x = a$ to $x = b$; that is, $A = A(b)$ or $A = A(b) = \int_a^b f(x)dx$.

We know from above that as x increases, the area $A(x)$ also increases. Suppose now that we increase x by a positive amount Δx. This implies that $A(x + \Delta x)$ is the area *amgd* in Figure 8.4.4(B). Therefore, $\Delta A = A(x + \Delta x) - A(x)$ gives the difference of the areas *axed* in Figure 8.4.4(A) and *amgd* in Figure 8.4.4(B).

Let us now magnify this change in area (ΔA) as illustrated in Figure 8.4.5(A). A visual inspection of the figure suggests that the change in area (ΔA) cannot be smaller than the area $\Delta x f(x)$ and cannot be larger than the area $\Delta x f(x + \Delta x)$. In other words, the change

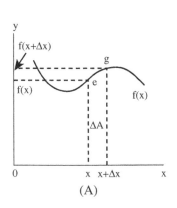
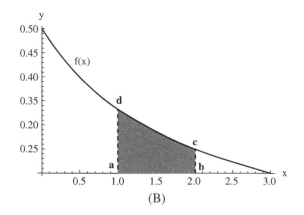

Figure 8.4.5

in area must lie between the areas $\Delta x f(x)$ and $\Delta x f(x + \Delta x)$. This, in symbols, becomes $\Delta x f(x) \leq A(x + \Delta x) - A(x) \leq \Delta x f(x + \Delta x)$. Notice that if $\Delta x < 0$, the last inequality will be reversed. Dividing all the three parts of the last inequality by Δx yields $f(x) \leq [A(x + \Delta x) - A(x)]/\Delta x \leq f(x + \Delta x)$. Taking the limit, as Δx tends to zero, of the last inequality we obtain an important result that $\lim\limits_{\Delta x \to 0} f(x) \leq \lim\limits_{\Delta x \to 0} [\{A(x + \Delta x) - A(x)\}/\Delta x] \leq \lim\limits_{\Delta x \to 0} f(x + \Delta x)$. This inequality can be written as $f(x) \leq \lim\limits_{\Delta x \to 0} [\{A(x + \Delta x) - A(x)\}/\Delta x] \leq f(x)$. What does the limit, as Δx tends to zero, of the expression inside the last square brackets stand for? It is, as we presented in Chapter 3, the derivative of the area function $A(x)$ with respect to its independent variable x; that is, $A'(x) = \lim\limits_{\Delta x \to 0} [\{A(x + \Delta x) - A(x)\}/\Delta x]$. Therefore, we have the result

$$A'(x) = f(x) \tag{8.4.13}$$

What does equation (8.4.13) imply? It implies that the derivative of the area function $A(x)$ (or $A'(x)$) is the graph's height function $f(x)$. In other words, the derivative of $A(x)$ is $f(x)$; or $A(x)$ is an anti-derivative of $f(x)$. Suppose now that $F(x)$ is another anti-derivative of $f(x)$. We know from Section 8.2.2 that the two anti-derivatives differ only by a constant C. Therefore, we can write $A(x) = F(x) + C$. We stated earlier in the present section that $A(a) = 0$. This implies that, when $x = a$ (or when x takes the value of the left-end of the closed interval), $A(a) = F(a) + C$, or $F(a) + C = 0$, or $C = -F(a)$. Substituting $C = -F(a)$ into $A(x) = F(x) + C$, we obtain $A(x) = F(x) - F(a)$. If $x = b$ or if x takes the right-end value of the closed interval, the previous equation can be recast as $A(b) = F(b) - F(a)$. But, we stated earlier that $A = A(b)$ or $A = A(b) = \int_a^b f(x)dx$. Therefore, we have the result

$$A = A(b) = \int_a^b f(x)dx = F(b) - F(a) = [F(x)]_a^b \tag{8.4.14}$$

Notice that to represent the expression $[F(x)]_a^b$ in equation (8.4.14), some writers use expressions such as $F(x)]_a^b$, $F(x)|_a^b$, $_a^b|F(x)$, etc.

A meticulous reader would have noticed by now the relationship between a definite integral and anti-differentiation. Equation (8.4.14) shows that to find the area under the graph of a continuous, positive function and over a closed interval $[a, b]$ on the function's domain (that is, to find $\int_a^b f(x)dx$), all one needs to do is to find an anti-derivative of the integrand $f(x)$, denoted by $F(x)$, and subtract the value of $F(x)$ when x takes the left-end value of the interval (that is, when $x = a$) from the value of $F(x)$ when x takes the right-end value of the interval (that is, when $x = b$); that is, find $F(b) - F(a)$. This result is called the *fundamental theorem of integral calculus* (FTIC).

Let us now state the FTIC formally: if a function $f(x) = F'(x)$ is positive and continuous on a closed interval $[a, b]$ and if $F(x)$ is an anti-derivative of $f(x)$, then the area bounded by the function and the closed interval on the domain of the function is given by $\int_a^b f(x)dx = F(b) - F(a)$. The reader would have noticed the distinction between a definite integral and an indefinite integral. We found in the last section that a definite integral, such as $\int_a^b f(x)dx$, is just a number and this number is the limit of a sum. But, the FTIC implies that the indefinite integral $\int f(x)dx$, which is a function of x and an anti-derivative of $F'(x) = f(x)$, can be used to find the limit of the sum in the definite integral. The FTIC also implies that all the rules of indefinite integral we exposed in Section 8.3 are applicable in the case of definite integral.

So far our exposition of the FTIC has been based on a positive-valued function. Many functions may be negative-valued or some portion(s) of a single function may be negative-valued. How do we find the area bounded by the graph of a negative-valued function and the closed interval $[a, b]$ on the domain of the function? Suppose that $f(x) = F'(x)$ is continuous on the closed interval $[a, b]$ with $f(x) \leq 0$ for all $x \in [a, b]$. Notice that the graph of the function $f(x) = F'(x)$, the x-axis, and the vertical lines $x = a$ and $x = b$ still enclose an area even though $f(x) = F'(x)$ is negative. Notice also that the area referred to here (or any other area) cannot be negative. Now suppose that $F(x)$ is an anti-derivative of $f(x)$. We can now define the area as the negative of the integral; that is, as $\int_a^b f(x)dx = -[F(b) - F(a)]$.

Let us now apply the FTIC to few specific examples. As the first example, consider the problem posed at the beginning of Section 8.4.2. This problem was to find the area bounded by the graph of the function $y = f(x) = 2x$ and the closed interval $[0, 1]$ on the x-axis. Notice that in the present problem $a = 0$ and $b = 1$. In the spirit of equation (8.4.14), the problem can be restated as $\int_{a=0}^{b=1} f(x)dx = \int_0^1 2xdx$. We know that the integral $\int f(x)dx = \int 2x\,dx = F(x) + C = (2x^2/2) + C$. Therefore, we can write the definite integral as $\int_0^1 2xdx = F(b) - F(a) = [F(x) + C]_a^b = [(2x^2/2) + C]_0^1 = [\{(2 \times 1^2)/2\} + C] - [\{(2 \times 0^2)/2\} + C] = 1 + C - 0 - C = 1$, which is exactly the result obtained at the beginning of Section 8.4.2 and with equations (8.4.10) and (8.4.11). The reader would have noticed that it is much easier to use the FTIC to evaluate a definite integral than to use the method of exhaustion.

As another example, consider evaluating $\int_{a=1}^{b=2} f(x)dx = \int_1^2 (2x + 1)dx$. We know that the integral $\int f(x)dx = \int (2x + 1)\,dx = (2x^2/2) + x + C = F(x) + C$. Therefore, we can write the definite integral as $\int_1^2 f(x)dx = \int_1^2 (2x + 1)dx = F(b) - F(a) = [F(x) + C]_a^b = [(2x^2/2) + x + C]_1^2 = 6 + C - 2 - C = 4$.

8.4.5 *Properties of definite integrals*

In this section we present, without proofs, some of the important properties of definite integrals. As we shall see later, these properties are highly useful in solving problems.

All the properties we present are concerned with continuous functions $f(x)$ and $g(x)$, and three points a, b, and c, such that $a < b < c$, on a closed interval $[a, c]$ on the domains of the functions.

Property I. Reversing the order of limits, or interchanging the limits, of integration changes only the sign and not the absolute value of the definite integral. In other words,

$$\int_a^b f(x)dx = -\int_b^a f(x)dx \qquad (8.4.15)$$

Property II. If the two limits of a definite integral are identical, then the definite integral is zero. That is,

$$\int_a^a f(x)dx = 0 \qquad (8.4.16)$$

Property III. The sum of the definite integral from point a to b and the definite integral from b to c is equal to the definite integral from a to c. This means that

$$\int_a^c f(x)dx = \int_a^b f(x)dx + \int_b^c f(x)dx \qquad (8.4.17)$$

Property IV. The definite integral of the sum or the difference of two functions, $f(x)$ and $g(x)$, is equal to the sum or difference of the integrals of the two functions. This means that

$$\int_a^b [f(x) \pm g(x)]dx = \int_a^b f(x)dx \pm \int_a^b g(x)dx \qquad (8.4.18)$$

Property V. The definite integral of a constant, k, times the function $f(x)$ is equal to the constant times the definite integral of the function $f(x)$. That is,

$$\int_a^b kf(x)dx = k\int_a^b f(x)dx \qquad (8.4.19)$$

Property VI. The definite integral of the negative of the function $f(x)$ is equal to the negative of the definite integral of $f(x)$. Symbolically, we have

$$\int_a^b -f(x)dx = -\int_a^b f(x)dx \qquad (8.4.20)$$

8.4.6 Indefinite integral from definite integral

We defined definite integral in equation (8.4.14). This definition was given symbolically as $\int_a^b f(x)dx = F(b) - F(a) = [F(x)]_a^b$. In fact, to find the definite integral we were appending the two limits, the lower limit a and the upper limit b, to the indefinite integral $\int f(x)dx$. In other words, we obtained the definite integral from the indefinite integral by appending the limits to the latter. A pertinent question at this stage is: can one derive the indefinite integral from the definite integral? The answer is yes. The procedure is outlined below.

Notice that in Figure 8.4.4(A) we represented the area as a variable $A(x)$. This representation of the area is based on the assumption that the lower limit a of the closed interval is a constant, while a point x in the interval is treated as a variable. This means that as x increases, the area $A(x)$ also increases. Let us now replace the upper limit b of the interval by the variable of integration x. Then the integral can be written as $\int_a^x f(x)dx$. If we evaluate this, we will obtain $\int_a^x f(x)dx = F(x) - F(a)$. Since a is a constant, $-F(a)$ must be a numerical value and can be treated as a constant C. Therefore, we can write the last equation as $\int_a^x f(x)dx = F(x)+C$. If we differentiate this integral with respect to x what we obtain is $d[\int_a^x f(x)dx]/dx = d[F(x)+C]/dx = F'(x) = f(x)$. Notice that the last term of this equation is nothing but the integrand of the definite integral. In other words, it is the indefinite integral. This shows the connection between the definite integral and indefinite integral.

8.4.7 Area under a curve

In Section 8.4.2 we defined the area of a region, using the method of exhaustion, as the limit of a sum and then found this limit by equation (8.4.12). We later found in Section 8.4.4 (equation (8.4.14)) that this sum could also be found by using the FTIC. In the present section we use the FTIC (equation (8.4.14)) to find the areas bounded by specific functions.

Suppose that we have a function of the form $y = f(x) = 1/(2+x)$ and a closed interval on its domain given by $[a = 1, b = 2]$. Our problem is to find the area $abcd$ bounded by the graph of the function and the closed interval, as illustrated in Figure 8.4.5(B). Applying equation (8.4.14) we can obtain this area by integrating the function $y = f(x) = 1/(2+x)$ from point a to point b; that is, by evaluating $\int_a^b f(x)dx = F(b) - F(a) = [F(x)]_a^b$. Since $\int f(x)dx = \int [1/(2+x)]dx = \ln(2+x)+C$, we can write $\int_a^b f(x)dx = F(b) - F(a) = [F(x)]_a^b = [\ln(2+x)+C]_a^b$. And since $a = 1$ and $b = 2$, the last equation can be written as $\int_1^2 f(x)dx = [F(x)]_1^2 = [\ln(2+x)+C]_1^2 = [\ln 4 + C] - [\ln 3 + C] = \ln 4 - \ln 3 = \ln[4/3] = 0.287$. Therefore, the area $abcd$ in Figure 8.4.5(B) is 0.287 units.

As another example, consider the function $y = f(x) = 4 - x$ with the closed interval given by $[a = 2, d = 6]$ on its domain. Our problem now is to find the sum of the areas abc and cde, which are bounded from above by the graph of the function and from below by the closed subinterval $[a = 2, c = 4]$ and bounded from above by the closed subinterval $[c = 4, d = 6]$ and from below by the same graph of the function, respectively, as illustrated in Figure 8.4.6(A).

We first split the closed interval $[a = 2, d = 6]$ into two: $[a = 2, c = 4]$ and $[c = 4, d = 6]$. We can then find the area abc under the graph and above the interval $[a = 2, c = 4]$ and then the area under the interval $[c = 4, b = 6]$ and above the graph separately by applying Property III in equation (8.4.17) of definite integral: $\int_a^d f(x)dx = \int_a^c f(x)dx + \int_c^d f(x)dx$. The area under the graph and above the interval $[a = 2, c = 4]$ is given by $\int_a^c f(x)dx = F(c) - F(a) = [F(x)]_a^c$. Since $\int f(x)dx = F(x)+C = \int (4-x)dx = 4x - (x^2/2)+C$, we can write $\int_a^c f(x)dx = F(c) - F(a) = [F(x)+C]_a^c = [4x - (x^2/2)+C]_a^c$. And since $a = 2$ and $c = 4$, the last equation can be written as $\int_{a=2}^{c=4} f(x)dx = [F(x)]_2^4 = [4x - (x^2/2)+C]_2^4 = [8 + C] - [6 + C] = 2$. Therefore, the area abc in Figure 8.4.6(A) is 2 units. This result is true because the area abc is the area of the triangle abc whose height (h) and length (l) are 2 units each. Therefore, the area of the triangle is $(1/2) \times h \times l = (1/2) \times 2 \times 2 = 2$ units.

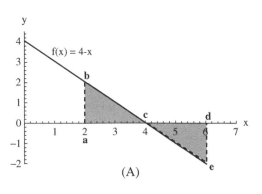

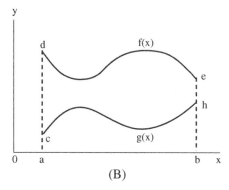

Figure 8.4.6

Let us now find the area *cde*, which is below the interval $[c = 4, d = 6]$ and above the graph of the function. Notice that the value of the function to the right of point $c = 4$ is negative. This implies that the integral of the function from $c = 4$ to $d = 6$ will be negative. Therefore, we shall multiply, as stated at the end of Section 8.4.4, the integral by -1 to make the area positive. The area above the graph and below the interval $[c = 4, d = 6]$ is given by $-\int_c^d f(x)dx$. Since $\int f(x)dx = [4x - (x^2/2) + C]$, we can write $-\int_c^d f(x)dx = -[F(d) - F(c)] = -[F(x) + C]_c^d = -[4x - (x^2/2) + c]_c^d$. And since $c = 4$ and $d = 6$, the last equation can be written as $-\int_{c=4}^{d=6} f(x)dx = -[F(x)]_4^6 = -[4x - (x^2/2) + C]_4^6 = -\{[6 + C] - [8 + C]\} = 2$. Therefore, the area *cde* in Figure 8.4.6(A) is 2. This result is also true because the area *cde* is the area of the triangle *cde* whose height (h) and length (l) are 2 units each. Therefore, the area of the triangle is, as before, $(1/2) \times h \times l = (1/2) \times 2 \times 2 = 2$ units.

Adding the two areas *abc* and *cde* we obtain $abc + cde = 2 + 2 = 4$ units. Therefore, the total area we required in Figure 8.4.6(A) is 4 units. This result is again true because adding the two triangles *abc* and *cde* we will obtain a square. The length of the square is $l = 2$ units and its height is $h = 2$ units. Therefore, the area of the square is $h \times l = 2 \times 2 = 4$ units.

8.4.8 Area between curves

So far we have been concerned with the determination of areas that are bounded by their respective graphs and the closed intervals on their respective domains. How can we determine the area bounded by two functions and a common closed interval on the domains of the two functions? Suppose that the two functions we have are $f(x)$ and $g(x)$ and their common closed interval is $[a, b]$. Our problem is to determine the area *cdeh* bounded by these two functions and the vertical lines that represent the end points of the functions' common closed interval, as illustrated in Figure 8.4.6(B).

We know how to find the area below the graph of the function $f(x)$ and above the interval $[a, b]$; that is, the area *adeb* in Figure 8.4.6(B). This is found by integrating the function $f(x)$ from point a to point b; that is, by evaluating $\int_a^b f(x)dx$. Similarly, the area below the graph of the function $g(x)$ and above the interval $[a, b]$ or the area *achb* can be found by evaluating $\int_a^b g(x)dx$. If we subtract the latter value from the former value we obtain the required area.

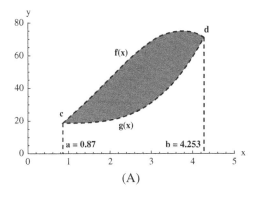

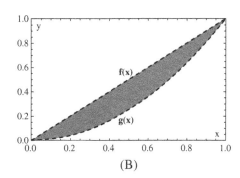

(A)   (B)

Figure 8.4.7

Therefore, applying Property IV in equation (8.4.18) of definite integral, we can write the area *cdeh* as $\int_a^b [f(x) - g(x)]dx = \int_a^b f(x)dx - \int_a^b g(x)dx$.

As an example, assume that we have two functions $f(x) = 10 + 14x^2 - 2.5x^3$ and $g(x) = 20 - 2.5x^2 + 1.25x^3$, the graphs of which are illustrated in Figure 8.4.7(A). Also assume that we need to find the shaded area between the two graphs. To find this area we can apply the above result: $\int_a^b [f(x) - g(x)]dx = \int_a^b f(x)dx - \int_a^b g(x)dx$. We know that $\int_{a=0.87}^{b=4.253} f(x)dx = \int_{0.874.253}^{4.253}(10 + 12x^2 - 2.25x^3)dx = [F(x) + C]_{0.87}^{4.253} = [10x + 4x^3 - 0.5625x^4 + C]_{0.87}^{4.253} \approx 185.629 - C$ and $\int_{a=0.87}^{b=4.253} g(x)dx = \int_{0.87}^{4.253}(5 - 2.25x^2 + 1.25x^3)dx = [G(x) + C]_{0.87}^{4.253} = [5x - 0.75x^3 + 0.3125x^4 + C]_{0.87}^{4.253} \approx 106.165 - C$. Therefore, we obtain the area as $\int_a^b f(x)dx - \int_a^b g(x)dx = [185.629 - C] - [106.165 - C] \approx 79.464$ units.

As another example, suppose that we have two functions $f(x) = x$ and $g(x) = x^2$, and their common closed interval $[a = 0, b = 1]$, as illustrated in Figure 8.4.7(B). Our problem is to find the shaded area bounded by these two functions above their common closed interval. To find this area, we can again apply Property IV in equation (8.4.18) of definite integral. Therefore, we can write the required area as $\int_a^b [f(x) - g(x)]dx = \int_a^b f(x)dx - \int_a^b g(x)dx$. We know that $\int_{a=0}^{b=1} f(x)dx = \int_0^1 f(x)dx = [F(x) + C]_{a=0}^{b=1} = [(x^2/2) + C]_0^1 = 1/2$ and $\int_{a=0}^{b=1} g(x)dx = \int_0^1 x^2 dx = [G(x) + C]_{a=0}^{b=1} = [(x^3/3) + C]_0^1 = 1/3$. Therefore, we obtain the required area as $\int_a^b [f(x) - g(x)]dx = \int_a^b f(x)dx - \int_a^b g(x)dx = (1/2) - (1/3) = 1/6$ units.

8.4.9 Average value of a function

We know from the solution to the problem in example 4 in Section 1.11.3 that the arithmetic mean or the average, denoted by $\bar{x}$, of n numbers x_i, where $i = 1, 2, \ldots, n$, is the sum of these numbers divided by n; that is, $\bar{x} = \sum_{i=1}^n x_i/n$. But, what about the average value of a function? Consider the function that represents the total profit of a firm, which sells a good, given by $\Pi = f(x) = 5x - x^2$, where Π denotes total profits in dollars and x denotes the quantity of the good sold. The graph of this function is illustrated in Figure 8.4.8(A). How do we find the average profit of the firm when the firm sells quantities in between 1 and 3 units of the good? In other words, what will be firm's average profit over the closed interval $[a = 1, b = 3]$ on

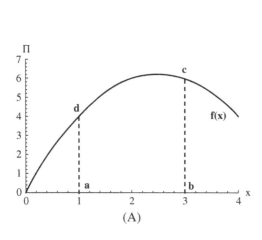
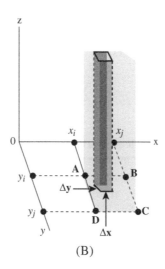

(A)   (B)

Figure 8.4.8

the domain of the firm's profit function? The answer to this question seems to be difficult as there are an infinite number of values in the closed interval. Then, how do we solve the problem? The method of solving problems like these is explained below.

As we split the interval into n subintervals of equal length in the method of exhaustion in Section 8.4.2, we can split the interval $[a = 1, b = 3]$ in Figure 8.4.8(A) into n subintervals of equal length. Therefore, each subinterval will have length equal to $\Delta x = (b - a)/n$. After this we can find the profit at a point in every subinterval. Then we can take the average of these sub-profits as an approximation to the average profit over the interval $[a = 1, b = 3]$. As we stated in the method of exhaustion, this approximation will improve as the number of subintervals increases; that is, the approximate average value of the function will tend to the true average value of the function as $n \to \infty$. In other words, the average value of the profit function can be written as

$$\bar{\Pi} = (1/n)[f(x_1) + f(x_2) + \cdots + f(x_i) + \cdots + f(x_n)] \tag{8.4.21}$$

where $\bar{\Pi}$ denotes the average profit or the average value of the function and x_i denotes the value of x in the ith subinterval. We know from above that $(b - a)$ measures the length of the closed interval in Figure 8.4.8(A). Let us multiply equation (8.4.21) by $[(b - a)/(b - a)]$ to obtain

$$\bar{\Pi} = (1/n)[f(x_1) + f(x_2) + \cdots + f(x_i) + \cdots + f(x_n)][(b - a)/(b - a)]$$

$$= \frac{1}{(b - a)}\left[\frac{(b - a)}{n}f(x_1) + \frac{(b - a)}{n}f(x_2) + \cdots + \frac{(b - a)}{n}f(x_i) + \cdots + \frac{(b - a)}{n}f(x_n)\right]$$

$$= \frac{1}{(b - a)}[f(x_1)\Delta x + f(x_2)\Delta x + \cdots + f(x_i)\Delta x + \cdots + f(x_n)\Delta x] = \frac{1}{(b - a)}\sum_{i=1}^{n}f(x_i)\Delta x$$

In the third step of the last equation we used our above result $\Delta x = (b-a)/n$. Now suppose that we increase the number of subintervals to infinity. Therefore, in the limit as $n \to \infty$, if the limit exists, equation (8.4.12) states that the sum on the RHS of the last equation will be replaced by $\int_a^b f(x) dx$. Therefore, we have the following result:

$$\bar{\Pi} = \lim_{n \to \infty} \frac{1}{(b-a)} \sum_{i-1}^n f(x_i)\Delta x = \frac{1}{(b-a)} \lim_{n \to \infty} \sum_{i-1}^n f(x_i)\Delta x = \frac{1}{(b-a)} \int_a^b f(x) dx$$

(8.4.22)

Let us now define the average value of a function formally. Suppose that we have a function $f(x)$. Then the average value of this function over a closed interval $[a, b]$ on its domain is given by $f(\bar{x}) = [1/(b-a)] \int_a^b f(x) dx$.

We are now ready to find the average value of the profit function: $\Pi = f(x) = 5x - x^2$. Therefore, the average profit over the closed interval $[a = 1, b = 3]$ on the function's domain can be found by applying equation (8.4.22). The result is $\bar{\Pi} = [1/(b-a)] \int_a^b f(x) dx = [1/(b-a)] \int_1^3 (5x - x^2) dx = F(b) - F(a) = [F(x) + C]_a^b = [F(x) + C]_1^3$. Since $F(x) + C = (5x^2/2) - (x^3/3) + C$, the last equation can be written as $\bar{\Pi} = [1/(3-1)]\{[F(x) + C]_1^3\} = [1/(3-1)] \times [(5x^2/2) - (x^3/3) + C]_1^3$, which can be simplified to obtain $\bar{\Pi} = (1/2)[(5x^2/2) - (x^3/3) + C]_1^3 = 5.66$. Therefore, the firm's average profit over the closed interval $[a = 1, b = 3]$ on the function's domain is $5.66.

8.4.10 Definite partial integrals

In Section 8.3.12, we saw how to find indefinite integrals where the integrands are multivariate functions such as $f(x_1, x_2)$, $f(x_1, x_2, x_3)$, etc. So far in the present section we have been concerned with evaluating definite integrals where the integrands are univariate functions. We now extend our exposition of definite integrals to the case where the integrands are multivariate functions.

Notice that we evaluated univariate definite integrands by first finding the primitive function using the rules of indefinite integrals, and then applying the lower and upper limit to the primitive function. We may follow the same procedure in evaluating definite integrals involving multivariate integrands. However, as in the case of finding indefinite integrals with multivariate integrands, we must treat all variables other than the variable of integration as constants when we evaluate a *definite partial integral* with multivariate integrands. Notice also that definite partial integrals are indefinite partial integrals appended with the limits of integration.

As an example, consider the first problem we solved in Section 8.3.12. This problem is to find the indefinite integral of the function $f(x_1, x_2) = 2x_1x_2 + x_2$ with respect to x_1. Let us now find the definite integral of the function when x_1 varies from 0 to 1. Notice that here only x_1 is varying and, therefore, x_1 is the variable of integration and x_2 is a constant. Following the symbol we used for definite integral, and using the result $\int (2x_1x_2 + x_2) dx_1 = x_2 x_1^2 + x_2 x_1 + C$, we may write the current problem symbolically as $\int_0^1 f(x_1, x_2) dx_1 = \int_0^1 (2x_1x_2 + x_2) dx_1$, which can be evaluated to obtain $\int_0^1 (x_1^2 x_2 + x_1 x_2) dx_1 = [F(x_1, x_2)]_0^1 = [x_1^2 x_2 + x_1 x_2 + C]_0^1 = 2x_2$.

As another example, consider the second problem we solved in Section 8.3.12. This problem is to find the indefinite integral of the function $f(x_1, x_2) = 2x_1x_2 + x_2$ with respect to x_2. Let us now find the definite integral of the function when x_2 varies from 0 to 1.

Notice that, unlike in the last example, x_2 is the variable of integration in the present problem and, therefore, x_1 is a constant. Following the symbol we used for definite integral, and using the result $\int F_{x1} dx_2 = x_1 x_2^2 + (x_2^2/2) + C$, we may write the current problem symbolically as $\int_0^1 f(x_1, x_2) dx_2 = \int_0^1 (2x_1 x_2 + x_2) dx_2$, which can be evaluated to obtain $\int_0^1 (2x_1 x_2 + x_2) dx_2 = [F(x_1, x_2)]_0^1 = [x_1 x_2^2 + (x_2^2/2) + C]_0^1 = (1/2) + x_1$.

As the final example, consider integrating the function $f(x_1, x_2, x_3) = x_1 + x_2 + x_3 + x_1 x_2 + x_1 x_3$ when x_1 varies from 0 to 2. Therefore, we have $\int_0^2 f(x_1, x_2, x_3) dx_1 = \int_0^2 (x_1 + x_2 + x_3 + x_1 x_2 + x_1 x_3) dx_1 = [F(x_1, x_2, x_3)]_0^2$. We know that the indefinite integral of $(x_1 + x_2 + x_3 + x_1 x_2 + x_1 x_3)$ with respect to x_1 is $\int (x_1 + x_2 + x_3 + x_1 x_2 + x_1 x_3) dx_1 = (x_1^2/2) + x_1 x_2 + x_1 x_3 + (x_1^2 x_2/2) + (x_1^2 x_3/2) + C$. Therefore, we have the result $\int_0^2 (x_1 + x_2 + x_3 + x_1 x_2 + x_1 x_3) dx_1 = [(x_1^2/2) + x_1 x_2 + x_1 x_3 + (x_1^2 x_2/2) + (x_1^2 x_3/2) + C]_0^2 = [2 + 2x_2 + 2x_3 + 2x_2 + 2x_3 + C] - C = 2[1 + 2x_2 + 2x_3]$.

8.4.11 Definite multiple integrals

In Section 8.3.13, we explored indefinite multiple integrals or iterated indefinite integration of multivariate functions such as $F_{x1} = f(x_1, x_2)$ and $F_{x1} = f(x_1, x_2, x_3)$. In the present section we present *definite multiple integrals* or definite integration of multivariate functions. Before we begin this, let us attempt to present a geometric representation of definite integrals of multivariate functions.

Notice that we have already interpreted the definite integrals of a univariate function as the area under the graph of the function and above the chosen closed interval on the domain of the function. The question is: can one give such a geometric interpretation to the definite integrals of multivariate functions? The answer is yes in the case of definite integrals of bivariate functions. Since a geometric representation of functions involving more than two independent variables is not possible, we restrict the geometric representation of definite integrals to the case of integrands with two independent variables.

Suppose that we have a multivariate function given by $z = f(x, y)$ with the closed intervals $[x_i, x_j]$ and $[y_i, y_j]$ on its domain. Suppose also that the graph of the function and the intervals are as illustrated in Figure 8.4.8(B). Our problem is to determine the *volume*, not the area, bounded by the intervals $[x_i, x_j]$ and $[y_i, y_j]$ and the function; that is, the lightly shaded area between the two-dimensional x, y-plane (or ABCD) defined by the intervals $[x_i, x_j]$ and $[y_i, y_j]$ and the three-dimensional plane or the *surface* defined by the function $f(x, y)$, as illustrated in the figure.

Recall that we used the method of exhaustion with rectangles to find the area bounded by a univariate function in Section 8.4.2. The same method can be used to determine the volume of the region bounded by a bivariate function and a two-dimensional space such as the one referred to above. For this first consider the object, like the one at the center of the two-dimensional space ABCD, with base area equal to $\Delta x \times \Delta y$ and height equal to the value of the function corresponding to the point at the top of the object. We know that the volume of this object is equal to its base times its height. We can now construct n similar objects such that the required region is exhausted and find the sum of the volumes of all these objects. But, notice that as $n \to \infty$, the sum of the volumes of the n objects will tend to the volume of the required region. And this sum is given by the multiple integral

$$\int_{y_i}^{y_j} \int_{x_i}^{x_j} f(x, y) \, dx \, dy \tag{8.4.23}$$

which is also called a *double-definite integral*. If the integrand involves three variables, such as $f(x_1, x_2, x_3)$, then the sum of the volumes of the n objects when $n \to \infty$ will be given by the definite multiple integral

$$\int_{z_i}^{z_j} \int_{y_i}^{y_j} \int_{x_i}^{x_j} f(x, y, z) \, dx \, dy \, dz \qquad (8.4.24)$$

which is also called a *triple-definite integral*.

Let us now apply the idea to find the definite integrals of multivariate functions. As an example, consider the problem in the first example of the last section. This problem was to find the partial definite integral of the function $f(x, y) = 2xy + y$, where $x = x_1$ and $y = x_2$, with respect to x. Let us now find the double-definite integral of the function when x varies from 0 to 1 and y varies from 0 to 1. The problem can be written as $\int_{y=0}^{y=1} \int_{x=0}^{x=1} f(x, y) \, dx \, dy = \int_0^1 \int_0^1 [2xy + y] \, dx \, dy$. First, we can integrate the function with respect to x treating y as constant, and then integrate the result with respect to y. We have already obtained the result for the first step as the solution to the problem in the first example in the last section. This result was $\int_0^1 f(x, y) \, dx = \int_0^1 (2xy + y) \, dx = 2y$. Therefore, we can write $\int_{y=0}^{y=1} \int_{x=0}^{x=1} f(x, y) \, dx \, dy = \int_0^1 [\int_0^1 (2xy + y) \, dx] \, dy = \int_0^1 2y \, dy$. We now integrate this result with respect to y. The result is $\int_0^1 2y \, dy = [y^2]_0^1 = 1$.

As another example, consider the problem in the last example of the last section. Treating $x = x_1$, $y = x_2$, and $z = x_3$, we can reformulate the integrand in that example as $f(x, y, z) = x + y + z + xy + xz$. Let us now find the triple-definite integral of the function when x, y, and z vary from 0 to 1. The problem can be written as $\int_0^1 [\int_0^1 [\int_0^1 [(x + y + z + xy + xz) \, dx] \, dy] \, dz$. Firstly, we can integrate the function with respect to x treating y and z as constants; secondly, integrate the result in the last step with respect to y treating z as constant; and, thirdly, integrate the result in the last step with respect to z. Therefore, following these steps, we obtain $\int_0^1 [\int_0^1 [(1/2) + y + z + (y/2) + (z/2)] \, dy] \, dz$, $\int_0^1 [(5/4) + (3z/2)] \, dz$, and $\int_0^1 [(5/4) + (3z/2)] \, dz = [(5z/4) + (3z^2/4) + C] = 2$.

8.4.12 Application examples

Example 1.   This example is concerned with *income distribution*. Let x denote the *cumulative percentage* of people, from the poorest to the richest, who receive income in an economy and that $y = f(x)$ denotes the cumulative percentage of income. This suggests that on the graph of the function $y = f(x) = x$ there exists perfect equality in the distribution of income, as illustrated in Figure 8.4.9(A). Suppose that we have another function $y = g(x) = (x/4) + (6x^2/8)$, where the variables possess the same interpretation as they do in the case of $y = f(x)$. The graph represented by this function, or the function itself, is called a *Lorentz curve*. The degree of deviation from equality is computed by the *coefficient of inequality* for a Lorentz curve, which is the area between $y = f(x) = x$ and $y = g(x) = (x/4) + (6x^2/8)$. Determine the coefficient of inequality for the Lorentz curve $y = g(x) = (x/4) + (6x^2/8)$.

Solution.   Our aim is to determine the area bounded by the equality curve (the graph of the function $y = f(x) = x$) and the Lorentz curve (the graph of the function $y = g(x) = (x/4) + (6x^2/8)$) over the common closed interval [0, 1]. These graphs are illustrated in Figure 8.4.9(A). As presented in Section 8.4.8, the area between $f(x) = x$ and

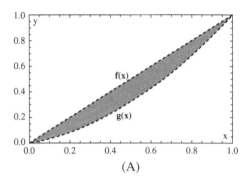

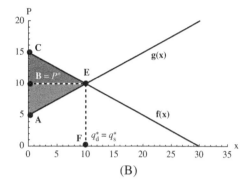

(A)   (B)

Figure 8.4.9

$y = g(x) = (x/4) + (6x^2/8)$ can be found by integrating the two functions and subtracting the latter from the former. Assume that an anti-derivative of $f(x)$ is $F(x) + C$ and that of $g(x)$ is $G(x) + C$. Then, the required area can be written as $\int_0^1 f(x)dx - \int_0^1 g(x)dx$. We know that $\int_0^1 f(x)dx = \int_0^1 xdx = [F(x) + C]_0^1 = [(x^2/2)]_0^1 = 1/2$. Similarly, we can obtain $\int_0^1 g(x)dx = \int_0^1 [(x/4) + (6x^2/8)]dx = [G(x) + C]_0^1 = [(x^2/8) + (6x^3/24) + C]_0^1 = (1/8) + (6/24) = 9/24 = 0.375$. Therefore, the required area is $\int_0^1 f(x)dx - \int_0^1 g(x)dx = \int_0^1 xdx - \int_0^1 [(x/4) + (6x^2/8)]dx = (1/2) - (9/24) = 0.5 - 0.375 = 0.125$. Therefore, the coefficient of inequality for the present Lorentz curve is 0.125.

Example 2. Assume that the inverse demand for and the inverse supply of a good are given by the functions $P = f(x) = 15 - 0.5x_d$ and $P = g(x) = 5 + 0.5x_s$, respectively, where P, x_d, and x_s denote the price of the good in dollars, the quantity demanded of the good, and the quantity supplied of the good, respectively. Find the *consumers' surplus* and *producers' surplus*.

Solution. Let us first plot the graphs of the two functions, as illustrated in Figure 8.4.9(B). Notice that the equilibrium price P^* and quantity $x_d^* = x_s^*$ of the good are determined by the intersection of the demand and supply functions as at point E in the figure. Equating the demand and supply functions, we obtain the equilibrium price P^* and quantity $x_d^* = x_s^*$ as $P^* = 10$ (point B) and $x_d^* = x_s^* = 10$ (point F), respectively. This means that in equilibrium (point E), the price of the good will be \$10 and 10 units of the good are demanded and supplied.

Consumers' surplus is defined as the extra amount of money over the equilibrium price that consumers are willing to pay to obtain the good. Therefore, consumers' surplus is the difference or the area between what the consumers are willing to pay (the area below the demand curve, or the area 0CEF) and what they are actually paying (the area 0BEF). This area is equal to the area of the triangle BEC. This area can be found by subtracting the area 0BEF (which equals the area below $P^* = 10$) from the area 0CEF. We know that definite integral can be used to approximate this area. Thus, in terms of definite integral, this area can be written, following our demonstration in Section 8.4.8, as $\int_0^{10} 15 - 0.5x - 10]dx$, where $x = x_d$. If we integrate the last integral we obtain $[15x - 0.25x^2 - 10x]_0^{10} = 25$. Therefore, the consumers' surplus in the present example is equal to $25.

Producers' surplus is defined as the extra amount of money over the *cost price* that a producer receives when it produces the equilibrium quantity of the good. Therefore, the difference or the area between what the producer is receiving when he produces the equilibrium quantity (the area 0BEF) and what it costs for him to produce the equilibrium quantity of the good (the area 0AEF). This area is equal to the area of the triangle AEB. To determine this area, we can subtract the area 0AEF from the area 0BEF (which equals the area below $P^* = 10$). This, in terms of definite integral, can be written as $\int_0^{10}[10 - 5 - 0.5x]\,dx$, where $x = x_s$. If we integrate the last integral we obtain $[10x - 5x - 0.25x^2]_0^{10} = 25$. Therefore, the producers' surplus in the present example is equal to $25.

Let us now present the general formulas for the consumers' and producers' surpluses. Suppose that the general inverse demand and supply functions of a good are given by $P = f(x_d)$ and $P = g(x_s)$, respectively, which can be equated and solved for the equilibrium price P^*. Substitution of P^* in the inverse demand and supply functions gives us the equilibrium quantity $x^* = x_d^* = x_s^*$. Then the consumers' surplus is given by

$$\int_0^{x^*} [(x_d) - P^*]\,dx_d \tag{8.4.25}$$

and the producers' surplus is given by

$$\int_0^{x^*} [P^* - g(x_s)]\,dx_s \tag{8.4.26}$$

Example 3.   Suppose that a firm's marginal revenue function is as that in example 1 in Section 8.3.14: $F'(x) = f(x) = 50 - x - x^2$. Find the change in the firm's total revenue when sales increase from 2 to 4 units of the good.

Solution.   We know that the integral of the marginal revenue function yields the total revenue function. If we integrate the function from 2 to 4, we will obtain the change in total revenue. We obtained in the solution to the problem in example 1 in Section 8.3.14 that, after ignoring the constant C, the total revenue of the firm as $R = F(x) = \int F'(x)dx = \int (50 - x - x^2)dx = 50x - (x^2/2) - (x^3/3)$. Let us now take the limits of integration to get $\Delta R = \Delta F(x) = \int_2^4 F'(x)dx = [F[x]]_2^4 = [50x - (x^2/2) - (x^3/3)]_2^4 \approx \$81.34$. This means that the firm's total revenue will increase by $81.34 if it increases production from 2 to 4 units of the good.

Example 4.   Suppose that $100 invested in the share of a company is expected to grow by the function $f(x) = 100e^{0.08x}$, where x denotes time in years. What will be the value of this share after 12 years?

Solution.   The value of the share at the time of investment (that is, at $x = 0$) is $100 because $f(0) = 100e^{0.08 \times 0} = 100$. To find the value of the investment after 12 years, we need to integrate the function from time $x = 0$ to $x = 12$. Therefore, integrating the function $f(x) = 100e^{0.08x}$ from $x = 0$ to $x = 12$ we obtain $\int_0^{12} f(x)dx = \int_0^{12} 100e^{0.08x}dx = [100(e^{0.08x}/0.08)]_0^{12} = 3264.62 - 1250 = 2014.62$. Therefore, the value of the investment after 12 years will be $2014.62.

Example 5. What will be the *present value* of receiving $100 each year for six years when the *discount rate r* = 0.08 is compounded continuously? What will be the accumulated sum of these receipts?

Solution. Suppose that a person invests P amount of money in a bank account at interest rate r when the bank compounds interest annually. The amount of money in the account after one year will be $F_t = P + r.P = P(1 + r)$. The amount of money in the account after the second year will be $F_t = P(1 + r)(1 + r) = P(1 + r)^2$. The amount of money in the account after t years will be

$$F_t = P(1 + r)^t \tag{8.4.27}$$

where $(1 + r)$ is called the *growth factor* and r is called the *rate of growth*. The *present value* of the future amount of money F_t received in t years from now can be obtained from equation (8.4.27) as

$$P = F_t(1 + r)^{-t} \tag{8.4.28}$$

where r is called the discount rate. Finding the present value of a future sum of money is called *discounting*. Suppose now that the bank applies compound interest, compounding the interest m times a year. Given compound interest, the amount in the person's account after t years will be

$$F_t = P[1 + (r/m)]^{m \times t} \tag{8.4.29}$$

Last, assume that the bank compounds interest infinitely; that is, $m \to \infty$. Then equation (8.4.29) can be written as $F_t = \lim_{m \to \infty} P[1 + (r/m)]^{m \times t}$. Multiplying the exponent by r/r and moving the limit operator beyond P, the last equation can be written as $F_t = P \lim_{m \to \infty} [1 + (r/m)]^{(m/r) \times r \times t}$. Letting $n = r/m$, the last equation can be written as $F_t = P\{\lim_{m \to \infty} [1 + (1/n)]^n\}^{r \times t}$. Using equation (1.10.14), this result can be written as

$$F_t = Pe^{r \times t} \tag{8.4.30}$$

where e denotes the base of natural logarithm. Therefore, the present value of F_t amount of money to be received in t years from now, when the discount rate is r, is

$$P = F_t e^{-r \times t} \tag{8.4.31}$$

The reader would have noticed that it is the constant discount rate r that is treated either discretely or continuously in equations (8.4.27)–(8.4.31). The independent variable (t) in these equations is still treated discretely. That is, the future sum(s) of money is received at specific point(s) in time, say, at the end of the first year, second year, and so on. But, what if the future sums are received as a continuous stream of receipts (at every point in time)? How can we find the present value of such continuous future streams of receipts?

Assume that an individual receives a sum of money every year and, therefore, we can consider the sum as a function of time; that is, we can treat it as $f(t)$. This sum is received between the closed time interval $[t = 0, t = T]$. Assume also that we divide this interval into n subintervals of equal length Δt. Then the sum received over the ith subinterval, where $i = 1, 2, \ldots, n$, of time will be $f(t_i)\Delta t e^{r \times t}$ and the present value of this sum will be $f(t_i)\Delta t e^{-r \times t}$. Therefore, the present value of the sums received over all other subintervals can be written as $\sum_{t=0}^{T} f(t_i)\Delta t e^{-r \times t}$. Treating $f(t)$ as F, we can write the last result as $\sum_{t=0}^{T} F \Delta t e^{-r \times t}$. We know that as the number of subintervals n increase, $\Delta t \to 0$. But, when $\Delta t \to 0$, we know from equation (8.4.12) that the sum can be approximated by the definite integral, which gives the formula to find the present value of continuous future stream of receipts as

$$\int_0^T F e^{-r \times t} dt = [F e^{-r \times t}/(-r)]_0^T = (F/r)[1 - e^{-r \times T}] \tag{8.4.32}$$

We defined above the present value of the sum of the continuous future stream of receipts over the entire subintervals as the definite integral $\int_0^T e^{-r \times t} f(t) dt$. One might wonder what the total value or the accumulated sum of this continuous future stream of receipts would be. We can find this *future value* or the *accumulated sum* of continuous future stream of receipts by applying the formula

$$\int_0^T F e^{r(T-t)} dt \tag{8.4.33}$$

Having developed the required tools, let us now attempt to solve the problem in the present example. In our example, $F = 100$, $r = 0.08$, and $t = 6$ years. Therefore, applying the formula for the present value of continuous future stream of receipts (equation (8.4.32)) we obtain $\int_0^T F e^{-r \times t} dt = \int_0^6 100 e^{-0.08t} dt = [100(e^{-0.08t})/(-0.08)]_0^6 = [(100 e^{-0.08 \times 6})/(-0.08)] - [(100 e^{-0.08 \times 6})/(-0.08) = 477$. Therefore, the present value of receiving \$100 each year for six years is \$477.

We can now find the accumulated sum of receiving \$100 each year for six years using the formula $\int_0^T F e^{r(T-t)} dt$ (equation (8.4.33)). Therefore, we have $\int_0^T F e^{r(T-t)} dt = \int_0^T 100 e^{0.08(6-t)} dt = \int_0^6 100 e^{0.08(6-t)} dt = \int_0^6 100 e^{0.48-0.08t} dt = 100 e^{0.48} \int_0^6 e^{-0.08 \times t} dt = [161.6 \times (e^{-0.08 \times t})/(-0.08)]_0^6 = [-2020 e^{-0.08 \times t}]_0^6 = -1250 + 2020 = 770$. Therefore, the accumulated sum after six years will be \$770.

Example 6.   Consider the problem in example 4 above. In this example, \$100 invested in the share of a company is expected to grow by the function $f(x) = 100 e^{0.08x}$, where x denotes time in years. What will be the average value of this share after 12 years?

Solution.   We can apply equation (8.4.22) to solve this problem. To apply this equation we define $a = 0$ and $b = 12$. Then we can write $f(\bar{x}) = [1/(b-a)] \int_a^b f(x) dx = [1/(12 - 0)] \int_0^{12} 100 e^{0.08x} dx = (100/12) \int_0^{12} e^{0.08x} dx = 8.33 \int_0^{12} e^{0.08x} dx = [8.33 e^{0.08x}/0.08]_0^{12} = [104.13 e^{0.08x}]_0^{12} = 272 - 104.13 = 167.87$. Therefore, the average value of this share after 12 years will be \$167.87.

Example 7.   Suppose that the marginal cost in dollars of a firm is given by the function $f(x) = 100 - 10x + x^2$, where x denotes the quantity of output produced. What will be total cost of increasing output from $x = 10$ to $x = 20$?

Solution. We know that we obtain the total cost function when we integrate the marginal cost function. Therefore, the total cost function is $TC = \int f(x)dx$. If we evaluate this integral using the limits $x = 10$ to $x = 20$, we obtain the total cost of increasing output from $x = 10$ to $x = 20$. Notice that $TC = \int f(x)dx = \int (100 - 10x + x^2)dx = [100x - 5x^2 + (x^3/3) + C]$. Therefore, evaluating this integral from $x = 10$ to $x = 20$ we obtain $\int_{10}^{20} f(x)dx = \int_{10}^{20} (100 - 10x + x^2)dx = [100x - 5x^2 + (x^3/3)]_{10}^{20} = 2666.67 - 833.33 = 1833.34$. Thus, the total cost of increasing output from 10 to 20 units is \$1833.34.

Example 8. Assume that the total investment, in billions of dollars, in an economy is a function of time (t) and is given by $I = f(t) = 10\,t^{3/4}$. Find the total *capital formation* for the first five years.

Solution. The total capital formation in the economy is given by $\int f(t)dt = K(t)$, where $K(t)$ denotes the level of capital at time t. Therefore, the total capital formation in the economy during the first five years (that is, from time $t = 0$ to $t = 5$) can be written as $\int_0^5 f(t)dt = \int_0^5 10t^{3/4}dt = [40t^{7/4}/7]_0^5 = 95.53$. Therefore, the level of capital formation in the economy is \$95.53 billion during the first five years.

Example 9. Assume that the Oil and Natural Gas Commission of India (ONGC) estimates that the yearly rate of oil extraction (in millions of barrels) from its offshore site Bombay High is given by the function $f(t) = 40e^{1-t}$. ONGC also believes that the stock of oil at the site will be exhausted in 50 years. What will be total quantity of oil that the company can extract from Bombay High if the above rate of extraction is continued for the next 50 years?

Solution. To find the total quantity of oil that can be extracted during the next 50 years if the rate of extraction is $f(t) = 40e^{1-t}$, we need to integrate the function with respect to time from $t = 0$ to $t = 50$. Therefore, we obtain $\int_0^{50} f(t)dt = \int_0^{50} 40e^{1-t}dt = [-40e^{1-t}]_0^{50} = -2.241 + 108.73 = 108.731$. This means that the total quantity of oil that ONGC can extract from Bombay High before oil deposits are exhausted is 108.731 million barrels.

Example 10. Assume that an individual has a choice of either going for work with only high school education or going for work after obtaining an undergraduate degree in economics, business, or finance. If the individual decides to go for work with only high school education, the salary (in dollars) over the years will be given by the function $g(t) = 8000e^{0.05\,t}$; and if the individual decides to go for work after obtaining a degree, the salary over the years will be given by the function $f(t) = 12000e^{(0.05+0.02)\,t}$. If the individual works for 25 years, what will be the difference in total salary earned over the entire period of work?

Solution. Notice that the individual's salary at the beginning of work (when $t = 0$) will be $g(0) = 8000e^{0.05\times0} = \$8000$ with high school education and will be $f(0) = 12\,000e^{(0.05+0.02)\times0} = \$12\,000$ with a degree. Notice also that the total salary from work after undergraduate education is given by the definite integral $\int_0^{25} f(t)dt = \int_0^{25} 12\,000e^{(0.05+0.02)\,t}$ $dt = \int_0^{25} 12\,000e^{0.07\,t}dt = \$815\,075$. Similarly, the total salary from work with only high school education is given by the definite integral $\int_0^{25} g(t)dt = \int_0^{25} 8000e^{0.05\,t}dt = \$398\,455$. If we subtract the latter value from the former value, we obtain \$416\,620, which is the difference between the total salaries for work for 25 years with undergraduate education and with only high school education.

Example 11. Assume that the total quantity of a good produced by a firm using capital (K) and labor (L) is given by the Cobb–Douglas production function $f(K, L) = K^{0.7} L^{0.3}$. What will be the total increase in output if the firm increases the quantities of capital and labor to 10 units each from the present levels of 5 units each?

Solution. Notice that the present integrand is a function of two variables, K and L, and, therefore, the integrand will make a three-dimensional plane when it is graphed. As we noticed in Section 8.4.11, the definite integral of this function over the intervals $[5 \leq K \leq 10]$ and $[5 \leq L \leq 10]$ will give us the volume of the region below the graph of the function and above the intervals. This area will be equal to the increase in output when the firm increases capital and labor by the stated amounts.

We can now apply the method of evaluating definite integrals of multivariate integrands covered in Section 8.4.11. Therefore, we can write the double-definite integral of the present problem as $\int_5^{10} [\int_5^{10} [f(K, L)] dK] dL = \int_5^{10} [\int_5^{10} [K^{0.7} L^{0.3}] dK] dL$. Evaluating this integral will yield $\int_5^{10} \{[K^{1.7} L^{0.3} / 1.7]_5^{10}\} dL = \int_5^{10} 20.41 L^{0.3} dL = [20.41 L^{1.3} / 1.3]_5^{10} = 186.013$. Therefore, the increase in output when the firm increases the quantities of capital and labor from 5 units each to 10 units each will be 186.013 units.

8.4.13 Exercises

1. Evaluate the following definite integrals:
   (i) $\int_0^1 5 dx$; (ii) $\int_0^1 5x dx$; (iii) $\int_0^1 x(1-x)^2 dx$; (iv) $\int_0^1 (1-x)^4 dx$; (v) $\int_0^1 (1/5x) dx$;
   (vi) $\int_{-1}^1 (x^2 + 1)^3 dx$; (vii) $\int_{-1}^1 e^{x^2} dx$; (viii) $\int_{-1}^1 6e^{3x} dx$; (ix) $\int_{-1}^1 [5/(5 + e^x)] dx$.

2. Evaluate the following definite integrals:
   (i) $\int_0^1 2x(x^2 + 1)^3 dx$; (ii) $\int_0^1 [2x/(x^2 + 1)^3] dx$; (iii) $\int_0^1 [4x/((x^2 - 1))] dx$;
   (iv) $\int_0^1 2xe^{x^2+4} dx$.

3. Evaluate the following definite integrals:
   (i) $\int_0^1 xe^x dx$; (ii) $\int_0^1 2xe^{-4x} dx$; (iii) $\int_0^1 5x(x-1)^{1/2} dx$; (iv) $\int_0^1 [x/(e^x)] dx$.

4. Evaluate the following definite integrals:
   (i) $\int_0^1 [(x+2)/(x^2 + 2x + 1)] dx$; (ii) $\int_0^1 [(x+2)/(x^2 - 2x + 1)] dx$;
   (iii) $\int_0^1 [(x+2)/(x^3 - 2x^2 + x)] dx$.

5. Evaluate the following definite integrals:
   (i) $\int_0^1 \int_0^1 x^2 y^2 dx dy$; (ii) $\int_0^1 \int_0^1 (x/y) dx dy$; (iii) $\int_0^1 \int_0^1 \int_0^1 x^2 y^2 z^2 dx dy dz$;
   (iv) $\int_0^1 \int_0^1 \int_0^1 (2x + 2y + 2z) dx dx dy dz$.

6. Find the volumes of the following regions defined by the given functions and the given intervals using definite integrals:
   (i) $f(x) = x^2 - 3x$, $1 \leq x \leq 4$; (ii) $f(x) = 5 + x - x^2$, $-2 \leq x \leq 2$; (iii) $f(x) = 2x^2 + x$, $-2 \leq x \leq 2$; (iv) $f(x, y) = xy$, $0 \leq x \leq 2$ and $0 \leq x \leq 2$; (v) $f(x, y) = x + y$, $0 \leq x \leq 2$ and $0 \leq x \leq 2$; (vi) $f(x, y) = x - y$, $0 \leq x \leq 2$ and $0 \leq x \leq 2$.

7. *Application exercise.* Suppose that a firm's marginal cost in dollars of producing x units of its output is given by $f(x) = dC/dx = 10 + 0.4x$, where C denotes the firm's total cost in dollars. Determine the change in the firm's total cost if it decides to increase output from $x = 50$ units to $x = 60$ units.

8. *Application exercise.* Assume that a firm's marginal revenue in dollars from the sale of x units of its output is given by $f(x) = dR/dx = 50/x^{1/2}$, where R denotes the firm's total revenue in dollars. Determine the change in the firm's total revenue if it decides to increase output from $x = 100$ units to $x = 150$ units.

9. *Application exercise.* Determine the coefficient of inequality for the Lorentz curve $f(x) = (5x^2 + x)/6$.

10. *Application exercise.* Assume that the inverse demand for a good is given by the function $f(x) = 25 - x^2$, where x denotes the quantity of the good demanded. Determine the consumers' surplus if the equilibrium price is $10.

11. *Application exercise.* Assume that the inverse supply of a good is given by the function $f(x) = 10 + x^2$, where x denotes the quantity of the good supplied. Determine the producers' surplus if the equilibrium price is $5.

12. *Application exercise.* Assume that the inverse demand for and supply of a good are given by the functions $f(x) = 10 - 0.5x$ and $g(x) = 5 + 0.5x$, respectively. Determine the consumers' surplus and producers' surplus when the market is in equilibrium.

13. *Application exercise.* Suppose that a person purchased a used car for $5000 the value of which is expected to decline by $f(x) = 5000e^{-0.06x}$ every year, where x denotes time in years. What will be the total decline in the value of the car after five years?

14. *Application exercise.* What will be the present value of receiving $1000 each year for five years when the interest rate $r = 0.1$ is compounded continuously? What will be the accumulated sum of these receipts?

15. *Application exercise.* Assume that the total quantity of a good produced by a firm using capital (K) and labor (L) is given by the Cobb–Douglas production function $f(K, L) = K^{0.5}L^{0.5}$. What will be total increase in output if the firm increases the quantity of capital from 10 units to 20 units?

 Web supplement: S8.4.14 Mathematica applications

8.5 Improper integrals

8.5.1 Introduction

In Section 8.4 we were primarily concerned with evaluating definite integrals whose limits (both lower and upper) were well defined. In other words, these limits were assumed to be *finite* numbers. But, one question emerges now is what happens if the limits tend to infinity. We will see in Section 8.5.2 how we can integrate functions where the limits of integration tend to infinity.

We have also integrated functions that could generate bounded regions. That is, we have dealt with integrals with which we are able to measure areas of bounded regions. Sometimes one may come across unbounded regions and want to determine the areas of such regions. We will present this topic in Section 8.5.3.

In our exposition of integral calculus we have maintained an important assumption that the integrands are *continuous* functions of some variable(s). But, as we saw in Section 1.8.4, one can find many functions, in the subjects of our interest, which are discontinuous. How can we integrate such functions? This topic will be presented in Section 8.5.4.

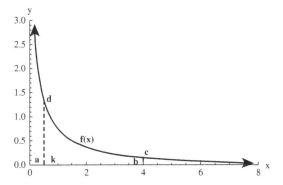

Figure 8.5.1

8.5.2 Improper integrals: integration with infinite limits

As we mentioned in the last section, we sometimes need to evaluate definite integrals with limits approaching infinity. This topic is taken up here. It is easier to understand the meaning of *improper integrals* (with infinite limits) with the help of the graph of the function $y = f(x)$ illustrated in Figure 8.5.1.

Suppose that we want to evaluate the definite integral $\int_a^b f(x)dx$. If the lower limit (a) of this integral approaches negative infinity, or the upper limit (b) approaches positive infinity, or both the lower and the upper limits approach infinity, negative and positive, respectively, then we may write the integral as

$$\int_{-\infty}^b f(x)dx = \lim_{a \to -\infty} \int_a^b f(x)dx \tag{8.5.1}$$

$$\int_a^{+\infty} f(x)dx = \lim_{b \to +\infty} \int_a^b f(x)dx \tag{8.5.2}$$

and

$$\int_{-\infty}^{+\infty} f(x)dx = \lim_{a \to -\infty, b \to +\infty} \int_a^b f(x)dx \tag{8.5.3}$$

respectively. In terms of the graph in Figure 8.5.1, the improper integrals in equations (8.5.1)–(8.5.3) give the areas under the graph of the function and above the intervals $[-\infty, b]$, $[a, \infty]$, and $[-\infty, +\infty]$, respectively. The integrals in equations (8.5.1)–(8.5.3) are called improper integrals or, more precisely, improper integrals with infinite limit(s). If the limits in equations (8.5.1)–(8.5.3) exist, then the improper integral is said to be convergent; if the limits do not exist, then the improper integral is said to be divergent.

As an example, consider the integral $\int_1^{+\infty} (1/x^4)dx$. Notice that this integral is an improper integral because the upper limit of the integral is infinity and that the limits of this integral are identical to those in equation (8.5.2) if we treat $a = 1$ and $b \to +\infty$. Therefore, we can write the integral in the form of equation (8.5.2) as $\int_1^{+\infty}(1/x^4)dx = \lim_{b\to+\infty}\int_1^b(1/x^4)dx =$ $\lim_{b\to+\infty}[-x^{-3}/3]_1^b = (-1/3b^3) + 1/3 = 0 + 1/3 = 1/3$. This result shows that the given improper integral converges to $1/3$.

As another example, consider the integral $\int_{-\infty}^0 e^{2x}dx$. Notice that this integral is an improper integral because the lower limit of the integral is infinity and that the limits of this integral are identical to those in equation (8.5.1) if we treat $a\to-\infty$ and $b=0$. Therefore, we can write the integral in the form of equation (8.5.1) as $\int_{-\infty}^0 e^{2x}dx = \lim_{a\to-\infty}\int_a^0 e^{2x}dx = \lim_{a\to-\infty}[e^{2x}/2]_a^0 = \lim_{a\to-\infty}[(1/2)-(e^{2a}/2)] = (1/2) - 0 = 1/2$. This result shows that the given improper integral converges to $1/2$.

As the last example, consider the improper integral $\int_{-\infty}^{+\infty}e^{2x}dx$. Notice that this integral is an improper integral because both the lower and the upper limits of the integral are infinity and that the limits of this integral are identical to those in equation (8.5.3) if we treat $a\to-\infty$ and $b\to+\infty$. Therefore, we can write the integral in the form of equation (8.5.3) as $\int_{-\infty}^{+\infty}e^{2x}dx = \lim_{a\to-\infty,b\to+\infty}\int_a^b e^{2x}dx$. We have already evaluated the integral with its lower limit tending to infinity in the last example and obtained $\int_{-\infty}^0 e^{2x}dx = \lim_{a\to-\infty}\int_a^0 e^{2x}dx = 1/2$. We can now evaluate the integral with its upper limit tending to infinity to obtain $\int_0^{+\infty}e^{2x}dx = \lim_{b\to+\infty}\int_0^b e^{2x}dx = \lim_{b\to+\infty}[e^{2x}/2]_0^b = \lim_{b\to+\infty}[(e^{2b}/2)-(1/2)] = \infty$. Although the part $\int_{-\infty}^0 e^{2x}dx$ of the integral $\int_{-\infty}^{+\infty}e^{2x}dx$ converges to the finite value of $1/2$, the other part $\int_0^{+\infty}e^{2x}dx$ does not converge to a finite value. Therefore the integral $\int_{-\infty}^{+\infty}e^{2x}dx$ does not converge; instead, it diverges.

8.5.3 Improper integrals: integration with infinite integrands

In all our examples so far in the case of definite integrals, most of the integrands or the functions we integrated were *finite integrands* or *finite functions*. What this means is that most of the functions we integrated could give us a specific value for the function when the variable of integration assumed any value within the interval over which integration was carried out. Because of this nature of the integrands, we were able to measure the regions under (above) the graph of the function and above (below) the specified interval. This implies that all the regions we considered so far were bounded regions or regions bounded by the function and the interval(s). But, some integrands can generate *unbounded regions*, even with finite limits, as the region illustrated in Figure 8.5.1.

Notice that the graph of the function $f(x)$ in Figure 8.5.1 is continuous and is not defined at $a=0$. When $x\to0$, $f(x)$ becomes infinite. The problem now is to evaluate the integral $\int_a^b f(x)dx$; that is, to determine the area under the graph of the function and over the half-open interval $(a, b]$. How do we proceed? The trick is as follows. Suppose that we pick a number k in $(a, b]$. Therefore, we can write the integral as

$$\int_a^b f(x)dx = \lim_{k\to0^+}\int_{a+k}^b f(x)dx \tag{8.5.4}$$

If the limit exists, then the integral in equation (8.5.4) is said to be convergent. The integral in equation (8.5.4) is another type of improper integral or, more specifically, improper integral with *infinite integrand*. In the spirit of equation (8.5.4), we can also write

$$\int_a^b f(x)dx = \lim_{k \to 0^+} \int_a^{b-k} f(x)dx \tag{8.5.5}$$

Sometimes, we may need to integrate functions that are not defined at both limits; that is, the integrand is not defined at both a and b. In this case, we can apply Property III (equation (8.4.17)) of definite integrals. Therefore, we can write, with a number c that lies between a and b,

$$\int_a^b f(x)dx = \int_a^c f(x)dx + \int_c^d f(x)dx \tag{8.5.6}$$

if the integrals on the RHS of equation (8.5.6) converge.

As an example, consider the integral $\int_k^4 f(x)dx = \int_k^4 (1/\sqrt{x})dx$. Therefore, following equation (8.5.4), we can write the integral as $\int_k^4 f(x)dx = \int_k^4 (1/\sqrt{x})dx = [2x^{1/2}]_k^4 = 4 - 2k^{1/2}$. This result approaches 4 when $k \to 0$. Therefore, the improper integral in the present example is convergent and the area below the graph of the function and the half-open interval $(a, b]$ is 4 units.

As another example, consider the integral $\int_{-1}^1 (1/x^2)dx$. Notice that this integral is an improper integral because the integrand approaches infinity as x approaches zero. Suppose that we write this integral as the sum of two integrals: $\int_{-1}^1 (1/x^2)dx = \int_{-1}^0 (1/x^2)dx + \int_0^1 (1/x^2)dx = \int_{-1}^0 x^{-2}dx + \int_0^1 x^{-2}dx$. Notice that the last equation is similar to equation (8.5.6). Let us now evaluate the last integral using equation (8.5.5). The first part of the last integral can be evaluated to obtain $\int_{-1}^0 x^{-2}dx = \lim_{k \to -0} \int_{-1}^k x^{-2}dx = \lim_{k \to -0} [-1/x]_{-1}^k = \lim_{k \to -0} [(-1/k)+1] = \infty$. Similarly, the second part of the integral can be evaluated to obtain $\int_0^1 x^{-2}dx = \lim_{k \to +0} \int_k^1 x^{-2}dx = \lim_{k \to +0} [-1/x]_k^1 = \lim_{k \to -0} [-1+(1/k)] = \infty$. These two results imply that $\int_{-1}^1 (1/x^2)dx = \int_{-1}^0 (1/x^2)dx + \int_0^1 (1/x^2)dx = \int_{-1}^0 x^{-2}dx + \int_0^1 x^{-2}dx = \infty + \infty = \infty$; that is, the integral is infinity and, therefore, it is divergent.

8.5.4 Integrals of discontinuous functions

As we mentioned in Section 8.5.1, in our exposition of integral calculus so far we have maintained an important assumption that the integrands are *continuous* functions of some variable(s). But, as we saw in Section 1.8.4, one can find many functions, in the subjects of our interest, which are discontinuous. As an example, consider the graph of the function in Figure 8.5.2(A). How do we integrate the discontinuous function $y = f(x)$ in this figure over the interval $[a, d]$?

Our problem is to integrate the discontinuous function $y = f(x)$ in Figure 8.5.2(A) over the interval $[a, d]$ or to determine the area under the graph $y = f(x)$ and above the interval $[a, d]$. For this we can proceed as follows. Assume that a function $g(x)$ is equal to $f(x)$ in the subinterval $[a, d]$ that has the same value as $f(x)$ when $x \to {}^-b$. Then, we can write $\int_a^b f(x)dx = \int_a^b g(x)dx$. The portions of the function $f(x)$ corresponding to other subintervals

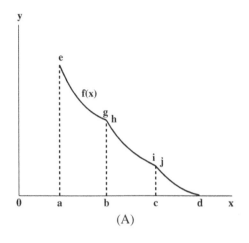
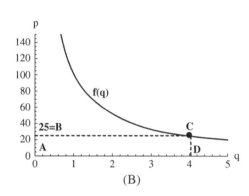

Figure 8.5.2

can be treated similarly. Therefore, we obtain the result

$$\int_a^d f(x)dx = \int_a^b g(x)dx + \int_b^c h(x)dx + \int_c^d j(x)dx \tag{8.5.7}$$

where $h(x)$ and $j(x)$ are functions that are considered to be equal to the function $f(x)$ in intervals $[b, c)$ and $[c, d]$, respectively. Therefore, equation (8.5.7) shows that the integral $\int_a^d f(x)dx$ is equal to the sum of the integrals $\int_a^b g(x)dx$, $\int_b^c h(x)dx$, and $\int_c^d j(x)dx$; or the area $aeghijd$ is equal to the sum of the areas $aegb$, $bhijc$, and cjd.

8.5.5 Application examples

Example 1.   Suppose that a bond possessed by an individual gives a *perpetual cash flow* of $1000 every year and that the discount rate is 8 percent. Determine the present value of this bond or the present value of the perpetual cash flow in this case.

Solution.   We saw in the solution to example 5 in Section 8.4.12 that the present value of a continuous stream of future receipts (F_t) with discount rate (r) for a finite period of time (T) is given by equation (8.4.32): $\int_0^T Fe^{-r \times t}dt = [Fe^{-rt}/(-r)]_0^T = (F/r).[1 - e^{-rt}]$. Since in the present example the receipts are perpetual, the last equation can be written, using equation (8.4.12), as

$$\int_0^{+\infty} Fe^{-r \times t}dt = \lim_{T \to +\infty} [\int_0^T Fe^{-r\,t}dt] = \lim_{T \to +\infty} [(F/r) \times (1 - e^{-r\,t})]_0^T = F/r \tag{8.5.8}$$

Notice that the last integral represents an improper integral. In the present example, $F = 1000$ and $r = 0.08$. Then applying the last result we obtain $\int_0^{+\infty} Fe^{-r \times t}dt = \lim_{T \to +\infty} [\int_0^T 1000e^{-0.08t}]dt = F/r = 12\,500$. Therefore, the present value of a bond that gives a perpetual cash flow of $1000 at discount rate of 8 percent is $12 500.

Example 2.   Suppose that the inverse demand function for a good is given by $p = 100/q^2$, where p denotes the price in dollars per unit of the good and q denotes the quantity of the good demanded. Determine the consumers' surplus when the price is $p = \$25$.

Solution.   We need to determine the consumers' surplus when the price is $25. Consumers' surplus is equal to the area under the graph of the demand function and above the price. But, notice that the area is unbounded because it tends to infinity as $q \to 0$ or as $p \to \infty$.

However, we know that the required area is equal to the integral $\int_{25}^{+\infty} f(q)dq = \int_{25}^{+\infty} (100/q^2)dq$. Notice that this integral is similar to the integral in equation (8.5.2) and, therefore, we can write it as $\int_{25}^{+\infty} f(q)dq = \lim_{b \to +\infty} \int_{25}^{b} f(q)dq$. This last integral can be evaluated to obtain $\int_{25}^{+\infty} f(q)dq = \lim_{b \to +\infty} \int_{25}^{b} f(q)dq = \lim_{b \to +\infty} \int_{25}^{b} (100/q^2)dq = \lim_{b \to +\infty} [-100/q]_{25}^{b}$
$= \lim_{b \to +\infty} [(-100/b) + 4] = 4$. Therefore, the consumers' surplus in the present example is equal to $4.

Example 3.   Suppose that the inverse demand function for a good is given by $p = 100/q$, where p denotes the price in dollars per unit of the good and q denotes the quantity of the good purchased. Determine the consumers' surplus when the price is $p = \$25$.

Solution.   Let us first plot the graph of the demand function $p = f(q) = 100/q$, which is illustrated in Figure 8.5.2(B). We need to determine the consumers' surplus when the price is $25. Consumers' surplus is equal to the area under the graph of the demand function and above the price. This implies that we need to determine the area under the graph of the function and above the line BC. But, notice that the area is unbounded because it tends to infinity as $q \to 0$ or as $p \to \infty$. But, we know that the required area is equal to the integral $\int_{25}^{+\infty} f(q)dq = \int_{25}^{+\infty} (100/q)dq$. Notice that this integral is similar to the integral in equation (8.5.2) and, therefore, we can write it as $\int_{25}^{+\infty} f(q)dq = \lim_{b \to +\infty} \int_{25}^{b} f(q)dq$. This last integral can be evaluated to obtain $\int_{25}^{+\infty} f(q)dq = \lim_{b \to +\infty} \int_{25}^{b} f(q)dq = \lim_{b \to +\infty} \int_{25}^{b} (100/q)dq = \lim_{b \to +\infty} [100 \times \ln q]_{25}^{b} = \lim_{b \to +\infty} [100 \times \ln b - 100 \times \ln 25] = \infty$. Therefore, the consumers' surplus in the present example is equal to infinity. This suggests that the integral in the present example is divergent.

8.5.6 Exercises

1. Evaluate the following integrals if they exist. Which of them are divergent or convergent?
    (i) $\int_0^{+\infty} (1/x)dx$; (ii) $\int_{-\infty}^{0} (1/x)dx$; (iii) $\int_1^{+\infty} (1/\sqrt{x})dx$; (iv) $\int_{-\infty}^{+\infty} (2 + 2x)dx$;
    (v) $\int_0^{+\infty} e^{1-x}dx$; (vi) $\int_0^{+\infty} e^{-rx}dx$; (vii) $\int_{-\infty}^{0} e^{-rx}dx$; (viii) $\int_0^{+\infty} e^{2x}dx$; (ix) $\int_{-\infty}^{0} e^{2x}dx$;
    (x) $\int_3^{+\infty} [1/(x-2)^2]dx$.
2. *Application exercise.* Suppose that a piece of land gives a perpetual cash flow of $1 000 every year and that the discount rate is 6 percent. Determine the present value of this piece of land.
3. *Application exercise.* Suppose that the inverse demand function for a good is given by $p = \sqrt[3]{125/q}$, where p denotes the price in dollars per unit of the good and q denotes the

quantity of the good purchased. Determine the change in consumers' surplus when the price declines from $27 to $8.

4. *Application exercise.* Assume that the per capita income in an economy is expected to grow at $f(t) = 80\,000/(2+t)^2$. What will be the long run (that is, $t \to \infty$) per capita income in the economy?

 Web supplement: S8.5.7 Integration and *probability distributions*

Notes

1 Review of basics

1 A phantasmagoria is a fast-changing and confused group of real or imagined images, one following the other, as in a dream.
2 A list of Greek letters is given at the end of the book.
3 Notice that the *radical* sign ($\sqrt{\,}$), by default, refers to the positive root. Therefore, $4^{1/2} = \pm 2$ but $\sqrt{4} = +2$.
4 Notice that, although Properties I through VII are expressed in natural logarithmic forms (ln) for convenience, all these properties (except III and IV) are equally applicable for common logarithms.
5 We present three methods here. The fourth method, using matrices, will be presented in Chapter 2.
6 Although we use mainly the $<$ sign, these properties are applicable to signs $\leq$, $\geq$, and $>$.
7 Notice that the quantity of goods cannot be negative. Therefore, the nonnegative constraint is implicitly assumed in this example.
8 Notice that we are at the moment considering univariate functions; that is, functions with one independent variable (say, x). Later we will consider multivariate functions; that is, functions with more than one independent variable.
9 A function is a one-to-one function if each value in the range corresponds to exactly one value in the domain. A function is an onto function if each value in the range corresponds to more than one value in the domain.

2 Linear algebra: vectors and matrices

1 Notice that, for the properties to hold, all the vectors must be of the same dimension.

4 Classical optimization

1 "Optima" is the plural of "optimum," and "extrema" is the plural of "extremum."
2 Although a relative minimum and a relative maximum occur at point F and G respectively in Figure 4.2.2, we have ruled out these cases as the first derivative is not defined at these points.
3 Notice that points A and B are inflection points.
4 Notice that the SOC or the sufficient conditions (4.2.3) and (4.2.4) are similar to the second parts of the conditions (4.2.1) and (4.2.2), respectively. Since it is much easier to work with the inequalities in (4.2.3) and (4.2.4), we will use them in our later presentations.
5 It is much easier to work with differentials when we have a multivariate objective function. This is the reason why we introduce the application of differentials in optimization problems. However, one may still find it easier to work with the FOC and SOC using the derivatives stated in the previous section.
6 Although one can use the discriminants, the Hessians, or the characteristic roots to determine the sign of the quadratic forms (as presented in Sections 2.8.2–2.8.4), we will use the discriminants.
7 The Lagrangian function can also be set up as $L = h(x_1, x_2, \lambda) = x_1^{1/2} + x_2^{1/2} - \lambda(2x_1 + 4x_2 - 40)$. We follow the form in equation (4.4.3) because we can use it to interpret the meaning of the Lagrangian multiplier.

8 All economic agents will be at some point on the budget constraint when the agents optimize their objective functions. This implies that, as in the present example, the agent utilizes the complete resource. Then the constraint is said to be a binding constraint. A binding constraint is a special feature of optimization with equality constraints. In optimization problems involving inequality constraints, the available resources may or may not be utilized completely. In this case, the constraint is said to be a *nonbinding constraint*.

5 Linear programming

1 An extreme point or a basic solution can be found by using the *basis theorem*. This theorem states that a solution in which at least $n - m$ variables equal zero is an extreme point. Then, one can find an extreme point or a basic solution by setting $n - m$ variables equal to zero and solving the m equations for the remaining m variables. The total number of basic solutions or extreme points can be found by applying the formula for combinations, $n!/(m!(n - m)!)$, given in equation (1.10.12).

2 The bottom row of the first column of the tableau contains the value $\Pi_j - \pi_j$. Although we will hereafter drop π_j from this position for convenience, it will still contain the value $\Pi_j - \pi_j$. See equations (5.3.11) and (5.3.15) for more on Π_j and $\Pi_j - \pi_j$.

Bibliography

Allen, R.G.D. (1966) *Mathematical Economics*. London: Macmillan.

Allen, R.G.D. (1974) *Mathematical Analysis for Economists*. London: Macmillan.

Almon, C. (1967) *Matrix Methods in Economics*. Reading, MA: Addison-Wesley.

Bailey, D. (1998) *Mathematics in Economics*. London: McGraw-Hill.

Barnett, R.A., Ziegler, M.R., and Byleen, K.E. (1999) *Calculus for Business, Economics, Life Sciences, and Social Sciences*, 8th edition. Upper Saddle River, NJ: Prentice Hall.

Birchenhall, C. and Grout, P. (1984) *Mathematics for Modern Economics*. New York: Philip Allan.

Bronson, R. and Naadimuthu, G. (1982) *Operations Research*, 2nd edition. New York: McGraw-Hill.

Chiang, A.C. and Wainwright, K. (2005) *Fundamental Methods of Mathematical Economics*, 4th edition. New York: McGraw-Hill.

Courant, R. and John, F. (1965) *Introduction to Calculus and Analysis*, Volume I. New York: Interscience Publishers.

Courant, R. and John, F. (1965) *Introduction to Calculus and Analysis*, Volume II. New York: Interscience Publishers.

Dhrymes, P.J. (2000) *Mathematics for Econometrics*, 3rd edition. New York: Springer-Verlag.

Dixit, A.K. (1990) *Optimization in Economic Theory*, 2nd edition. Oxford: Oxford University Press.

Dixit, A.K. and Skeath, S. (2004) *Games of Strategy*, 2nd edition. New York: W.W. Norton.

Dorfman, R., Samuelson, P., and Solow, R. (1958) *Linear Programming and Economic Analysis*, New York: McGraw-Hill.

Dowling, E.T. (1992) *Introduction to Mathematical Economics*, 2nd edition. New York: McGraw-Hill.

Dutta, P.K. (1999) *Strategies and Games: Theory and Practice*. Cambridge, MA: MIT Press.

Eichberger, J. (1993) *Game Theory for Economists*. New York: Academic Press.

Fisher, I. (1925) *Mathematical Investigations in the Theory of Value and Prices*. New Haven, CT: Yale University Press.

Franklin, J. (1980) *Methods of Mathematical Economics: Linear and Nonlinear Programming, Fixed-Point Theorems*. New York: Springer-Verlag.

Friedman, J.W. (1989) *Game Theory with Applications to Economics*. Oxford: Oxford University Press.

Frisch, R. (1966) *Maxima and Minima: Theory and Economic Applications*. Amsterdam: Rand McNally.

Fudenberg, D. and Tirole, J. (2005) *Game Theory*. New Delhi: Ane Books.

Gardner, R. (1995) *Games for Business and Economics*. New York: John Wiley.

Gibbons, R. (1992) *A Primer in Game Theory*. Upper Saddle River, NJ: Prentice Hall.

Glaister, S. (1984) *Mathematical Methods for Economists*, 3rd edition. Oxford: Blackwell.

Goldstein, L.J., Lay, D.C., and Schneider, D.I. (2001) *Calculus and Its Applications*, 9th edition, Upper Saddle River, NJ: Prentice Hall.

Hadley, G. (1962) *Linear Algebra*. Reading, MA: Addison-Wesley.

Hadley, G. (1962) *Linear Programming*. Reading, MA: Addison-Wesley.

Hess, P. (2002) *Using Mathematics in Economic Analysis*. Upper Saddle River, NJ: Prentice Hall.

Hillier, F.S. and Lieberman, G.J. (1995) *Introduction to Operations Research*, 6th edition. New York: McGraw-Hill.

Hoy, M., Livernois, J., McKenna, C., Rees, R., and Stengos, T. (2004) *Mathematics for Economics*, 2nd edition. New Delhi: Prentice Hall of India.

Huang, C.J. and Crooke, P.S. (1997) *Mathematics and Mathematica for Economists*. Oxford: Blackwell.

Intriligator, M.D. (1971) *Mathematical Optimization and Economic Theory*. Englewood Cliffs, NJ: Prentice Hall.

Klein, M.W. (2002) *Mathematical Methods for Economics*, 2nd edition. New York: Addison-Wesley.

Koo, D. (1977) *Elements of Optimization: With Applications in Economics and Business*. New York: Springer-Verlag.

Kreps, D.M. (1990) *Game Theory and Economic Modeling*. Oxford: Oxford University Press.

Lambert, P.J. (1985) *Advanced Mathematics for Economists: Static and Dynamic Optimization*. Oxford: Blackwell.

Luenberger, D.G. (2003) *Linear and Nonlinear Programming*. Boston, MA: Kluwer Academic.

Mas-Collel, A., Whinston, M.D., and Green, J.R. (1995) *Microeconomic Theory*. Oxford: Oxford University Press.

Novshek, W. (1993) *Mathematics for Economists*. London: Academic Press.

Osborne, M.J. (2004) *An Introduction to Game Theory*. Oxford: Oxford University Press.

Osborne, M.J. and Rubinstein, A. (1994) *A Course in Game Theory*. Cambridge, MA: MIT Press.

Ostaszewski, A. (1993) *Mathematical Methods in Economics: Models and Methods*. Oxford: Blackwell.

Read, R.C. (1972) *A Mathematical Background for Economists and Social Scientists*. Englewood Cliffs, NJ: Prentice Hall.

Roberts, B. and Schulze, D.L. (1973) *Modern Mathematics and Economic Analysis*. New York: W.W. Norton.

Romp, G. (1997) *Game Theory: Introduction and Applications*. Oxford: Oxford University Press.

Scott Bierman, H. and Fernandez, L. (1998) *Game Theory with Economic Application*, 2nd edition. Reading, MA: Addison-Wesley.

Simon, K.P. and Blume, L. (1994) *Mathematics for Economists*. New York: W.W. Norton.

Sydsaeter, K. and Hammond, P. (1995) *Mathematics for Economic Analysis*. Englewood Cliffs, NJ: Prentice Hall.

Sydsaeter, K., Hammond, P., Seierstad, A., and Strom, A. (2005) *Further Mathematics for Economic Analysis*. London: Financial Times/Prentice Hall.

Sydsaeter, K., Strom, A., and Berck, P. (1999) *Economists' Mathematical Manual*, 3rd edition. New York: Springer-Verlag.

Taha, H.A. (2003) *Operations Research: An Introduction*, 7th edition. Upper Saddle River, NJ: Prentice Hall.

Takayama, A. (1985) *Mathematical Economics*, 2nd edition. Cambridge: Cambridge University Press.

Thie, P.R. (1989) *An Introduction to Linear Programming and Game Theory*, 2nd edition. New York: John Wiley.

Thomas, Jr., G.B. and Finney, R.L. (1996) *Calculus and Analytic Geometry*, 9th edition. New Delhi: Pearson Education.

Varian, H.R. (1992) *Microeconomic Analysis*, 3rd edition. New York: W.W. Norton.

Watson, J. (2002) *Strategy: An Introduction to Game Theory*. New York: W.W. Norton.

Webster, T.J. (2009) *Introduction to Game Theory in Business and Economics*. New York: M.E. Sharpe.

Weintraub, E.Y. (1982) *Mathematics for Economists: An Integrated Approach*. Cambridge: Cambridge University Press.

Werner, F. and Sotskov, Y.N. (2006) *Mathematics of Economics and Business*. London: Routledge.

Yamane, T. (1968) *Mathematics for Economists: An Elementary Survey*, 2nd edition. New Delhi: Prentice Hall.

Index

Greek alphabet

Name	Lowercase	Uppercase	Name	Lowercase	Uppercase
Alpha	α	A	**Omega**	ω	Ω
Beta	β	B	**Omicron**	o	O
Chi	χ	X	**Pi**	π	Π
Delta	δ	Δ	**Phi**	ϕ	Φ
Epsilon	ε	E	**Psi**	ψ	Ψ
Eta	η	H	**Rho**	ρ	P
Gamma	γ	Γ	**Sigma**	σ	Σ
Iota	ι	I	**Tau**	τ	T
Kappa	κ	K	**Theta**	θ	Θ
Lambda	λ	Λ	**Upsilon**	υ	Υ
Mu	μ	M	**Xi**	ξ	Ξ
Nu	ν	N	**Zeta**	ζ	Z

Printed and bound by CPI Group (UK) Ltd, Croydon, CR0 4YY

01/11/2024

01782609-0004